Graduate Texts in Mathematics

Graduate Texts in Mathematics bridge the gap between passive study and creative understanding, offering graduate-level introductions to advanced topics in mathematics. The volumes are carefully written as teaching aids and highlight characteristic features of the theory. Although these books are frequently used as textbooks in graduate courses, they are also suitable for individual study.

For further volumes:
http://www.springer.com/series/136

Daniel Bump

Lie Groups

Second Edition

 Springer

Daniel Bump
Department of Mathematics
Stanford University
Stanford, CA, USA

ISSN 0072-5285
ISBN 978-1-4939-3842-1   ISBN 978-1-4614-8024-2 (eBook)
DOI 10.1007/978-1-4614-8024-2
Springer New York Heidelberg Dordrecht London

Mathematics Subject Classification: 22Exx, 17Bxx

Springer is part of Springer Science+Business Media (www.springer.com)

Preface

This book aims to be both a graduate text and a study resource for Lie groups. It tries to strike a compromise between accessibility and getting enough depth to communicate important insights. In discussing the literature, often secondary sources are preferred: cited works are usually recommended ones.

There are four parts. Parts I, II or IV are all "starting points" where one could begin reading or lecturing. On the other hand, Part III assumes familiarity with Part II. The following chart indicates the approximate dependencies of the chapters. There are other dependencies, where a result is used from a chapter that is not a prerequisite according to this chart: but these are relatively minor. The dashed lines from Chaps. 1 and 2 to the opening chapters of Parts II and IV indicate that the reader will benefit from knowledge of Schur orthogonality but may skip or postpone Chaps. 1 and 2 before starting Part II or Part IV. The other dashed line indicates that the Bruhat decomposition (Chap. 27) is assumed in the last few chapters of Part IV.

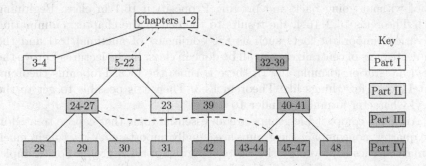

The two lines of development in Parts II–IV were kept independent because it was possible to do so. This has the obvious advantage that one may start reading with Part IV for an alternative course. This should not obscure the fact that these two lines are complementary, and shed light on each other. We hope the reader will study the whole book.

Part I treats two basic topics in the analysis of compact Lie groups: Schur orthogonality and the Peter–Weyl theorem, which says that the irreducible unitary representations of a compact group are all finite-dimensional.

Usually the study of Lie groups begins with compact Lie groups. It is attractive to make this the complete content of a short course because it can be treated as a self-contained subject with well-defined goals, about the right size for a 10-week class. Indeed, Part II, which covers this theory, could be used as a traditional course culminating in the Weyl character formula. It covers the basic facts about compact Lie groups: the fundamental group, conjugacy of maximal tori, roots and weights, the Weyl group, the Weyl integration formula, and the Weyl character formula. These are basic tools, and a short course in Lie theory might end up with the Weyl character formula, though usually I try to do a bit more in a 10-week course, even at the expense of skipping a few proofs in the lectures. The last chapter in Part II introduces the affine Weyl group and computes the fundamental group. It can be skipped since Part III does not depend on it.

Sage, the free mathematical software system, is capable of doing typical Lie theory calculations. The student of Part II may want to learn to use it. An appendix illustrates its use.

The goal of Part I is the Peter–Weyl theorem, but Part II does not depend on this. Therefore one could skip Part I and start with Part II. Usually when I teach this material, I do spend one or two lectures on Part I, proving Schur orthogonality but not the Peter–Weyl formula. In the interests of speed I tend to skip a few proofs in the lectures. For example, the conjugacy of maximal tori needs to be proved, and this depends in turn on the surjectivity of the exponential map for compact groups, that is, Theorem 16.3. This is proved completely in the text, and I think it should be proved in class but some of the differential geometry details behind it can be replaced by intuitive explanations. So in lecturing, I try to explain the intuitive content of the proof without going back and proving Proposition 16.1 in class. Beginning with Theorems 16.2–16.4, the results to the end of the chapter, culminating in various important facts such as the conjugacy of maximal tori and the connectedness of centralizers can all be done in class. In the lectures I prove the Weyl integration formula and (if there is time) the local Frobenius theorem. But I skip a few things like Theorem 13.3. Then it is possible to get to the Weyl Character formula in under 10 weeks.

Although compact Lie groups are an essential topic that can be treated in one quarter, noncompact Lie groups are equally important. A key role in much of mathematics is played by the Borel subgroup of a Lie group. For example, if $G = \mathrm{GL}(n, \mathbb{R})$ or $\mathrm{GL}(n, \mathbb{C})$, the Borel subgroup is the subgroup of upper triangular matrices, or any conjugate of this subgroup. It is involved in two important results, the Bruhat and Iwasawa decompositions. A noncompact Lie group has two important classes of homogeneous spaces, namely symmetric spaces and flag varieties, which are at the heart of a great deal of important modern mathematics. Therefore, noncompact Lie groups cannot be ignored, and we tried hard to include them.

In Part III we first introduce a class of noncompact groups, the complex reductive groups, that are obtained from compact Lie groups by "complexification." These are studied in several chapters before eventually taking on general noncompact Lie groups. This allows us to introduce key topics such as the Iwasawa and Bruhat decompositions without getting too caught up in technicalities. Then we look at the Weyl group and affine Weyl group, already introduced in Part II, as Coxeter groups. There are two important facts about them to be proved: that they have Coxeter group presentations, and the theorem of Matsumoto and Tits that any two reduced words for the same element may be related by applications of the braid relations.

For these two facts we give geometric proofs, based on properties of the complexes on which they act. These complexes are the system of Weyl chambers in the first case, and of alcoves in the second. Applications are given, such as Demazure characters and the Bruhat order. For complex reductive groups, we prove the Iwasawa and Bruhat decompositions, digressing to discuss some of the implications of the Bruhat decomposition for the flag manifold. In particular the Schubert and Bott–Samelson varieties, the Borel-Weil theorem and the Bruhat order are introduced. Then we look at symmetric spaces, in a chapter that alternates examples with theory. Symmetric spaces occur in pairs, a compact space matched with a noncompact one. We see how some symmetric spaces, the Hermitian ones, have complex structures and are important in the theory of functions of several complex variables. Others are convex cones. We take a look at Freudenthal's "magic square." We discuss the embedding of a noncompact symmetric space in its compact dual, the boundary components and Bergman–Shilov boundary of a symmetric tube domain, and Cartan's classification. By now we are dealing with arbitrary noncompact Lie groups, where before we limited ourselves to the complex analytic ones. Another chapter constructs the relative root system, explains Satake diagrams and gives examples illustrating the various phenomena that can occur. The Iwasawa decomposition, formerly obtained for complex analytic groups, is reproved in this more general context. Another chapter surveys the different ways Lie groups can be embedded in one another. Part III ends with a somewhat lengthy discussion of the spin representations of the double covers of orthogonal groups. First, we consider what can be deduced from the Weyl theory. Second, as an alternative, we construct the spin representations using Clifford algebras. Instead of following the approach (due to Chevalley) often taken in embedding the spin group into the multiplicative group of the Clifford algebra, we take a different approach suggested by the point of view in Howe [75, 77].

This approach obtains the spin representation as a projective representation from the fact that the orthogonal group acts by automorphisms on a ring having a unique representation. The existence of the spin group is a byproduct of the projective representation. This is the same way that the Weil representation is usually constructed from the Stone–von Neumann theorem, with the Clifford algebra replacing the Heisenberg group.

Part IV, we have already mentioned, is largely independent of the earlier parts. Much of it concerned with *correspondences* which were emphasized by Howe, though important examples occur in older work of Frobenius and Schur, Weyl, Weil and others. Following Howe, a *correspondence* is a bijection between a set of representations of a group G with a set of representations of another group H which arise as follows. There is a representation Ω of $G \times H$ with the following property. Let $\pi_i \otimes \pi_i'$ be the irreducible representations of $G \times H$ that occur in the restriction. It is assumed that each occurs with multiplicity one, and moreover, that there are no repetitions among the π_i, and none among the π_i'. This gives a bijection between the representations π_i of G and the representations π_i' of H. Often Ω has an explicit description with special properties that allow us to transfer calculation from one group to the other. Sometimes Ω arises by restriction of a "small" representation of a big group W that contains $G \times H$ as a subgroup.

The first example is the Frobenius–Schur duality. This is the correspondence between the irreducible representations of the symmetric group and the general linear groups. The correspondence comes from decomposing tensor spaces over both groups simultaneously. Another correspondence, for the groups $GL(n)$ and $GL(m)$, is embodied in the Cauchy identity. We will focus on these two correspondences, giving examples of how they can be used to transfer calculations from one group to the other.

Frobenius–Schur duality is very often called "Schur–Weyl duality," and indeed Weyl emphasized this theory both in his book on the classical groups and in his book on quantum mechanics. However Weyl was much younger than Schur and did not begin working on Lie groups until the 1920s, while the duality is already mature in Schur's 1901 dissertation. Regarding Frobenius' contribution, Frobenius invented character theory before the relationship between characters and representations was clarified by his student Schur. With great insight Frobenius showed in 1900 that the characters of the symmetric group could be computed using symmetric functions. This very profound idea justifies attaching Frobenius' name with Schur's to this phenomenon. Now Green has pointed out that the 1892 work of Deruyts in invariant theory also contains results almost equivalent to this duality. This came a few years too soon to fully take the point of view of group representation theory. Deruyt's work is prescient but less historically influential than that of Frobenius and Schur since it was overlooked for many years, and in particular Schur was apparently not aware of it. For these reasons we feel the term "Frobenius–Schur duality" is most accurate. See the excellent history of Curtis [39].

Frobenius–Schur duality allows us to simultaneously develop the representation theories of $GL(n, \mathbb{C})$ and S_k. For $GL(n, \mathbb{C})$, this means a proof of the Weyl character formula that is independent of the arguments in Part II. For the symmetric group, this means that (following Frobenius) we may use symmetric functions to describe the characters of the irreducible representations of S_k. This gives us a double view of symmetric function theory that sheds light on a great many things. The double view is encoded in the structure of a graded algebra (actually a Hopf algebra) $\mathcal{R}$ whose homogeneous part of degree k consists of the characters of representations of S_k. This is isomorphic to the ring of Λ of symmetric polynomials, and familiarity with this equivalence is the key to understanding a great many things.

One very instructive example of using Frobenius–Schur duality is the computation by Diaconis and Shahshahani of the moments of the traces of unitary matrices. The result has an interesting interpretation in terms of random matrix theory, and it also serves as an example of how the duality can be used: directly computing the moments in question is feasible but leads to a difficult combinatorial problem. Instead, one translates the problem from the unitary group to an equivalent but easier question on the symmetric group.

The $\mathrm{GL}(n) \times \mathrm{GL}(m)$ duality, like the Frobenius–Schur duality, can be used to translate a calculation from one context to another, where it may be easier. As an example, we consider a result of Keating and Snaith, also from random matrix theory, which had significant consequences in understanding the distribution of the values of the Riemann zeta function. The computation in question is that of the $2k$-th moment of the characteristic polynomial of $\mathrm{U}(n)$. Using the duality, it is possible to transfer the computation from $\mathrm{U}(n)$ to $\mathrm{U}(2k)$, where it becomes easy.

Other types of problems that may be handled this way are branching rules: a branching rule describes how an irreducible representation of a group G decomposes into irreducibles when restricted to a subgroup H. We will see instances where one uses a duality to transfer a calculation from one pair (G, H) to another, (G', H'). For example, we may take G and H to be $\mathrm{GL}(p+q)$ and its subgroup $\mathrm{GL}(p) \times \mathrm{GL}(q)$, and G' and H' to be $\mathrm{GL}(n) \times \mathrm{GL}(n)$ and its diagonal subgroup $\mathrm{GL}(n)$.

Chapter 42 shows how the Jacobi–Trudi identity from the representation theory of the symmetric group can be translated using Frobenius–Schur duality to compute minors of Toeplitz matrices. Then we look at involution models for the symmetric group, showing how it is possible to find a set of induced representations whose union contains every irreducible representation exactly once. Translated by Frobenius–Schur duality, this gives some decompositions of symmetric algebras over the symmetric and exterior square representations, a topic that is also treated by a different method in Part II.

Towards the end of Part IV, we discuss several other ways that the graded ring $\mathcal{R}$ occurs. First, the representation theory of the symmetric group has a deformation in the Iwahori Hecke algebra, which is ubiquitous in mathematics, from the representation theory of p-adic groups to the K-theory of flag varieties and developments in mathematical physics related to the Yang–Baxter equation. Second, the Hopf algebra $\mathcal{R}$ has an analog in which the representation theory of $\mathrm{GL}(k)$ (say over a finite field) replaces the representation theory of S_k; the multiplication and comultiplication are parabolic induction and its adjoint (the Jacquet functor). The ground field may be replaced by a p-adic field or an adele ring, and ultimately this "philosophy of cusp forms" leads to the theory of automorphic forms. Thirdly, the ring $\mathcal{R}$ has as a homomorphic image the cohomology rings of flag varieties, leading to the Schubert calculus. These topics are surveyed in the final chapters.

What's New? I felt that the plan of the first edition was a good one, but that substantial improvements were needed. Some material has been removed, and a fair amount of new material has been added. Some old material has been streamlined or rewritten, sometimes extensively. In places what was implicit in the first edition but not explained well is now carefully explained with attention to the underlying principles. There are more exercises. A few chapters are little changed, but the majority have some revisions, so the changes are too numerous to list completely. Highlights in the newly added material include the affine Weyl group, new material about Coxeter groups, Demazure characters, Bruhat order, Schubert and Bott–Samelson varieties, the Borel-Weil theorem the appendix on Sage, Clifford algebras, the Keating–Snaith theorem, and more.

Notation. The notations $GL(n, F)$ and $GL_n(F)$ are interchangeable for the group of $n \times n$ matrices with coefficients in F. By $Mat_n(F)$ we denote the ring of $n \times n$ matrices, and $Mat_{n \times m}(F)$ denotes the vector space of $n \times m$ matrices. In $GL(n)$, I or I_n denotes the $n \times n$ identity matrix and if g is any matrix, ${}^t g$ denotes its transpose. Omitted entries in a matrix are zero. Thus, for example,

$$\begin{pmatrix} & 1 \\ -1 & \end{pmatrix} = \begin{pmatrix} 0 & 1 \\ -1 & 0 \end{pmatrix}.$$

The identity element of a group is usually denoted 1 but also as I, if the group is $GL(n)$ (or a subgroup), and occasionally as e when it seemed the other notations could be confusing. The notations $\subset$ and $\subseteq$ are synonymous, but we mostly use $X \subset Y$ if X and Y are known to be unequal, although we make no guarantee that we are completely consistent in this. If X is a finite set, $|X|$ denotes its cardinality.

Acknowledgements The proofs of the Jacobi–Trudi identity were worked out years ago with Karl Rumelhart when he was still an undergraduate at Stanford. Chapters 39 and 42 owe a great deal to Persi Diaconis and (for the Keating–Snaith result) Alex Gamburd. For the second edition, I thank the many people who informed me of typos; I cannot list them all but I especially thank Yunjiang (John) Jiang for his careful reading of Chap. 18. And thanks in advance to all who will report typos in this edition.

This work was supported in part by NSF grants DMS-9970841 and DMS-1001079.

Stanford, CA, USA   Daniel Bump

Contents

Part III Noncompact Lie Groups

Part IV Duality and Other Topics

Compact Groups

1

Haar Measure

If G is a locally compact group, there is, up to a constant multiple, a unique regular Borel measure μ_L that is invariant under left translation. Here *left translation invariance* means that $\mu(X) = \mu(gX)$ for all measurable sets X. *Regularity* means that

$$\mu(X) = \inf\{\mu(U) \mid U \supseteq X, U \text{ open}\} = \sup\{\mu(K) \mid K \subseteq X, K \text{ compact}\}.$$

Such a measure is called a *left Haar measure*. It has the properties that any compact set has finite measure and any nonempty open set has measure > 0.

We will not prove the existence and uniqueness of the Haar measure. See for example Halmos [61], Hewitt and Ross [69], Chap. IV, or Loomis [121] for a proof of this. Left-invariance of the measure amounts to left-invariance of the corresponding integral,

$$\int_G f(\gamma g)\,\mathrm{d}\mu_L(g) = \int_G f(g)\,\mathrm{d}\mu_L(g), \tag{1.1}$$

for any Haar integrable function f on G.

There is also a right-invariant measure, μ_R, unique up to constant multiple, called a *right Haar measure*. Left and right Haar measures may or may not coincide. For example, if

$$G = \left\{ \begin{pmatrix} y & x \\ 0 & 1 \end{pmatrix} \,\middle|\, x, y \in \mathbb{R}, y > 0 \right\},$$

then it is easy to see that the left- and right-invariant measures are, respectively,

$$\mathrm{d}\mu_L = y^{-2}\,\mathrm{d}x\,\mathrm{d}y, \qquad \mathrm{d}\mu_R = y^{-1}\,\mathrm{d}x\,\mathrm{d}y.$$

They are not the same. However, there are many cases where they do coincide, and if the left Haar measure is also right-invariant, we call G *unimodular*.

D. Bump, *Lie Groups*, Graduate Texts in Mathematics 225,
DOI 10.1007/978-1-4614-8024-2_1, © Springer Science+Business Media New York 2013

Conjugation is an automorphism of G, and so it takes a left Haar measure to another left Haar measure, which must be a constant multiple of the first. Thus, if $g \in G$, there exists a constant $\delta(g) > 0$ such that

$$\int_G f(g^{-1}hg)\,\mathrm{d}\mu_L(h) = \delta(g) \int_G f(h)\,\mathrm{d}\mu_L(h).$$

If G is a topological group, a *quasicharacter* is a continuous homomorphism $\chi : G \longrightarrow \mathbb{C}^\times$. If $|\chi(g)| = 1$ for all $g \in G$, then χ is a (linear) *character* or *unitary quasicharacter*.

Proposition 1.1. *The function* $\delta : G \longrightarrow \mathbb{R}_+^\times$ *is a quasicharacter. The measure* $\delta(h)\mu_L(h)$ *is right-invariant.*

The measure $\delta(h)\mu_L(h)$ is a right Haar measure, and we may write $\mu_R(h) = \delta(h)\mu_L(h)$. The quasicharacter δ is called the *modular quasicharacter*.

Proof. Conjugation by first g_1 and then g_2 is the same as conjugation by $g_1 g_2$ in one step. Thus $\delta(g_1 g_2) = \delta(g_1)\,\delta(g_2)$, so δ is a quasicharacter. Using (1.1),

$$\delta(g) \int_G f(h)\,\mathrm{d}\mu_L(h) = \int_G f(g \cdot g^{-1}hg)\,\mathrm{d}\mu_L(h) = \int_G f(hg)\,\mathrm{d}\mu_L(h).$$

Replace f by $f\delta$ in this identity and then divide both sides by $\delta(g)$ to find that

$$\int_G f(h)\,\delta(h)\,\mathrm{d}\mu_L(h) = \int_G f(hg)\,\delta(h)\,\mathrm{d}\mu_L(h).$$

Thus, the measure $\delta(h)\,\mathrm{d}\mu_L(h)$ is right-invariant.   □

Proposition 1.2. *If G is compact, then G is unimodular and $\mu_L(G) < \infty$.*

Proof. Since δ is a homomorphism, the image of δ is a subgroup of $\mathbb{R}_+^\times$. Since G is compact, $\delta(G)$ is also compact, and the only compact subgroup of $\mathbb{R}_+^\times$ is just $\{1\}$. Thus δ is trivial, so a left Haar measure is right-invariant. We have mentioned as an assumed fact that the Haar volume of any compact subset of a locally compact group is finite, so if G is finite, its Haar volume is finite.   □

If G is compact, then it is natural to normalize the Haar measure so that G has volume 1.

To simplify our notation, we will denote $\int_G f(g)\,\mathrm{d}\mu_L(g)$ by $\int_G f(g)\,\mathrm{d}g$.

Proposition 1.3. *If G is unimodular, then the map* $g \longrightarrow g^{-1}$ *is an isometry.*

Proof. It is easy to see that $g \longrightarrow g^{-1}$ turns a left Haar measure into a right Haar measure. If left and right Haar measures agree, then $g \longrightarrow g^{-1}$ multiplies the left Haar measure by a positive constant, which must be 1 since the map has order 2.   □

Exercises

Exercise 1.1. Let $d_{\mathbf{a}}X$ denote the Lebesgue measure on $\mathrm{Mat}_n(\mathbb{R})$. It is of course a Haar measure for the additive group $\mathrm{Mat}_n(\mathbb{R})$. Show that $|\det(X)|^{-n}d_{\mathbf{a}}X$ is both a left and a right Haar measure on $\mathrm{GL}(n,\mathbb{R})$.

Exercise 1.2. Let P be the subgroup of $\mathrm{GL}(r+s,\mathbb{R})$ consisting of matrices of the form

$$p = \begin{pmatrix} g_1 & X \\ & g_2 \end{pmatrix}, \qquad g_1 \in \mathrm{GL}(r,\mathbb{R}),\ g_2 \in \mathrm{GL}(s,\mathbb{R}), \quad X \in \mathrm{Mat}_{r\times s}(\mathbb{R}).$$

Let dg_1 and dg_2 denote Haar measures on $\mathrm{GL}(r,\mathbb{R})$ and $\mathrm{GL}(s,\mathbb{R})$, and let $d_{\mathbf{a}}X$ denote an additive Haar measure on $\mathrm{Mat}_{r\times s}(\mathbb{R})$. Show that

$$d_L p = |\det(g_1)|^{-s}\, dg_1\, dg_2\, d_{\mathbf{a}}X, \qquad d_R p = |\det(g_2)|^{-r}\, dg_1\, dg_2\, d_{\mathbf{a}}X,$$

are (respectively) left and right Haar measures on P, and conclude that the modular quasicharacter of P is

$$\delta(p) = |\det(g_1)|^s |\det(g_2)|^{-r}.$$

2

Schur Orthogonality

In this chapter and the next two, we will consider the representation theory of compact groups. Let us begin with a few observations about this theory and its relationship to some related theories.

If V is a finite-dimensional complex vector space, or more generally a Banach space, and $\pi : G \longrightarrow \mathrm{GL}(V)$ a continuous homomorphism, then (π, V) is called a *representation*. Assuming $\dim(V) < \infty$, the function $\chi_\pi(g) = \mathrm{tr}\,\pi(g)$ is called the *character* of π. Also assuming $\dim(V) < \infty$, the representation (π, V) is called *irreducible* if V has no proper nonzero invariant subspaces, and a character is called *irreducible* if it is a character of an irreducible representation.

[If V is an infinite-dimensional topological vector space, then (π, V) is called irreducible if it has no proper nonzero invariant *closed* subspaces.]

A quasicharacter χ is a character in this sense since we can take $V = \mathbb{C}$ and $\pi(g)v = \chi(g)v$ to obtain a representation whose character is χ.

The archetypal compact Abelian group is the circle $\mathbb{T} = \{z \in \mathbb{C}^\times \mid |z| = 1\}$. We normalize the Haar measure on $\mathbb{T}$ so that it has volume 1. Its characters are the functions $\chi_n : \mathbb{T} \longrightarrow \mathbb{C}^\times$, $\chi_n(z) = z^n$. The important properties of the χ_n are that they form an orthonormal system and (deeper) an orthonormal basis of $L^2(\mathbb{T})$.

More generally, if G is a compact Abelian group, the characters of G form an orthonormal basis of $L^2(G)$. If $f \in L^2(G)$, we have a Fourier expansion,

$$f(g) = \sum_\chi a_\chi\, \chi(g), \qquad a_\chi = \int_G f(g)\overline{\chi(g)}\,\mathrm{d}g, \qquad (2.1)$$

and the Plancherel formula is the identity:

$$\int_G |f(g)|^2\,\mathrm{d}g = \sum_\chi |a_\chi|^2. \qquad (2.2)$$

These facts can be directly generalized in two ways. First, Fourier analysis on locally compact Abelian groups, including Pontriagin duality, Fourier

D. Bump, *Lie Groups*, Graduate Texts in Mathematics 225,
DOI 10.1007/978-1-4614-8024-2_2, © Springer Science+Business Media New York 2013

inversion, the Plancherel formula, etc. is an important and complete theory due to Weil [169] and discussed, for example, in Rudin [140] or Loomis [121]. The most important difference from the compact case is that the characters can vary continuously. The characters themselves form a group, the *dual group* $\hat{G}$, whose topology is that of uniform convergence on compact sets. The Fourier expansion (2.1) is replaced by the *Fourier inversion formula*

$$f(g) = \int_{\hat{G}} \hat{f}(\chi)\,\chi(g)\,\mathrm{d}\chi, \qquad \hat{f}(\chi) = \int_{G} f(g)\,\overline{\chi(g)}\,\mathrm{d}g.$$

The symmetry between G and $\hat{G}$ is now evident. Similarly in the Plancherel formula (2.2) the sum on the right is replaced by an integral.

The second generalization, to arbitrary *compact* groups, is the subject of this chapter and the next two. In summary, group representation theory gives a orthonormal basis of $L^2(G)$ in the matrix coefficients of irreducible representations of G and a (more important and very canonical) orthonormal basis of the subspace of $L^2(G)$ consisting of class functions in terms of the characters of the irreducible representations. Most importantly, the irreducible representations are all finite-dimensional. The orthonormality of these sets is Schur orthogonality; the completeness is the Peter–Weyl theorem.

These two directions of generalization can be unified. Harmonic analysis on locally compact groups agrees with representation theory. The Fourier inversion formula and the Plancherel formula now involve the matrix coefficients of the irreducible unitary representations, which may occur in continuous families and are usually infinite-dimensional. This field of mathematics, largely created by Harish-Chandra, is fundamental but beyond the scope of this book. See Knapp [104] for an extended introduction, and Gelfand, Graev and Piatetski-Shapiro [55] and Varadarajan [165] for the Plancherel formula for $\mathrm{SL}(2, \mathbb{R})$.

Although *infinite-dimensional* representations are thus essential in harmonic analysis on a noncompact group such as $\mathrm{SL}(n, \mathbb{R})$, noncompact Lie groups also have irreducible *finite-dimensional* representations, which are important in their own right. They are seldom unitary and hence not relevant to the Plancherel formula. The scope of this book includes finite-dimensional representations of Lie groups but not infinite-dimensional ones.

In this chapter and the next two, we will be mainly concerned with compact groups. In this chapter, all representations will be complex and finite-dimensional except when explicitly noted otherwise.

By an *inner product* on a complex vector space, we mean a positive definite Hermitian form, denoted $\langle\ ,\ \rangle$. Thus, $\langle v, w \rangle$ is linear in v, conjugate linear in w, satisfies $\langle w, v \rangle = \overline{\langle v, w \rangle}$, and $\langle v, v \rangle > 0$ if $v \neq 0$. We will also use the term *inner product* for real vector spaces—an inner product on a real vector space is a positive definite symmetric bilinear form. Given a group G and a real or complex representation $\pi : G \longrightarrow \mathrm{GL}(V)$, we say the inner product $\langle\ ,\ \rangle$ on V is *invariant* or *G-equivariant* if it satisfies the identity

$$\langle \pi(g)v, \pi(g)w \rangle = \langle v, w \rangle.$$

Proposition 2.1. *If G is compact and (π, V) is any finite-dimensional complex representation, then V admits a G-equivariant inner product.*

Proof. Start with an arbitrary inner product $\langle\!\langle\,,\,\rangle\!\rangle$. Averaging it gives another inner product,

$$\langle v, w \rangle = \int_G \langle\!\langle \pi(g)v, \pi(g)w \rangle\!\rangle \, dg,$$

for it is easy to see that this inner product is Hermitian and positive definite. It is G-invariant by construction. $\qquad\qquad\square$

Proposition 2.2. *If G is compact, then each finite-dimensional representation is the direct sum of irreducible representations.*

Proof. Let (π, V) be given. Let V_1 be a nonzero invariant subspace of minimal dimension. It is clearly irreducible. Let $V_1^\perp$ be the orthogonal complement of V_1 with respect to a G-invariant inner product. It is easily checked to be invariant and is of lower dimension than V. By induction $V_1^\perp = V_2 \oplus \cdots \oplus V_n$ is a direct sum of invariant subspaces and so $V = V_1 \oplus \cdots \oplus V_n$ is also. $\quad\square$

A function of the form $\phi(g) = L\big(\pi(g)\,v\big)$, where (π, V) is a finite-dimensional representation of G, $v \in V$ and $L : V \longrightarrow \mathbb{C}$ is a linear functional, is called a *matrix coefficient* on G. This terminology is natural, because if we choose a basis $e_1, \ldots, e_n$, of V, we can identify V with $\mathbb{C}^n$ and represent g by matrices:

$$\pi(g)v = \begin{pmatrix} \pi_{11}(g) & \cdots & \pi_{1n}(g) \\ \vdots & & \vdots \\ \pi_{n1}(g) & \cdots & \pi_{nn}(g) \end{pmatrix} \begin{pmatrix} v_1 \\ \vdots \\ v_n \end{pmatrix}, \qquad v = \begin{pmatrix} v_1 \\ \vdots \\ v_n \end{pmatrix} = \sum_{j=1}^n v_j e_j.$$

Then each of the n^2 functions π_{ij} is a matrix coefficient. Indeed

$$\pi_{ij}(g) = L_i\big(\pi(g)e_j\big),$$

where $L_i(\sum_j v_j e_j) = v_i$.

Proposition 2.3. *The matrix coefficients of G are continuous functions. The pointwise sum or product of two matrix coefficients is a matrix coefficient, so they form a ring.*

Proof. If $v \in V$, then $g \longrightarrow \pi(g)v$ is continuous since by definition a representation $\pi : G \longrightarrow \mathrm{GL}(V)$ is continuous and so a matrix coefficient $L\big(\pi(g)\,v\big)$ is continuous.

If (π_1, V_1) and (π_2, V_2) are representations, $v_i \in V_i$ are vectors and $L_i : V_i \longrightarrow \mathbb{C}$ are linear functionals, then we have representations $\pi_1 \oplus \pi_2$ and $\pi_1 \otimes \pi_2$ on $V_1 \oplus V_2$ and $V_1 \otimes V_2$, respectively. Given vectors $v_i \in V_i$ and functionals $L_i \in V_i^*$, then $L_1\big(\pi(g)v_1\big) \pm L_2\big(\pi(g)v_2\big)$ can be expressed as

$L\big((\pi_1 \oplus \pi_2)(g)(v_1, v_2)\big)$ where $L : V_1 \oplus V_2 \longrightarrow \mathbb{C}$ is $L(x_1, x_2) = L_1(x_1) \pm L_2(x_2)$, so the matrix coefficients are closed under addition and subtraction.

Similarly, we have a linear functional $L_1 \otimes L_2$ on $V_1 \otimes V_2$ satisfying

$$(L_1 \otimes L_2)(x_1 \otimes x_2) = L_1(x_1)L_2(x_2)$$

and

$$(L_1 \otimes L_2)\big((\pi_1 \otimes \pi_2)(g)(v_1 \otimes v_2)\big) = L_1\big(\pi_1(g)v_1\big) L_2\big(\pi_2(g)v_2\big),$$

proving that the product of two matrix coefficients is a matrix coefficient.   □

If (π, V) is a representation, let V^* be the dual space of V. To emphasize the symmetry between V and V^*, let us write the dual pairing $V \times V^* \longrightarrow \mathbb{C}$ in the symmetrical form $L(v) = [\![v, L]\!]$. We have a representation $(\hat{\pi}, V^*)$, called the *contragredient* of π, defined by

$$[\![v, \hat{\pi}(g)L]\!] = [\![\pi(g^{-1})v, L]\!] . \tag{2.3}$$

Note that the inverse is needed here so that $\hat{\pi}(g_1 g_2) = \hat{\pi}(g_1)\hat{\pi}(g_2)$.

If (π, V) is a representation, then by Proposition 2.3 any linear combination of functions of the form $L\big(\pi(g)v\big)$ with $v \in V$, $L \in V^*$ is a matrix coefficient, though it may be a function $L'\big(\pi'(g)v'\big)$ where (π', V') is not (π, V), but a larger representation. Nevertheless, we call any linear combination of functions of the form $L\big(\pi(g)v\big)$ a *matrix coefficient of the representation* (π, V). Thus, the matrix coefficients of π form a vector space, which we will denote by $\mathcal{M}_\pi$. Clearly, $\dim(\mathcal{M}_\pi) \leqslant \dim(V)^2$.

Proposition 2.4. *If f is a matrix coefficient of (π, V), then $\check{f}(g) = f(g^{-1})$ is a matrix coefficient of $(\hat{\pi}, V^*)$.*

Proof. This is clear from (2.3), regarding v as a linear functional on V^*.   □

We have actions of G on the space of functions on G by left and right translation. Thus if f is a function and $g \in G$, the left and right translates are

$$(\lambda(g)f)(x) = f(g^{-1}x), \qquad (\rho(g)f)(x) = f(xg).$$

Theorem 2.1. *Let f be a function on G. The following are equivalent.*

(i) The functions $\lambda(g)f$ span a finite-dimensional vector space.
(ii) The functions $\rho(g)f$ span a finite-dimensional vector space.
(iii) The function f is a matrix coefficient of a finite-dimensional representation.

Proof. It is easy to check that if f is a matrix coefficient of a particular representation V, then so are $\lambda(g)f$ and $\rho(g)f$ for any $g \in G$. Since V is finite-dimensional, its matrix coefficients span a finite-dimensional vector space; in fact, a space of dimension at most $\dim(V)^2$. Thus, (iii) implies (i) and (ii).

Suppose that the functions $\rho(g)f$ span a finite-dimensional vector space V. Then (ρ, V) is a finite-dimensional representation of G, and we claim that f is a matrix coefficient. Indeed, define a functional $L : V \longrightarrow \mathbb{C}$ by $L(\phi) = \phi(1)$. Clearly, $L(\rho(g)f) = f(g)$, so f is a matrix coefficient, as required. Thus (ii) implies (iii).

Finally, if the functions $\lambda(g)f$ span a finite-dimensional space, composing these functions with $g \longrightarrow g^{-1}$ gives another finite-dimensional space which is closed under right translation, and $\check{f}$ defined as in Proposition 2.4 is an element of this space; hence $\check{f}$ is a matrix coefficient by the case just considered. By Proposition 2.4, f is also a matrix coefficient, so (i) implies (iii). $\square$

If (π_1, V_1) and (π_2, V_2) are representations, an *intertwining operator*, also known as a *G-equivariant map* $T : V_1 \longrightarrow V_2$ or (since V_1 and V_2 are sometimes called *G-modules*) a *G-module homomorphism*, is a linear transformation $T : V_1 \longrightarrow V_2$ such that

$$T \circ \pi_1(g) = \pi_2(g) \circ T$$

for $g \in G$. We will denote by $\mathrm{Hom}_{\mathbb{C}}(V_1, V_2)$ the space of all linear transformations $V_1 \longrightarrow V_2$ and by $\mathrm{Hom}_G(V_1, V_2)$ the subspace of those that are intertwining maps.

For the remainder of this chapter, unless otherwise stated, G will denote a compact group.

Theorem 2.2 (Schur's lemma).

(i) Let (π_1, V_1) and (π_2, V_2) be irreducible representations, and let $T : V_1 \longrightarrow V_2$ be an intertwining operator. Then either T is zero or it is an isomorphism.

(ii) Suppose that (π, V) is an irreducible representation of G and $T : V \longrightarrow V$ is an intertwining operator. Then there exists a scalar $\lambda \in \mathbb{C}$ such that $T(v) = \lambda v$ for all $v \in V$.

Proof. For (i), the kernel of T is an invariant subspace of V_1, which is assumed irreducible, so if T is not zero, $\ker(T) = 0$. Thus, T is injective. Also, the image of T is an invariant subspace of V_2. Since V_2 is irreducible, if T is not zero, then $\mathrm{im}(T) = V_2$. Therefore T is bijective, so it is an isomorphism.

For (ii), let λ be any eigenvalue of T. Let $I : V \longrightarrow V$ denote the identity map. The linear transformation $T - \lambda I$ is an intertwining operator that is not an isomorphism, so it is the zero map by (i). $\square$

We are assuming that G is compact. The Haar volume of G is therefore finite, and we normalize the Haar measure so that the volume of G is 1.

We will consider the space $L^2(G)$ of functions on G that are square-integrable with respect to the Haar measure. This is a Hilbert space with the inner product

$$\langle f_1, f_2 \rangle_{L^2} = \int_G f_1(g)\, \overline{f_2(g)}\, dg.$$

Schur orthogonality will give us an orthonormal basis for this space.

If (π, V) is a representation and $\langle\, ,\, \rangle$ is an invariant inner product on V, then every linear functional is of the form $x \longrightarrow \langle x, v \rangle$ for some $v \in V$. Thus a matrix coefficient may be written in the form $g \longrightarrow \langle \pi(g)w, v \rangle$, and such a representation will be useful to us in our discussion of Schur orthogonality.

Lemma 2.1. *Suppose that (π_1, V_1) and (π_2, V_2) are complex representations of the compact group G. Let $\langle\, ,\, \rangle$ be any inner product on V_1. If $v_i, w_i \in V_i$, then the map $T : V_1 \longrightarrow V_2$ given by*

$$T(w) = \int_G \langle \pi_1(g)w, v_1 \rangle \, \pi_2(g^{-1})v_2\, dg \qquad (2.4)$$

is G-equivariant.

Proof. We have

$$T\big(\pi_1(h)w\big) = \int_G \langle \pi_1(gh)w, v_1 \rangle \, \pi_2(g^{-1})v_2\, dg.$$

The variable change $g \longrightarrow gh^{-1}$ shows that this equals $\pi_2(h)T(w)$, as required.
□

Theorem 2.3 (Schur orthogonality). *Suppose that (π_1, V_1) and (π_2, V_2) are irreducible representations of the compact group G. Either every matrix coefficient of π_1 is orthogonal in $L^2(G)$ to every matrix coefficient of π_2, or the representations are isomorphic.*

Proof. We must show that if there exist matrix coefficients $f_i : G \longrightarrow \mathbb{C}$ of π_i that are *not* orthogonal, then there is an isomorphism $T : V_1 \longrightarrow V_2$. We may assume that the f_i have the form $f_i(g) = \langle \pi_i(g)w_i, v_i \rangle$ since functions of that form span the spaces of matrix coefficients of the representations π_i. Here we use the notation $\langle\, ,\, \rangle$ to denote invariant bilinear forms on both V_1 and V_2, and $v_i, w_i \in V_i$. Then our assumption is that

$$\int_G \langle \pi_1(g)w_1, v_1 \rangle \langle \pi_2(g^{-1})v_2, w_2 \rangle \, dg = \int_G \langle \pi_1(g)w_1, v_1 \rangle \overline{\langle \pi_2(g)w_2, v_2 \rangle} \, dg \neq 0.$$

Define $T : V_1 \longrightarrow V_2$ by (2.4). The map is nonzero since the last inequality can be written $\langle T(w_1), w_2 \rangle \neq 0$. It is an isomorphism by Schur's lemma.   □

This gives orthogonality for matrix coefficients coming from *nonisomorphic* irreducible representations. But what about matrix coefficients from the same representation? (If the representations are isomorphic, we may as well assume they are equal.) The following result gives us an answer to this question.

Theorem 2.4 (Schur orthogonality). *Let (π, V) be an irreducible representation of the compact group G, with invariant inner product $\langle \, , \, \rangle$. Then there exists a constant $d > 0$ such that*

$$\int_G \langle \pi(g)w_1, v_1 \rangle \overline{\langle \pi(g)w_2, v_2 \rangle} \, dg = d^{-1} \langle w_1, w_2 \rangle \langle v_2, v_1 \rangle. \tag{2.5}$$

Later, in Proposition 2.9, we will show that $d = \dim(V)$.

Proof. We will show that if v_1 and v_2 are fixed, there exists a constant $c(v_1, v_2)$ such that

$$\int_G \langle \pi(g)w_1, v_1 \rangle \overline{\langle \pi(g)w_2, v_2 \rangle} \, dg = c(v_1, v_2) \langle w_1, w_2 \rangle. \tag{2.6}$$

Indeed, T given by (2.4) is G-equivariant, so by Schur's lemma it is a scalar. Thus, there is a constant $c = c(v_1, v_2)$ depending only on v_1 and v_2 such that $T(w) = cw$. In particular, $T(w_1) = cw_1$, and so the right-hand side of (2.6) equals

$$\langle T(w_1), w_2 \rangle = \int_G \langle \pi(g)w_1, v_1 \rangle \langle \pi(g^{-1})v_2, w_2 \rangle \, dg,$$

Now the variable change $g \longrightarrow g^{-1}$ and the properties of the inner product show that this equals the left-hand side of (2.6), proving the identity. The same argument shows that there exists another constant $c'(w_1, w_2)$ such that for all v_1 and v_2 we have

$$\int_G \langle \pi(g)w_1, v_1 \rangle \overline{\langle \pi(g)w_2, v_2 \rangle} \, dg = c'(w_1, w_2) \langle v_2, v_1 \rangle.$$

Combining this with (2.6), we get (2.5). We will compute d later in Proposition 2.9, but for now we simply note that it is positive since, taking $w_1 = w_2$ and $v_1 = v_2$, both the left-hand side of (2.5) and the two inner products on the right-hand side are positive. $\qquad\square$

Before we turn to the evaluation of the constant d, we will prove a different orthogonality for the characters of irreducible representations (Theorem 2.5). This will require some preparations.

Proposition 2.5. *The character χ of a representation (π, V) is a matrix coefficient of V.*

Proof. If $v_1, \ldots, v_n$ is a matrix of V, and $L_1, \ldots, L_n$ is the dual basis of V^*, then $\chi(g) = \sum_{i=1}^{n} L_i(\pi(g)v_i)$. $\qquad\square$

Proposition 2.6. *Suppose that (π, V) is a representation of G. Let χ be the character of π.*

(i) If $g \in V$ then $\chi(g^{-1}) = \overline{\chi(g)}$.

(ii) Let $(\hat{\pi}, V^*)$ be the contragredient representation of π. Then the character of $\hat{\pi}$ is the complex conjugate $\overline{\chi}$ of the character χ of G.

Proof. Since $\pi(g)$ is unitary with respect to an invariant inner product $\langle\,,\,\rangle$, its eigenvalues $t_1, \ldots, t_n$ all have absolute value 1, and so

$$\operatorname{tr} \pi(g)^{-1} = \sum_i t_i^{-1} = \sum_i \overline{t_i} = \overline{\chi(g)}.$$

This proves (i). As for (ii), referring to (2.3), $\hat{\pi}(g)$ is the adjoint of $\pi(g)^{-1}$ with respect to the dual pairing $[\![\,,\,]\!]$, so its trace equals the trace of $\pi(g)^{-1}$.   □

The *trivial representation* of any group G is the representation on a one-dimensional vector space V with $\pi(g)v = v$ being the trivial action.

Proposition 2.7. *If (π, V) is an irreducible representation and χ its character, then*

$$\int_G \chi(g)\,\mathrm{d}g = \begin{cases} 1 \text{ if } \pi \text{ is the trivial representation;} \\ 0 \text{ otherwise.} \end{cases}$$

Proof. The character of the trivial representation is just the constant function 1, and since we normalized the Haar measure so that G has volume 1, this integral is 1 if π is trivial. In general, we may regard $\int_G \chi(g)\,\mathrm{d}g$ as the inner product of χ with the character 1 of the trivial representation, and if π is nontrivial, these are matrix coefficients of different irreducible representations and hence orthogonal by Theorem 2.3.   □

If (π, V) is a representation, let V^G be the subspace of G-*invariants*, that is,

$$V^G = \{v \in V \mid \pi(g)v = v \text{ for all } g \in G\}.$$

Proposition 2.8. *If (π, V) is a representation of G and χ its character, then*

$$\int_G \chi(g)\,\mathrm{d}g = \dim(V^G).$$

Proof. Decompose $V = \oplus_i V_i$ into a direct sum of irreducible invariant subspaces, and let χ_i be the character of the restriction π_i of π to V_i. By Proposition 2.7, $\int_G \chi_i(g)\,\mathrm{d}g = 1$ if and only if π_i is trivial. Hence $\int_G \chi(g)\,\mathrm{d}g$ is the number of trivial π_i. The direct sum of the V_i with π_i trivial is V^G, and the statement follows.   □

If (π_1, V_1) and (π_2, V_2) are irreducible representations, and χ_1 and χ_2 are their characters, we have already noted in proving Proposition 2.3 that we may form representations $\pi_1 \oplus \pi_2$ and $\pi_1 \otimes \pi_2$ on $V_1 \oplus V_2$ and $V_1 \otimes V_2$. It is easy to see that $\chi_{\pi_1 \oplus \pi_2} = \chi_{\pi_1} + \chi_{\pi_2}$ and $\chi_{\pi_1 \otimes \pi_2} = \chi_{\pi_1} \chi_{\pi_2}$. It is not quite true that the characters form a ring. Certainly the negative of a matrix coefficient is a matrix coefficient, yet the negative of a character is not a character. The set of characters is closed under addition and multiplication but not subtraction. We define a *generalized* (or *virtual*) *character* to be a function of the form $\chi_1 - \chi_2$, where χ_1 and χ_2 are characters. It is now clear that the generalized characters form a ring.

Lemma 2.2. *Define a representation* $\Psi : \mathrm{GL}(n,\mathbb{C}) \times \mathrm{GL}(m,\mathbb{C}) \longrightarrow \mathrm{GL}(\Omega)$ *where* $\Omega = \mathrm{Mat}_{n\times m}(\mathbb{C})$ *by* $\Psi(g_1, g_2) : X \longrightarrow g_2 X g_1^{-1}$. *Then the trace of* $\Psi(g_1, g_2)$ *is* $\mathrm{tr}(g_1^{-1})\,\mathrm{tr}(g_2)$.

Proof. Both $\mathrm{tr}\,\Psi(g_1, g_2)$ and $\mathrm{tr}(g_1^{-1})\,\mathrm{tr}(g_2)$ are continuous, and since diagonalizable matrices are dense in $\mathrm{GL}(n,\mathbb{C})$ we may assume that both g_1 and g_2 are diagonalizable. Also if γ is invertible we have $\Psi(\gamma g_1 \gamma^{-1}, g_2) = \Psi(\gamma, 1)\Psi(g_1, g_2)\Psi(\gamma, 1)^{-1}$ so the trace of both $\mathrm{tr}\,\Psi(g_1, g_2)$ and $\mathrm{tr}(g_1^{-1})\mathrm{tr}(g_2)$ are unchanged if g_1 is replaced by $\gamma g_1 \gamma^{-1}$. So we may assume that g_1 is diagonal, and similarly g_2. Now if $\alpha_1, \ldots, \alpha_n$ and $\beta_1, \ldots, \beta_m$ are the diagonal entries of g_1 and g_2^{-1}, the effect of $\Psi(g_1, g_2)$ on $X \in \Omega$ is to multiply the columns by the α_i^{-1} and the rows by the β_j. So the trace is $\mathrm{tr}(g_1^{-1})\mathrm{tr}(g_2)$. □

Theorem 2.5 (Schur orthogonality). *Let* (π_1, V_1) *and* (π_2, V_2) *be representations of* G *with characters* χ_1 *and* χ_2. *Then*

$$\int_G \chi_1(g)\,\overline{\chi_2(g)}\,\mathrm{d}g = \dim\mathrm{Hom}_G(V_1, V_2). \tag{2.7}$$

If π_1 *and* π_2 *are irreducible, then*

$$\int_G \chi_1(g)\overline{\chi_2(g)}\,\mathrm{d}g = \begin{cases} 1 & \text{if } \pi_1 \cong \pi_2; \\ 0 & \text{otherwise.} \end{cases}$$

Proof. Define a representation Π of G on the space $\Omega = \mathrm{Hom}_{\mathbb{C}}(V_1, V_2)$ of all linear transformations $T : V_1 \longrightarrow V_2$ by

$$\Pi(g)T = \pi_2(g) \circ T \circ \pi_1(g)^{-1}.$$

By lemma 2.2 and Proposition 2.6, the character of $\Pi(g)$ is $\chi_2(g)\overline{\chi_1(g)}$. The space of invariants Ω^G exactly of the T which are G-module homomorphisms, so by Proposition 2.8 we get

$$\int_G \overline{\chi_1(g)}\,\chi_2(g)\,\mathrm{d}g = \dim\mathrm{Hom}_G(V_1, V_2).$$

Since this is real, we may conjugate to obtain (2.7). □

Proposition 2.9. *The constant d in Theorem 2.4 equals* $\dim(V)$.

Proof. Let $v_1, \ldots, v_n$ be an orthonormal basis of V, $n = \dim(V)$. We have

$$\chi(g) = \sum_i \langle \pi_i(g) v_i, v_i \rangle$$

since $\langle \pi(g) v_j, v_i \rangle$ is the i, j component of the matrix of $\pi(g)$ with respect to this basis. Now

$$1 = \int_G |\chi(g)|^2 \, dg = \sum_{i,j} \int_G \langle \pi(g) v_i, v_i \rangle \overline{\langle \pi(g) v_j, v_j \rangle} \, dg.$$

There are n^2 terms on the right, but by (2.5) only the terms with $i = j$ are nonzero, and those equal d^{-1}. Thus, $d = n$. $\qquad\square$

We now return to the matrix coefficients $\mathcal{M}_\pi$ of an irreducible representation (π, V). We define a representation Θ of $G \times G$ on $\mathcal{M}_\pi$ by

$$\Theta(g_1, g_2) f(x) = f(g_2^{-1} x g_1).$$

We also have a representation Π of $G \times G$ on $\mathrm{End}_\mathbb{C}(V)$ by

$$\Pi(g_1, g_2) T = \pi(g_2)^{-1} T \pi(g_1).$$

Proposition 2.10. *If $f \in \mathcal{M}_\pi$ then so is $\Theta(g_1, g_2) f$. The representations Θ and Π are equivalent.*

Proof. Let $L \in V^*$ and $v \in V$. Define $f_{L,v}(g) = L(\pi(g) v)$. The map $L, v \longmapsto f_{L,v}$ is bilinear, hence induces a linear map $\sigma : V^* \otimes V \longrightarrow \mathcal{M}_\pi$. It is surjective by the definition of $\mathcal{M}_\pi$, and it follows from Proposition 2.4 that if L_i and v_j run through orthonormal bases, then f_{L_i, v_j} are orthonormal, hence linearly independent. Therefore, σ is a vector space isomorphism. We have

$$\Theta(g_1, g_2) f_{L,v}(g) = L(g_2^{-1} g g_1 v) = f_{\hat{\pi}(g_2) L, \pi(g_1) v}(x),$$

where we recall that $(\hat{\pi}, V^*)$ is the contragredient representation. This means that σ is a $G \times G$-module homomorphism and so $\mathcal{M}_\pi \cong V^* \otimes V$ as $G \times G$-modules. On the other hand we also have a bilinear map $V^* \times V \longrightarrow \mathrm{End}_\mathbb{C}(V)$ that associates with (L, v) the rank-one linear map $T_{L,v}(u) = L(u) v$. This induces an isomorphism $V^* \otimes V \longrightarrow \mathrm{End}_\mathbb{C}(V)$ which is $G \times G$ equivariant. We see that $\mathcal{M}_\pi \cong V^* \otimes V \cong \mathrm{End}_\mathbb{C}(V)$. $\qquad\square$

A function f on G is called a *class function* if it is constant on conjugacy classes, that is, if it satisfies the equation $f(hgh^{-1}) = f(g)$. The character of a representation is a class function since the trace of a linear transformation is unchanged by conjugation.

Proposition 2.11. *If f is the matrix coefficient of an irreducible representation (π, V), and if f is a class function, then f is a constant multiple of χ_π.*

Proof. By Schur's lemma, there is a unique G-invariant vector in $\mathrm{Hom}_{\mathbb{C}}(V, V)$; hence. by Proposition 2.10, the same is true of $\mathcal{M}_{\pi}$ in the action of G by conjugation. This matrix coefficient is of course χ_{π}. $\qquad\square$

Theorem 2.6. *If f is a matrix coefficient and also a class function, then f is a finite linear combination of characters of irreducible representations.*

Proof. Write $f = \sum_{i=1}^{n} f_i$, where each f_i is a class function of a distinct irreducible representation (π_i, V_i). Since f is conjugation-invariant, and since the f_i live in spaces $\mathcal{M}_{\pi_i}$, which are conjugation-invariant and mutually orthogonal, each f_i is itself a class function and hence a constant multiple of χ_{π_i} by Proposition 2.11. $\qquad\square$

Exercises

Exercise 2.1. Suppose that G is a compact Abelian group and $\pi : G \longrightarrow \mathrm{GL}(n, \mathbb{C})$ an irreducible representation. Prove that $n = 1$.

Exercise 2.2. Suppose that G is compact group and $f : G \longrightarrow \mathbb{C}$ is the matrix coefficient of an irreducible representation π. Show that $g \longmapsto f(g^{-1})$ is a matrix coefficient of the same representation π.

Exercise 2.3. Suppose that G is compact group. Let $C(G)$ be the space of continuous functions on G. If f_1 and $f_2 \in C(G)$, define the convolution $f_1 * f_2$ of f_1 and f_2 by

$$(f_1 * f_2)(g) = \int_G f_1(gh^{-1}) \, f_2(h) \, \mathrm{d}h = \int_G f_1(h) \, f_2(h^{-1}g) \, \mathrm{d}h.$$

(i) Use the variable change $h \longrightarrow h^{-1}g$ to prove the identity of the last two terms. Prove that this operation is associative, and so $C(G)$ is a ring (without unit) with respect to covolution.

(ii) Let π be an irreducible representation. Show that the space $\mathcal{M}_{\pi}$ of matrix coefficients of π is a 2-sided ideal in $C(G)$, and explain how this fact implies Theorem 2.3.

Exercise 2.4. Let G be a compact group, and let $G \times G$ act on the space $\mathcal{M}_{\pi}$ by left and right translation: $(g, h)f(x) = f(g^{-1}xh)$. Show that $\mathcal{M}_{\pi} \cong \hat{\pi} \otimes \pi$ as $(G \times G)$-modules.

Exercise 2.5. Let G be a compact group and let $g, h \in G$. Show that g and h are conjugate if and only if $\chi(g) = \chi(h)$ for every irreducible character χ. Show also that every character is real-valued if and only if every element is conjugate to its inverse.

Exercise 2.6. Let G be a compact group, and let V, W be irreducible G-modules. An *invariant* bilinear form $B : V \times W \to \mathbb{C}$ is one that satisfies $B(g \cdot v, g \cdot w) = B(v, w)$ for $g \in G$, $v \in V$, $w \in W$. Show that the space of invariant bilinear forms is at most one-dimensional, and is one-dimensional if and only if V and W are contragredient.

3

Compact Operators

If $\mathfrak{H}$ is a normed vector space, a linear operator $T : \mathfrak{H} \to \mathfrak{H}$ is called *bounded* if there exists a constant C such that $|Tx| \leqslant C|x|$ for all $x \in \mathfrak{H}$. In this case, the smallest such C is called the *operator norm* of T, and is denoted $|T|$. The boundedness of the operator T is equivalent to its continuity. If $\mathfrak{H}$ is a Hilbert space, then a bounded operator T is *self-adjoint* if

$$\langle Tf, g \rangle = \langle f, Tg \rangle$$

for all $f, g \in \mathfrak{H}$. As usual, we call f an *eigenvector* with *eigenvalue* λ if $f \neq 0$ and $Tf = \lambda f$. Given λ, the set of eigenvectors with eigenvalue λ (together with 0, which is not an eigenvector) is called the λ-*eigenspace*. It follows from elementary and well-known arguments that if T is a self-adjoint bounded operator, then its eigenvalues are real, and the eigenspaces corresponding to distinct eigenvalues are orthogonal. Moreover, if $V \subset \mathfrak{H}$ is a subspace such that $T(V) \subset V$, it is easy to see that also $T(V^{\perp}) \subset V^{\perp}$.

A bounded operator $T : \mathfrak{H} \to \mathfrak{H}$ is *compact* if whenever $\{x_1, x_2, x_3, \ldots\}$ is any bounded sequence in $\mathfrak{H}$, the sequence $\{Tx_1, Tx_2, \ldots\}$ has a convergent subsequence.

Theorem 3.1 (Spectral theorem for compact operators). *Let T be a compact self-adjoint operator on a Hilbert space $\mathfrak{H}$. Let $\mathfrak{N}$ be the nullspace of T. Then the Hilbert space dimension of $\mathfrak{N}^{\perp}$ is at most countable. $\mathfrak{N}^{\perp}$ has an orthonormal basis ϕ_i $(i = 1, 2, 3, \ldots)$ of eigenvectors of T so that $T\phi_i = \lambda_i \phi_i$. If $\mathfrak{N}^{\perp}$ is not finite-dimensional, the eigenvalues $\lambda_i \to 0$ as $i \to \infty$.*

Since the eigenvalues $\lambda_i \to 0$, if λ is any nonzero eigenvalue, it follows from this statement that the λ-eigenspace is finite-dimensional.

Proof. This depends upon the equality

$$|T| = \sup_{0 \neq x \in \mathfrak{H}} \frac{|\langle Tx, x \rangle|}{\langle x, x \rangle}. \tag{3.1}$$

D. Bump, *Lie Groups*, Graduate Texts in Mathematics 225, DOI 10.1007/978-1-4614-8024-2_3, © Springer Science+Business Media New York 2013

To prove this, let B denote the right-hand side. If $0 \neq x \in \mathfrak{H}$,

$$| \langle Tx, x \rangle | \leqslant |Tx| \cdot |x| \leqslant |T| \cdot |x|^2 = |T| \cdot \langle x, x \rangle \, ,$$

so $B \leqslant |T|$. We must prove the converse. Let $\lambda > 0$ be a constant, to be determined later. Using $\langle T^2 x, x \rangle = \langle Tx, Tx \rangle$, we have

$$
\begin{aligned}
&\langle Tx, Tx \rangle \\
&= \tfrac{1}{4} \left| \langle T(\lambda x + \lambda^{-1} Tx), \lambda x + \lambda^{-1} Tx \rangle - \langle T(\lambda x - \lambda^{-1} Tx), \lambda x - \lambda^{-1} Tx \rangle \right| \\
&\leqslant \tfrac{1}{4} \left| \langle T(\lambda x + \lambda^{-1} Tx), \lambda x + \lambda^{-1} Tx \rangle \right| + \left| \langle T(\lambda x - \lambda^{-1} Tx), \lambda x - \lambda^{-1} Tx \rangle \right| \\
&\leqslant \tfrac{1}{4} \left[B \langle \lambda x + \lambda^{-1} Tx, \lambda x + \lambda^{-1} Tx \rangle + B \langle \lambda x - \lambda^{-1} Tx, \lambda x - \lambda^{-1} Tx \rangle \right] \\
&= \tfrac{B}{2} \left[\lambda^2 \langle x, x \rangle + \lambda^{-2} \langle Tx, Tx \rangle \right] .
\end{aligned}
$$

Now taking $\lambda = \sqrt{|Tx|/|x|}$, we obtain

$$|Tx|^2 = \langle Tx, Tx \rangle \leqslant B |x| \, |Tx|,$$

so $|Tx| \leqslant B|x|$, which implies that $|T| \leqslant B$, whence (3.1).

We now prove that $\mathfrak{N}^\perp$ has an orthonormal basis consisting of eigenvectors of T. It is an easy consequence of self-adjointness that $\mathfrak{N}^\perp$ is T-stable. Let Σ be the set of all orthonormal subsets of $\mathfrak{N}^\perp$ whose elements are eigenvectors of T. Ordering Σ by inclusion, Zorn's lemma implies that it has a maximal element S. Let V be the closure of the linear span of S. We must prove that $V = \mathfrak{N}^\perp$. Let $\mathfrak{H}_0 = V^\perp$. We wish to show $\mathfrak{H}_0 = \mathfrak{N}$. It is obvious that $\mathfrak{N} \subseteq \mathfrak{H}_0$. To prove the opposite inclusion, note that $\mathfrak{H}_0$ is stable under T, and T induces a compact self-adjoint operator on $\mathfrak{H}_0$. What we must show is that $T|\mathfrak{H}_0 = 0$. If T has a nonzero eigenvector in $\mathfrak{H}_0$, this will contradict the maximality of Σ. It is therefore sufficient to show that *a compact self-adjoint operator on a nonzero Hilbert space has an eigenvector.*

Replacing $\mathfrak{H}$ by $\mathfrak{H}_0$, we are therefore reduced to the easier problem of showing that if $T \neq 0$, then T has a nonzero eigenvector. By (3.1), there is a sequence $x_1, x_2, x_3, \ldots$ of unit vectors such that $| \langle Tx_i, x_i \rangle | \to |T|$. Observe that if $x \in \mathfrak{H}$, we have

$$\langle Tx, x \rangle = \langle x, Tx \rangle = \overline{\langle Tx, x \rangle}$$

so the $\langle Tx_i, x_i \rangle$ are real; we may therefore replace the sequence by a subsequence such that $\langle Tx_i, x_i \rangle \to \lambda$, where $\lambda = \pm |T|$. Since $T \neq 0$, $\lambda \neq 0$. Since T is compact, there exists a further subsequence $\{x_i\}$ such that Tx_i converges to a vector v. We will show that $x_i \to \lambda^{-1} v$.

Observe first that

$$| \langle Tx_i, x_i \rangle | \leqslant |Tx_i| \, |x_i| = |Tx_i| \leqslant |T| \, |x_i| = |\lambda|,$$

and since $\langle Tx_i, x_i \rangle \to \lambda$, it follows that $|Tx_i| \to |\lambda|$. Now

$$|\lambda\, x_i - Tx_i|^2 = \langle \lambda\, x_i - Tx_i, \lambda\, x_i - Tx_i \rangle = \lambda^2 |x_i|^2 + |Tx_i|^2 - 2\lambda \langle Tx_i, x_i \rangle,$$

and since $|x_i| = 1$, $|Tx_i| \to |\lambda|$, and $\langle Tx_i, x_i \rangle \to \lambda$, this converges to 0. Since $Tx_i \to v$, the sequence λx_i therefore also converges to v, and $x_i \to \lambda^{-1} v$. Now, by continuity, $Tx_i \to \lambda^{-1} Tv$, so $v = \lambda^{-1} Tv$. This proves that v is an eigenvector with eigenvalue λ. This completes the proof that $\mathfrak{N}^{\perp}$ has an orthonormal basis consisting of eigenvectors.

Now let $\{\phi_i\}$ be this orthonormal basis and let λ_i be the corresponding eigenvalues. If $\epsilon > 0$ is given, only finitely many $|\lambda_i| > \epsilon$ since otherwise we can find an infinite sequence of ϕ_i with $|T\phi_i| > \epsilon$. Such a sequence will have no convergent subsequence, contradicting the compactness of T. Thus, $\mathfrak{N}^{\perp}$ is countable-dimensional, and we may arrange the $\{\phi_i\}$ in a sequence. If it is infinite, we see the $\lambda_i \longrightarrow 0$. $\qquad\square$

Proposition 3.1. *Let X and Y be compact topological spaces with Y a metric space with distance function d. Let U be a set of continuous maps $X \longrightarrow Y$ such that for every $x \in X$ and every $\epsilon > 0$ there exists a neighborhood N of x such that $d\big(f(x), f(x')\big) < \epsilon$ for all $x' \in N$ and for all $f \in U$. Then every sequence in U has a uniformly convergent subsequence.*

We refer to the hypothesis on U as *equicontinuity*.

Proof. Let $S_0 = \{f_1, f_2, f_3, \ldots\}$ be a sequence in U. We will show that it has a convergent subsequence. We will construct a subsequence that is uniformly Cauchy and hence has a limit. For every $n > 1$, we will construct a subsequence $S_n = \{f_{n1}, f_{n2}, f_{n3}, \ldots\}$ of S_{n-1} such that $\sup_{x \in X} d(f_{ni}(x), f_{nj}(x)) \leqslant 1/n$.

Assume that S_{n-1} is constructed. For each $x \in X$, equicontinuity guarantees the existence of an open neighborhood N_x of x such that $d\big(f(y), f(x)\big) \leqslant \frac{1}{3n}$ for all $y \in N_x$ and all $f \in X$. Since X is compact, we can cover X by a finite number of these sets, say $N_{x_1}, \ldots, N_{x_m}$. Since the $f_{n-1,i}$ take values in the compact space Y, the m-tuples $\big(f_{n-1,i}(x_1), \ldots, f_{n-1,i}(x_m)\big)$ have an accumulation point, and we may therefore select the subsequence $\{f_{ni}\}$ such that $d\big(f_{ni}(x_k), f_{nj}(x_k)\big) \leqslant \frac{1}{3n}$ for all i, j and $1 \leqslant k \leqslant m$. Then for any y, there exists x_k such that $y \in N_{x_k}$ and

$$d\big(f_{ni}(y), f_{nj}(y)\big) \leqslant d\big(f_{ni}(y), f_{ni}(x_k)\big) + d\big(f_{ni}(x_k), f_{nj}(x_k)\big)$$
$$+ d\big(f_{nj}(y), f_{nj}(x_k)\big) \leqslant \tfrac{1}{3n} + \tfrac{1}{3n} + \tfrac{1}{3n} = \tfrac{1}{n}.$$

This completes the construction of the sequences $\{f_{ni}\}$.

The diagonal sequence $\{f_{11}, f_{22}, f_{33}, \ldots\}$ is uniformly Cauchy. Since Y is a compact metric space, it is complete, and so this sequence is uniformly convergent. $\qquad\square$

We topologize $C(X)$ by giving it the L^∞ norm $|\ |_\infty$ (sup norm).

Proposition 3.2 (Ascoli and Arzela). *Suppose that X is a compact space and that $U \subset C(X)$ is a bounded subset such that for each $x \in X$ and $\epsilon > 0$ there is a neighborhood N of x such that $|f(x) - f(y)| \leqslant \epsilon$ for all $y \in N$ and all $f \in U$. Then every sequence in U has a uniformly convergent subsequence.*

Again, the hypothesis on U is called *equicontinuity*.

Proof. Since U is bounded, there is a compact interval $Y \subset \mathbb{R}$ such that all functions in U take values in Y. The result follows from Proposition 3.1. $\square$

Exercises

Exercise 3.1. Suppose that T is a bounded operator on the Hilbert space $\mathfrak{H}$, and suppose that for each $\epsilon > 0$ there exists a compact operator T_ϵ such that $|T - T_\epsilon| < \epsilon$. Show that T is compact. (Use a diagonal argument like the proof of Proposition 3.1.)

Exercise 3.2 (Hilbert–Schmidt operators). Let X be a locally compact Hausdorff space with a positive Borel measure μ. Assume that $L^2(X)$ has a countable basis. Let $K \in L^2(X \times X)$. Consider the operator on $L^2(X)$ with kernel K defined by

$$Tf(x) = \int_X K(x, y) \, f(y) \, d\mu(y).$$

Let ϕ_i be an orthonormal basis of $L^2(X)$. Expand K in a Fourier expansion:

$$K(x, y) = \sum_{i=1}^{\infty} \psi_i(x) \, \overline{\phi_i(y)}, \qquad \psi_i = T\phi_i.$$

Show that $\sum |\psi_i|^2 = \int \int |K(x, y)|^2 d\mu(x) \, d\mu(y) < \infty$. Consider the operator T_N with kernel

$$K_N(x, y) = \sum_{i=1}^{N} \psi_i(x) \, \overline{\phi_i(y)}.$$

Show that T_N is compact, and deduce that T is compact.

4

The Peter–Weyl Theorem

In this chapter, we assume that G is a compact group. Let $C(G)$ be the convolution ring of continuous functions on G. It is a ring (without unit unless G is finite) under the multiplication of *convolution*:

$$(f_1 * f_2)(g) = \int_G f_1(gh^{-1}) \, f_2(h) \, dh = \int_G f_1(h) \, f_2(h^{-1}g) \, dh.$$

(Use the variable change $h \longrightarrow h^{-1}g$ to prove the identity of the last two terms. See Exercise 2.3.) We will sometimes define $f_1 * f_2$ by this formula even if f_1 and f_2 are not assumed continuous. For example, we will make use of the convolution defined this way if $f_1 \in L^\infty(G)$ and $f_2 \in L^1(G)$, or vice versa.

Since G has total volume 1, we have inequalities (where $|\ |_p$ denotes the L^p norm, $1 \leqslant p \leqslant \infty$)

$$|f|_1 \leqslant |f|_2 \leqslant |f|_\infty. \tag{4.1}$$

The second inequality is trivial, and the first is Cauchy–Schwarz:

$$|f|_1 = \langle |f|, 1 \rangle \leqslant |f|_2 \cdot |1|_2 = |f|_2.$$

(Here $|f|$ means the function $|f|(x) = |f(x)|$.)

If $\phi \in C(G)$ let T_ϕ be left convolution with ϕ. Thus,

$$(T_\phi f)(g) = \int_G \phi(gh^{-1}) \, f(h) \, dh.$$

Proposition 4.1. *If $\phi \in C(G)$, then T_ϕ is a bounded operator on $L^1(G)$. If $f \in L^1(G)$, then $T_\phi f \in L^\infty(G)$ and*

$$|T_\phi f|_\infty \leqslant |\phi|_\infty |f|_1. \tag{4.2}$$

D. Bump, *Lie Groups*, Graduate Texts in Mathematics 225,
DOI 10.1007/978-1-4614-8024-2_4, © Springer Science+Business Media New York 2013

Proof. If $f \in L^1(G)$, then

$$|T_\phi f|_\infty = \sup_{g \in G} \left| \int_G \phi(gh^{-1}) f(h) \, dh \right| \leqslant |\phi|_\infty \int_G |f(h)| \, dh,$$

proving (4.2). Using (4.1), it follows that the operator T_ϕ is bounded. In fact, (4.1) shows that it is bounded in each of the three metrics $|\ |_1$, $|\ |_2$, $|\ |_\infty$. $\square$

Proposition 4.2. *If $\phi \in C(G)$, then convolution with ϕ is a bounded operator T_ϕ on $L^2(G)$ and $|T_\phi| \leqslant |\phi|_\infty$. The operator T_ϕ is compact, and if $\phi(g^{-1}) = \overline{\phi(g)}$, it is self-adjoint.*

Proof. Using (4.1), $L^\infty(G) \subset L^2(G) \subset L^1(G)$, and by (4.2), $|T_\phi f|_2 \leqslant |T_\phi f|_\infty \leqslant |\phi|_\infty |f|_1 \leqslant |\phi|_\infty |f|_2$, so the operator norm $|T_\phi| \leqslant |\phi|_\infty$.

By (4.1), the unit ball in $L^2(G)$ is contained in the unit ball in $L^1(G)$, so it is sufficient to show that $\mathfrak{B} = \{T_\phi f \,|\, f \in L^1(G), |f|_1 \leqslant 1\}$ is sequentially compact in $L^2(G)$. Also, by (4.1), it is sufficient to show that it is sequentially compact in $L^\infty(G)$, that is, in $C(G)$, whose topology is induced by the $L^\infty(G)$ norm. It follows from (4.2) that $\mathfrak{B}$ is bounded. We show that it is equicontinuous. Since ϕ is continuous and G is compact, ϕ is uniformly continuous. This means that given $\epsilon > 0$ there is a neighborhood N of the identity such that $|\phi(kg) - \phi(g)| < \epsilon$ for all g when $k \in N$. Now, if $f \in L^1(G)$ and $|f|_1 \leqslant 1$, we have, for all g,

$$|(\phi * f)(kg) - (\phi * f)(g)| = \left| \int_G \left[\phi(kgh^{-1}) - \phi(gh^{-1}) \right] f(h) \, dh \right|$$

$$\leqslant \int_G |\phi(kgh^{-1}) - \phi(gh^{-1})| \, |f(h)| \, dh \leqslant \epsilon |f|_1 \leqslant \epsilon.$$

This proves equicontinuity, and sequential compactness of $\mathfrak{B}$ now follows by the Ascoli–Arzela lemma (Proposition 3.2).

If $\phi(g^{-1}) = \overline{\phi(g)}$, then

$$\langle T_\phi f_1, f_2 \rangle = \int_G \int_G \phi(gh^{-1}) f_1(h) \, \overline{f_2(g)} \, dg \, dh$$

while

$$\langle f_1, T_\phi f_2 \rangle = \int_G \int_G \overline{\phi(hg^{-1})} f_1(h) \, \overline{f_2(g)} \, dg \, dh.$$

These are equal, so T is self-adjoint. $\square$

If $g \in G$, let $(\rho(g)f)(x) = f(xg)$ be the right translate of f by g.

Proposition 4.3. *If $\phi \in C(G)$, and $\lambda \in \mathbb{C}$, the λ-eigenspace*

$$V(\lambda) = \{f \in L^2(G) \,|\, T_\phi f = \lambda f\}$$

is invariant under $\rho(g)$ for all $g \in G$.

Proof. Suppose $T_\phi f = \lambda f$. Then

$$(T_\phi \rho(g)f)(x) = \int_G \phi(xh^{-1})\, f(hg)\, \mathrm{d}h.$$

After the change of variables $h \longrightarrow hg^{-1}$, this equals

$$\int_G \phi(xgh^{-1})\, f(h)\, \mathrm{d}h = \rho(g)(T_\phi f)(x) = \lambda\rho(g)f(x).$$

$\square$

Theorem 4.1 (Peter and Weyl). *The matrix coefficients of G are dense in $C(G)$.*

Proof. Let $f \in C(G)$. We will prove that there exists a matrix coefficient f' such that $|f - f'|_\infty < \epsilon$ for any given $\epsilon > 0$.

Since G is compact, f is uniformly continuous. This means that there exists an open neighborhood U of the identity such that if $g \in U$, then $|\lambda(g)f - f|_\infty < \epsilon/2$, where $\lambda : G \to \mathrm{End}(C(G))$ is the action by left translation: $(\lambda(g)f)(h) = f(g^{-1}h)$. Let ϕ be a nonnegative function supported in U such that $\int_G \phi(g)\, \mathrm{d}g = 1$. We may arrange that $\phi(g) = \phi(g^{-1})$ so that the operator T_ϕ is self-adjoint as well as compact. We claim that $|T_\phi f - f|_\infty < \epsilon/2$. Indeed, if $h \in G$,

$$\begin{aligned}
\left|(\phi * f)(h) - f(h)\right| &= \left|\int_G [\phi(g)\, f(g^{-1}h) - \phi(g)f(h)]\, \mathrm{d}g\right| \\
&\leqslant \int_U \phi(g)\, \left|f(g^{-1}h) - f(h)\right| \mathrm{d}g \\
&\leqslant \int_U \phi(g)\, |\lambda(g)f - f|_\infty\, \mathrm{d}g \\
&\leqslant \int_U \phi(g)\, (\epsilon/2)\, \mathrm{d}g = \frac{\epsilon}{2}.
\end{aligned}$$

By Proposition 4.1, T_ϕ is a compact operator on $L^2(G)$. If λ is an eigenvalue of T_ϕ, let $V(\lambda)$ be the λ-eigenspace. By the spectral theorem, the spaces $V(\lambda)$ are finite-dimensional [except perhaps $V(0)$], mutually orthogonal, and they span $L^2(G)$ as a Hilbert space. By Proposition 4.3 they are T_ϕ-invariant. Let f_λ be the projection of f on $V(\lambda)$. Orthogonality of the f_λ implies that

$$\sum_\lambda |f_\lambda|_2^2 = |f|_2^2 < \infty. \tag{4.3}$$

Let

$$f' = T_\phi(f''), \qquad f'' = \sum_{|\lambda| > q} f_\lambda,$$

where $q > 0$ remains to be chosen. We note that f' and f'' are both contained in $\bigoplus_{|\lambda|>q} V(\lambda)$, which is a finite-dimensional vector space, and closed under right translation by Proposition 4.3, and by Theorem 2.1, it follows that they are matrix coefficients.

By (4.3), we may choose q so that $\sum_{0<q<|\lambda|} |f_\lambda|_2^2$ is as small as we like. Using (4.1) may thus arrange that

$$\left| \sum_{0<|\lambda|<q} f_\lambda \right|_1 \leqslant \left| \sum_{0<|\lambda|<q} f_\lambda \right|_2 = \sqrt{\sum_{0<|\lambda|<q} |f_\lambda|_2^2} < \frac{\epsilon}{2|\phi|_\infty}. \qquad (4.4)$$

We have

$$T_\phi(f - f'') = T_\phi\left(f_0 + \sum_{0<|\lambda|<q} f_\lambda \right) = T_\phi\left(\sum_{0<|\lambda|<q} f_\lambda \right).$$

Using (4.2) and (4.4) we have $|T_\phi(f - f'')|_\infty \leqslant \epsilon/2$. Now

$$|f - f'|_\infty = |f - T_\phi f + T_\phi(f - f'')| \leqslant |f - T_\phi f| + |T_\phi f - T_\phi f''|$$
$$\leqslant \tfrac{\epsilon}{2} + \tfrac{\epsilon}{2} = \epsilon.$$

$\square$

Corollary 4.1. *The matrix coefficients of G are dense in $L^2(G)$.*

Proof. Since $C(G)$ is dense in $L^2(G)$, this follows from the Peter–Weyl theorem and (4.1). $\square$

We say that a topological group G has *no small subgroups* if it has a neighborhood U of the identity such that the only subgroup of G contained in U is just $\{1\}$. For example, we will see that Lie groups have no small subgroups. On the other hand, some groups, such as $\mathrm{GL}(n, \mathbb{Z}_p)$ where $\mathbb{Z}_p$ is the ring of p-adic integers, have a neighborhood basis at the identity consisting of open subgroups. Such a group is called *totally disconnected*, and for such a group the no small subgroups property fails very strongly.

A representation is called *faithful* if its kernel is trivial.

Theorem 4.2. *Let G be a compact group that has no small subgroups. Then G has a faithful finite-dimensional representation.*

Proof. Let U be a neighborhood of the identity that contains no subgroup but $\{1\}$. By the Peter–Weyl theorem, we can find a finite-dimensional representation π and a matrix coefficient f such that $f(1) = 0$ but $f(g) > 1$ when $g \notin U$. The function f is constant on the kernel of π, so that kernel is contained in U. It follows that the kernel is trivial. $\square$

We will now prove a fact about infinite-dimensional representations of a compact group G. The Peter–Weyl Theorem amounts to a "completeness" of the finite-dimensional representations from the point of view of harmonic analysis. One aspect of this is the L^2 completeness asserted in Corollary 4.1. Another aspect, which we now prove, is that there are no irreducible unitary infinite-dimensional representations. From the point of view of harmonic analysis, these two statements are closely related and are in fact equivalent. Representation theory and Fourier analysis on groups are essentially the same thing.

If H is a Hilbert space, a representation $\pi : G \longrightarrow \mathrm{End}(H)$ is called *unitary* if $\langle \pi(g)v, \pi(g)w \rangle = \langle v, w \rangle$ for all $v, w \in H$, $g \in G$. It is also assumed that the map $(g, v) \longmapsto \pi(g)v$ from $G \times H \longrightarrow H$ is continuous.

Theorem 4.3 (Peter and Weyl). *Let H be a Hilbert space and G be a compact group. Let $\pi : G \longrightarrow \mathrm{End}(H)$ be a unitary representation. Then H is a direct sum of finite-dimensional irreducible representations.*

Proof. We first show that if H is nonzero then it has an irreducible finite-dimensional invariant subspace. We choose a nonzero vector $v \in H$. Let N be a neighborhood of the identity of G such that if $g \in N$ then $|\pi(g)v - v| \leqslant |v|/2$. We can find a nonnegative continuous function ϕ on G supported in N such that $\int_G \phi(g)\, dg = 1$.

We claim that $\int_G \phi(g)\, \pi(g)v\, dg \neq 0$. This can be proved by taking the inner product with v. Indeed

$$\left\langle \int_G \phi(g)\, \pi(g)v\, dg, v \right\rangle = \langle v, v \rangle - \left\langle \int_N \phi(g)\big(v - \pi(g)v\big)\, dg, v \right\rangle \qquad (4.5)$$

and

$$\left| \left\langle \int_N \phi(g)\big(v - \pi(g)v\big)\, dg, v \right\rangle \right| \leqslant \int_N |v - \pi(g)v|\, dg \cdot |v| \leqslant |v|^2/2.$$

Thus, the two terms in (4.5) differ in absolute value and cannot cancel.

Next, using the Peter–Weyl theorem, we may find a matrix coefficient f such that $|f - \phi|_\infty < \epsilon$, where ϵ can be chosen arbitrarily. We have

$$\left| \int_G (f - \phi)(g)\, \pi(g)v\, dg \right| \leqslant \epsilon |v|,$$

so if ϵ is sufficiently small we have $\int_G f(g)\, \pi(g)v\, dg \neq 0$.

Since f is a matrix coefficient, so is the function $g \longmapsto f(g^{-1})$ by Proposition 2.4. Thus, let (ρ, W) be a finite-dimensional representation with $w \in W$ and $L : W \longrightarrow \mathbb{C}$ a linear functional such that $f(g^{-1}) = L(\rho(g)w)$. Define a map $T : W \longrightarrow H$ by

$$T(x) = \int_G L\big(\rho(g^{-1})x\big)\, \pi(g)v\, dg.$$

This is an intertwining map by the same argument used to prove (2.4). It is nonzero since $T(w) = \int f(g)\,\pi(g)v\,dg \neq 0$. Since W is finite-dimensional, the image of T is a nonzero finite-dimensional invariant subspace.

We have proven that every nonzero unitary representation of G has a nonzero finite-dimensional invariant subspace, which we may obviously assume to be irreducible. From this we deduce the stated result. Let (π, H) be a unitary representation of G. Let Σ be the set of all sets of orthogonal finite-dimensional irreducible invariant subspaces of H, ordered by inclusion. Thus, if $S \in \Sigma$ and $U, V \in S$, then U and V are finite-dimensional irreducible invariant subspaces, If $U \neq V$. then $U \perp V$. By Zorn's lemma, Σ has a maximal element S and we are done if S spans H as a Hilbert space. Otherwise, let H' be the orthogonal complement of the span of S. By what we have shown, H' contains an invariant irreducible subspace. We may append this subspace to S, contradicting its maximality. $\square$

Exercises

Exercise 4.1. Let G be totally disconnected, and let $\pi : G \longrightarrow \mathrm{GL}(n, \mathbb{C})$ be a finite-dimensional representation. Show that the kernel of π is open. (**Hint:** Use the fact that $\mathrm{GL}(n, \mathbb{C})$ has no small subgroups.) Conclude (in contrast with Theorem 4.2) that the compact group $\mathrm{GL}(n, \mathbb{Z}_p)$ has no faithful finite-dimensional representation.

Exercise 4.2. Suppose that G is a compact Abelian group and $H \subset G$ a closed subgroup. Let $\chi : H \longrightarrow \mathbb{C}^\times$ be a character. Show that χ can be extended to a character of G. (**Hint:** Apply Theorem 4.3 to the space $V = \{f \in L^2(G) \mid f(hg) = \chi(h)\,f(g)\}$. To show that V is nonzero, note that if $\phi \in C(G)$ then $f(g) = \int \phi(hg)\,\chi(h)^{-1}\,dh$ defines an element of V. Use Urysohn's lemma to construct ϕ such that $f \neq 0$.)

Compact Lie Groups

5

Lie Subgroups of $\mathrm{GL}(n, \mathbb{C})$

If U is an open subset of $\mathbb{R}^n$, we say that a map $\phi : U \longrightarrow \mathbb{R}^m$ is *smooth* if it has continuous partial derivatives of all orders. More generally, if $X \subset \mathbb{R}^n$ is not necessarily open, we say that a map $\phi : X \longrightarrow \mathbb{R}^n$ is *smooth* if for each $x \in X$ there exists an open set U of $\mathbb{R}^n$ containing x such that ϕ can be extended to a smooth map on U. A *diffeomorphism* of $X \subseteq \mathbb{R}^n$ with $Y \subseteq \mathbb{R}^m$ is a homeomorphism $F : X \longrightarrow Y$ such that both F and F^{-1} are smooth. We will assume as known the following useful criterion.

Inverse Function Theorem. *If $U \subset \mathbb{R}^d$ is open and $u \in U$, if $F : U \longrightarrow \mathbb{R}^n$ is a smooth map, with $d < n$, and if the matrix of partial derivatives $(\partial F_i / \partial x_j)$ has rank d at u, then u has a neighborhood N such that F induces a diffeomorphism of N onto its image.*

A subset X of a topological space Y is *locally closed* (in Y) if for all $x \in X$ there exists an open neighborhood U of x in Y such that $X \cap U$ is closed in U. This is equivalent to saying that X is the intersection of an open set and a closed set. We say that X is a *submanifold of $\mathbb{R}^n$ of dimension d* if it is a locally closed subset and every point of X has a neighborhood that is diffeomorphic to an open set in $\mathbb{R}^d$.

Let us identify $\mathrm{Mat}_n(\mathbb{C})$ with the Euclidean space $\mathbb{C}^{n^2} \cong \mathbb{R}^{2n^2}$. The subset $\mathrm{GL}(n, \mathbb{C})$ is open, and if a closed subgroup G of $\mathrm{GL}(n, \mathbb{C})$ is a submanifold of $\mathbb{R}^{2n^2}$ in this identification, we say that G is a *closed Lie subgroup* of $\mathrm{GL}(n, \mathbb{C})$. It may be shown that any closed subgroup of $\mathrm{GL}(n, \mathbb{C})$ is a closed Lie subgroup. See Remarks 7.1 and 7.2 for some subtleties behind the innocent term "closed Lie subgroup."

More generally, a *Lie group* is a topological group G that is a differentiable manifold such that the multiplication and inverse maps $G \times G \longrightarrow G$ and $G \longrightarrow G$ are smooth. We will give a proper definition of a differentiable manifold in the next chapter. In this chapter, we will restrict ourselves to closed Lie subgroups of $\mathrm{GL}(n, \mathbb{C})$.

D. Bump, *Lie Groups*, Graduate Texts in Mathematics 225,
DOI 10.1007/978-1-4614-8024-2_5, © Springer Science+Business Media New York 2013

Example 5.1. If F is a field, then the *general linear group* GL(n, F) is the group of invertible $n \times n$ matrices with coefficients in F. It is a Lie group. Assuming that $F = \mathbb{R}$ or $\mathbb{C}$, the group GL(n, F) is an open set in Mat$_n(F)$ and hence a manifold of dimension n^2 if $F = \mathbb{R}$ or $2n^2$ if $F = \mathbb{C}$. The *special linear group* is the subgroup SL(n, F) of matrices with determinant 1. It is a closed Lie subgroup of GL(n, F) of dimension $n^2 - 1$ or $2(n^2 - 1)$.

Example 5.2. If $F = \mathbb{R}$ or $\mathbb{C}$, let O(n, F) $= \{g \in$ GL(n, F) $| \, g \cdot {}^t g = I\}$. This is the $n \times n$ *orthogonal group*. More geometrically, O(n, F) is the group of linear transformations preserving the quadratic form $Q(x_1, \ldots, x_n) = x_1^2 + x_2^2 + \cdots + x_n^2$. To see this, if $(x) = {}^t(x_1, \ldots, x_n)$ is represented as a column vector, we have $Q(x) = Q(x_1, \ldots, x_n) = {}^t x \cdot x$, and it is clear that $Q(gx) = Q(x)$ if $g \cdot {}^t g = I$. The group O($n, \mathbb{R}$) is compact and is usually denoted simply O(n). The group O(n) contains elements of determinants ± 1. The subgroup of elements of determinant 1 is the *special orthogonal group* SO(n). The dimension of O(n) and its subgroup SO(n) of index 2 is $\frac{1}{2}(n^2 - n)$. This will be seen in Proposition 5.6 when we compute their Lie algebra (which is the same for both groups).

Example 5.3. More generally, over any field, a vector space V on which there is given a quadratic form q is called a *quadratic space*, and the set O(V, q) of linear transformations of V preserving q is an *orthogonal group*. Over the complex numbers, it is not hard to prove that all orthogonal groups are isomorphic (Exercise 5.4), but over the real numbers, some orthogonal groups are not isomorphic to O(n). If $k + r = n$, let O(k, r) be the subgroup of GL($n, \mathbb{R}$) preserving the indefinite quadratic form $x_1^2 + \cdots + x_k^2 - x_{k+1}^2 - \cdots - x_n^2$. If $r = 0$, this is O(n), but otherwise this group is noncompact. The dimensions of these Lie groups are, like SO(n), equal to $\frac{1}{2}(n^2 - n)$.

Example 5.4. The *unitary group* U(n) $= \{g \in$ GL($n, \mathbb{C}$) $| \, g \cdot {}^{\overline{t}} g = I\}$. If $g \in$ U(n) then $|\det(g)| = 1$, and every complex number of absolute value 1 is a possible determinant of $g \in$ U(n). The *special unitary group* SU(n) $=$ U(n) $\cap$ SL($n, \mathbb{C}$). The dimensions of U(n) and SU(n) are n^2 and $n^2 - 1$, just like GL($n, \mathbb{R}$) and SL($n, \mathbb{R}$).

Example 5.5. If $F = \mathbb{R}$ or $\mathbb{C}$, let Sp($2n, F$) $= \{g \in$ GL($2n, F$) $| \, g \cdot J \cdot {}^t g = J\}$, where

$$J = \begin{pmatrix} 0 & -I_n \\ I_n & 0 \end{pmatrix}.$$

This is the *symplectic group*. The compact group Sp($2n, \mathbb{C}$) $\cap$ U($2n$) will be denoted as simply Sp($2n$).

A *Lie algebra* over a field F is a vector space $\mathfrak{g}$ over F endowed with a bilinear map, the *Lie bracket*, denoted $(X, Y) \longrightarrow [X, Y]$ for $X, Y \in \mathfrak{g}$, that satisfies $[X, Y] = -[Y, X]$ and the *Jacobi identity*

$$[X, [Y, Z]] + [Y, [Z, X]] + [Z, [X, Y]] = 0. \tag{5.1}$$

The identity $[X, Y] = -[Y, X]$ implies that $[X, X] = 0$.

We will show that it is possible to associate a Lie algebra with any Lie group. We will show this for closed Lie subgroups of GL(n, $\mathbb{C}$) in this chapter and for arbitrary lie groups in Chap. 7.

First we give two purely algebraic examples of Lie algebras.

Example 5.6. Let A be an associative algebra. Define a bilinear operation on A by $[X, Y] = XY - YX$. With this definition, A becomes a Lie algebra.

If $A = \text{Mat}_n(F)$, where F is a field, we will denote the Lie algebra associated with A by the previous example as $\mathfrak{gl}(n, F)$. After Proposition 5.5 it will become clear that this is the Lie algebra of GL(n, F) when $F = \mathbb{R}$ or $\mathbb{C}$. Similarly, if V is a vector space over F, then the space End(V) of F-linear transformations $V \longrightarrow V$ is an associative algebra and hence a Lie algebra, denoted $\mathfrak{gl}(V)$.

Example 5.7. Let F be a field and let A be an F-algebra. By a *derivation* of A we mean a map $D : A \longrightarrow A$ that is F-linear, and satisfies $D(fg) = fD(g) + D(f)g$. We have $D(1 \cdot 1) = 2D(1)$, which implies that $D(1) = 0$, and therefore $D(c) = 0$ for any $c \in F \subset A$. It is easy to check that if D_1 and D_2 are derivations, then so is $[D_1, D_2] = D_1 D_2 - D_2 D_1$. However, $D_1 D_2$ and $D_2 D_1$ are themselves not derivations. It is easy to check that the derivations of A form a Lie algebra.

The *exponential map* $\exp : \text{Mat}_n(\mathbb{C}) \longrightarrow$ GL(n, $\mathbb{C}$) is defined by

$$\exp(X) = I + X + \tfrac{1}{2}X^2 + \tfrac{1}{6}X^3 + \cdots . \tag{5.2}$$

This series is convergent for all matrices X.

Remark 5.1. If X and Y commute, then $\exp(X + Y) = \exp(X) \exp(Y)$. If they do not commute, this is not true.

A *one-parameter subgroup* of a Lie group G is a continuous homomorphism $\mathbb{R} \longrightarrow G$. We denote this by $t \mapsto g_t$. Since tX and uX commute, for $X \in \text{Mat}_n(\mathbb{C})$, the map $t \longrightarrow \exp(tX)$ is a one-parameter subgroup. We will also denote $\exp(X) = e^X$.

Proposition 5.1. *Let U be an open subset of $\mathbb{R}^n$, and let $x \in U$. Then we may find a smooth function f with compact support contained in U that does not vanish at x.*

Proof. We may assume $x = (x_1, \ldots, x_n)$ is the origin. Define

$$f(x_1, \ldots, x_n) = \begin{cases} e^{-(1-|x|^2/r^2)^{-1}} & \text{if } |x| \leqslant r, \\ 0 & \text{otherwise.} \end{cases}$$

This function is smooth and has support in the ball $\{|x| \leqslant r\}$. Taking r sufficiently small, we can make this vanish outside U. $\square$

Proposition 5.2. *Let G be a closed Lie subgroup of* GL($n, \mathbb{C}$), *and let $X \in$* Mat$_n(\mathbb{C})$. *Then the path $t \longrightarrow \exp(tX)$ is tangent to the submanifold G of* GL($n, \mathbb{C}$) *at $t = 0$ if and only if it is contained in G for all t.*

Proof. If $\exp(tX)$ is contained in G for all t, then clearly it is tangent to G at $t = 0$. We must prove the converse. Suppose that $\exp(t_0 X) \notin G$ for some $t_0 > 0$. Using Proposition 5.1, Let ϕ_0 be a smooth compactly supported function on GL($n, \mathbb{C}$) such that $\phi_0(g) = 0$ for all $g \in G$, $\phi_0 \geqslant 0$, and $\phi_0\big(\exp(t_0 X)\big) \neq 0$. Let

$$f(t) = \phi\big(\exp(tX)\big), \qquad \phi(h) = \int_G \phi_0(hg)\, \mathrm{d}g, \qquad t \in \mathbb{R},$$

in terms of a left Haar measure on G. Clearly, ϕ is constant on the cosets hG of G, vanishes on G, but is nonzero at $\exp(t_0 X)$. For any t,

$$f'(t) = \frac{\mathrm{d}}{\mathrm{d}u} \phi\big(\exp(tX)\exp(uX)\big)\big|_{u=0} = 0$$

since the path $u \longrightarrow \exp(tX)\exp(uX)$ is tangent to the coset $\exp(tX)G$ and ϕ is constant on such cosets. Moreover, $f(0) = 0$. Therefore, $f(t) = 0$ for all t, which is a contradiction since $f(t_0) \neq 0$.   □

Proposition 5.3. *Let G be a closed Lie subgroup of* GL($n, \mathbb{C}$). *The set* Lie(G) *of all $X \in$* Mat$_n(\mathbb{C})$ *such that $\exp(tX) \subset G$ is a vector space whose dimension is equal to the dimension of G as a manifold.*

Proof. This is clear from the characterization of Proposition 5.2.   □

Proposition 5.4. *Let G be a closed Lie subgroup of* GL($n, \mathbb{C}$). *The map*

$$X \longrightarrow \exp(X)$$

gives a diffeomorphism of a neighborhood of the identity in Lie(G) *onto a neighborhood of the identity in G.*

Proof. First we note that since $\exp(X) = I + X + \frac{1}{2}X^2 + \cdots$, the Jacobian of exp at the identity is 1, so exp induces a diffeomorphism of an open neighborhood U of the identity in Mat$_n(\mathbb{C})$ onto a neighborhood of the identity in GL$_n(\mathbb{C}) \subset$ Mat$_n(\mathbb{C})$. Now, since by Proposition 5.3 Lie(H) is a vector subspace of dimension equal to the dimension of H as a manifold, the Inverse Function Theorem implies that the image of Lie(H) $\cap\, U$ must be mapped onto an open neighborhood of the identity in H.   □

Proposition 5.5. *If G is a closed Lie subgroup of* GL($n, \mathbb{C}$), *and if $X, Y \in$* Lie(G), *then $[X, Y] \in$* Lie(G).

Proof. It is evident that Lie(G) is mapped to itself under conjugation by elements of G. Thus, Lie(G) contains

$$\tfrac{1}{t}\left(\mathrm{e}^{tX}Y\mathrm{e}^{-tX} - Y\right) = XY - YX + \tfrac{t}{2}(X^2Y - 2XYX + YX^2) + \cdots.$$

Because this is true for all t, passing to the limit $t \longrightarrow 0$ shows that $[X, Y] \in$ Lie(G).   □

We see that Lie(G) is a Lie subalgebra of $\mathfrak{gl}(n, \mathbb{C})$. Thus, we are able to associate a Lie algebra with a Lie group.

Example 5.8. The Lie algebra of GL(n, F) with $F = \mathbb{R}$ or $\mathbb{C}$ is $\mathfrak{gl}(n, F)$.

Example 5.9. Let $\mathfrak{sl}(n, F)$ be the subspace of $X \in \mathfrak{gl}(n, F)$ such that $\text{tr}(X) = 0$. This is a Lie subalgebra, and it is the Lie algebra of SL(n, F) when $F = \mathbb{R}$ or $\mathbb{C}$. This follows immediately from the fact that $\det(e^X) = e^{\text{tr}(X)}$ for any matrix X because if $x_1, \ldots, x_n$ are the eigenvalues of X, then $e^{x_1}, \ldots, e^{x_n}$ are the eigenvalues of e^X.

Example 5.10. Let $\mathfrak{o}(n, F)$ be the set of $X \in \mathfrak{gl}(n, F)$ that are skew-symmetric, in other words, that satisfy $X + {}^t X = 0$. It is easy to check that $\mathfrak{o}(n, F)$ is closed under the Lie bracket and hence is a Lie subalgebra.

Proposition 5.6. *If $F = \mathbb{R}$ or $\mathbb{C}$, the Lie algebra of O(n, F) is $\mathfrak{o}(n, F)$. The dimension of O(n) is $\frac{1}{2}(n^2 - n)$, and the dimension of O($n, \mathbb{C}$) is $n^2 - n$.*

Proof. Let $G = \text{O}(n, F)$, $\mathfrak{g} = \text{Lie}(G)$. Suppose $X \in \mathfrak{o}(n, F)$. Exponentiate the identity $-tX = t^t X$ to get

$$\exp(tX)^{-1} = {}^t \exp(tX),$$

whence $\exp(tX) \in \text{O}(n, F)$ for all $t \in \mathbb{R}$. Thus, $\mathfrak{o}(n, F) \subseteq \mathfrak{g}$. To prove the converse, suppose that $X \in \mathfrak{g}$. Then, for all t,

$$\begin{aligned}
I &= \exp(tX) \cdot {}^t \exp(tX) \\
&= (I + tX + \tfrac{1}{2}t^2 X^2 + \cdots)(I + t\,{}^t X + \tfrac{1}{2}t^2 \cdot {}^t X^2 + \cdots) \\
&= I + t(X + {}^t X) + \tfrac{1}{2}t^2(X^2 + 2X \cdot {}^t X + {}^t X^2) + \cdots.
\end{aligned}$$

Since this is true for all t, each coefficient in this Taylor series must vanish (except of course the constant one). In particular, $X + {}^t X = 0$. This proves that $\mathfrak{g} = \mathfrak{o}(n, F)$.

The dimensions of O(n) and O($n, \mathbb{C}$) are most easily calculated by computing the dimension of the Lie algebras. A skew-symmetric matrix is determined by its upper triangular entries, and there are $\frac{1}{2}(n^2 - n)$ of these. $\square$

Example 5.11. Let $\mathfrak{u}(n)$ be the set of $X \in \text{GL}(n, \mathbb{C})$ such that $X + {}^t \overline{X} = 0$. One checks easily that this is closed under the $\mathfrak{gl}(n, \mathbb{C})$ Lie bracket $[X, Y] = XY - YX$. Despite the fact that these matrices have complex entries, this is a *real* Lie algebra, for it is only a real vector space, not a complex one. (It is not closed under multiplication by complex scalars.) It may be checked along the lines of Proposition 5.6 that $\mathfrak{u}(n)$ is the Lie algebra of U(n), and similarly $\mathfrak{su}(n) = \{X \in \mathfrak{u}(n) \mid \text{tr}(X) = 0\}$ is the Lie algebra of SU(n).

Example 5.12. Let $\mathfrak{sp}(2n, F)$ be the set of matrices $X \in \text{Mat}_{2n}(F)$ that satisfy $XJ + J\,{}^t X = 0$, where

$$J = \begin{pmatrix} 0 & -I_n \\ I_n & 0 \end{pmatrix}.$$

This is the Lie algebra of Sp($2n, F$).

Exercises

Exercise 5.1. Show that $O(n, m)$ is the group of $g \in GL(n + m, \mathbb{R})$ such that $g J_1 {}^t g = J_1$, where

$$J_1 = \begin{pmatrix} I_n & \\ & -I_m \end{pmatrix}.$$

Exercise 5.2. If $F = \mathbb{R}$ or $\mathbb{C}$, let $O_J(F)$ be the group of all $g \in GL(N, F)$ such that $g J {}^t g = J$, where J is the $N \times N$ matrix

$$J = \begin{pmatrix} & & 1 \\ & \cdot{}^{\cdot{}^\cdot} & \\ 1 & & \end{pmatrix}. \tag{5.3}$$

Show that $O_J(\mathbb{R})$ is conjugate in $GL(N, \mathbb{R})$ to $O(n, n)$ if $N = 2n$ and to $O(n + 1, n)$ if $N = 2n + 1$. [**Hint:** Find a matrix $\sigma \in GL(N, \mathbb{R})$ such that $\sigma J {}^t \sigma = J_1$, where J is as in the previous exercise.]

Exercise 5.3. Let J be as in the previous exercise, and let

$$\sigma = \begin{pmatrix} \frac{1}{\sqrt{2i}} & & \cdots & & -\frac{i}{\sqrt{2i}} \\ & \frac{1}{\sqrt{2i}} & & -\frac{i}{\sqrt{2i}} & \\ \vdots & & \cdot{}^{\cdot{}^\cdot} \cdot{}_{\cdot{}_\cdot} & & \vdots \\ & \frac{i}{\sqrt{2i}} & & -\frac{1}{\sqrt{2i}} & \\ \frac{i}{\sqrt{2i}} & & \cdots & & -\frac{1}{\sqrt{2i}} \end{pmatrix},$$

with all entries not on one of the two diagonals equal to zero. If N is odd, the middle element of this matrix is 1.

(i) Check that $\sigma {}^t \sigma = J$, with J as in (5.3). With $O_J(F)$ as in Example 5.2, deduce that $\sigma^{-1} O_J(\mathbb{C}) \sigma = O(N, \mathbb{C})$. Why does the same argument *not* prove that $\sigma^{-1} O_J(\mathbb{R}) \sigma = O(n, \mathbb{R})$?

(ii) Check that σ is unitary. Show that if $g \in O_J(\mathbb{C})$ and $h = \sigma^{-1} g \sigma$, then h is real if and only if g is unitary.

(iii) Show that the group $O_J(\mathbb{C}) \cap U(N)$ is conjugate in $GL(N, \mathbb{C})$ to $O(N)$.

Exercise 5.4. Let V_1 and V_2 be vector spaces over a field F, and let q_i be a quadratic form on V_i for $i = 1, 2$. The quadratic spaces are called *equivalent* if there exists an isomorphism $l : V_1 \longrightarrow V_2$ such that $q_1 = q_2 \circ l$.

(i) Show that over a field of characteristic not equal to 2, any quadratic form is equivalent to $\sum a_i x_i^2$ for some constants a_i.

(ii) Show that, if $F = \mathbb{C}$, then any quadratic space of dimension n is equivalent to $\mathbb{C}^n$ with the quadratic form $x_1^2 + \cdots + x_n^2$.

(iii) Show that, if $F = \mathbb{R}$, then any quadratic space of dimension n is equivalent to $\mathbb{R}^n$ with the quadratic form $x_1^2 + \cdots + x_r^2 - x_{r+1}^2 - \cdots - x_n^2$ for some r.

Exercise 5.5. Compute the Lie algebra of $Sp(2n, \mathbb{R})$ and the dimension of the group.

Let $\mathbb{H} = \mathbb{R} \oplus \mathbb{R}i \oplus \mathbb{R}j \oplus \mathbb{R}k$ be the ring of quaternions, where $i^2 = j^2 = k^2 = -1$ and $ij = -ji = k$, $jk = -kj = i$, $ki = -ik = j$. Then $\mathbb{H} = \mathbb{C} \oplus \mathbb{C}j$. If $x = a + bi + cj + dk \in \mathbb{H}$ with a, b, c, d real, let $\bar{x} = a - bi - cj - dk$. If $u \in \mathbb{C}$, then $juj^{-1} = \bar{u}$. The group GL$(n, \mathbb{H})$ consists of all $n \times n$ invertible quaternion matrices.

Exercise 5.6. Show that there is a ring isomorphism $\mathrm{Mat}_n(\mathbb{H}) \longrightarrow \mathrm{Mat}_{2n}(\mathbb{C})$ with the following description. Any $A \in \mathrm{Mat}_n(\mathbb{H})$ may be written uniquely as $A_1 + A_2 j$ with $A_1, A_2 \in \mathrm{Mat}_n(\mathbb{C})$. The isomorphism in question maps

$$A_1 + A_2 j \longmapsto \begin{pmatrix} A_1 & A_2 \\ -\bar{A}_2 & \bar{A}_1 \end{pmatrix}.$$

Exercise 5.7. Show that if $A \in \mathrm{Mat}_n(\mathbb{H})$, then $A \cdot {}^t\bar{A} = I$ if and only if the complex $2n \times 2n$ matrix

$$\begin{pmatrix} A_1 & A_2 \\ -\bar{A}_2 & \bar{A}_1 \end{pmatrix}$$

is in both Sp$(2n, \mathbb{C})$ and U$(2n)$. Recall that the intersection of these two groups was the group denoted Sp$(2n)$.

Exercise 5.8. Show that the groups SO(2) and SU(2) may be identified with the groups of matrices

$$\left\{ \begin{pmatrix} a & b \\ -\bar{b} & \bar{a} \end{pmatrix} \,\middle|\, a, b \in F, \ |a|^2 + |b|^2 = 1 \right\},$$

where $F = \mathbb{R}$ or $\mathbb{C}$, respectively.

Exercise 5.9. The group SU$(1, 1)$ is by definition the group of $g \in$ SL$(2, \mathbb{C})$ such that

$$g \cdot J \cdot {}^t g = J, \qquad J = \begin{pmatrix} 1 & \\ & -1 \end{pmatrix}.$$

(i) Show that SU$(1, 1)$ consists of all elements of SL$(2, \mathbb{C})$ of the form

$$\begin{pmatrix} a & b \\ \bar{b} & \bar{a} \end{pmatrix}, \qquad |a|^2 - |b|^2 = 1.$$

(ii) Show that the Lie algebra $\mathfrak{su}(1, 1)$ of SU$(1, 1)$ consists of all matrices of the form

$$\begin{pmatrix} ai & b \\ \bar{b} & -ai \end{pmatrix}$$

with a real.

(iii) Let $C = \frac{1}{\sqrt{2i}} \begin{pmatrix} 1 & -i \\ 1 & i \end{pmatrix} \in$ SL$(2, \mathbb{C})$. This element is sometimes called the *Cayley transform*. Show that $C \cdot$ SL$(2, \mathbb{R}) \cdot C^{-1} =$ SU$(1, 1)$ and $C \cdot \mathfrak{sl}(2, \mathbb{R}) \cdot C^{-1} = \mathfrak{su}(1, 1)$.

6

Vector Fields

A *smooth premanifold* of dimension n is a Hausdorff topological space M together with a set $\mathcal{U}$ of pairs (U, ϕ), where the set of U such that $(U, \phi) \in \mathcal{U}$ for some ϕ is an open cover of M and such that, for each $(U, \phi) \in \mathcal{U}$, the image $\phi(U)$ of ϕ is an open subset of $\mathbb{R}^n$ and ϕ is a homeomorphism of U onto $\phi(U)$. We assume that if $U, V \in \mathcal{U}$, then $\phi_V \circ \phi_U^{-1}$ is a diffeomorphism from $\phi_U(U \cap V)$ onto $\phi_V(U \cap V)$. The set $\mathcal{U}$ is called a *preatlas*.

If M and N are premanifolds, a continuous map $f : M \longrightarrow N$ is *smooth* if whenever (U, ϕ) and (V, ψ) are charts of M and N, respectively, the map $\psi \circ f \circ \phi^{-1}$ is a smooth map from $\phi\big(U \cap f^{-1}(V)\big) \longrightarrow \psi(V)$. Smooth maps are the morphisms in the category of smooth premanifolds. The smooth map f is a *diffeomorphism* if it is a bijection and has a smooth inverse. Open subsets of $\mathbb{R}^n$ are naturally premanifolds, and the definitions of smooth maps and diffeomorphisms are consistent with the definitions already given in that special case.

If M is a premanifold with preatlas $\mathcal{U}$, and if we replace $\mathcal{U}$ by the larger set $\mathcal{U}'$ of all pairs (U, ϕ), where U is an open subset of M and ϕ is a diffeomorphism of U onto an open subset of $\mathbb{R}^n$, then the set of smooth maps $M \longrightarrow N$ or $N \longrightarrow M$, where N is another premanifold, is unchanged. If $\mathcal{U} = \mathcal{U}'$, then we call $\mathcal{U}'$ an *atlas* and M a *smooth manifold*.

Suppose that M is a smooth manifold and $m \in M$. If U is a neighborhood of x and (ϕ, U) is a chart such that $\phi(x)$ is the origin in $\mathbb{R}^n$, then we may write $\phi(u) = (x_1(u), \ldots, x_n(u))$, where $x_1, \ldots, x_m : U \longrightarrow \mathbb{R}$ are smooth functions. Composing ϕ with a translation in $\mathbb{R}^n$, we may arrange that $x_i(m) = 0$, and it is often advantageous to do so. We call $x_1, \ldots, x_m$ a set of *local coordinates at* m or *coordinate functions on* U. The set U itself may be called a *coordinate neighborhood*.

Let $m \in M$, and let $F = \mathbb{R}$ or $\mathbb{C}$. A *germ* of an F-valued function is an equivalence class of pairs (U, f_U), where U is an open neighborhood of x and $f : U \longrightarrow F$ is a function. The equivalence relation is that (U, f_U) and (V, f_V) are equivalent if f_U and f_V are equal on some open neighborhood W of x contained in $U \cap V$. Let $\mathcal{O}_m$ be the set of germs of smooth

D. Bump, *Lie Groups*, Graduate Texts in Mathematics 225,
DOI 10.1007/978-1-4614-8024-2_6, © Springer Science+Business Media New York 2013

real-valued functions. It is a ring in an obvious way, and evaluation at m induces a surjective homomorphism $\mathcal{O}_m \longrightarrow \mathbb{R}$, the *evaluation map*. We will denote the evaluation map $f \mapsto f(m)$, a slight abuse of notation since f is a germ, not a function. Let $\mathcal{M}_m$ be the kernel of this homomorphism; that is, the ideal of germs of smooth functions vanishing at m. Then $\mathcal{O}_m$ is a local ring and $\mathcal{M}_m$ is its maximal ideal.

Lemma 6.1. *Suppose that f is a smooth function on a neighborhood U of the origin in $\mathbb{R}^n$, and $f(0, x_2, \ldots, x_n) = 0$ for $(0, x_2, \ldots, x_n) \in U$. Then*

$$g(x_1, x_2, \ldots, x_n) = \begin{cases} x_1^{-1} f(x_1, \ldots, x_n) & \text{if } x_1 \neq 0 , \\ (\partial f / \partial x_1)(0, x_2, \ldots, x_n) & \text{if } x_1 = 0 , \end{cases}$$

defines a smooth function on U.

Proof. We show first that g is continuous. Indeed, with $x_2, \ldots, x_n$ fixed,

$$\lim_{x_1 \to 0} x_1^{-1} f(x_1, \ldots, x_n) = (\partial f / \partial x_1)(0, x_2, \ldots, x_n)$$

by the definition of the derivative. Convergence is uniform on compact sets in $x_2, \ldots, x_n$ since by the remainder form of Taylor's theorem

$$\left| x_1^{-1} f(x_1, \ldots, x_n) - (\partial f / \partial x_1)(0, x_2, \ldots, x_n) \right| \leqslant \tfrac{B}{2} x_1,$$

where B is an upper bound for $|\partial^2 f / \partial x_1|$. Since $\partial f / \partial x_1(0, x_2, \ldots, x_n)$ is continuous by the smoothness of f, it follows that g is continuous.

A similar argument based on Taylor's theorem shows that the higher partial derivatives $\partial^n g / \partial x_1^n$ are also continuous.

Finally, the two functions

$$\frac{\partial^{k_2 + \cdots + k_n} f}{\partial x_2^{k_2} \cdots \partial x_n^{k_n}} \quad \text{and} \quad \frac{\partial^{k_2 + \cdots + k_n} g}{\partial x_2^{k_2} \cdots \partial x_n^{k_n}}$$

bear the same relationship to each other as f and g, so we obtain similarly continuity of the mixed partials $\partial^{k_1 + k_2 + \cdots + k_n} g / \partial x_1^{k_1} \partial x_2^{k_2} \cdots \partial x_n^{k_n}$. $\qquad \square$

Proposition 6.1. *Let $m \in M$, where M is a smooth manifold of dimension n. Let $\mathcal{O} = \mathcal{O}_m$ and $\mathcal{M} = \mathcal{M}_m$. Let $x_1, \ldots, x_n$ be the germs of a set of local coordinates at m. Then $x_1, \ldots, x_n$ generate the ideal $\mathcal{M}$. Moreover, $\mathcal{M}/\mathcal{M}^2$ is a vector space of dimension n generated by the images of $x_1, \ldots, x_n$.*

Proof. Although this is really a statement about germs of functions, we will work with representative functions defined in some neighborhood of m.

If $f \in \mathcal{M}$, we write $f = f_1 + f_2$, where $f_1(x_1, \ldots, x_n) = f(0, x_2, \ldots, x_n)$ and $f_2 = f - f_1$. Then $f_2 \in x_1 \mathcal{O}$ by Lemma 6.1, while f_1 is the germ of a function in $x_2, \ldots, x_n$ vanishing at m and lies in $x_2 \mathcal{O} + \cdots + x_n \mathcal{O}$ by induction on n.

As for the last assertion, if $f \in \mathcal{M}$, let $a_i = (\partial f/\partial x_i)(m)$. Then $f - \sum_i a_i x_i$ vanishes to order 2 at m. We need to show that it lies in $\mathcal{M}^2$. Thus, what we must prove is that if f and $\partial f/\partial x_i$ vanish at m, then f is in $\mathcal{M}^2$. To prove this, write $f = f_1 + f_2 + f_3$, where

$$f_1(x_1, x_2, \ldots, x_n) = f(x_1, \ldots, x_n) - f(0, x_2, \ldots, x_n) - x_1 \frac{\partial f}{\partial x_1}(0, x_2, \ldots, x_n),$$

$$f_2(x_1, \ldots, x_n) = f(0, x_2, \ldots, x_n),$$

$$f_3(x_1, x_2, \ldots, x_n) = x_1 \frac{\partial f}{\partial x_1}(0, x_2, \ldots, x_n).$$

Two applications of Lemma 6.1 show that $f_1 = x_1^{-2} h$ where h is smooth, so $f_1 \in \mathcal{M}^2$. The function f_2 also vanishes, with its first-order partial derivatives at m, but is a function in one fewer variables, so by induction it is in $\mathcal{M}^2$. Lastly, $\partial f/\partial x_1$ vanishes at m and hence is in $\mathcal{M}$ by the part of this proposition that is already proved, so multiplying by x_1 gives an element of $\mathcal{M}^2$.   □

A *local derivation* of $\mathcal{O}_m$ is a map $X : \mathcal{O}_m \longrightarrow \mathbb{R}$ that is $\mathbb{R}$-linear and such that

$$X(fg) = f(m)X(g) + g(m)X(f). \tag{6.1}$$

Taking $f = g = 1$ gives $X(1 \cdot 1) = 2X(1)$ so X annihilates constant functions.

For example, if $x_1, \ldots, x_n$ are a set of local coordinates and $a_1, \ldots, a_n \in \mathbb{R}$, then

$$Xf = \sum_{i=1}^{n} a_i \frac{\partial f}{\partial x_i}(m) \tag{6.2}$$

is a local derivation.

Proposition 6.2. *Let m be a point on an n-dimensional smooth manifold M. Every local derivation of $\mathcal{O}_m$ is of the form (6.2). The set $T_m(M)$ of such local derivations is an n-dimensional real vector space.*

Proof. If f and g both vanish at m, then (6.1) implies that a local derivation X vanishes on $\mathcal{M}^2$, and by Proposition 6.1 it is therefore determined by its values on $x_1, \ldots, x_n$. If these are $a_1, \ldots, a_n$, then X agrees with the right-hand side of (6.2).   □

We now define the *tangent space* $T_m(M)$ to be the space of local derivations of $\mathcal{O}_m$. We will call elements of $T_m(M)$ *tangent vectors*. Thus, a tangent vector at m is the same thing as a local derivation of the ring $\mathcal{O}_m$.

This definition of tangent vector and tangent space has the advantage that it is intrinsic. Proposition 6.2 allows us to relate this definition to the intuitive notion of a tangent vector. Intuitively, a tangent vector should be an equivalence class of paths through m: two paths are equivalent if they are tangent.

By a path we mean a smooth map $u : (-\epsilon, \epsilon) \longrightarrow M$ such that $u(0) = m$ for some $\epsilon > 0$. Given a function, or the germ of a function at m, we can use the path to define a local derivation

$$Xf = \frac{\mathrm{d}}{\mathrm{d}t} f\big(u(t)\big)\Big|_{t=0}. \tag{6.3}$$

Using the chain rule, this equals (6.2) with $a_i = (\mathrm{d}/\mathrm{d}t)(x_i(u(t)))\big|_{t=0}$.

Let M and N be smooth manifolds with a smooth map $f : M \to N$. Let $m \in M$ and $n \in N$ such that $f(m) = n$. Then we have a map $\mathrm{d}f : T_m(M) \to T_n(N)$ defined as follows. Note that f induces a map from $\mathcal{O}_n(N)$ to $\mathcal{O}_m(M)$. Now $X \in T_m(M)$ then X is a local derivation of $\mathcal{O}_m(M)$, and composition with f produces a local derivation of $\mathcal{O}_n(N)$. This is the tangent vector we will denote $\mathrm{d}f(X)$. The map $\mathrm{d}f : T_m(M) \to T_n(N)$ is called the *differential* of f.

We will use the notation

$$X = \sum_{i=1}^{n} a_i \frac{\partial}{\partial x_i}$$

to denote the element (6.2) of $T_m(M)$. By a *vector field* X on M we mean a rule that assigns to each point $m \in M$ an element $X_m \in T_m(M)$. The assignment $m \longrightarrow X_m$ must be smooth. This means that if $x_1, \ldots, x_n$ are local coordinates on an open set $U \subseteq M$, then there exist smooth functions $a_1, \ldots, a_n$ on U such that

$$X_m = \sum_{i=1}^{n} a_i(m) \frac{\partial}{\partial x_i}. \tag{6.4}$$

It follows from the chain rule that this definition is independent of the choice of local coordinates x_i.

Now let $A = C^\infty(M, \mathbb{R})$ be the ring of smooth real-valued functions on M. Given a vector field X on M, we may obtain a derivation of A as follows. If $f \in A$, let $X(f)$ be the smooth function that assigns to $m \in M$ the value $X_m(f)$, where we are of course applying X_m to the germ of f at m. For example, if $M = U$ is an open set on $\mathbb{R}^n$ with coordinate functions $x_1, \ldots, x_n$ on U, given smooth functions $a_i : U \longrightarrow \mathbb{R}$, we may associate a derivation of A with the vector field (6.4) by

$$(Xf)(m) = \sum_{i=1}^{n} a_i(m) \frac{\partial f}{\partial x_i}(m). \tag{6.5}$$

The content of the next theorem is that every derivation of A is associated with a vector field in this way.

Proposition 6.3. *There is a one-to-one correspondence between vector fields on a smooth manifold M and derivations of $C^\infty(M, \mathbb{R})$. Specifically, if D is any derivation of $C^\infty(M, \mathbb{R})$, there is a unique vector field X on M such that $Df = Xf$ for all f.*

Proof. We show first that if $m \in M$, and if $f \in A = C^\infty(M, \mathbb{R})$ has germ zero at m, then the function Df vanishes at m. This implies that D induces a well-defined map $X_m : \mathcal{O}_m \longrightarrow \mathbb{R}$ that is a local derivation. Our assumption means that f vanishes in a neighborhood of m, so there is another smooth function g such that $gf = f$, yet $g(m) = 0$. Now $D(f)(m) = g(m)D(f) + f(m)D(g)$. Since both f and g vanish at m, we see that $D(f)(m) = 0$.

Now let x_i be local coordinates on an open set U of M. For each $m \in U$ there are real numbers $a_i(m)$ such that (6.4) is true. We need to know that the $a_i(m)$ are smooth functions. Indeed, we have $a_i(m) = D(x_i)$, so it is smooth. $\square$

Now let X and Y be vector fields on M. By Proposition 6.3, we may regard these as derivations of $C^\infty(M, \mathbb{R})$. As we have noted in Example 5.7, derivations of an arbitrary ring form a Lie algebra. Thus $[X, Y] = XY - YX$ defines a derivation:

$$[X, Y]f = X(Yf) - Y(Xf). \tag{6.6}$$

By Proposition 6.3 this derivation $[X, Y]$ corresponds to a vector field. Let us see this again concretely by computing its effect in local coordinates. If $X = \sum a_i \frac{\partial}{\partial x_i}$ and $Y = \sum b_i \frac{\partial}{\partial x_i}$, we have $X(Yf) = \sum_{i,j} \left[a_j \frac{\partial b_i}{\partial x_j} \frac{\partial f}{\partial x_i} + a_i b_j \frac{\partial^2 f}{\partial x_i \partial x_j} \right]$. This is not a derivation, but if we subtract $Y(Xf)$ to cancel the unwanted mixed partials, we see that

$$[X, Y] = \sum_{i,j} \left[a_j \frac{\partial b_i}{\partial x_j} - b_j \frac{\partial a_i}{\partial x_j} \right] \frac{\partial}{\partial x_i}.$$

Exercises

The following exercise requires some knowledge of topology.

Exercise 6.1. Let X be a vector field on the sphere S^k. Assume that X is nowhere zero, i.e., $X_m \neq 0$ for all $m \in S^k$. Show that the *antipodal map* $a : S^k \longrightarrow S^k$ and the identity map $S^k \longrightarrow S^k$ are homotopic. Deduce that k is odd.

Hint: Normalize the vector field so that X_m is a unit tangent vector for all m. If $m \in S^k$ consider the great circle $\theta_m : [0, 2\pi] \longrightarrow S^k$ tangent to X_m. Then $\theta_m(0) = \theta_m(2\pi) = m$, but $m \longmapsto \theta_m(\pi)$ is the antipodal map. Also, think about the effect of the antipodal map on $H^k(S^k)$.

7

Left-Invariant Vector Fields

To recapitulate, a *Lie group* is a differentiable manifold with a group structure in which the multiplication and inversion maps $G \times G \longrightarrow G$ and $G \longrightarrow G$ are smooth. A homomorphism of Lie groups is a group homomorphism that is also a smooth map.

Remark 7.1. There is a subtlety in the definition of a Lie subgroup. A *Lie subgroup of G* is best defined as a Lie group H with an injective homomorphism $i : H \longrightarrow G$. With this definition, the image of i in G is not closed, however, as the following example shows. Let G be $\mathbb{T} \times \mathbb{T}$, where $\mathbb{T}$ is the circle $\mathbb{R}/\mathbb{Z}$. Let H be $\mathbb{R}$, and let $i : H \longrightarrow G$ be the map $i(t) = (\alpha t, \beta t)$ modulo 1, where the ratio α/β is irrational. This is a Lie subgroup, but the image of H is not closed. To require a closed image in the definition of a Lie subgroup would invalidate a theorem of Chevalley that subalgebras of the Lie algebra of a Lie group correspond to Lie subgroups. If we wish to exclude this type of example, we will explicitly describe a Lie subgroup of G as a *closed* Lie subgroup.

Remark 7.2. On the other hand, in the expression "closed Lie subgroup," the term "Lie" is redundant. It may be shown that a closed subgroup of a Lie group is a submanifold and hence a Lie group. See Bröcker and Tom Dieck [25], Theorem 3.11 on p. 28; Knapp [106] Chap. I Sect. 4; or Knapp [105], Theorem 1.5 on p. 20. We will only prove this for the special case of an abelian subgroup in Theorem 15.2 below.

Suppose that M and N are smooth manifolds and $\phi : M \longrightarrow N$ is a smooth map. As we explained in Chap. 6, if $m \in M$ and $n = \phi(m)$, we get a map $d\phi : T_m(M) \longrightarrow T_n(N)$, called the *differential* of f. If ϕ is a diffeomorphism of M onto N, then we can push a vector field X on M forward this way to obtain a vector field on N. This vector field may be denoted $\phi_* X$, defined by $(\phi_* X)_n = d\phi(X_m)$ when $f(m) = n$. If ϕ is *not* a diffeomorphism, this may not work because some points in N may not even be in the image of ϕ, while others may be in the image of two different points m_1 and m_2 with no guarantee that $d\phi X_{m_1} = d\phi X_{m_2}$.

D. Bump, *Lie Groups*, Graduate Texts in Mathematics 225,
DOI 10.1007/978-1-4614-8024-2_7, © Springer Science+Business Media New York 2013

Now let G be a Lie group. If $g \in G$, then $L_g : G \longrightarrow G$ defined by $L_g(h) = gh$ is a diffeomorphism and hence induces maps $L_{g,*} : T_h(G) \longrightarrow T_{gh}(G)$. A vector field X on G is *left-invariant* if $L_{g,*}(X_h) = X_{gh}$.

Proposition 7.1. *The vector space of left-invariant vector fields is closed under* $[\,,\,]$ *and is a Lie algebra of dimension* $\dim(G)$. *If* $X_e \in T_e(G)$, *there is a unique left-invariant vector field* X *on* G *with the prescribed tangent vector at the identity.*

Proof. Given a tangent vector X_e at the identity element e of G, we may define a left-invariant vector field by $X_g = L_{g,*}(X_e)$, and conversely any left-invariant vector field must satisfy this identity, so the space of left-invariant vector fields is isomorphic to the tangent space of G at the identity. Therefore, its vector space dimension equals the dimension of G. □

Let $\mathrm{Lie}(G)$ be the vector space of left-invariant vector fields, which we may identify with the $T_e(G)$. It is clearly closed under $[\,,\,]$.

Suppose now that $G = \mathrm{GL}(n, \mathbb{C})$. We have defined two different Lie algebras for G: first, in Chap. 5, we defined the Lie algebra $\mathfrak{gl}(n, \mathbb{C})$ of G to be $\mathrm{Mat}_n(\mathbb{C})$ with the commutation relation $[X, Y] = XY - YX$ (matrix multiplication); and second, we have defined the Lie algebra to be the Lie algebra of left-invariant vector fields with the bracket (6.6). We want to see that these two definitions are the same. We will accomplish this in Proposition 7.2 below.

If $X \in \mathrm{Mat}_n(\mathbb{C})$, we begin by associating with X a left-invariant vector field. Since G is an open subset of the real vector space $V = \mathrm{Mat}_n(\mathbb{C})$, we may identify the tangent space to G at the identity with V. With this identification, an element $X \in V$ is the local derivation at I [see (6.3)] defined by

$$f \longmapsto \frac{\mathrm{d}}{\mathrm{d}t} f(I + tX) \Big|_{t=0},$$

where f is the germ of a smooth function at I. The two paths $t \longrightarrow I + tX$ and $t \longrightarrow \exp(tX) = I + tX + \cdots$ are tangent when $t = 0$, so this is the same as

$$f \longrightarrow \frac{\mathrm{d}}{\mathrm{d}t} f\big(\exp(tX)\big) \Big|_{t=0},$$

which is a better definition. Indeed, if H is a Lie subgroup of $\mathrm{GL}(n, \mathbb{C})$ and X is in the Lie algebra of H, then by Proposition 5.2, the second path $\exp(tX)$ stays within H, so this definition still makes sense.

It is clear how to extrapolate this local derivation to a left-invariant global derivation of $C^\infty(G, \mathbb{R})$. We must define

$$(\mathrm{d}X)f(g) = \frac{\mathrm{d}}{\mathrm{d}t} f\big(g \exp(tX)\big) \Big|_{t=0}. \tag{7.1}$$

By Proposition 2.8, the left-invariant derivation $\mathrm{d}X$ of $C^\infty(G, \mathbb{R})$ corresponds to a left-invariant vector field. To distinguish this derivation from the element X of $\mathrm{Mat}_n(\mathbb{C})$, we will resist the temptation to denote this derivation also as X and denote it by $\mathrm{d}X$.

Lemma 7.1. *Let f be a smooth map from a neighborhood of the origin in $\mathbb{R}^n$ into a finite-dimensional vector space. We may write*

$$f(x) = c_0 + c_1(x) + B(x, x) + r(x), \tag{7.2}$$

where $c_1 : \mathbb{R}^n \longrightarrow V$ is linear, $B : \mathbb{R}^n \times \mathbb{R}^n \longrightarrow V$ is symmetric and bilinear, and r vanishes to order 3.

Proof. This is just the familiar Taylor expansion. Denoting $u = (u_1, \ldots, u_n)$, let $c_0 = f(0)$,

$$c_1(u) = \sum_i \frac{\partial f}{\partial x_i}(0) \, u_i,$$

and

$$B(u, v) = \frac{1}{2} \sum_{i,j} \frac{\partial^2 f}{\partial x_i \partial x_j}(0) \, u_i v_j.$$

Both $f(x)$ and $c_0 + c_1(x) + B(x, x)$ have the same partial derivatives of order $\leqslant 2$, so the difference $r(x)$ vanishes to order 3. The fact that B is symmetric follows from the equality of mixed partials:

$$\frac{\partial^2 f}{\partial x_i \partial x_j}(0) = \frac{\partial^2 f}{\partial x_j \partial x_i}(0).$$

$\square$

Proposition 7.2. *If X, $Y \in \mathrm{Mat}_n(\mathbb{C})$, and if f is a smooth function on $G = \mathrm{GL}(n, \mathbb{C})$, then $\mathrm{d}[X, Y]f = \mathrm{d}X(\mathrm{d}Y f) - \mathrm{d}Y(\mathrm{d}X f)$.*

Here $[X, Y]$ means $XY - YX$ computed using matrix operations; that is, the bracket computed as in Chap. 5. This proposition shows that if $X \in \mathrm{Mat}_n(\mathbb{C})$, and if we associate with X a derivation of $C^\infty(G, \mathbb{R})$, where $G = \mathrm{GL}(n, \mathbb{C})$, using the formula (7.1), then this bracket operation gives the same result as the bracket operation (6.6) for left-invariant vector fields.

Proof. We fix a function $f \in C^\infty(G)$ and an element $g \in G$. By Lemma 7.1, we may write, for X near 0,

$$f\big(g(I + X)\big) = c_0 + c_1(X) + B(X, X) + r(X),$$

where c_1 is linear in X, B is symmetric and bilinear, and r vanishes to order 3 at $X = 0$. We will show that

$$(\mathrm{d}X \, f)(g) = c_1(X) \tag{7.3}$$

and

$$(\mathrm{d}X \circ \mathrm{d}Y \, f)(g) = c_1(XY) + 2B(X, Y). \tag{7.4}$$

Indeed,

$$(\mathrm{d}X\,f)(g) = \frac{\mathrm{d}}{\mathrm{d}t}\,f\big(g(I+tX)\big)|_{t=0}$$

$$= \frac{\mathrm{d}}{\mathrm{d}t}\,\big(c_0 + c_1(tX) + B(tX,tX) + r(tX)\big)\Big|_{t=0}.$$

We may ignore the B and r terms because they vanish to order $\geqslant 2$, and since c_1 is linear, this is just $c_1(X)$ proving (7.3). Also

$$(\mathrm{d}X \circ \mathrm{d}Y\,f)(g) = \frac{\mathrm{d}}{\mathrm{d}t}\left((\mathrm{d}Y\,f)\big(g(I+tX)\big)\right)\Big|_{u=0}$$

$$= \frac{\partial}{\partial t}\frac{\partial}{\partial u}\,f\big(g(I+tX)(I+uY)\big)\Big|_{t=u=0}$$

$$= \frac{\partial}{\partial t}\frac{\partial}{\partial u}[c_0 + c_1(tX + uY + tuXY)$$

$$+ B(tX + uY + tuXY, tX + uY + tuXY)$$

$$+ r(tX + uY + tuXY)]\,|_{t=u=0}.$$

We may omit r from this computation since it vanishes to third order. Expanding the linear and bilinear maps c_1 and B, we obtain (7.4).

Similarly,

$$(\mathrm{d}Y \circ \mathrm{d}X\,f)(g) = c_1(YX) + 2B(X,Y).$$

Subtracting this from (7.4) to kill the unwanted B term, we obtain

$$\big((\mathrm{d}X \circ \mathrm{d}Y - \mathrm{d}Y \circ \mathrm{d}X)\,f\big)(g) = c_1(XY - YX) = (\mathrm{d}[X,Y]\,f)\,(g)$$

by (7.3).   □

If $\phi : G \longrightarrow H$ is a homomorphism of Lie groups, there is an induced map of Lie algebras, as we will now explain. Let X be a left-invariant vector field on G. We have induced a map $\mathrm{d}\phi : T_e(G) \longrightarrow T_e(H)$, and by Proposition 7.1 applied to H there is a unique left-invariant vector field Y on H such that $\mathrm{d}\phi(X_e) = Y_e$. It is easy to see that for any $g \in G$ we have $\mathrm{d}\phi(X_g) = Y_{\phi(g)}$. We regard Y as an element of $\mathrm{Lie}(H)$, and $X \longmapsto Y$ is a map $\mathrm{Lie}(G) \longrightarrow \mathrm{Lie}(H)$, which we denote $\mathrm{Lie}(\phi)$ or, more simply, $\mathrm{d}\phi$. The Lie algebra homomorphism $\mathrm{d}\phi = \mathrm{Lie}(\phi)$ is called the *differential* of ϕ. A map $f : \mathfrak{g} \longrightarrow \mathfrak{h}$ of Lie algebras is naturally called a *homomorphism* if $f([X,Y]) = [f(X), f(Y)]$.

Proposition 7.3. *If* $\phi : G \longrightarrow H$ *is a Lie group homomorphism, then* $\mathrm{Lie}(\phi) :$ $\mathrm{Lie}(G) \longrightarrow \mathrm{Lie}(H)$ *is a Lie algebra homomorphism.*

Proof. If $X, Y \in G$, then X_e and Y_e are local derivations of $\mathcal{O}_e(G)$, and it is clear from the definitions that $\phi_*([X_e, Y_e]) = [\phi_*(X_e), \phi_*(Y_e)]$. Consequently, $[\mathrm{Lie}(\phi)X, \mathrm{Lie}(\phi)Y]$ and $\mathrm{Lie}(\phi)([X,Y])$ are left-invariant vector fields on H that agree at the identity, so they are the same by Proposition 7.1.   □

We may ask to what extent the Lie algebra homomorphism Lie(ϕ) contains complete information about ϕ. For example, given Lie groups G and H with Lie algebras $\mathfrak{g}$ and $\mathfrak{h}$, and a homomorphism $f : \mathfrak{g} \longrightarrow \mathfrak{h}$, is there a homomorphism $G \longrightarrow H$ with Lie(ϕ) = f?

In general, the answer is no, as the following example will show.

Example 7.1. Let $H = \mathrm{SU}(2)$ and let $G = \mathrm{SO}(3)$. H acts on the three-dimensional space V of Hermitian matrices $\xi = \begin{pmatrix} x & y+iz \\ y-iz & -x \end{pmatrix}$ of trace zero by $h : \xi \mapsto h\xi h^{-1} = h\xi^{t}\bar{h}$, and

$$\xi \mapsto -\det(\xi) = x^2 + y^2 + z^2$$

is an invariant positive definite quadratic form on V invariant under this action. Thus, the transformation $\xi \mapsto h\xi h^{-1}$ of V is orthogonal, and we have a homomorphism $\psi : \mathrm{SU}(2) \longrightarrow \mathrm{SO}(3)$. Both groups are three-dimensional, and ψ is a local homeomorphism at the identity. The differential Lie(ψ) : $\mathfrak{su}(2) \longrightarrow \mathfrak{so}(3)$ is therefore an isomorphism and has an inverse, which is a Lie algebra homomorphism $\mathfrak{so}(3) \longrightarrow \mathfrak{su}(2)$. However, ψ itself does not have an inverse since it has a nontrivial element in its kernel, $-I$. Therefore, Lie(ψ)$^{-1}$: $\mathfrak{so}(3) \longrightarrow \mathfrak{su}(2)$ is an example of a Lie algebra homomorphism that does not correspond to a Lie group homomorphism $\mathrm{SO}(3) \longrightarrow \mathrm{SU}(2)$.

Nevertheless, we will see later (Proposition 14.2) that if G is *simply connected*, then any Lie algebra homomorphism $\mathfrak{g} \longrightarrow \mathfrak{h}$ corresponds to a Lie group homomorphism $G \longrightarrow H$. Thus, the obstruction to lifting the Lie algebra homomorphism $\mathfrak{so}(3) \longrightarrow \mathfrak{su}(2)$ to a Lie group homomorphism is topological and corresponds to the fact that $\mathrm{SO}(3)$ is not simply connected.

Exercises

Exercise 7.1. Compute the Lie algebra homomorphism Lie(ψ) : $\mathfrak{su}(2) \longrightarrow \mathfrak{so}(3)$ of Example 7.1 explicitly.

Exercise 7.2. Show that no Lie group can be homeomorphic to the sphere S^k if k is even. On the other hand, show that $\mathrm{SU}(2) \cong S^3$. (**Hint**: Use Exercise 6.1.)

Exercise 7.3. Let J be the matrix (5.3). Let $\mathfrak{o}(N, \mathbb{C})$ and $\mathfrak{o}_J(\mathbb{C})$ be the complexified Lie algebras of the groups $\mathrm{O}(N)$ and $\mathrm{O}_J(\mathbb{C})$ in Exercise 5.9. Show that these complex Lie algebras are isomorphic. Describe $\mathfrak{o}(N, \mathbb{C})$ explicitly, i.e., write down a typical matrix.

8

The Exponential Map

The exponential map, introduced for closed Lie subgroups of $\mathrm{GL}(n, \mathbb{C})$ in Chap. 5, can be defined for a general Lie group G as a map $\mathrm{Lie}(G) \longrightarrow G$.

We may consider a vector field (6.5) that is allowed to vary smoothly. By this we mean that we introduce a real parameter $\lambda \in (-\epsilon, \epsilon)$ for some $\epsilon > 0$ and smooth functions $a_i : M \times (-\epsilon, \epsilon) \longrightarrow \mathbb{C}$ and consider a vector field, which in local coordinates is given by

$$(Xf)(m) = \sum_{i=1}^{n} a_i(m, \lambda) \frac{\partial f}{\partial x_i}(m). \tag{8.1}$$

Proposition 8.1. *Suppose that M is a smooth manifold, $m \in M$, and X is a vector field on M. Then, for sufficiently small $\epsilon > 0$, there exists a path $p : (-\epsilon, \epsilon) \longrightarrow M$ such that $p(0) = m$ and $p_*(d/dt)(t) = X_{p(t)}$ for $t \in (-\epsilon, \epsilon)$. Such a curve, on whatever interval it is defined, is uniquely determined. If the vector field X is allowed to depend on a parameter λ as in (8.1), then for small values of t, $p(t)$ depends smoothly on λ.*

Here we are regarding the interval $(-\epsilon, \epsilon)$ as a manifold, and $p_*(d/dt)$ is the image of the tangent vector d/dt. We call such a curve an *integral curve* for the vector field.

Proof. In terms of local coordinates $x_1, \ldots, x_n$ on M, the vector field X is

$$\sum a_i(x_1, \ldots, x_n) \frac{\partial}{\partial x_i},$$

where the a_i are smooth functions in the coordinate neighborhood. If a path $p(t)$ is specified, let us write $x_i(t)$ for the x_i component of $p(t)$, with the coordinates of m being $x_1 = \cdots = x_n = 0$. Applying the tangent vector $p_*(t)(d/dt)(t)$ to a function $f \in C^\infty(G)$ gives

$$\frac{d}{dt} f(x_1(t), \ldots, x_n(t)) = \sum x_i'(t) \frac{\partial f}{\partial x_i}(x_1(t), \ldots, x_n(t)).$$

D. Bump, *Lie Groups*, Graduate Texts in Mathematics 225,
DOI 10.1007/978-1-4614-8024-2_8, © Springer Science+Business Media New York 2013

On the other hand, applying $X_{p(t)}$ to the same f gives

$$\sum_i a_i\big(x_1(t),\ldots,x_n(t)\big) \frac{\partial f}{\partial x_i}\big(x_1(t),\ldots,x_n(t)\big),$$

so we need a solution to the first-order system

$$x_i'(t) = a_i\big(x_1(t),\ldots,x_n(t)\big), \qquad x_i(0) = 0, \qquad (i = 1,\ldots,n).$$

The existence of such a solution for sufficiently small $|t|$, and its uniqueness on whatever interval it does exist, is guaranteed by a standard result in the theory of ordinary differential equations, which may be found in most texts. See, for example, Ince [81], Chap. 3, particularly Sect. 3.3, for a rigorous treatment. The required Lipschitz condition follows from smoothness of the a_i. For the statement about continuously varying vector fields, one needs to know the corresponding fact about first-order systems, which is discussed in Sect. 3.31 of [81]. Here Ince imposes an assumption of analyticity on the dependence of the differential equation on λ, which he allows to be a complex parameter, because he wants to conclude analyticity of the solutions; if one weakens this assumption of analyticity to smoothness, one still gets smoothness of the solution. □

In general, the existence of the integral curve of a vector field is only guaranteed in a small segment $(-\epsilon, \epsilon)$, as in Proposition 8.1. However, we will now see that, for left-invariant vector fields on a Lie group, the integral curve extends to all $\mathbb{R}$. This fact underlies the construction of the exponential map.

Theorem 8.1. *Let G be a Lie group and $\mathfrak{g}$ its Lie algebra. There exists a map* $\exp : \mathfrak{g} \longrightarrow G$ *that is a local homeomorphism in a neighborhood of the origin in $\mathfrak{g}$ such that, for any $X \in \mathfrak{g}$, $t \longrightarrow \exp(tX)$ is an integral curve for the left-invariant vector field X. Moreover, $\exp\big((t + u)X\big) = \exp(tX)\exp(uX)$.*

Proof. Let $X \in \mathfrak{g}$. We know that for sufficiently small $\epsilon > 0$ there exists an integral curve $p : (-\epsilon, \epsilon) \longrightarrow G$ for the left-invariant vector field X with $p(0) = 1$. We show first that if $p : (a, b) \longrightarrow G$ is any integral curve for an open interval (a, b) containing 0, then

$$p(s)\,p(t) = p(s + t) \text{ when } s, t, s + t \in (a, b). \tag{8.2}$$

Indeed, since X is invariant under left-translation, left-translation by $p(s)$ takes an integral curve for the vector field into another integral curve. Thus, $t \longrightarrow p(s)\,p(t)$ and $t \longrightarrow p(s + t)$ are both integral curves, with the same initial condition $0 \longrightarrow p(s)$. They are thus the same.

 With this in mind, we show next that if $p : (-a, a) \longrightarrow G$ is an integral curve for the left-invariant vector field X, then we may extend it to all of $\mathbb{R}$. Of course, it is sufficient to show that we may extend it to $(-\frac{3}{2}a, \frac{3}{2}a)$. We extend it by the rule $p(t) = p(a/2)\,p(t - a/2)$ when $-a/2 \leqslant t \leqslant 3a/2$ and

$p(t) = p(-a/2)\, p(t + a/2)$ when $-3a/2 \leqslant t \leqslant a/2$, and it follows from (8.2) that this definition is consistent on regions of overlap.

Now define $\exp : \mathfrak{g} \longrightarrow G$ as follows. Let $X \in \mathfrak{g}$, and let $p : \mathbb{R} \longrightarrow G$ be an integral curve for the left-invariant vector field X with $p(0) = 0$. We define $\exp(X) = p(1)$. We note that if $u \in \mathbb{R}$, then $t \mapsto p(tu)$ is an integral curve for uX, so $\exp(uX) = p(u)$.

The exponential map is a smooth map, at least for X near the origin in $\mathfrak{g}$, by the last statement in Proposition 8.1. Identifying the tangent space at the origin in the vector space $\mathfrak{g}$ with $\mathfrak{g}$ itself, $\exp$ induces a map $T_0(\mathfrak{g}) \longrightarrow T_e(G)$ (that is $\mathfrak{g} \longrightarrow \mathfrak{g}$), and this map is the identity map by construction. Thus, the Jacobian of $\exp$ is nonzero and, by the Inverse Function Theorem, $\exp$ is a local homeomorphism near 0. $\square$

We also denote $\exp(X)$ as e^X for $X \in \mathfrak{g}$.

Remark 8.1. If $G = \mathrm{GL}(n, \mathbb{C})$, then as we explained in Chap. 7, Proposition 7.2 allows us to identify the Lie algebra of G with $\mathrm{Mat}_n(\mathbb{C})$. We observe that the definition of $\exp : \mathrm{Mat}_n(\mathbb{C}) \longrightarrow \mathrm{GL}(n, \mathbb{C})$ by a series in (5.2) agrees with the definition in Theorem 8.1. This is because $t \longmapsto \exp(tX)$ with either definition is an integral curve for the same left-invariant vector field, and the uniqueness of such an integral curve follows from Proposition 8.1.

Proposition 8.2. *Let G, H be Lie groups and let $\mathfrak{g}$, $\mathfrak{h}$ be their respective Lie algebras. Let $f : G \to H$ be a homomorphism. Then the following diagram is commutative:*

$$
\begin{array}{ccc}
\mathfrak{g} & \xrightarrow{\; df \;} & \mathfrak{h} \\
\downarrow{\scriptstyle \exp} & & \downarrow{\scriptstyle \exp} \\
G & \xrightarrow{\; f \;} & H
\end{array}
$$

Proof. It is clear from the definitions that f takes an integral curve for a left-invariant vector field X on G to an integral curve for $df(X)$, and the statement follows. $\square$

A *representation* of a Lie algebra $\mathfrak{g}$ over a field F is a Lie algebra homomorphism $\pi : \mathfrak{g} \longrightarrow \mathrm{End}(V)$, where V is an F-vector space, or more generally a vector space over a field E containing F, and $\mathrm{End}(V)$ is given the Lie algebra structure that it inherits from its structure as an associative algebra. Thus,

$$\pi([x, y]) = \pi(x)\, \pi(y) - \pi(y)\, \pi(x).$$

We may sometimes find it convenient to denote $\pi(x)v$ as just xv for $x \in \mathfrak{g}$ and $v \in V$. We may think of $(x, v) \mapsto xv = \pi(x)v$ as a multiplication. If V is a vector space, given a map $\mathfrak{g} \times V \longrightarrow V$ denoted $(x, v) \mapsto xv$ such that $x \mapsto \pi(x)$ is a representation, where $\pi(x) : V \longrightarrow V$ is the endomorphism $v \longrightarrow xv$, then we call V a $\mathfrak{g}$-*module*. A *homomorphism* $\phi : U \longrightarrow V$ of $\mathfrak{g}$-modules is an F-linear map satisfying $\phi(xv) = x\phi(v)$.

Example 8.1. If $\pi : G \longrightarrow \mathrm{GL}(V)$ is a representation, where V is a real or complex vector space, then the Lie algebra of $\mathrm{GL}(V)$ is $\mathrm{End}(V)$, so the differential $\mathrm{Lie}(\pi) : \mathrm{Lie}(G) \longrightarrow \mathrm{End}(V)$, defined by Proposition 7.3, is a Lie algebra representation.

By the universal property of $U(\mathfrak{g})$ in Theorem 10.1, A Lie algebra representation $\pi : \mathfrak{g} \longrightarrow \mathrm{End}(V)$ extends to a ring homomorphism $U(\mathfrak{g}) \longrightarrow \mathrm{End}(V)$, which we continue to denote as π.

If $\mathfrak{g}$ is a Lie algebra over a field F, we get a homomorphism $\mathrm{ad} : \mathfrak{g} \longrightarrow \mathrm{End}(\mathfrak{g})$, called the *adjoint map*, defined by $\mathrm{ad}(x)y = [x, y]$. We give $\mathrm{End}(\mathfrak{g})$ the Lie algebra structure it inherits as an associative ring. We have

$$\mathrm{ad}(x)([y, z]) = [\mathrm{ad}(x)(y), z] + [y, \mathrm{ad}(x)(z)] \tag{8.3}$$

since, by the Jacobi identity, both sides equal $[x, [y, z]] = [[x, y], z] + [y, [x, z]]$. This means that $\mathrm{ad}(x)$ is a derivation of $\mathfrak{g}$.

Also

$$\mathrm{ad}(x)\,\mathrm{ad}(y) - \mathrm{ad}(y)\,\mathrm{ad}(x) = \mathrm{ad}([x, y]) \tag{8.4}$$

since applying either side to $z \in \mathfrak{g}$ gives $[x, [y, z]] - [y, [x, z]] = [[x, y], z]$ by the Jacobi identity. So $\mathrm{ad} : \mathfrak{g} \longrightarrow \mathrm{End}(\mathfrak{g})$ is a Lie algebra representation.

We next explain the geometric origin of ad. To begin with, representations of Lie algebras arise naturally from representations of Lie groups. Suppose that G is a Lie group and $\mathfrak{g}$ is its Lie algebra. If V is a vector space over $\mathbb{R}$ or $\mathbb{C}$, any Lie group homomorphism $\pi : G \longrightarrow \mathrm{GL}(V)$ induces a Lie algebra homomorphism $\mathfrak{g} \longrightarrow \mathrm{End}(V)$ by Proposition 7.3; that is, a real or complex representation.

In particular, G acts on itself by conjugation, and so it acts on $\mathfrak{g} = T_e(G)$. This representation is called the *adjoint representation* and is denoted $\mathrm{Ad} : G \longrightarrow \mathrm{GL}(\mathfrak{g})$. We show next that the differential of Ad is ad. That is:

Theorem 8.2. *Let G be a Lie group, $\mathfrak{g}$ its Lie algebra, and $\mathrm{Ad} : G \longrightarrow \mathrm{GL}(\mathfrak{g})$ the adjoint representation. Then the Lie group representation $\mathfrak{g} \longrightarrow \mathrm{End}(\mathfrak{g})$ corresponding to Ad by Proposition 7.3 is ad.*

Proof. It will be most convenient for us to think of elements of the Lie algebra as tangent vectors at the identity or as local derivations of the local ring there. Let $X, Y \in \mathfrak{g}$. If $f \in C^\infty(G)$, define $c(g)f(h) = f(g^{-1}hg)$. Then our definitions of the adjoint representation amount to

$$\big(\mathrm{Ad}(g)Y\big)f = Y\big(c(g^{-1})f\big).$$

To compute the differential of Ad, note that the path $t \longrightarrow \exp(tX)$ in G is tangent to the identity at $t = 0$ with tangent vector X. Therefore, under the representation of $\mathfrak{g}$ in Proposition 7.3, X maps Y to the local derivation at the identity

$$f \longmapsto \frac{\mathrm{d}}{\mathrm{d}t}\big(\mathrm{Ad}(\mathrm{e}^{tX})Y\big)f\,\Big|_{t=0} = \frac{\mathrm{d}}{\mathrm{d}t}\frac{\mathrm{d}}{\mathrm{d}u}f(\mathrm{e}^{tX}\mathrm{e}^{uY}\mathrm{e}^{-tX})\,\Big|_{t=u=0}.$$

By the chain rule, if $F(t_1, t_2)$ is a function of two real variables,

$$\frac{\mathrm{d}}{\mathrm{d}t} F(t,t) \Big|_{t=0} = \frac{\partial F}{\partial t_1}(0,0) + \frac{\partial F}{\partial t_2}(0,0). \tag{8.5}$$

Applying this, with u fixed to $F(t_1, t_2) = f(e^{t_1 X} e^{uY} e^{-t_2 X})$, our last expression equals

$$\frac{\mathrm{d}}{\mathrm{d}u} \frac{\mathrm{d}}{\mathrm{d}t} f(e^{tX} e^{uY}) \Big|_{t=u=0} - \frac{\mathrm{d}}{\mathrm{d}u} \frac{\mathrm{d}}{\mathrm{d}t} f(e^{uY} e^{tX}) \Big|_{t=u=0} = XYf(1) - YXf(1).$$

This is, of course, the same as the effect of $[X, Y] = \mathrm{ad}(X)Y$.   □

Exercises

Exercise 8.1. Show that the exponential map $\mathfrak{su}(2) \to \mathrm{SU}(2)$ is surjective, but the exponential map $\mathfrak{sl}(2, \mathbb{R}) \to \mathrm{SL}(2, \mathbb{R})$ is not.

9

Tensors and Universal Properties

We will review the basic properties of the tensor product and use them to illustrate the basic notion of a *universal property*, which we will see repeatedly.

If R is a commutative ring and M, N, and P are R-modules, then a *bilinear map* $f : M \times N \longrightarrow P$ is a map satisfying

$$f(r_1 m_1 + r_2 m_2, n) = r_1 f(m_1, n) + r_2 f(m_2, n), \qquad r_i \in R, m_i \in M, n \in N,$$

$$f(m, r_1 n_1 + r_2 n_2) = r_1 f(m, n_1) + r_2 f(m, n_2), \qquad r_i \in R, n_i \in N, m \in M.$$

More generally, if $M_1, \ldots, M_k$ are R-modules, the notion of a *k-linear map* $M_1 \times \cdots \times M_k \longrightarrow P$ is defined similarly: the map must be linear in each variable.

The *tensor product* $M \otimes_R N$ is an R-module together with a bilinear map $\otimes : M \times N \longrightarrow M \otimes_R N$ satisfying the following property.

Universal Property of the Tensor Product. *If P is any R-module and $p : M \times N \longrightarrow P$ is a bilinear map, there exists a unique R-module homomorphism $F : M \otimes N \longrightarrow P$ such that $p = F \circ \otimes$.*

Why do we call this a universal property? It says that $\otimes : M \times N \longrightarrow M \otimes N$ is a "universal" bilinear map in the sense that any bilinear map of $M \times N$ factors through it. As we will explain, the module $M \otimes_R N$ is uniquely determined by the universal property. This is important beyond the immediate example because often objects are described by universal properties. Before we explain this point (which is obvious if one thinks about it correctly), let us make a categorical observation.

If $\mathcal{C}$ is a category, an *initial object* in $\mathcal{C}$ is an object X_0 such that, for each object Y, the Hom set $\text{Hom}_{\mathcal{C}}(X_0, Y)$ consists of a single element. A *terminal object* is an object X_∞ such that, for each object Y, the Hom set $\text{Hom}_{\mathcal{C}}(Y, X_\infty)$ consists of a single element. For example, in the category of sets, the empty set is an initial object and a set consisting of one element is a terminal object.

Lemma 9.1. *In any category, any two initial objects are isomorphic. Any two terminal objects are isomorphic.*

Proof. If X_0 and X_1 are initial objects, there exist unique morphisms $f : X_0 \longrightarrow X_1$ (since X_0 is initial) and $g : X_1 \longrightarrow X_0$ (since X_1 is initial). Then $g \circ f : X_0 \longrightarrow X_0$ and $1_{X_0} : X_0 \longrightarrow X_0$ must coincide since X_0 is initial, and similarly $f \circ g = 1_{X_1}$. Thus f and g are inverse isomorphisms. Similarly, terminal objects are isomorphic. $\square$

Theorem 9.1. *The tensor product $M \otimes_R N$, if it exists, is determined up to isomorphism by the universal property.*

Proof. Let $\mathcal{C}$ be the following category. An object in $\mathcal{C}$ is an ordered pair (P, p), where P is an R-module and $p : M \times N \longrightarrow P$ is a bilinear map. If $X = (P, p)$ and $Y = (Q, q)$ are objects, then a morphism $X \longrightarrow Y$ consists of an R-module homomorphism $f : P \longrightarrow Q$ such that $q = f \circ p$. The universal property of the tensor product means that $\otimes : M \times N \longrightarrow M \otimes N$ is an initial object in this category and therefore determined up to isomorphism. $\square$

Of course, we usually denote $\otimes(m, n)$ as $m \otimes n$ in $M \otimes_R N$. We have not proved that $M \otimes_R N$ exists. We refer to any text on algebra for this fact, such as Lang [116], Chap. XVI.

In general, by a *universal property* we mean *any characterization of a mathematical object that can be expressed by saying that some associated object is an initial or terminal object in some category.* The basic paradigm is that *a universal property characterizes an object up to isomorphism.*

A typical application of the universal property of the tensor product is to make $M \otimes_R N$ into a functor. Specifically, if $\mu : M \longrightarrow M'$ and $\nu : N \longrightarrow N'$ are R-module homomorphisms, then there is a unique R-module homomorphism $\mu \otimes \nu : M \otimes_R N \longrightarrow M' \otimes_R N'$ such that $(\mu \otimes \nu)(m \otimes n) = \mu(m) \otimes \nu(n)$. We get this by applying the universal property to the R-bilinear map $M \times N \longrightarrow M' \otimes N'$ defined by $(m, n) \longmapsto \mu(m) \otimes \nu(n)$.

As another example of an object that can be defined by a universal property, let V be a vector space over a field F. Let us ask for an F-algebra $\bigotimes V$ together with an F-linear map $i : V \longrightarrow \bigotimes V$ satisfying the following condition.

Universal Property of the Tensor Algebra. *If A is any F-algebra and $\phi : V \longrightarrow A$ is an F-linear map then there exists a unique F-algebra homomorphism $\Phi : \bigotimes V \longrightarrow A$ such that $r = \rho \circ i$.*

It should be clear from the previous discussion that this universal property characterizes the tensor algebra up to isomorphism. To prove existence, we can construct a ring with this exact property as follows. Let unadorned $\otimes$ mean $\otimes_F$ in what follows. By $\otimes^k V$ we mean the k-fold tensor product $V \otimes \cdots \otimes V$ (k times); if $k = 0$, then it is natural to take $\otimes^0 V = F$ while $\otimes^1 V = V$. If V has finite dimension d, then $\otimes^k V$ has dimension d^k. Let

$$\bigotimes V = \bigoplus_{k=0}^{\infty} \left(\otimes^k V \right).$$

Then $\bigotimes V$ has the natural structure of a graded F-algebra in which the multiplication $\otimes^k V \times \otimes^l V \longrightarrow \otimes^{k+l} V$ sends

$$(v_1 \otimes \cdots \otimes v_k, u_1 \otimes \cdots \otimes u_l) \longrightarrow v_1 \otimes \cdots \otimes v_k \otimes u_1 \otimes \cdots \otimes u_l.$$

We regard V as a subset of $\bigotimes V$ embedded onto $\otimes^1 V = V$.

Proposition 9.1. *The universal property of the tensor algebra is satisfied.*

Proof. If $\phi : V \longrightarrow A$ is any linear map of V into an F-algebra, define a map $\Phi : \bigotimes V \longrightarrow A$ by $\Phi(v_1 \otimes \cdots \otimes v_k) = \phi(v_1) \cdots \phi(v_k)$ on $\otimes^k V$. It is easy to see that Φ is a ring homomorphism. It is unique since V generates $\bigotimes V$ as an F-algebra. $\qquad\square$

A *graded algebra* over the field F is an F-algebra A with a direct sum decomposition

$$A = \bigoplus_{k=0}^{\infty} A_k$$

such that $A_k A_l \subseteq A_{k+l}$. In most examples we will have $A_0 = F$. Elements of A_k are called *homogeneous* of degree k. The tensor algebra is a graded algebra, with $\otimes^k V$ being the homogeneous part of degree k.

Next we define the *symmetric* and *exterior powers* of a vector space V over the field F. Let V^k denote $V \times \cdots \times V$ (k times). A k-linear map $f : V^k \longrightarrow U$ into another vector space is called *symmetric* if for any $\sigma \in S_k$ it satisfies $f(v_{\sigma(1)}, \ldots, v_{\sigma(k)}) = f(v_1, \ldots, v_k)$ and *alternating* if $f(v_{\sigma(1)}, \ldots, v_{\sigma(k)}) = \varepsilon(\sigma) f(v_1, \ldots, v_k)$, where $\varepsilon : S_k \longrightarrow \{\pm 1\}$ is the alternating (sign) character. The kth symmetric and exterior powers of V, denoted $\vee^k V$ and $\wedge^k V$, are F-vector spaces, together with k-linear maps $\vee : V^k \longrightarrow \vee^k V$ and $\wedge : V^k \longrightarrow \wedge^k V$. The map $\vee$ is symmetric, and the map $\wedge$ is alternating. We normally denote $\vee(v_1, \ldots, v_k) = v_1 \vee \cdots \vee v_k$ and similarly for $\wedge$. The following universal properties are required.

Universal Properties of the Symmetric and Exterior Powers: *Let $f : V^k \longrightarrow U$ be any symmetric (resp. alternating) k-linear map. Then there exists a unique F-linear map $\phi : \vee^k V \longrightarrow U$ (resp. $\wedge^k V \longrightarrow U$) such that $f = \phi \circ \vee$ (resp. $f = \phi \circ \wedge$).*

As usual, the symmetric and exterior algebras are characterized up to isomorphism by the universal property. We may construct $\vee^k V$ as a quotient of $\otimes^k V$, dividing by the subspace W generated by elements of the form $v_1 \otimes \cdots \otimes v_k - v_{\sigma(1)} \otimes \cdots \otimes v_{\sigma(k)}$, with a similar construction for $\wedge^k$. The universal property of $\vee^k V$ then follows from the universal property of the tensor product. Indeed, if $f : V^k \longrightarrow U$ is any symmetric k-linear map, then there is induced a linear map $\psi : \otimes^k V \longrightarrow U$ such that $f = \psi \circ \otimes$. Since f is symmetric, ψ vanishes on W, so ψ induces a map $\vee^k V = \otimes^k V / W \longrightarrow U$ and the universal property follows.

If V has dimension d, then $\vee^k V$ has dimension $\binom{d+k-1}{k}$, for if $x_1, \ldots, x_d$ is a basis of V, then $\{x_{i_1} \vee \cdots \vee x_{i_k} \mid 1 \leqslant i_1 \leqslant i_2 \leqslant \cdots \leqslant i_k \leqslant d\}$ is a basis for $\vee^k V$. On the other hand, the exterior power vanishes unless $k \leqslant d$, in which case it has dimension $\binom{d}{k}$. A basis consists of $\{x_{i_1} \wedge \cdots \wedge x_{i_k} \mid 1 \leqslant i_1 < i_2 < \cdots < i_k \leqslant d\}$. The vector spaces $\vee^k V$ may be collected together to make a commutative graded algebra:

$$\bigvee V = \bigoplus_{k=0}^{\infty} \vee^k V.$$

This is the *symmetric algebra*. The exterior algebra $\bigwedge V = \bigoplus_k \wedge^k V$ is constructed similarly. The spaces $\vee^0 V$ and $\wedge^0 V$ are one-dimensional and it is natural to take $\vee^0 V = \wedge^0 V = F$.

Exercises

Exercise 9.1. Let V be a finite-dimensional vector space over a field F that may be assumed to be infinite. Let $\mathcal{P}(V)$ be the ring of polynomial functions on V. Note that an element of the dual space V^* is a function on V, so regarding this function as a polynomial gives an injection $V^* \longrightarrow \mathcal{P}(V)$. Show that this linear map extends to a ring isomorphism $\bigvee V^* \longrightarrow \mathcal{P}(V)$.

Exercise 9.2. Prove that if V is a vector space, then $V \otimes V \cong (V \wedge V) \oplus (V \vee V)$.

Exercise 9.3. Use the universal properties of the symmetric and exterior power to show that if V and W are vector spaces, then there are maps $\vee^k f : \vee^k V \longrightarrow \vee^k W$ and $\wedge^k f : \wedge^k V \longrightarrow \wedge^k W$ such that

$$\vee^k f(v_1 \vee \cdots \vee v_k) = f(v_1) \vee \cdots \vee f(v_k), \qquad \wedge^k f(v_1 \wedge \cdots \wedge v_k) = f(v_1) \wedge \cdots \wedge f(v_k).$$

Exercise 9.4. Suppose that $V = F^4$. Let $f : V \longrightarrow V$ be the linear transformation with eigenvalues a, b, c, d. Compute the traces of the linear transformations $\vee^2 f$ and $\wedge^2 f$ on $\vee^2 V$ and $\wedge^2 V$ as polynomials in a, b, c, d.

Exercise 9.5. Let A and B be algebras over the field F. Then $A \otimes B$ is also an algebra, with multiplication $(a \otimes b)(a' \otimes b') = aa' \otimes bb'$. Show that there are ring homomorphisms $i : A \to A \otimes B$ and $j : B \to A \otimes B$ such that if $f : A \to C$ and $g : B \to C$ are ring homomorphisms into a ring C satisfying $f(a) g(b) = g(b) f(a)$ for $a \in A$ and $b \in B$, then there exists a unique ring homomorphism $\phi : A \otimes B \to C$ such that $\phi \circ i = f$ and $\phi \circ j = g$.

Exercise 9.6. Show that if U and V are finite-dimensional vector spaces over F then show that

$$\bigvee(U \oplus V) \cong \left(\bigvee U\right) \otimes \left(\bigvee U\right)$$

and

$$\bigwedge(U \oplus V) \cong \left(\bigwedge U\right) \otimes \left(\bigwedge U\right).$$

10

The Universal Enveloping Algebra

We have seen that elements of the Lie algebra of a Lie group G are derivations of $C^\infty(G)$. They are thus first-order differential operators that are left-invariant. The universal enveloping algebra is a purely algebraically defined ring that may be identified with the ring of all left-invariant differential operators, including higher-order ones.

We recall from Example 5.6 that if A is an associative algebra, then A may be regarded as a Lie algebra by the rule $[a, b] = ab - ba$ for $a, b \in A$. We will denote this Lie algebra by $\mathrm{Lie}(A)$.

Theorem 10.1. *Let $\mathfrak{g}$ be a Lie algebra over a field F. There exists an associative F-algebra $U(\mathfrak{g})$ with a Lie algebra homomorphism $i : \mathfrak{g} \longrightarrow \mathrm{Lie}(U(\mathfrak{g}))$ such that if A is any F-algebra, and $\phi : \mathfrak{g} \longrightarrow \mathrm{Lie}(A)$ is a Lie algebra homomorphism, then there exists a unique F-algebra homomorphism $\Phi : U(\mathfrak{g}) \longrightarrow A$ such that $\phi = \Phi \circ i$.*

As always, an object [in this case $U(\mathfrak{g})$] defined by a universal property is characterized up to isomorphism by that property.

Proof. Let $\mathcal{K}$ be the ideal in $\bigotimes \mathfrak{g}$ generated by elements of the form $[x, y] - (x \otimes y - y \otimes x)$ for $x, y \in \mathfrak{g}$, and let $U(\mathfrak{g})$ be the quotient $\bigotimes V/\mathcal{K}$. Let $\phi : \mathfrak{g} \longrightarrow \mathrm{Lie}(A)$ be a Lie algebra homomorphism. This means that ϕ is an F-linear map such that $\phi([x, y]) = \phi(x)\phi(y) - \phi(y)\phi(x)$. Then ϕ extends to a ring homomorphism $\bigotimes \mathfrak{g} \longrightarrow A$ by Proposition 9.1. Our assumption implies that $\mathcal{K}$ is in the kernel of this homomorphism, and so there is induced a ring homomorphism $U(\mathfrak{g}) \longrightarrow A$. Clearly, $U(\mathfrak{g})$ is generated by the image of $\mathfrak{g}$, so this homomorphism is uniquely determined. $\square$

Suppose that $\mathfrak{g}$ is the Lie algebra of a Lie group G. Consider the ring A of vector space endomorphisms of $C^\infty(G)$ that commute with left translation by elements of G. As we have already seen, elements of $\mathfrak{g}$ are left-invariant differential operators, by means of the action

$$X f(g) = \frac{\mathrm{d}}{\mathrm{d}t} f(g e^{tX})|_{t=0}. \tag{10.1}$$

By the universal property of the universal enveloping algebra, this action extends to a ring homomorphism $U(\mathfrak{g}) \longrightarrow A$, the image of which consists of left-invariant differential operators [Exercise 10.2 (i)]. Let us apply this observation to give a quick analytic proof of a fact that has a longer purely algebraic proof.

Proposition 10.1. *If $\mathfrak{g}$ is the Lie algebra of a Lie group G, then the natural map $i : \mathfrak{g} \longrightarrow U(\mathfrak{g})$ is injective.*

It is a consequence of the Poincaré–Birkhoff–Witt theorem, a standard and purely algebraic theorem, that $i : \mathfrak{g} \longrightarrow U(\mathfrak{g})$ is injective for *any* Lie algebra. Instead of proving the Poincaré–Birkhoff–Witt theorem, we give a short proof of this weaker statement.

Proof. Let A be the ring of endomorphisms of $C^\infty(G)$. Regarding $X \in \mathfrak{g}$ as a derivation of $C^\infty(G)$ acting by (10.1), we have a Lie algebra homomorphism $\mathfrak{g} \longrightarrow \mathrm{Lie}(A)$, which by Theorem 10.1 induces a map $U(\mathfrak{g}) \longrightarrow A$. If $X \in \mathfrak{g}$ had zero image in $U(\mathfrak{g})$, it would have zero image in A. It would therefore be zero.   □

The center of $U(\mathfrak{g})$ is very important. One reason for this is that while elements of $U(\mathfrak{g})$ are realized as differential operators that are invariant under left-translation, elements of the center are invariant under both left and right translation. [Exercise 10.2 (ii)]. Moreover, the center acts by scalars on any irreducible subspace, as we see in the following version of Schur's lemma.

A representation (π, V) of a Lie algebra $\mathfrak{g}$ is *irreducible* if there is no proper nonzero subspace $U \subset V$ such that $\pi(x)U \subseteq U$ for all $x \in \mathfrak{g}$.

Proposition 10.2. *Let $\pi : \mathfrak{g} \longrightarrow \mathrm{End}(V)$ be an irreducible representation of the Lie algebra $\mathfrak{g}$. If c is in the center of $U(\mathfrak{g})$, then there exists a scalar λ such that $\pi(c) = \lambda I_V$.*

Proof. Let λ be any eigenvalue of $\pi(c)$. Let U be the λ-eigenspace of $\pi(c)$. Since $\pi(c)$ commutes with $\pi(x)$ for all $x \in \mathfrak{g}$, we see that $\pi(x)U \subseteq U$ for all $x \in \mathfrak{g}$. By the definition of irreducibility, $U = V$, so $\pi(c)$ acts by the scalar λ.   □

Thus, the center of $U(\mathfrak{g})$ is extremely important. One particular element, the *Casimir element*, is especially important. To give two examples of its significance, the Casimir element gives rise to the Laplace–Beltrami operator, the spectral theory for which is very important in noneuclidean geometry. It is also fundamental in the theory of Kac–Moody Lie algebras. This theory generalizes the theory of finite-dimensional Lie algebras to an infinite-dimensional setting in which (remarkably) all the main theorems remain valid. One of the key features of this theory is how the Casimir element becomes the key ingredient in many proofs (such as that of the Weyl character formula) where other tools are no longer available. See [92].

Our next task will be to construct the Casimir elements. This requires a discussion of invariant bilinear forms. If V is a vector space over F and $\pi : \mathfrak{g} \longrightarrow \operatorname{End}(V)$ is a representation, then we call a bilinear form B on V *invariant* if

$$B(\pi(X)v, w) + B(v, \pi(X)w) = 0 \tag{10.2}$$

for $X \in \mathfrak{g}$, $v, w \in V$. The following proposition shows that this notion of invariance is the Lie algebra analog of the more intuitive corresponding notion for Lie groups.

Proposition 10.3. *Suppose that G is a Lie group, $\mathfrak{g}$ its Lie algebra, and $\pi : G \longrightarrow \operatorname{GL}(V)$ a representation admitting an invariant bilinear form B. Then B is invariant for the differential of π.*

Proof. Invariance under π means that

$$B(\pi(e^{tX})v, \pi(e^{tX})w) = B(v, w).$$

The derivative of this with respect to t is zero. By (8.5), this derivative is

$$B(\pi(X)v, w) + B(v, \pi(X)w).$$

We see that (10.2) is satisfied.   □

If (π, V) is a representation of $\mathfrak{g}$, define a bilinear form $B_V : \mathfrak{g} \times \mathfrak{g} \longrightarrow \mathbb{C}$ by $B_V(X, Y) = \operatorname{tr}(\pi(X)\pi(Y))$. This is the *trace bilinear form* on $\mathfrak{g}$ with respect to V. In the special case where $V = \mathfrak{g}$ and π is the adjoint representation, the trace bilinear form is called the *Killing form*.

Proposition 10.4. *Suppose that (π, V) is a representation of $\mathfrak{g}$. Then the trace bilinear form on $\mathfrak{g}$ is invariant for the adjoint representation $\operatorname{ad} : \mathfrak{g} \longrightarrow \operatorname{End}(\mathfrak{g})$.*

Proof. Invariance under ad means

$$B([x, y], z) + B(y, [x, z]) = 0. \tag{10.3}$$

Since π is a representation, $\pi([x, y]) = \pi(x)\pi(y) - \pi(y)\pi(x)$, so $B([x, y], z)$ is the trace of

$$\pi(x)\,\pi(y)\,\pi(z) - \pi(y)\,\pi(x)\,\pi(z)$$

while $B(y, [x, z])$ is the trace of

$$\pi(y)\pi(x)\pi(z) - \pi(y)\pi(z)\pi(x).$$

Using the property of endomorphisms A and B of a vector space that $\operatorname{tr}(AB) = \operatorname{tr}(BA)$, these sum to zero. This same fact implies that $B(x, y) = B(y, x)$.   □

Now given an invariant bilinear form on $\mathfrak{g}$, we may construct an element of the center, provided the bilinear form is nondegenerate.

Theorem 10.2. *Suppose that the Lie algebra $\mathfrak{g}$ admits a nondegenerate invariant bilinear form B. Let $x_1, \ldots, x_d$ be a basis of $\mathfrak{g}$, and let $y_1, \ldots, y_d$ be the dual basis, so that $B(x_i, y_j) = \delta_{ij}$ (Kronecker δ). Then the element $\Delta = \sum_i x_i y_i$ of $U(\mathfrak{g})$ is in the center of $U(\mathfrak{g})$. The element Δ is independent of the choice of basis $x_1, \ldots, x_d$.*

The element Δ is called the *Casimir element* of $U(\mathfrak{g})$ (with respect to B).

Proof. Let $z \in \mathfrak{g}$. There exist constants α_{ij} and β_{ij} such that $[z, x_i] = \sum_j \alpha_{ij} x_j$ and $[z, y_i] = \sum_j \beta_{ij} y_j$. Since B is invariant, we have

$$0 = B([z, x_i], y_j) + B(x_i, [z, y_j]) = \alpha_{ij} + \beta_{ji}.$$

Now

$$z \sum_i x_i y_i = \sum_i ([z, x_i] y_i + x_i z y_i) = \left(\sum_{i,j} \alpha_{ij} x_j y_i \right) + \sum_i x_i z y_i,$$

while

$$\sum_i x_i y_i z = \sum_i (-x_i [z, y_i] + x_i z y_i) = -\left(\sum_{i,j} \beta_{ij} x_i y_j \right) + \sum_i x_i z y_i,$$

and since $\beta_{ij} = -\alpha_{ji}$, these are equal. Thus Δ commutes with $\mathfrak{g}$, and since $\mathfrak{g}$ generates $U(\mathfrak{g})$ as a ring, it is in the center.

It remains to be shown that Δ is independent of the choice of basis $x_1, \ldots, x_d$. Suppose that $x_1', \ldots, x_d'$ is another basis. Write $x_i' = \sum_j c_{ij} x_j$, and if $y_1', \ldots, y_d'$ is the corresponding dual basis, let $y_i' = \sum_j d_{ij} y_j$. The condition that $B(x_i', y_j') = \delta_{ij}$ (Kronecker δ) implies that $\sum_k c_{ik} d_{jk} = \delta_{ij}$. Therefore, the matrices (c_{ij}) and (d_{ij}) are transpose inverses of each other and so we have also $\sum_k c_{ki} d_{kj} = \delta_{ij}$. Now $\sum_k x_k' y_k' = \sum_{i,j,k} c_{ki} d_{kj} x_i y_j = \sum_k x_k y_k = \Delta$. $\square$

Although Proposition 10.4 provides us with a supply of invariant bilinear forms, there is no guarantee that they are nonzero, which is required by Theorem 10.2. We will not address this point now.

One might wonder, since there may be many irreducible representations, whether the invariant bilinear forms produced by Proposition 10.4 are all distinct. Also, since these invariant bilinear forms are all symmetric, one might wonder if we are missing some invariant bilinear forms that are not symmetric. The following proposition shows that for simple Lie algebras, there is essentially a unique invariant bilinear form, and that it is symmetric.

A Lie algebra $\mathfrak{g}$ is called *simple* if it has no proper nonzero ideals. An ideal is just an invariant subspace of $\mathfrak{g}$ for the adjoint representation, so another way of saying the same thing is that $\mathrm{ad} : \mathfrak{g} \longrightarrow \mathrm{End}(\mathfrak{g})$ is irreducible. For example, it is not hard to see that for any field F, the Lie algebra $\mathfrak{sl}(n, F)$ is simple.

Proposition 10.5. *Let $\mathfrak{g}$ be a finite-dimensional simple Lie algebra over a field F. Then there exists, up to scalar, at most one invariant bilinear form on $\mathfrak{g}$. If a nonzero invariant bilinear form exists it is nondegenerate and symmetric.*

Proof. Let $\mathfrak{g}^*$ be the dual space to $\mathfrak{g}$. If $\lambda \in \mathfrak{g}^*$ and $x \in \mathfrak{g}$ we will use the notation $\langle x, \lambda \rangle$ for $\lambda(x)$. Let $\alpha : \mathfrak{g} \longrightarrow \operatorname{End}(\mathfrak{g}^*)$ be defined by the rule $\langle x, \alpha(y)\lambda \rangle = -\langle [y, x], \lambda \rangle$. It is easy to check using the Jacobi identity that this α is a representation. We will regard $\mathfrak{g}^*$ as a $\mathfrak{g}$-module by means of α.

Every bilinear for $B : \mathfrak{g} \times \mathfrak{g} \longrightarrow F$ is of the form $B(x, y) = \langle x, \theta(y) \rangle$ for some linear map $\theta : \mathfrak{g} \longrightarrow \mathfrak{g}^*$. We claim that the condition for B to be invariant is equivalent to θ being a homomorphism of $\mathfrak{g}$-modules. Indeed, for θ to be a $\mathfrak{g}$-module homomorphism we need $\alpha(x)\theta(z) = \theta(\operatorname{ad}(x)z)$. Applying these linear functionals to $y \in \mathfrak{g}$, this condition is equivalent to $-B([x, y], z) = B([y, [x, z]])$ for all y.

Thus, the vector space of invariant bilinear forms is isomorphic to the space of $\mathfrak{g}$-module homomorphisms $\theta : \mathfrak{g} \longrightarrow \mathfrak{g}^*$. Since $\mathfrak{g}$ is simple, any such homomorphism is either zero or injective; if it is nonzero, it is bijective since $\mathfrak{g}$ and $\mathfrak{g}^*$ have the same finite dimension. By Schur's lemma (Exercise 10.5) the space of such θ is at most one-dimensional, and so, therefore, is the space of invariant bilinear forms.

We must show that if B is nonzero and invariant it is symmetric and nondegenerate. Since θ is injective, $B(x, y) = 0$ for all x implies that $y = 0$, and so it is nondegenerate. To see that it is symmetric, it is unique up to a scalar, so $B(x, y) = cB(y, x)$ for some scalar c. Applying this twice, $c^2 = 1$, and we need to show $c = 1$. If the characteristic of F is two, then $c^2 = 1$ implies $c = 1$, so assume the characteristic is not two. Then we show that $c \neq -1$. Arguing by contradiction, $c = -1$ implies that

$$B([x, y], z) = -B(z, [x, y]) = B([x, z], y) = -B([z, x], y).$$

Applying this identity three times, $B([x, y], z) = -B([x, y], z)$ and because the characteristic is not two, we have $B([x, y], z) = 0$ for all x, y, z. Now we may assume that $\mathfrak{g}$ is non-Abelian since otherwise it is one-dimensional and any bilinear form is symmetric. Then $[\mathfrak{g}, \mathfrak{g}]$ is an ideal of $\mathfrak{g}$ (as follows from the Jacobi identity) and is nonzero since $\mathfrak{g}$ is non-Abelian. Since $\mathfrak{g}$ is simple $[\mathfrak{g}, \mathfrak{g}] = \mathfrak{g}$ and we have proved that $B = 0$, a contradiction. $\square$

Exercises

Exercise 10.1. Let $X_{ij} \in \mathfrak{gl}(n, \mathbb{R})$ $(1 \leqslant i, j \leqslant n)$ be the $n \times n$ matrix with a 1 in the i, j position and 0's elsewhere. Show that $[X_{ij}, X_{kl}] = \delta_{jk} X_{il} - \delta_{il} X_{kj}$, where δ_{jk} is the Kronecker δ. From this, show for any positive integer d that

$$\sum_{i_1=1}^{n} \cdots \sum_{i_r=1}^{n} X_{i_1 i_2} X_{i_2 i_3} \cdots X_{i_d i_1}$$

is in the center of $U(\mathfrak{gl}(n,\mathbb{R}))$.

Exercise 10.2. Let G be a connected Lie group and $\mathfrak{g}$ its Lie algebra. Define an action of $\mathfrak{g}$ on the space $C^\infty(G)$ of smooth functions on G by (10.1).

(i) Show that this is a representation of G. Explain why Theorem 10.1 implies that this action of $\mathfrak{g}$ on $C^\infty(G)$ can be extended to a representation of the associative algebra $U(\mathfrak{g})$ on $C^\infty(G)$.

(ii) If $h \in G$, let $\rho(h)$ and $\lambda(h)$ be the endomorphisms of G given by left and right translation. Thus

$$\rho(h)f(g) = f(gh), \qquad \lambda(h)f(g) = f(h^{-1}g).$$

Show that if $h \in G$ and $D \in U(\mathfrak{g})$, then $\lambda(h) \circ D = D \circ \lambda(h)$. If D is in the center of $U(\mathfrak{g})$ then prove that $\rho(h) \circ D = D \circ \rho(h)$. (**Hint:** Prove this first if h is of the form e^X for some $X \in G$, and recall that G was assumed to be connected, so it is generated by a neighborhood of the identity.)

Exercise 10.3. Let $G = \mathrm{GL}(n,\mathbb{R})$. Let B be the "Borel subgroup" of upper triangular matrices with positive diagonal entries, and let B_0 be the connected component of the identity, whose matrices have positive diagonal entries. Let $K = \mathrm{O}(n)$.

(i) Show that every element of $g \in G$ has a unique decomposition as $g = bk$ with $b \in B_0$ and $k \in K$.

(ii) Let $s = (s_1, \ldots, s_n) \in \mathbb{C}^n$. By (i), we may define an element $\phi = \phi_s$ of $C^\infty(G)$ by

$$\phi_s\left(\begin{pmatrix} y_1 & * & \cdots & * \\ 0 & y_2 & \cdots & * \\ \vdots & \vdots & \ddots & \vdots \\ 0 & 0 & \cdots & y_n \end{pmatrix} k\right) = \prod_{i=1}^{n} y_i^{s_i}, \qquad y_i > 0, \, k \in K.$$

Show that ϕ is an eigenfunction of the center of $U(\mathfrak{g})$. That is, if D is in the center of $U(\mathfrak{g})$, then $D\phi = \lambda\phi$ for some complex number λ. [**Hint:** Characterize ϕ by properties of left and right translation and use Exercise 10.2 (ii).]

(iii) Define $\sigma_s(g) = \int_K \phi_s(kg) \, dk$. Clearly σ_s satisfies $\sigma_s(kgk') = \sigma(g)$ for $k, k' \in K$. Show that σ is an eigenfunction of the center of $U(\mathfrak{g})$. This is the *spherical function*.

Exercise 10.4. Give a construction similar to that in Exercise 10.3 for eigenfunctions of the center of $U(\mathfrak{g})$ when $\mathfrak{g} = \mathfrak{gl}_n(\mathbb{C})$.

Exercise 10.5 (Schur's lemma). Let $\mathfrak{g}$ be a Lie algebra, and let V, W be $\mathfrak{g}$-modules.

(i) Show that the space of $\mathfrak{g}$-module homomorphisms $\phi : V \to W$ is at most one-dimensional.

(ii) Show that the space of invariant bilinear forms $V \times W \to \mathbb{C}$ is at most one-dimensional.

11

Extension of Scalars

We will be interested in *complex* representations of both real and complex Lie algebras. There is an important distinction to be made. If $\mathfrak{g}$ is a real Lie algebra, then a complex representation is an $\mathbb{R}$-linear homomorphism $\mathfrak{g} \longrightarrow \mathrm{End}(V)$, where V is a complex vector space. On the other hand, if $\mathfrak{g}$ is a *complex* Lie algebra, we require that the homomorphism be $\mathbb{C}$-linear. The reader should note that we ask more of a complex representation of a complex Lie algebra than we do of a complex representation of a real Lie algebra.

The interplay between real and complex Lie groups and Lie algebras will prove important to us. We begin this theme right here with some generalities about extension of scalars.

If R is a commutative ring and S is a larger commutative ring containing R, we may think of S as an R-algebra. In this case, there are functors between the categories of R-modules and S-modules. Namely, if N is an S-module, we may regard it as an R-module. On the other hand, if M is an R-module, then thinking of S as an R-module, we may form the R-module $M_S = S \otimes_R M$. This has an S-module structure such that $t(s \otimes m) = ts \otimes m$ for $t, s \in S$, and $m \in M$. We call this the S-module obtained by *extension of scalars*. If $\phi : M \longrightarrow N$ is an R-module homomorphism, $1 \otimes \phi : M_S \longrightarrow N_S$ is an S-module homomorphism, so extension of scalars is a functor.

Of the properties of extension of scalars, we note the following:

Proposition 11.1. *Let $S \supseteq R$ be commutative rings.*

(i) If M_1 and M_2 are R-modules, we have the following natural isomorphisms of S-modules:

$$S \otimes_R R \cong S, \tag{11.1}$$

$$S \otimes_R (M_1 \oplus M_2) \cong (S \otimes_R M_1) \oplus (S \otimes_R M_2), \tag{11.2}$$

$$(S \otimes_R M_1) \otimes_S (S \otimes_R M_2) \cong S \otimes_R (M_1 \otimes_R M_2). \tag{11.3}$$

(ii) If M is an R-module and N is an S-module, we have a natural isomorphism

$$\mathrm{Hom}_R(M, N) \cong \mathrm{Hom}_S(S \otimes_R M, N). \tag{11.4}$$

D. Bump, *Lie Groups*, Graduate Texts in Mathematics 225,
DOI 10.1007/978-1-4614-8024-2_11, © Springer Science+Business Media New York 2013

Proof. To prove (11.1), note that the multiplication $S \times R \longrightarrow S$ is an R-bilinear map hence by the universal property of the tensor product induces an R-module homomorphism $S \otimes_R R \longrightarrow S$. On the other hand, $s \longrightarrow s \otimes 1$ is an R-module homomorphism $S \longrightarrow S \otimes_R R$, and these maps are inverses of each other. With our definition of the S-module structure on $S \otimes_R R$, they are S-module isomorphisms.

To prove (11.2), one may characterize the direct sum $M_1 \oplus M_2$ as follows: given an R-module M with maps $j_i : M_i \longrightarrow M$, $p_i : M \longrightarrow M_i$ $(i = 1, 2)$ such that $p_i \circ j_i = 1_{M_i}$ and $j_1 \circ p_1 + j_2 \circ p_2 = 1_M$, then there are maps

$$M \longrightarrow M_1 \oplus M_2, \qquad m \longmapsto (p_1(m), p_2(m)),$$

$$M_1 \oplus M_2 \longrightarrow M, \qquad (m_1, m_2) \longmapsto i_1 m_1 + i_2 m_2.$$

These are easily checked to be inverses of each other, and so $M \cong M_1 \oplus M_2$. For example, if $M = M_1 \oplus M_2$, such maps exist—take the inclusion and projection maps in and out of the direct sum. Now applying the functor $M \mapsto S \otimes_R M$ to the maps j_1, j_2, p_1, p_2 gives corresponding maps for $S \otimes_R (M_1 \otimes_R M_2)$ showing that it is isomorphic to the left-hand side of (11.2).

To prove (11.3), one has an S-bilinear map

$$(S \otimes_R M_1) \times (S \otimes_R M_2) \longrightarrow S \otimes_R (M_1 \otimes_R M_2) \qquad (11.5)$$

such that $((s_1 \otimes m_1), (s_2 \otimes m_2)) \mapsto s_1 s_2 \otimes (m_1 \otimes m_2)$. This map is S-bilinear, so it induces a homomorphism

$$(S \otimes_R M_1) \otimes_S (S \otimes_R M_2) \longrightarrow S \otimes_R (M_1 \otimes_R M_2). \qquad (11.6)$$

Similarly, there is an R-bilinear map

$$S \times (M_1 \otimes_R M_2) \longrightarrow (S \otimes_R M_1) \otimes_S (S \otimes_R M_2)$$

such that $(s, m_1 \otimes m_2) \mapsto (s \otimes m_1) \otimes (1 \otimes m_2) = (1 \otimes m_1) \otimes (s \otimes m_2)$. This induces an S-module homomorphism that is the inverse to (11.6).

To prove (11.4), we describe the correspondence explicitly. If

$$\phi \in \mathrm{Hom}_R(M, N) \quad \text{and} \quad \Phi \in \mathrm{Hom}_S(S \otimes M, N),$$

then ϕ and Φ correspond if $\phi(m) = \Phi(1 \otimes m)$ and $\Phi(s \otimes m) = s\phi(m)$. It is easily checked that $\phi \mapsto \Phi$ and $\Phi \mapsto \phi$ are well-defined inverse isomorphisms. $\qquad \square$

If V is a d-dimensional real vector space, then the complex vector space $V_{\mathbb{C}} = \mathbb{C} \otimes_{\mathbb{R}} V$ is a d-dimensional complex vector space. This follows from Proposition 11.1 because if $V \cong \mathbb{R} \oplus \ldots \oplus \mathbb{R}$ (d copies), then (11.1) and (11.2) imply that $V_{\mathbb{C}} \cong \mathbb{C} \oplus \cdots \oplus \mathbb{C}$ (d copies). We call $V_{\mathbb{C}}$ the *complexification* of V. The natural map $V \longrightarrow V_{\mathbb{C}}$ given by $v \mapsto 1 \otimes v$ is injective, so we may think of V as a real vector subspace of $V_{\mathbb{C}}$.

Proposition 11.2.

(i) *If V is a real vector space and W is a complex vector space, any $\mathbb{R}$-linear transformation $V \longrightarrow W$ extends uniquely to a $\mathbb{C}$-linear transformation $V_\mathbb{C} \longrightarrow W$.*

(ii) *If V and U are real vector spaces, any $\mathbb{R}$-linear transformation $V \longrightarrow U$ extends uniquely to a $\mathbb{C}$-linear map $V_\mathbb{C} \longrightarrow U_\mathbb{C}$.*

(iii) *If V and U are real vector spaces, any $\mathbb{R}$-bilinear map $V \times V \longrightarrow U$ extends uniquely to a $\mathbb{C}$-bilinear map $V_\mathbb{C} \times V_\mathbb{C} \longrightarrow U_\mathbb{C}$.*

Proof. Part (i) is a special case of (ii) of Proposition 11.1. Part (ii) follows by taking $W = U_\mathbb{C}$ in part (i) after composing the given linear map $V \longrightarrow U$ with the inclusion $U \longrightarrow W$. As for (iii), an $\mathbb{R}$-bilinear map $V \times V \longrightarrow U$ induces an $\mathbb{R}$-linear map $V \otimes_\mathbb{R} V \longrightarrow U$ and hence by (ii) a $\mathbb{C}$-linear map $(V \otimes_\mathbb{R} V)_\mathbb{C} \longrightarrow U_\mathbb{C}$. But by (11.3), $(V \otimes_\mathbb{R} V)_\mathbb{C}$ is $V_\mathbb{C} \otimes_\mathbb{C} V_\mathbb{C}$, and a $\mathbb{C}$-linear map $V_\mathbb{C} \otimes_\mathbb{C} V_\mathbb{C} \longrightarrow U_\mathbb{C}$ is the same thing as a $\mathbb{C}$-bilinear map $V_\mathbb{C} \times V_\mathbb{C} \longrightarrow U_\mathbb{C}$. $\square$

Proposition 11.3.

(i) *The complexification $\mathfrak{g}_\mathbb{C}$ of a real Lie algebra $\mathfrak{g}$ with the bracket extended as in Proposition 11.2 (iii) is a Lie algebra.*

(ii) *If $\mathfrak{g}$ is a real Lie algebra, $\mathfrak{h}$ is a complex Lie algebra, and $\rho : \mathfrak{g} \longrightarrow \mathfrak{h}$ is a real Lie algebra homomorphism, then ρ extends uniquely to a homomorphism $\rho_\mathbb{C} : \mathfrak{g}_\mathbb{C} \longrightarrow \mathfrak{h}$ of complex Lie algebras. In particular, any complex representation of $\mathfrak{g}$ extends uniquely to a complex representation of $\mathfrak{g}_\mathbb{C}$.*

(iii) *If $\mathfrak{g}$ is a real Lie subalgebra of the complex Lie algebra $\mathfrak{h}$, and if $\mathfrak{h} = \mathfrak{g} \oplus i\mathfrak{g}$ (i.e., if $\mathfrak{g}$ and $i\mathfrak{g}$ span $\mathfrak{h}$ but $\mathfrak{g} \cap i\mathfrak{g} = \{0\}$), then $\mathfrak{h} \cong \mathfrak{g}_\mathbb{C}$ as complex Lie algebras.*

Proof. For (i), the extended bracket satisfies the Jacobi identity since both sides of (5.1) are trilinear maps on $\mathfrak{g}_\mathbb{C} \times \mathfrak{g}_\mathbb{C} \times \mathfrak{g}_\mathbb{C} \longrightarrow \mathfrak{g}_\mathbb{C}$, which by assumption vanish on $\mathfrak{g} \times \mathfrak{g} \times \mathfrak{g}$. Since $\mathfrak{g}$ generates $\mathfrak{g}_\mathbb{C}$ over the complex numbers, (5.1) is therefore true on $\mathfrak{g}_\mathbb{C}$.

For (ii), the extension is given by Proposition 11.2 (i), taking $W = \mathfrak{h}$. To see that the extension is a Lie algebra homomorphism, note that both $\rho([x, y])$ and $\rho(x)\rho(y) - \rho(y)\rho(x)$ are bilinear maps $\mathfrak{g}_\mathbb{C} \times \mathfrak{g}_\mathbb{C} \longrightarrow \mathfrak{h}$ that agree on $\mathfrak{g} \times \mathfrak{g}$. Since $\mathfrak{g}$ generates $\mathfrak{g}_\mathbb{C}$ over $\mathbb{C}$, they are equal for all $x, y \in \mathfrak{g}_\mathbb{C}$.

For (iii), by Proposition 11.2 (i), it will be least confusing to distinguish between $\mathfrak{g}$ and its image in $\mathfrak{h}$, so we prove instead the following equivalent statement: if $\mathfrak{g}$ is a real Lie algebra, $\mathfrak{h}$ is a complex Lie algebra, $f : \mathfrak{g} \longrightarrow \mathfrak{h}$ is an injective homomorphism, and if $\mathfrak{h} = f(\mathfrak{g}) \oplus i f(\mathfrak{g})$, then f extends to an isomorphism $\mathfrak{g}_\mathbb{C} \longrightarrow \mathfrak{h}$ of complex Lie algebras. Now f extends to a Lie algebra homomorphism $f_\mathbb{C} : \mathfrak{g}_\mathbb{C} \longrightarrow \mathfrak{h}$ by part (ii). To see that this is an isomorphism, note that it is surjective since $f(\mathfrak{g})$ spans $\mathfrak{h}$. To prove that it is injective, if $f_\mathbb{C}(X + iY) = 0$ with $X, Y \in \mathfrak{g}$, then $f(X) + if(Y) = 0$. Now $f(X) = f(Y) = 0$ because $f(\mathfrak{g}) \cap i f(\mathfrak{g}) = 0$. Since f is injective, $X = Y = 0$. $\square$

Of course, given any complex representation of $\mathfrak{g}_{\mathbb{C}}$, we may also restrict it to $\mathfrak{g}$, so Proposition 11.3 implies that complex representations of $\mathfrak{g}$ and complex representations of $\mathfrak{g}_{\mathbb{C}}$ are really the same thing. (They are equivalent categories.)

As an example, let us consider the complexification of $\mathfrak{u}(n)$.

Proposition 11.4.

(i) *Every $n \times n$ complex matrix X can be written uniquely as $X_1 + iX_2$, where X_1 and X_2 are $n \times n$ complex matrices satisfying $X_1 = -{}^t X_1$ and $X_2 = {}^t X_2$.*

(ii) *The complexification of the real Lie algebra $\mathfrak{u}(n)$ is isomorphic to $\mathfrak{gl}(n, \mathbb{C})$.*

(iii) *The complexification of the real Lie algebra $\mathfrak{su}(n)$ is isomorphic to $\mathfrak{sl}(n, \mathbb{C})$.*

Proof. For (i), the unique solution is clearly

$$X_1 = \tfrac{1}{2}(X - {}^t X), \qquad X_2 = \tfrac{1}{2i}(X + {}^t X).$$

For (ii), we will use the criterion of Proposition 11.3 (iii). We recall that $\mathfrak{u}(n)$ is the *real* Lie algebra consisting of *complex* $n \times n$ matrices satisfying $X = \overline{-{}^t X}$. We want to get the complex conjugation out of the picture before we try to complexify it, so we write $X = X_1 + iX_2$, where X_1 and X_2 are real $n \times n$ matrices. We must have $X_1 = -{}^t X_1$ and $X_2 = {}^t X_2$. Thus, as a vector space, we may identify $\mathfrak{u}(2)$ with the real vector space of pairs $(X_1, X_2) \in \mathrm{Mat}_n(\mathbb{R}) \oplus \mathrm{Mat}_n(\mathbb{R})$, where X_1 is skew-symmetric and X_2 symmetric. The Lie bracket operation, required by the condition that

$$[X, Y] = XY - YX \text{ when } X = X_1 + iX_2 \text{ and } Y = Y_1 + iY_2, \qquad (11.7)$$

amounts to the rule

$$[(X_1, X_2), (Y_1, Y_2)]$$
$$= (X_1 Y_1 - X_2 Y_2 - Y_1 X_1 + Y_2 X_2, X_1 Y_2 + X_2 Y_1 - Y_2 X_1 - Y_1 X_2). \quad (11.8)$$

Now (i) shows that the complexification of this vector space (allowing X_1 and X_2 to be complex) can be identified with $\mathrm{Mat}_n(\mathbb{C})$. Of course, (11.7) and (11.8) are still equivalent if X_1, X_2, Y_1, and Y_2 are allowed to be complex, so with the Lie bracket in (11.8), this Lie algebra is $\mathrm{Mat}_n(\mathbb{C})$ with the usual bracket.

(iii) is similar to (ii), and we leave it to the reader. $\square$

Theorem 11.1. *Every complex representation of the Lie algebra $\mathfrak{u}(n)$ or the Lie algebra $\mathfrak{gl}(n, \mathbb{R})$ extends uniquely to a complex representation of $\mathfrak{gl}(n, \mathbb{C})$. Every complex representation of the Lie algebra $\mathfrak{su}(n)$ or the Lie algebra $\mathfrak{sl}(n, \mathbb{R})$ extends uniquely to a complex representation of $\mathfrak{sl}(n, \mathbb{C})$.*

Proof. This follows from Proposition 11.3 since the complexification of $\mathfrak{u}(n)$ or $\mathfrak{gl}(n, \mathbb{R})$ is $\mathfrak{gl}(n, \mathbb{C})$, while the complexification of $\mathfrak{su}(n)$ or $\mathfrak{sl}(n, \mathbb{R})$ is $\mathfrak{sl}(n, \mathbb{C})$. For $\mathfrak{gl}(2, \mathbb{R})$ or $\mathfrak{sl}(2, \mathbb{R})$, this is obvious. For $\mathfrak{u}(n)$ and $\mathfrak{su}(n)$, this is Proposition 11.4. $\square$

12

Representations of $\mathfrak{sl}(2, \mathbb{C})$

Unless otherwise indicated, in this chapter a *representation* of a Lie group or Lie algebra is a complex representation. We remind the reader that if $\mathfrak{g}$ is a complex Lie algebra [e.g. $\mathfrak{sl}(2, \mathbb{C})$], then a complex representation $\pi : \mathfrak{g} \to \mathrm{End}(V)$ is assumed to be complex linear, while if $\mathfrak{g}$ is a real Lie algebra [e.g. $\mathfrak{su}(2)$ or $\mathfrak{sl}(2, \mathbb{R})$] then there is no such assumption.

Let us exhibit some representations of the group $\mathrm{SL}(2, \mathbb{C})$. We start with the standard representation on $\mathbb{C}^2$, with $\mathrm{SL}(2, \mathbb{C})$ acting by matrix multiplication on column vectors. Due to the functoriality of $\vee^k$, there is induced a representation of $\mathrm{SL}(2, \mathbb{C})$ on $\vee^k \mathbb{C}^2$. The dimension of this vector space is $k + 1$. In short, $\vee^k$ gives us a representation $\mathrm{SL}(2, \mathbb{C}) \longrightarrow \mathrm{GL}(k + 1, \mathbb{C})$. There is an induced map of Lie algebras $\mathfrak{sl}(2, \mathbb{C}) \longrightarrow \mathfrak{gl}(k + 1, \mathbb{C})$ by Proposition 7.3, and it is not hard to see that this is a complex Lie algebra homomorphism. We have corresponding representations of the real subalgebras $\mathfrak{sl}(2, \mathbb{R})$ and $\mathfrak{su}(2)$, and we will eventually see that these are all the irreducible representations of these groups.

Let us make these symmetric power representations more explicit for the algebra $\mathfrak{g} = \mathfrak{sl}(2, \mathbb{R})$. A basis of $\mathfrak{g}$ consists of the three matrices

$$H = \begin{pmatrix} 1 & 0 \\ 0 & -1 \end{pmatrix}, \qquad R = \begin{pmatrix} 0 & 1 \\ 0 & 0 \end{pmatrix}, \qquad L = \begin{pmatrix} 0 & 0 \\ 1 & 0 \end{pmatrix}.$$

They satisfy the commutation relations

$$[H, R] = 2R, \qquad [H, L] = -2L, \qquad [R, L] = H. \tag{12.1}$$

Let

$$\mathbf{x} = \begin{pmatrix} 1 \\ 0 \end{pmatrix}, \qquad \mathbf{y} = \begin{pmatrix} 0 \\ 1 \end{pmatrix},$$

be the standard basis of $\mathbb{C}^2$. We have a corresponding basis of $k + 1$ elements in $\vee^k \mathbb{C}^2$, which we will label by integers $k, k-2, k-4, \ldots, -k$ for reasons that will become clear presently. Thus, we let

D. Bump, *Lie Groups*, Graduate Texts in Mathematics 225,
DOI 10.1007/978-1-4614-8024-2_12, © Springer Science+Business Media New York 2013

$$v_{k-2l} = \mathbf{x} \vee \cdots \vee \mathbf{x} \vee \mathbf{y} \vee \cdots \vee \mathbf{y} \qquad (k - l \text{ copies of } x, \, l \text{ copies of } y).$$

Since $\vee^k$ is a functor, if $f : \mathbb{C}^2 \longrightarrow \mathbb{C}^2$ is a linear transformation, there is induced a linear transformation $\vee^k f$ of $\vee^k \mathbb{C}^2$. (See Exercise 9.3.) For simplicity, if $X \in \mathfrak{g}$ and $v \in \vee^k \mathbb{C}^2$ we will denote write $X \cdot v$ or Xv instead of $(\vee^k X)v$.

Proposition 12.1. *We have*

$$H \cdot v_{k-2l} = (k - 2l)v_{k-2l}, \qquad (0 \leqslant l \leqslant k), \tag{12.2}$$

$$R \cdot v_{k-2l} = \begin{cases} lv_{k-2l+2} & \text{if } l > 0, \\ 0 & \text{if } l = 0, \end{cases} \tag{12.3}$$

and

$$L \cdot v_{k-2l} = \begin{cases} (k - l)v_{k-2l-2} & \text{if } l < k, \\ 0 & \text{if } l = k, \end{cases} \tag{12.4}$$

The first identity is the reason for the labeling of the vectors v_{k-2l}: each v_{k-2l} is an eigenvector of H, and the subscript is the eigenvalue. We may visualize the effects of R and L as in Fig. 12.1. Each dot represents a one-dimensional eigenspace of H, called a *weight space*.

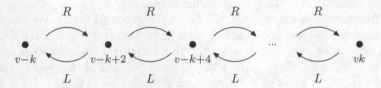

Fig. 12.1. Effects of R and L on weight vectors

What this diagram means is that the operator R maps v_j to a multiple of v_{j+2}, while L maps v_j to a multiple of v_{j-2}. The operators R and L shift between the various weight spaces. The only exceptions are that R kills v_k and L kills v_{-k}.

Proof. For example, let us compute the effect of $\vee^k R$ on v_i. In $\mathbb{C}^2$,

$$\exp(tR) : \begin{cases} \mathbf{x} \longmapsto \mathbf{x}, \\ \mathbf{y} \longmapsto \mathbf{y} + t\mathbf{x}. \end{cases}$$

So

$$R \cdot v_{k-2l} = \frac{d}{dt} \exp(tR)v_{k-2l}|_{t=0}.$$

Therefore, in $\vee^k V$, remembering that the $\vee$ operation is symmetric (commutative), we see that $\exp(tR)$ maps v_{k-2l} to

$$v_{k-2l} + tlv_{k-2l+2} + t^2 \binom{l}{2} v_{k-2l+4} + \cdots \quad .$$

Differentiating with respect to t, then letting $t = 0$ gives (12.3). We leave the reader to compute the effects of H and L. □

For example, if $k = 3$, then with respect to the basis v_3, v_1, v_{-1}, v_{-3}, we find that

$$\vee^3 R = \begin{pmatrix} 0 & 1 & 0 & 0 \\ 0 & 0 & 2 & 0 \\ 0 & 0 & 0 & 3 \\ 0 & 0 & 0 & 0 \end{pmatrix},$$

$$\vee^3 L = \begin{pmatrix} 0 & 0 & 0 & 0 \\ 3 & 0 & 0 & 0 \\ 0 & 2 & 0 & 0 \\ 0 & 0 & 1 & 0 \end{pmatrix}, \qquad \vee^3 H = \begin{pmatrix} 3 & 0 & 0 & 0 \\ 0 & 1 & 0 & 0 \\ 0 & 0 & -1 & 0 \\ 0 & 0 & 0 & -3 \end{pmatrix}.$$

It may be checked directly that these matrices satisfy the commutation relations (12.1).

Proposition 12.2. *The representation $\vee^k \mathbb{C}^2$ of $\mathfrak{sl}(2,\mathbb{R})$ is irreducible.*

Proof. Suppose that U is a nonzero invariant subspace. Choose a nonzero element $\sum a_{k-2l} v_{k-2l}$ of U. Let $k - 2l$ be the smallest integer such that $a_{k-2l} \neq 0$. Applying R to this vector l times shifts each $v_r \longrightarrow v_{r+2}$ times a nonzero constant, except for v_k, which it kills. Consequently, this operation R^l will kill every vector v_r with $r \geqslant k - 2l$, leaving only a nonzero constant times v_k. Thus $v_k \in U$. Now applying L repeatedly shows that $v_{k-2}, v_{k-4}, \ldots \in U$, so U contains a basis of $\vee^k \mathbb{C}^2$. We see that any nonzero invariant subspace of $\vee^k \mathbb{C}^2$ is the whole space, so the representation is irreducible. $\qquad\square$

If $k = 0$, we reiterate that $\vee^0 \mathbb{C}^2 = \mathbb{C}$. It is a *trivial* $\mathfrak{sl}(2,\mathbb{R})$-module, meaning that $\pi(X)$ acts as zero on it for all $X \in \mathfrak{sl}(2,\mathbb{R})$.

Now we need an element of the center of $U(\mathfrak{sl}(2,\mathbb{R}))$. An invariant bilinear form on $\mathfrak{g}$ is given by $B(x,y) = \frac{1}{2}\operatorname{tr}(xy)$, where the trace is the usual trace of a matrix, and xy is the product of two matrices, *not* multiplication in $U(\mathfrak{sl}(2,\mathbb{R}))$. The invariance of this bilinear form follows from the property of the trace that $\operatorname{tr}(xy) = \operatorname{tr}(yx)$ since

$$B([x,y],z) + B(y,[x,z]) = \tfrac{1}{2}\bigl(\operatorname{tr}(xyz) - \operatorname{tr}(yxz) + \operatorname{tr}(yxz) - \operatorname{tr}(yzx)\bigr) = 0\,,$$

proving (10.3). Dual to the basis H, R, L of $\mathfrak{sl}(2,\mathbb{R})$ is the basis H, $2L$, $2R$, and it follows from Theorem 10.2 that the Casimir element

$$\Delta = H^2 + 2RL + 2LR$$

is an element of the center of $U(\mathfrak{sl}(2,\mathbb{R}))$.

Proposition 12.3. *Suppose that (π, V) is an irreducible representation of $\mathfrak{sl}(2,\mathbb{R})$. Assume that there exists a vector v_k in V such that $v_k \neq 0$ but $R v_k = 0$. Then $\Delta v = \lambda v$ for all $v \in V$, where $\lambda = k^2 + 2k$.*

Proof. By Proposition 10.2 there exists λ such that $\Delta v = \lambda v$ for all v. To calculate λ, we use the identity $[R, L] = H$ to write

$$\Delta = H^2 + 2H + 4LR. \tag{12.5}$$

Using $Rv_k = 0$ and $Hv_k = kv_k$ we have $\Delta v_k = (k^2 + 2k)v_k$ so $\lambda = k^2 + 2k$. $\quad\square$

Proposition 12.4. *The element Δ acts by the scalar $\lambda = k^2 + 2k$ on $\vee^k\mathbb{C}^2$.*

Proof. This follows from Proposition 12.3. $\qquad\qquad\qquad\qquad\qquad\qquad\quad\square$

The following fact, though trivial to prove, is very important. It may be visualized as in Fig. 12.1.

Lemma 12.1. *Suppose v is an H-eigenvector in some module for $\mathfrak{sl}(2,\mathbb{R})$ with eigenvalue k. Then Rv (if nonzero) is also an eigenvector with eigenvalue $k+2$, and Lv (if nonzero) is an eigenvector with eigenvalue $k - 2$.*

Proof. In the enveloping algebra, we have $HR - RH = [H, R] = 2R$, so $HRv = RHv + 2Rv = (r + 2)Rv$. This proves the statement for R, and L is handled similarly. $\qquad\qquad\qquad\qquad\qquad\qquad\qquad\qquad\qquad\qquad\quad\square$

Proposition 12.5. *Let V be a finite-dimensional representation of $\mathfrak{sl}(2,\mathbb{R})$. Let $v_k \in V$ be an H-eigenvector with eigenvalue k maximal. Then k is a positive integer and v_k is contained in an irreducible subspace of V isomorphic to $\vee^k\mathbb{C}^2$.*

Proof. We have $Rv_k = 0$ by Lemma 12.1 and the maximality of k. Then $\Delta v_k = (k^2 + 2k)v_k$ follows from (12.5). Consider the submodule U generated by v_k. Every element of U is of the form ξv_k where ξ is in the universal enveloping algebra, and since Δ is in the center, it follows that $\Delta \xi v_k = \lambda \xi v_k$ with $\lambda = k^2 + 2k$. It remains to be shown that U is isomorphic to $\vee^k\mathbb{C}^2$.

Define $v_{k-2}, v_{k-4}, \ldots, v_{-k}$ by

$$v_{k-2l-2} = \frac{1}{k-l}Lv_{k-2l}.$$

Then (12.2) is satisfied by Lemma 12.1, and (12.4) is also satisfied by construction. To prove (12.3), the case $l = 0$ is known, so assume $l \geqslant 1$. Writing $\Delta = H^2 - 2H + 4RL$, the relation $\Delta v_{k-2l+2} = (k^2 + 2k)v_{k-2l+2}$ applied to v_{k-2l+2} gives

$$(k^2 + 2k)v_{k-2l+2} = [(k - 2l + 2)^2 - 2(k - 2l + 2)]v_{k-2l+2} + 4(k - l + 1)Rv_{k-2l}.$$

This can be simplified, giving (12.3). It is now clear that U is isomorphic to $\vee^k\mathbb{C}^2$. $\qquad\qquad\qquad\qquad\qquad\qquad\qquad\qquad\qquad\qquad\qquad\qquad\quad\square$

Proposition 12.6. *Let (π, V) be an irreducible complex representation of the Lie algebra $\mathfrak{sl}(2,\mathbb{R})$. Then Δ acts by a scalar λ on V, and $\lambda = k^2 + 2k$ for some nonnegative integer k. The representation π is isomorphic to $\vee^k\mathbb{C}^2$.*

Proof. Let v_k be an eigenvector for H with eigenvalue k maximal. Then $Rv_k = 0$ since otherwise Rv_k is an eigenvector with eigenvalue $k + 2$. By Proposition 12.5 v_k generates an irreducible subspace isomorphic to $\vee^k \mathbb{C}^2$. Since V is irreducible, the result follows. □

Theorem 12.1. *Let (π, V) be any irreducible complex representation of $\mathfrak{sl}(2, \mathbb{R})$, $\mathfrak{su}(2)$ or $\mathfrak{sl}(2, \mathbb{C})$. Then π is isomorphic to $\vee^k \mathbb{C}^2$ for some k.*

Proof. By Theorem 11.1, it is sufficient to show this for $\mathfrak{sl}(2, \mathbb{R})$, in which case the statement follows from Proposition 12.6. □

We can't quite say yet that the finite-dimensional representations of $\mathfrak{sl}(2, \mathbb{R})$, $\mathfrak{su}(2)$, and $\mathfrak{sl}(2, \mathbb{C})$ are now classified. We know the irreducible representations of these three Lie algebras. What we haven't yet proved is the theorem of Weyl that says that every irreducible representation is *completely reducible*, that is, a direct sum of irreducible representations. We will prove this next. Another proof will be given later in Theorem 14.4. Therefore, the reader may skip the rest of this chapter with no loss of continuity.

The proof below in Theorem 14.4 is not purely algebraic. So even though it is not needed, it is instructive to give a purely algebraic proof of complete reducibility. The following proof depends on only two facts about $\mathfrak{g}$ and the Casimir element Δ. First, we have $[\mathfrak{g}, \mathfrak{g}] = \mathfrak{g}$, and second, that if V is an irreducible module then $\Delta v = \lambda v$ for $v \in V$ where the scalar λ is zero if and only if V is trivial.

It may be shown that these properties are true for an arbitrary semisimple Lie algebra, so the following arguments are applicable in that generality. The exercises give an indication of how to extend the proof to other Lie algebras. But in the special case where $\mathfrak{g}$ is $\mathfrak{sl}(2, \mathbb{R})$, $\mathfrak{su}(2)$ or $\mathfrak{sl}(2, \mathbb{C})$, the first statement, that $[\mathfrak{g}, \mathfrak{g}] = \mathfrak{g}$ follows from (12.1), and the second statement, that the only irreducible module annihilated by Δ is the trivial module, follows from Proposition 12.4 and our classification of the irreducible modules.

If $\mathfrak{g} = \mathfrak{su}(2)$, we haven't proven that Δ is an element of $U(\mathfrak{g})$. This can be checked by direct computation, but we don't really need it—it is an element of $U(\mathfrak{g}_{\mathbb{C}}) \cong U(\mathfrak{g})_{\mathbb{C}}$ and as such acts as a scalar on any complex representation of $\mathfrak{g}$.

Proposition 12.7. *Let $\mathfrak{g} = \mathfrak{sl}(2, \mathbb{R})$, $\mathfrak{su}(2)$ or $\mathfrak{sl}(2, \mathbb{C})$. Let (π, V) be a finite-dimensional complex representation of $\mathfrak{g}$. If there exists $k \geqslant 1$ such that $\pi(\Delta^k)v = 0$ for all $v \in V$, then $\pi(X)v = 0$ for all $X \in \mathfrak{g}$, $v \in V$.*

Proof. There is nothing to do if $V = \{0\}$. Assume therefore that U is a maximal proper invariant subspace of U. By induction on $\dim(V)$, $\mathfrak{g}$ acts trivially on U. Now V/U is irreducible by the maximality of U, and Δ annihilates V/U, so by the classification of the irreducible representations of $\mathfrak{g}$ in Theorem 12.1, $\mathfrak{g}$ acts trivially on V/U. This means that if $Y \in \mathfrak{g}$ and $v \in V$, then $\pi(Y)v \in U$. Since $\mathfrak{g}$ acts trivially on U, if X is another element of $\mathfrak{g}$, we have $\pi(X)\pi(Y)v = 0$ and similarly $\pi(Y)\pi(X) = 0$. Thus, $\pi([X, Y])v = \pi(X)\pi(Y)v - \pi(Y)\pi(X)v = 0$, and since by (12.1) elements of the form $[X, Y]$ span $\mathfrak{g}$, it follows that $\mathfrak{g}$ acts trivially on V. □

Proposition 12.8. *Let* $\mathfrak{g} = \mathfrak{sl}(2, \mathbb{R})$, $\mathfrak{su}(2)$, *or* $\mathfrak{sl}(2, \mathbb{C})$. *Let* (π, V) *be a finite-dimensional complex representation of* $\mathfrak{g}$.

(i) If $v \in V$ *and* $\Delta^2 v = 0$, *then* $\Delta v = 0$.
(ii) We have $V = V_0 \oplus V_1$, *where* V_0 *is the kernel of* Δ *and* V_1 *is the image of* Δ. *Both are invariant subspaces. If* $X \in \mathfrak{g}$ *and* $v \in V_0$, *then* $\pi(X)v = 0$.
(iii) The subspace $V_0 = \{v \in V \mid \pi(X) = 0 \text{ for all } X \in \mathfrak{g}\}$.
(iv) If $0 \longrightarrow V \longrightarrow W \longrightarrow Q \longrightarrow 0$ *is an exact sequence of* $\mathfrak{g}$-*modules, then there is an exact sequence* $0 \longrightarrow V_0 \longrightarrow W_0 \longrightarrow Q_0 \longrightarrow 0$.

Proof. Since Δ commutes with the action of $\mathfrak{g}$, the kernel W of Δ^k is an invariant subspace. Now (i) follows from Proposition 12.7.

It follows from (i) that $V_0 \cap V_1 = \{0\}$. Now for any linear endomorphism of a vector space, the dimension of the image equals the codimension of the kernel, so $\dim(V_0) + \dim(V_1) = \dim(V)$. It follows that $V_0 + V_1 = V$ and this sum is direct. Since Δ commutes with the action of $\mathfrak{g}$, both V_0 and V_1 are invariant subspaces.

It follows from Proposition 12.7 that $\mathfrak{g}$ acts trivially on V_0. This proves (ii) and also (iii) since it is obvious that $\{v \in V \mid \pi(X)v = 0\} \subseteq V_0$, and we have proved the other inclusion.

For (iv), any homomorphism $V \longrightarrow W$ of $\mathfrak{g}$-modules maps V_0 into W_0, so $V \longrightarrow V_0$ is a functor. Given a short exact sequence $0 \longrightarrow V \longrightarrow W \longrightarrow Q \longrightarrow 0$, consider

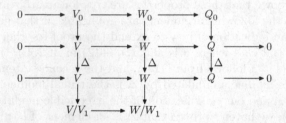

Exactness of the two middle rows implies exactness of the top row. We must show that $W_0 \longrightarrow Q_0$ is surjective. We will deduce this from the Snake Lemma. The cokernel of $\Delta : V \longrightarrow V$ is $V/V_1 \cong V_0$, and similarly the cokernel of $\Delta : W \longrightarrow W$ is $W/W_1 \cong W_0$, so the Snake lemma gives us a long exact sequence:

$$0 \longrightarrow V_0 \longrightarrow W_0 \longrightarrow Q_0 \longrightarrow V_0 \longrightarrow W_0.$$

Since the last map is injective, the map $Q_0 \longrightarrow V_0$ is zero, and hence $W_0 \longrightarrow Q_0$ is surjective. $\qquad\square$

If V is a $\mathfrak{g}$-module, we call $V_0 = \{v \in V \mid Xv = 0 \text{ for all } X \in \mathfrak{g}\}$ the *module of invariants*. The proposition shows that it is an exact functor.

If $\mathfrak{g}$ is a Lie algebra and V, W are $\mathfrak{g}$-modules, we can make the space $\mathrm{Hom}(V, W)$ of all $\mathbb{C}$-linear transformations $V \longrightarrow W$ into a $\mathfrak{g}$-module by:

$$(X\phi)v = X\phi(v) - \phi(Xv).$$

It is straightforward to check that Π is a Lie algebra representation. The module of invariants is the space

$$\text{Hom}_{\mathfrak{g}}(V,W) = \{\phi : V \longrightarrow W \mid \phi(Xv) = X\phi(v) \text{ for all } X \in \mathfrak{g}\}$$

of all $\mathfrak{g}$-module homomorphisms.

Proposition 12.9. *Let U, V, W, Q be $\mathfrak{g}$-modules, where $\mathfrak{g}$ is one of $\mathfrak{sl}(2,\mathbb{R})$, $\mathfrak{su}(2)$, or $\mathfrak{sl}(2,\mathbb{C})$, and let*

$$0 \longrightarrow V \longrightarrow W \longrightarrow Q \longrightarrow 0$$

be an exact sequence of $\mathfrak{g}$-modules. Composition with these maps gives an exact sequence:

$$0 \longrightarrow \text{Hom}_{\mathfrak{g}}(U,V) \longrightarrow \text{Hom}_{\mathfrak{g}}(U,W) \longrightarrow \text{Hom}_{\mathfrak{g}}(U,Q) \longrightarrow 0.$$

Proof. Composition with these maps gives a short exact sequence:

$$0 \longrightarrow \text{Hom}(U,V) \longrightarrow \text{Hom}(U,W) \longrightarrow \text{Hom}(U,Q) \longrightarrow 0.$$

Here, of course, $\text{Hom}(U,V)$ is just the space of all linear transformations of complex vector spaces. Taking the spaces of invariants gives the exact sequence of $\text{Hom}_{\mathfrak{g}}$ spaces, and by Proposition 12.8 it is exact.   □

Theorem 12.2. *Let $\mathfrak{g} = \mathfrak{sl}(2,\mathbb{R})$, $\mathfrak{su}(2)$, or $\mathfrak{sl}(2,\mathbb{C})$. Any finite-dimensional complex representation of $\mathfrak{g}$ is a direct sum of irreducible representations.*

Proof. Let W be a $\mathfrak{g}$-module. If W is zero or irreducible, there is nothing to check. Otherwise, let V be a proper nonzero submodule and let $Q = W/V$. We have an exact sequence

$$0 \longrightarrow V \longrightarrow W \longrightarrow Q \longrightarrow 0$$

and by induction on $\dim(W)$ both V and Q decompose as direct sums of irreducible submodules. By Proposition 12.9, composition with these maps produces an exact sequence

$$0 \longrightarrow \text{Hom}(Q,V)_{\mathfrak{g}} \longrightarrow \text{Hom}(Q,W)_{\mathfrak{g}} \longrightarrow \text{Hom}(Q,Q)_{\mathfrak{g}} \longrightarrow 0.$$

The surjectivity of the map $\text{Hom}(Q,W)_{\mathfrak{g}} \longrightarrow \text{Hom}(Q,Q)_{\mathfrak{g}}$ means that there is a map $i : Q \longrightarrow W$ which has a composition $p \circ i$ with the projection $p : W \longrightarrow Q$ that is the identity map on Q.

Now V and $i(Q)$ are submodules of W such that $V \cap i(Q) = \{0\}$ and $W = V + i(Q)$. Indeed, if $x \in V \cap i(Q)$, then $p(x) = 0$ since $p(V) = \{0\}$, and writing $x = i(q)$ with $q \in Q$, we have $q = (p \circ i)(q) = p(x) = 0$; so $x = 0$ and if $w \in W$ we can write $w = v + q$, where $v = w - ip(w)$ and $q = ip(w)$ and, since $p(v) = p(w) - p(w) = 0$, $v \in \ker(p) = V$ and $q \in i(Q)$.

We see that $W = V \oplus i(Q)$, and since V and Q are direct sums of irreducible submodules, so is W.   □

Exercises

Exercise 12.1. If (π, V) is a representation of $SL(2,\mathbb{R})$, $SU(2)$ or $SL(2,\mathbb{C})$, then we may restrict the character of π to the diagonal subgroup. This gives

$$\xi_\pi(t) = \text{tr } \pi \begin{pmatrix} t & \\ & t^{-1} \end{pmatrix},$$

which is a Laurent polynomial, that is, a polynomial in t and t^{-1}.

(i) Compute $\xi_\pi(t)$ for the symmetric power representations. Show that the polynomials $\xi_\pi(t)$ are linearly independent and determine the representation π.

(ii) Show that if $\Pi = \pi \otimes \pi'$, then $\xi_\Pi = \xi_\pi \xi_{\pi'}$. Use this observation to compute the decomposition of $\pi \otimes \pi'$ into irreducibles when $\pi = \vee^n \mathbb{C}^2$ and $\pi' = \vee^m \mathbb{C}^2$.

Exercise 12.2. Show that each representation of $\mathfrak{sl}(2,\mathbb{R})$ comes from a representation of $SL(2,\mathbb{R})$.

Exercise 12.3. Let $\mathfrak{g} = \mathfrak{sl}(3,\mathbb{R})$. Let Δ be the Casimir element with respect to the invariant bilinear form $\frac{1}{2}\text{tr}(xy)$ on $\mathfrak{g}$. Show that if (π, V) is an irreducible representation with $\Delta \cdot V = 0$, then V is trivial.

[**Hint:** Here are some suggestions for a direct approach. Let

$$H_1 = \begin{pmatrix} 1 & & \\ & -1 & \\ & & 0 \end{pmatrix}, \qquad H_2 = \begin{pmatrix} 0 & & \\ & 1 & \\ & & -1 \end{pmatrix},$$

and (denoting by E_{ij} the matrix with 1 in the i,j position, 0 elsewhere) let $R_1 = E_{12}$, $R_2 = E_{23}$, $R_3 = E_{13}$, $L_1 = E_{21}$, $L_2 = E_{32}$, $L_3 = E_{31}$. These eight elements are a basis. Since $[H_1, H_2] = 0$ there exists a vector v_λ that is a simultaneous eigenvector, so that $H_1 v_\lambda = (\lambda_1 - \lambda_2)v_\lambda$ and $H_2 v_\lambda = (\lambda_2 - \lambda_3)v_\lambda$ for some triple $(\lambda_1, \lambda_2, \lambda_3)$ of real numbers. (We may normalize them so $\lambda_1 + \lambda_2 + \lambda_3 = 0$, and it may then be shown that $\lambda_i \in \frac{1}{3}\mathbb{Z}$, though you may not need that fact.) Let V_λ be the space of such vectors. Show that R_1 maps V_λ into $V_{\lambda + \alpha_1}$ and R_2 maps V_λ into $V_{\lambda + \alpha_2}$ where $\alpha_1 = (1, -1, 0)$ and $\alpha_2 = (0, 1, -1)$. (What does R_3 do to V_λ, and what do the L_i do?) Conclude that there is a nonzero vector v_λ in some V_λ that is annihilated by R_1, R_2 and R_3. Show that $\lambda_1 \geqslant \lambda_2 \geqslant \lambda_3$. For this, it may be useful to observe that there are two copies of $\mathfrak{sl}_2(\mathbb{R})$ in $\mathfrak{sl}_3(\mathbb{R})$ spanned by H_i, R_i, L_i with $i = 1, 2$, so you may restrict the representation to these and make use of the theory in the text. Compute the eigenvector of Δ and show that $\Delta v_\lambda = 0$ implies $\lambda = (0,0,0)$.]

Exercise 12.4. Show that complex representations of $\mathfrak{su}(3)$, $\mathfrak{sl}(3,\mathbb{R})$, and $\mathfrak{sl}(3,\mathbb{C})$ are completely reducible.

Exercise 12.5. Show that if (π, V) is a faithful representation of $\mathfrak{sl}(2,\mathbb{R})$, then the trace bilinear form $B_V : \mathfrak{g} \times \mathfrak{g} \to \mathbb{C}$ defined by $B_V(X, Y) = \text{tr}\big(\pi(X)\pi(Y)\big)$ is nonzero.

Exercise 12.6. Let $\mathfrak{g}$ be a simple Lie algebra. Assume that $\mathfrak{g}$ contains a subalgebra isomorphic to $\mathfrak{sl}(2,\mathbb{R})$. Let $\pi : \mathfrak{g} \to \text{End}(V)$ be an irreducible representation. Assume that π is not the trivial representation.

(i) Show that π is faithful.

(ii) Show that the trace bilinear form B_V on $\mathfrak{g}$ defined by $B(X, Y) = \mathrm{tr}(\pi(X)\pi(Y))$ is nondegenerate. (**Hint**: First show that it is nonzero.)

(iii) By (ii) there exists a Casimir element Δ in the center of $U(\mathfrak{g})$ as in Theorem 10.2. Show that the eigenvalue of Δ on V is $1/\dim(V)$. (**Hint**: Take traces.)

(iv) Show that representations of $\mathfrak{g}$ are completely reducible. (**Hint**: Use Proposition 10.5.)

Exercise 12.7. Show that complex representations of $\mathfrak{su}(n)$, $\mathfrak{sl}(n, \mathbb{R})$, and $\mathfrak{sl}(n, \mathbb{C})$ are completely reducible.

13

The Universal Cover

If U is a Hausdorff topological space, a *path* is a continuous map $p : [0,1] \longrightarrow U$. The path is *closed* if the endpoints coincide: $p(0) = p(1)$. A closed path is also called a *loop*.

An object in the category of *pointed* topological spaces consists of a pair (X, x_0), where X is a topological space and $x_0 \in X$. The chosen point $x_0 \in X$ is called the *base point*. A morphism in this category is a continuous map taking base point to base point.

If U and V are topological spaces and $\phi, \psi : U \longrightarrow V$ are continuous maps, a *homotopy* $h : \phi \rightsquigarrow \psi$ is a continuous map $h : U \times [0,1] \longrightarrow V$ such that $h(u,0) = \phi(u)$ and $h(u,1) = \psi(1)$. To simplify the notation, we will denote $h(u,t)$ as $h_t(u)$ in a homotopy. Two maps ϕ and ψ are called *homotopic* if there exists a homotopy $\phi \rightsquigarrow \psi$. Homotopy is an equivalence relation.

If $p : [0,1] \longrightarrow U$ and $p' : [0,1] \longrightarrow U$ are two paths, we say that p and p' are *path-homotopic* if there is a homotopy $h : p \rightsquigarrow p'$ that *does not move the endpoints*. This means that $h_t(0) = p(0) = p'(0)$ and $h_t(1) = p(1) = p'(1)$ for all t. We call h a *path-homotopy*, and we write $p \approx p'$ if a path-homotopy exists.

Suppose there exists a continuous function $f : [0,1] \longrightarrow [0,1]$ such that $f(0) = 0$ and $f(1) = 1$ and that $p' = p \circ f$. Then we say that p' is a *reparametrization* of p. The paths are path-homotopic since we can consider $p_t(u) = p((1-t)u + tf(u))$. Because the interval $[0,1]$ is convex, $(1-t)u + tf(u) \in [0,1]$ and $p_t : p \rightsquigarrow p'$.

Let us say that a map of topological spaces is *trivial* if it is constant, mapping the entire domain to a single point. A topological space U is *contractible* if the identity map $U \longrightarrow U$ is homotopic to a trivial map. A space U is *path-connected* if for all $x, y \in U$ there exists a path $p : [0,1] \longrightarrow U$ such that $p(0) = x$ and $p(1) = y$.

Suppose that $p : [0,1] \longrightarrow U$ and $q : [0,1] \longrightarrow U$ are two paths in the space U such that the right endpoint of p coincides with the left endpoint of q; that is, $p(1) = q(0)$. Then we can *concatenate* the paths to form the path $p \star q$:

D. Bump, *Lie Groups*, Graduate Texts in Mathematics 225,
DOI 10.1007/978-1-4614-8024-2_13, © Springer Science+Business Media New York 2013

$$(p \star q)(t) = \begin{cases} p(2t) & \text{if } 0 \leqslant t \leqslant \frac{1}{2}, \\ q(2t-1) & \text{if } \frac{1}{2} \leqslant t \leqslant 1. \end{cases}$$

We may also *reverse* a path: $-p$ is the path $(-p)(t) = p(1-t)$. These operations are compatible with path-homotopy, and the path $p \star (-p)$ is homotopic to the trivial path $p_0(t) = p(0)$. To see this, define

$$p_t(u) = \begin{cases} p(2tu) & \text{if } 0 \leqslant u \leqslant 1/2, \\ p\big(2t(1-u)\big) & \text{if } 1/2 \leqslant u \leqslant 1. \end{cases}$$

This is a path-homotopy $p_0 \rightsquigarrow p \star (-p)$. Also $(p \star q) \star r \approx p \star (q \star r)$ if $p(1) = q(0)$ and $q(1) = r(0)$, since these paths differ by a reparametrization.

The space U is *simply connected* if it is path-connected and given any closed path [that is, any $p : [0,1] \longrightarrow U$ such that $p(0) = p(1)$], there exists a path-homotopy $f : p \rightsquigarrow p_0$, where p_0 is a trivial loop mapping $[0,1]$ onto a single point. Visually, the space is simply connected if every closed path can be shrunk to a point. It may be convenient to fix a base point $x_0 \in U$. In this case, to check whether U is simply-connected or not, it is sufficient to consider loops $p : [0,1] \longrightarrow U$ such that $p(0) = p(1) = x_0$. Indeed, we have:

Proposition 13.1. *Suppose the space U is path-connected. The following are equivalent.*

(i) *Every loop in U is path-homotopic to a trivial loop.*
(ii) *Every loop p in U with $p(0) = p(1) = x_0$ is path-homotopic to a trivial loop.*
(iii) *Every continuous map of the circle $S^1 \longrightarrow U$ is homotopic to a trivial map.*

Thus, any one of these conditions is a criterion for simple connectedness.

Proof. Clearly, (i) implies (ii). Assuming (ii), if p is a loop in U, let x be the endpoint $p(0) = p(1)$ and (using path-connectedness) let q be a path from x_0 to x. Then $q \star p \star (-q)$ is a loop beginning and ending at x_0, so using (ii) it is path-homotopic to the trivial path $p_0(t) = x_0$ for all $t \in [0,1]$. Since $p_0 \approx q \star p \star (-q)$, $p \approx (-q) \star p_0 \star q$, which is path homotopic to the trivial loop $t \longmapsto x$. Thus, (ii) implies (i).

As for (iii), a continuous map of the circle $S^1 \longrightarrow U$ is equivalent to a path $p : [0,1] \longrightarrow U$ with $p(0) = p(1)$. To say that this path is homotopic to a trivial path is not quite the same as saying it is path-homotopic to a trivial path because in deforming p we need $p_t(0) = p_t(1)$ (so that it extends to a continuous map of the circle), but we do not require that $p_t(0) = p(0)$ for all t. Thus, it may not be a path-homotopy. However, we may modify it to obtain a path-homotopy as follows: let

$$q_t(u) = \begin{cases} p_{3tu}(0) & \text{if } 0 \leqslant u \leqslant 1/3, \\ p_t(3u-1) & \text{if } 1/3 \leqslant u \leqslant 2/3, \\ p_{(3-3u)t}(1) & \text{if } 2/3 \leqslant u \leqslant 1. \end{cases}$$

Then q_t is a path-homotopy. When $t = 0$, it is a reparametrization of the original path, and when $t = 1$, since p_1 is trivial, q_1 is path-homotopic to a trivial path. Thus, (iii) implies (i), and the converse is obvious. $\qquad\square$

A map $\pi : N \longrightarrow M$ is called a *covering map* if the fibers $\pi^{-1}(x)$ are discrete for $x \in M$, and every point $m \in M$ has a neighborhood U such that $\pi^{-1}(U)$ is homeomorphic to $U \times \pi^{-1}(x)$ in such a way that the composition

$$\pi^{-1}(U) \cong U \times \pi^{-1}(x) \longrightarrow U,$$

where the second map is the projection, coincides with the given map π. We say that the cover is *trivial* if N is homeomorphic to $M \times F$, where the space F is discrete, in such a way that π is the composition $N \cong M \times F \longrightarrow M$ (where the second map is the projection). Thus, each $m \in M$ has a neighborhood U such that the restricted covering map $\pi^{-1}(U) \longrightarrow U$ is trivial, a property we will cite as *local triviality* of the cover.

Proposition 13.2. *Let $\pi : N \longrightarrow M$ be a covering map.*

(i) *If $p : [0,1] \longrightarrow M$ is a path, and if $y \in \pi^{-1}(p(0))$, then there exists a unique path $\tilde{p} : [0,1] \longrightarrow N$ such that $\pi \circ \tilde{p} = p$ and $\tilde{p}(0) = y$.*

(ii) *If $\tilde{p}, \tilde{p}' : [0,1] \longrightarrow N$ are paths with $\tilde{p}(0) = \tilde{p}'(0)$, and if the paths $\pi \circ \tilde{p}$ and $\pi \circ \tilde{p}'$ are path-homotopic, then the paths $\tilde{p}$ and $\tilde{p}'$ are path-homotopic.*

We refer to (i) as the *path lifting property* of the covering space. We refer to (ii) as the *homotopy lifting property*.

Proof. If the cover is trivial, then we may assume that $N = M \times F$ where F is discrete, and if $y = (x, f)$, where $x = p(0)$ and $f \in F$, then the unique solution to this problem is $\tilde{p}(t) = (p(t), f)$.

Since $p([0,1])$ is compact, and since the cover is locally trivial, there are a finite number of open sets $U_1, U_2, \ldots, U_n$ and points $x_0 = 0 < x_1 < \cdots < x_n = 1$ such that $p([x_{i-1}, x_i]) \subset U_i$ and such that the restriction of the cover to U_i is trivial. On each interval $[x_{i-1}, x]$, there is a unique solution, and patching these together gives the unique general solution. This proves (i).

For (ii), since $p = \pi \circ \tilde{p}$ and $p' = \pi \circ \tilde{p}'$ are path-homotopic, there exists a continuous map $(u, t) \mapsto p_t(u)$ from $[0,1] \times [0,1] \longrightarrow M$ such that $p_0(u) = p(u)$ and $p_1(u) = p'(u)$. For each t, using (i) there is a unique path $\widetilde{p_t} : [0,1] \longrightarrow M$ such that $p_t = \pi \circ \tilde{p}_t$ and $\tilde{p}_t(0) = p(0)$. One may check that $(u, t) \mapsto \tilde{p}_t(u)$ is continuous, and $\widetilde{p_0} = \tilde{p}$ and $\widetilde{p_1} = \tilde{p}'$, so $\tilde{p}$ and $\tilde{p}'$ are path-homotopic. $\qquad\square$

Covering spaces of a fixed space M form a category: if $\pi : N \longrightarrow M$ and $\pi' : N' \longrightarrow M$ are covering maps, a *morphism* is a covering map $f : N \longrightarrow N'$ such that $\pi = \pi' \circ f$. If M is a pointed space, we are actually interested in the subcategory of pointed covering maps: if x_0 is the base point of M, the base point of N must lie in the fiber $\pi^{-1}(x_0)$, and in this category the morphism f must preserve base points. We call this category the category of *pointed covering maps* or *pointed covers* of M.

Let M be a path-connected space with a fixed base point x_0. We assume that every point has a contractible neighborhood. The *fundamental group* $\pi_1(M)$ consists of the set of homotopy classes of loops in M with left and right endpoints equal to x_0. The multiplication in $\pi_1(M)$ is concatenation, and the inverse operation is path-reversal. Clearly, $\pi_1(M) = 1$ if and only if M is simply connected. Changing the base point replaces $\pi_1(M)$ by an isomorphic group, but not canonically so. Thus, $\pi_1(M)$ is a functor from the category of pointed spaces to the category of groups—*not* a functor on the category of topological spaces. If M happens to be a topological group, we will always take the base point to be the identity element.

Proposition 13.3. *If M is simply connected, is N path-connected, and $\pi : N \longrightarrow M$ is a covering map, then π is a homeomorphism.*

Proof. Since a covering map is always a local homeomorphism, what we need to show is that π is bijective. It is, of course, surjective. Suppose that $n, n' \in N$ have the same image in M. Since N is path-connected, let $\tilde{p} : [0, 1] \longrightarrow N$ be a path with $\tilde{p}(0) = n$ and $\tilde{p}(1) = n'$. Because M is simply connected and $\tilde{\pi} \circ p(0) = \tilde{\pi} \circ p(1)$, the path $\tilde{\pi} \circ p$ is path-homotopic to a trivial path. By Proposition 13.2 (ii), so is p. Therefore $n = n'$. $\qquad\square$

Theorem 13.1. *Let M be a path-connected space with base point x_0 in which every point has a contractible neighborhood. Then there exists a simply connected space $\tilde{M}$ with a covering map $\tilde{\pi} : \tilde{M} \longrightarrow M$. If $\pi : N \longrightarrow M$ is any pointed covering map, there is a unique morphism $\tilde{M} \longrightarrow N$ of pointed covers of M. If N is simply connected, this map is an isomorphism. Thus, M has a unique simply connected cover.*

Note that this is a universal property. Therefore it characterizes $\tilde{M}$ up to isomorphism. The space $\tilde{M}$ is called the *universal covering space* of M.

Proof. To construct $\tilde{M}$, let $\tilde{M}$ as a set be the set of all paths $p : [0, 1] \longrightarrow M$ such that $p(0) = x_0$ modulo the equivalence relation of path-homotopy. We define the covering map $\tilde{\pi} : \tilde{M} \longrightarrow M$ by $\tilde{\pi}(p) = p(1)$. To topologize $\tilde{M}$, let $x \in M$ and let U be a contractible neighborhood of x. Let $F = \tilde{\pi}^{-1}(x)$. It is a set of path-homotopy classes of paths from x_0 to x. Using the contractibility of U, it is straightforward to show that, given $p \in \pi^{-1}(U)$ with $y = \pi(p) \in U$, there is a unique element F represented by a path p' such that $p \approx p' \star q$, where q is a path from x to y lying entirely within U. We topologize $\tilde{\pi}^{-1}(U)$ in the unique way such that the map $p \mapsto (p', y)$ is a homeomorphism $\tilde{\pi}^{-1}(U) \longrightarrow F \times U$.

We must show that, given a pointed covering map $\pi : N \longrightarrow M$, there exists a unique morphism $\tilde{M} \longrightarrow N$ of pointed covers of M. Let y_0 be the base point of N. An element of $\tilde{\pi}^{-1}(x)$, for $x \in M$, is an equivalence class under the relation of path-homotopy of paths $p : [0, 1] \longrightarrow M$ with $x_0 = p(0)$. By Proposition 13.2 (i), there is a unique path $q : [0, 1] \longrightarrow N$ lifting this with

$q(0) = y_0$, and Proposition 13.2 (ii) shows that the path-homotopy class of q depends only on the path-homotopy class of p. Then mapping $p \mapsto q(1)$ is the unique morphism $\tilde{M} \longrightarrow N$ of pointed covers of M.

If N is simply connected, any covering map $M \longrightarrow N$ is an isomorphism by Proposition 13.3. $\square$

Proposition 13.4. *Let M, N and N' be topological spaces such that every point has a contractible neighborhood. Assume that M is simply-connected. Let $\pi : N' \longrightarrow N$ be a covering map, and let $f : M \longrightarrow N$ be continuous. Then there exists a continuous map $f' : M \longrightarrow N'$ such that $\pi \circ f' = f$.*

This result shows that the universal cover is a functor: if $\tilde{M}$ and $\tilde{N}$ are the universal covers of M and N, then this proposition implies that a continuous map $\phi : M \to N$ induces a map $\tilde{\phi} : \tilde{M} \to \tilde{N}$.

Proof. Let x_0 be a base point for M, and let y_0' be an element of N' such that $\pi(y_0') = y_0$ where $y_0 = f(x_0)$. If $x \in M$, we may find a path $p : [0, 1] \longrightarrow M$ such that $p(0) = x_0$ and $p(1) = x$. By Proposition 13.2 (i) we may then find a path $\tilde{p} : [0, 1] \longrightarrow N'$ such that $\pi \circ \tilde{p} = f \circ p$ and $\tilde{p}(0) = y_0'$. We will define $f'(x) = \tilde{p}(1)$, but first we must check that this is well-defined. If q is another path with $q(0) = x_0$ and $q(1) = x$, and if $\tilde{q} : [0, 1] \longrightarrow N'$ is the corresponding lift of $f \circ p'$ with $\tilde{q}(0) = y_0'$, then we must show $\tilde{q}(1) = \tilde{p}(1)$. The paths p' and p are homotopic because M is simply connected. That is, the concatenation of p with the inverse path to p' is a loop, hence contractible, and this implies that p and p' are homotopic. It follows that the paths $\tilde{q}$ and $\tilde{p}$ are path-homotopic, and in particular they have the same right endpoint $\tilde{q}(1) = \tilde{p}(1)$. Hence we may define $f'(x) = \tilde{p}(1)$ and this is the required map. $\square$

If M is a pointed space and x_0 is its base point, then the fiber $\tilde{\pi}^{-1}(x_0)$ coincides with its fundamental group $\pi_1(M)$. We are interested in the case where $M = G$ is a Lie group. We take the base point to be the origin.

Theorem 13.2. *Suppose that G is a path-connected group in which every point has a contractible neighborhood. Then the universal covering space $\tilde{G}$ admits a group structure in which both the natural inclusion map $\pi_1(G) \hookrightarrow \tilde{G}$ and the projection $\tilde{\pi} : \tilde{G} \longrightarrow G$ are homomorphisms. The kernel of $\tilde{\pi}$ is $\pi_1(G)$.*

Proof. If $p : [0, 1] \longrightarrow G$ and $q : [0, 1] \longrightarrow G$ are paths, so is $t \mapsto p \cdot q(t) = p(t)q(t)$. If $p(0) = q(0) = 1_G$, the identity element in G, then $p \cdot q(0) = 1_G$ also. If p and p' are path-homotopic and q, q' are another pair of path-homotopic paths, then $p \cdot q$ and $p' \cdot q'$ are path-homotopic, for if $t \mapsto p_t$ is a path-homotopy $p \rightsquigarrow p'$ and $t \mapsto q_t$ is a path-homotopy $q \rightsquigarrow q'$, then $t \mapsto p_t \cdot q_t$ is a path-homotopy $p \cdot q \rightsquigarrow p' \cdot q'$.

It is straightforward to see that the projection $\tilde{\pi}$ is a group homomorphism. To see that the inclusion of the fundamental group as the fiber over the identity in $\tilde{G}$ is a group homomorphism, let p and q be loops with $p(0) = p(1) = q(0) = q(1) = 1_G$. There is a continuous map $f : [0, 1] \times [0, 1] \longrightarrow G$ given by $(t, u) \longrightarrow p(t)q(u)$. Taking different routes from $(0,0)$ to $(1,1)$ will give path-homotopic paths. Going directly via $t \mapsto f(t,t) = p(t)q(t)$ gives $p \cdot q$, while going indirectly via

$$
t \mapsto \begin{cases} f(2t, 0) = p(2t) & \text{if } 0 \leqslant t \leqslant \frac{1}{2}, \\ f(1, 2t - 1) = q(2t - 1) & \text{if } \frac{1}{2} \leqslant t \leqslant 1, \end{cases}
$$

gives the concatenated path $p \star q$. Thus, $p \star q$ and $p \cdot q$ are path-homotopic, so the multiplication in $\pi_1(G)$ is compatible with the multiplication in $\tilde{G}$.

The last statement, that the kernel of $\tilde{\pi}$ is $\pi_1(G)$, is true by definition. $\quad\square$

Proposition 13.5. *Let S^r denote the r-sphere. Then $\pi_1(S^1) \cong \mathbb{Z}$, while S^r is simply-connected if $r \geqslant 2$.*

Proof. We may identify the circle S^1 with the unit circle in $\mathbb{C}$. Then $x \mapsto e^{2\pi i x}$ is a covering map $\mathbb{R} \longrightarrow S^1$. The space $\mathbb{R}$ is contractible and hence simply-connected, so it is the universal covering space. If we give $S^1 \subset \mathbb{C}^\times$ the group structure it inherits from $\mathbb{C}^\times$, then this map $\mathbb{R} \longrightarrow S^1$ is a group homomorphism, so by Theorem 13.2 we may identify the kernel $\mathbb{Z}$ with $\pi_1(S^1)$.

To see that S^r is simply connected for $r \geqslant 2$, let $p : [0,1] \longrightarrow S^r$ be a path. Since it is a mapping from a lower-dimensional manifold, perturbing the path slightly if necessary, we may assume that p is not surjective. If it omits one point $P \in S^r$, its image is contained in $S^r - \{P\}$, which is homeomorphic to $\mathbb{R}^r$ and hence contractible. Therefore p, is path-homotopic to a trivial path.

$\square$

Proposition 13.6. *The group $\mathrm{SU}(2)$ is simply-connected. The group $\mathrm{SO}(3)$ is not. In fact $\pi_1(\mathrm{SO}(3)) \cong \mathbb{Z}/2\mathbb{Z}$.*

Proof. Note that $\mathrm{SU}(2) = \left\{ \begin{pmatrix} a & b \\ -\bar{b} & \bar{a} \end{pmatrix} \middle| \, |a|^2 + |b|^2 = 1 \right\}$ is homeomorphic to the 3 sphere in $\mathbb{C}^2$. As such, it is simply connected. We have a homomorphism $\mathrm{SU}(2) \longrightarrow \mathrm{SO}(3)$, which we constructed in Example 7.1. Since this mapping induced an isomorphism of Lie algebras, its image is an open subgroup of $\mathrm{SO}(3)$, and since $\mathrm{SO}(3)$ is connected, this homomorphism is surjective. The kernel $\{\pm I\}$ of this homomorphism is finite, so this is a covering map. Because $\mathrm{SU}(2)$ is simply connected, it follows from the uniqueness of the simply connected covering group that it is the universal covering group of $\mathrm{SO}(3)$. The kernel of this homomorphism $\mathrm{SU}(2) \longrightarrow \mathrm{SO}(3)$ is therefore the fundamental group, and it has order 2. $\quad\square$

Let G and H be topological groups. By a *local homomorphism* $G \longrightarrow H$ we mean the following data: a neighborhood U of the identity and a continuous map $\phi : U \longrightarrow H$ such that $\phi(uv) = \phi(u)\phi(v)$ whenever u, v, and $uv \in U$. This implies that $\phi(1_G) = 1_H$, so if $u, u^{-1} \in U$ we have $\phi(u^{-1}) = \phi(u)$. We may as well replace U by $U \cap U^{-1}$ so this is true for all $u \in U$.

Theorem 13.3. *Let G and H be topological groups, and assume that G is simply connected. Let U be a neighborhood of the identity in G. Then any local homomorphism $U \longrightarrow H$ can be extended to a homomorphism $G \longrightarrow H$.*

Proof. Let $g \in G$. Let $p : [0,1] \longrightarrow G$ be a path with $p(0) = 1_G$, $p(1) = g$. (Such a path exists because G is path-connected.) We first show that there exists a unique path $q : [0,1] \longrightarrow H$ such that $q(0) = 1_H$, and

$$q(v) \, q(u)^{-1} = \phi\big(p(v) \, p(u)^{-1}\big) \tag{13.1}$$

when $u, v \in [0,1]$ and $|u - v|$ is sufficiently small. We note that when u and v are sufficiently close, $p(v)p(u)^{-1} \in U$, so this makes sense. To construct a path q with this property, find $0 = x_0 < x_1 < \cdots < x_n = 1$ such that when u and v lie in an interval $[x_{i-1}, x_{i+1}]$, we have $p(v)p(u)^{-1} \in U$ $(1 \leqslant i < n)$. Define $q(x_0) = 1_H$, and if $v \in [x_i, x_{i+1}]$ define

$$q(v) = \phi\big(p(v) \, p(x_i)^{-1}\big) q(x_i). \tag{13.2}$$

This definition is recursive because here $q(x_i)$ is defined by (13.2) with i replaced by $i - 1$ if $i > 0$. With this definition, (13.2) is actually true for $v \in [x_{i-1}, x_{i+1}]$ if $i \geqslant 1$. Indeed, if $v \in [x_{i-1}, x_i]$ (the subinterval for which this is not a definition), we have

$$q(v) = \phi\big(p(v) \, p(x_{i-1})^{-1}\big) \, q(x_{i-1}),$$

so what we need to show is that

$$q(x_i) \, q(x_{i-1})^{-1} = \phi\big(p(v) \, p(x_i)^{-1}\big)^{-1} \phi\big(p(v)p(x_{i-1})^{-1}\big).$$

It follows from the fact that ϕ is a local homomorphism that the right-hand side is

$$\phi\big(p(x_i) \, p(x_{i-1})^{-1}\big).$$

Replacing i by $i - 1$ in (13.2) and taking $v = x_i$, this equals $q(x_i)q(x_{i-1})^{-1}$. Now (13.1) follows for this path by noting that if $\epsilon = \frac{1}{2} \min |x_{i+1} - x_i|$, then when $|u - v| < \epsilon$, $u, v \in [0,1]$, there exists an i such that $u, v \in [x_{i-1}, x_{i+1}]$, and (13.1) follows from (13.2) and the fact that ϕ is a local homomorphism. This proves that the path q exists. To show that it is unique, assume that (13.1) is valid for $|u-v| < \epsilon$, and choose the x_i so that $|x_i - x_{i+1}| < \epsilon$; then for $v \in [x_i, x_{i+1}]$, (13.2) is true, and the values of q are determined by this property.

Next we indicate how one can show that if p and p' are path-homotopic, and if q and q' are the corresponding paths in H, then $q(1) = q'(1)$. It is sufficient to prove this in the special case of a path-homotopy $t \mapsto p_t$, where $p_0 = p$ and $p_1 = p'$, such that there exists a sequence $0 = x_1 \leqslant \cdots \leqslant x_n = 1$ with $p_t(u)p_{t'}(v)^{-1} \in U$ when $u, v \in [x_{i-1}, x_{i+1}]$ and t and $t' \in [0,1]$. For although a general path-homotopy may not satisfy this assumption, it can be broken into steps, each of which does. In this case, we define

$$q_t(v) = \phi\big(p_t(v)\,p(x_i)^{-1}\big)q(x_i)$$

when $v \in [x_i, x_{i+1}]$ and verify that this q_t satisfies

$$q_t(v)q_t(u)^{-1} = \phi\big(p_t(v)\,p_t(u)^{-1}\big)$$

when $|u - v|$ is small. In particular, this is satisfied when $t = 1$ and $p_1 = p'$, so $q_1 = q'$ by definition. Now $q'(1) = \phi\big(p'(1)\,p(1)^{-1}\big)q(1) = q(1)$ since $p(1) = p'(1)$, as required.

We now define $\phi(g) = q(1)$. Since G is simply connected, any two paths from the identity to g are path-homotopic, so this is well-defined. It is straightforward to see that it agrees with ϕ on U. We must show that it is a homomorphism. Given g and g' in G, let p be a path from the identity to g, and let p' be a path from the identity to g', and let q and q' be the corresponding paths in H defined by (13.1). We construct a path p'' from the identity to gg' by

$$p''(t) = \begin{cases} p'(2t) & \text{if } 0 \leqslant t \leqslant 1/2, \\ p(2t-1)g' & \text{if } 1/2 \leqslant t \leqslant 1. \end{cases}$$

Let

$$q''(t) = \begin{cases} q'(2t) & \text{if } 0 \leqslant t \leqslant 1/2, \\ q(2t-1)q'(1) & \text{if } 1/2 \leqslant t \leqslant 1. \end{cases}$$

Then it is easy to check that q'' is related to p'' by (13.1), and taking $t = 1$, we see that $\phi(gg') = q''(1) = q(1)q'(1) = \phi(g)\phi(g')$. □

We turn next to the computation of the fundamental groups of some *noncompact* Lie groups.

As usual, we call a square complex matrix g *Hermitian* if $g = {}^t\overline{g}$. The eigenvalues of a Hermitian matrix are real, and it is called *positive definite* if these eigenvalues are positive. If g is Hermitian, so are g^2 and $e^g = I + g + \frac{1}{2}g^2 + \cdots$. According to the spectral theorem, the Hermitian matrix g can be written kak^{-1}, where a is real and diagonal and k is unitary. We have $g^2 = ka^2k^{-1}$ and ke^ak^{-1}, so g^2 and e^g are positive definite.

Proposition 13.7.

(i) *If g_1 and g_2 are positive definite Hermitian matrices, and if $g_1^2 = g_2^2$, then $g_1 = g_2$.*

(ii) *If g_1 and g_2 are Hermitian matrices and $e^{g_1} = e^{g_2}$, then $g_1 = g_2$.*

Proof. To prove (i), assume that the g_i are positive definite and that $g_1^2 = g_2^2$. We may write $g_i = k_i a_i k_i^{-1}$, where a_i is diagonal with positive entries, and we may arrange it so the entries in a_i are in descending order. Since a_1^2 and a_2^2 are similar diagonal matrices with their entries in descending order, they are equal, and since the squaring map on the positive reals is injective, $a_1 = a_2$. Denote

$a = a_1 = a_2$. It is not necessarily true that $k_1 = k_2$, but denoting $k = k_1^{-1}k_2$, k commutes with a^2. Let $\lambda_1 > \lambda_2 > \cdots$ be the distinct eigenvalues of a with multiplicities $d_1, d_2, \ldots$. Since k commutes with

$$a^2 = \begin{pmatrix} \lambda_1^2 I_{d_1} & & \\ & \lambda_2^2 I_{d_2} & \\ & & \ddots \end{pmatrix},$$

it has the form

$$k = \begin{pmatrix} K_1 & & \\ & K_2 & \\ & & \ddots \end{pmatrix},$$

where K_i is a $d_i \times d_i$ block. This implies that k commutes with a, and so $g_2 = kak^{-1} = g_1$.

The proof assuming $e^{g_1} = e^{g_2}$ is similar. It is no longer necessary to assume that g_1 and g_2 are positive definite because (unlike the squaring map) the exponential map is injective on all of $\mathbb{R}$.   □

Theorem 13.4. *Let P be the space of positive definite Hermitian matrices. If $g \in \mathrm{GL}(n, \mathbb{C})$, then g may be written uniquely as pk, where $k \in \mathrm{U}(n)$ and $p \in P$. Moreover, the multiplication map $P \times \mathrm{U}(n) \longrightarrow \mathrm{GL}(n, \mathbb{C})$ is a diffeomorphism.*

This is one of several related decompositions referred to as the *Cartan decomposition*. See Chap. 28 for related material.

Proof. The matrix $g \cdot {}^t\overline{g}$ is positive definite and Hermitian, so by the spectral theorem it can be diagonalized by a unitary matrix. This means we can write $g \cdot {}^t\overline{g} = \kappa a \kappa^{-1}$, where κ is unitary and a is a diagonal matrix with positive real entries. We may take the square root of a, writing $a = d^2$, where d is another diagonal matrix with positive real entries. Let $p = \kappa d \kappa^{-1}$. Since ${}^t\overline{\kappa} = \kappa^{-1}$, we have $g \cdot {}^t\overline{g} = \kappa d \kappa^{-1} \cdot {}^t\overline{(\kappa d \kappa^{-1})} = p \cdot {}^t\overline{p}$, which implies that $k = p^{-1}g$ is unitary.

The existence of the decomposition is now proved. To see that it is unique, suppose that $pk = p'k'$, where p and p' are positive definite Hermitian matrices, and k and k' are unitary. To show that $p = p'$ and $k = k'$, we may move the k' to the other side, so it is sufficient to show that if $pk = p'$, then $p = p'$. Taking the conjugate transpose, $k^{-1}p\,{}^t\overline{k}p = p'$, so $(p')^2 = pkk^{-1}p = p^2$. The uniqueness now follows from Proposition 13.7.

We now know that the multiplication map $P \times \mathrm{U}(n) \longrightarrow \mathrm{GL}(n, \mathbb{C})$ is a bijection. To see that it is a diffeomorphism, we can use the inverse function theorem. One must check that the Jacobian of the map is nonzero near any given point $(p_0, k_0) \in P \times \mathrm{U}(n)$. Let X_0 be a fixed Hermitian matrix such that $\exp(X_0) = p_0$. Parametrize P by elements of the vector space $\mathfrak{p}$ of Hermitian matrices, which we map to P by the map $\mathfrak{p} \ni X \longmapsto \exp(X_0 + X)$, and parametrize $U(n)$ by elements of $\mathfrak{u}(n)$ by means of the map $\mathfrak{u}(n) \ni Y \longmapsto$ $\exp(Y)p_0$. Noting that $\mathfrak{p}$ and $\mathfrak{u}(n)$ are complementary subspaces of $\mathfrak{gl}(n,\mathbb{C})$, it is clear using this parametrization of a neighborhood of (p_0, k_0) that the Jacobian is nonzero there, and so the multiplication map is a diffeomorphism. $\quad\square$

Theorem 13.5. *We have*

$$\pi_1\big(\mathrm{GL}(n,\mathbb{C})\big) \cong \pi_1\big(\mathrm{U}(n)\big), \qquad \pi_1\big(\mathrm{SL}(n,\mathbb{C})\big) \cong \pi_1\big(\mathrm{SU}(n)\big),$$

and

$$\pi_1\big(\mathrm{SL}(n,\mathbb{R})\big) \cong \pi_1\big(\mathrm{SO}(n)\big).$$

We have omitted $\mathrm{GL}(n,\mathbb{R})$ from this list because it is not connected. There is a general principle here: the fundamental group of a connected Lie group is often the same as the fundamental group of a maximal compact subgroup.

Proof. First, let $G = \mathrm{GL}(n,\mathbb{C})$, $K = \mathrm{U}(n)$, and P be the space of positive definite Hermitian matrices. By the Cartan decomposition, multiplication $K \times P \longrightarrow G$ is a bijection, and in fact, a homeomorphism, so it will follow that $\pi_1(K) \cong \pi_1(G)$ if we can show that P is contractible. However, the exponential map from the space $\mathfrak{p}$ of Hermitian matrices to P is bijective (in fact, a homeomorphism) by Proposition 13.7, and the space $\mathfrak{p}$ is a real vector space and hence contractible.

For $G = \mathrm{SL}(n,\mathbb{C})$, one argues similarly, with $K = \mathrm{SU}(n)$ and P the space of positive definite Hermitian matrices of determinant one. The exponential map from the space $\mathfrak{p}$ of Hermitian matrices of trace zero is again a homeomorphism of a real vector space onto P.

Finally, for $G = \mathrm{SL}(n,\mathbb{R})$, one takes $K = \mathrm{SO}(n)$, P to be the space of positive definite real matrices of determinant one, and $\mathfrak{p}$ to be the space of real symmetric matrices of trace zero. $\quad\square$

The remainder of this chapter will be less self-contained, but can be skipped with no loss of continuity. We will calculate the fundamental groups of $\mathrm{SO}(n)$ and $\mathrm{SU}(n)$, making use of some facts from algebraic topology that we do not prove. (These fundamental groups can alternatively be computed using the method of Chap. 23. See Exercise 23.4.)

If G is a Hausdorff topological group and H is a closed subgroup, then the coset space G/H is a Hausdorff space with the quotient topology. Such a quotient is called a *homogeneous space*.

Proposition 13.8. *Let G be a Lie group and H a closed subgroup. If the homogeneous space G/H is homeomorphic to a sphere S^r where $r \geqslant 3$, then $\pi_1(G) \cong \pi_1(H)$.*

Proof. The map $G \longrightarrow G/H$ is a fibration (Spanier [149], Example 4 on p. 91 and Corollary 14 on p. 96). It follows that there is an exact sequence

$$\pi_2(G/H) \longrightarrow \pi_1(H) \longrightarrow \pi_1(G) \longrightarrow \pi_1(G/H)$$

(Spanier [149], Theorem 10 on p. 377). Since G/H is a sphere of dimension $\geqslant 3$, its first and second homotopy groups are trivial and the result follows.

$\square$

Theorem 13.6. *The groups* $\mathrm{SU}(n)$ *are simply connected for all* n. *On the other hand,*

$$\pi_1(\mathrm{SO}(n)) \cong \begin{cases} \mathbb{Z} & \text{if } n = 2, \\ \mathbb{Z}/2\mathbb{Z} & \text{if } n > 2. \end{cases}$$

Proof. Since $\mathrm{SO}(2)$ is a circle, its fundamental group is $\mathbb{Z}$. By Proposition 13.6 $\pi_1(\mathrm{SO}(3)) \cong \mathbb{Z}/2\mathbb{Z}$ and $\pi_1(\mathrm{SU}(2))$ is trivial. The group $\mathrm{SO}(n)$ acts transitively on the unit sphere S^{n-1} in $\mathbb{R}^n$, and the isotropy subgroup is $\mathrm{SO}(n-1)$, so $\mathrm{SO}(n)/\mathrm{SO}(n-1)$ is homeomorphic to S^{n-1}. By Proposition 13.8, we see that $\pi_1(\mathrm{SO}(n)) \cong \pi_1(\mathrm{SO}(n-1))$ if $n \geqslant 4$. Similarly, $\mathrm{SU}(n)$ acts on the unit sphere S^{2n-1} in $\mathbb{C}^n$, and so $\mathrm{SU}(n)/\mathrm{SU}(n-1) \cong S^{2n-1}$, whence $\mathrm{SU}(n) \cong \mathrm{SU}(n-1)$ for $n \geqslant 2$. $\square$

If $n \geqslant$, the universal covering group of $\mathrm{SO}(n)$ is called the *spin* group and is denoted Spin(n). We will take a closer look at it in Chap. 31.

Exercises

Exercise 13.1. Let $\widetilde{\mathrm{SL}}(2, \mathbb{R})$ be the universal covering group of $\mathrm{SL}(2, \mathbb{R})$. Let $\pi : \widetilde{\mathrm{SL}}(2, \mathbb{R}) \longrightarrow \mathrm{GL}(V)$ be any finite-dimensional irreducible representation. Show that π factors through $\mathrm{SL}(2, \mathbb{R})$ and is hence not a faithful representation. (**Hint:** Use Exercise 12.2.)

14

The Local Frobenius Theorem

Let M be an n-dimensional smooth manifold. The *tangent bundle* TM of M is the disjoint union of all tangent spaces of points of M. It can be given the structure of a manifold of dimension $2\dim(M)$ as follows. If U is a coordinate neighborhood and $x_1, \ldots, x_n$ are local coordinates on U, then $T(U) = \{T_x M \mid x \in U\}$ can be taken to be a coordinate neighborhood of TM. Every element of $T_x M$ with $x \in U$ can be written uniquely as

$$\sum_{i=1}^n a_i \frac{\partial}{\partial x_i},$$

and mapping this tangent vector to $(x_1, \ldots, x_n, a_1, \ldots, a_n) \in \mathbb{R}^{2n}$ gives a chart on $T(U)$, making TM into a manifold.

By a d-dimensional *family* D in the tangent bundle of M we mean a rule that associates with each $x \in M$ a d-dimensional subspace $D_x \subset T_x(M)$. We ask that the family be *smooth*. By this we mean that in a neighborhood U of any given point x there are smooth vector fields $X_1, \ldots, X_d$ such that for $u \in U$ the vectors $X_{i,u} \in T_u(M)$ span D_u.

We say that a vector field X is *subordinate* to the family D if $X_x \in D_x$ for all $x \in U$. The family is called *involutory* if whenever X and Y are vector fields subordinate to D then so is $[X, Y]$. This definition is motivated by the following considerations.

An *integral manifold* of the family D is a d-dimensional submanifold N such that, for each point $x \in N$, the tangent space $T_x(N)$, identified with its image in $T_x(M)$, is D_x. We may ask whether it is possible, at least locally in a neighborhood of every point, to pass an integral manifold. This is surely a natural question.

Let us observe that if it is true, then the family D is involutory. To see this (at least plausibly), let U be an open set in M that is small enough that through each point in U there is an integral submanifold that is closed in U. Let J be the subspace of $C^\infty(U)$ consisting of functions that are constant on these integral submanifolds. Then the restriction of a vector field X to U is

D. Bump, *Lie Groups*, Graduate Texts in Mathematics 225,
DOI 10.1007/978-1-4614-8024-2_14, © Springer Science+Business Media New York 2013

subordinate to D if and only if it annihilates J. It is clear from (6.6) that if X and Y have this property, then so does $[X, Y]$.

The Frobenius theorem is a converse to this observation. A global version may be found in Chevalley [35]. We will content ourselves with the local theorem.

Lemma 14.1. *If $X_1, \ldots, X_d$ are vector fields on M such that $[X_i, X_j]$ lies in the $C^\infty(M)$ span of $X_1, \ldots, X_d$, and if for each $x \in M$ we define D_x to be the span of $X_{1x}, \ldots, X_{dx}$, then D is an involutory family.*

Proof. Any vector field subordinate to D has the form (locally near x) $\sum_i f_i X_i$, where f_i are smooth functions. To check that the commutator of two such vector fields is also of the same form amounts to using the formula

$$[fX, gY] = fg[X, Y] + fX(g)Y - gY(f)X,$$

which follows easily on applying both sides to a function h and using the fact that X and Y are derivations of $C^\infty(M)$. □

Theorem 14.1 (Frobenius). *Let D be a smooth involutory d-dimensional family in the tangent bundle of M. Then for each point $x \in M$ there exists a neighborhood U of x and an integral manifold N of D through x in U. If N' is another integral manifold through x, then N and N' coincide near x. That is, there exists a neighborhood V of x such that $V \cap N = V \cap N'$.*

Proof. Since this is a strictly local statement, it is sufficient to prove this when M is an open set in $\mathbb{R}^n$ and x is the origin.

We show first that if X is a vector field that does not vanish at x, then we may find a system $y_1, \ldots, y_n$ of coordinates in which $X = \partial/\partial y_n$. Let $x_1, \ldots, x_n$ be the standard Cartesian functions. Since X does not vanish at the origin, the function $X(x_i)$ does not vanish at the origin for some i, so after permuting the variables if necessary, we may assume that $X(x_n) \neq 0$. Write

$$X = a_1 \frac{\partial}{\partial x_1} + \cdots + a_n \frac{\partial}{\partial x_n}$$

in terms of smooth functions $a_i = a_i(x_1, \ldots, x_n)$. Then $a_n(0, \ldots, 0) \neq 0$.

The new coordinate system $y_1, \ldots, y_n$ will have the property that

$$(y_1, \ldots, y_{n-1}, 0) = (x_1, \ldots, x_{n-1}, 0).$$

To describe $(y_1, \ldots, y_n)$ when $y_n \neq 0$, let us fix small numbers $u_1, \ldots, u_{n-1}$. Then we will describe the path which is, in the y coordinates,

$$t \longmapsto (u_1, \ldots, u_{n-1}, t).$$

This path is to be an integral curve for the vector field through the point $(u_1, \ldots, u_{n-1}, 0)$. By Proposition 8.1 a unique such path exists (for t small).

Thus, we have a path that is (in the x coordinates) $t \longrightarrow (x_1(t), \ldots, x_n(t))$, satisfying the first-order system

$$x_i'(t) = a_i(x_1(t), \ldots, x_n(t)), \tag{14.1}$$
$$(x_i(0), \ldots, x_n(0)) = (u_1, \ldots, u_{n-1}, 0).$$

For $u_1, \ldots, u_{n-1}$ sufficiently small, we have $a_n(u_1, \ldots, u_{n-1}, 0) \neq 0$ and so this integral curve is transverse to the hyperplane $x_n = 0$. We choose our coordinate system $y_1, \ldots, y_n$ so that

$$y_i(x_1(t), \ldots, x_n(t)) = u_i, \qquad (i = 1, 2, 3, \ldots, n-1),$$
$$y_n(x_1(t), \ldots, x_n(t)) = t.$$

Now $\partial x_i / \partial y_n = a_i$ because the partial derivative is the derivative along one of the paths (14.1). Thus

$$\frac{\partial}{\partial y_n} = \sum_i \frac{\partial x_i}{\partial y_n} \frac{\partial}{\partial x_i} = \sum_i a_i \frac{\partial}{\partial x_i} = X.$$

This proves that there exists a coordinate system in which $X = \partial/\partial y_n$.

If $d = 1$, the result is proved by this. We will assume that $d > 1$ and that the existence of integral manifolds is known for lower-dimensional involutory families. Let $X_1, \ldots, X_d$ be smooth vector fields such that $X_{i,u}$ span D_u for u near the origin. We have just shown that we may assume that $X = X_d = \partial/\partial y_n$. Since D is involutory, $[X_d, X_i] = \sum_j g_{ij} X_j$ for smooth functions g_{ij}. We will show that we can arrange things so that $g_{id} = 0$ when $i < d$; that is,

$$[X_d, X_i] = \sum_{j=1}^{d-1} g_{ij} X_j, \qquad (i < d). \tag{14.2}$$

Indeed, writing

$$X_i = \sum_{k=1}^{n} h_{ik} \frac{\partial}{\partial y_k}, \qquad (i = 1, \ldots, d-1), \tag{14.3}$$

we will still have a spanning set if we subtract $h_{in} X_d$ from X_i. We may therefore assume that $h_{in} = 0$ for $i < d$. Thus

$$X_i = \sum_{k=1}^{n-1} h_{ik} \frac{\partial}{\partial y_k}, \qquad (i = 1, \ldots, d-1). \tag{14.4}$$

In other words, we may assume that X_i does not involve $\partial/\partial y_n$ for $i < d$. Now

$$[X_d, X_i] = \sum_{k=1}^{n-1} \frac{\partial h_{ik}}{\partial y_n} \frac{\partial}{\partial y_j}. \tag{14.5}$$

On the other hand, we have

$$[X_d, X_i] = \sum_{j=1}^{d-1} g_{ij} X_j + g_{id} X_d = \sum_{j=1}^{d-1} \sum_{k=1}^{n-1} g_{ij} h_{jk} \frac{\partial}{\partial y_k} + g_{id} \frac{\partial}{\partial y_n}.$$

Comparing the coefficients of $\partial/\partial y_n$ in this expression with that in (14.5) shows that $g_{id} = 0$, proving (14.2).

Next we show that if $(c_1, \ldots, c_{d-1})$ are real constants, then there exist smooth functions $f_1, \ldots, f_{d-1}$ such that for small $y_1, \ldots, y_{n-1}$ we have

$$f_i(y_1, y_2, \ldots, y_{n-1}, 0) = c_i, \qquad (i = 1, \ldots, d-1), \tag{14.6}$$

and

$$\left[X_d, \sum_{i=1}^{d-1} f_i X_i \right] = 0.$$

Indeed,

$$\left[X_d, \sum_{i=1}^{d-1} f_i X_i \right] = \sum_{i=1}^{d=1} \frac{\partial f_i}{\partial y_n} X_i + \sum_{i,j=1}^{d-1} f_i g_{ij} X_j.$$

For this to be zero, we need the f_i to be solutions to the first-order system

$$\frac{\partial f_j}{\partial y_n} + \sum_{i=1}^{d-1} g_{ij} f_i = 0, \qquad j = 1, \ldots, d-1.$$

This first-order system has a solution locally with the prescribed initial condition.

Since the c_i can be arbitrary, we may choose

$$c_i = \begin{cases} 1 \text{ if } i = 1, \\ 0 \text{ otherwise.} \end{cases}$$

Then the vector field $\sum f_i X_i$ agrees with X_1 on the hyperplane $y_n = 0$. Replacing X_1 by $\sum f_i X_i$, we may therefore assume that $[X_d, X_1] = 0$. Repeating this process, we may similarly assume that $[X_d, X_i] = 0$ for all $i < d$. Now with the h_{ij} as in (14.3), this means that $\partial h_{ij}/\partial y_n = 0$, so the h_{ij} are independent of y_n.

Since the h_{ij} are independent of y_n, we may interpret (14.4) as defining $d-1$ vector fields on $\mathbb{R}^{n-1}$. They span a $(d-1)$-dimensional involutory family of tangent vectors in $\mathbb{R}^{n-1}$ and by induction there exists an integral manifold for this vector field. If this manifold is $N_0 \subset \mathbb{R}^{n-1}$, then it is clear that

$$N = \{(y_1, \ldots, y_n) \mid (y_1, \ldots, y_{n-1}) \in N_0\}$$

is an integral manifold for D.

We have established the existence of an integral submanifold. The local uniqueness of the integral submanifold can also be proved now. In fact, if we repeat the process by which we selected the coordinate system $y_1, \ldots, y_n$ so that the vector field $\partial/\partial y_n$ was subordinate to the involutory family D, we eventually arrive at a system in which D is spanned by $\partial/\partial y_{n-d+1}, \ldots, \partial/\partial y_n$. Then the integral manifold is given by the equations $y_1 = \cdots = y_{n-d} = 0$. $\square$

If G is a Lie group, a *local subgroup* of G consists of an open neighborhood U of the identity and a closed subset K of U such that $1_G \in K$, and if $x, y \in K$ such that $xy \in U$, then $xy \in K$, and if $x \in K$ such that $x^{-1} \in U$, then $x^{-1} \in K$. For example, if H is a closed subgroup of G and U is any open set, then $U \cap H$ is a local subgroup.

Proposition 14.1. *Let G be a Lie group with Lie algebra $\mathfrak{g}$, and let $\mathfrak{k}$ be a Lie subalgebra of $\mathfrak{g}$. Then there exists a local subgroup K of G with a tangent space at the identity that is $\mathfrak{k}$. The exponential map sends a neighborhood of the identity in $\mathfrak{k}$ onto a neighborhood of the identity in K.*

Proof. The Lie algebra $\mathfrak{g}$ of G has two incarnations: as the tangent space to the identity of G and as the set of left-invariant vector fields. For definiteness, we identify $\mathfrak{g} = T_e(G)$ and recall how the left-invariant vector field arises.

If $g \in G$, let $\lambda_g : G \longrightarrow G$ be left translation by g, so that $\lambda_g(x) = gx$. Let $\lambda_{g*} : T_e(G) \longrightarrow T_x(G)$ be the induced map of tangent spaces. Then the left-invariant vector field associated with $X_e \in \mathfrak{g}$ has $X_g = \lambda_{g*}(X_e)$.

Let $d = \dim(\mathfrak{k})$ and let D be the d-dimensional family of tangent vectors such that $D_g = \lambda_{g*}(\mathfrak{k})$. Since $\mathfrak{k}$ is closed under the bracket, it follows from Lemma 14.1 that D is involutory, so there exists an integral submanifold K in a neighborhood U of the identity. We will show that if U is sufficiently small, then K is a local group.

Indeed, let x and y be elements of K such that $xy \in U$. Since the vector fields associated with elements of $\mathfrak{k}$ are left-invariant, the involutory family D is invariant under left translation. The image of K under right translation by x is also an integral submanifold of D through x, so this submanifold is K itself. These submanifolds therefore coincide near x and, since y is in K, its left translate xy by x is also in K.

Since the one-parameter subgroups $\exp(tX)$ with $X \in \mathfrak{k}$ are tangent to the left-invariant vector field at every point, they are contained in the integral submanifold K near the identity, and the image of a neighborhood of the identity under $\exp$ is a manifold of the same dimension as K, so the last statement is clear. $\square$

We recall that the notion of a *local homomorphism* was defined in Chap. 13 before Theorem 13.3.

Proposition 14.2. *Let G and H be Lie groups with Lie algebras $\mathfrak{g}$ and $\mathfrak{h}$, respectively, and let $\pi : \mathfrak{g} \longrightarrow \mathfrak{h}$ be a Lie algebra homomorphism. Then there exists a neighborhood U of G and a local homomorphism $\pi : U \longrightarrow H$ whose differential is π.*

Proof. The tangent space to $G \times H$ at the identity is $\mathfrak{g} \oplus \mathfrak{h}$. Let

$$\mathfrak{k} = \{(X, \pi(X)) \mid X \in \mathfrak{g}\} \subset \mathfrak{g} \oplus \mathfrak{h}.$$

It is a Lie subalgebra, corresponding by Proposition 14.1 to a local subgroup K of $G \times H$. The tangent space to the identity of K is thus its Lie algebra $\mathfrak{k}$, which intersects $\mathfrak{h}$ in $\mathfrak{g} \oplus \mathfrak{h}$ transversally in a single point. Thus $\mathfrak{g}$ is the direct sum of $\mathfrak{k}$ and $\mathfrak{h}$. Concretely, this reflects the fact that $\mathfrak{k}$ is the graph of a map $\pi : \mathfrak{g} \longrightarrow \mathfrak{h}$. Using the inverse function theorem, the same is true locally of K: since its tangent space at the identity is a direct sum complement of the tangent space of H in the tangent space of $G \times H$, it is, locally, the graph of a mapping. Thus, there exists a map $\pi : U \longrightarrow H$ of a sufficiently small neighborhood of the identity in G such that if $(g, h) \in G \times H$, $g \in U$, and $h \in \pi(U)$, then $(g, h) \in K$ if and only if $h = \pi(g)$. Because K is a local subgroup, this implies that π is a local homomorphism. $\square$

Theorem 14.2. *Let G and H be Lie groups with Lie algebras $\mathfrak{g}$ and $\mathfrak{h}$, respectively, and let $\pi : \mathfrak{g} \longrightarrow \mathfrak{h}$ be a Lie algebra homomorphism. Assume that G is simply connected. Then there exists a Lie group homomorphism $\pi : G \longrightarrow H$ with differential π.*

Proof. This follows from Proposition 14.2 and Theorem 13.3. $\square$

We can now give another proof of Theorem 12.2. The basic idea here is to use a compact subgroup to prove the complete reducibility of some class of representations of a noncompact group. This idea was called the "Unitarian Trick" by Hermann Weyl. We will extend the validity of Theorem 12.2, though the algebraic method would work as well for this.

Theorem 14.3. *Let G and K be Lie groups with Lie algebras $\mathfrak{g}$ and $\mathfrak{k}$. Assume K is compact and simply connected. Suppose that $\mathfrak{g}$ and $\mathfrak{k}$ have isomorphic complexifications. Then every finite-dimensional irreducible complex representation of $\mathfrak{g}$ is completely reducible. If G is connected, then every irreducible complex representation of G is completely reducible.*

Proof. Let (π, V) be a finite-dimensional representation of G, and let W be a proper nonzero invariant subspace. We will show that there is another invariant subspace W' such that $V = W \oplus W'$. By induction on $\dim(V)$, it will follow that both W and W' are direct sums of irreducible representations.

The differential of π is a complex representation of $\mathfrak{g}$. As in Proposition 11.3, we may extend it to a representation of $\mathfrak{g}_{\mathbb{C}} \cong \mathfrak{k}_{\mathbb{C}}$ and then restrict it to $\mathfrak{k}$. Since K is simply connected, the resulting Lie algebra homomorphism $\mathfrak{k} \longrightarrow \mathfrak{gl}(V)$ is the differential of a Lie group homomorphism $\pi_K : K \longrightarrow \mathrm{GL}(V)$.

Now, because K is compact, this representation of K is completely reducible (Proposition 2.2). Thus, there exists a K-invariant subspace W' such that $V = W \oplus W'$. Of course, W' is also invariant with respect to $\mathfrak{k}$

and hence $\mathfrak{k}_{\mathbb{C}} \cong \mathfrak{g}_{\mathbb{C}}$, and hence $\mathfrak{g}$. It is therefore invariant under $\exp(\mathfrak{g})$. If G is connected, it is generated by a neighborhood of the identity, and so W' is G-invariant.   □

Theorem 14.4. *Let* (π, V) *be a finite-dimensional irreducible complex representation of* $\mathfrak{g} = \mathfrak{sl}(n, \mathbb{R})$, $\mathfrak{su}(n)$, *or* $\mathfrak{sl}(n, \mathbb{C})$. *If* $\mathfrak{g}$ *is* $\mathfrak{sl}(n, \mathbb{C})$ *then assume that* $\pi : \mathfrak{g} \longrightarrow \mathrm{End}(V)$ *is complex linear. Then* π *is completely reducible.*

Proof. We will prove this for $\mathfrak{sl}(n, \mathbb{R})$ and $\mathfrak{su}(n)$. By Theorem 13.6, K is simply-connected and the hypotheses of Theorem 14.3 are satisfied. For $\mathfrak{sl}(n, \mathbb{R})$, we can take $G = \mathrm{SL}(n, \mathbb{R})$, $K = \mathrm{SU}(n)$. For $\mathfrak{su}(n)$, we can take $G = K = \mathrm{SU}(n)$.

The case of $\mathfrak{sl}(n, \mathbb{C})$ requires a minor modification to Theorem 14.3 and is left to the reader.   □

Theorem 14.5. *Let* (π, V) *be a finite-dimensional irreducible complex representation of* $\mathrm{SL}(n, \mathbb{R})$. *Then* π *is completely reducible.*

Proof. We take $G = \mathrm{SL}(n, \mathbb{R})$, $K = \mathrm{SU}(n)$.   □

Exercises

Exercise 14.1. Let G be a connected complex analytic Lie group, and let $K \subset G$ be a compact Lie subgroup. Let $\mathfrak{g}$ and $\mathfrak{k} \subset \mathfrak{g}$ be the Lie algebras of G and K, respectively. Assume that $\mathfrak{g}$ is the complexification of $\mathfrak{k}$ and that K is simply-connected. Prove that every finite-dimensional irreducible complex representation of $\mathfrak{g}$ is completely reducible. If G is connected, then every irreducible complex analytic representation of G is completely reducible.

15

Tori

A *complex manifold* M is constructed analogously to a smooth manifold. We specify an atlas $\mathcal{U} = \{(U, \phi)\}$, where each chart $U \subset M$ is an open set and $\phi : U \longrightarrow \mathbb{C}^m$ is a homeomorphism of U onto its image that is assumed to be open in $\mathbb{C}^m$. It is assumed that the transition functions $\psi \circ \phi^{-1} : \phi(U \cap V) \longrightarrow \psi(U \cap V)$ are holomorphic for any two charts (U, ϕ) and (V, ψ). A *complex Lie group* (or *complex analytic group*) is a Hausdorff topological group that is a complex manifold in which the multiplication and inversion maps $G \times G \longrightarrow G$ and $G \longrightarrow G$ are holomorphic. The Lie algebra of a complex Lie group is a complex Lie algebra. For example, $\mathrm{GL}(n, \mathbb{C})$ is a complex Lie group.

If $\mathfrak{g}$ is a Lie algebra and $X, Y \in \mathfrak{g}$, we say that X and Y *commute* if $[X, Y] = 0$. We call the Lie algebra $\mathfrak{g}$ *Abelian* if $[X, Y] = 0$ for all $X, Y \in \mathfrak{g}$.

Proposition 15.1. *The Lie algebra of an Abelian Lie group is Abelian.*

Proof. The action of G on itself by conjugation is trivial, so the induced action Ad of G on its Lie algebra is trivial. By Theorem 8.2, it follows that $\mathrm{ad} : \mathrm{Lie}(G) \longrightarrow \mathrm{End}(\mathrm{Lie}(G))$ is the zero map, so $[X, Y] = \mathrm{ad}(X)Y = 0$. $\qquad\square$

Proposition 15.2. *If G is a Lie group, and X and Y are commuting elements of $\mathrm{Lie}(G)$, then $\mathrm{e}^{X+Y} = \mathrm{e}^X \mathrm{e}^Y$. In particular, $\mathrm{e}^X \mathrm{e}^Y = \mathrm{e}^Y \mathrm{e}^X$.*

Proof. First note that, since the differential of Ad is ad (Theorem 8.2), $\mathrm{Ad}(\mathrm{e}^{tX})Y = Y$ for all t. Recalling that $\mathrm{Ad}(\mathrm{e}^{tX})$ is the endomorphism of $\mathrm{Lie}(G)$ induced by conjugation, this means that conjugation by e^{tX} takes the one-parameter subgroup $u \longrightarrow \mathrm{e}^{uY}$ to itself, so $\mathrm{e}^{tX} \mathrm{e}^{uY} \mathrm{e}^{-tX} = \mathrm{e}^{uY}$. Thus, e^{tX} and e^{uY} commute for all real t and u.

We recall from Chap. 8 that the path $p(t) = \mathrm{e}^{tY}$ is characterized by the fact that $p(0) = 1_G$, while $p_*(\mathrm{d}/\mathrm{d}t) = Y_{p(t)}$. The latter condition means that if $f \in C^\infty(G)$ we have

$$\frac{\mathrm{d}}{\mathrm{d}t} f(p(t)) = (Yf)(p(t)).$$

D. Bump, *Lie Groups*, Graduate Texts in Mathematics 225,
DOI 10.1007/978-1-4614-8024-2_15, © Springer Science+Business Media New York 2013

Let $q(t, u) = e^{tX} e^{uY}$. The vector field Y is invariant under left translation, in particular left translation by e^{tX}, so

$$\frac{\partial}{\partial u} f\big(q(t, u)\big) = (Yf)(e^{tX} e^{uY}).$$

Similarly (making use of $e^{tX} e^{uY} = e^{uY} e^{tX}$),

$$\frac{\partial}{\partial t} f\big(q(t, u)\big) = (Xf)(e^{tX} e^{uY}).$$

Now, by the chain rule,

$$\frac{\mathrm{d}}{\mathrm{d}v} f\big(q(v, v)\big) = \frac{\partial}{\partial t} f\big(q(t, u)\big)\Big|_{t=u=v} + \frac{\partial}{\partial u} f\big(q(t, u)\big)\Big|_{t=u=v}$$
$$= (Yf + Xf)\big(q(v, v)\big).$$

This means that the path $v \longrightarrow r(v) = q(v, v)$ satisfies $r_*(\mathrm{d}/\mathrm{d}v) = (X+Y)_{r(v)}$ whence $e^{v(X+Y)} = e^{vX} e^{vY}$. Taking $v = 1$, the result is proved.   □

A *compact torus* is a compact connected Lie group that is Abelian. In the context of Lie group theory a compact torus is usually just called a *torus*, though in the context of algebraic groups the term "torus" is used slightly differently.

For example, $\mathbb{T} = \{z \in \mathbb{C}^\times \mid |z| = 1\}$ is a torus. This group is isomorphic to $\mathbb{R}/\mathbb{Z}$. Even though $\mathbb{R}$ and $\mathbb{Z}$ are additive groups, we may, during the following discussion, sometimes write the group law in $\mathbb{R}/\mathbb{Z}$ multiplicatively.

Proposition 15.3. *Let T be a torus, and let $\mathfrak{t}$ be its Lie algebra. Then* $\exp :$ $\mathfrak{t} \longrightarrow T$ *is a homomorphism, and its kernel is a lattice. We have* $T \cong (\mathbb{R}/\mathbb{Z})^r \cong \mathbb{T}^r$, *where r is the dimension of T.*

Proof. Let $\mathfrak{t}$ be the Lie algebra of T. Since T is Abelian, so is $\mathfrak{t}$, and by Proposition 15.2, exp is a homomorphism from the additive group $\mathfrak{t}$ to T. The kernel $\Lambda \subset \mathfrak{t}$ is discrete since exp is a local homeomorphism, and Λ is cocompact since T is compact. Thus, Λ is a lattice and $T \cong \mathfrak{t}/\Lambda \cong (\mathbb{R}/\mathbb{Z})^r \cong \mathbb{T}^r$.   □

A character of $\mathbb{R}^r$ of the form

$$(x_1, \ldots, x_r) \mapsto e^{2\pi i \left(\sum k_j x_j\right)}, \tag{15.1}$$

where $(k_1, \ldots, k_r) \in \mathbb{Z}^r$, induces a character on $(\mathbb{R}/\mathbb{Z})^r$.

Proposition 15.4. *Every irreducible complex representation of $(\mathbb{R}/\mathbb{Z})^r$ coincides with (15.1) for suitable $k_i \in \mathbb{Z}$.*

Proof. By classical Fourier analysis, these characters span $L^2((\mathbb{R}/\mathbb{Z})^r)$. Thus, the character χ of any complex representation π is not orthogonal to (15.1) for some $(k_1, \ldots, k_r) \in \mathbb{Z}^r$. By Schur orthogonality, χ agrees with this character.   □

We also want to know the irreducible *real* representations of $(\mathbb{Z}/\mathbb{R})^r$. Let $k_1, \ldots, k_r \in \mathbb{Z}$ be given. Assume that they are not all zero. The complex character (15.1) is not a real representation. However, regarding it as a homomorphism $(\mathbb{Z}/\mathbb{R})^r \longrightarrow \mathbb{T}$, we may compose it with the real representation $\mathbb{T} \ni t = e^{2\pi i \theta} \mapsto \begin{pmatrix} \cos(2\pi\theta) & \sin(2\pi\theta) \\ -\sin(2\pi\theta) & \cos(2\pi\theta) \end{pmatrix}$ of $\mathbb{T}$. We obtain a real representation

$$(x_1, \ldots, x_r) \mapsto \begin{pmatrix} \cos(2\pi \sum k_i x_i) & \sin(2\pi \sum k_i x_i) \\ -\sin(2\pi \sum k_i x_i) & \cos(2\pi \sum k_i x_i) \end{pmatrix}. \tag{15.2}$$

Proposition 15.5. *Let $T = (\mathbb{Z}/\mathbb{R})^r$ and let (π, V) be an irreducible real representation. Then either π is trivial or π is two-dimensional and is one of the irreducible representations (15.2) with $k_i \in \mathbb{Z}$ not all zero. In the two-dimensional case the complexified module $V_{\mathbb{C}} = \mathbb{C} \otimes V$ decomposes into two one-dimensional representations corresponding to a character and its inverse.*

Proof. It is straightforward to see that the real representation (15.2) is irreducible. The completeness of this set of irreducible real representations follows from the corresponding classification of the irreducible complex characters (Proposition 15.4). It is also easy to see that the complexified representation is equivalent to

$$(x_1, \ldots, x_r) \mapsto \begin{pmatrix} e^{2\pi i \sum k_i x_i} & \\ & e^{-2\pi i \sum k_i x_i} \end{pmatrix}.$$

$\square$

If T is a compact torus, we will associate with T a complex analytic group $T_{\mathbb{C}}$, which we call the *complexification* of T. Let $\mathfrak{t}_{\mathbb{C}} = \mathbb{C} \otimes \mathfrak{t}$ be the complexification of the Lie algebra, and let $T_{\mathbb{C}} = \mathfrak{t}_{\mathbb{C}}/\Lambda$, where $\Lambda \subset \mathfrak{t}$ is the kernel of $\exp : \mathfrak{t} \longrightarrow T$. It is easy to see that this construction is functorial: given a homomorphism $\phi : T \longrightarrow U$ of compact tori, the differential $\phi_* : \text{Lie}(T) \longrightarrow \text{Lie}(U)$ commutes with the exponential map, so ϕ_* kills the kernel Λ of $\exp : \mathfrak{t} \longrightarrow T$. Therefore, there is an induced map $T_{\mathbb{C}} \longrightarrow U_{\mathbb{C}}$.

If we identify $T = (\mathbb{R}/\mathbb{Z})^r$, the complexification $T_{\mathbb{C}} \cong (\mathbb{C}/\mathbb{Z})^r$. Since $x \longrightarrow e^{2\pi i x}$ induces an isomorphism of the additive group $\mathbb{C}/\mathbb{Z}$ with the multiplicative group $\mathbb{C}^\times$, we see that $T_{\mathbb{C}} \cong (\mathbb{C}^\times)^r$. We call any complex Lie group isomorphic to $(\mathbb{C}^\times)^r$ for some r a *complex torus*.

By a *linear character* χ of a compact torus T, we mean a continuous homomorphism $T \longrightarrow \mathbb{C}^\times$. These are just the characters of irreducible representations, known explicitly by (15.1). They take values in $\mathbb{T}$, as we may see from (15.1), or by noting that the image is a compact subgroup of $\mathbb{C}^\times$.

By a *rational character* χ of a complex torus T, we mean an analytic homomorphism $T \longrightarrow \mathbb{C}^\times$.

Proposition 15.6. *Let T be a compact torus. Then any linear character χ of T extends uniquely to a rational character of $T_{\mathbb{C}}$.*

Proof. Without loss of generality, we may assume that $T = (\mathbb{R}/\mathbb{Z})^r$ and that $T_{\mathbb{C}} = (\mathbb{C}^{\times})^r$, where the embedding $T \longrightarrow T_{\mathbb{C}}$ is the map $(x_1, \ldots, x_r) \longrightarrow (e^{2\pi i x_1}, \ldots, e^{2\pi i x_r})$. Every linear character of T is given by (15.1) for suitable $k_i \in \mathbb{Z}$, and this extends to the rational character $(t_1, \ldots, t_r) \longrightarrow \prod t_i^{k_i}$ of $T_{\mathbb{C}}$. Since a rational character is holomorphic, it is determined by its values on the image $\mathbb{T}^r$ of T. $\qquad\square$

We will denote the group of characters of a compact torus T as $X^*(T)$. We will denote its group law *additively*: if χ_1 and χ_2 are characters, then $(\chi_1 + \chi_2)(t) = \chi_1(t)\chi_2(t)$. We may identify $X^*(T)$ with the group of rational characters of $T_{\mathbb{C}}$.

A *(topological) generator* of a compact torus T is an element t such that the smallest closed subgroup of T containing t is T itself.

Theorem 15.1 (Kronecker). *Let $(t_1, \ldots, t_r) \in \mathbb{R}^r$, and let t be the image of this point in $T = (\mathbb{R}/\mathbb{Z})^r$. Then t is a generator of T if and only if $1, t_1, \ldots, t_r$ are linearly independent over $\mathbb{Q}$.*

Proof. Let H be the closure of the group $\langle t \rangle$ generated by t in $T = (\mathbb{R}/\mathbb{Z})^r$. Then T/H is a compact Abelian group, and if it is not reduced to the identity it has a character χ. We may regard this as a character of T that is trivial on H, and as such it has the form (15.1) for suitable $k_i \in \mathbb{Z}$. Since t itself is in H, this means that $\sum k_j t_j \in \mathbb{Z}$, so $1, t_1, \ldots, t_r$ are linearly dependent. The existence of nontrivial characters of T/H is thus equivalent to the linear dependence of $1, t_1, \ldots, t_r$ and the result follows. $\qquad\square$

Corollary 15.1. *Each compact torus T has a generator. Indeed, generators are dense in T.*

Proof. We may assume that $T = (\mathbb{R}/\mathbb{Z})^r$. By Kronecker's Theorem 15.1, what we must show is that r-tuples $(t_1, \ldots, t_r)$ such that $1, t_1, \ldots, t_r$ are linearly independent over $\mathbb{Q}$ are dense in $\mathbb{R}^r$. If $1, t_1, \ldots, t_{i-1}$ are linearly independent, then linear independence of $1, t_1, \ldots, t_i$ excludes only countably many t_i, and the result follows from the uncountability of $\mathbb{R}$. $\qquad\square$

Proposition 15.7. *Let $T = (\mathbb{R}/\mathbb{Z})^r$.*

(i) *Each automorphism of T is of the form $t \longrightarrow Mt \pmod{\mathbb{Z}^r}$, where $M \in \mathrm{GL}(r, \mathbb{Z})$. Thus, $\mathrm{Aut}(T) \cong \mathrm{GL}(r, \mathbb{Z})$.*

(ii) *If H is a connected topological space and $f : H \longrightarrow \mathrm{Aut}(T)$ is a map such that $(h, t) \longrightarrow f(h)t$ is a continuous map $H \times T \longrightarrow T$, then f is constant.*

We can express (ii) by saying that $\mathrm{Aut}(T)$ is discrete since if it is given the discrete topology, then $(h, t) \longrightarrow f(h)t$ is continuous if and only if f is locally constant.

Proof. If $\phi : T \longrightarrow T$ is an automorphism, then ϕ induces an invertible linear transformation M of the Lie algebra $\mathfrak{t}$ of T that commutes with the exponential map. Because T is Abelian, the exponential map $\exp : \mathfrak{t} \rightarrow T$ is a group homomorphism, and ϕ must preserve its kernel Λ. We may identify $\mathfrak{t} = \mathbb{R}^r$ in such a way that Λ is identified with $\mathbb{Z}^r$, in which case the matrix of M must lie in $\mathrm{GL}(r, \mathbb{Z})$. Part (i) is now clear.

For part (ii), since T is compact and f is continuous, as $h \longrightarrow h_1$, $f(h)t \longrightarrow f(h_1)t$ uniformly for $t \in T$. It is easy to see from (i) that this is impossible unless f is locally constant. $\qquad\square$

In the remainder of this chapter, we will consider tori embedded in Lie groups. First, we prove a general statement that implies the existence of tori.

Theorem 15.2. *Let G be a Lie group and H a closed Abelian subgroup. Then H is a Lie subgroup of G. If G is compact, then the connected component of the identity in H is a torus.*

The assumption that H is Abelian is unnecessary. See Remark 7.2 for references to a result without this assumption.

Proof. Let $\mathfrak{g} = \mathrm{Lie}(G)$. The exponential map $\mathfrak{g} \longrightarrow G$ is a local homeomorphism near the origin. Let U be a neighborhood of $0 \in \mathfrak{g}$ such that exp has a smooth inverse $\log : \exp(U) \longrightarrow U$. Let

$$\mathfrak{h} = \{X \in \mathfrak{g} \mid \exp(tX) \in H \text{ for all } t \in \mathbb{R}\}.$$

Lemma 15.1. *If $X \in \mathfrak{h}$ and $Y \in U$, and if $e^Y \in H$ then $[X, Y] = 0$.*

To prove the lemma, note that for any $t > 0$ both e^{tX} and $e^Y \in H$ commute, so $e^Y = e^{tX} e^Y e^{-tX} = \exp(\mathrm{Ad}(tX)Y)$. If t is small enough, both Y and $\mathrm{Ad}(tX)Y$ are in U, so applying log we have $\mathrm{Ad}(tX)Y = Y$. By Theorem 8.2, it follows that $\mathrm{ad}(X)Y = 0$, proving the lemma.

Let us now show that $\mathfrak{h}$ is an Abelian Lie algebra. It is clearly closed under scalar multiplication. If X and Y are in $\mathfrak{h}$, then $e^{tY} \in H$ and $tY \in U$ for small enough t, so by the lemma $[X, tY] = 0$. Thus, $[X, Y] = 0$. By Proposition 15.2 we have $e^{t(X+Y)} = e^{tX} e^{tY}$ for all t, so $X + Y \in \mathfrak{h}$.

Now we will show that there exists a neighborhood V of the identity in G such that $V \subseteq \exp(U)$ and $V \cap H = \{\exp(X) \mid X \in \mathfrak{h} \cap \log(V)\}$. This will show that $V \cap H$ is a smooth locally closed submanifold of G. Since every point of H has a neighborhood diffeomorphic to this neighborhood of the identity, it will follow that H is a submanifold of G and hence a Lie subgroup.

It is clear that, for each open neighborhood of V contained in $\exp(U)$, we have $V \cap H \supseteq \{\exp(X) \mid X \in \mathfrak{h} \cap \log(V)\}$. If this inclusion is proper for every V, then there exists a sequence $\{h_n\} \subset H \cap \exp(U)$ such that $h_n \longrightarrow 1$ but $\log(h_n) \notin \mathfrak{h}$. We write $\log(h_n) = X_n$. Thus, $X_n \to 0$.

Let us write $\mathfrak{g} = \mathfrak{h} \oplus \mathfrak{p}$, where $\mathfrak{p}$ is a vector subspace. We will show that we may choose $X_n \in \mathfrak{p}$. Write $X_n = Y_n + Z_n$, where $Y_n \in \mathfrak{h}$ and $Z_n \in \mathfrak{p}$. By the lemma, $[X_n, Y_n] = 0$, so $e^{Z_n} = e^{X_n} e^{-Y_n} \in H$. We may replace X_n by Z_n

and h_n by e^{Z_n}, and we still have $h_n \longrightarrow 1$, but $\log(h_n) \notin \mathfrak{h}$, and after this substitution we have $X_n \in \mathfrak{p}$.

Let us put an inner product on $\mathfrak{g}$. We choose it so that the unit ball is contained in U. The vectors $X_n/|X_n|$ lie on the unit ball in $\mathfrak{p}$, which is compact, so they have an accumulation point. Passing to a subsequence, we may assume that $X_n/|X_n| \longrightarrow X_\infty$, where X_∞ lies in the unit ball in $\mathfrak{p}$. We will show that $X_\infty \in \mathfrak{h}$, which is a contradiction since $\mathfrak{h} \cap \mathfrak{p} = \{0\}$.

To show that $X_\infty \in \mathfrak{h}$, we must show that $e^{tX_\infty} \in H$. It is sufficient to show this for $t < 1$. With t fixed, let r_n be the smallest integer greater than $t/|X_n|$. Since $X_n \to 0$ we have $r_n|X_n| \to t$. Thus, $r_n X_n \longrightarrow tX_\infty$ and $e^{r_n X_n} = (e^{X_n})^{r_n} \in H$ since $e^{X_n} \in H$. Since H is closed, $e^{tX_\infty} \in H$ and the proof that H is a Lie group is complete.

If G is compact, then so is H. The connected component of the identity in H is a connected compact Abelian Lie group and hence a torus. $\square$

If G is a group and H a subgroup, we will denote by $N_G(H)$ and $C_G(H)$ the normalizer and centralizers of H. If no confusion is possible, we will denote them as simply $N(H)$ and $C(H)$.

Let G be a compact, connected Lie group. It contains tori, for example $\{1\}$, and an ascending chain $T_1 \subsetneq T_2 \subsetneq T_3 \subsetneq \cdots$ has length bounded by the dimension of G. Therefore, G contains maximal tori. Let T be a maximal torus.

The normalizer $N(T) = \{g \in G \mid gTg^{-1} = T\}$. It is a closed subgroup since if $t \in T$ is a generator, $N(T)$ is the inverse image of T under the continuous map $g \longrightarrow gtg^{-1}$.

Proposition 15.8. *Let G be a compact Lie group and T a maximal torus. Then $N(T)$ is a closed subgroup of G. The connected component $N(T)^\circ$ of the identity in $N(T)$ is T itself. The quotient $N(T)/T$ is a finite group.*

Proof. We have a homomorphism $N(T) \longrightarrow \mathrm{Aut}(T)$ in which the action is by conjugation. By Proposition 15.7, $\mathrm{Aut}(T) \cong \mathrm{GL}(r, \mathbb{Z})$ is discrete, so any connected group of automorphisms must act trivially. Thus, if $n \in N(T)^\circ$, n commutes with T. If $N(T)^\circ \neq T$, then it contains a one-parameter subgroup $\mathbb{R} \ni t \longrightarrow n(t)$, and the closure of the group generated by T and $n(t)$ is a closed commutative subgroup strictly larger than T. By Theorem 15.2, it is a torus, contradicting the maximality of T. It follows that $T = N(T)^\circ$.

The quotient group $N(T)^\circ/T$ is both discrete and compact and hence finite. $\square$

The quotient $N(T)/T$ is called the *Weyl group* of G with respect to T.

Example 15.1. Suppose that $G = \mathrm{U}(n)$. A maximal torus is

$$T = \left\{ \begin{pmatrix} t_1 & & \\ & \ddots & \\ & & t_n \end{pmatrix} \,\middle|\, |t_1| = \cdots = |t_n| = 1 \right\}.$$

Its normalizer $N(T)$ consists of all monomial matrices (matrices with a single nonzero entry in each row and column) so the quotient $N(T)/T \cong S_n$.

Proposition 15.9. *Let T be a maximal torus in the compact connected Lie group G, and let $\mathfrak{t}$, $\mathfrak{g}$ be the Lie algebras of T and G, respectively.*

(i) Any vector in $\mathfrak{g}$ fixed by $\mathrm{Ad}(T)$ is in $\mathfrak{t}$.

(ii) We have $\mathfrak{g} = \mathfrak{t} \oplus \mathfrak{p}$, where $\mathfrak{p}$ is invariant under $\mathrm{Ad}(T)$. Under the restriction of Ad to T, $\mathfrak{p}$ decomposes into a direct sum of two-dimensional irreducible representations of T of the form (15.2).

Proof. For (i), if $X \in \mathfrak{g}$ is fixed by $\mathrm{Ad}(T)$, then by Proposition 15.2, $\exp(tX)$ is a one-parameter subgroup that is not contained in T but that commutes with T, and unless $X \in \mathfrak{t}$, the closure of the group it generates with T will be a torus strictly larger than T, which is a contradiction.

Since G is compact, there exists a positive definite symmetric bilinear form on the real vector space that is $\mathfrak{g}$-invariant under the real representation $\mathrm{Ad} : G \longrightarrow \mathrm{GL}(\mathfrak{g})$. The orthogonal complement $\mathfrak{p}$ of $\mathfrak{t}$ is invariant under $\mathrm{Ad}(T)$. It contains no $\mathrm{Ad}(T)$-fixed vectors by (i). Since every nontrivial irreducible real representation of T is of the form (15.2), (ii) follows.   □

Corollary 15.2. *If G is a compact connected Lie group and T a maximal torus, then $\dim(G) - \dim(T)$ is even.*

Proof. This follows since $\dim(G/T) = \dim(\mathfrak{p})$, and $\mathfrak{p}$ decomposes as a direct sum of two-dimensional irreducible representations.   □

We review the notion of an orientation. Let M be a manifold of dimension n. The *orientation bundle* of M is a certain twofold cover that we now describe. One way of constructing $\tilde{M}$ begins with the n-fold exterior power of the tangent bundle: the fiber over $x \in M$ is $\wedge^n T_x(M)$. This is a one-dimensional real vector space. Omitting the origin and dividing by the equivalence relation $v \sim w$ if $v = \lambda w$ for $0 < \lambda \in \mathbb{R}$, when v, w are elements of $\wedge^n T_x(M)$, produces a set $F(x)$ with two points. The disjoint union $\tilde{M} = \bigcup_{x \in M} F(x)$ is topologized as follows. Let $\pi : \tilde{M} \longrightarrow M$ be the map sending $F(x)$ to x. If $X_1, \ldots, X_n$ are vector fields that are linearly independent on an open set U, then $X_1 \wedge \cdots \wedge X_n$ determines, for each $x \in U$, an element $s(x)$ of $\pi^{-1}(x)$. We topologize $\tilde{M}$ by requiring that $s : U \longrightarrow \tilde{M}$ be a local homeomorphism.

Now an *orientation* of the manifold M is a *global section* of the orientation bundle, that is, a continuous map $s : M \longrightarrow \tilde{M}$ such that $p \circ s(x) = x$ for all $x \in M$. If an orientation exists, then $\tilde{M}$ is a trivial cover, and $\tilde{M} \cong M \times (\mathbb{Z}/2\mathbb{Z})$. In this case, the bundle M is called *orientable*. Any complex manifold is orientable. On the other hand, a Möbius strip is not orientable.

If M and N are manifolds of dimension n and $f : M \longrightarrow N$ is a diffeomorphism, there is induced for each $x \in M$ an isomorphism $\wedge^n T_x(M) \longrightarrow \wedge^n T_{f(x)}(N)$ and so there is induced a canonical map $\tilde{f} : \tilde{M} \longrightarrow \tilde{N}$ covering f.

Proposition 15.10. *Let G be a connected Lie group and H a connected closed Lie subgroup. Then the quotient space G/H is a connected orientable manifold.*

The manifold G/T is called a *flag manifold.*

Proof. To make G/H a manifold, choose a subspace $\mathfrak{p}$ of $\mathfrak{g} = \mathrm{Lie}(G)$ complementary to $\mathfrak{h} = \mathrm{Lie}(H)$. Then $X \longrightarrow \exp(X)gH$ is a local homeomorphism of a neighborhood of the identity in $\mathfrak{p}$ with a neighborhood of the coset gH in G/H.

To see that $M = G/H$ is orientable, let $\pi : \tilde{M} \longrightarrow M$ be the orientation bundle, and let ω be an element of $\pi^{-1}(H)$. If $g \in G$ then g acts by left translation on M and hence induces an automorphism $\tilde{g}$ of $\tilde{M}$. We can define a global section s of $\tilde{M}$ by $s(gH) = \tilde{g}(\omega)$ if we can check that this is well-defined. Thus, if $gH = g'H$, we must show that $\tilde{g}(\omega) = \tilde{g}'(\omega)$ in the fiber of $\tilde{M}$ above gH. We will show that the map $\tilde{g} : M \longrightarrow M$ can be deformed into $\tilde{g}'$ through a sequence of maps $\tilde{g}_t$, each of them mapping $H \longrightarrow gH$, so that $\tilde{g}_0 = \tilde{g}$ and $\tilde{g}'_1 = \tilde{g}'$. This is sufficient because the fiber of $\tilde{M}$ above gH is a discrete set consisting of two elements, and $t \longrightarrow \tilde{g}_t(\omega)$ is then a continuous map from $[0, 1]$ into this discrete set.

The existence of $\tilde{g}_t$ will follow from the connectedness of H. Note that if $\gamma \in G$ we have

$$\gamma gH = gH \quad \Longleftrightarrow \quad \gamma \in gHg^{-1}. \tag{15.3}$$

In particular, $g'g^{-1} \in gHg^{-1}$. Since H is connected, so is gHg^{-1}, and there is a path $t \longmapsto \gamma_t$ from the identity to $g'g^{-1}$ within gHg^{-1}. Then $xH \longmapsto \gamma_t gxH$ is a diffeomorphism of M that agrees with left translation by g when $t = 0$ and left translation by g' when $t = 1$, and by (15.3), each canonical lifting $\tilde{g}_t$ takes $H \longrightarrow gH$, as required. $\qquad \square$

We have seen in Corollary 15.2 that the flag manifold X is even-dimensional, and by Proposition 15.10 it is orientable. These facts will be explained by Theorem 26.4, where we will see that X is actually a complex analytic manifold.

Exercises

Exercise 15.1. Compute the dimensions of the flag manifolds for $\mathfrak{su}(n)$, $\mathfrak{sp}(2n)$ and $\mathfrak{so}(n)$.

16

Geodesics and Maximal Tori

An important theorem of Cartan asserts that any two maximal tori in a compact Lie group are conjugate. There are different ways of proving this. We will deduce it from the surjectivity of the exponential map, which we will prove by showing that a geodesic between the origin and an arbitrary point of the group has the form $t \mapsto e^{tX}$ for some X in the Lie algebra.

We begin by establishing the properties of geodesics that we will need. These properties are rather well-known, though they do require proof. Some readers may want to start reading with Theorem 16.1.

A *Riemannian manifold* consists of a smooth manifold M and for each $x \in M$ an inner product on the tangent space T_x. Since T_x is a real vector space and not a complex one, an *inner product* in this context is a positive definite symmetric real-valued bilinear form. We also describe this family of inner products on the tangent spaces as a *Riemannian structure* on the manifold M. We will denote the inner product of $X, Y \in T_x$ by $\langle X, Y \rangle$ and the length $\sqrt{\langle X, X \rangle} = |X|$. As part of the definition, the inner product must vary smoothly with x. To make this condition precise, we choose a system of coordinates $x_1, \ldots, x_n$ on some open set U of M, where $n = \dim(M)$. Then, at each point $x \in U$, a basis of $T_x(M)$ consists of $\partial/\partial x_1, \ldots, \partial/\partial x_n$. Let

$$g_{ij} = \left\langle \frac{\partial}{\partial x_i}, \frac{\partial}{\partial x_j} \right\rangle. \tag{16.1}$$

Thus, the matrix (g_{ij}) representing the inner product is positive definite symmetric. Smoothness of the inner product means that the g_{ij} are smooth functions of $x \in U$.

We also define (g^{ij}) to be the inverse matrix to (g_{ij}). Thus, the functions g^{ij} satisfy

$$\sum_j g_{ij} g^{jk} = \delta_i^k, \quad \text{where} \quad \delta_i^k = \begin{cases} 1 \text{ if } i = k, \\ 0 \text{ otherwise}, \end{cases} \tag{16.2}$$

D. Bump, *Lie Groups*, Graduate Texts in Mathematics 225,
DOI 10.1007/978-1-4614-8024-2_16, © Springer Science+Business Media New York 2013

and of course

$$g_{ij} = g_{ji}, \quad g^{ij} = g^{ji}.$$

Suppose that $p : [0, 1] \longrightarrow M$ is a path in the Riemannian manifold M. We say p is *admissible* if it is smooth, and moreover the movement along the path never "stops," that is, the tangent vector $p_*(\mathrm{d}/\mathrm{d}t)$, where t is the coordinate function on $[0, 1]$, is never zero. The *length* or *arclength* of p is

$$|p| = \int_0^1 \left| p_* \left(\frac{\mathrm{d}}{\mathrm{d}t} \right) \right| \mathrm{d}t. \tag{16.3}$$

In terms of local coordinates, if we write $x_i(t) = x_i(p(t))$ the integrand is

$$\left| p_* \left(\frac{\mathrm{d}}{\mathrm{d}t} \right) \right| = \sqrt{\sum_{i,j} g_{ij} \frac{\partial x_i}{\partial t} \frac{\partial x_j}{\partial t}}.$$

We call the path *well-paced* if

$$\int_0^a \left| p_* \left(\frac{\mathrm{d}}{\mathrm{d}t} \right) \right| \mathrm{d}t = |p|a$$

for all $0 \leqslant a \leqslant 1$. Intuitively, this means that the point $p(t)$ moves along the path at a constant "velocity."

It is an easy application of the chain rule that the arclength of p is unchanged under reparametrization. Moreover, each path has a unique reparametrization that is well-paced.

A Riemannian manifold becomes a complete metric space by defining the distance between two points a and b as the infimum of the lengths of the paths connecting them. It is not immediately obvious that there will be a shortest path, and indeed there may not be for some Riemannian manifolds, but it is easy to check that this definition satisfies the triangle inequality and induces the standard topology.

We will encounter various quantities indexed by $1 \leqslant i, j, k, \cdots \leqslant n$, where n is the dimension of the manifold M under consideration. We will make use of Einstein's *summation convention* (in this chapter only). According to this convention, if any index is repeated in a term, it is summed. For example, suppose that $p : [0, 1] \longrightarrow M$ is a path lying entirely in a single chart $U \subset M$ with coordinate functions $x_1, \ldots, x_n$. Then we may regard $x_1, \ldots, x_n$ as functions of $t \in [0, 1]$, namely $x_i(t) = x_i(p(t))$. If $f : U \longrightarrow \mathbb{C}$ is a smooth function, then according to the chain rule

$$\frac{\mathrm{d}f}{\mathrm{d}t} (x_1(t), \ldots, x_n(t)) = \sum_{i=1}^n \frac{\mathrm{d}x_i}{\mathrm{d}t} \frac{\partial f}{\partial x_i} (x_1(t), \ldots, x_n(t)).$$

According to the summation convention, we can write this as simply

$$\frac{\mathrm{d}f}{\mathrm{d}t} = \frac{\mathrm{d}x_i}{\mathrm{d}t} \frac{\partial f}{\partial x_i},$$

and the summation over i is understood because it is a repeated index.

If for each smooth curve $q : [0,1] \longrightarrow M$ with the same endpoints as p we have $|p| \leqslant |q|$, then we say that p is a *path of shortest length*. We will presently define geodesics by means of a differential equation, but for the moment we may provisionally describe a *geodesic* as a well-paced path along a manifold M that *on short intervals* is a path of shortest length.

An example will explain the qualification "on short intervals" in this definition. On a sphere, a geodesic is a great circle. The path in Fig. 16.1 is a geodesic. It is obviously *not* the path of shortest length between a and b.

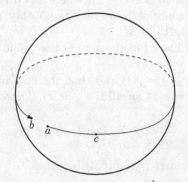

Fig. 16.1. A geodesic on a sphere

Although the indicated geodesic is not a path of shortest length, if we break it into smaller segments, we may still hope that these shorter paths may be paths of shortest length. Indeed they *will* be paths of shortest length if they are not too long, and this is the content of Proposition 16.4 below. For example, the segment from a to c is a path of shortest length.

Let $p : [0,1] \longrightarrow M$ be an admissible path. We can consider deformations of p, namely we can consider a smooth family of paths $u \longrightarrow p_u$, where, for each $u \in (-\epsilon, \epsilon)$, p_u is a path from a to b and $p_0 = p$. Note that, as with the definition of path-homotopy, we require that the endpoints be fixed as the path is deformed. We consider the function $f(u) = |p_u|$. We say the path is of *stationary length* if $f'(0) = 0$ for each such deformation.

If p is a path of shortest length, then 0 will be a minimum of f so $f'(0) = 0$. As for the example in Fig. 16.1, the path from a to b may be deformed by raising it up above the equator and simultaneously shrinking it, but even under such a deformation we will have $f'(0) = 0$. So although this path is not a path of shortest length, it is still a path of stationary length.

Let $x_1, \ldots, x_n$ be coordinate functions on some open set U on M. Relative to this coordinate system, let g_{ij} and g^{ij} be as in (16.1) and (16.2). We define the *Christoffel symbols*

$$[ij, k] = \frac{1}{2} \left(\frac{\partial g_{ik}}{\partial x_j} + \frac{\partial g_{jk}}{\partial x_i} - \frac{\partial g_{ij}}{\partial x_k} \right), \qquad \{ij, k\} = g^{kl}[ij, l].$$

In the last expression, l is summed by the summation convention.

Proposition 16.1. *Suppose that $p : [0,1] \longrightarrow M$ is a well-paced admissible path. If the path lies within an open set U on which $x_1, \ldots, x_n$ is a system of coordinates, then writing $x_i(t) = x_i(p(t))$, the path is of stationary length if and only if it satisfies the differential equation*

$$\frac{\mathrm{d}^2 x_k}{\mathrm{d}t^2} = -\{ij, k\} \frac{\mathrm{d}x_i}{\mathrm{d}t} \frac{\mathrm{d}x_j}{\mathrm{d}t}. \tag{16.4}$$

Proof. Let us consider the effect of deforming the path. We consider a family p_u of paths parametrized by $u \in (-\epsilon, \epsilon)$, where $\epsilon > 0$ is a small real number. It is assumed that the family of paths varies smoothly, so $(t, u) \longmapsto p_u(t)$ is a smooth map $(-\epsilon, \epsilon) \times [0,1] \longrightarrow M$.

We regard the coordinate functions x_i of the point $x = p_u(t)$ to be functions of u and t.

It is assumed that $p_0(t) = p(t)$ and that the endpoints are fixed, so that $p_u(0) = p(u)$ and $p_u(1) = p(1)$ for all $u \in (-\epsilon, \epsilon)$. Therefore,

$$\frac{\partial x_i}{\partial u} = 0 \quad \text{when } t = 0 \text{ or } 1. \tag{16.5}$$

In local coordinates, the arclength (16.3) becomes

$$|p_u| = \int_0^1 \sqrt{g_{ij} \frac{\partial x_i}{\partial t} \frac{\partial x_j}{\partial t}} \, \mathrm{d}t. \tag{16.6}$$

Because the path $p(t) = p_0(t)$ is well-paced, the integrand is constant (independent of t) when $u = 0$, so

$$\frac{\partial}{\partial t} \sqrt{g_{ij} \frac{\partial x_i}{\partial t} \frac{\partial x_j}{\partial t}} = 0 \quad \text{when } u = 0. \tag{16.7}$$

We do not need to assume that the deformed path $p(t, u)$ is well-paced for any $u \neq 0$.

Let $f(u) = |p_u|$. We have

$$f'(u) = \frac{\partial}{\partial u} \int_0^1 \sqrt{g_{ij} \frac{\partial x_i}{\partial t} \frac{\partial x_j}{\partial t}} \, \mathrm{d}t \,.$$

This equals

$$\int_0^1 \left(g_{ij} \frac{\partial x_i}{\partial t} \frac{\partial x_j}{\partial t} \right)^{-\frac{1}{2}} \left[\frac{1}{2} \frac{\partial g_{ij}}{\partial u} \frac{\partial x_i}{\partial t} \frac{\partial x_j}{\partial t} + \frac{1}{2} g_{ij} \frac{\partial^2 x_i}{\partial u \partial t} \frac{\partial x_j}{\partial t} + \frac{1}{2} g_{ij} \frac{\partial x_i}{\partial t} \frac{\partial^2 x_j}{\partial u \partial t} \right] \mathrm{d}t$$

$$= \int_0^1 \left(g_{ij} \frac{\partial x_i}{\partial t} \frac{\partial x_j}{\partial t} \right)^{-\frac{1}{2}} \left[\frac{1}{2} \frac{\partial g_{ij}}{\partial x_l} \frac{\partial x_l}{\partial u} \frac{\partial x_i}{\partial t} \frac{\partial x_j}{\partial t} + g_{ij} \frac{\partial^2 x_i}{\partial u \partial t} \frac{\partial x_j}{\partial t} \right] \mathrm{d}t,$$

where we have used the chain rule, and combined two terms that are equal. (The variables i and j are summed by the summation convention, so we may interchange them, and using $g_{ij} = g_{ji}$, the last two terms on the left-hand side are equal.) We integrate the second term by parts with respect to t, making use of (16.5) and (16.7) to obtain

$$f'(0) = \int_0^1 \left(g_{ij} \frac{\partial x_i}{\partial t} \frac{\partial x_j}{\partial t} \right)^{-\frac{1}{2}} \left[\frac{1}{2} \frac{\partial g_{ij}}{\partial x_l} \frac{\partial x_l}{\partial u} \frac{\partial x_i}{\partial t} \frac{\partial x_j}{\partial t} - \frac{\partial x_i}{\partial u} \frac{\partial}{\partial t} \left(g_{ij} \frac{\partial x_j}{\partial t} \right) \right] dt$$

$$= \int_0^1 \left(g_{ij} \frac{\partial x_i}{\partial t} \frac{\partial x_j}{\partial t} \right)^{-\frac{1}{2}} \left[\frac{1}{2} \frac{\partial g_{ij}}{\partial x_l} \frac{\partial x_i}{\partial t} \frac{\partial x_j}{\partial t} - \frac{\partial}{\partial t} \left(g_{lj} \frac{\partial x_j}{\partial t} \right) \right] \frac{\partial x_l}{\partial u} dt.$$

Now all the partial derivatives are evaluated when $u = 0$. The last step is just a relabeling of a summed index.

We observe that the displacements $\partial x_l / \partial u$ are arbitrary except that they must vanish when $t = 0$ and $t = 1$. (We did not assume the deformed path to be well-paced except when $u = 0$.) Thus, the path is of stationary length if and only if

$$0 = \frac{1}{2} \frac{\partial g_{ij}}{\partial x_l} \frac{\partial x_i}{\partial t} \frac{\partial x_j}{\partial t} - \frac{\partial}{\partial t} \left(g_{lj} \frac{\partial x_j}{\partial t} \right),$$

so the condition is

$$g_{lj} \frac{\partial^2 x_j}{\partial t^2} = \frac{1}{2} \frac{\partial g_{ij}}{\partial x_l} \frac{\partial x_i}{\partial t} \frac{\partial x_j}{\partial t} - \frac{\partial g_{lj}}{\partial t} \frac{\partial x_j}{\partial t}.$$

Now

$$\frac{\partial g_{lj}}{\partial t} \frac{\partial x_j}{\partial t} = \frac{\partial g_{lj}}{\partial x_i} \frac{\partial x_i}{dt} \frac{\partial x_j}{\partial t} = \frac{1}{2} \left[\frac{\partial g_{lj}}{\partial x_i} + \frac{\partial g_{li}}{\partial x_j} \right] \frac{\partial x_i}{dt} \frac{\partial x_j}{\partial t}.$$

The two terms on the right-hand side are of course equal since both i and j are summed indices. We obtain in terms of the Christoffel symbols

$$g_{lj} \frac{\partial^2 x_j}{\partial t^2} = -[ij, l] \frac{\partial x_i}{\partial t} \frac{\partial x_j}{\partial t}.$$

Multiplying by g^{kl}, summing the repeated index l, and using (16.2), we obtain (16.4). □

We define a *geodesic* to be a solution to the differential equation (16.4). This definition does not depend upon the choice of coordinate systems because the differential equation (16.4) arose from a variational problem that was formulated without reference to coordinates. Naturally, one may alternatively confirm by direct computation that the differential equation (16.4) is stable under coordinate changes.

Proposition 16.2. *Let x be a point on the Riemannian manifold M, and let $X \in T_x(M)$. Then, for sufficiently small ϵ, there is a unique geodesic $p : (-\epsilon, \epsilon) \longrightarrow M$ such that $p(0) = x$ and $p_*(\mathrm{d}/\mathrm{d}t) = X$.*

Proof. Let $x_1, \ldots, x_n$ be coordinate functions. Let $y_1, \ldots, y_n$ be a set of new variables, and rewrite (16.4) as a first-order system

$$\frac{\mathrm{d}x_i}{\mathrm{d}t} = y_i,$$

$$\frac{\mathrm{d}y_k}{\mathrm{d}t} = -\{ij, k\}\, y_i y_j.$$

The conditions $p(0) = x$ and $p_*(\mathrm{d}/\mathrm{d}t) = X$ amount to initial conditions for this first-order system, and the existence and uniqueness of the solution follow from the general theory of first-order systems.   □

We now come to a property of geodesics that may be less intuitive. Let U be a smooth submanifold of M, homeomorphic to a disk, of codimension 1. If $x \in U$, we consider the geodesic $t \longmapsto p_x(t)$ such that $p_x(0) = x$ and such that $p_{x,*}(\mathrm{d}/\mathrm{d}t)$ is the unit normal vector to M at x in a fixed direction. For small $\epsilon > 0$, let $U' = \{p_x(\epsilon) | x \in U\}$. In other words, U' is a translation of the disk U along the family of geodesics normal to U.

It is obvious that U is normal to each of the geodesic curves p_x. What is less obvious, and will be proved in the next proposition, is that U' is also normal to the geodesics p_x.

In order to prove this, we will work with a particular set of coordinates. Let $x_2, \ldots, x_n$ be local coordinates on U. At each point $x = (x_2, \ldots, x_n) \in U$, we choose the unit normal vector in a fixed direction and construct the geodesic path through the point with that tangent vector. We prescribe a coordinate system on M near U by asking that $(0, x_2, \ldots, x_n)$ agree with the point $x \in \dot{U}$ and that the path $t \longmapsto (t, x_2, \ldots, x_n)$ agree with p_x. We describe such a coordinate system as *geodesic coordinates*.

Proposition 16.3. *In geodesic coordinates, $g_{1i} = 0$ for $2 \leqslant i \leqslant n$. Also $g_{11} = 1$.*

In view of (16.1), this amounts to saying that the geodesic curves (having tangent vector $\partial/\partial x_1$) are orthogonal to the level hypersurfaces $x_1 = $ constant (having tangent spaces spanned by $\partial/\partial x_2, \ldots, \partial/\partial x_n$), such as U and U' in Fig. 16.2.

Proof. Having chosen coordinates so that the path $t \longmapsto (t, x_2, \ldots, x_n)$ is a geodesic, we see that if all $\mathrm{d}x_i/\mathrm{d}t = 0$ in (16.4), for $i \neq 1$, then $\mathrm{d}^2 x_k/\mathrm{d}t^2 = 0$ for all k. This means that $\{11, k\} = 0$. Since the matrix (g_{kl}) is invertible, it follows that $[11, k] = 0$, so

$$\frac{\partial g_{1k}}{\partial x_1} = \frac{1}{2} \frac{\partial g_{11}}{\partial x_k}. \tag{16.8}$$

First, take $k = 1$ in (16.8). We see that $\partial g_{11}/\partial x_1 = 0$, so if $x_2, \ldots, x_n$ are held constant, g_{11} is constant. When $x_1 = 0$, the initial condition of the geodesic curve p_x through $(0, x_2, \ldots, x_n)$ is that it is tangent to the unit normal to the surface, that is, its tangent vector $\partial/\partial x_1$ has length one, and by (16.1) it follows that $g_{11} = 1$ when $x_1 = 0$, so $g_{11} = 1$ throughout the geodesic coordinate neighborhood.

Now let $2 \leqslant k \leqslant n$ in (16.8). Since g_{11} is constant, $\partial g_{1k}/\partial x_1 = 0$, and so g_{1k} is also constant when $x_2, \ldots, x_n$ are held constant. When $x_1 = 0$, our assumption that the geodesic curve p_x is normal to the surface means that $\partial/\partial x_1$ and $\partial/\partial x_k$ are orthogonal, so by (16.1), g_{1k} vanishes when $x_1 = 0$ and so it vanishes for all x_1.   □

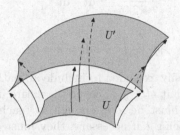

Fig. 16.2. Hypersurface remains perpendicular to geodesics on parallel translation

With these preparations, we may now prove that short geodesics are paths of shortest length.

Proposition 16.4.

(i) Let $p : [0, 1] \longrightarrow M$ be a geodesic. Then there exists an $\epsilon > 0$ such that the restriction of p to $[0, \epsilon]$ is the unique path of shortest length from $p(0)$ to $p(\epsilon)$.

(ii) Let $x \in M$. There exists a neighborhood N of x such that for all $y \in N$ there exists a unique path of shortest distance from x to y, and that path is a geodesic.

Proof. We choose a hypersurface U orthogonal to p at $t = 0$ and construct geodesic coordinates as explained before Proposition 16.3. We choose ϵ and B so small that the set N of points with coordinates $\{x_1 \in [0, \epsilon], 0 \leqslant |x_2|, \ldots, |x_n| \leqslant B\}$ is contained within the interior of this geodesic coordinate neighborhood. We can assume that the coordinates of $p(0)$ are $(0, \ldots, 0)$, so by construction $p(t) = (t, 0, \ldots, 0)$. Then $|p| = \epsilon$, where now $|p|$ denotes the length of the restriction of the path to the interval from 0 to ϵ.

We will show that if $q : [0, \epsilon] \longrightarrow M$ is any path with $q(0) = p(0)$ and $q(\epsilon) = p(\epsilon)$, then $|q| \geqslant |p|$.

First, we consider paths $q : [0, \epsilon] \longrightarrow M$ that lie entirely within the set N and such that the x_1-coordinate of $q(t)$ is monotonically increasing. Reparametrizing q, we may arrange that $q(t)$ and $p(t)$ have the same x_1-coordinate, which equals t. Let us write $q(t) = (t, x_2(t), \ldots, x_n(t))$. We also denote $x_1(t) = t$. Since $g_{1k} = g_{k1} = 0$ when $k \geqslant 2$ and $g_{11} = 1$, we have

$$|q| = \int_0^\epsilon \sqrt{\sum_{i,j} g_{ij} \frac{dx_i}{dt} \frac{dx_j}{dt}} \, dt$$

$$= \int_0^\epsilon \sqrt{1 + \sum_{2 \leqslant i,j \leqslant n} g_{ij} \frac{dx_i}{dt} \frac{dx_j}{dt}} \, dt.$$

Now since the matrix $(g_{ij})_{1 \leqslant i,j \leqslant n}$ is positive definite, its principal minor $(g_{ij})_{2 \leqslant i,j \leqslant n}$ is also positive definite, so

$$\sum_{2 \leqslant i,j \leqslant n} g_{ij} \frac{dx_i}{dt} \frac{dx_j}{dt} \geqslant 0$$

and

$$|q| \geqslant \int_0^\epsilon \sqrt{1} \, dt = \epsilon = |p|.$$

This argument is easily extended to include all paths such that the values of x_1 for those t such that $q(t) \in N$ cover the entire interval $[0, \epsilon]$. Paths for which this is not true must be long enough to reach the edges of the box $x_i > B$, and after reducing ϵ if necessary, they must be longer than ϵ. This completes our discussion of (i).

For (ii), given each unit tangent vector $X \in T_x(M)$, there is a unique geodesic $p_X : [0, \epsilon_X] \longrightarrow M$ through x tangent to X, and $\epsilon_X > 0$ may be chosen so that this geodesic is a path of shortest length. We assert that ϵ_X may be chosen so that the same value ϵ_X is valid for nearby unit tangent vectors Y. We leave this point to the reader except to remark that it is perhaps easiest to see this by applying a diffeomorphism of M that moves X to Y and regarding X as fixed while the metric g_{ij} varies; if Y is sufficiently near X, the variation of g_{ij} will be small and the ϵ in part (i) can be chosen to work for small variations of the g_{ij}. So for each unit tangent vector $X \in T_x(M)$ there exists an $\epsilon_X > 0$ and a neighborhood N_X of X in the unit ball of $T_x(M)$ such that $p_Y : [0, \epsilon_X] \longrightarrow M$ is a path of shortest length for all $Y \in N_X$. Since the unit tangent ball in $T_x(M)$ is compact, a finite number of N_X suffice to cover it, and if ϵ is the minimum of the corresponding ϵ_X, then we can take N to be the set of all points connected to x by a geodesic of length $< \epsilon$. $\square$

If M is a connected Riemannian manifold, we make M into a metric space by defining $d(x, y)$ to be the infimum of $|p|$, where p is a smooth path from x to y.

Theorem 16.1. *Let M be a compact connected Riemannian manifold, and let x and y be points of M. Then there is a geodesic $p : [0, 1] \longrightarrow M$ with $p(0) = x$ and $p(1) = y$.*

A more precise statement may be found in Kobayashi and Nomizu [110], Theorem 4.2 on p. 172. It is proved there that if M is connected and *geodesically complete*, meaning that any well-paced geodesic can be extended to $(-\infty, \infty)$, then the conclusion of the theorem is true. (It is not hard to see that a compact manifold is geodesically complete.)

Proof. Let $\{p_i\}$ be a sequence of well-paced paths from x to y such that $|p_i| \longrightarrow d(x, y)$. Because they are well-paced, if $0 \leqslant a < b \leqslant 1$ we have $d\big(p_i(a), p_i(b)\big) = (b - a)|p_i|$, and it follows that $\{p_i\}$ are equicontinuous. Thus by Proposition 3.1 there is a subsequence that converges uniformly to a path p. It is not immediately evident that p is smooth, but it is clearly continuous. So we can partition $[0, 1]$ into short intervals. On each sufficiently short interval $0 \leqslant a < b \leqslant 1$, $p(b)$ is near enough to $p(a)$ that the unique path of shortest distance between them is a geodesic by Proposition 16.4. It follows that p is a geodesic. □

Theorem 16.2. *Let G be a compact Lie group. There exists on G a Riemannian metric that is invariant under both left and right translation. In this metric, a geodesic is a translate (either left or right) of a map $t \longrightarrow \exp(tX)$ for some $X \in \mathrm{Lie}(G)$.*

Proof. Let $\mathfrak{g} = \mathrm{Lie}(G)$. Since G is a compact group acting by Ad on the real vector space $\mathfrak{g}$, there exists an $\mathrm{Ad}(G)$-invariant inner product on $\mathfrak{g}$. Regarding G as the tangent space to G at the identity, if $g \in G$, left translation induces an isomorphism $\mathfrak{g} = T_e(G) \longrightarrow T_g(G)$ and we may transfer this inner product to $T_g(G)$. This gives us an inner product on $T_g(G)$ and hence a Riemannian structure on G, which is invariant under left translation. Right translation by g induces a different isomorphism $\mathfrak{g} = T_e(G) \longrightarrow T_g(G)$, but these two isomorphisms differ by $\mathrm{Ad}(g) : \mathfrak{g} \longrightarrow \mathfrak{g}$, and since the original inner product is invariant under $\mathrm{Ad}(g)$, we see that the Riemannian structure we have obtained is invariant under both left and right translation.

It remains to be shown that a geodesic is a translate of the exponential map. This is essentially a local statement. Indeed, it is sufficient to show that any short segment of a geodesic is of the form $t \longmapsto g \cdot \exp(tX)$ since any path that is of such a form on every short interval is globally of the same form. Moreover, since the Riemannian metric is translation-invariant, it is sufficient to show that a geodesic near the origin is of the form $t \longrightarrow \exp(tX)$.

First, we consider the case where G is a torus. In this case, $G \cong \mathbb{R}^n/\Lambda$, where Λ is a lattice. We identify the tangent space to $\mathbb{R}^n$ at any point with $\mathbb{R}^n$ itself. By a linear change of variables, we may assume that the inner product on $\mathbb{R}^n = T_e(G)$ corresponding to the Riemannian structure is the standard Euclidean inner product. Since the Riemannian structure is invariant under translation it follows that $G \cong \mathbb{R}^n/\Lambda$ is a Riemannian manifold as well as a group. Geodesics are straight lines and so are translates of the exponential map.

We turn now to the general case. If $X \in \mathfrak{g}$, let $E_X : (-\epsilon, \epsilon) \longrightarrow G$ denote the geodesic through the origin tangent to $X \in \mathfrak{g}$. It is defined for sufficiently small ϵ (depending on X). If $\lambda \in \mathbb{R}$, then $t \longmapsto E_X(\lambda t)$ is the geodesic through the origin tangent to λX, so $E_X(\lambda t) = E_{\lambda X}(t)$. Thus, there is a neighborhood U of the origin in $\mathfrak{g}$ and a map $E : U \longrightarrow G$ such that $E_X(t) = E(tX)$ for $X, tX \in U$. We must show that E coincides with the exponential map.

If $g \in G$, then translating $E(tX)$ on the left by g and on the right by g^{-1} gives another geodesic, which is tangent to $\mathrm{Ad}(g)X$. Thus, if $tX \in U$,

$$g\, E(tX)\, g^{-1} = E(t\, \mathrm{Ad}(g)X). \tag{16.9}$$

We now fix $X \in \mathfrak{g}$. Let T be a maximal torus containing the one-parameter subgroup $\{e^{tX} \mid t \in \mathbb{R}\}$. It follows from (16.9) that $E(tX)$ commutes with $g \in H$ when $tX \in U$. Thus the path $t \longmapsto E(tX)$ runs through the centralizer $C(T)$ and *a fortiori* through $N(T)$. By Proposition 15.8, it follows that $E(tX) \in T$.

Now the translation-invariant Riemannian structure on G induces a translation-invariant Riemannian structure on T, and since the geodesic path $t \longmapsto E(tX)$ of G is contained in T, it is a geodesic path in T also. The result therefore follows from the special case of the torus, which we have already handled. $\qquad\square$

Theorem 16.3. *Let G be a compact Lie group and $\mathfrak{g}$ its Lie algebra. Then the exponential map $\mathfrak{g} \longrightarrow G$ is surjective.*

Proof. Put a Riemannian structure on G as in Theorem 16.2. By Theorem 16.1, given $g \in G$, there exists a geodesic path from the identity to g. By Theorem 16.2, this path is of the form $t \longmapsto e^{tX}$ for some $X \in \mathfrak{g}$, so $g = e^X$. $\qquad\square$

Theorem 16.4. *Let G be a compact connected Lie group, and let T be a maximal torus. Let $g \in G$. Then there exists $k \in G$ such that $g \in kTk^{-1}$.*

Proof. Let $\mathfrak{g}$ and $\mathfrak{t}$ be the Lie algebras of G and T, respectively. Let t_0 be a generator of T. Using Theorem 16.3, find $X \in \mathfrak{g}$ and $H_0 \in \mathfrak{t}$ such that $e^X = g$ and $e^{H_0} = t_0$.

Since G is a compact group acting by Ad on the real vector space $\mathfrak{g}$, there exists on $\mathfrak{g}$ an $\mathrm{Ad}(G)$-invariant inner product for which we will denote the corresponding symmetric bilinear form as $\langle\ ,\ \rangle$. Choose $k \in G$ so that the real value $\langle X, \mathrm{Ad}(k)H_0 \rangle$ is maximal, and let $H = \mathrm{Ad}(k)H_0$. Thus, $\exp(H) = kt_0k^{-1}$ generates kTk^{-1}.

If $Y \in \mathfrak{g}$ is arbitrary, then $\langle X, \mathrm{Ad}(e^{tY})H \rangle$ has a maximum when $t = 0$, so using Theorem 8.2 we have

$$0 = \frac{\mathrm{d}}{\mathrm{d}t}\left\langle X, \mathrm{Ad}(e^{tY})H \right\rangle\Big|_{t=0} = \langle X, \mathrm{ad}(Y)H \rangle = -\langle X, [H, Y] \rangle .$$

By Proposition 10.3, this means that

$$\langle [H, X], Y \rangle = 0$$

for all Y. Since an inner product is by definition positive definite, the bilinear form $\langle\ ,\ \rangle$ is nondegenerate, which implies that $[H, X] = 0$. Now, by Proposition 15.2, e^H commutes with e^{tX} for all $t \in \mathbb{R}$. Since e^H generates the maximal torus kTk^{-1}, it follows that the one-parameter subgroup $\{e^{tX}\}$ is contained in the centralizer of kTk^{-1}, and since kTk^{-1} is a maximal torus, it follows that $\{e^{tX}\} \subset kTk^{-1}$. In particular, $g = e^X \in kTk^{-1}$. □

Theorem 16.5 (E. Cartan). *Let G be a compact connected Lie group, and let T be a maximal torus. Then every maximal torus is conjugate to T, and every element of G is contained in a conjugate of T.*

Proof. The second statement is contained in Theorem 16.4. As for the first statement, let T' be another maximal torus, and let t be a generator. Then t' is contained in kTk^{-1} for some k, so $T' \subseteq kTk^{-1}$. Since both are maximal tori, they are equal. □

Proposition 16.5. *Let G be a compact connected Lie group, $S \subset G$ a torus (not necessarily maximal), and $g \in C_G(S)$ an element of its centralizer. Let H be the closure of the group generated by S and g. Then H has a topological generator. That is, there exists $h \in H$ such that the subgroup generated by h is dense in H.*

Proof. Since H is closed and Abelian, its connected component H° of the identity is a torus by Proposition 15.2. Let h_0 be a topological generator.

The group H/H° is compact and discrete and hence finite. Since $S \subseteq H^\circ$, and since S and g generate a dense subgroup of H, the finite group H/H° is cyclic and generated by gH°. Let r be the order of H/H°. Then $g^r \in H^\circ$. Since the rth power map $H^\circ \longrightarrow H^\circ$ is surjective, we can find $u \in H^\circ$ such that $(gu)^r = h_0$. Then the group generated by $h = ug$ contains both a generator h_0 of H° and a generator $gH^\circ = (gu)H^\circ$ of H/H°. Clearly, it is a topological generator of H. □

Proposition 16.6. *If G is a Lie group and $u \in G$, then the centralizer $C_G(u)$ is a closed Lie subgroup, and its Lie algebra is $\{X \in \mathrm{Lie}(G) \,|\, \mathrm{Ad}(u)X = X\}$.*

Proof. To show that $H = C_G(u)$ is a closed submanifold of G, it is sufficient to show that its intersection with a small neighborhood of the identity is a closed submanifold since translation by an element h of H will give a diffeomorphism of that neighborhood onto a neighborhood of h. In a neighborhood N of the origin in $\mathrm{Lie}(G)$, the exponential map is a diffeomorphism onto $\exp(N)$, and we see that the preimage of $C_G(u)$ in N is a vector subspace by recalling that conjugation by u corresponds to the linear transformation $\mathrm{Ad}(u)$ of N. Particularly, $u e^{tX} u^{-1} = e^{t\,\mathrm{ad}(u)X}$, so $e^{tX} \in C_G(u)$ for all t if and only if $\mathrm{Ad}(u)X = X$. □

Theorem 16.6. *Let G be a compact connected Lie group and $S \subset G$ a torus (not necessarily maximal). Then the centralizer $C_G(S)$ is a closed connected Lie subgroup of G.*

Proof. We first prove that $C_G(S)$ is connected. Let $g \in C_G(S)$. By Proposition 16.5, there exists an element h of $C_G(S)$ that generates the closure H of the group generated by S and g. Let T be a maximal torus in G containing h. Then T centralizes S, so the closure of TS is a connected compact Abelian group and hence a torus, and by the maximality of T it follows that $S \subseteq T$. Now clearly $T \subseteq C_G(S)$, and since T is connected, $T \subseteq C_G(S)^\circ$. Now $g \in H \subseteq T \subset C_G(S)^\circ$. We have shown that $C_G(S)^\circ = C_G(S)$, so $C_G(S)$ is connected.

To show that $C_G(S)$ is a closed Lie subgroup, let $u \in S$ be a generator. Then $C_G(S) = C_G(u)$, and the statement follows by Proposition 16.6. $\qquad\square$

Exercises

Exercise 16.1. Give an example of a connected Riemannian manifold with two points P and Q such that no geodesic connects P and Q.

Exercise 16.2. Let G be a compact connected Lie group and let $g \in G$. Show that the centralizer $C_G(g)$ of g is connected.

Exercise 16.3. Show that the conclusion of Exercise 16.2 fails for the connected *noncompact* Lie group $\mathrm{SL}(2, \mathbb{R})$ by exhibiting an element with a centralizer that is not connected.

If M and N are Riemannian manifolds of the same dimension, and if $f : M \longrightarrow N$ is a diffeomorphism, then f is called a *conformal map* if there exists a positive function ϕ on M such that if $x \in M$ and $y = f(x)$, and if we use the notation $\langle\,,\,\rangle$ to denote the inner products in both $T_x(M)$ and $T_y(N)$, then

$$\langle f_* X, f_* Y \rangle = \phi(x) \langle X, Y \rangle, \qquad X, Y \in T_x(M),$$

where $f_* : T_x(M) \longrightarrow T_y(N)$ is the induced map. Intuitively, a conformal map is one that preserves angles. If the function $\phi = 1$, then f is called *isometric*.

Exercise 16.4. Show that if M and N are open subsets in $\mathbb{C}$ and $f : M \longrightarrow N$ is a holomorphic map such that the inverse map $f^{-1} : N \longrightarrow M$ exists and is holomorphic (so f' is never zero), then f is a conformal map.

The next exercises describe the geodesics for some familiar homogeneous spaces. Let $\mathfrak{D} = \{z \in \mathbb{C} \,|\, |z| < 1\}$ be the complex disk in $\mathbb{C}$, and let $\mathfrak{R} = \mathbb{C} \cup \{\infty\}$ be the Riemann sphere. The group $\mathrm{SL}(2, \mathbb{C})$ acts on $\mathfrak{R}$ by linear fractional transformations:

$$\begin{pmatrix} a & b \\ c & d \end{pmatrix} : z \longmapsto \frac{az + b}{cz + d}.$$

In this action, it is understood that ∞ is mapped to a/c and z is mapped to ∞ if $cz + d = 0$. The map $z \longmapsto -1/z$ is a chart near zero, and $\mathfrak{R}$ is a complex analytic manifold. Let

$$A = \left\{ \begin{pmatrix} a & b \\ 0 & \bar{a} \end{pmatrix} \,\Big|\, a, b \in \mathbb{C}, \, |a|^2 = 1 \right\},$$

$$\mathrm{SU}(2) = \left\{ \begin{pmatrix} a & b \\ -\bar{b} & \bar{a} \end{pmatrix} \Big| \, a,b \in \mathbb{C}, \ |a|^2 + |b|^2 = 1 \right\},$$

$$\mathrm{SU}(1,1) = \left\{ \begin{pmatrix} a & b \\ \bar{b} & \bar{a} \end{pmatrix} \Big| \, a,b \in \mathbb{C}, \ |a|^2 - |b|^2 = 1 \right\},$$

and

$$K = \left\{ \begin{pmatrix} a & 0 \\ 0 & \bar{a} \end{pmatrix} \Big| \, |a|^2 = 1 \right\} \cong \mathrm{U}(1).$$

It will be shown in Chap. 28 that the group $\mathrm{SU}(1,1)$ is conjugate in $\mathrm{SL}(2,\mathbb{C})$ to $\mathrm{SL}(2,\mathbb{R})$. Let G be one of the groups $\mathrm{SU}(2)$, A, or $\mathrm{SU}(1,1)$. The stabilizer of $0 \in \mathfrak{R}$ is the group K, so we may identify the orbit of $0 \in \mathfrak{R}$ with the homogeneous space G/H by the bijection $g(0) \longleftrightarrow gH$. The orbit of 0 is given in the following table.

G	K	orbit of $0 \in \mathfrak{R}$
$\mathrm{SU}(1,1)$	$\mathrm{U}(1)$	$\mathfrak{D}$
A	$\mathrm{U}(1)$	$\mathbb{C}$
$\mathrm{SU}(2)$	$\mathrm{U}(1)$	$\mathfrak{H}$

Exercise 16.5. Show that if G is one of the groups $\mathrm{SU}(1,1)$, A, or $\mathrm{SU}(2)$, then the quotient G/K, which we may identify with $\mathfrak{D}$, $\mathbb{C}$, or $\mathfrak{H}$, has a unique G-invariant Riemannian structure.

Exercise 16.6. Show that the inclusions $\mathfrak{D} \longrightarrow \mathbb{C} \longrightarrow \mathfrak{R}$ are conformal maps but are not isometric.

A subset C of $\mathfrak{R}$ is called a *circle* if either $C \subset \mathbb{C}$ and C is a circle in the Euclidean sense. In other words, C is the set of all solutions z to the equation $|z - \alpha| = r$ for $\alpha \in \mathbb{C}$, or else $C = L \cup \{\infty\}$, where L is a straight line. Let $\partial\mathfrak{D} = \{z \, | \, |z| = 1\}$ be the unit circle.

Exercise 16.7.

(i) Show that the group $\mathrm{SL}(n,\mathbb{C})$ preserves the set of circles. Show, however, that a linear fractional transformation $g \in \mathrm{SL}(n,\mathbb{C})$ may take a circle with center α to a circle with center different from $g(\alpha)$.

(ii) Show that if $M = \mathfrak{D}$, $\mathbb{C}$ or $\mathfrak{R}$, then each geodesic is a circle, but not each circle is a geodesic.

(iii) Show that the geodesics in $\mathbb{C}$ are the straight lines and that the geodesics in $\mathfrak{D}$ are the curves $C \cap \mathfrak{D}$, where C is a circle in $\mathbb{C}$ perpendicular to $\partial\mathfrak{D}$.

(iv) Show that $\partial\mathfrak{D}$ is a geodesic in $\mathfrak{R}$.

The Weyl Integration Formula

Let G be a compact, connected Lie group, and let T be a maximal torus. Theorem 16.5 implies that every conjugacy class meets T. Thus, we should be able to compute the Haar integral of a class function (e.g., the inner product of two characters) as an integral over the torus. The formula that allows this, the *Weyl integration formula*, is therefore fundamental in representation theory and in other areas, such as random matrix theory.

If G is a locally compact group and H a closed subgroup, then the quotient space G/H consisting of all cosets gH with $g \in G$, given the quotient topology, is a locally compact Hausdorff space. (See Hewitt and Ross [69, Theorems 5.21 and 5.22 on p. 38].) Such a coset space is called a *homogeneous space*.

If X is a locally compact Haudorff space let $C_c(X)$ be the space of continuous, compactly supported functions on X. If X is a locally compact Hausdorff space, a linear functional I on $C_c(X)$ is called *positive* if $I(f) \geqslant 0$ if f is nonnegative. According to the *Riesz representation theorem*, each such I is of the form

$$I(f) = \int_X f \, \mathrm{d}\mu$$

for some regular Borel measure $\mathrm{d}\mu$. See Halmos [61, Sect. 56], or Hewitt and Ross [69, Corollary 11.37 on p. 129]. (Regularity of the measure is discussed after Definition 11.34 on p. 127.)

Proposition 17.1. *Let G be a locally compact group, and let H be a compact subgroup. Let $\mathrm{d}\mu_G$ and $\mathrm{d}\mu_H$ be left Haar measures on G and H, respectively. Then there exists a regular Borel measure $\mathrm{d}\mu_{G/H}$ on G/H which is invariant under the action of G by left translation. The measure $\mathrm{d}\mu_{G/H}$ may be normalized so that, for $f \in C_c(G)$, we have*

$$\int_{G/H} \int_H f(gh) \, \mathrm{d}\mu_H(h) \, \mathrm{d}\mu_{G/H}(gH).$$

Here the function $g \longmapsto \int_H f(gh) \, \mathrm{d}\mu_H$ is constant on the cosets gH, and we are therefore identifying it with a function on G/H.

D. Bump, *Lie Groups*, Graduate Texts in Mathematics 225,
DOI 10.1007/978-1-4614-8024-2_17, © Springer Science+Business Media New York 2013

Proof. We may choose the normalization of $\mathrm{d}\mu_H$ so that H has total volume 1. We define a map $\lambda : C_c(G) \longrightarrow C_c(G/H)$ by

$$(\lambda f)(g) = \int_H f(gh) \, \mathrm{d}\mu_H(h).$$

Note that λf is a function on G which is right invariant under translation by elements of H, so it may be regarded as a function on G/H. Since H is compact, λf is compactly supported. If $\phi \in C_c(G/H)$, regarding ϕ as a function on G, we have $\lambda\phi = \phi$ because

$$(\lambda\phi)(g) = \int_H \phi(gh) \, \mathrm{d}\mu_H(h) = \int_H \phi(g) \, \mathrm{d}\mu_H(h) = \phi(g).$$

This shows that λ is surjective. We may therefore define a linear functional I on $C_c(G/H)$ by

$$I(\lambda f) = \int_G f(g) \, \mathrm{d}\mu_G(g), \qquad f \in C_c(G)$$

provided we check that this is well defined. We must show that if $\lambda f = 0$ then

$$\int_G f(g) \, \mathrm{d}\mu_G(g) = 0. \tag{17.1}$$

We note that the function $(g, h) \longmapsto f(gh)$ is compactly supported and continuous on $G \times H$, so if $\lambda f = 0$ we may use Fubini's theorem to write

$$0 = \int_G (\lambda f)(g) \, \mathrm{d}\mu_G(g) = \int_H \int_G f(gh) \, \mathrm{d}\mu_G(g) \, \mathrm{d}\mu_H(h).$$

In the inner integral on the right-hand side we make the variable change $g \longmapsto gh^{-1}$. Recalling that $\mathrm{d}\mu_G(g)$ is *left* Haar measure, this produces a factor of $\delta_G(h)$, where δ_G is the modular quasicharacter on G. Thus,

$$0 = \int_H \delta_G(h) \int_G f(g) \, \mathrm{d}\mu_G(g) \, \mathrm{d}\mu_H(h).$$

Now the group H is compact, so its image under δ_G is a compact subgroup of $\mathbb{R}_+^\times$, which must be just $\{1\}$. Thus, $\delta_G(h) = 1$ for all $h \in H$ and we obtain (17.1), justifying the definition of the functional I. The existence of the measure on G/H now follows from the Riesz representation theorem. $\square$

We have seen in Proposition 15.9 that in the adjoint action on $\mathfrak{g} = \mathrm{Lie}(G)$, restricted to T, the Lie algebra $\mathfrak{t}$ is an invariant subspace, complemented by a space $\mathfrak{p}$, which decomposes as the direct sum of nontrivial two-dimensional irreducible real representations as described in Proposition 15.5.

Let $W = N(T)/T$ be the Weyl group of G. The Weyl group acts on T by conjugation. Indeed, the elements of the Weyl group are cosets $w = nT$ for $n \in N(T)$. If $t \in T$, the element ntn^{-1} depends only on w so by abuse of notation we denote it wtw^{-1}.

Theorem 17.1.

(i) Two elements of T are conjugate in G if and only if they are conjugate in $N(T)$.

(ii) The inclusion $T \longrightarrow G$ induces a bijection between the orbits of W on T and the conjugacy classes of G.

Proof. Suppose that $t, u \in T$ are conjugate in G, say $gtg^{-1} = u$. Let $H = C_G(u)^\circ$ be the connected component of the identity in the centralizer of u. It is a closed Lie subgroup of G by Proposition 16.6. Both T and gTg^{-1} are contained in H since they are connected commutative groups containing u. As they are maximal tori in G, they are maximal tori in H, and so they are conjugate in the compact connected group H. If $h \in H$ such that $hTh^{-1} = gTg^{-1}$, then $w = h^{-1}g \in N(T)$. Since $wtw^{-1} = h^{-1}uh = u$, we see that t and u are conjugate in $N(T)$.

Since G is the union of the conjugates of T, (ii) is a restatement of (i).   □

Proposition 17.2. *The centralizer $C(T) = T$.*

Proof. Since $C(T) \subset N(T)$, T is of finite index in $C(T)$ by Proposition 15.8. Thus, if $x \in C(T)$, we have $x^n \in T$ for some n. Let t_0 be a generator of T. Since the nth power map $T \longrightarrow T$ is surjective, there exists $t \in T$ such that $(xt)^n = t_0$. Now xt is contained in a maximal torus T', which contains t_0 and hence $T \subset T'$. Since T is maximal, $T' = T$ and $x \in T$.   □

Proposition 17.3. *There exists a dense open set Ω of T such that the $|W|$ elements wtw^{-1} ($w \in W$) are all distinct for $t \in \Omega$.*

See Proposition 23.4 for a more precise result.

Proof. If $w \in W$, let $\Omega_w = \{t \in T \,|\, wtw^{-1} \neq t\}$. It is an open subset of T since its complement is evidently closed. If $w \neq 1$ and t is a generator of T, then $t \in \Omega_w$ because otherwise if $n \in N(T)$ represents w, then $n \in C(t) = C(T)$, so $n \in T$ by Proposition 17.2. This is a contradiction since $w \neq 1$. The finite intersection $\Omega = \bigcap_{w \neq 1} \Omega_w$ is dense by Kronecker's Theorem 15.1. It thus fits our requirements.   □

Theorem 17.2 (Weyl). *Let G be a compact connected Lie group, and let $\mathfrak{p}$ be as in Proposition 15.9. If f is a class function, and if dg and dt are Haar measures on G and T (normalized so that G and T have volume 1), then*

$$\int_G f(g)\,\mathrm{d}g = \frac{1}{|W|} \int_T f(t) \det\left([\mathrm{Ad}(t^{-1}) - I_{\mathfrak{p}}] \,|\, \mathfrak{p}\right) \mathrm{d}t.$$

Proof. Let $X = G/T$. We give X the measure d_X invariant under left translation by G such that X has volume 1. Consider the map

$$\phi : X \times T \longrightarrow G, \quad \phi(xT, t) = xtx^{-1}.$$

Both $X \times T$ and G are orientable manifolds of the same dimension. Of course, G and T both are given the Haar measures such that G and T have volume 1.

We choose volume elements on the Lie algebras $\mathfrak{g}$ and $\mathfrak{t}$ of G and T, respectively, so that the Jacobians of the exponential maps $\mathfrak{g} \longrightarrow G$ and $\mathfrak{t} \longrightarrow T$ at the identity are 1.

We compute the Jacobian $J\phi$ of ϕ: Parametrize a neighborhood of xT in X by a chart based on a neighborhood of the origin in $\mathfrak{p}$. This chart is the map

$$\mathfrak{p} \ni U \mapsto xe^U T .$$

We also make use of the exponential map to parametrize a neighborhood of $t \in T$. This is the chart $\mathfrak{t} \ni V \mapsto te^V$. We therefore have the chart near the point (xT, t) in $X \times T$ mapping

$$\mathfrak{p} \times \mathfrak{t} \ni (U, V) \longrightarrow (xe^U T, te^V) \in X \times T$$

and, in these coordinates, ϕ is the map

$$(U, V) \mapsto xe^U te^V e^{-U} x^{-1}.$$

To compute the Jacobian of this map, we translate on the left by $t^{-1}x^{-1}$ and on the right by x. There is no harm in this because these maps are Haar isometries. We are reduced to computing the Jacobian of the map

$$(U, V) \mapsto t^{-1}e^U te^V e^{-U} = e^{\mathrm{Ad}(t^{-1})U} e^V e^{-U}.$$

Identifying the tangent space of the real vector space $\mathfrak{p} \times \mathfrak{t}$ with itself (that is, with $\mathfrak{g} = \mathfrak{p} \oplus \mathfrak{t}$), the differential of this map is

$$U + V \mapsto \left(\mathrm{Ad}(t^{-1}) - I_{\mathfrak{p}}\right) U + V.$$

The Jacobian is the determinant of the differential, so

$$(J\phi)(xT, t) = \det\left([\mathrm{Ad}(t^{-1}) - I_{\mathfrak{p}}] \,|\, \mathfrak{p}\right). \tag{17.2}$$

By Proposition 17.3, the map $\phi : X \times T \longrightarrow G$ is a $|W|$-fold cover over a dense open set and so, for any function f on G, we have

$$\int_G f(g) \, \mathrm{d}g = \frac{1}{|W|} \int_{X \times T} f\big(\phi(xT, t)\big) \, J\big(\phi(xT, t)\big) \, \mathrm{d}x \times \mathrm{d}t.$$

The integrand $f\big(\phi(xT, t)\big) \, J\big(\phi(xT, t)\big) = f(t) \det\left([\mathrm{Ad}(t^{-1}) - I_{\mathfrak{p}}] \,|\, \mathfrak{p}\right)$ is independent of x since f is a class function, and the result follows.   □

An example may help make this result more concrete.

Proposition 17.4. *Let* $G = U(n)$, *and let* T *be the diagonal torus. Writing*

$$t = \begin{pmatrix} t_1 & & \\ & \ddots & \\ & & t_n \end{pmatrix} \in T,$$

and letting $\int_T dt$ *be the Haar measure on* T *normalized so that its volume is* 1, *we have*

$$\int_G f(g)\,dg = \frac{1}{n!} \int_T f \begin{pmatrix} t_1 & & \\ & \ddots & \\ & & t_n \end{pmatrix} \prod_{i<j} |t_i - t_j|^2 dt. \tag{17.3}$$

Proof. This will follow from Theorem 17.2 once we check that

$$\det\left([\mathrm{Ad}(t^{-1}) - I_{\mathfrak{p}}]\,|\,\mathfrak{p}\right) = \prod_{i<j} |t_i - t_j|^2.$$

To compute this determinant, we may as well consider the linear transformation induced by $\mathrm{Ad}(t^{-1}) - I_{\mathfrak{p}}$ on the complexified vector space $\mathbb{C} \otimes \mathfrak{p}$. As in Proposition 11.4, we may identify $\mathbb{C} \otimes \mathfrak{u}(n)$ with $\mathfrak{gl}(n,\mathbb{C}) = \mathrm{Mat}_n(\mathbb{C})$. We recall that $\mathbb{C} \otimes \mathfrak{p}$ is spanned by the T-eigenspaces in $\mathbb{C} \otimes \mathfrak{u}(n)$ corresponding to nontrivial characters of T. These are spanned by the elementary matrices E_{ij} with a 1 in the i,jth position and zeros elsewhere, where $1 \leqslant i, j \leqslant n$ and $i \neq j$. The eigenvalue of t on E_{ij} is $t_i t_j^{-1}$. Hence

$$\det\left([\mathrm{Ad}(t^{-1}) - I_{\mathfrak{p}}]\,|\,\mathfrak{p}\right) = \prod_{i \neq j}(t_i t_j^{-1} - 1) = \prod_{i<j}(t_i t_j^{-1} - 1)(t_j t_i^{-1} - 1).$$

Since $|t_i| = |t_j| = 1$, we have $(t_i t_j^{-1} - 1)(t_j t_i^{-1} - 1) = (t_i - t_j)(t_i^{-1} - t_j^{-1}) = |t_i - t_j|^2$, proving (17.3). $\qquad\qquad\square$

Exercises

Exercise 17.1. Let $G = SO(2n + 1)$. Choose the realization of Exercise 5.3. Show that

$$\int_{SO(2n+1)} f(g)\,dg = \frac{1}{2^n n!} \int_{\mathbb{T}^n} f \begin{pmatrix} t_1 & & & & & & \\ & \ddots & & & & & \\ & & t_n & & & & \\ & & & 1 & & & \\ & & & & t_n^{-1} & & \\ & & & & & \ddots & \\ & & & & & & t_1^{-1} \end{pmatrix}$$

$$\times \prod_{i<j}\left\{|t_i - t_j|^2\,|t_i - t_j^{-1}|^2\right\} \prod_i |t_i - 1|^2\,dt_1 \cdots dt_n\,.$$

Exercise 17.2. Let $G = SO(2n)$. Choose the realization of Exercise 5.3. Show that

$$\int_{SO(2n)} f(g)\,dg = \frac{1}{2^{n-1}n!} \int_{\mathbb{T}^n} f \begin{pmatrix} t_1 & & & & & \\ & \ddots & & & & \\ & & t_n & & & \\ & & & t_n^{-1} & & \\ & & & & \ddots & \\ & & & & & t_1^{-1} \end{pmatrix}$$

$$\times \prod_{i<j} \left\{ |t_i - t_j|^2 \, |t_i - t_j^{-1}|^2 \right\}\, dt_1 \cdots dt_n\,.$$

Exercise 17.3. Describe the Haar measure on $Sp(2n)$ as an integral over the diagonal maximal torus.

Exercise 17.4. Let f be a class function on $SU(2)$. Suppose that

$$f\begin{pmatrix} z & \\ & z^{-1} \end{pmatrix} = \sum_n a(n)\, z^n.$$

Give at least two proofs that

$$\int_{SU(2)} f(g)\,dg = a(0) - a(2).$$

For the first proof, check that this is true for every irreducible character. For the second proof, show that $a(n) = a(-n)$. Then use the Weyl integration formula and make use of the fact that $a(2) = a(-2)$.

Exercise 17.5. Prove that

$$\int_{SU(2)} |\text{tr}(g)|^{2k}\,dg = \frac{1}{k+1}\binom{2k}{k}.$$

The moments of trace are thus the *Catalan numbers*.

18

The Root System

A *Euclidean space* is a real vector space $\mathcal{V}$ endowed with an inner product, that is, a positive definite symmetric bilinear form. We denote this inner product by $\langle\ ,\ \rangle$. If $0 \neq \alpha \in \mathcal{V}$, consider the transformation $s_\alpha : \mathcal{V} \longrightarrow \mathcal{V}$ given by

$$s_\alpha(x) = x - \frac{2\langle x, \alpha\rangle}{\langle \alpha, \alpha\rangle}\alpha. \tag{18.1}$$

This is the *reflection* attached to α. Geometrically, it is the reflection in the plane perpendicular to α. We have $s_\alpha(\alpha) = -\alpha$, while any element of that plane (with $\langle x, \alpha\rangle = 0$) is unchanged by s_α.

Definition 18.1. *Let $\mathcal{V}$ be a finite-dimensional real Euclidean space, $\Phi \subset \mathcal{V}$ a finite subset of nonzero vectors. Then Φ is called a* root system *if for all $\alpha \in \Phi$, $s_\alpha(\Phi) = \Phi$, and if $\alpha, \beta \in \Phi$ then $2\langle \alpha, \beta\rangle / \langle \alpha, \alpha\rangle \in \mathbb{Z}$. The root system is called* reduced *if $\alpha, \lambda\alpha \in \Phi$, $\lambda \in \mathbb{R}$ implies that $\lambda = \pm 1$.*

There is another, more modern notion which was introduced in Demazure [10] (Exposé XXI). This notion is known as a *root datum*. We will give the definition, then discuss the relationship between the two notions. We will find both structures in a compact Lie group.

A *root datum* consists of a quadruple $(\Lambda, \Phi, \Lambda^\vee, \Phi^\vee)$ of data which are to be as follows. First, Λ is a lattice, that is, a free $\mathbb{Z}$-module, and let $\Lambda^\vee = \mathrm{Hom}(\Lambda, \mathbb{Z})$ is the dual lattice. Inside each lattice there is given a finite set of nonzero vectors, denoted $\Phi \subset \Lambda$ and $\hat{\Phi} \subset \Lambda^\vee$, together with a bijection $\alpha \to \alpha^\vee$ from Φ to $\hat{\Phi}$. It is required that $\alpha^\vee(\alpha) = 2$ and that $\alpha^\vee(\Phi) \subset \mathbb{Z}$. Using these we may define, for each $\alpha \in \Phi$, linear maps $s_\alpha : \Lambda \to \Lambda$ and $s_{\alpha^\vee} : \Lambda^\vee \to \Lambda^\vee$ of order 2. These are defined by the formulas

$$s_\alpha(v) = v - \alpha^\vee(v)\alpha, \qquad s_{\alpha^\vee}(v^*) = v^* - v^*(\alpha)\alpha^\vee.$$

It is easy to see that $s_{\alpha^\vee}$ is the adjoint of s_α, that is,

$$s_{\alpha^\vee}(v^*)(v) = v^*(s_\alpha^{-1}v) = v^*(s_\alpha v).$$

D. Bump, *Lie Groups*, Graduate Texts in Mathematics 225,
DOI 10.1007/978-1-4614-8024-2_18, © Springer Science+Business Media New York 2013

Let us now explain the relationship between the root system and the root datum. We will always obtain the root system with another piece of data: a lattice Λ that spans $\mathcal{V}$ such that $\Phi \subset \Lambda$. It will have the property of being invariant under the s_α. Let $\mathcal{V}^*$ be the real dual space of $\mathcal{V}$. The *dual lattice* $\Lambda^\vee$ is the set of linear functionals $v^* : \mathcal{V} \to \mathbb{R}$ such that $v^*(L) \subset \mathbb{Z}$. It can be identified with $\mathrm{Hom}(\Lambda, \mathbb{Z})$. If $\alpha \in \Phi$ the linear functional

$$\alpha^\vee(x) = \frac{2\langle x, \alpha \rangle}{\langle \alpha, \alpha \rangle} \tag{18.2}$$

is in $L^\vee$ by the definition of a root system. If α is a root, then $\alpha^\vee$ is the called the *associated coroot*. Now if $\Phi^\vee$ is the set of coroots, then $(\Lambda, \Phi, \Lambda^\vee, \Phi^\vee)$ is a root datum.

The root datum notion has several advantages. First, the root datum gives complete information sufficient to uniquely determine the group G. This is perhaps less important if G is semisimple, for then one may specify the group by describing its root system and its fundamental group. However, if G is reductive but not semisimple, the root system is not enough data. Another, more subtle value of the root datum is that if $(\Lambda, \Phi, \Lambda^\vee, \Phi^\vee)$ is a root datum then so is $(\Lambda^\vee, \Phi^\vee, \Lambda, \Phi)$. This root datum describes another group $\hat{G}$, usually taken in its complexified form, as a complex analytic group. This is the *Langlands L-group*, which plays an important role in the representation theory of both Lie groups and p-adic groups. See Springer [152] and Borel [19]. In the root system, we are making use of the Euclidean inner product structure to identify the ambient vector space $\mathcal{V}$ with its dual. This has the psychological advantage of allowing us to envision s_α as reflection in the hyperplane perpendicular to the root α. On the other hand from a purely mathematical point of view, the identification of $\mathcal{V}$ with its dual is a somewhat artificial procedure.

The goal of this chapter is to associate a reduced root system with an arbitrary compact connected Lie group G. The lattice Λ will be $X^*(T)$, where T is a maximal torus, and the vector space $\mathcal{V}$ will be $\mathbb{R} \otimes \Lambda$. Elements of Λ will be called *weights*, and Λ will be called the *weight lattice*.

Let G be a compact connected Lie group and T a maximal torus. The dimension r of T is called the *rank* of G. We note that this terminology is not completely standard, for if $Z(G)$ is not finite, the term *rank* might refer to $\dim(T) - \dim(Z(G))$. We will refer to the latter statistic as the *semisimple rank* of G.

Let $\mathfrak{g} = \mathrm{Lie}(G)$ and $\mathfrak{t} = \mathrm{Lie}(T)$. Recall that $\mathbb{T}$ is the Lie group of complex numbers of absolute value 1. If we identify the Lie algebra of $\mathbb{C}^\times$ with $\mathbb{C}$ then the Lie algebra of $\mathbb{T}$ is $i\mathbb{R}$. Thus, if $\lambda : T \longrightarrow \mathbb{T}$ is a character, let $\mathrm{d}\lambda : \mathfrak{t} \longrightarrow i\mathbb{R}$ be the differential of λ, defined as usual by

$$\mathrm{d}\lambda(H) = \frac{\mathrm{d}}{\mathrm{d}t}\lambda(e^{tH})\Big|_{t=0}, \qquad H \in \mathfrak{t}. \tag{18.3}$$

Then $\mathrm{d}\lambda$ takes purely imaginary values.

Remark 18.1. Since $T \cong (\mathbb{R}/\mathbb{Z})^r$, its character group $X^*(T) \cong \mathbb{Z}^r$. We want to embed $X^*(T)$ into a real vector space $\mathcal{V} \cong \mathbb{R}^r$. There are two natural ways of doing this. First, we may note that $X^*(T) \cong \mathbb{Z}^r$, so we can take $\mathcal{V} = \mathbb{R} \otimes_{\mathbb{Z}} X^*(T)$. Alternatively, as we have just explained, if λ is a character of T, then $d\lambda \in \mathrm{Hom}(\mathfrak{t}, i\mathbb{R})$. Extending $d\lambda$ to a complex linear map $\mathfrak{t}_{\mathbb{C}}$, we see that $d\lambda$ also maps $i\mathfrak{t} \to \mathbb{R}$. Part of the construction will be to produce elements $H_\alpha \in i\mathfrak{t}$ such that for $\lambda \in X^*(T)$ we have

$$d\lambda(H_\alpha) = \alpha^\vee(\lambda) \tag{18.4}$$

(See Proposition 18.13.) In view of this close relationship both the $\alpha^\vee$ and the H_α may be referred to as *coroots*.

The Weyl group $W = N(T)/T$ acts on T by conjugation and hence on $\mathcal{V}$, and it will be convenient to give $\mathcal{V}$ an inner product (that is, a positive definite symmetric bilinear form) that is W-invariant. We may, of course, do this for any finite group acting on a real vector space.

If $\pi : G \to \mathrm{GL}(V)$ is a complex representation, then we may restrict π to T, where it will decompose into one-dimensional characters. The elements of $\Lambda = X^*(T)$ that occur in π restricted to T are called the *weights* of the representation. A *root* of G with respect to T is a nonzero weight of the adjoint representation. We recall from Chap. 17 that $\mathfrak{g} = \mathfrak{t} \oplus \mathfrak{p}$ where $\mathfrak{p}$ is the direct sum of nontrivial two-dimensional real subspaces that are irreducible T-modules. Then $\mathfrak{g}_{\mathbb{C}}$ decomposes as $\mathfrak{t}_{\mathbb{C}} \oplus \mathfrak{p}_{\mathbb{C}}$. The space $\mathfrak{p}_{\mathbb{C}}$ will further decompose into one-dimensional $\mathfrak{t}$-invariant complex subspaces. More precisely, if U is one of the irreducible two-dimensional $\mathfrak{t}$-invariant subspaces of the real vector space $\mathfrak{p}$, then by Proposition 15.5, $U_{\mathbb{C}}$ is the direct sum of two one-dimensional invariant complex vector spaces, each corresponding to a root α and its negative $-\alpha$. So we may say that a root is a character of T that occurs in the adjoint representation on $\mathfrak{p}_{\mathbb{C}}$. If α is a root, let $\mathfrak{X}_\alpha \subset \mathfrak{p}_{\mathbb{C}}$ be the α-eigenspace. We will denote by $\Phi \subset \mathcal{V}$ the set of roots of G with respect to T. We will show in Theorem 18.2 that Φ is a root system.

Because the proofs are somewhat long, it may be useful to have a couple of examples in mind. First, SU(2) will play a role in the sequel, so we review it. The Lie algebra $\mathfrak{g}$ consists of 2×2 skew-Hermitian matrices of trace 0. Every element of the Lie algebra $\mathfrak{sl}(2, \mathbb{C})$ of SL(2, $\mathbb{C}$) may be written uniquely as $X + iY$ with X and Y in $\mathfrak{g}$, so $\mathfrak{sl}(2, \mathbb{C}) = \mathfrak{g} \oplus i\mathfrak{g}$. In other words the complexified Lie algebra $\mathfrak{g}_{\mathbb{C}} = \mathfrak{sl}(2, \mathbb{C})$, the Lie algebra with representations that were studied in Chap. 12.

Let T be the group of diagonal matrices in SU(2). A character of T has the form

$$\lambda_k \begin{pmatrix} t & \\ & t^{-1} \end{pmatrix} = t^k. \tag{18.5}$$

Define

$$H_\alpha = \begin{pmatrix} 1 & \\ & -1 \end{pmatrix}, \qquad X_\alpha = \begin{pmatrix} 0 & 1 \\ 0 & 0 \end{pmatrix}, \qquad X_{-\alpha} = \begin{pmatrix} 0 & 0 \\ 1 & 0 \end{pmatrix}. \tag{18.6}$$

We will see that $H_\alpha \in i\mathfrak{t}$ is the coroot, and that X_α and $X_{-\alpha}$ span the one-dimensional weight spaces $\mathfrak{X}_\alpha$ and $\mathfrak{X}_{-\alpha}$. Thus, the root system $\Phi = \{\alpha, -\alpha\}$.

Let us say that λ_k is the *highest* weight in an irreducible representation V if k is maximal such that λ_k occurs in the restriction of the representation to V. This *ad hoc* definition is a special case of a partial order on the weights for general compact Lie groups in Chap. 21.

Proposition 18.1. *If $k \in \mathbb{Z}$ then $d\lambda_k(H_\alpha) = k$. The roots of $\mathrm{SU}(2)$ are $\alpha = \lambda_2$ and $-\alpha = \lambda_{-2}$. If k is a nonnegative integer then $\mathrm{SU}(2)$ has a unique irreducible representation with highest weight λ_k. The weights of this representation are λ_l with $-k \leqslant l \leqslant k$ and $l \equiv k$ modulo 2.*

Proof. Although H_α is not in $\mathfrak{t}$, iH_α is and we find that

$$d\lambda_k(iH_\alpha) = \frac{d}{dt}\lambda_k \left(\begin{matrix} e^{it} & \\ & e^{-it} \end{matrix} \right) \Big|_{t=0} = \frac{d}{dt}e^{ikt}|_{t=0} = ik$$

so $d\lambda_k(H_\alpha) = k$. We have

$$\mathrm{Ad}\begin{pmatrix} t & \\ & t^{-1} \end{pmatrix} X_\alpha = t^2 X_\alpha,$$

so X_α spans a T-eigenspace affording the character λ_2, which is thus a root α. If k is a nonnegative integer, then we proved in Chap. 12 that $\mathfrak{sl}(2, \mathbb{C})$ has an irreducible representation $\vee^k \mathbb{C}^2$ and the weights are seen from (12.2) to be the integers between $-k$ and k with the same parity as k. $\qquad \square$

To give a higher-rank example, let us consider the group $G = \mathrm{Sp}(4)$. This is a maximal compact subgroup of $\mathrm{Sp}(4, \mathbb{C})$, which we will take to be the group of $g \in \mathrm{GL}(4, \mathbb{C})$ that satisfy $g\,J\,{}^t g = J$, where

$$J = \begin{pmatrix} & & & -1 \\ & & -1 & \\ & 1 & & \\ 1 & & & \end{pmatrix}.$$

This is not the same as the group introduced in Example 5.5, but it is conjugate to that group in $\mathrm{GL}(4, \mathbb{C})$. The subgroup $\mathrm{Sp}(4)$ is the intersection of $\mathrm{Sp}(4, \mathbb{C})$ with $\mathrm{U}(4)$. A maximal torus T can be taken to be the group of diagonal elements, and the we will show that the roots are the eight characters

$$
T \ni t = \begin{pmatrix} t_1 & & & \\ & t_2 & & \\ & & t_2^{-1} & \\ & & & t_1^{-1} \end{pmatrix} \longmapsto \begin{cases} \alpha_1(t) = t_1 t_2^{-1}, \\ \alpha_2(t) = t_2^2, \\ (\alpha_1 + \alpha_2)(t) = t_1 t_2, \\ (2\alpha_1 + \alpha_2)(t) = t_1^2, \\ -\alpha_1(t) = t_1^{-1} t_2, \\ -\alpha_2(t) = t_2^{-2}, \\ -(\alpha_1 + \alpha_2)(t) = t_1^{-1} t_2^{-1}, \\ -(2\alpha_1 + \alpha_2)(t) = t_1^{-2}. \end{cases}
$$

They form a configuration in $\mathcal{V}$ that can be seen in Fig. 19.4 of the next chapter. The reader can check that this forms a root system.

The complexified Lie algebra $\mathfrak{g}_{\mathbb{C}}$ consists of matrices of the form

$$
\begin{pmatrix} t_1 & x_{12} & x_{13} & x_{14} \\ x_{21} & t_2 & x_{23} & x_{13} \\ x_{31} & x_{32} & -t_2 & -x_{12} \\ x_{41} & x_{31} & -x_{21} & -t_1 \end{pmatrix}. \tag{18.7}
$$

The spaces $\mathfrak{X}_{\alpha_1}$ and $\mathfrak{X}_{-\alpha_1}$ are spanned by the vectors

$$
X_{\alpha_1} = \begin{pmatrix} 0 & 1 & 0 & 0 \\ 0 & 0 & 0 & 0 \\ 0 & 0 & 0 & -1 \\ 0 & 0 & 0 & 0 \end{pmatrix}, \qquad X_{-\alpha_1} = \begin{pmatrix} 0 & 0 & 0 & 0 \\ 1 & 0 & 0 & 0 \\ 0 & 0 & 0 & 0 \\ 0 & 0 & -1 & 0 \end{pmatrix}.
$$

Similarly the spaces $\mathfrak{X}_{\alpha_2}$ and $\mathfrak{X}_{-\alpha_2}$ are spanned by

$$
X_{\alpha_2} = \begin{pmatrix} 0 & 0 & 0 & 0 \\ 0 & 0 & 1 & 0 \\ 0 & 0 & 0 & 0 \\ 0 & 0 & 0 & 0 \end{pmatrix}, \qquad X_{-\alpha_2} = \begin{pmatrix} 0 & 0 & 0 & 0 \\ 0 & 0 & 0 & 0 \\ 0 & 1 & 0 & 0 \\ 0 & 0 & 0 & 0 \end{pmatrix}.
$$

As you can see, $\operatorname{Ad}(t) X_\alpha = \alpha(t) X_\alpha$ when $\alpha = \alpha_1$ or α_2. This proves that α_1 and α_2 are roots, and the four others are handled similarly. Note that these X_α are elements not of $\mathfrak{g}$ but of its complexification $\mathfrak{g}_{\mathbb{C}}$, since to be in $\mathfrak{g}$, the element (18.7) must be skew-Hermitian, which means that the t_i are purely imaginary, and $x_{ij} = -\overline{x_{ji}}$.

As we have mentioned, the proof that the set of roots of a compact Lie group form a root system involves constructing certain elements H_α of it. In this example

$$
H_{\alpha_1} = \begin{pmatrix} 1 & & & \\ & -1 & & \\ & & 1 & \\ & & & -1 \end{pmatrix}, \qquad H_{\alpha_2} = \begin{pmatrix} 0 & & & \\ & 1 & & \\ & & -1 & \\ & & & 0 \end{pmatrix}.
$$

Note that $H_\alpha \notin \mathfrak{t}$, but $-iH_\alpha \in \mathfrak{t}$, since the elements of $\mathfrak{t}$ are diagonal and purely imaginary. The H_α satisfy

$$[H_\alpha, X_\alpha] = 2X_\alpha, \qquad [H_\alpha, X_{-\alpha}] = -2X_{-\alpha},$$

and they are elements of the intersection of it with the complex Lie algebra generated by X_α and $X_{-\alpha}$. We note that X_α and $X_{-\alpha}$ are only determined up to constant multiples by the description we have given, but H_α is fully characterized. The H_α will be constructed in Proposition 18.8 below. They form a root system that is dual to the one we want to construct—if α is a long root, then H_α is short, and conversely, in root systems where not all the roots have the same length. (See Exercise 18.1.)

A key step will be to construct an element w_α of the Weyl group $W = N(T)/T$ corresponding to the reflection s_α in (18.1). In order to produce this, we will construct a homomorphism $i_\alpha : \mathrm{SU}(2) \longrightarrow G$. Then $w_\alpha \in N(T)$ will then be the image of $\begin{pmatrix} 0 & -1 \\ 1 & 0 \end{pmatrix}$ under i_α.

Let us offer a word about how one can get a grip on i_α. The centralizer $C(T_\alpha)$ of the kernel T_α of the homomorphism $\alpha : T \longrightarrow \mathbb{C}^\times$ is a close relative of this group $i_\alpha(\mathrm{SU}(2))$. In fact, $C(T_\alpha) = i_\alpha(\mathrm{SU}(2)) \cdot T$. Later, in Proposition 18.6 we will use this circumstance to show that $\mathfrak{X}_\alpha$ is one-dimensional, after which the structure of $C(T_\alpha)$ will become clear: since this group has only two roots α and $-\alpha$, it is itself a close relative of $SU(2)$. Its Lie algebra contains a copy of $\mathfrak{su}(2)$ (Proposition 18.8) and using this fact we will be able to construct the homomorphism i_α in Theorem 18.1.

Let us take a look at the groups $C(T_\alpha)$ and the homomorphisms i_α in the example of $\mathrm{Sp}(4)$. The subgroup T_{α_1} of T is characterized by $t_1 = t_2$, so its centralizer consists of elements of the form

$$\begin{pmatrix} a & b & & \\ c & d & & \\ & & * & * \\ & & * & * \end{pmatrix}, \qquad \begin{pmatrix} a & b \\ c & d \end{pmatrix} \in \mathrm{U}(2),$$

where the elements marked $*$ are determined by the requirement that the matrix be in $\mathrm{Sp}(4)$. The homomorphism i_{α_1} is given by

$$i_{\alpha_1} \begin{pmatrix} a & b \\ c & d \end{pmatrix} = \begin{pmatrix} a & b & & \\ c & d & & \\ & & a & -b \\ & & -c & d \end{pmatrix}, \qquad \begin{pmatrix} a & b \\ c & d \end{pmatrix} \in \mathrm{SU}(2).$$

Similarly, T_{α_2} is characterized by $t_2 = \{\pm 1\}$, and

$$i_{\alpha_2} \begin{pmatrix} a & b \\ c & d \end{pmatrix} = \begin{pmatrix} 1 & & & \\ & a & b & \\ & c & d & \\ & & & 1 \end{pmatrix}.$$

We turn now to the general case and to the proofs.

Proposition 18.2. *A maximal Abelian subalgebra $\mathfrak{h}$ of $\mathfrak{g}$ is the Lie algebra of a conjugate of T. Its dimension is the rank r of G.*

Proof. By Proposition 15.2, $\exp(\mathfrak{h})$ is a commutative group that is connected since it is the continuous image of a connected space. By Theorem 15.2 its closure H is a Lie subgroup of G, closed, connected and Abelian and therefore a torus. It is therefore contained in a maximal torus H'. By maximality of $\mathfrak{h} \subseteq \mathrm{Lie}(H')$ we must have $\mathfrak{h} = \mathrm{Lie}(H')$ and $H' = H$. By Cartan's Theorem 16.5, H is a conjugate of T. $\qquad\square$

Lemma 18.1. *Suppose that G is a compact Lie group with Lie algebra $\mathfrak{g}$, $\pi : G \longrightarrow \mathrm{GL}(V)$ a representation, and $d\pi : \mathfrak{g} \longrightarrow \mathrm{End}(V)$ the differential. If $v \in V$ and $X \in \mathfrak{g}$ such that $d\pi(X)^n v = 0$ for any $n > 1$, then $d\pi(X)v = 0$.*

Proof. We may put a G-invariant positive definite inner product $\langle\,,\,\rangle$ on V. The inner product is then $\mathfrak{g}$-invariant, which means that $\langle d\pi(X)v, w\rangle = -\langle v, d\pi(X)w\rangle$. Thus, $d\pi(X)$ is skew-Hermitian, which by the spectral theorem implies that V has a basis with respect to which its matrix is diagonal. It is clear that, for a diagonal matrix M, $M^n v = 0$ implies that $Mv = 0$. $\qquad\square$

Let (π, V) be any finite-dimensional complex representation of G. If $\lambda \in X^*(T)$, let $V(\lambda) = \{v \in V \mid \pi(t)v = \lambda(t)v\}$. Then V is the direct sum of the $V(\lambda)$. If $(\pi, V) = (\mathrm{Ad}, \mathfrak{g}_{\mathbb{C}})$ and $\lambda = \alpha$ is a root, then $V(\lambda) = \mathfrak{X}_\alpha$.

Proposition 18.3. *Let (π, V) be any irreducible representation of G, and let α be a root.*

(i) If $d\pi : \mathfrak{g} \longrightarrow \mathfrak{gl}(V)$ is the differential of π, then

$$d\pi(H)v = d\lambda(H)v, \qquad H \in \mathfrak{t},\, v \in V(\lambda). \tag{18.8}$$

(ii) We have

$$[H, X_\alpha] = \mathrm{ad}(H)X_\alpha = d\alpha(H)X_\alpha, \qquad H \in \mathfrak{t},\, X_\alpha \in \mathfrak{X}_\alpha. \tag{18.9}$$

(iii) If (π, V) is a finite-dimensional complex representation of G and $v \in V(\lambda)$ for some $\lambda \in X^*(T)$, then $d\pi(X_\alpha)v \in V(\lambda + \alpha)$.

Proof. For (i), if $H \in \mathfrak{t}$ and $t \in \mathbb{R}$, then for $v \in V(\lambda)$ we have

$$\pi(e^{tH})v = \lambda(e^{tH})v = e^{t d\lambda(H)}v.$$

Taking the derivative and setting $t = 0$, using (18.3) we obtain (18.8). When $V = \mathfrak{g}_{\mathbb{C}}$ and $\pi = \mathrm{Ad}$, we have $\mathfrak{X}_\alpha = V(\lambda)$. Remembering that the differential of Ad is ad (Theorem 8.2), we see that (18.9) is a special case of (18.8), and (ii) follows.

For (iii), we have, by (18.9),

$$d\pi(H)\, d\pi(X_\alpha) - d\pi(X_\alpha)\, d\pi(H) = d\pi[H, X_\alpha] = d\alpha(H)d\pi(X_\alpha).$$

Applying this to v and using (18.8) gives, with $w = d\pi(X_\alpha)v$,

$$d\pi(H)w = \big(d\lambda(H) + d\alpha(H)\big)w,$$

so $w \in V(\lambda + \alpha)$. □

We may write $\mathfrak{g}_\mathbb{C} = \mathfrak{g} + i\mathfrak{g}$. Let $c : \mathfrak{g}_\mathbb{C} \longrightarrow \mathfrak{g}_\mathbb{C}$ be the conjugation with respect to $\mathfrak{g}$, that is, the real linear transformation $X + iY \longrightarrow X - iY$ ($X, Y \in \mathfrak{g}$). Although c is not complex linear, it is an automorphism of $\mathfrak{g}_\mathbb{C}$ as a real Lie algebra. We have $c(aZ) = \bar{a} \cdot c(Z)$, so c is complex antilinear.

Proposition 18.4.

(i) We have $c(\mathfrak{X}_\alpha) = \mathfrak{X}_{-\alpha}$.
(ii) If $X_\alpha \in \mathfrak{X}_\alpha$, $X_\beta \in \mathfrak{X}_\beta$, $\alpha, \beta \in \Phi$, then

$$[X_\alpha, X_\beta] \in \left\{ \begin{array}{ll} \mathfrak{t}_\mathbb{C} & \text{if } \beta = -\alpha, \\ \mathfrak{X}_{\alpha+\beta} & \text{if } \alpha + \beta \in \Phi. \end{array} \right.$$

while $[X_\alpha, X_\beta] = 0$ if $\beta \neq -\alpha$ and $\alpha + \beta \notin \Phi$.
(iii) If $0 \neq X_\alpha \in \mathfrak{X}_\alpha$, then $[X_\alpha, c(X_\alpha)]$ is a nonzero element of it, and $d\alpha([X_\alpha, c(X_\alpha)]) \neq 0$.

In case (ii), if $\alpha + \beta \in \Phi$, we will eventually show that $[X_\alpha, X_\beta]$ is a *nonzero* element of $\mathfrak{X}_{\alpha+\beta}$. See Corollary 18.1.

Proof. For (i), apply c to (18.9) using the complex antilinearity of c, and the fact that $d\alpha(H)$ is purely imaginary to obtain, for $H \in \mathfrak{t}$

$$[H, c(X_\alpha)] = [c(H), c(X_\alpha)] = c[H, X_\alpha] = c(d\alpha(H)X_\alpha) = -d\alpha(H)c(X_\alpha).$$

This shows that $c(X_\alpha) \in \mathfrak{X}_{-\alpha}$.

Part (ii) is the special case of Proposition 18.3 (iii) when $\pi = \mathrm{Ad}$ and $V = \mathfrak{g}_\mathbb{C}$ since $\mathfrak{t}_\mathbb{C} = V(0)$ while $\mathfrak{X}_\alpha = V(\alpha)$ when $\alpha \in \Phi$.

Next we prove (iii). By (i) and (ii), $[X_\alpha, c(X_\alpha)] \in \mathfrak{t}_\mathbb{C}$. Applying c to this element,

$$c\big([X_\alpha, c(X_\alpha)]\big) = [c(X_\alpha), X_\alpha] = -[X_\alpha, c(X_\alpha)],$$

so $[X_\alpha, c(X_\alpha)] \in i\mathfrak{t}$. We show that $[X_\alpha, c(X_\alpha)] \neq 0$. Let $\mathfrak{t}_\alpha \subset \mathfrak{t}$ be the kernel of $d\alpha$. It is of course a subspace of codimension 1. Let $H_1, \ldots, H_{r-1}$ be a basis. If $[X_\alpha, c(X_\alpha)] = 0$, then denoting

$$Y_\alpha = \tfrac{1}{2}\big(X_\alpha + c(X_\alpha)\big), \qquad Z_\alpha = \tfrac{1}{2i}\big(X_\alpha - c(X_\alpha)\big), \tag{18.10}$$

Y_α and Z_α are c-invariant and hence in $\mathfrak{g}$, and

$$H_1, \ldots, H_{r-1}, Y_\alpha, Z_\alpha$$

are $r + 1$ commuting elements of $\mathfrak{t}$ that are linearly independent over $\mathbb{R}$. This contradicts Proposition 18.2, so $[X_\alpha, c(X_\alpha)] \neq 0$.

It remains to be shown that $d\alpha([X_\alpha, c(X_\alpha)]) \neq 0$. If on the contrary this vanishes, then $[H_0, X_\alpha] = [H_0, c(X_\alpha)] = 0$ by (18.9), where $H_0 = -i[X_\alpha, c(X_\alpha)] \in \mathfrak{t}$. With Y_α and Z_α as in (18.10), this implies that $[H_0, Y_\alpha] = [H_0, Z_\alpha] = 0$. Now

$$[Y_\alpha, Z_\alpha] = \tfrac{1}{2}H_0, \qquad [Y_\alpha, H_0] = 0.$$

Thus, $\mathrm{ad}(Y_\alpha)^2 Z_\alpha = 0$, yet $\mathrm{ad}(Y_\alpha)Z_\alpha \neq 0$, contradicting Lemma 18.1.   □

Proposition 18.5. *If* $\dim(T) = 1$, *then either* $G = T$ *or* $\dim(G) = 3$. *If* α *is any root, then* $\mathfrak{X}_\alpha$ *is one-dimensional, and* $\alpha, -\alpha$ *are the only roots.*

Proof. Since $\mathfrak{t}_{\mathbb{C}}$ is one-dimensional, let H be a basis vector. Assuming $G \neq T$, Φ is nonempty. The spaces $\mathfrak{X}_\alpha$ are just the eigenspaces of H on $\mathfrak{p}_{\mathbb{C}}$. Since T is one-dimensional, so is $\mathcal{V}$. Thus, if $\alpha \in \Phi$, every $\beta \in \Phi$ is of the form $\lambda\alpha$ for a nonzero constant λ. We choose α so that all $|\lambda| \geqslant 1$. Let $0 \neq X_\alpha \in \mathfrak{X}_\alpha$, and let $X_{-\alpha} = -c(X_\alpha)$. We consider the complex vector space

$$V = \mathbb{C}X_{-\alpha} \oplus \mathfrak{t}_{\mathbb{C}} \oplus \bigoplus_{\substack{\lambda\alpha \in \Phi \\ \lambda > 0}} \mathfrak{X}_{\lambda\alpha}.$$

By Proposition 18.4, each component space is mapped into another by $\mathrm{ad}(X_\alpha)$ and $\mathrm{ad}(X_{-\alpha})$. Indeed, $\mathrm{ad}(X_{-\alpha})$ kills $X_{-\alpha}$, shifts $\mathfrak{t}_{\mathbb{C}}$ into $\mathbb{C}X_{-\alpha}$, and shifts $\mathfrak{X}_{\lambda\alpha}$ into $\mathfrak{t}_{\mathbb{C}}$ if $\lambda = 1$ or $\mathfrak{X}_{(\lambda-1)\alpha}$ if $\lambda \neq 1$. Note that $\lambda > 0$ implies $\lambda > 1$, so indeed V is stable under $\mathrm{ad}(X_{-\alpha})$. The case of $\mathrm{ad}(X_\alpha)$ is similar. Moreover, $[X_\alpha, X_{-\alpha}]$ is a nonzero multiple of H by Proposition 18.4. Since the commutator of two linear transformations on a finite-dimensional vector space has trace zero, the trace of H on V is therefore zero.

On the other hand, denoting $C = d\alpha(H)$, the trace of $\mathrm{ad}(H)$ on $\mathfrak{X}_{\lambda\alpha}$ equals $\lambda C \dim(\mathfrak{X}_{\lambda\alpha})$, while the trace of $\mathrm{ad}(H)$ on $\mathbb{C}X_{-\alpha}$ is $-C$, and the trace of $\mathrm{ad}(H)$ on $\mathfrak{t}_{\mathbb{C}}$ is zero. We see that the trace is $-C + \sum_{\lambda \geqslant 1} \lambda C \dim(\mathfrak{X}_{\lambda\alpha})$. Since this is zero, there can be only one $\mathfrak{X}_{\lambda\alpha}$ with $\lambda > 0$, namely $\mathfrak{X}_\alpha$, and $\dim(\mathfrak{X}_\alpha) = 1$. Now $\mathfrak{g} = \mathbb{C}H \oplus \mathbb{C}X_\alpha \oplus \mathbb{C}X_{-\alpha}$ is three-dimensional.   □

We return now to the general case. If $\alpha \in \Phi$, let $T_\alpha \subset T$ be the kernel of α. This closed subgroup of T may or may not be connected. Its Lie algebra is the kernel $\mathfrak{t}_\alpha$ of $d\alpha$.

Proposition 18.6.

(i) If $\alpha \in \Phi$, *then* $\dim(\mathfrak{X}_\alpha) = 1$.
(ii) If $\alpha, \beta \in \Phi$ *and* $\alpha = \lambda\beta$, $\lambda \in \mathbb{R}$, *then* $\lambda = \pm 1$.

Proof. The group $H = C_G(T_\alpha)$ is a closed connected Lie subgroup by Theorem 16.6. It has T_α as a normal subgroup. The Lie algebra of H is the centralizer $\mathfrak{h}$ in $\mathfrak{g}$ of $\mathfrak{t}_\alpha$, so

$$\mathfrak{h}_\mathbb{C} = \mathfrak{t}_\mathbb{C} \oplus \bigoplus_{\substack{\lambda\alpha \in \Phi \\ \lambda \in \mathbb{R}}} \mathfrak{X}_{\lambda\alpha}.$$

Thus H/T_α is a rank 1 group with maximal torus T/T_α. Its complexified Lie algebra is therefore three-dimensional by Proposition 18.5. However, $\bigoplus \mathfrak{X}_{\lambda\alpha}$ is embedded injectively in this complexified Lie algebra, so $\lambda = \pm 1$ are the only λ, and $\mathfrak{X}_{\pm\alpha}$ are one-dimensional. $\square$

Proposition 18.7.

(i) *Let $\mathfrak{g}$ be the Lie algebra of a compact Lie group. If $X, Y \in \mathfrak{g}$ such that $[X, Y] = cY$ with c a nonzero real constant, then $Y = 0$.*

(ii) *There does not exist any embedding of $\mathfrak{sl}(2, \mathbb{R})$ into the Lie algebra of a compact Lie group.*

Proof. Let G be a compact Lie group with Lie algebra $\mathfrak{g}$. Then given any finite-dimensional representation $\pi : G \longrightarrow \mathrm{GL}(V)$ on a real vector space V, there exists a positive-definite symmetric bilinear form B on V such that $B(\pi(g)v, \pi(g)w) = B(v, w)$ for $g \in G$, $v, w \in V$. By Proposition 10.3, we have $B(d\pi(X)v, w) = -B(v, d\pi(X)w)$ for $X \in \mathfrak{g}$. Now let us apply this with $V = \mathfrak{g}$ and $\pi = \mathrm{Ad}$, so by Theorem 8.2 we have $B([X, Y], Z) = -B(Y, [X, Z])$. If X and Y such that $[X, Y] = cY$ with c a nonzero constant, then

$$cB(Y, Y) = B([X, Y], Y) = -B(Y, [X, Y]) = -cB(Y, Y).$$

Since $c \neq 0$ and B is positive definite, it follows that $Y = 0$. This proves (i).

As for (ii), if $\mathfrak{g}$ contains a subalgebra isomorphic to $\mathfrak{sl}(2, \mathbb{R})$ then we may take X and Y to be the images of $\begin{pmatrix} 1 & \\ & -1 \end{pmatrix}$ and $\begin{pmatrix} 0 & 1 \\ 0 & 0 \end{pmatrix}$ and obtain a contradiction to (i). $\square$

We remind the reader that the Lie algebra $\mathfrak{su}(2)$ of $\mathrm{SU}(2)$ consists of the trace zero skew-Hermitian matrices in $\mathrm{Mat}(2, \mathbb{C})$. The Lie algebra $\mathfrak{sl}(2, \mathbb{C})$ of $\mathrm{SL}(2, \mathbb{C})$ consists of all trace zero matrices. Any trace zero matrix may X may be uniquely written as $X_1 + iX_2$ where X_1 and X_2 are in $\mathfrak{su}(2)$, so $\mathfrak{sl}(2, \mathbb{C})$ is the complexification of $\mathfrak{su}(2)$.

Proposition 18.8. *Let $\alpha \in \Phi$ and let $0 \neq X_\alpha \in \mathfrak{X}_\alpha$. Let $X_{-\alpha} = -c(X_\alpha) \in \mathfrak{X}_{-\alpha}$. Then X_α and $X_{-\alpha}$ generate a complex Lie subalgebra $\mathfrak{g}_{\alpha,\mathbb{C}}$ of $\mathfrak{g}_\mathbb{C}$ isomorphic to $\mathfrak{sl}(2, \mathbb{C})$. Its intersection $\mathfrak{g}_\alpha = \mathfrak{g} \cap \mathfrak{g}_{\alpha,\mathbb{C}}$ is isomorphic to $\mathfrak{su}(2)$. We may choose X_α and the isomorphism $i_\alpha : \mathfrak{sl}(2, \mathbb{C}) \longrightarrow \mathfrak{g}_{\alpha,\mathbb{C}}$ so that*

$$i_\alpha\begin{pmatrix} 1 & \\ & -1 \end{pmatrix} = H_\alpha, \qquad i_\alpha\begin{pmatrix} 0 & 1 \\ 0 & 0 \end{pmatrix} = X_\alpha, \qquad i_\alpha\begin{pmatrix} 0 & 0 \\ 1 & 0 \end{pmatrix} = X_{-\alpha}, \quad (18.11)$$

where $H_\alpha = [X_\alpha, X_{-\alpha}]$. In this case, $H_\alpha \in i\mathfrak{t}$ and

$$[X_\alpha, X_{-\alpha}] = H_\alpha, \qquad [H_\alpha, X_\alpha] = 2X_\alpha, \qquad [H_\alpha, X_{-\alpha}] = -2X_{-\alpha}. \quad (18.12)$$

Proof. Let $H_\alpha = [X_\alpha, X_{-\alpha}]$. By Proposition 18.4(iii), H_α is a nonzero element of $i\mathfrak{t}$ not in $i\mathfrak{t}_\alpha$. By Proposition 18.4(iii) and (18.9), we have $[H_\alpha, X_\alpha] = 2\lambda X_\alpha$, where λ is a nonzero real constant. Applying c and using $c(H_\alpha) = -H_\alpha$, we have $[H_\alpha, X_{-\alpha}] = -2\lambda X_{-\alpha}$.

We will show later that $\lambda > 0$. For now, assume this. Replacing X_α, $X_{-\alpha}$ and H_α by $\lambda^{-1/2} X_\alpha$, $\lambda^{-1/2} X_{-\alpha}$, and $\lambda^{-1} H_\alpha$, we may arrange that (18.12) is satisfied. Since the three matrices in $\mathfrak{sl}(2, \mathbb{C})$ in (18.11) satisfy the same relations, we have an isomorphism $i_\alpha : \mathfrak{sl}(2, \mathbb{C}) \longrightarrow \mathfrak{g}_{\alpha, \mathbb{C}}$ such that (18.11) is true. Now the real subalgebra $\mathfrak{g}_\alpha$ fixed by the conjugation c is spanned as a real vector space by iH, $i(X_\alpha + X_{-\alpha})$ and $X_\alpha - X_{-\alpha}$. (Here $i = \sqrt{-1}$, not to be confused with i_α.) These are the image under i_α of

$$\begin{pmatrix} i & \\ & -i \end{pmatrix}, \qquad \begin{pmatrix} 0 & i \\ i & 0 \end{pmatrix}, \qquad \begin{pmatrix} & 1 \\ -1 & \end{pmatrix},$$

which span $\mathfrak{su}(2)$.

It remains to be shown that $\lambda > 0$. If not, we will obtain a contradiction. Replacing X_α, $X_{-\alpha}$ and H_α by $|\lambda|^{-1/2} X_\alpha$, $|\lambda|^{-1/2} X_{-\alpha}$, and $\lambda^{-1} H_\alpha$ gives

$$[X_\alpha, X_{-\alpha}] = H_\alpha, \qquad [H_\alpha, X_\alpha] = -2X_\alpha, \qquad [H_\alpha, X_{-\alpha}] = 2X_{-\alpha}.$$

We may now obtain an isomorphism i'_α of $\mathfrak{sl}(2, \mathbb{C})$ with $\mathfrak{g}_\mathbb{C}$ by

$$i'_\alpha \begin{pmatrix} 1 & \\ & -1 \end{pmatrix} = H_\alpha, \qquad i'_\alpha \begin{pmatrix} 0 & 0 \\ i & 0 \end{pmatrix} = X_\alpha, \qquad i'_\alpha \begin{pmatrix} 0 & i \\ 0 & 0 \end{pmatrix} = X_{-\alpha},$$

The real subalgebra $\mathfrak{g}_\alpha$ fixed by the conjugation c is generated by iH, $i(X_\alpha + X_{-\alpha})$ and $X_\alpha - X_{-\alpha}$, and these correspond to

$$\begin{pmatrix} i & \\ & -i \end{pmatrix}, \qquad \begin{pmatrix} & -1 \\ -1 & \end{pmatrix}, \qquad \begin{pmatrix} & -i \\ i & \end{pmatrix},$$

and these generate the Lie algebra $\mathfrak{su}(1, 1)$, which is isomorphic to $\mathfrak{sl}(2, \mathbb{R})$ by Exercise 5.9. This is a contradiction because $\mathfrak{sl}(2, \mathbb{R})$ cannot be embedded in the Lie algebra of a compact Lie group by Proposition 18.7. $\square$

Since $\mathfrak{X}_\alpha$ is one-dimensional, the group $\mathfrak{g}_\alpha$ does not depend on the choice of X_α.

Proposition 18.9. *If $H \in \mathfrak{t}_\alpha = \ker(d\alpha)$, then $[H, \mathfrak{g}_\alpha] = 0$.*

Proof. H centralizes X_α and $X_{-\alpha}$ by (18.9); that is, $[H, X_\alpha] = [H, X_{-\alpha}] = 0$, and it follows that $[H, X] = 0$ for all $X \in \mathfrak{g}_\alpha$. $\square$

We gave the ambient vector space $\mathcal{V}$ of the set Φ of roots an inner product (Euclidean structure) invariant under W. The Weyl group acts on T by conjugation and hence it acts on $X^*(T)$. It acts on $\mathfrak{p}$ by the adjoint representation (induced from conjugation) so it permutes the roots. The Weyl group elements are realized as orthogonal motions with respect to this metric.

We may now give a method of constructing Weyl group elements. Let $\alpha \in \Phi$. Let $T_\alpha = \{t \in T \mid \alpha(t) = 1\}$.

Theorem 18.1. *Let $\alpha \in \Phi$. There exists a homomorphism $i_\alpha : \mathrm{SU}(2) \longrightarrow C(T_\alpha)^\circ \subset G$ such that the image of the differential $di_\alpha : \mathfrak{su}(2) \longrightarrow \mathfrak{g}$ is the Lie algebra homomorphism of Proposition 18.8. If*

$$w_\alpha = i_\alpha \begin{pmatrix} & -1 \\ 1 & \end{pmatrix}, \tag{18.13}$$

then $w_\alpha \in N(T)$ and w_α induces s_α in its action on $X^*(T)$.

Proof. Since $\mathrm{SU}(2)$ is simply connected, it follows from Theorem 14.2 that the Lie algebra homomorphism $\mathfrak{su}(2) \longrightarrow \mathfrak{g}$ of Proposition 18.8 is the differential of a homomorphism $i_\alpha : \mathrm{SU}(2) \longrightarrow G$. By Proposition 18.9, $\mathfrak{g}_\alpha$ centralizes $\mathfrak{t}_\alpha$, and since $\mathrm{SU}(2)$ is connected, it follows that $i_\alpha(\mathrm{SU}(2)) \subseteq C(T_\alpha)^\circ$.

By Proposition 18.4, $-iH_\alpha \notin \mathfrak{t}_\alpha$, so $\mathfrak{t}$ is generated by its codimension-one subspace $\mathfrak{t}_\alpha$ and $i_\alpha(\mathfrak{su}(2)) \cap \mathfrak{t}$. Since $\mathrm{Lie}(T_\alpha) = \mathfrak{t}_\alpha$, it follows that T is generated by T_α and $T \cap i_\alpha(\mathrm{SU}(2))$. By construction, w_α normalizes

$$T \cap i_\alpha(\mathrm{SU}(2)) = i_\alpha \left\{ \begin{pmatrix} y & \\ & y^{-1} \end{pmatrix} \;\middle|\; y \in \mathbb{C}, |y| = 1 \right\},$$

and since $i_\alpha(\mathrm{SU}(2)) \subseteq C(T_\alpha)^\circ$, w_α also normalizes T_α.

Since we chose a W-invariant inner product, any element of the Weyl group acts by a Euclidean motion. Since w_α centralizes T_α, it acts trivially on $\mathfrak{t}_\alpha$ and thus fixes a codimension-one subspace in $\mathcal{V}$. It also maps $\alpha \longrightarrow -\alpha$, and these two properties characterize s_α. □

Proposition 18.10. *Let (π, V) be a finite-dimensional representation of G, and let $\lambda \in X^*(T)$ such that $V(\lambda) \neq 0$. Then $2 \langle \lambda, \alpha \rangle / \langle \alpha, \alpha \rangle \in \mathbb{Z}$ for all $\alpha \in \Phi$.*

Proof. Let

$$W = \bigoplus_{k \in \mathbb{Z}} V(\lambda + k\alpha).$$

By Proposition 18.4, this subspace is stable under $d\pi(X_\alpha)$ and $d\pi(X_{-\alpha})$. It is therefore invariant under the Lie algebra $\mathfrak{g}_{\alpha,\mathbb{C}}$ that they generate and its subalgebra $\mathfrak{g}_\alpha$. Thus, it is invariant under $i_\alpha(\mathrm{SU}(2))$, in particular by w_α in Theorem 18.1. Thus, $w_\alpha V(\lambda) = V(\lambda + k\alpha)$ for some $k \in \mathbb{Z}$ and by (18.1) we have $k = -2\langle \lambda, \alpha \rangle / \langle \alpha, \alpha \rangle$. That proves that this is an integer. □

Theorem 18.2. *If Φ is the set of roots associated with a compact Lie group and its maximal torus T, then Φ is a reduced root system.*

Proof. Clearly, Φ is a set of nonzero vectors in a Euclidean space $\mathcal{V}$. The fact that Φ is invariant under s_α, $\alpha \in \Phi$ follows from the construction of $w_\alpha \in N(T)$, the conjugation of which induces s_α in Theorem 18.1. The fact that the integers $2\langle\beta,\alpha\rangle / \langle\alpha,\alpha\rangle \in \mathbb{Z}$ for $\alpha,\beta \in \Phi$ follows from applying Proposition 18.10 to $(\mathrm{Ad}, \mathfrak{g}_\mathbb{C})$. Thus Φ is a root system. It is reduced by Proposition 18.6. $\qquad\square$

Proposition 18.11. *Let $\lambda \in X^*(T)$. Then there exists a finite-dimensional complex representation (π, V) of G such that $V(\lambda) \neq 0$.*

Proof. Consider the subspace $L(\lambda)$ of $L^2(G)$ of functions f satisfying

$$f(tg) = \lambda(t)f(g)$$

for $t \in T$. Let G act on $L(\lambda)$ by right translation: $\rho : G \longrightarrow \mathrm{End}(V)$ is the map $\rho(g)f(x) = f(xg)$. Clearly, $L(\lambda)$ is an invariant subspace under this action, and by Theorem 4.3 it decomposes into a direct sum of finite-dimensional irreducible invariant subspaces. Let V be one of these subspaces, and let π be the representation of G on V. Every linear functional on V has the form $x \longrightarrow \langle x, f_0\rangle$, where f_0 is a vector and $\langle\,,\,\rangle$ is the L^2 inner product. Thus, there exists an $f_0 \in V$ such that $f(1) = \langle f, f_0\rangle$ for all $f \in V$. Clearly, $f_0 \neq 0$. We have

$$\langle f, \pi(t)f_0\rangle = \langle\pi(t^{-1})f, f_0\rangle = \pi(t^{-1})f(1) = f(t^{-1}) = \lambda(t)^{-1}f(1) = \langle f, \lambda(t)f_0\rangle.$$

Therefore $\pi(t)f_0 = \lambda(t)f_0$ and so $V(\lambda) \neq 0$. $\qquad\square$

Proposition 18.12. *If H_α is as in Proposition 18.8 and $w_\alpha \in N(T)$ is as in Theorem 18.1, then $\mathrm{ad}(w_\alpha)H_\alpha = -H_\alpha$.*

Proof. Since w_α lies in $i_\alpha(\mathrm{SU}(2))$, and since by Proposition 18.8 the element $-iH_\alpha$ lies in the image of the Lie algebra of $\mathrm{SU}(2)$ under the differential of i_α, we may work in $\mathrm{SU}(2)$ to confirm this. The result follows from (18.11) and (18.13). $\qquad\square$

We now check the identity (18.4).

Proposition 18.13. *Let $\lambda \in \mathcal{V}$ and $\alpha \in \Phi$. Then $\mathrm{d}\lambda(H_\alpha) = \alpha^\vee(\lambda)$.*

(See Remark 18.1 about the notation $\mathrm{d}\lambda$.)

Proof. First let us show that λ and α are orthogonal if and only if $\mathrm{d}\lambda(H_\alpha) = 0$ with H_α as in Proposition 18.4. It is sufficient to show that the orthogonal complement of α is contained in the kernel of this functional since both are subspaces of codimension 1. Assuming therefore that α and λ are orthogonal, $s_\alpha(\lambda) = \lambda$, and since the action of W on $X^*(T)$ and $\mathcal{V} = \mathbb{R}\otimes X^*(T)$ is induced by the action of W on T by conjugation, whose differential is the action of W on $\mathfrak{t}$ via Ad, we have

$$d\lambda(H_\alpha) = d\lambda(\mathrm{Ad}(w_\alpha)H_\alpha) = -d\lambda(H_\alpha)$$

by Proposition 18.12.

The result is now proved in the case where λ and α are orthogonal. Therefore $d\lambda(H_\alpha) = c\alpha^\vee(\lambda)$ for some constant c. To show that $c = 1$, we take $\lambda = \alpha$ and (remembering that $iH_\alpha \in \mathfrak{t}$) check that $d\alpha(iH_\alpha) = 2i$. Indeed we have

$$d\alpha\, i_\alpha \begin{pmatrix} i & \\ & -i \end{pmatrix} = \frac{d}{dt}\alpha\left(i_\alpha \begin{pmatrix} e^{it} & \\ & e^{-it} \end{pmatrix}\right)\bigg|_{t=0} = \frac{d}{dt}e^{2it}\big|_{t=0} = 2i.$$

$\square$

We recall that if α is a root of G, then $T_\alpha \subset T$ is the kernel of α. An element of is called *regular* if it is contained in a unique maximal torus. Otherwise, it is called *singular*.

Proposition 18.14.

(i) $\bigcap_{\alpha \in \Phi} T_\alpha$ is the center $Z(G)$.
(ii) $\bigcup_{\alpha \in \Phi} T_\alpha$ is the set of singular elements of T.

Of course, $T_\alpha = T_{-\alpha}$, so we could equally well write $Z(G) = \bigcap_{\alpha \in \Phi^+} T_\alpha$.

Proof. For (i), any element of G is conjugate to an element of T. If it is in $Z(G)$, conjugation does not move it, so $Z(G) \subset T$. Now G is generated by T together with the subgroups $i_\alpha(\mathrm{SU}(2))$ as α runs through the roots of G because the Lie algebras of these groups generate the Lie algebra of $\mathfrak{g}$, and G is connected. Hence $x \in T$ is in $Z(G)$ if and only if it commutes with each of these subgroups. From the construction of the groups $i_\alpha(\mathrm{SU}(2))$, this is true if and only if x is in the kernel of the representation induced by Ad on the two-dimensional T-invariant subspace $\mathfrak{X}_\alpha \oplus \mathfrak{X}_{-\alpha}$. This kernel is T_α, for every root α. Thus, the center of G is the intersection of the T_α.

For (ii), suppose that T and T' are distinct maximal tori containing t. Then both are contained in the connected centralizer $C(t)^\circ$, and so by Theorem 16.5 applied to this connected Lie group, they are conjugate in $C(t)^\circ$. The complexified Lie algebra of $C(t)^\circ$ must contain $\mathfrak{X}_\alpha$ for some α since otherwise $C(t)^\circ$ would be a compact connected Lie group with no roots and hence a torus, contradicting the assumption that $T \neq T'$. Thus, $t \in T_\alpha$. Conversely, if $t \in T_\alpha$, it is contained in every maximal torus in $C(T_\alpha)^\circ$, which is non-Abelian, so there are more than one of these. $\square$

Theorem 18.3. *The Weyl group $W = N(T)/T$ is generated by the w_α with $\alpha \in \Phi$.*

Proof. Arguing by contradiction, choose $w \in N(T)/T$ that is not in the subgroup generated by the w_α. If $\alpha \in \Phi$ let $\mathfrak{t}_\alpha$ be the Lie algebra of the group T_α which is the kernel of α. They are hyperplanes in $\mathfrak{t}$, the kernels of the linear functionals $d\alpha$. Let us partition $\mathfrak{t}$ into open chambers which are the complements of the $\mathfrak{t}_\alpha$. Let $\mathfrak{C}$ be one of these. Choose the counterexample w to minimize the number of hyperplanes $\mathfrak{t}_\alpha$ separating the chambers $\mathfrak{C}$ and $w\mathfrak{C}$. Since w_α reflects in the hyperplane $\mathfrak{t}_\alpha$, we must have $w(\mathfrak{C}) = \mathfrak{C}$. We will argue that $w = 1$, which will be a contradiction. Let $n \in N(T)$ represent w. What we need to show is that $n \in T$.

Since w has finite order and maps $\mathfrak{C}$ to itself, and since $\mathfrak{C}$ is convex, we may find an element H of $\mathfrak{C}$ such that $w(H) = H$; simply averaging any element over its orbit under powers of w will produce such an H. Since H does not lie in any of the $\mathfrak{t}_\alpha$, the one-parameter subgroup $S = \{\exp(tH) \,|\, t \in \mathbb{R}\} \subset T$ contains regular elements. Since $\mathrm{Ad}(n)$ fixes H, n is in the centralizer $C_G(S)$. We claim that $C_G(S) = T$. First note that if $g \in C_G(S)$ then gTg^{-1} contains regular elements of T, so $gTg^{-1} = T$. Thus $C_G(S) \subset N_G(T)$. But $C_G(S)$ is connected by Theorem 16.6, so $n \in C_G(S) \subseteq N_G(T)^\circ = T$ by Proposition 15.8. Therefore $n \in T$, as required. $\square$

Proposition 18.15. *Suppose that $\alpha \in \Phi$. Let $\beta \neq \pm\alpha$ be another root. Let*

$$W = \bigoplus_{\substack{k \,\in\, \mathbb{Z} \\ \beta + k\alpha \,\in\, \Phi}} \mathfrak{X}_{\beta+k\alpha}. \tag{18.14}$$

Then W is an irreducible module for $i_\alpha(\mathfrak{sl}(2,\mathbb{C}))$ in the adjoint representation.

Proof. Denote $\mathfrak{g}_\alpha = i_\alpha(\mathfrak{sl}(2,\mathbb{C}))$. First we note that W is an $\mathfrak{sl}(2,\mathbb{C})$-module, since by Proposition 18.4 it is closed under the Lie bracket with X_α and $X_{-\alpha}$ (which generate $\mathfrak{g}_\alpha$). Therefore, it is a module for $i_\alpha(\mathfrak{sl}(2,\mathbb{C}))$.

We must show that it is irreducible. Let $T_{\mathrm{SU}(2)}$ be the maximal torus

$$T_{\mathrm{SU}(2)} = \left\{ \begin{pmatrix} t & \\ & t^{-1} \end{pmatrix} |t \in \mathbb{T} \right\}$$

of $\mathrm{SU}(2)$. The inclusion $i_\alpha : T_{\mathrm{SU}(2)} \longrightarrow T$ induces a homomorphism $X^*(T) \longrightarrow X^*(T_{\mathrm{SU}(2)})$. The image of α is the positive root α' of $\mathrm{SU}(2)$, and the image of β is a weight β'. In the notation (18.5) we have $\alpha' = \lambda_2$ and $\beta' = \lambda_m$ for some m. All of the weights $\beta' + k\alpha'$ of $i_\alpha(\mathrm{SU}(2))$, or of its complexified Lie algebra $i_\alpha(\mathfrak{sl}(2,\mathbb{C}))$ in W are of the form λ_{m+2k} where the indices $m + 2k$ have the same parity as m. Thus, decomposing into irreducibles, W is a direct sum of modules $\vee^{n_i}\mathbb{C}^2$ $(i = 1, 2, \ldots)$ where the n_i have the same parity as m. If there is more than one of these, then without loss of generality we may assume that $n_1 \geqslant n_2$, in which case λ_{n_2} occurs as a weight of both $\vee^{n_1}\mathbb{C}^2$ and $\vee^{n_2}\mathbb{C}^2$, so λ_{n_2} occurs with multiplicity two in W. Writing $n_2 = m + 2k$, this means that $\mathfrak{X}_{\beta+k\alpha}$ is more than one-dimensional, contradicting Proposition 18.6. Therefore W is irreducible. $\square$

Corollary 18.1. *Suppose that α, β and $\alpha+\beta \in \Phi$. Let X_α and X_β be nonzero elements of $\mathfrak{X}_\alpha$ and $\mathfrak{X}_\beta$. Then $[X_\alpha, X_\beta]$ is a nonzero element of $\mathfrak{X}_{\alpha+\beta}$.*

Proof. We may identify the decomposition (18.14) with the irreducible module described in (12.1). Now X_α is $i_\alpha(R)$ in the notation of that Proposition. Since $\alpha + \beta$ is a root, X_β is v_{k-2l} with $l > 0$, and the nonvanishing of $[X_\alpha, X_\beta] = \mathrm{ad}(X_\alpha)X_\beta$ follows from (12.3).   □

Exercises

Exercise 18.1.

(i) Let Φ be a root system in a Euclidean space $\mathcal{V}$ and for $\alpha \in \Phi$ let

$$\alpha^\vee = \frac{2\alpha}{\langle \alpha, \alpha \rangle}.$$

Show that the $\alpha^\vee$ also form a root system. Note that long roots in Φ correspond to short vectors in $\Phi^\vee$. (**Hint**: Prove this first for rank two root systems, then note that if $\alpha, \beta \in \Phi$ are linearly independent roots the intersection of Φ with their span is a rank two root system.)

(ii) Explain why this implies that the H_α form a root system in *it*.

Exercise 18.2. Analyze the root system of SO(5) similarly to the case of Sp(4) in the text. It may be helpful to use Exercise 7.3.

19

Examples of Root Systems

It may be easiest to read the next chapter with examples in mind. In this chapter we will describe various root systems and in particular we will illustrate the rank 2 root systems. Since the purpose of this chapter is to give examples, we will state various facts here without proof. The proofs will come in later chapters.

A root system Φ in a Euclidean space $\mathcal{V}$ is called *reducible* if we can decompose $\mathcal{V} = \mathcal{V}_1 \oplus \mathcal{V}_2$ into orthogonal subspaces with $\Phi = \Phi_1 \cup \Phi_2$, with both $\Phi_i = \mathcal{V}_i \cap \Phi$ nonempty. Then the Φ_i are themselves smaller root systems. In classifying root systems, one may clearly restrict to irreducible root systems, and these were classified by Killing and Cartan. The irreducible root systems are classified by a *Cartan type* which can be one of the *classical Cartan types* A_r $(r \geqslant 1)$, B_r $(r \geqslant 2)$, C_r $(r \geqslant 2)$ D_r $(r \geqslant 4)$, or one of the five *exceptional types* G_2, F_4, E_6, E_7 and E_8. The subscript is the (semisimple) rank of the corresponding Lie groups. We have an accidental isomorphism $B_2 \cong C_2$. The Cartan types D_2 and D_3 are usually excluded, but it may be helpful to consider D_2 as a synonym for the reducible Cartan type $A_1 \times A_1$ (that is, A_1 is the Cartan type of both Φ_1 and Φ_2 in the orthogonal decomposition); and D_3 as a synonym for A_3.

In the last chapter we saw how to associate a root system Φ with a compact Lie group G. The Euclidean $\mathcal{V}$ containing Φ is $\mathbb{R} \otimes X^*(T)$ where T is a maximal torus. The group G is called *semisimple* if the root system Φ spans $\mathcal{V} = \mathbb{R} \otimes \Lambda$ where $\Lambda = X^*(T)$ is the group of rational characters of a maximal torus T. We will denote by $\mathfrak{g}$ the Lie algebra of G and other notations will be as in Chap. 18.

Within each Cartan type there may be several Lie groups to consider, but in each case there is a unique semisimple simply connected group. There is also a unique simple semisimple group, which is isomorphic to the simply connected group modulo its finite center. This is called the *adjoint group* since it is isomorphic to its image in $\mathrm{GL}(\mathfrak{g})$ under the adjoint representation. Here is a table giving the simply connected and adjoint groups for each of the classical Cartan types.

D. Bump, *Lie Groups*, Graduate Texts in Mathematics 225,
DOI 10.1007/978-1-4614-8024-2_19, © Springer Science+Business Media New York 2013

Cartan type	Simply connected G	Adjoint group	Other common instance
A_r	$\mathrm{SU}(r+1)$	$\mathrm{U}(r+1)/\text{center}$	$\mathrm{U}(r+1)$ (not semisimple)
B_r	$\mathrm{Spin}(2r+1)$	$\mathrm{SO}(2r+1)$	
C_r	$\mathrm{Sp}(2r)$	$\mathrm{Sp}(2r)/\text{center}$	
D_r	$\mathrm{Spin}(2r)$	$\mathrm{SO}(2r)/\{\pm I\}$	$\mathrm{SO}(2r)$

Let us consider first the Cartan type A_r. We will describe three distinct groups, $\mathrm{U}(r+1)$, $\mathrm{SU}(r+1)$ and $\mathrm{PU}(r+1)$, which is $U(r+1)$ modulo its one-dimensional center. These have the same root system, but the ambient vector space $\mathcal{V}$ is different in each case.

The group $\mathrm{U}(r+1)$ is not semisimple. Its rank is $r+1$ but its semisimple rank is r. The maximal torus T consists of diagonal matrices t with eigenvalues $t_1, \ldots, t_{r+1}$ in $\mathbb{T}$, the group of complex numbers of absolute value 1. We may identify $\Lambda = X^*(T) \cong \mathbb{Z}^{r+1}$, in which $\lambda = (\lambda_1, \ldots, \lambda_{r+1})$ with $\lambda_i \in \mathbb{Z}$ represents the character $t \mapsto \prod t_i^{\lambda_i}$. So $\mathcal{V} = \mathbb{R} \otimes X^*(T)$ may be identified with $\mathbb{R}^{r+1}$ with the usual Euclidean inner product. Let $\mathbf{e}_i = (0, \ldots, 0, 1, 0, \ldots, 0)$ be the standard basis of $\mathbb{R}^{r+1}$. The root system consists of the $r(r-1)$ vectors

$$\mathbf{e}_i - \mathbf{e}_j, \qquad i \neq j, \tag{19.1}$$

having exactly two nonzero entries, one being 1 and the other -1. To see that this is the root system, we recall that the complexified Lie algebra is $\mathbb{C} \otimes \mathfrak{g} \cong \mathfrak{gl}_{r+1}(\mathbb{C}) = \mathrm{Mat}_{r+1}(\mathbb{C})$, since $\mathrm{Mat}_{r+1}(\mathbb{C}) = \mathfrak{g} \oplus i\mathfrak{g}$. (Every complex matrix can be written uniquely as $X + iY$ with X and Y skew-Hermitian.) If $\alpha = \mathbf{e}_i - \mathbf{e}_j$, the one-dimensional vector space $\mathfrak{X}_\alpha$ of $\mathrm{Mat}_{r+1}(\mathbb{C})$ spanned by the matrix E_{ij} with a 1 in the i, j-position and 0's everywhere else is an eigenspace for T affording the character α, and these eigenspaces, together with the Lie algebra of T, span V. So the $\mathbf{e}_i - \mathbf{e}_j$ are precisely the roots of $\mathrm{U}(r+1)$. The group $\mathrm{U}(r+1)$ has semisimple rank r, since that is the dimension of the space spanned by these vectors.

Next consider the group $\mathrm{SU}(r+1)$. This is the semisimple and simply connected group with the same root system. The ambient space $\mathcal{V}$ is one dimension smaller than for $\mathrm{U}(r+1)$, because the t_i are subject to the equation $\prod t_i = \det(t) = 1$. Therefore, the character represented by $\lambda \in \mathbb{Z}^{r+1}$ is trivial if λ is in the diagonal lattice $\Delta = \mathbb{Z}(1, \ldots, 1)$. Thus, for this group, the weight lattice, which we will denote $\Lambda_{\mathrm{SU}(r+1)}$, is $\mathbb{Z}^{r+1}/\Delta$, and the space $\mathcal{V}$ is r-dimensional. It is spanned by the roots, so this group is semisimple.

The group $\mathrm{PU}(r+1)$ is $\mathrm{U}(r+1)$ modulo its one-dimensional central torus. It is the adjoint group for the Cartan type A_r. It is isomorphic to $\mathrm{SU}(r+1)$ modulo its finite center of order $r+1$. A character of $\mathrm{U}(r+1)$ parametrized by $\lambda \in \mathbb{Z}^{r+1}$ is well-defined if and only if it is trivial on the center of $\mathrm{U}(r+1)$, which requires $\sum \lambda_i = 0$. So the lattice $\Lambda_{\mathrm{PU}(r+1)}$ is isomorphic to the sublattice of $\mathbb{Z}^{r+1}$ determined by this condition. The composition

$$\Lambda_{\mathrm{PU}(r+1)} \longrightarrow \mathbb{Z}^{r+1} \longrightarrow \mathbb{Z}^{r+1}/\Delta = \Lambda_{\mathrm{SU}(r+1)}$$

where the first map is the inclusion and the second the projection is injective, so we may regard $\Lambda_{\mathrm{PU}(r+1)}$ as a sublattice of $\Lambda_{\mathrm{SU}(r+1)}$. Its index is $r + 1$.

Turning now to the general case, the set Φ of roots will be partitioned into two parts, called Φ^+ and Φ^-. Exactly half the roots will be in Φ^+ and the other half in Φ^-. This is accomplished by choosing a hyperplane through the origin in $\mathcal{V}$ that does not pass through any root, and taking Φ^+ to be the roots on one side of the hyperplane, Φ^- the roots on the other side. Although the choice of the hyperplane is arbitrary, if another such decomposition $\Phi = \Phi_1^+ \cup \Phi_1^-$ is found by choosing a different hyperplane, a Weyl group element w can be found such that $w(\Phi^+) = \Phi_1^+$ and $w(\Phi^-) = \Phi_1^-$, so the procedure is not as arbitrary as one might think. The roots in Φ^+ will be called *positive*. In the figures of this chapter, the positive roots are labeled $\bullet$, and the negative roots are labeled $\circ$.

If G is a semisimple compact connected Lie group, then its universal cover $\tilde{G}$ is a cover of finite degree, and as in the last example (where $G = \mathrm{PU}(r+1)$ and $\tilde{G} = \mathrm{SU}(r + 1)$) the weight lattice of G is a sublattice of $\tilde{G}$. Moreover, if $\pi : G \longrightarrow \mathrm{GL}(V)$ is an irreducible representation, then we may compose it with the canonical map $\tilde{G} \longrightarrow G$ and get a representation of $\tilde{G}$. So if we understand the representation theory of $\tilde{G}$ we understand the representation theory of G. For this reason, we will consider mainly the case where G is simply connected and semisimple in the remaining examples of this chapter.

Assuming G is semisimple, so the α_i span $\mathcal{V}$, we will define certain special elements of $\mathcal{V}$ as follows. If $\Sigma = \{\alpha_1, \ldots, \alpha_r\}$ are the simple positive roots then let $\{\alpha_1^\vee \ldots, \alpha_r^\vee\}$ be the corresponding coroots. In the semisimple case, the coroots span $\mathcal{V}^*$, and the *fundamental dominant weights* ϖ_i are the dual basis of $\mathcal{V}$. Thus,

$$\alpha_j^\vee(\varpi_i) = \delta_{ij} \qquad \text{(Kronecker } \delta\text{)}.$$

We will show later that if G is simply connected, then the ϖ_i are in the weight lattice $\Lambda = X^*(T)$, though if G is not simply connected, they may not all be in Λ.

Another important particular vector is ρ, sometimes called the *Weyl vector*. It may be characterized as half the sum of the positive roots, and in the semisimple case it may also be characterized as the sum of the fundamental weights. (See Proposition 20.17.)

For example, the root system of type A_2, pictured in Fig. 19.1, consists of

$$\alpha_1 = (1, -1, 0), \qquad \alpha_2 = (0, 1, -1), \qquad (1, 0, -1),$$
$$(-1, 1, 0), \qquad\qquad (0, -1, 1), \qquad (-1, 0, 1).$$

With $G = \mathrm{SU}(3)$, we really mean the images of these vectors in $\mathbb{Z}^3/\Delta$, as explained above. Taking T to be the diagonal torus of $\mathrm{SU}(3)$, α_1 and $\alpha_2 \in X^*(T)$ are the roots

$$\alpha_1(t) = t_1 t_2^{-1}, \qquad \alpha_2(t) = t_2 t_3^{-1}, \qquad t = \begin{pmatrix} t_1 & & \\ & t_2 & \\ & & t_3 \end{pmatrix} \in T.$$

The corresponding eigenspaces are spanned by

$$E_{12} = \begin{pmatrix} 0 & 1 & 0 \\ 0 & 0 & 0 \\ 0 & 0 & 0 \end{pmatrix} \in \mathfrak{X}_{\alpha_1}, \qquad E_{23} = \begin{pmatrix} 0 & 0 & 0 \\ 0 & 0 & 1 \\ 0 & 0 & 0 \end{pmatrix} \in \mathfrak{X}_{\alpha_2}.$$

The fundamental dominant weights ϖ_1 and ϖ_2 are, respectively, $\varpi_1(t) = t_1$ and $\varpi_2(t) = t_3^{-1}$. Let $v_0 = (1,1,1)$, so in our previous notation $\Delta = \mathbb{Z}v_0$. The vector space $\mathcal{V}$ is $\mathbb{R}^3 / \mathbb{R}v_0$, but we may identify this with the codimension one vector subspace of $\mathbb{R}^3$ consisting of (x_1, x_2, x_3) with $\sum x_i = 0$. The fundamental weights are represented by the cosets in $\mathbb{Z}^3 / \mathbb{Z}v_0$ of the vectors $(1,0,0)$ and $(1,1,0)$, or in the subspace of codimension one in $\mathbb{R}^3$ consisting of vectors (x_0, x_1, x_2) satisfying $\sum_i x_i = 0$ by $(\frac{2}{3}, -\frac{1}{3}, -\frac{1}{3})$ and $(\frac{1}{3}, \frac{1}{3}, -\frac{2}{3})$, respectively.

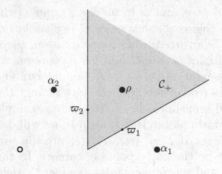

Fig. 19.1. The root system of type A_2

Figure 19.1 shows the root system of type A_2 associated with the Lie group SU(3). The shaded region in Fig. 19.1 is the *positive Weyl chamber* $\mathcal{C}_+$, which consists of $\{x \in \mathcal{V} \mid \langle x, \alpha \rangle \geqslant 0 \text{ for all } \alpha \in \Phi^+\}$. It is a fundamental domain for the Weyl group.

A role will also be played by a *partial order* on $\mathcal{V}$. We define $x \succcurlyeq y$ if $x - y \succcurlyeq 0$, where $x \succcurlyeq 0$ if x is a linear combination, with nonnegative coefficients, of the elements of Σ. The shaded region in Fig. 19.2 is the set of x such that $x \succcurlyeq 0$ for the root system of type A_2.

Next we turn to the remaining classical root systems. The root system of type B_n is associated with the odd orthogonal group SO$(2n+1)$ or with its double cover spin$(2n+1)$. The root system of type C_n is associated with the symplectic group Sp$(2n)$. Finally, the root system of type D_n is associated with the even orthogonal group SO$(2n)$ or its double cover spin$(2n)$. We will now describe these root systems. Let $\mathbf{e}_i = (0, \ldots, 0, 1, 0, \ldots, 0)$ be the standard basis of $\mathbb{R}^n$.

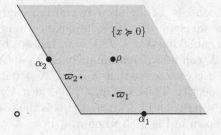

Fig. 19.2. The partial order

The root system of type B_n can be embedded in $\mathbb{R}^n$. The roots are not all of the same length. There are $2n$ short roots

$$\pm\mathbf{e}_i \qquad (1 \leqslant i \leqslant n)$$

and $2(n^2 - n)$ long roots

$$\pm\mathbf{e}_i \pm \mathbf{e}_j \qquad (i \neq j).$$

The simple positive roots are

$$\alpha_1 = \mathbf{e}_1 - \mathbf{e}_2, \quad \alpha_2 = \mathbf{e}_2 - \mathbf{e}_3, \quad \ldots \quad \alpha_{n-1} = \mathbf{e}_{n-1} - \mathbf{e}_n, \quad \alpha_n = \mathbf{e}_n.$$

To see that this is the root system of $\mathrm{SO}(2n+1)$, it is most convenient to use the representation of $\mathrm{SO}(2n+1)$ in Exercise 5.3. Thus, we replace the usual realization of $\mathrm{SO}(2n+1)$ as a group of real matrices by the subgroup of all $g \in \mathrm{U}(2n+1)$ that satisfy $g\,J\,{}^t g = J$, where

$$J = \begin{pmatrix} & & 1 \\ & \cdot^{\cdot^{\cdot}} & \\ 1 & & \end{pmatrix}.$$

A maximal torus consists of all diagonal elements, which have the form (when $n = 4$, for example)

$$t = \begin{pmatrix} t_1 \\ & t_2 \\ & & t_3 \\ & & & t_4 \\ & & & & 1 \\ & & & & & t_4^{-1} \\ & & & & & & t_3^{-1} \\ & & & & & & & t_2^{-1} \\ & & & & & & & & t_1^{-1} \end{pmatrix}.$$

The Lie algebra $\mathfrak{g}$ consists of all skew-Hermitian matrices X satisfying $X J + J{}^t X = 0$. Now we claim that the complexification of $\mathfrak{g}$ just consists of *all* complex matrices satisfying $X J + J{}^t X = 0$. Indeed, by Proposition 11.4, any complex matrix X can be written uniquely as $X_1 + i X_2$ with X_1 and X_2 skew-Hermitian, and it is easy to see that $X J + J{}^t X = 0$ if and only if X_1 and X_2 satisfy the same identity. Thus, $\mathfrak{g} \oplus i\mathfrak{g} = \{ X \in \mathfrak{gl}(n, \mathbb{C}) \,|\, X J + J{}^t X = 0 \}$. It now follows from Proposition 11.3(iii) that this is the complexification of $\mathfrak{g}$. This Lie algebra is shown in Fig. 19.3 when $n = 4$.

$$\begin{pmatrix}
t_1 & x_{12} & x_{13} & x_{14} & x_{15} & x_{16} & x_{17} & x_{18} & 0 \\
x_{21} & t_2 & x_{23} & x_{24} & x_{25} & x_{26} & x_{27} & 0 & -x_{18} \\
x_{31} & x_{32} & t_3 & x_{34} & x_{35} & x_{36} & 0 & -x_{27} & -x_{17} \\
x_{41} & x_{42} & x_{43} & t_4 & x_{45} & 0 & -x_{36} & -x_{26} & -x_{16} \\
x_{51} & x_{52} & x_{53} & x_{54} & 0 & -x_{45} & -x_{35} & -x_{25} & -x_{15} \\
x_{61} & x_{62} & x_{63} & 0 & -x_{54} & -t_4 & -x_{34} & -x_{24} & -x_{14} \\
x_{71} & x_{72} & 0 & -x_{63} & -x_{53} & -x_{43} & -t_3 & -x_{23} & -x_{13} \\
x_{81} & 0 & -x_{72} & -x_{62} & -x_{52} & -x_{42} & -x_{32} & -t_2 & -x_{12} \\
0 & -x_{81} & -x_{71} & -x_{61} & -x_{51} & -x_{41} & -x_{31} & -x_{21} & -t_1
\end{pmatrix}$$

Fig. 19.3. The Lie algebra $\mathfrak{so}(9)$. The Dynkin diagram, which will be explained in Chap. 25, has been superimposed on top of the Lie algebra

We order the roots so the root spaces $\mathfrak{X}_\alpha$ with $\alpha \in \Phi^+$ are upper triangular. In particular, the simple roots are $\alpha_1(t) = t_1 t_2^{-1}$, acting on $\mathfrak{X}_{\alpha_1}$, the space of matrices in which all entries are zero except x_{12}; $\alpha_2(t) = t_2 t_3^{-1}$, with root space corresponding to x_{23}; $\alpha_3(t) = t_3 t_4^{-1}$ corresponding to x_{34}; and $\alpha_4(t) = t_4$, corresponding to x_{45}. We have circled these positions. Note, however that (for example) x_{12} appears in a second place which has not been circled. The lines connecting the circles, one of them double, map out the Dynkin diagram, which will explained in greater detail Chap. 25. Briefly, the Dynkin diagram is a graph whose vertices correspond to the simple roots; simple roots are connected in the Dynkin diagram if they are not perpendicular. We have drawn the nodes corresponding to each simple root on top of the variable x_{ij} for the corresponding eigenspace.

We have drawn a double bond with an arrow pointing from a long root to a short root, which is the convention when two nonadjacent roots have different lengths.

If we take $\mathbf{e}_i \in X^*(T)$ to be the character $\mathbf{e}_i(t) = t_i$, then it is clear that the root system consists of the $2n^2$ roots $\pm\mathbf{e}_i$ and $\pm\mathbf{e}_i \pm e_j$ $(i \neq j)$, as claimed.

The root system of type C_n is similar, but the long and short roots are reversed. Now there are $2n$ long roots

$$\pm 2\mathbf{e}_i \qquad (1 \leqslant i \leqslant n)$$

and $2(n^2 - n)$ short roots

$$\pm\mathbf{e}_i \pm \mathbf{e}_j \qquad (i \neq j).$$

The simple positive roots are

$$\alpha_1 = \mathbf{e}_1 - \mathbf{e}_2, \quad \alpha_2 = \mathbf{e}_2 - \mathbf{e}_3, \quad \ldots \quad \alpha_{n-1} = \mathbf{e}_{n-1} - \mathbf{e}_n, \quad \alpha_n = 2\mathbf{e}_n.$$

We leave it to the reader to show that C_n is the root system of $\mathrm{Sp}(2n)$ in Exercise 19.2. (Fig. 30.15 may help with this.)

The root system of type D_n consists of just the long roots in the root system of type B_n. There are $2(n^2 - n)$ roots, all of the same length:

$$\pm\mathbf{e}_i \pm \mathbf{e}_j \qquad (i \neq j).$$

The simple positive roots are

$$\alpha_1 = \mathbf{e}_1 - \mathbf{e}_2, \quad \alpha_2 = \mathbf{e}_2 - \mathbf{e}_3, \quad \ldots \quad \alpha_{n-1} = \mathbf{e}_{n-1} - \mathbf{e}_n, \quad \alpha_n = \mathbf{e}_{n-1} + \mathbf{e}_n.$$

To see that D_n is the root system of $\mathrm{SO}(2n)$, one may again use the realization of Exercise 5.3. We leave this verification to the reader in Exercise 19.2. (Fig. 30.1 may help with this.)

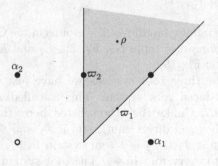

Fig. 19.4. The root system of type C_2, which coincides with type B_2

It happens that $\mathrm{spin}(5) \cong \mathrm{Sp}(4)$, so the root systems of types B_2 and C_2 coincide. These are shown in Fig. 19.4. The shaded region is the positive

Weyl chamber. (We have labeled the roots so that the order coincides with the root system C_2 in the notations of Bourbaki [23], in the appendix at the back of the book. For type B_2, the roots α_1 and α_2 would be switched.)

There is a nonreduced root system whose type is called BC_n. The root system of type BC_n can be realized as all elements of the form

$$\pm\mathbf{e}_i \pm \mathbf{e}_j (i < j), \qquad \pm\mathbf{e}_i, \qquad \pm 2\mathbf{e}_i,$$

where $\mathbf{e}_i$ are standard basis vectors of $\mathbb{R}^n$. Nonreduced root systems do not occur as root systems of compact Lie groups, but they occur as relative root systems (Chap. 29). The root system of type BC_2 may be found in Fig. 19.5.

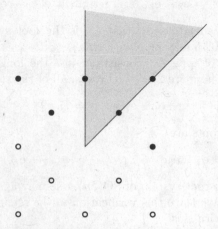

Fig. 19.5. The nonreduced root system BC_2

In addition to the infinite families of Lie groups in the Cartan classification are five *exceptional groups*, of types G_2, F_4, E_6, E_7 and E_8. The root system of type G_2 is shown in Fig. 19.6.

In addition to the three root systems we have just considered there is another rank two reduced root system. This is called $A_1 \times A_1$, and it is illustrated in Fig. 19.7. Unlike the others listed here, this one is *reducible*. If $\mathcal{V} = \mathcal{V}_1 \oplus \mathcal{V}_2$ (orthogonal direct sum), and if Φ_1 and Φ_2 are root systems in $\mathcal{V}_1$ and $\mathcal{V}_2$, then $\Phi = \Phi_1 \cup \Phi_2$ is a root system in $\mathcal{V}$ such that every root in Φ_1 is orthogonal to every root in Φ_2. The root system Φ is reducible if it decomposes in this way.

We leave two other rank 2 root systems, which are neither reduced nor irreducible, to the imagination of the reader. Their types are $A_1 \times BC_1$ and $BC_1 \times BC_1$.

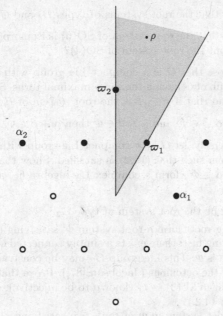

Fig. 19.6. The root system of type G_2

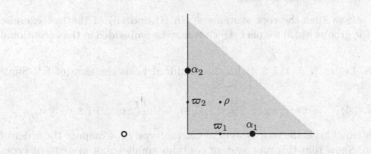

Fig. 19.7. The reducible root system $A_1 \times A_1$

Exercises

Exercise 19.1. Show that any irreducible rank 2 root system is isomorphic to one of those described in this chapter, of type A_2, B_2, G_2 or BC_2.

Exercise 19.2. Verify, as we did for type $SO(2n+1)$, that the root system of the Lie group $SO(2n)$ is of type D_n and that the root system of $Sp(2n)$ is of type C_n.

Exercise 19.3. Show that the root systems of types B_2 and C_2 are isomorphic.

Exercise 19.4. Show that the root system of SO(6) is isomorphic to that of SU(4). What can you say about the root system of SO(4)?

Exercise 19.5. Suppose that G is a compact Lie group with root system Φ, and that H is a Lie subgroup of G having the same maximal torus. Show that every root of H is a root of G, and that if $\Phi' \subseteq \Phi$ is the root system of H, then

$$\text{If } \alpha, \beta \in \Phi' \text{ and } \alpha + \beta \in \Phi \text{ then } \alpha + \beta \in \Phi'. \tag{19.2}$$

Exercise 19.6. Conversely, let G be a compact Lie group with root system Φ. Let $\Phi' \subseteq \Phi$ be a root system such that (19.2) is satisfied. Show that in the notation of Chap. 18, $\mathfrak{t}_\mathbb{C}$ and $\mathfrak{X}_\alpha$ ($\alpha \in \Phi'$) form a complex Lie algebra $\mathfrak{h}_\mathbb{C}$, and that $\mathfrak{h} = \mathfrak{h}_\mathbb{C}$ is a Lie subalgebra of $\mathfrak{g}$.

Exercise 19.7. Let Φ be the root system of type G_2.

(i) Show that the long roots form a root system Φ' satisfying (19.2).
(ii) Assume the following fact: there exists a simply connected compact Lie group G whose root system is Φ. This Lie group G_2 may be constructed as the group of automorphisms of the octonions (Jacobson [87]). Prove that there exists a non-trivial homomorphism SU(3) $\rightarrow G$ (known to be injective). (**Hint**: Use Exercise 19.6 and Theorem 14.2.)
(ii) Exhibit another root system in Φ of rank two satisfying (19.2). Note that you cannot use the short roots for this.

It may be shown that the root systems Φ' in (i) and (ii) of the last exercise correspond to Lie groups [$\mathfrak{su}(3)$ for part (i)] that may be embedded in the exceptional group G_2.

Exercise 19.8. Let $\mathbf{e}_i$ ($i = 1, 2, 3, 4$) be the standard basis elements of $\mathbb{R}^4$. Show that the 48 vectors

$$\pm \mathbf{e}_i \ (1 \leqslant i \leqslant 4), \qquad \pm \mathbf{e}_i \pm \mathbf{e}_j \ (1 \leqslant i < j \leqslant 4), \qquad \frac{1}{2}(\pm \mathbf{e}_1 \pm \mathbf{e}_2 \pm \mathbf{e}_3 \pm \mathbf{e}_4),$$

form a root system. This is the root system of Cartan type F_4. Compute the order of the Weyl group. Show that this root system contains smaller root systems of types B_3 and C_3.

Exercise 19.9. Let Φ_8 consist of the following vectors in $\mathbb{R}^8$. First, the 112 vectors

$$\pm \mathbf{e}_i \pm \mathbf{e}_j \qquad 1 \leqslant i < j \leqslant 8. \tag{19.3}$$

Second, the 128 vectors

$$\frac{1}{2}(\pm \mathbf{e}_1 \pm \mathbf{e}_2 \pm \mathbf{e}_3 \pm \mathbf{e}_4 \pm \mathbf{e}_5 \pm \mathbf{e}_6 \pm \mathbf{e}_7 \pm \mathbf{e}_8) \tag{19.4}$$

where the number of $-$ signs is even. We will refer to the vectors (19.3) as *integral roots* and the vectors (19.4) as *half-integral roots*. Prove that Φ_8 is a root system. This is the exceptional root system of type E_8. Note that the integral roots form a root system of type D_8.

Hint: To show that if α and β are roots then $s_\alpha(\beta) \in \Phi_8$, observe that the D_8 Weyl group permutes the roots, and using this action we may assume that $\alpha = \mathbf{e}_1 + \mathbf{e}_2$ or $\alpha = \frac{1}{2}\sum \mathbf{e}_i$. The first case is easy so assume $\alpha = \frac{1}{2}\sum \mathbf{e}_i$. We may then use the action of the symmetric group on β and there are only a few cases to check.

Exercise 19.10. Let Φ_7 consist of the vectors in Φ_8 that are orthogonal to $e_7 + e_8$. Show that Φ_7 is a root system containing 126 roots. This is the exceptional root system of type E_7.

Exercise 19.11. Let Φ_6 consist of the vectors in Φ_7 that are orthogonal to $e_6 - e_7$. Show that Φ_6 is a root system containing 72 roots. This is the exceptional root system of type E_6.

Abstract Weyl Groups

In this chapter, we will associate a Weyl group with an abstract root system, and develop some of its properties.

Let V be a Euclidean space and $\Phi \subset V$ a reduced root system. (At the end of the chapter we will remove the assumption that Φ is reduced, but many of the results of this chapter are false without it.)

Since Φ is a finite set of nonzero vectors, we may choose $\rho_0 \in V$ such that $\langle \alpha, \rho_0 \rangle \neq 0$ for all $\alpha \in \Phi$. Let Φ^+ be the set of roots α such that $\langle \alpha, \rho_0 \rangle > 0$. This consists of exactly half the roots since evidently a root $\alpha \in \Phi^+$ if and only if $-\alpha \notin \Phi^+$. Elements of Φ^+ are called *positive roots*. Elements of set $\Phi^- = \Phi - \Phi^+$ are called *negative roots*.

If α, $\beta \in \Phi^+$ and $\alpha + \beta \in \Phi$, then evidently $\alpha + \beta \in \Phi^+$. Let Σ be the set of elements in Φ^+ that cannot be expressed as a sum of other elements of Φ^+. If $\alpha \in \Sigma$, then we call α a *simple positive root*, or sometimes just a *simple root* and we call s_α defined by (18.1) a *simple reflection*.

Proposition 20.1.

(i) *The elements of Σ are linearly independent.*

(ii) *If $\alpha \in \Sigma$ and $\beta \in \Phi^+$, then either $\beta = \alpha$ or $s_\alpha(\beta) \in \Phi^+$.*

(iii) *If α and β are distinct elements of Σ, then $\langle \alpha, \beta \rangle \leqslant 0$.*

(iv) *Each element $\alpha \in \Phi$ can be expressed uniquely as a linear combination*

$$\alpha = \sum_{\beta \in \Sigma} n_\beta \cdot \beta$$

in which each $n_\beta \in \mathbb{Z}$ and either all $n_\beta \geqslant 0$ (if $\beta \in \Phi^+$) or all $n_\beta \leqslant 0$ (if $\beta \in \Phi^-$).

Proof. Let Σ' be a subset of Φ^+ that is minimal with respect to the property that every element of Φ^+ is a linear combination with nonnegative coefficients of elements of Σ'. (Subsets with this property clearly exist—e.g., Σ' itself.) We will eventually show that $\Sigma' = \Sigma$.

D. Bump, *Lie Groups*, Graduate Texts in Mathematics 225,
DOI 10.1007/978-1-4614-8024-2_20, © Springer Science+Business Media New York 2013

First, we show that if $\alpha \in \Sigma'$ and $\beta \in \Phi^+$, then either $\beta = \alpha$ or $s_\alpha(\beta) \in \Phi^+$. Otherwise $-s_\alpha(\beta) \in \Phi^+$, and

$$\alpha^\vee(\beta)\,\alpha = \beta + \big(-s_\alpha(\beta) \big)$$

is a sum of two positive roots β and $-s_\beta(\alpha)$. Therefore, we have

$$\beta = \sum_{\gamma \in \Sigma'} n'_\gamma \cdot \gamma, \qquad -s_\alpha(\beta) = \sum_{\gamma \in \Sigma'} n''_\gamma \cdot \gamma,$$

where $n'_\gamma, n''_\gamma \geqslant 0$ and so

$$\alpha^\vee(\beta)\,\alpha = \sum_{\gamma \in \Sigma'} n_\gamma \cdot \gamma, \qquad n_\gamma \geqslant 0,$$

where $n_\gamma = n'_\gamma + n''_\gamma$. There exists some $\gamma \in \Sigma'$ such that $n'_\gamma \neq 0$, because β is not α, and (the root system being reduced), it follows that β is not a multiple of α. Therefore,

$$(\alpha^\vee(\beta) - n_\alpha)\,\alpha = \sum_{\substack{\gamma \in \Sigma' \\ \gamma \neq \alpha}} n_\gamma \cdot \gamma,$$

and the right-hand side is not zero. Taking the inner product with ρ_0 shows that the coefficient on the left-hand side is strictly positive; dividing by this positive constant, we see that α may be expressed as a linear combination of the elements $\gamma \in \Sigma'$ distinct from α, and so α may be omitted from Σ', contradicting its assumed minimality. This contradiction shows that $s_\alpha(\beta) \in \Phi^+$.

Next we show that if α and β are distinct elements of Σ', then $\langle \alpha, \beta \rangle \leqslant 0$. We have already shown that $s_\alpha(\beta) \in \Phi^+$. If $\langle \alpha, \beta \rangle > 0$, then by (18.2) we have $\alpha^\vee(\beta) > 0$. Write

$$\beta = s_\alpha(\beta) + \alpha^\vee(\beta)\,\alpha. \tag{20.1}$$

We have already shown that $s_\alpha(\beta) \in \Phi^+$. Writing $s_\alpha(\beta)$ as a linear combination with nonnegative coefficients of the elements of Σ', and noting that the coefficient of α on the right-hand side of (20.1) is strictly positive, we may write

$$\beta = \sum_{\gamma \in \Sigma'} n_\gamma \cdot \gamma,$$

where $n_\alpha > 0$. We rewrite this

$$(1 - n_\beta) \cdot \beta = \sum_{\substack{\gamma \in \Sigma' \\ \gamma \neq \beta}} n_\gamma \cdot \gamma.$$

At least one coefficient $n_\alpha > 0$ on the right, so taking the inner product with ρ_0 we see that $1 - n_\beta > 0$. Thus, β is a linear combination with nonnegative coefficients of other elements of Σ' and hence may be omitted, contradicting the minimality of Σ'.

Now let us show that the elements of Σ' are $\mathbb{R}$-linearly independent. In a relation of algebraic dependence, we move all the negative coefficients to the other side of the identity and obtain a relation of the form

$$\sum_{\alpha \in \Sigma_1} c_\alpha \cdot \alpha = \sum_{\beta \in \Sigma_2} d_\beta \cdot \beta, \tag{20.2}$$

where Σ_1 and Σ_2 are disjoint subsets of Σ' and the coefficients c_α, d_β are all positive. Call this vector v. We have

$$\langle v, v \rangle = \sum_{\substack{\alpha \in \Sigma_1 \\ \beta \in \Sigma_2}} c_\alpha d_\beta \langle \alpha, \beta \rangle \leqslant 0$$

since we have already shown that the inner products $\langle \alpha, \beta \rangle \leqslant 0$. Therefore, $v = 0$. Now taking the inner product of the left-hand side in (20.2) with ρ_0 gives

$$0 = \sum_{\alpha \in \Sigma_1} c_\alpha \langle \alpha, \rho_0 \rangle.$$

Since $\langle \alpha, \rho_0 \rangle > 0$ and $c_\alpha > 0$, this is a contradiction. This proves the linear independence of the elements of Σ'.

Next let us show that every element of Φ^+ may be expressed as a linear combination of elements of Σ' with *integer* coefficients. We define a function h from Φ^+ to the positive real numbers as follows. If $\alpha \in \Phi^+$ we may write

$$\alpha = \sum_{\beta \in \Sigma'} n_\beta \cdot \beta, \qquad n_\beta \geqslant 0.$$

The coefficients n_β are uniquely determined since the elements of Σ' are linearly independent. We define

$$h(\alpha) = \sum n_\beta. \tag{20.3}$$

Evidently $h(\alpha) > 0$. We want to show that the coefficients n_β are integers. Assume a counterexample with $h(\alpha)$ minimal. Evidently, $\alpha \notin \Sigma'$ since if $\alpha \in \Sigma'$, then $n_\alpha = 1$ while all other $n_\beta = 0$, so such an α has all $n_\beta \in \mathbb{Z}$. Since

$$0 < \langle \alpha, \alpha \rangle = \sum_{\beta \in \Sigma'} n_\beta \langle \alpha, \beta \rangle, \tag{20.4}$$

it is impossible that $\langle \alpha, \beta \rangle \leqslant 0$ for all $\beta \in \Sigma'$. Thus, there exists $\gamma \in \Sigma'$ such that $\langle \alpha, \gamma \rangle > 0$. Then by what we have already proved, $\alpha' = s_\gamma(\alpha) \in \Phi^+$, and by (18.1) we see that

$$\alpha' = \sum_{\beta \in \Sigma'} n'_\beta \cdot \beta,$$

where

$$n'_\beta = \begin{cases} n_\beta & \text{if } \beta \neq \gamma, \\ n_\gamma - \gamma^\vee(\alpha) & \text{if } \beta = \gamma. \end{cases}$$

Since $\langle \gamma, \alpha \rangle > 0$, we have

$$h(\alpha') < h(\alpha),$$

so by induction we have $n'_\beta \in \mathbb{Z}$. Since Φ is a root system, $\gamma^\vee(\alpha) \in \mathbb{Z}$, so $n_\beta \in \mathbb{Z}$ for all $\beta \in \Sigma'$. This is a contradiction.

Finally, let us show that $\Sigma = \Sigma'$.

If $\alpha \in \Sigma$, then by definition of Σ, α cannot be expressed as a linear combination with integer coefficients of other elements of Φ^+. Hence α cannot be omitted from Σ'. Thus, $\Sigma \subset \Sigma'$.

On the other hand, if $\alpha \in \Sigma'$, then we claim that $\alpha \in \Sigma$. Otherwise, we may write $\alpha = \beta + \gamma$ with $\beta, \gamma \in \Phi^+$, and β and γ may both be written as linear combinations of elements of Σ' with positive integer coefficients, and thus $h(\beta), h(\gamma) \geqslant 1$, so $h(\alpha) = h(\beta) + h(\gamma) > 1$. But evidently $h(\alpha) = 1$ since $\alpha \in \Sigma'$. This contradiction shows that $\Sigma' \subset \Sigma$.   □

Let W be the group generated by the simple reflections s_α with $\alpha \in \Sigma$. If $w \in W$, let the length $l(w)$ be defined to be the smallest k such that w admits a factorization $w = s_1 \cdots s_k$ into simple reflections, or $l(w) = 0$ if $w = 1$. Let $l'(w)$ be the number of $\alpha \in \Phi^+$ such that $w(\alpha) \in \Phi^-$. We will eventually show that the functions l and l' are the same.

Proposition 20.2. *Let $s = s_\alpha$ ($\alpha \in \Sigma$) be a simple reflection, and let $w \in W$. We have*

$$l'(sw) = \begin{cases} l'(w) + 1 & \text{if } w^{-1}(\alpha) \in \Phi^+, \\ l'(w) - 1 & \text{if } w^{-1}(\alpha) \in \Phi^-, \end{cases} \tag{20.5}$$

and

$$l'(ws) = \begin{cases} l'(w) + 1 & \text{if } w(\alpha) \in \Phi^+, \\ l'(w) - 1 & \text{if } w(\alpha) \in \Phi^-, \end{cases} \tag{20.6}$$

Proof. Since $s(\Phi^-)$ is obtained from Φ^- by deleting $-\alpha$ and adding α, we see that $(sw)^{-1}\Phi^- = w^{-1}(s\Phi^-)$ is obtained from $w^{-1}\Phi^-$ by deleting $-w^{-1}(\alpha)$ and adding $w^{-1}(\alpha)$. Since $l'(w)$ is the cardinality of $\Phi^+ \cap w^{-1}\Phi^-$, we obtain (20.5). To prove (20.6), we note that $l'(ws)$ is the cardinality of $\Phi^+ \cap (ws)^{-1}\Phi^-$, which equals the cardinality of $s(\Phi^+ \cap (ws)^{-1}\Phi^-) = s\Phi^+ \cap w^{-1}\Phi^-$, and since $s\Phi^+$ is obtained from Φ^+ by deleting the element α and adjoining $-\alpha$, (20.6) is evident.   □

If w is any orthogonal linear endomorphism of $\mathcal{V}$, then evidently $ws_\alpha w^{-1}$ is the reflection in the hyperplane perpendicular to $w(\alpha)$, so

$$ws_\alpha w^{-1} = s_{w(\alpha)}. \tag{20.7}$$

Proposition 20.3. *Suppose that $\alpha_1, \ldots, \alpha_k$ and α are elements of Σ, and let $s_i = s_{\alpha_i}$. Suppose that*

$$s_1 s_2 \cdots s_k(\alpha) \in \Phi^-.$$

Then there exists a $1 \leqslant j \leqslant k$ such that

$$s_1 s_2 \cdots s_k = s_1 s_2 \cdots \hat{s}_j \cdots s_k s_\alpha, \qquad (20.8)$$

where the "hat" on the right signifies the omission of the single element s_j.

Proof. Let $1 \leqslant j \leqslant k$ be minimal such that $s_{j+1} \cdots s_k(\alpha) \in \Phi^+$. Then $s_j s_{j+1} \cdots s_k(\alpha) \in \Phi^-$. Since α_j is the unique element of Φ^+ mapped into Φ^- by s_j, we have

$$s_{j+1} \cdots s_k(\alpha) = \alpha_j,$$

and by (20.7) we have

$$(s_{j+1} \cdots s_k) s_\alpha (s_{j+1} \cdots s_k)^{-1} = s_j$$

or

$$s_{j+1} \cdots s_k s_\alpha = s_j s_{j+1} \cdots s_k.$$

This implies (20.8). □

Proposition 20.4. *Suppose that $\alpha_1, \ldots, \alpha_k$ are elements of Σ, and let $s_i = s_{\alpha_i}$. Suppose that $l'(s_1 s_2 \cdots s_k) < k$. Then there exist $1 \leqslant i < j \leqslant k$ such that*

$$s_1 s_2 \cdots s_k = s_1 s_2 \cdots \hat{s}_i \cdots \hat{s}_j \cdots s_k, \qquad (20.9)$$

where the "hats" on the right signify omission of the elements s_i and s_j.

Proof. Evidently there is a first j such that $l'(s_1 s_2 \cdots s_j) < j$, and [since $l'(s_1) = 1$] we have $j > 1$. Then $l'(s_1 s_2 \cdots s_{j-1}) = j - 1$, and by Proposition 20.2 we have $s_1 s_2 \cdots s_{j-1}(\alpha_j) \in \Phi^-$. The existence of i satisfying $s_1 \cdots s_{j-1} = s_1 \cdots \hat{s}_i \cdots s_{j-1} s_j$ now follows from Proposition 20.3, which implies (20.9). □

Proposition 20.5. *If $w \in W$, then $l(w) = l'(w)$.*

Proof. The inequality

$$l'(w) \leqslant l(w)$$

follows from Proposition 20.2 because we may write $w = s w_1$, where s is a simple reflection and $l(w_1) = l(w) - 1$, and by induction on $l(w_1)$ we may assume that $l'(w_1) \leqslant l(w_1)$, so $l'(w) \leqslant l'(w_1) + 1 \leqslant l(w_1) + 1 = l(w)$.

Let us show that

$$l'(w) \geqslant l(w).$$

Indeed, let $w = s_1 \cdots s_k$ be a counterexample with $l(w) = k$, where each $s_i = s_{\alpha_i}$ with $\alpha_i \in \Sigma$. Thus, $l'(s_1 \cdots s_k) < k$. Then, by Proposition 20.4 there exist i and j such that

$$w = s_1 s_2 \cdots \hat{s}_i \cdots \hat{s}_j \cdots s_k.$$

This expression for w as a product of $k - 2$ simple reflections contradicts our assumption that $l(w) = k$.   □

Proposition 20.6. *If* $w(\Phi^+) = \Phi^+$, *then* $w = 1$.

Proof. If $w(\Phi^+) = \Phi^+$, then $l'(w) = 0$, so $l(w) = 0$, that is, $w = 1$.   □

Proposition 20.7. *If* $\alpha \in \Phi$, *there exists an element* $w \in W$ *such that* $w(\alpha) \in \Sigma$.

Proof. First, assume that $\alpha \in \Phi^+$. We will argue by induction on $h(\alpha)$, which is defined by (20.3). In view of Proposition 20.1(iv), we know that $h(\alpha)$ is a positive integer, and if $\alpha \notin \Sigma$ (which we may as well assume), then $h(\alpha) > 1$. As in the proof of Proposition 20.1, (20.4) implies that $\langle \alpha, \beta \rangle > 0$ for some $\beta \in \Sigma$, and then with $\alpha' = s_\beta(\alpha)$ we have $h(\alpha') < h(\alpha)$. On the other hand, $\alpha' \in \Phi^+$ since $\alpha \neq \beta$ by Proposition 20.1(ii). By our inductive hypothesis, $w'(\alpha') \in \Sigma$ for some $w' \in W$. Then $w(\alpha) = w'(\alpha')$ with $w = w' s_\beta \in W$. This shows that if $\alpha \in \Phi^+$, then there exists $w \in W$ such that $w(\alpha) \in \Sigma$.

If, on the other hand, $\alpha \in \Phi^-$, then $-\alpha \in \Phi^+$ so we may find $w_1 \in W$ such that $w_1(-\alpha) \in \Sigma$. Letting $w_1(-\alpha) = \beta$ we have $w(\alpha) = \beta$ with $w = s_\beta w_1$.

In both cases, $w(\alpha) \in \Sigma$ for some $w \in W$.   □

Proposition 20.8. *The group* W *contains* s_α *for each* $\alpha \in \Phi$.

Proof. Indeed, $w(\alpha) \in \Sigma$ for some $w \in W$, so $s_{w(\alpha)} \in W$ and s_α is conjugate in W to $s_{w(\alpha)}$ by (20.7). Therefore, $s_\alpha \in W$.   □

Proposition 20.9. *The group* W *is finite.*

Proof. By Proposition 20.6, $w \in W$ is determined by $w(\Phi^+) \subset \Phi$. Since Φ is finite, W is finite.   □

Proposition 20.10. *Suppose that* $w \in W$ *such that* $l(w) = k$. *Write* $w = s_1 \cdots s_k$, *where* $s_i = s_{\alpha_i}$, $\alpha_1, \ldots, \alpha_k \in \Sigma$. *Then*

$$\{\alpha \in \Phi^+ | w(\alpha) \in \Phi^-\} = \{\alpha_k, s_k(\alpha_{k-1}), s_k s_{k-1}(\alpha_{k-2}), \ldots, s_k s_{k-1} \cdots s_2(\alpha_1)\}.$$

Proof. By Proposition 20.5, the cardinality of $\{\alpha \in \Phi^+ | w(\alpha) \in \Phi^-\}$ is k, so the result will be established if we show that the described elements are distinct and in the set. Let $w = s_1 w_1$, where $w_1 = s_2 \cdots s_k$, so that $l(w_1) = l(w) - 1$. By induction, we have

$$\{\alpha \in \Phi^+ | w_1(\alpha) \in \Phi^-\} = \{\alpha_k, s_k(\alpha_{k-1}), s_k s_{k-1}(\alpha_{k-2}), \ldots, s_k s_{k-1} \cdots s_3(\alpha_2)\},$$

and the elements on the right are distinct. We claim that

$$\{\alpha \in \Phi^+ | w_1(\alpha) \in \Phi^-\} \subset \{\alpha \in \Phi^+ | s_1 w_1(\alpha) \in \Phi^-\}. \tag{20.10}$$

Otherwise, let $\alpha \in \Phi^+$ such that $w_1(\alpha) \in \Phi^-$, while $s_1 w_1(\alpha) \in \Phi^+$. Let $\beta = -w_1(\alpha)$. Then $\beta \in \Phi^+$, while $s_1(\beta) \in \Phi^-$. By Proposition 20.1(ii), this implies that $\beta = \alpha_1$. Therefore, $\alpha = -w_1^{-1}(\alpha_1)$. By Proposition 20.2, since $l(s_1 w_1) = k = l(w_1) + 1$, we have $-\alpha = w_1^{-1}(\alpha_1) \in \Phi^+$. This contradiction proves (20.10).

We will be done if we show that the last remaining element $s_k \cdots s_2(\alpha_1)$ is in $\{\alpha \in \Phi^+ | s_1 w_1(\alpha) \in \Phi^-\}$ but not $\{\alpha \in \Phi^+ | w_1(\alpha) \in \Phi^-\}$ since that will guarantee that it is distinct from the other elements listed. This is clear since if $\alpha = s_k \cdots s_2(\alpha_1)$ we have $w_1(\alpha) = \alpha_1 \notin \Phi^-$, while $s_1 w_1(\alpha) = -\alpha_1 \in \Phi^-$. $\qquad\square$

A connected component of the complement of the union of the hyperplanes

$$\{x \in \mathcal{V} \mid \langle x, \alpha \rangle = 0 \text{ for all } \alpha \in \Phi\}$$

is called an *open Weyl chamber*. The closure of an open Weyl chamber is called a *Weyl chamber*. For example, $\mathcal{C}_+ = \{x \in \mathcal{V} \mid \langle x, \alpha \rangle \geqslant 0 \text{ for all } \alpha \in \Sigma\}$ is called the *positive Weyl chamber*. Since every element of Φ^+ is a linear combination of elements of $\mathcal{C}$ with positive coefficients, $\mathcal{C}_+ = \{x \in \mathcal{V} \mid \langle x, \alpha \rangle \geqslant 0 \text{ for all } \alpha \in \Phi^+\}$. The interior

$$\mathcal{C}_+^\circ = \{x \in \mathcal{V} \mid \langle x, \alpha \rangle > 0 \text{ for all } \alpha \in \Sigma\} = \{x \in \mathcal{V} \mid \langle x, \alpha \rangle > 0 \text{ for all } \alpha \in \Phi^+\}$$

is an open Weyl chamber.

If $y \in \mathcal{V}$, let $W(y)$ be the stabilizer $\{w \in W | w(y) = y\}$.

Proposition 20.11. *Suppose that $w \in W$ such that $l(w) = k$. Write $w = s_1 \cdots s_k$, where $s_i = s_{\alpha_i}$, $\alpha_1, \ldots, \alpha_k \in \Sigma$. Assume that $x \in \mathcal{C}_+$ such that $wx \in \mathcal{C}_+$ also.*

(i) We have $\langle x, \alpha_i \rangle = 0$ for $1 \leqslant i \leqslant k$.
(ii) Each $s_i \in W(x)$.
(iii) We have $w(x) = x$.

Proof. If $\alpha \in \Phi^+$ and $w\alpha \in \Phi^-$, then we have $\langle x, \alpha \rangle = 0$. Indeed, $\langle x, \alpha \rangle \geqslant 0$ since $\alpha \in \Phi^+$ and $x \in \mathcal{C}_+$, and $\langle x, \alpha \rangle = \langle wx, w\alpha \rangle \leqslant 0$ since $wx \in \mathcal{C}_+$ and $w\alpha \in \Phi^-$.

The elements of $\{\alpha \in \Phi^+ | w\alpha \in \Phi^-\}$ are listed in Proposition 20.10. Since α_k is in this set, we have $s_k(x) = x - (2 \langle x, \alpha_k \rangle / \langle \alpha_k, \alpha_k \rangle) \alpha_k = x$. Thus, $s_k \in W(x)$. Now since $s_k(\alpha_{k-1}) \in \{\alpha \in \Phi^+ | w\alpha \in \Phi^-\}$, we have $0 = \langle x, s_k(\alpha_{k-1}) \rangle = \langle s_k(x), \alpha_{k-1} \rangle = \langle x, \alpha_{k-1} \rangle$, which implies $s_{k-1}(x) = x - 2 \langle x, \alpha_{k-1} \rangle / \langle \alpha_{k-1}, \alpha_{k-1} \rangle = x$. Proceeding in this way, we prove (i) and (ii) simultaneously. Of course, (ii) implies (iii). $\qquad\square$

Theorem 20.1. *The positive Weyl chamber $\mathcal{C}_+$ is a fundamental domain for the action of W on $\mathcal{V}$. More precisely, let $x \in \mathcal{V}$.*

(i) There exists $w \in W$ such that $w(x) \in \mathcal{C}_+$.

(ii) If $w, w' \in W$ and $w(x) \in \mathcal{C}_+, w'(x) \in \mathcal{C}_+^\circ$, then $w = w'$.
(iii) If $w, w' \in W$ and $w(x) \in \mathcal{C}_+, w'(x) \in \mathcal{C}_+$, then $w(x) = w'(x)$.

Proof. Let $w \in W$ be chosen so that the cardinality of

$$S = \{\alpha \in \Phi^+ \mid \langle w(x), \alpha \rangle < 0\}$$

is as small as possible. We claim that S is empty. If not, then there exists an element of $\beta \in \Sigma \cap S$. We have $\langle w(x), -\beta \rangle > 0$, and since s_β preserves Φ^+ except for β, which it maps to $-\beta$, the set

$$S' = \{\alpha \in \Phi^+ \mid \langle w(x), s_\beta(\alpha) \rangle < 0\}$$

is smaller than S by one. Since $S' = \{\alpha \in \Phi^+ \mid \langle s_\beta w(x), \alpha \rangle < 0\}$ this contradicts the minimality of $|S|$. Clearly, $w(x) \in \mathcal{C}_+$. This proves (i).

We prove (ii). We may assume that $w' = 1$, so $x \in \mathcal{C}_+^\circ$. Since $\langle x, \alpha \rangle > 0$ for all $\alpha \in \Phi^+$, we have $\Phi^+ = \{\alpha \in \Phi \mid \langle x, \alpha \rangle > 0\} = \{\alpha \in \Phi \mid \langle x, \alpha \rangle \geqslant 0\}$. Since $w'(x) \in \mathcal{C}_+$, if $\alpha \in \Phi^+$, we have $\langle w^{-1}(\alpha), x \rangle = \langle \alpha, w(x) \rangle \geqslant 0$ so $w^{-1}(\alpha) \in \Phi^+$. By Proposition 20.6, this implies that $w^{-1} = 1$, whence (ii).

Part (iii) follows from Proposition 20.11(iii).   □

Proposition 20.12. *The function $w \longmapsto (-1)^{l(w)} \in \{\pm 1\}$ is a character of W. If $\alpha \in \Phi$, then $(-1)^{l(s_\alpha)} = -1$.*

Proof. If $l(w) = k$ and $l(w') = k'$, write $w = s_1 \cdots s_k$ and $w' = s_1' \cdots s_{k'}'$ as products of simple reflections. It follows from Proposition 20.4 that we may obtain a decomposition of ww' into a product of simple reflections of minimal length from $ww' = s_1 \cdots s_k s_1' \cdots s_{k'}'$ by discarding elements in pairs until the result is reduced. Therefore, $l(ww') \equiv l(w) + l(w')$ modulo 2, so $w \longmapsto (-1)^{l(w)}$ is a character. (One may argue alternatively by showing that $(-1)^{l(w)}$ is the determinant of w in its action on $\mathcal{V}$.)

If $\alpha \in \Phi$, then by Proposition 20.7 there exists $w \in W$ such that $w(\alpha) \in \Sigma$. By (20.7), we have $w s_\alpha w^{-1} = s_{w(\alpha)}$, and $l(s_{w(\alpha)}) = 1$. It follows that $(-1)^{s_\alpha} = -1$.   □

Proposition 20.13. *Let $\tilde{w}$ be a linear transformation of $\mathcal{V}$ that maps Φ to itself. Then there exists $w \in W$ such that $\tilde{w}(\mathcal{C}_+) = w\mathcal{C}_+$. The transformation $w^{-1}\tilde{w}$ of $\mathcal{V}$ permutes the elements of Φ^+ and of Σ.*

It is possible that $w^{-1}\tilde{w}$ is not the identity. (See Exercise 25.2.)

Proof. It is sufficient to show that $w^{-1}\tilde{w}(\mathcal{C}_+^\circ) = \mathcal{C}_+^\circ$. Let $x \in \mathcal{C}_+^\circ$. Since the open Weyl chambers are defined to be the connected components of the complement of the set of hyperplanes perpendicular to the roots, and since $\tilde{w}$ permutes the roots, $\tilde{w}(\mathcal{C}_+^\circ)$ is an open Weyl chamber. By Theorem 20.1 there is an element $w \in W$ such that $w^{-1}\tilde{w}(x) \in \mathcal{C}_+$, and $w^{-1}\tilde{w}(x)$ must be in the interior $\mathcal{C}_+^\circ$ since x lies in an open Weyl chamber, and these are permuted by W as well as by

$\tilde{w}$. Now $w^{-1}\tilde{w}(\mathcal{C}^\circ_+)$ and $\mathcal{C}^\circ_+$ are open Weyl chambers intersecting nontrivially in x, so they are equal.

The positive roots are characterized by the condition that $\alpha \in \Phi^+$ if and only if $\langle \alpha, x \rangle > 0$ for $x \in \mathcal{C}^\circ_+$. It follows that $w^{-1}\tilde{w}$ permutes the elements of Φ^+. Since the Σ are determined by Φ^+, these too are permuted by $w^{-1}\tilde{w}$. $\square$

Proposition 20.14. *If C is any Weyl chamber then there is a unique element w of W such that $C = wC_+$. In particular, let w_0 be the unique element such that $-C_+ = w_0 C$. Then $w_0 \Phi^+ = \Phi^-$ and w_0 is the longest element of W.*

The element w_0 is often called the *long element* of the Weyl group.

Proof. It is clear that W permutes the Weyl chambers transitively. The uniqueness of w of W such that $C = wC_+$ follows from Theorem 20.1.

Regarding w_0, since

$$\mathcal{C}_+ = \{x | \{\alpha, x\} \text{ for } \alpha \in \Phi^+\},$$

the element w_0 such that $wC_+ = -C_+$ sends positive roots to negative roots. Thus, its length equals the number of positive roots, and is maximal. $\square$

An important particular element of $\mathcal{V}$ is the *Weyl vector*

$$\rho = \frac{1}{2} \sum_{\alpha \in \Phi^+} \alpha.$$

Proposition 20.15. *If α is a simple root, then*

$$s_\alpha(\rho) = \rho - \alpha, \qquad \alpha \in \Sigma. \tag{20.11}$$

Proof. This follows since s_α changes the sign of α and permutes the remaining positive roots. $\square$

Let there be given a lattice Λ contained in $\mathcal{V}$ that contains a basis of $\mathcal{V}$. Then $\mathcal{V}$ may be identified with $\mathbb{R} \otimes_{\mathbb{Z}} \Lambda$. We will assume that $\alpha^\vee(\Lambda) \subseteq \mathbb{Z}$, and that every root α is in Λ. For example if Φ is the root system of a compact Lie group G with maximal torus T as in Chap. 18, then by Proposition 18.10 we may take $\Lambda = X^*(T)$. Elements of Λ are to be called *weights*, and our assumptions are satisfies by Proposition 18.10. A weight λ is called *dominant* if $\lambda \in \mathcal{C}^+$. By Theorem 20.1, every weight is equivalent by the action of W to a unique dominant weight.

Proposition 20.16. *If $\lambda \in \Lambda$, then $\lambda - w(\lambda) \in \Lambda_{\text{root}}$.*

Proof. This is true if w is a simple reflection by (18.1). The general case follows, since if $w = s_1 \cdots s_r$, where the s_i are simple reflections, we may write $\lambda - w(\lambda) = (\lambda - s_r(\lambda)) + (s_r(\lambda) - s_{r-1}(s_r(\lambda)) + \cdots$. $\square$

Now let us assume that Φ spans $\mathcal{V}$. This will be true if G is semisimple. Let $\tilde{\Lambda}$ to be the set of vectors v such that $\alpha^{\vee}(v) \in \mathbb{Z}$ for $\alpha^{\vee} \in \Phi^{\vee}$. In the semisimple case the $\alpha^{\vee}$ span $\mathcal{V}^*$, $\tilde{\Lambda}$ is a lattice. We have $\tilde{\Lambda} \supseteq \Lambda \supseteq \Lambda_{\text{root}}$, and all three lattices span $\mathcal{V}$, so $[\tilde{\Lambda} : \Lambda_{\text{root}}] < \infty$. The $\alpha_i^{\vee}$ are linearly independent, and in the semisimple case they are a basis of $\mathcal{V}$, so let ϖ_i be the dual basis of $\mathcal{V}$. In other words, these vectors are defined by $\alpha_i^{\vee}(\varpi_j) = \delta_{ij}$ (Kronecker delta). The ϖ_i are called the *fundamental dominant weights*. Strictly speaking, because in our usage only elements of Λ will be called weights, the ϖ_i might not be weights by our conventions. However, we will call them the fundamental weights because this terminology is standard. Clearly the ϖ_i span $\tilde{\Lambda}$ as a $\mathbb{Z}$-module.

Proposition 20.17. *In the semisimple case $\rho = \varpi_1 + \cdots + \varpi_h$. In particular, ρ is a dominant weight. It lies in $\mathcal{C}_+^{\circ}$.*

Proof. Let $\alpha = \alpha_i \in \Sigma$. By (20.11), we have $\alpha^{\vee}(\rho)\,\alpha = \rho - s_{\alpha}(\rho) = \alpha$. Thus, $\alpha_i^{\vee}(\rho) = 1$ for each $\alpha_i \in \Sigma$. It follows that ρ is the sum of the fundamental dominant weights. Since $\langle \rho, \alpha_i \rangle > 0$, ρ lies in the interior of $\mathcal{C}_+$. □

Up until now we have assumed that Φ is a reduced root system, and much of the foregoing is false without this assumption. In Chap. 18, and indeed most of the book, the root systems are reduced, so this is enough for now. In Chap. 29, however, we will encounter *relative root systems*, which may *not* be reduced, so let us say a few words about them. If $\Phi \subset \mathcal{V}$ is not reduced, then we may still choose v_0 and partition Φ into positive and negative roots. We call a positive root *simple* if it cannot be expressed as a linear combination (with nonnegative coefficients) of other positive roots.

Proposition 20.18. *Let $(\Phi, \mathcal{V})$ be a root system that is not necessarily reduced. If α and $\lambda\alpha \in \Phi$ with $\lambda > 0$, then $\lambda = 1, 2$ or $\frac{1}{2}$. Partition Φ into positive and negative roots, and let Σ be the set of simple roots. The elements of Σ are linearly independent. Any positive root may be expressed as a linear combination of elements of Σ with nonnegative integer coefficients.*

Proof. If α and β are proportional roots, say $\beta = \lambda\alpha$, then $2\langle \beta, \alpha \rangle / \langle \alpha, \alpha \rangle \in \mathbb{Z}$ implies that 2λ is an integer and, by symmetry, so is $2\lambda^{-1}$. The first assertion is therefore clear. Let Ψ be the set of all roots that are *not* the double of another root. Then it is clear that Ψ is another root system with the same Weyl group as Φ. Let $\Psi^+ = \Phi^+ \cap \Psi$. With our definitions, the set Σ of simple positive roots of Ψ^+ is precisely the set of simple positive roots of Φ. They are linearly independent by Proposition 20.1. If $\alpha \in \Phi^+$, we need to know that α can be expressed as a linear combination, with integer coefficients, of the elements of Σ. If $\alpha \in \Psi$, this follows from Proposition 20.1, applied to Ψ. Otherwise, $\alpha/2 \in \Psi$, so $\alpha/2$ is a linear combination of the elements of Σ with integer coefficients, and therefore so is α. □

Exercises

Exercise 20.1. Suppose that S is any subset of Φ such that if $\alpha \in \Phi$, then either $\alpha \in S$ or $-\alpha \in S$. Assume further more that if $\alpha, \beta \in S$ and if $\alpha + \beta \in \Phi$ then $\alpha + \beta \in S$. Show that there exists $w \in W$ such that $w(S) \supseteq \Phi^+$. If either for every $\alpha \in \Phi$ either $\alpha \in S$ or $-\alpha \in W$ but never both, then w is unique.

Exercise 20.2. Generalize (20.11) by proving, for $w \in W$:

$$w(\rho) = \rho - \sum_{\substack{\alpha \in \Phi^+ \\ w^{-1}(\alpha) \in \Phi^-}} \alpha. \tag{20.12}$$

Highest Weight Vectors

If G is a compact connected Lie group, we will show in Chap. 22 that its irreducible representations are parametrized uniquely by their *highest weight vectors*. In this chapter, we will explain what this means and give some illustrative examples. This chapter is to some extent a continuation of the example Chap. 19. As in that chapter, we will make many assertions that will only be proved in later chapters, mostly Chap. 22.

We return to the figures in Chap. 19 (which the reader should review). Let T be a maximal torus in G, with $\Lambda = X^*(T)$ embedded as a lattice in the Euclidean space $\mathcal{V} = \mathbb{R} \otimes X^*(T)$. Let $\Lambda_{\text{root}} \subseteq \Lambda$ be the lattice spanned by the roots.

If G is semisimple, then Λ_{root} spans $\mathcal{V}$ and has finite codimension in Λ. In this case, the coroots also span $\mathcal{V}^*$, so we may ask for the dual basis of $\mathcal{V}$. These are elements called ϖ_i such that $\alpha_i^\vee(\varpi_j) = \delta_{ij}$. These are the *fundamental dominant weights*. They are not necessarily in Λ, however: they are in Λ if G is simply connected as well as semisimple. We only will call elements of $\mathcal{V}$ *weights* if they are in Λ, so if G is not connected, the term "fundamental dominant weight" is a misnomer. But if G is semisimple and simply connected, the ϖ_i are uniquely defined and span the weight lattice Λ. The fundamental dominant weights do not play a major role in the general theory but they give a convenient parametrization of Λ when G is semisimple, since then every element of Λ is of the form $\sum n_i \varpi_i$ with n_i nonnegative integers. (This is true even if G is not simply connected.) Since our examples will be semisimple, we will make use of the fundamental dominant weights.

Our first example is $G = \mathrm{SU}(3)$. The lattices Λ and its sublattice Λ_{root} (of index 3) are marked in Fig. 21.1. The positive Weyl chamber $\mathcal{C}^+$ is the shaded cone. It is a fundamental group for the Weyl group W, acting by simple reflections, which are the reflections in the two walls of $\mathcal{C}^+$. The weight lattice Λ is marked with light dots and the root sublattice with darker ones. In this case G is semisimple and simply connected, so the fundamental dominant weights ϖ_1 and ϖ_2 are defined and span the weight lattice. The root lattice is of codimension 3 in Λ.

D. Bump, *Lie Groups*, Graduate Texts in Mathematics 225,
DOI 10.1007/978-1-4614-8024-2_21, © Springer Science+Business Media New York 2013

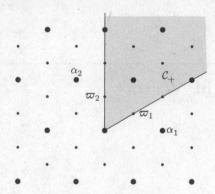

Fig. 21.1. The weight and root lattices for SU(2)

Let (π, V) be an irreducible complex representation of G. Then the restriction of π to T is a representation of T that will *not* be irreducible if π is not one-dimensional (since the irreducible representations of T are one-dimensional). It can be decomposed into a direct sum of one-dimensional irreducible subspaces of T corresponding to the characters of T. Some characters may occur with multiplicity greater than one. If $\mu \in X^*(T)$, let $m(\mu)$ be the multiplicity of μ in the decomposition of π over T. Thus, $m(\mu)$ is the dimension of $V(\mu) = \{v \in V | \pi(t)v = \lambda(t)v \text{ for } t \in T\}$. If $m(\lambda) \neq 0$, we say that λ is a *weight* of the representation π.

For example, let $G = \mathrm{SU}(3)$, and let T be the diagonal torus. Let $\varpi_1, \varpi_2 : T \longrightarrow \mathbb{C}$ be the fundamental dominant weights, labeled as in Chap. 19. They are the characters $\varpi_1(t) = t_1$ and $\varpi_2(t) = t_1 t_2 = t_3^{-1}$ where t_1, t_2, t_3 are the entries in the diagonal matrix t.

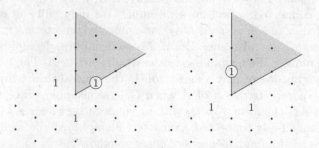

Fig. 21.2. Left: The standard representation; **Right**: its dual

The *standard representation* of SU(3) is just the usual embedding $\mathrm{SU}(3) \longrightarrow \mathrm{GL}(3, \mathbb{C})$. The three one-dimensional subspaces spanned by the standard basis vectors of $\mathbb{C}^3$ afford the characters ϖ_1, $-\varpi_1 + \varpi_2$, and $-\varpi_2$. These are the weights of the standard representation. Each occurs with multiplicity one. On the other hand, the contragredient of the standard representation is its composition with the transpose-inverse automorphism of $GL(3, \mathbb{C})$. The standard basis vectors in this dual representation afford the characters $-\varpi_1$, $\varpi_1 - \varpi_2$, and ϖ_2.

In Fig. 21.2 (left), we have labeled the three weights in the standard representation with their multiplicities. (For this example each multiplicity is one.) In Fig. 21.2 (right), we have labeled the three weights of the dual of the standard representation. Such a diagram, illustrating the weights of an irreducible representation, is called a *weight diagram*.

In each irreducible representation π, there is always a weight λ of π in the positive Weyl chamber such that if μ is another weight of π then $\lambda \succeq \mu$ in the partial order. This weight is called the *highest weight* of the representation. We always have $m(\lambda) = 1$, so $V(\lambda)$ is one-dimensional, and we call an element of $V(\lambda)$ a *highest weight vector*. We have circled the highest weight vectors of the standard representation and its dual in Fig. 21.2.

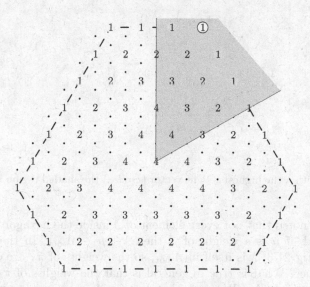

Fig. 21.3. The irreducible representation $\pi_{3\varpi_1 + 6\varpi_2}$ of SU(2). The *shaded region* is the positive Weyl chamber, and $\lambda = 3\varpi_1 + 6\omega_2$ is *circled*

The highest weight can be any element of $\Lambda \cap C^+$. In fact, there is a bijection between $\Lambda \cap C^+$ and the isomorphism classes of irreducible representations of G. Since there is a unique irreducible representation with the given highest weight λ, we will denote it by π_λ. For example, if $\lambda = 3\varpi_1 + 6\varpi_2$, the weight diagram of π_λ is shown in Fig. 21.3. Note that $\lambda = 3\varpi_1 + 6\varpi_2$ is marked with a circle.

From this we can see several features of the general situation. The set of weights of π_λ can be characterized as follows. First, if μ is a weight of π_λ then $\lambda \succeq \mu$ in the partial order. This puts μ in the translate by λ of the cone $\{\mu \preccurlyeq 0\}$. This is the shaded region in Fig. 21.4. Moreover since the set of weights is invariant under the Weyl group W, we can actually say that $\lambda \succcurlyeq w(\mu)$ for all $w \in W$. In Fig. 21.3, this puts μ in the hexagonal region that is the convex hull of the W-orbit $W\lambda = \{w(\lambda) \mid w \in W\}$. This region is marked with dashed lines.

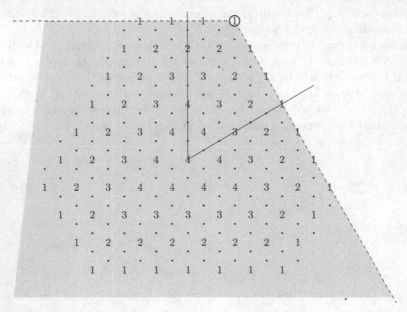

Fig. 21.4. With λ the highest weight vector (*circled*) the shaded region is $\{\mu \mid \mu \preccurlyeq \lambda\}$

It will be noted that not every element of Λ inside the hexagon is a weight of π_λ. Indeed, if μ is a weight of Λ then $\lambda - \mu \in \Lambda_{\text{root}}$. In the particular example of Fig. 21.3, λ is itself in Λ_{root}, so the weights of π_λ are elements of the root lattice. What is true in general is that the weights of π_λ are the μ inside the convex hull of $W\lambda$ such that $\lambda - \mu \in \Lambda_{\text{root}}$.

Next let $G = \text{Sp}(4)$. The root system is of type C_2. This group is also simply connected, so again the fundamental dominant weights ϖ_1 and ϖ_2 are in the weight lattice. The weight lattice and root lattice are illustrated in Fig. 21.5.

As in Fig. 21.1, the weight lattice Λ is marked with light dots and the root sublattice with darker ones. We have also marked the positive Weyl chamber, which is a fundamental group for the Weyl group W, acting by simple reflections.

The group $\text{Sp}(4)$ admits a homomorphism $\text{Sp}(4) \longrightarrow \text{SO}(5)$, so it has both a four-dimensional and a five-dimensional irreducible representation. These are π_{ϖ_1} and π_{ϖ_2}, respectively. Their root diagrams may be found in Fig. 21.6.

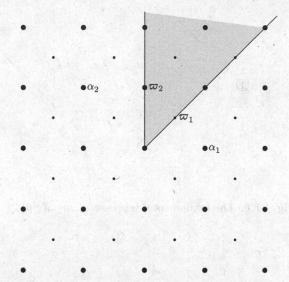

Fig. 21.5. The root and weight lattices of the C_2 root system

The weight diagram of the irreducible representation $\pi_{2\varpi_1+3\varpi_2}$ of Sp(4) is shown in Fig. 21.7.

If we considered SO(5), the group would not be simply connected, so we do not expect the fundamental weights to both lie in the root lattice. Let us see what happens. As explained in Chap. 19, the weight lattice is $\mathbb{Z}^2$ and the simple roots are $\mathbf{e}_1 - \mathbf{e}_2$ and $\mathbf{e}_2$, that is, $(1, -1)$ and $(0, 1)$. From this, the fundamental dominant weights are $\varpi_1 = (1, 0)$ and $\varpi_2 = (\frac{1}{2}, \frac{1}{2})$. The first is in the weight lattice but the second, being fractional, is not. So even though we call ϖ_2 a "fundamental dominant weight" it is *not* a weight of SO(5). It is, however, a weight of the universal covering group Spin(5). Indeed, ϖ_2 is the highest weight of a four-dimensional irreducible representation of Spin(5), the spin representation. Let t be an element of the maximal torus of Spin(5) that projects onto

$$\begin{pmatrix} t_1 & & & & \\ & t_2 & & & \\ & & 1 & & \\ & & & t_2^{-1} & \\ & & & & t_1^{-1} \end{pmatrix} \in \mathrm{SO}(5).$$

Then the four eigenvalues of t in the spin representation are $t_1^{\pm 1/2} t_2^{\pm 1/2}$, where the signs of the square roots depend on which of the two elements in the preimage of the above orthogonal matrix is chosen. The highest weight is $\sqrt{t_1 t_2}$, the character corresponding to $\varpi_2 = (\frac{1}{2}, \frac{1}{2})$.

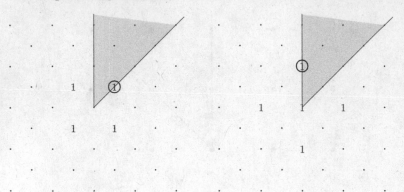

Fig. 21.6. The fundamental representations of Sp(4)

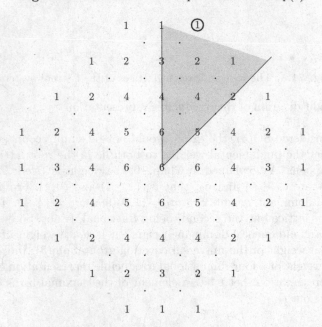

Fig. 21.7. The irreducible representation $\pi_{2\varpi_1 + 3\varpi_2}$ of Sp(4)

Exercises

Exercise 21.1. Consider the adjoint representation of SU(3) acting on the eight-dimensional Lie algebra $\mathfrak{g}$ of SU(3). (It may be shown to be irreducible.) Show that the highest weight vector is $\varpi_1 + \varpi_2$, and construct a weight diagram.

Exercise 21.2. Construct a weight diagram for the adjoint representation of $Sp(4)$ or, equivalently, $SO(5)$.

Exercise 21.3. Consider the symmetric square of the standard representation of $SU(3)$. This is an irreducible representation. Show that it has dimension six, and that its highest weight vector is $2\varpi_1$. Construct its weight diagram.

Exercise 21.4. Consider the tensor product of the contragredient of the standard representation of $SU(3)$, having highest weight vector ϖ_2, with the adjoint representation, having highest weight vector $\varpi_1 + \varpi_2$. We will see later in Exercise 22.4 that this tensor product has three irreducible constituents. They are the contragredient of the standard representation, the symmetric square of the standard representation, and another piece, which we will call $\pi_{\varpi_1 + 2\varpi_2}$. The first two pieces are known, and the third can be obtained by subtracting the two others. Accepting for now the validity of this decomposition, construct the weight diagram for the irreducible representation $\pi_{\varpi_1 + 2\varpi_2}$.

Exercise 21.5. The Lie group G_2 has an irreducible seven-dimensional representation. This information, together with the root system, described in Chap. 19, is enough to determine the weight diagram. Give the weight diagram for this representation, and for the 14-dimensional adjoint representation.

The Weyl Character Formula

The character formula of Weyl [174] is the gem of the representation theory of compact Lie groups.

Let G be a compact connected Lie group and T a maximal torus. Let $\Lambda = X^*(T)$, and let Λ_{root} be the lattice spanned by the roots. Then $\Lambda \supseteq \Lambda_{\text{root}}$. The index $[\Lambda : \Lambda_{\text{root}}]$ may be finite (e.g. if $G = \mathrm{SU}(n)$) or infinite (e.g. if $G = \mathrm{U}(n)$). If the index is finite, then we say G is *semisimple*, and this corresponds to the semisimple case in Chap. 20. Elements of Λ will be called *weights*.

We have written the characters of T additively. Sometimes we want to write them multiplicatively, however, so we introduce symbols e^λ for $\lambda \in \mathcal{V}$ subject to the rule $e^\lambda e^\mu = e^{\lambda+\mu}$. More formally, let $\mathcal{E}(R)$ denote the free R-module on the set of symbols $\{e^\lambda | \lambda \in \Lambda\}$. It consists of all formal sums $\sum_{\lambda \in \Lambda} n_\lambda e^\lambda$ with $n_\lambda \in R$ such that $n_\lambda = 0$ for all but finitely many λ. It is a ring with the multiplication

$$\left(\sum_{\lambda \in \Lambda} n_\lambda \cdot e^\lambda \right) \left(\sum_{\mu \in \Lambda} m_\mu \cdot e^\mu \right) = \sum_{\nu \in \Lambda} \left(\sum_{\lambda+\mu=\nu} n_\lambda m_\mu \right) \cdot e^\nu. \qquad (22.1)$$

This makes sense because only finitely many n_λ and only finitely many m_μ are nonzero. Of course, $\mathcal{E}(R)$ is just the group algebra over R of Λ. The Weyl group acts on $\mathcal{E}(R)$, and we will denote by $\mathcal{E}(R)^W$ the subring of W-invariant elements. Usually, we are interested in the case $R = \mathbb{Z}$, and we will denote $\mathcal{E} = \mathcal{E}(\mathbb{Z})$, $\mathcal{E}^W = \mathcal{E}(\mathbb{Z})^W$. We will find it sometimes convenient to work in the larger ring $\mathcal{E}_2$, which is the free Abelian group on $\frac{1}{2}\Lambda$.

If $\xi = \sum_\lambda n_\lambda \cdot e^\lambda$, we will sometimes denote $m(\xi, \lambda) = n_\lambda$, the *multiplicity* of λ in ξ. We will denote by $\overline{\xi} = \sum_\lambda n_\lambda \cdot e^{-\lambda}$ the *conjugate* of ξ.

By Theorem 17.1, class functions on G are the same thing as W-invariant functions on T. In particular, if χ is the character of a representation of G, then its restriction to T is a sum of characters of T and is invariant under the action of W. Thus, if $\lambda \in \Lambda$, let $n_\lambda(\chi)$ denote the multiplicity of λ in this restriction. We associate with χ the element

D. Bump, *Lie Groups*, Graduate Texts in Mathematics 225,
DOI 10.1007/978-1-4614-8024-2_22, © Springer Science+Business Media New York 2013

$$\sum_\lambda n_\lambda(\chi)\,\mathrm{e}^\lambda \in \mathcal{E}^W.$$

We will identify χ with this expression. We thus regard characters χ as elements of $\mathcal{E}^W$. The operation of conjugation that we have defined corresponds to the conjugation of characters. The conjugate of a character is a character by Proposition 2.6.

If $\mu_1, \mu_2, \ldots, \mu_n$ is a basis of the free $\mathbb{Z}$-module Λ, then $\mathcal{E}$ is the Laurent polynomial ring

$$\mathcal{E} = \mathbb{Z}[\mu_1, \ldots, \mu_r, \mu_1^{-1}, \ldots \mu_r^{-1}].$$

It is the localization $S^{-1}\mathbb{Z}[\mu_1, \ldots, \mu_r]$, where S is the multiplicative subset of $\mathbb{Z}[\mu_1, \ldots, \mu_r]$ generated by $\{\mu_1^{-1}, \ldots, \mu_r^{-1}\}$. As such, it is a unique factorization domain. (See Lang [116], Exercise 5 on p. 115.)

Let $\Sigma = \{\alpha_1, \ldots, \alpha_r\}$ be the simple roots, and Φ the set of positive roots, partitioned into Φ^+ and Φ^- as usual. We will denote by $\Delta \in \mathcal{E}$ the element

$$\mathrm{e}^{-\rho} \prod_{\alpha \in \Phi^+} (\mathrm{e}^\alpha - 1) = \mathrm{e}^\rho \prod_{\alpha \in \Phi^+} (1 - \mathrm{e}^{-\alpha}). \tag{22.2}$$

The equivalence of the two expressions follows easily from the fact that 2ρ is the sum of the positive roots.

Proposition 22.1. *We have* $w(\Delta) = (-1)^{l(w)}\Delta$ *for all* $w \in W$.

Proof. It is sufficient to check that $s_\beta(\Delta) = -\Delta$ for every simple root β. We recall that s_β changes the sign of β and permutes the remaining simple roots. Of the factors in the first expression for Δ in (22.2), only two are changed: $\mathrm{e}^{-\rho}$ and $(\mathrm{e}^\beta - 1)$. These become [see (20.11)] $\mathrm{e}^{-\rho+\beta}$ and $(\mathrm{e}^{-\beta} - 1)$. The net effect is that Δ changes sign.   □

An alternative way of explaining the same proof begins with the equation

$$\Delta = \prod_{\alpha \in \Phi^+} (\mathrm{e}^{\alpha/2} - \mathrm{e}^{-\alpha/2}). \tag{22.3}$$

Here $\alpha/2$ may not be an element of Λ, so each individual factor on the right is not really an element of $\mathcal{E}$ but of the larger ring $\mathcal{E}_2$. Proposition 22.1 follows by noting that by Proposition 20.1(ii) each simple reflection alters the sign of exactly one term in (22.3), and the result follows.

Proposition 22.2. *If* $\xi \in \mathcal{E}$ *satisfies* $w(\xi) = (-1)^{l(w)}\xi$ *for all* $w \in W$, *then* ξ *is divisible by* Δ *in* $\mathcal{E}$.

Proof. In the ring $\mathcal{E}$, by Proposition 22.1, Δ is a product of distinct irreducible elements $1 - \mathrm{e}^\alpha$, where α runs through Φ^+, times a unit $\mathrm{e}^{-\rho}$. It is therefore sufficient to show that ξ is divisible by each $1 - \mathrm{e}^\alpha$. By Proposition 20.12, we have $s_\alpha(\xi) = -\xi$. Write $\xi = \sum_{\lambda \in \Lambda} n_\lambda \cdot l$. Since $s_\alpha(\xi) = -\xi$, we have $n_{s_\alpha(\lambda)} = -n_\lambda$. Noting that $s_\alpha(\lambda) = \lambda - k\alpha$ where $k = \alpha^\vee(\lambda) \in \mathbb{Z}$, we see that

$$\xi = \sum_{\substack{\lambda \in \Lambda \\ \lambda \bmod \langle s_\alpha \rangle}} n_\lambda (e^\lambda - e^{\lambda - k\alpha}).$$

The notation means that we choose only one representative for each s_α orbit of Λ. (If $s_\alpha(\lambda) = \lambda$, then $n_\lambda = 0$.) Since

$$e^\lambda - e^{\lambda - k\alpha} = (1 - e^\alpha)(-e^{\lambda - \alpha} - e^{\lambda - 2\alpha} - \cdots - e^{\lambda - k\alpha}),$$

this is divisible by Δ. $\qquad\qquad\qquad\qquad\qquad\qquad\qquad\qquad\qquad\square$

If $\lambda \in \Lambda \cap \mathcal{C}_+$, let

$$\chi(\lambda) = \Delta^{-1} \sum_{w \in W} (-1)^{l(w)} e^{w(\lambda + \rho)}. \qquad\qquad (22.4)$$

By Proposition 22.2, $\chi(\lambda) \in \mathcal{E}$. Moreover, applying $w \in W$ multiplies both $\sum_{w \in W} (-1)^w e^{w(\lambda + \rho)}$ and Δ by $(-1)^w$, so $\chi(\lambda)$ is actually in $\mathcal{E}^W$.

We will eventually prove that if $\lambda \in \Lambda \cap \mathcal{C}_+$ this is an irreducible character of G. Then (22.4) is called the *Weyl character formula*.

If $\xi = \sum n_\lambda e^\lambda \in \mathcal{E}$, we define the *support* of ξ to be the finite set $\mathrm{supp}(\xi) = \{\lambda \in L \mid n_\lambda \neq 0\}$. We define a partial order on $\mathcal{V}$ by $\lambda \preccurlyeq \mu$ if $\lambda = \mu + \sum_{\alpha \in \Sigma} c_\alpha \alpha$, where $c_\alpha \geqslant 0$.

Proposition 22.3. *If $\lambda \in \mathcal{C}_+$, then $\lambda \succcurlyeq w(\lambda)$ for $w \in W$. If $\lambda \in \mathcal{C}_+^\circ$ and $w \neq 1$, then $w(\lambda) \succ \lambda$.*

Proof. It is easy to see that, for $x \in \mathcal{V}$, $x \succcurlyeq 0$ if and only if $\langle x, v \rangle \geqslant 0$ for all $v \in \mathcal{C}_+^\circ$. So if $\lambda \in \mathcal{C}_+$ and $\lambda \not\succcurlyeq w(\lambda)$, then there exists $v \in \mathcal{C}_+^\circ$ such that $\langle \lambda - w(\lambda), v \rangle < 0$. We choose w to maximize $\langle w(\lambda), v \rangle$. Since $w(\lambda) \neq \lambda$ and $\lambda \in \mathcal{C}_+$, it follows from Theorem 20.1 that $w(\lambda) \notin \mathcal{C}_+$. Therefore, there exists $\alpha \in \Sigma$ such that $\langle w(\lambda), \alpha \rangle < 0$, or equivalently, $\alpha^\vee(w(\lambda)) < 0$. Now

$$\langle s_\alpha w(\lambda), v \rangle = \left\langle w(\lambda) - 2 \frac{\langle w(\lambda), \alpha \rangle}{\langle \alpha, \alpha \rangle} \alpha, v \right\rangle$$

$$= \langle w(\lambda), v \rangle - 2 \frac{\langle w(\lambda), \alpha \rangle}{\langle \alpha, \alpha \rangle} \langle \alpha, v \rangle > \langle w(\lambda), v \rangle.$$

The maximality of $\langle w(\lambda), v \rangle$ is contradicted. $\qquad\qquad\qquad\qquad\square$

Proposition 22.4. *Let $\lambda \in \mathcal{C}_+$. Then $\lambda \in \mathrm{supp}\, \chi(\lambda)$. Indeed, writing $\chi(\lambda) = \sum_\mu n_\mu \cdot e^\mu$, we have $n_\lambda = 1$. Moreover, if $\mu \in \mathrm{supp}\, \chi(\lambda)$, then $\lambda \succcurlyeq \mu$, and $\lambda - \mu \in \Lambda_{\mathrm{root}}$. In particular, λ is the largest weight in the support of $\chi(\lambda)$.*

Proof. We enlarge the ring $\mathcal{E}$ as follows. Let $\hat{\mathcal{E}}$ be the "completion" consisting of all formal sums $\sum_{\lambda \in \Lambda} n_\lambda \cdot \lambda$, where we now allow $n_\lambda \neq 0$ for an infinite number of λ. However, we ask that there be a $v \in \mathcal{V}$ such that $n_\lambda \neq 0$ implies that $\lambda \preccurlyeq v$. This means that, in the product (22.1), only finitely many terms will be nonzero, so $\hat{\mathcal{E}}$ is a ring. We can write

$$\Delta = e^\rho \prod_{\alpha \in \Phi^+} (1 - e^{-\alpha}),$$

so in $\hat{\mathcal{E}}$ we have

$$\Delta^{-1} = e^{-\rho} \prod_{\alpha \in \Phi^+} (1 + e^{-\alpha} + e^{-2\alpha} + \cdots).$$

Therefore,

$$\chi(\lambda) = e^\lambda \prod_{\alpha \in \Phi^+} (1 + e^{-\alpha} + e^{-2\alpha} + \cdots) \sum_{w \in W} (-1)^{l(w)} e^{w(\lambda+\rho)-(\lambda+\rho)}. \qquad (22.5)$$

Each factor in the product is $\prec 0$ except 1, and by Proposition 22.3 each term in the sum is $\prec 0$ except that corresponding to $w = 1$. Hence, each term in the expansion is $\preccurlyeq \lambda$, and exactly one term contributes λ itself.

It remains to be seen that if e^μ appears in the expansion of the right-hand side of (22.5), then $\lambda - \mu$ is an element of Λ_{root}. We note that $w(\lambda+\rho)-(\lambda+\rho) \in \Lambda_{\text{root}}$ by Proposition 20.16, and of course all the contributions coming from the product over $\alpha \in \Phi^+$ are roots, and the result follows. $\square$

Now let us write the Weyl integration formula in terms of Δ.

Theorem 22.1. *If f is a class function on G, we have*

$$\int_G f(g)\, dg = \frac{1}{|W|} \int_T f(t)\, |\Delta(t)|^2\, dt. \qquad (22.6)$$

Here there is an abuse of notation since Δ is itself only an element of $\mathcal{E}$, not even W-invariant, so it is not identifiable as a function on the group. However, it will follow from the proof that $\Delta\bar{\Delta}$ is always a function on the group, and we will naturally denote $\Delta\bar{\Delta}$ as $|\Delta|^2$.

Proof. We will show that, in the notation of Theorem 17.2,

$$\det\left([\text{Ad}(t^{-1}) - I_{\mathfrak{p}}]\,|\,\mathfrak{p}\right) = \Delta\bar{\Delta}. \qquad (22.7)$$

Indeed, since the complexification of $\mathfrak{p}$ is the direct sum of the spaces $\mathfrak{X}_\alpha$ on each of which $t \in T$ acts by $\alpha(t)$ in the adjoint representation,

$$\det\left([\text{Ad}(t^{-1}) - I_{\mathfrak{p}}]\,|\,\mathfrak{p}\right) = \prod_{\alpha \in \Phi} (\alpha(t)^{-1} - 1) = |\prod_{\alpha \in \Phi^+} (\alpha(t) - 1)|^2.$$

In $\mathcal{E}$, this becomes the element

$$\left[e^{-\rho} \prod_{\alpha \in \Phi^+} (e^\alpha - 1)\right] \overline{\left[e^{-\rho} \prod_{\alpha \in \Phi^+} (e^\alpha - 1)\right]} = \Delta\bar{\Delta}.$$

Now (22.6) is just the Weyl integration formula, Theorem 17.2. $\square$

We now introduce an inner product on $\mathcal{E}^W$. If $\xi, \eta \in \mathcal{E}^W$, let

$$\langle \xi, \eta \rangle = \frac{1}{|W|} m((\xi\Delta)\overline{(\eta\Delta)}, 0). \tag{22.8}$$

That is, it is the multiplicity of the zero weight in $(\xi\Delta)\overline{(\eta\Delta)}$ divided by $|W|$.

Theorem 22.2. *If ξ and η are characters of G, identified with elements of $\mathcal{E}$, then the inner product (22.8) agrees with the L^2 inner product of the characters.*

Proof. The L^2 inner product of ξ and η is just the integral of $\xi \cdot \overline{\eta}$ over the group and, using (22.6), this is just W^{-1} times the multiplicity of 0 in $(\xi\Delta)\overline{(\eta\Delta)}$.   □

Proposition 22.5. *If λ and μ are weights in $\mathcal{C}_+$, we have*

$$\langle \chi(\lambda), \chi(\mu) \rangle = \begin{cases} 1 & \text{if } \lambda = \mu, \\ 0 & \text{otherwise.} \end{cases}$$

Proof. Using (22.8), this inner product is the multiplicity of 0 in

$$\frac{1}{|W|} \left[\sum_{w \in W} (-1)^w e^{w(\rho+\lambda)} \right] \left[\sum_{w' \in W} (-1)^{w'} e^{w'(\rho+\mu)} \right]$$

$$= \frac{1}{|W|} \left[\sum_{w,w' \in W} (-1)^{w+w'} e^{w(\rho+\lambda)-w'(\rho+\mu)} \right].$$

We must therefore ask, with both λ and $\mu \in \mathcal{C}_+$, under what circumstances $w(\rho + \lambda) - w'(\rho + \mu) = 0$ can vanish. Then $\rho + \lambda = w^{-1}w'(\rho + \mu)$. Since both $\rho + \lambda$ and $\rho + \mu$ are in $\mathcal{C}_+^\circ$, it follows from Theorem 20.1 that w must equal w' and so λ must equal μ. The number of solutions is thus $|W|$ if $\lambda = \mu$ and zero otherwise.   □

Proposition 22.6. *The set of $\chi(\lambda)$ with $\lambda \in \Lambda \cap \mathcal{C}_+$ is a basis of the free $\mathbb{Z}$-module $\mathcal{E}^W$.*

Proof. The linear independence of the $\chi(\lambda)$ follows from their orthogonality. We must show that they span. Clearly, $\mathcal{E}^W$ is spanned by elements of the form

$$B(\lambda) = \sum_{\mu \in W \cdot \lambda} e^\mu, \qquad l \in \Lambda \cap \mathcal{C}_+,$$

where $W \cdot \lambda$ is the orbit of λ under the action of W. It is sufficient to show that $B(\lambda)$ is in the $\mathbb{Z}$-linear span of the $\chi(\lambda)$. It follows from Proposition 22.4 that when we expand $B(\lambda) - \chi(\lambda)$ in terms of the $B(\mu)$, only $\mu \in \Lambda$ with $\mu \prec \lambda$ can occur and, by induction, these are in the span of the $\chi(\mu)$.   □

Theorem 22.3. (Weyl) *Assume that G is semisimple. If $\lambda \in \Lambda \cap \mathcal{C}_+$, then $\chi(\lambda)$ is the character of an irreducible representation of G, and each irreducible representation is obtained this way.*

We will denote by $\pi(\lambda)$ the irreducible representation of G with character χ_λ.

Proof. Let χ be an irreducible representation of G. Regarding χ as an element of $\mathcal{E}^W$, we may expand χ in terms of the $\chi(\lambda)$ by Proposition 22.6. We write

$$\chi = \sum_{\lambda \in \Lambda \cap \mathcal{C}_+} n_\lambda \cdot \chi(\lambda), \qquad n_\lambda \in \mathbb{Z}.$$

We have

$$1 = \langle \chi, \chi \rangle = \sum_\lambda n_\lambda^2.$$

Therefore, exactly one n_λ is nonzero, and that has value ± 1. Thus, either $\chi(\lambda)$ or its negative is an irreducible character of G. To see that $-\chi(\lambda)$ is not a character, consider its restriction to T. By Proposition 22.4, the multiplicity of the character λ in $-\chi(\lambda)$ is -1, which is impossible if $-\chi(\lambda)$ is a character. Hence, $\chi(\lambda) = \chi$ is an irreducible character of G.

We have shown that every irreducible character of G is a $\chi(\lambda)$. It remains to be shown that every $\chi(\lambda)$ is a character. Since the class functions on G are identical to the W-invariant functions on T, the closure in $L^2(G)$ of $\mathcal{E}(\mathbb{C})^W$ is identified with the space of all class functions on G. By Proposition 22.6, the $\chi(\lambda)$ form an L^2-basis of $\mathcal{E}(\mathbb{C})^W$. Since by the Peter–Weyl theorem the set of irreducible characters of G are an L^2 basis of the space of class functions, the characters of G cannot be a proper subset of the set of $\chi(\lambda)$. $\qquad\square$

Now let us step back and see what we have established. We know that in group representation theory there is a duality between the irreducible characters of a group and its conjugacy classes. We can study both the conjugacy classes and the irreducible representations of a compact Lie group by restricting them to T. We find that the conjugacy classes of G are in one-to-one correspondence with the W-orbits of T. Dually, the irreducible representations of G are parametrized by the orbits of W on $\Lambda = X^*(T)$.

We study these orbits by embedding $X^*(T)$ in a Euclidean space $\mathcal{V}$. The positive Weyl chamber $\mathcal{C}_+$ is a fundamental domain for the action of W on $\mathcal{V}$, and so the dominant weights—those in $\mathcal{C}_+$—are thus used to parametrize the irreducible representations. Of the weights that appear in the parametrized representation $\chi(\lambda)$, the parametrizing weight $\lambda \in \mathcal{C}_+ \cap X^*(T)$ is maximal with respect to the partial order. We therefore call it the *highest weight vector* of the representation.

Proposition 22.7. *We have*

$$\Delta = \sum_{w \in W} (-1)^{l(w)} e^{w(\rho)}. \tag{22.9}$$

Proof. The irreducible representation $\chi(0)$ with highest weight vector 0 is obviously the trivial representation. Therefore, $\chi(0) = e^0 = 1$. The formula now follows from (22.4).   □

Weyl gave a formula for the dimension of the irreducible representation with character χ_λ. Of course, this is the value χ_λ at the identity element of G, but we cannot simply plug the identity into the Weyl character formula since the numerator and denominator both vanish there. Naturally, the solution is to use L'Hôpital's rule, which can be formulated purely algebraically in this context.

Theorem 22.4 (Weyl). *The dimension of* $\pi(\lambda)$ *is*

$$\frac{\prod_{\alpha \in \Phi^+} \langle \lambda + \rho, \alpha \rangle}{\prod_{\alpha \in \Phi^+} \langle \rho, \alpha \rangle}. \tag{22.10}$$

Proof. Let $\Omega : \mathcal{E}_2 \longrightarrow \mathbb{Z}$ be the map

$$\Omega\left(\sum_{\lambda \in \Lambda} n_\lambda \cdot e^\lambda\right) = \sum_{\lambda \in \Lambda} n_\lambda.$$

The dimension we wish to compute is $\Omega(\chi_\lambda)$.

If $\alpha \in \Phi$, let $\partial_\alpha : \mathcal{E}_2 \longrightarrow \mathcal{E}_2$ be the map

$$\partial_\alpha\left(\sum_{\lambda \in \Lambda} n_\lambda \cdot e^\lambda\right) = \sum_{\lambda \in \Lambda} n_\lambda \langle \lambda, \alpha \rangle \cdot e^\lambda.$$

It is straightforward to check that ∂_α is a derivation and that the operators ∂_α commute with each other. Let $\partial = \prod_{\alpha \in \Phi^+} \partial_\alpha$.

We show that if $w \in W$ and $f \in \mathcal{E}_2$, we have

$$w\partial(f) = (-1)^{l(w)}\partial w(f). \tag{22.11}$$

We note first that

$$\partial_{w(\alpha)} \circ w = w \circ \partial_\alpha \tag{22.12}$$

since applying the operator on the left-hand side to e^λ gives $\langle w(\lambda), w(\alpha) \rangle \, e^{w(\lambda)}$, while the second gives $\langle \lambda, \alpha \rangle \, e^{w\lambda}$, and these are equal. Now, to prove (22.11), we may assume that $w = s_\beta$ is a simple reflection. By (22.12), we have

$$w \circ \left(\prod_{\alpha \in \Phi^+} \partial_{w(\alpha)}\right) = \partial \circ w.$$

But by Proposition 20.1(ii), the set of $w(\alpha)$ consists of Φ^+ with just one element, namely β, replaced by its negative. So (22.11) is proved.

We consider now what happens when we apply $\Omega \circ \partial$ to both sides of the identity

$$\sum_{w \in W} (-1)^\lambda e^{w(\lambda + \rho)} = \chi_\lambda \cdot \prod_{\alpha \in \Phi^+} \left(e^{\alpha/2} - e^{-\alpha/2}\right). \qquad (22.13)$$

On the left-hand side, by (22.11), applying ∂ gives

$$\sum_{w \in W} w \left(\partial e^{\lambda + \rho}\right) = \sum_{w \in W} w \left(\prod_{\alpha \in \Phi^+} \langle \lambda + \rho, \alpha \rangle \, e^{\lambda + \rho}\right).$$

Now applying Ω gives $|W| \prod_{\alpha \in \Phi^+} \langle \lambda + \rho, \alpha \rangle$.

On the other hand, we apply $\partial = \prod \partial_\beta$ one derivation at a time to the right-hand side of (22.13), expanding by the Leibnitz product rule to obtain a sum of terms, each of which is a product of χ_λ and the terms $e^{\alpha/2} - e^{-\alpha/2}$, with various subsets of the ∂_β applied to each factor. When we apply Ω, any term in which a $e^{\alpha/2} - e^{-\alpha/2}$ is not hit by at least one ∂_β will be killed. Since the number of operators ∂_β and the number of factors $e^{\alpha/2} - e^{-\alpha/2}$ are equal, only the terms in which each $e^{\alpha/2} - e^{-\alpha/2}$ is hit by exactly one ∂_β survive. Of course, χ_λ is not hit by a ∂_β in any such term. In other words,

$$\Omega \circ \partial \left(\chi_\lambda \cdot \prod_{\alpha \in \Phi^+} \left(e^{\alpha/2} - e^{-\alpha/2}\right)\right) = \theta \cdot \Omega(\chi_\lambda),$$

where

$$\theta = \Omega \circ \partial \left(\prod_{\alpha \in \Phi^+} \left(e^{\alpha/2} - e^{-\alpha/2}\right)\right)$$

is independent of λ. We have proved that

$$|W| \prod_{\alpha \in \Phi^+} \langle \lambda + \rho, \alpha \rangle = \theta \cdot \Omega(\chi_\lambda).$$

To evaluate θ, we take $\lambda = 0$, so that χ_λ is the character of the trivial representation, and $\Omega(\chi_\lambda) = 1$. We see that $\theta = |W| \prod_{\alpha \in \Phi^+} \langle \rho, \alpha \rangle$. Dividing by this, we obtain (22.10). $\qquad \square$

Proposition 22.8. *Let λ be a dominant weight, and let (π, V) be an irreducible representation with highest weight λ. Let w_0 be the longest Weyl group element (Proposition 20.14). Then the highest weight of the contragredient representation is $-w_0 \lambda$.*

Proof. We recall that the character of the contragredient is the complex conjugate of the character of π (Proposition 2.6). The weights that occur in the contragredient are therefore the negatives of the weights that occur in π. It follows that $-\lambda$ is the lowest weight of $\hat{\pi}$. The highest weight is therefore in the same Weyl group element as $-\lambda$, and the unique dominant weight in that orbit is $-w_0 \lambda$. $\qquad \square$

In 1967 Klimyk described a method of decomposing tensor products that is very efficient for computation. It is based on a simple idea and in retrospect it was found that the same idea appeared in a much earlier paper of Brauer (1937). The same idea appears in Steinberg [154]. We will prove a special case, leaving the general case for the exercises.

Proposition 22.9 (Brauer, Klimyk). *Suppose that λ and μ are in $X^*(T) \cap$ $\mathcal{C}_+$. Decompose χ_μ into a sum of weights $\nu \in X^*(T)$ with multiplicities $m(\nu)$:*

$$\chi_\mu = \sum_\nu m(\nu)\, e^\nu.$$

Suppose that for each ν with $m(\nu) \neq 0$ the weight $\lambda + \nu$ is dominant. Then

$$\chi_\lambda \chi_\mu = \sum_\nu m(\nu)\chi_{\lambda+\nu}. \tag{22.14}$$

Since $\chi_\lambda\chi_\mu$ is the character of the tensor product representation, this gives the decomposition of this tensor product into irreducibles. The method of proof can be extended to the case where $\lambda + \nu$ is *not* dominant for all ν, though the answer is a bit more complicated to state (Exercise 22.5).

Proof. By the Weyl character formula, we may write

$$\chi_\lambda \chi_\mu = \Delta^{-1} \sum_\nu m(\nu)\, e^\nu \sum_w (-1)^{l(w)}\, e^{w(\lambda+\rho)}.$$

Interchange the order of summation, so that the sum over ν is the inner sum, and make the variable change $\nu \longrightarrow w(\nu)$. Since $m(\nu) = m(w\nu)$, we get

$$\Delta^{-1} \sum_w \sum_\nu m(\nu)\, (-1)^{l(w)}\, e^{w(\lambda+\nu+\rho)}.$$

Now we may interchange the order of summation again and apply the Weyl character formula to obtain (22.14). □

It is sometimes convenient to shift ρ by a W-invariant element of $\mathbb{R} \otimes X^*(T)$ so that it is in $X^*(T)$. Such a shift is harmless in the Weyl character formula and the Weyl dimension formula provided we shift by a vector that is orthogonal to all the roots, since ρ only appears in inner product with the roots. This is only possible if G is not semisimple, for if G is semisimple, there are no nonzero vectors orthogonal to the roots. Let us illustrate this trick with $G = \mathrm{U}(n)$. We identify $X^*(T)$ with $\mathbb{Z}^n$ by mapping the character

$$\begin{pmatrix} t_1 & & \\ & \ddots & \\ & & t_n \end{pmatrix} \longmapsto \prod t_i^{k_i} \tag{22.15}$$

to $(k_1, \ldots, k_n) \in \mathbb{Z}^n$. Then ρ is $\frac{1}{2}(n-1, n-3, \ldots, 1-n)$. If n is even, it is an element of $\mathbb{R} \otimes X^*(T)$ but not of $X^*(T)$. However, if we add to it the W-invariant element $\frac{1}{2}(n-1, \ldots, n-1)$, we get

$$\delta = (n-1, n-2, \ldots, 1, 0) \in X^*(T). \tag{22.16}$$

We can now write the Weyl character formula in the form

$$\chi(\lambda) = \Delta_0^{-1} \sum_{w \in W} (-1)^{l(w)} e^{w(\lambda+\delta)}, \tag{22.17}$$

where

$$\Delta_0 = \sum_{w \in W} (-1)^{l(w)} e^{w(\delta)}.$$

We have simply multiplied the numerator and the denominator by the same W-invariant element so that both the numerator and the denominator are in $X^*(T)$.

In (22.7), we write the factor $|\Delta|^2 = |\Delta_0|^2$ since $(\Delta_0/\Delta)^2 = e^{2(\delta-\rho)}$. As a function on the group, this is just $\det(g)^{n-1}$, which has absolute value 1. Therefore, we may write the Weyl integration formula in the form

$$\int_G f(g) \, dg = \frac{1}{|W|} \int_T f(t) \, |\Delta_0(t)|^2 \, dt. \tag{22.18}$$

Exercises

In the first batch of exercises, $G = \mathrm{SU}(3)$ and, as usual, ϖ_1 and ϖ_2 are the fundamental dominant weights.

Exercise 22.1. By Proposition 22.4, all the weights in χ_λ lie in the set

$$S(\lambda) = \{\mu \in \Lambda \mid \lambda \succcurlyeq w(\mu) \text{ for all } w \in W, \lambda - \mu \in \Lambda_{\text{root}}\}.$$

Confirm by examining the weights that this is true for all the examples in Chap. 21—in fact, for all these examples, $S(\mu)$ is *exactly* the set of weights.

Exercise 22.2. Use the Weyl dimension formula to compute the dimension of $\chi_{2\varpi_1}$. Deduce from this that the symmetric square of the standard representation is irreducible.

Exercise 22.3. Use the Weyl dimension formula to compute the dimension of $\chi_{\varpi_1+2\varpi_2}$. Deduce from this that the symmetric square of the standard representation is irreducible.

Exercise 22.4. Use the Brauer–Klimyk method (Proposition 22.9) to compute the tensor product of the contragredient of the standard representation (with character χ_{ϖ_2}) and the adjoint representation (with character $\chi_{\varpi_1+\varpi_2}$).

Exercise 22.5. Prove the following extension of Proposition 22.9. Suppose that λ is dominant and that ν is any weight. By Proposition 20.1, there exists a Weyl group element such that $w(\nu+\lambda+\rho) \in \mathcal{C}_+$. The point $w(\nu+\lambda+\rho)$ is uniquely determined, even though w may not be. If $w(\nu+\lambda+\rho)$ is on the boundary of $\mathcal{C}_+$, define $\xi(\nu,\lambda) = 0$. If $w(\nu+\lambda+\rho)$ is *not* on the boundary of $\mathcal{C}_+$, explain why $w(\nu+\lambda+\rho) - \rho \in \mathcal{C}_+$ and w is uniquely determined. In this case, define $\xi(\nu,\lambda) = (-1)^{l(w)}\chi_{w(\nu+\lambda+\rho)-\rho}$. Prove that if μ is a dominant weight, and $\chi_\mu = \sum m(\nu)e^\nu$, then

$$\chi_\mu\chi_\lambda = \sum_\nu m(\nu)\xi(\nu,\lambda).$$

Exercise 22.6. Use the last exercise to compute the decomposition of $\chi^2_{\varpi_1}$ into irreducibles, and obtain another proof that the symmetric square of the standard representation is irreducible.

Exercise 22.7. Let μ be an element of the root lattice. A *vector partition* of μ is a decomposition of μ into a linear combination, with nonnegative integer coefficients, of positive weights. In other words, it is an assignment of nonnegative integers n_α to $\alpha \in \Phi^+$ such that

$$\mu = \sum_{\alpha \in \Phi^+} n_\alpha\alpha.$$

The *Kostant partition function* $\mathfrak{P}(\mu)$ is defined to be the number of vector partitions of μ. Note that this is zero unless $\mu \succcurlyeq 0$. Let $\hat{\mathcal{E}}$ be the completion of $\mathcal{E}$ defined in the proof of Proposition 22.4. Show that in $\hat{\mathcal{E}}$

$$\prod_{\alpha \in \Phi}(1 - e^{-\alpha})^{-1} = \sum_{\substack{\mu \in \Lambda_{\text{root}} \\ \mu \succcurlyeq 0}} \mathfrak{P}(\mu)e^{-\mu},$$

and from (22.5) deduce the *Kostant multiplicity formula*, for λ a dominant weight: the multiplicity of μ in $\chi(\lambda)$ is

$$\sum_w(-1)^{l(w)}\sum_\mu \mathfrak{P}(w(\lambda + \rho) - \rho - \mu).$$

Exercise 22.8. Let $G = \mathrm{SU}(3)$ and let $\varpi_1, \varpi_2 : T \longrightarrow \mathbb{C}$ be the fundamental dominant weights, labeled as in Chap. 19. Use the Kostant multiplicity formula to compute the weights of $\chi(\varpi_1 + 2\varpi_2)$. Note that you need only need to consider weights in supp $\chi(\lambda)$ as computed in Proposition 22.4. Do you observe a shortcut for this type of calculation?

Exercise 22.9. Show that if $-w_0\lambda = \lambda$ for all weights in G, then every element of G is conjugate to its inverse.

Exercise 22.10. The nine-dimensional adjoint representation of $\mathrm{U}(3)$ has as an invariant subspace the eight-dimensional Lie algebra of $\mathrm{SU}(3)$.

(i) Identifying the weight lattice of $\mathrm{U}(3)$ or $GL(3,\mathbb{C})$ with $\mathbb{Z}^3$ as in Chap. 19, what is the highest weight vector in this eight-dimensional module?

(ii) Decompose the tensor square of this representations into irreducibles by computing the square of the character and finding irreducible representations whose character adds up to the character in question.

(iii) Compute the symmetric and exterior squares of the character.

Exercise 22.11. Let V be the ten-dimensional adjoint representation of $Sp(4)$. What is the decomposition of the symmetric and exterior squares of this representation into irreducibles?

Exercise 22.12. Generalize the last exercise as follows. Let V be the adjoint representation of $Sp(2n)$. Its degree is $n + 2n^2$. What is the decomposition of the symmetric and exterior squares of this representation into irreducibles? You might want to use a computer program such as **Sage** to collect some data but prove your answer.

Exercise 22.13. Let the weight lattices of $Sp(2n)$ and $SO(2n + 1)$ be identified with $\mathbb{Z}^n$ as in Chap. 19. Denote by ρ_C and ρ_B the Weyl vectors of these two groups. Show that

$$\rho_C = (n, n - 1, \ldots, 1), \qquad \rho_B = \left(n - \frac{1}{2}, n - \frac{3}{2}, \ldots, \frac{1}{2}\right).$$

Recall that $\mathrm{Spin}(2n + 1)$ is the double cover of $SO(2n + 1)$. The root lattice of $\mathrm{Spin}(2n+1)$ is naturally embedded in that of $SO(2n+1)$. Show that the root lattice of $\mathrm{Spin}(2n+1)$ consists of tuples $(\mu_1, \ldots, \mu_n)$ such that $2\mu_i \in \mathbb{Z}$, and the μ_i are either all integers or all half integers, that is, the $2\mu_i$ are either all even or odd. (**Hint:** Use Proposition 18.10 and look ahead to Proposition 31.2 if you need help.) Now let $\lambda = (\lambda_1, \ldots, \lambda_n) \in \mathbb{Z}^n$ such that $\lambda_1 \geqslant \cdots \geqslant \lambda_n$, and let $\mu = (\mu_1, \ldots, \mu_n)$ where $\mu_i = \lambda_i + \frac{1}{2}$. Show that λ and μ are dominant weights for $Sp(2n)$ and $SO(2n+1)$. Let V_λ and W_μ be irreducible representations of $Sp(2n)$ and $SO(2n + 1)$, respectively. Show that

$$\dim(W_\mu) = 2^n \cdot \dim(V_\lambda).$$

[**Hint:** It may be easiest to show that $\dim(W_\mu)/\dim(V_\lambda)$ is constant, then take $l = 0$ to determine the constant.]

The next exercise treats the *Frobenius–Schur indicator* of an irreducible representation. This will be covered (in a slightly different form) in Theorem 43.1. If (π, V) is a representation of the compact group G, and $B : V \times V \to \mathbb{C}$ is an invariant bilinear form, then B is unique up to scalar multiple (by a version of Schur's lemma), so $B(x, y) = c\,B(y, x)$ for some constant c. Since $B(x, y) = c^2\,B(x, y)$, $c = \pm 1$. Thus the form B is either symmetric or skew-symmetric. If it is symmetric, then $\pi(G)$ is contained in the orthogonal group of the form, in which case we say π is *orthogonal*. If B is skew-symmetric, then $\pi(G)$ is contained in the symplectic group of the form, in which case $\dim(V)$ is even, and we say that π is *symplectic*. Every self-contragredient representation is either orthogonal or symplectic but not both. See Chap. 43 for further details.

Exercise 22.14. Let χ be the character of an irreducible representation $\pi : G \to GL(V)$ of the compact group G.

(i) Show that $\chi(g^2) = \vee^2\chi(g) - \wedge^2\chi(g)$ and $\chi(g)^2 = \vee^2\chi(g) + \wedge^2\chi(g)$, where $\vee^2\chi$ and $\wedge^2\chi$ are the characters of the symmetric and exterior square representations.

(ii) Show that $\chi(g^2)$ is a generalized character, and that when χ is expanded in terms of irreducible characters, the coefficient of the trivial character is the *Frobenius–Schur indicator*

$$\varepsilon(\pi) = \begin{cases} 1 & \text{if } \pi \text{ is orthogonal,} \\ -1 & \text{if } \pi \text{ is symplectic,} \\ 0 & \text{if } \pi \text{ is not self-contragredient.} \end{cases}$$

The generalized $\chi(g^2)$ is (in the language of lambda rings) the second *Adams operation* applied to χ. More generally, for $k \geqslant 0$, the Adams operation $\psi^k \chi(g) = \chi(g^k)$. We will return to the Adams operations in Chap. 33, and in particular we will see that $\psi^k \chi$ is a generalized character for all k; for $k = 2$ this follows from Exercise 22.14. Suppose that for μ in the weight lattice Λ of G, $m(\mu)$ is the weight multiplicity for χ, so that in the notation of Chap. 22, we have

$$\chi = \sum_{\mu \in \Lambda} m(\mu)\, \mathrm{e}^\mu.$$

Then clearly

$$\psi^k \chi = \sum_{\mu \in \Lambda} m(\mu)\, \mathrm{e}^{k\mu}.$$

A method of computing the Frobenius–Schur indicator is simply to decompose $\psi^2 \chi$ into irreducibles and note the coefficient of the trivial representation. A better method is to use a result of Steinberg [155] (modestly called Lemma 79) that there exists an element η of order $\leqslant 2$ in the center of G such that if π is self-dual, then the central character of π applied to η is the Frobenius–Schur indicator. In Sage, irreducible representations (as WeylCharacterRing elements) have a method to compute the Frobenius–Schur indicator.

Exercise 22.15.

(i) Let $G = \mathrm{SU}(2)$. If k is a nonnegative integer, let χ_k be the character of the irreducible representation on $\vee^k \mathbb{C}^2$. Show that if $\chi = \chi_k$, then the generalized character $g \mapsto \chi(g^2)$ equals

$$\sum_{l=0}^{k} (-1)^l \chi_{k-l},$$

and deduce that the Frobenius–Schur indicator of χ_k is $(-1)^k$.

(ii) Show that the image of $\mathrm{SL}(2, \mathbb{C})$ under the kth symmetric power homomorphism to $\mathrm{GL}(k+1, \mathbb{C})$ is contained in $\mathrm{SO}(k+1)$ if k is even, and $\mathrm{Sp}(k+1)$ if k is odd.

Exercise 22.16. Let k be a positive integer, and let χ be an irreducible character of the compact connected Lie group G. If λ is a dominant weight, let χ_λ denote the irreducible character with highest weight λ. Let ρ be the Weyl vector, half the sum of the positive roots. Prove that

$$\chi_{k-1} \cdot \psi^k \chi_\lambda = \chi_{k\lambda + (k-1)\rho}.$$

(**Hint**: Use the Weyl character formula.)

Exercise 22.17. Let α_1 and α_2 be the simple roots for $\mathrm{SU}(3)$. The aim of this exercise is to compute $\psi^2 \chi_{n\rho}$. Note that $\rho = \alpha_1 + \alpha_2$.

(i) Show that

$$\chi_\rho \sum_{\substack{i,j \leqslant m \\ i = 0 \text{ or } j = 0}} (-1)^{i+j} \chi_{2m\rho - i\alpha_1 - j\alpha_2} = \begin{cases} \chi_{(2m+1)\rho} - \chi_{(2m-1)\rho} & \text{if } m > 0, \\ \chi_\rho & \text{if } m = 0. \end{cases}$$

(**Hint**: One way to prove this is to use the Brauer-Klimyk method.)

(ii) Show that

$$\psi^2 \chi_{k\rho} = \sum_{m=0}^{k} \sum_{\substack{i,j \leqslant m \\ i = 0 \text{ or } j = 0}} (-1)^{i+j} \chi_{2m\rho - i\alpha_1 - j\alpha_2}.$$

Exercise 22.18. Let χ be a character of the group G. Let $\vee^k \chi$, $\wedge^k \chi$ and $\psi^k \chi$ be the symmetric power, exterior power and Adams operations applied to χ. Prove that

$$\vee^k \chi = \frac{1}{k} \sum_{r=1}^{k} (\psi^r \chi)(\vee^k \chi),$$

$$\wedge^k \chi = \frac{1}{k} \sum_{r=1}^{k} (-1)^r (\psi^r \chi)(\wedge^k \chi).$$

Hint: The symmetric polynomial identity (37.3) below may be of use.

If G is a Lie group, using the Brauer–Klimyk method to compute the right-hand side in these identities, it is not necessary to decompose $\psi^k \chi$ into irreducibles. So this gives an efficient recursive way to compute the symmetric and exterior powers of a character.

Exercise 22.19. Let ϖ_1 and ϖ_2 be the fundamental dominant weights for SU(3). Show that the irreducible representation with highest weight $k\varpi_1 + l\varpi_2$ is self-contragredient if and only if $k = l$, and in this case, it is orthogonal.

Exercise 22.20. Let G be a semisimple Lie group. Show that the adjoint representation is orthogonal. (**Hint**: You may assume that G is simple. Use the Killing form.)

23

The Fundamental Group

In this chapter, we will look more closely at the fundamental group of a compact Lie group G. We will show that it is a finitely generated Abelian group and that each loop in G can be deformed into any given maximal torus. Then we will show how to calculate the fundamental group. Along the way we will encounter another important Coxeter group, the *affine Weyl group*. The key arguments in this chapter are topological and are adapted from Adams [2].

Proposition 23.1. *Let G be a connected topological group and Γ a discrete normal subgroup. Then $\Gamma \subset Z(G)$.*

Proof. Let $\gamma \in \Gamma$. Then $g \longrightarrow g\gamma g^{-1}$ is a continuous map $G \longrightarrow \Gamma$. Since G is connected and Γ discrete, it is constant, so $g\gamma g^{-1} = \gamma$ for all g. Therefore, $\gamma \in Z(G)$. $\qquad\square$

Proposition 23.2. *If G is a connected Lie group, then the fundamental group $\pi_1(G)$ is Abelian.*

Proof. Let $p : \tilde{G} \longrightarrow G$ be the universal cover. We identify the kernel $\ker(p)$ with $\pi_1(G)$. This is a discrete normal subgroup of $\tilde{G}$ and hence is central in $\tilde{G}$ by Proposition 23.1. In particular, it is Abelian. $\qquad\square$

We remind the reader that an element of G is *regular* if it is contained in a unique maximal torus. Clearly, a generator of a maximal torus is regular. An element of G is *singular* if it is not regular. Let G_{reg} and G_{sing} be the subsets of regular and singular elements of G, respectively.

Proposition 23.3. *The set G_{sing} is a finite union of submanifolds of G, each of codimension at least 3.*

Proof. By Proposition 18.14, the singular elements of G are the conjugates of the kernels T_α of the roots. We first show that the union of the set of conjugates of T_α is the image of a manifold of codimension 3 under a smooth map. Let $\alpha \in \Phi$. The set of conjugates of T_α is the image of $G/C_G(T_\alpha) \times T_\alpha$ under the

D. Bump, *Lie Groups*, Graduate Texts in Mathematics 225,
DOI 10.1007/978-1-4614-8024-2_23, © Springer Science+Business Media New York 2013

smooth map $(gC_G(T_\alpha), u) \mapsto gug^{-1}$. Let $r = \dim(T)$, so $r - 1 = \dim(T_\alpha)$. The dimension of $C_G(T_\alpha)$ is at least $r+2$ since its complexified Lie algebra contains $\mathfrak{t}_\mathbb{C}$, $\mathfrak{X}_\alpha$, and $\mathfrak{X}_{-\alpha}$. Thus, the dimension of the manifold $G/C_G(T_\alpha) \times T_\alpha$ is at most $\dim(G) - (r+2) + (r-1) = \dim(G) - 3$.

However, we have asserted more precisely that G_{sing} is a union of submanifolds of codimension $\geqslant 3$. This more precise statement requires a bit more work. If $S \subset \Phi$ is any nonempty subset, let $U_S = \bigcap \{T_\alpha | \alpha \in S\}$. Let V_S be the open subset of U_S consisting of elements not contained in $U_{S'}$ for any larger S'. It is easily checked along the lines of (17.2) that the Jacobian of the map

$$(g\, C_G(U_S), u) \mapsto gug^{-1}, \qquad G/C_G(U_S) \times V_S \longrightarrow G,$$

is nonvanishing, so its image is a submanifold of G by the inverse function theorem. The union of these submanifolds is G_{sing}, and each has dimension $\leqslant \dim(G) - 3$.   □

Lemma 23.1. *Let X and Y be Hausdorff topological spaces and $f : X \longrightarrow Y$ a local homeomorphism. Suppose that $U \in X$ is a dense open set and that the restriction of f to U is injective. Then f is injective.*

Proof. If $x_1 \neq x_2$ are elements of X such that $f(x_1) = f(x_2)$, find open neighborhoods V_1 and V_2 of x_1 and x_2, respectively, that are disjoint, and such that f induces a homeomorphism $V_i \longrightarrow f(V_i)$. Note that $U \cap V_i$ is a dense open subset of V_i, so $f(U \cap V_i)$ is a dense open subset of $f(V_i)$. Since $f(V_1) \cap f(V_2) \neq \varnothing$, it follows that $f(U \cap V_1 \cap V_2)$ is nonempty. If $z \in f(U \cap V_1 \cap V_2)$, then there exist elements $y_i \in U \cap V_i$ such that $f(y_i) = z$. Since V_i are disjoint $y_1 \neq y_2$; yet $f(y_1) = f(y_2)$, a contradiction since $f|U$ is injective.   □

We define a map $\phi : G/T \times T_{\text{reg}} \longrightarrow G_{\text{reg}}$ by $\phi(gT, t) = gtg^{-1}$. It is the restriction to the regular elements of the map studied in Chap. 17.

Proposition 23.4.

(i) The map ϕ is a covering map of degree $|W|$.
(ii) If $t \in T_{\text{reg}}$, then the $|W|$ elements wtw^{-1}, $w \in W$ are all distinct.

Proof. For $t \in T_{\text{reg}}$, the Jacobian of this map, computed in (17.2), is nonzero. Thus the map ϕ is a local homeomorphism.

We define an action of $W = N(T)/T$ on $G/T \times T_{\text{reg}}$ by

$$w : (gT, t) \longrightarrow (gn^{-1}T, ngn^{-1}), \qquad w = nT \in W.$$

W acts freely on G/T, so the quotient map $G/T \times T_{\text{reg}} \longrightarrow W \backslash (G/T \times T_{\text{reg}})$ is a covering map of degree $|W|$. The map ϕ factors through $W \backslash (G/T \times T_{\text{reg}})$. Consider the induced map $\psi : W \backslash (G/T \times T_{\text{reg}}) \longrightarrow G_{\text{reg}}$. We have a commutative diagram:

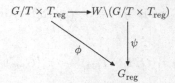

Both ϕ and the horizontal arrow are local homeomorphisms, so ψ is a local homeomorphism. By Proposition 17.3, the elements wtw^{-1} are all distinct for t in a dense open subset of T_{reg}. Thus, ψ is injective on a dense open subset of $W\backslash(G/T \times T_{\text{reg}})$, and since it is a local homeomorphism, it is therefore injective by Lemma 23.1. This proves both (i) and (ii).   □

Proposition 23.5. *Let* $p : X \longrightarrow Y$ *be a covering map. The map* $\pi_1(X) \longrightarrow \pi_1(Y)$ *induced by inclusion* $X \to Y$ *is injective.*

Proof. Suppose that p_0 and p_1 are loops in X with the same endpoints whose images in Y are path-homotopic. It is an immediate consequence of Proposition 13.2 that p_0 and p_1 are themselves path-homotopic.   □

Proposition 23.6. *The inclusion* $G_{\text{reg}} \longrightarrow G$ *induces an isomorphism of fundamental groups:* $\pi_1(G_{\text{reg}}) \cong \pi_1(G)$.

Proof. Of course, we usually take the base point of G to be the identity, but that is not in G_{reg}. Since G is connected, the isomorphism class of its fundamental group does not change if we move the base point P into G_{reg}.

If $p : [0,1] \longrightarrow G$ is a loop beginning and ending at P, the path may intersect G_{sing}. We may replace the path by a smooth path. Since G_{sing} is a finite union of submanifolds of codimension at least 3, we may move the path slightly and avoid G_{sing}. (For this we only need codimension 2.) Therefore, the induced map $\pi_1(G_{\text{reg}}) \longrightarrow \pi_1(G)$ is surjective.

Now suppose that p_0 and p_1 are two paths in G_{reg} that are path-homotopic in G. We may assume that both the paths and the homotopy are smooth. Since G_{sing} is a finite union of submanifolds of codimension at least 3, we may perturb the homotopy to avoid it, so p_0 and p_1 are homotopic in G_{reg}. Thus, the map $\pi_1(G_{\text{reg}}) \longrightarrow \pi_1(G)$ is injective.   □

Proposition 23.7. *We have* $\pi_1(G/T) = 1$.

In Exercise 27.4 we will see that the Bruhat decomposition gives an alternative proof of this fact.

Proof. Let $t_0 \in T_{\text{reg}}$ and consider the map $f_0 : G/T \longrightarrow G$, $f_0(gT) = gt_0g^{-1}$. We will show that the map $\pi_1(G/T) \longrightarrow \pi_1(G)$ induced by f_0 is injective. We may factor f_0 as

$$G/T \xrightarrow{v} G/T \times T_{\text{reg}} \xrightarrow{\phi} G_{\text{reg}} \longrightarrow G,$$

where the first map v sends $gT \longrightarrow (gT, t_0)$. We will show that each induced map

$$\pi_1(G/T) \xrightarrow{v} \pi_1(G/T \times T_{\text{reg}}) \xrightarrow{\phi} \pi_1(G_{\text{reg}}) \longrightarrow \pi_1(G) \qquad (23.1)$$

is injective. It should be noted that T_{reg} might not be connected, so $G/T \times T_{\text{reg}}$ might not be connected, and $\pi_1(G/T \times T_{\text{reg}})$ depends on the choice of a connected component for its base point. We choose the base point to be (T, t_0).

We can factor the identity map G/T as $G/T \xrightarrow{v} G/T \times T_{\text{reg}} \longrightarrow G/T$, where the second map is the projection. Applying the functor π_1, we see that $\pi_1(v)$ has a left inverse and is therefore injective. Also $\pi_1(\phi)$ is injective by Propositions 23.4 and 23.5, and the third map is injective by Proposition 23.6. This proves that the map induced by f_0 injects $\pi_1(G/T) \longrightarrow \pi_1(G)$.

However, the map $f_0 : G/T \to G$ is homotopic in G to the trivial map mapping G/t to a single point, as we can see by moving t_0 to $1 \in G$. Thus f_0 induces the trivial map $\pi_1(G/T) \longrightarrow \pi_1(G)$ and so $\pi_1(G/T) = 1$.   □

Theorem 23.1. *The induced map $\pi_1(T) \longrightarrow \pi_1(G)$ is surjective. The group $\pi_1(G)$ is finitely generated and Abelian.*

Proof. One way to see this is to use have the exact sequence

$$\pi_1(T) \longrightarrow \pi_1(G) \longrightarrow \pi_1(G/T)$$

of the fibration $G \longrightarrow G/T$ (Spanier [149, Theorem 10 on p. 377]). It follows using Proposition 23.7 that $\pi_1(T) \longrightarrow \pi_1(G)$ is surjective. Alternatively, we avoid recourse to the exact sequence to recall more directly why $\pi_1(G/T)$ implies that $\pi_1(T) \to \pi_1(G)$ is surjective. Given any loop in G, its image in G/T can be deformed to the identity, and lifting this homotopy to G deforms the original path to a path lying entirely in T.

As a quotient of the finitely generated Abelian group $\pi_1(T)$, the group $\pi_1(G)$ is finitely generated and Abelian.   □

Lemma 23.2. *Let $H \in \mathfrak{t}$.*

(i) Let $\lambda \in \Lambda$. Then $\lambda(e^H) = 1$ if and only if $\frac{1}{2\pi i} d\lambda(H) \in \mathbb{Z}$.
(ii) We have $e^H = 1$ if and only if $\frac{1}{2\pi i} d\lambda(H) \in \mathbb{Z}$ for all $\lambda \in L$.

Proof. Since $t \mapsto \lambda(e^{tH})$ is a character of $\mathbb{R}$ we have $\lambda(e^{tH}) = e^{2\pi i \theta t}$ for some $\theta = \theta(\lambda, H)$. Then $\theta = \frac{1}{2\pi i} \frac{d}{dt} \lambda(e^{tH})|_{t=0} = \frac{1}{2\pi i} d\lambda(H)$. On the other hand

$$\lambda(e^H) = \int_0^1 \frac{d}{dt} \lambda(e^{tH}) \, dt = \frac{1}{2\pi i \theta} \left[e^{2\pi i \theta} - 1 \right],$$

so $\lambda(e^H) = 1$ if and only if $\frac{1}{2\pi i} d\lambda(H) \in \mathbb{Z}$. And $\lambda(e^H) = 1$ for all $\lambda \in X^*(T)$ if and only if $e^H = 1$.   □

Since the map $\pi_1(T) \longrightarrow \pi_1(G)$ is surjective, we may study the fundamental group of G by determining the kernel of this homomorphism. The group $\pi_1(T)$ is easy to understand: the Lie algebra $\mathfrak{t}$ is simply-connected, and the exponential map $\exp : \mathfrak{t} \longrightarrow T$ is a surjective group homomorphism that is a covering map. Thus we may identify $\mathfrak{t}$ with the universal cover of T, and with this identification $\pi_1(T)$ is just the kernel of the exponential map $\mathfrak{t} \to T$. Moreover, the next result shows how we may further identify $\pi_1(T)$ with the *coweight lattice*, which is the lattice $\Lambda^\vee$ of linear functionals on $\mathcal{V}$ that map Λ into $\mathbb{Z}$.

Proposition 23.8. *Define* $\tau : \mathfrak{t} \to \mathcal{V}^*$ *by letting* $\tau(H) \in \mathcal{V}^*$ *be the linear functional that sends* λ *to* $\frac{1}{2\pi i}\mathrm{d}\lambda(H)$. *Then* τ *is a linear isomorphism, and* τ *maps the kernel of* $\exp : \mathfrak{t} \to T$ *to* $\Lambda^\vee$. *If* $\alpha^\vee$ *is a coroot, then*

$$\alpha^\vee = \tau(2\pi i H_\alpha). \tag{23.2}$$

Proof. It is clear that τ is a linear isomorphism. It follows from Lemma 23.2(ii) that it maps the kernel of $\exp$ onto the coweight lattice. The identity (23.2) follows from Proposition 18.13.   □

For each $\alpha \in \Phi$ and each $k \in \mathbb{Z}$ define the hyperplane $\mathfrak{H}_{\alpha,k} \subset \mathcal{V}$ to be

$$\{H \in \mathfrak{t} | \tau(H)(\alpha) = k\}.$$

By Lemma 23.2, the preimage of T_α in $\mathfrak{t}$ under the exponential map is the union of the $\mathfrak{H}_{\alpha,k}$.

The geometry of these hyperplanes leads to the affine Weyl group (Bourbaki [23, Chap. IV Sect. 2]). This structure goes back to Stiefel [156] and has subsequently proved very important. Adams [2] astutely based his discussion of the fundamental group on the affine Weyl group. The fundamental group is also discussed in Bourbaki [24, Chap. IX]. The affine Weyl group was used by Iwahori and Matsumoto [86] to introduce a Bruhat decomposition into a reductive p-adic group. The geometry introduced by Stiefel also reappears as the "apartment" in the Bruhat-Tits building. Iwahori and Matsumoto also introduced the affine Hecke algebra as a convolution algebra of functions on a p-adic group, but the affine Hecke algebra has an importance in other areas of mathematics for example because of its role in Kazhdan-Lusztig theory.

In addition to the simple positive roots $\alpha_1, \ldots, \alpha_r$, let $-\alpha_0$ be the highest weight in the adjoint representation. It is a positive root, so α_0 is a negative root, the so-called *affine root* which will appear later in Chap. 30. We call a connected component of the complement of the hyperplanes $\mathfrak{H}_{\alpha,k}$ an *alcove*. To identify a particular one, there is a unique alcove that is contained in the positive Weyl chamber which contains the origin in its closure. This alcove is the region bounded by the hyperplanes $\mathfrak{H}_{a_1,0}, \ldots, \mathfrak{H}_{a_r,0}$ and $\mathfrak{H}_{-\alpha_0,1} = \mathfrak{H}_{\alpha_0,-1}$. It will be called the *fundamental alcove* $\mathcal{F}$.

We have seen (by Lemma 23.2) that the lattice $\Lambda^\vee$ where the weights take integer values may be identified with the kernel of the exponential map

$\mathfrak{t} \longrightarrow T$. Thus $\Lambda^\vee$ may be identified with the fundamental group of T, which (by Proposition 23.1) maps surjectively onto $\pi_1(G)$. Therefore we need to compute the kernel of this homomorphism $\Lambda^\vee \longrightarrow \pi_1(G)$. We will show that it is the coroot lattice $\Lambda^\vee_{\text{coroot}}$, which is the sublattice generated by the coroots $\alpha^\vee$ ($\alpha \in \Phi$).

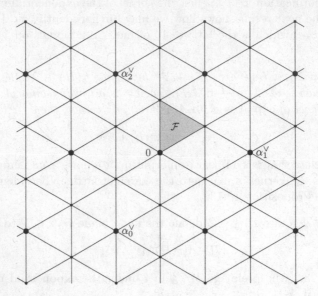

Fig. 23.1. The Cartan subalgebra $\mathfrak{t}$, partitioned into alcoves, when $G = \mathrm{SU}(3)$ or $\mathrm{PU}(3)$. We are identifying $\mathfrak{t} = \mathcal{V}^*$ via the isomorphism τ, so the coroots are in $\mathfrak{t}$

Before turning to the proofs, let us consider an example. Let $G = \mathrm{PU}(3)$, illustrated in Fig. 23.1. The hyperplanes $\mathfrak{H}_{\alpha,k}$ are labeled, subdividing $\mathfrak{t}$ into alcoves, and the fundamental alcove $\mathcal{F}$ is shaded. The coweight lattice $\Lambda^\vee \cong \pi_1(T)$ consists of the vertices of the alcoves, which are the smaller dots. The heavier dots mark the coroot lattice.

If we consider instead $G = \mathrm{SU}(3)$, the diagram would be the same with only one difference: now not all the vertices of alcoves are in $\Lambda^\vee$. In this example, $\Lambda^\vee = \Lambda_{\text{coroot}}$ consists of only the heavier dots. Not every vertex of the fundamental alcove is in the coweight lattice.

If $\alpha \in \Phi$ and $k \in \mathbb{Z}$ let $s_{\alpha,k}$ be the reflection in the hyperplane $\mathfrak{H}_{\alpha,k}$. Thus

$$s_{\alpha,k}(x) = x - (\alpha(x) - k)\,\alpha^\vee.$$

Let W_{aff} be the group of transformations of $\mathfrak{t}$ generated by the reflections in the hyperplanes $\mathfrak{H}_{\alpha,k}$. This is the *affine Weyl group*. In particular we will label the reflections in the walls of the fundamental alcove $s_0, s_1, \ldots, s_r$, where if $1 \leqslant i \leqslant r$ then s_i is the reflection in $\mathfrak{H}_{\alpha_i,0}$, and s_0 is the reflection in $\mathfrak{H}_{\alpha_0,-1}$.

Proposition 23.9.

(i) *The affine Weyl group acts transitively on the alcoves.*

(ii) *The group W_{aff} is generated by $s_0, s_1, \ldots, s_r$.*

(iii) *The group W_{aff} contains the group of translations by elements of $\Lambda_{\text{coroot}}^{\vee}$ as a normal subgroup and is the semidirect product of W and the translation group of $\Lambda_{\text{coroot}}^{\vee}$.*

Proof. Let W'_{aff} be the subgroup $s_0, s_1, \ldots, s_r$. Consider the orbit $W'_{\text{aff}} \mathcal{F}$ of alcoves. If $\mathcal{F}_1$ is in this orbit, and $\mathcal{F}_2$ is another alcove adjacent to $\mathcal{F}_1$, then $\mathcal{F} = w\mathcal{F}_1$ for some $w \in W'_F$ and so $w\mathcal{F}_2$ is adjacent to $\mathcal{F}$, i.e. $w\mathcal{F}_2 = s_i\mathcal{F}$ for some s_i. It follows that $\mathcal{F}_2$ is in $W'_{\text{aff}} \mathcal{F}$ also. Since every alcove adjacent to an alcove in $W'_{\text{aff}} \mathcal{F}$ is also in it, it is clear that $W'_{\text{aff}} \mathcal{F}$ consists of all alcoves, proving (i).

We may now show that $W_{\text{aff}} = W'_{\text{aff}}$. It is enough to show that the reflection $r = s_{\alpha,k}$ in the hyperplane $\mathfrak{H}_{\alpha,k}$ is in W'_{aff}. Let $\mathcal{A}$ be an alcove that is adjacent to $\mathfrak{H}_{\alpha,k}$, and find $w \in W'_{\text{aff}}$ such that $w(\mathcal{A}) = \mathcal{F}$. Then w maps $\mathfrak{H}_{\alpha,k}$ to one of the walls of $\mathcal{F}$, so wrw^{-1} is the reflection in that wall. This means $wrw^{-1} = s_i$ for some i and so $r = w^{-1}s_i w \in W'_{\text{aff}}$. This completes the proof that W_{aff} is generated by the s_i.

Now we recall that the group $\mathfrak{G}$ of all affine linear maps of the real vector space $\mathfrak{t}$ is the semidirect product of the group of linear transformations (which fix the origin) by the group of translations, a normal subgroup. If $v \in \mathfrak{t}$ we will denote by $T(v)$ the translation $x \mapsto x + v$. So $T(\mathfrak{t})$ is normal in $\mathfrak{G}$ and $\mathfrak{G} = \text{GL}(\mathfrak{t}) \cdot T(\mathfrak{t})$. This means that $\mathfrak{G}/T(\mathfrak{t}) \cong \text{GL}(\mathfrak{t})$. The homomorphism $\mathfrak{G} \longrightarrow \text{GL}(\mathfrak{t})$ maps the reflection in $\mathfrak{H}_{\alpha,k}$ to the reflection in the parallel hyperplane $\mathfrak{H}_{\alpha,0}$ through the origin. This induces a homomorphism from W_{aff} to W, and we wish to determine the kernel $\mathfrak{K}$, which is the group of translations in W_{aff}.

First observe $T(\alpha^{\vee})$ is in W_{aff}, since it is the reflection in $\mathfrak{H}_{\alpha,0}$ followed by the reflection in $\mathfrak{H}_{\alpha,1}$. Let us check that $\Lambda_{\text{coroot}}^{\vee}$ is normalized by W. Indeed we check easily that $s_{\alpha,0}T(\beta^{\vee})s_{\alpha_0}$ is translation by $s_{\alpha,0}(\beta^{\vee}) = \beta^{\vee} - \alpha(\beta^{\vee})\alpha^{\vee} \in \Lambda_{\text{coroot}}^{\vee}$. Therefore $W \cdot T(\Lambda_{\text{coroot}}^{\vee})$ is a subgroup of W'_{aff}. Finally, we note that $s_{\alpha,k} = T(k\alpha^{\vee})s_{\alpha,0}$, so $W \cdot T(\Lambda_{\text{coroot}}^{\vee})$ contains generators for W_{aff}, and the proof is complete. $\square$

The group W_{extended} generated by W_{aff} and translations by Λ may equal W_{aff} or it may be larger. This often larger group is called the *extended affine Weyl group*. We will not need it, but it is important and we mention it for completeness.

We have constructed a surjective homomorphism $\Lambda^{\vee} \longrightarrow \pi_1(G)$. To recapitulate, the exponential map $\mathfrak{t} \longrightarrow T$ is a covering by a simply-connected group, so its kernel $\Lambda^{\vee}$ may be identified with the fundamental group $\pi_1(T)$, and we have seen that the map $\pi_1(T) \longrightarrow \pi_1(G)$ induced by inclusion is surjective.

We recall that if X is a topological space then a path $p : [0,1] \longrightarrow X$ is called a *loop* if $p(0) = p(1)$.

Lemma 23.3. *Let $\psi : Y \longrightarrow X$ be a covering map, and let $p : [0,1] \longrightarrow Y$ be a path. If $\psi \circ p : [0,1] \longrightarrow X$ is a loop that is contractible in X, then p is a loop.*

Proof. Let $q = \phi \circ p$, and let $x = q(0)$. Let $y = p(0)$, so $\phi(y) = x$. What we know is that $q(1) = x$ and what we need to prove is that $p(1) = y$.

Since q is contractible in X, we may find a family q_u of paths indexed by $u \in [0,1]$ such that $q_0 = q$ while q_1 is the constant path $q_1(t) = x$. We may choose the deformation so that neither end point moves, that is, $q_u(0) = q_u(1) = x$ for all u. For each u, by the path lifting property of covering maps [Proposition 13.2(i)] we may find a unique path $p_u : [0,1] \longrightarrow Y$ such that $\psi \circ p_u = q_u$ and $p_u(0) = y$. In particular, $p_0 = p$. It is clear that $p_u(t)$ varies continuously as a function of both t and u. Since q_1 is constant, so is p_1 and therefore $p_1(1) = p_1(0) = y$. Now $u \mapsto p_u(1)$ is a path covering a constant path, hence it too is constant, so $p_0(1) = p_1(1) = y$. □

Proposition 23.10. *The kernel of the surjective homomorphism $\Lambda^\vee \longrightarrow \pi_1(G)$ that we have described is $\Lambda^\vee_{\mathrm{coroot}}$. Thus the fundamental group $\pi_1(G)$ is isomorphic to $\Lambda^\vee / \Lambda^\vee_{\mathrm{coroot}}$.*

Proof. First let us show that a coroot $\alpha^\vee$ is in the kernel of this homomorphism. In view of (23.2) this means, concretely that if we take a path from 0 to $2\pi i H_\alpha$ in $\mathfrak{t}$ then the exponential of this map (which has both left and right endpoint at the identity) is contractible in G. We may modify the path so that it is the straight-line path, passing through $\pi i H_\alpha$. Then we write it as the concatenation of two paths, $p \star q$ in the notation of Chap. 13, where $p(0) = 0$ and $p(1) = \pi i H_\alpha$ while $q(0) = \pi i H_\alpha$ and $q(1) = 2\pi i H_\alpha$.

The exponential of this path $e^{p \star q} = e^p \star e^q$ is a loop. In fact, we have

$$e^{p(0)} = e^{q(1)} = i_\alpha \begin{pmatrix} 1 & \\ & 1 \end{pmatrix}, \qquad e^{p(1)} = e^{q(0)} = i_\alpha \begin{pmatrix} -1 & \\ & -1 \end{pmatrix}.$$

We will deform the path q, leaving p unchanged. Let

$$g(u) = i_\alpha \begin{pmatrix} \cos(\pi u/2) & -\sin(\pi u/2) \\ \sin(\pi u/2) & \cos(\pi u/2) \end{pmatrix},$$

and consider $q_u = \mathrm{Ad}(g(u))q$. The endpoints of q_u do change as u goes from 0 to 1, but the endpoints of e^{q_u} do not. Indeed $e^{q_u(0)} = g(u) i_\alpha(-I) g(u)^{-1} = i_\alpha(-I)$ and similarly $e^{q_u(1)} = i_\alpha(-1)$. Thus the path $e^{p \star q}$ is homotopic to $e^p \star e^{q_1}$. Now $e^{q_1} = w_\alpha e^q w_\alpha^{-1}$, which is the negative of the path e^p, being the exponential of the straight line path from $-\pi i H_\alpha$ to $-2\pi i H_\alpha$. This proves that $e^{p \star q}$ is path-homotopic to the identity.

Thus far we have shown that $\Lambda_{\mathrm{coroot}}$ is in the kernel of the homomorphism $\Lambda^\vee \longrightarrow \pi_1(G)$. To complete the proof, we will now show that if $K \in \Lambda^\vee$ maps to the identity in $\pi_1(G)$, then $K \in \Lambda^\vee_{\mathrm{coroot}}$. We note that there are $|W|$ alcoves that are adjacent to the origin; these are the alcoves $w\mathcal{F}$ with $w \in W$. We will show that we may assume that K lies in the closure of one of these alcoves. Indeed $\mathcal{F} + K$ is an alcove, so there is some $w' \in W_{\text{aff}}$ such that $\mathcal{F} + K = w'\mathcal{F}$. Moreover, w' may be represented as $T(K')w$ with $K' \in \Lambda^{\vee}_{\text{coroot}}$ and $w \in W$. Thus $\mathcal{F} + K - K' = w\mathcal{F}$ and since we have already shown that K' maps to the identity in $\pi_1(G)$, we may replace K by $K - K'$ and assume that K is in the closure of $w\mathcal{F}$.

Our goal is to prove that $K = 0$, which will of course establish that establish $K \in \Lambda^{\vee}_{\text{coroot}}$, as required. Since K and the origin are in the same alcove, we may find a path $p : [0, 1] \longrightarrow \mathfrak{t}$ from 0 to K such that $p(u)$ is in the interior of the alcove for $p(u) \neq 0, 1$, while $p(0) = 0$ and $p(1) = K$ are vertices of the alcove.

Let $\mathfrak{t}_{\text{reg}}$ be the preimage of T_{reg} in $\mathfrak{t}$ under the exponential map. It is the complement of the hyperplanes $\mathfrak{H}_{\alpha,k}$, or equivalently, the union of the alcove interiors. We will make use of the map $\psi : G/T \times \mathfrak{t}_{\text{reg}} \longrightarrow G_{\text{reg}}$ defined by $\phi(gT, H) = ge^H g^{-1}$. It is the composition of the covering map ϕ in Proposition 23.4 with the exponential map $\mathfrak{t}_{\text{reg}} \longrightarrow T_{\text{reg}}$, which is also a covering, so ψ is a covering map.

Let N be a connected neighborhood of the identity in G that is closed under conjugation such that the preimage under exp of N in $\mathfrak{t}$ consists of disjoint open sets, each containing an a single element of the kernel $\Lambda^{\vee}$ of the exponential map on $\mathfrak{t}$. Let $N_{\mathfrak{t}} = \mathfrak{t} \cap \exp^{-1}(N)$ be this reimage. Each connected component of $N_{\mathfrak{t}}$ contains a unique element of $\Lambda^{\vee}$.

We will modify p so that it is outside $\mathfrak{t}_{\text{reg}}$ only near $t = 1$. Let $H \in \mathfrak{t}$ be a small vector in the interior of $w\mathcal{F} \cap N$ and let $p' : [0, 1] \longrightarrow \mathfrak{t}$ be the path shifted by H. The vector H can be chosen in the connected component of $N_{\mathfrak{t}}$ that contains 0 but no other element of $\Lambda^{\vee}$. So $p'(0) = H$ and $p'(1) = K + H$. When t is near 1 the path p' may cross some of the $\mathfrak{H}_{\alpha,k}$ but only inside N.

The exponentiated path $e^{p'(t)}$ will be a loop close to $e^{p(t)}$ hence contractible. And $e^{p'(t)}$ will be near the identity $1 = e^K$ in G at the end of the path where $p'(t)$ may not be in the interior of $w\mathcal{F}$. Because, by Proposition 23.3, G_{reg} is a union of codimension $\geqslant 3$ submanifolds of G, we may find a loop $q'' : [0, 1] \longrightarrow G_{\text{reg}}$ that coincides with $e^{p'}$ until near the end where $p'(t)$ is near $v + H$ and $e^{p'(t)}$ reenters N. At the end of the path, q'' dodges out of T to avoid the singular subset of G, but stays near the identity. More precisely, if q'' is close enough to $e^{p'}$ the paths will agree until $e^{p'(t)}$ and q'' are both inside N.

The loop q'' will have the same endpoints as $e^{p'}$. It is still contractible by Proposition 23.6. Therefore we may use the path lifting property of covering maps [Proposition 13.2(i)] to lift the path q'' to a path $p'' : [0, 1] \longrightarrow G/T \times \mathfrak{t}_{\text{reg}}$ by means of ψ, with $p''(0) = (1 \cdot T, H)$, the lifted path is a loop by Lemma 23.3. Thus $p''(1) = p''(0) = (1 \cdot T, H)$. Now consider the projection p''' of p'' onto $\mathfrak{t}_{\text{reg}}$. At the end of the path, the paths $e^{p'}$ and $\psi \circ p''$ are both within N, so p' and p''' can only vary within $N_{\mathfrak{t}}$. In particular their endpoints, which are $H + K$ and H respectively, must be the same, so $K = 0$ as required. $\square$

Proposition 23.11. *Let G be a compact connected Lie group. The following are equivalent.*

(i) The root system Φ spans $\mathcal{V} = \mathbb{R} \otimes X^*(T)$.
(ii) The fundamental group $\pi_1(G)$ is finite.
(iii) The center $Z(G)$ is finite.

If these conditions are satisfied, G is called *semisimple*.

Proof. The root lattice spans $\mathcal{V}$ if and only if the coroot lattice spans $\mathcal{V}^*$, which we are identifying with $\mathfrak{t}$. Since $\Lambda^\vee$ is a lattice in $\mathfrak{t}$ of rank equal to $\dim(\mathcal{V}) = \dim(T)$, the coroot lattice spans $\mathcal{V}^*$ if and only if $\Lambda^\vee / \Lambda_{\text{coroot}}$ is finite, and this, we have seen, is isomorphic to the fundamental group. Thus (i) and (ii) are equivalent. The center $Z(G)$ is the intersection of the T_α by Proposition 18.14. Thus the Lie algebra $\mathfrak{z}$ of $Z(G)$ is zero if and only if the $\mathfrak{t}_\alpha$, which are the kernels of the roots interpreted as linear functionals on $\mathfrak{t} = \mathcal{V}^*$. So $Z(G)$ is finite if and only if $\mathfrak{z} = 0$, if and only if the roots span $\mathcal{V}$. The equivalence of all three statements is now proved. $\square$

Let us now assume that G is semisimple. Let $\tilde{\Lambda}$ be the set of $\lambda \in \mathcal{V}$ such that $\alpha^\vee(\lambda) \in \mathbb{Z}$ for all coroots α, and let $\tilde{\Lambda}^\vee$ be the set of $H \in \mathcal{V}^* = \mathfrak{t}$ such that $\alpha(H) \in \mathbb{Z}$ for all $\alpha \in \Phi$.

The following result gives a complete determination of both the center and the fundamental group.

Theorem 23.2. *Assume that G is semisimple. Then*

$$\tilde{\Lambda} \supseteq \Lambda \supseteq \Lambda_{\text{root}}, \qquad \tilde{\Lambda}^\vee \supseteq \Lambda^\vee \supseteq \Lambda^\vee_{\text{coroot}}.$$

Regarding these as lattices in the dual real vector spaces $\mathcal{V}$ and $\mathcal{V}^*$, which we have identified with $\mathbb{R} \otimes X^*(T)$ and with $\mathfrak{t}$, respectively, $\tilde{\Lambda}$ is the dual lattice of $\Lambda^\vee_{\text{coroot}}$, Λ is the dual lattice of $\Lambda^\vee$ and Λ_{root} is the dual lattice of $\tilde{\Lambda}^\vee$. Both $\pi_1(G)$ and $Z(G)$ are finite Abelian groups and

$$\pi_1(G) \cong \tilde{\Lambda}/\Lambda \cong \Lambda^\vee/\Lambda^\vee_{\text{coroot}}, \qquad Z(G) \cong \tilde{\Lambda}^\vee/\Lambda^\vee \cong \Lambda/\Lambda_{\text{root}}. \qquad (23.3)$$

Proof. By Proposition 18.10 we have $\tilde{\Lambda} \supseteq \Lambda$ and $\Lambda \supseteq \Lambda_{\text{root}}$ is clear since roots are characters of $X^*(T)$. That Λ and $\Lambda^\vee$ are dual lattices is Lemma 23.2. That Λ_{root} and $\tilde{\Lambda}^\vee$ are dual lattices and that $\tilde{\Lambda}$ and $\Lambda^\vee_{\text{coroot}}$ are dual lattices are both by definition. The inclusions $\tilde{\Lambda} \supseteq \Lambda \supseteq \Lambda_{\text{root}}$ then imply $\tilde{\Lambda}^\vee \supseteq \Lambda^\vee \supseteq \Lambda^\vee_{\text{root}}$. Moreover, $\tilde{\Lambda}/\Lambda \cong \Lambda^\vee/\Lambda^\vee_{\text{root}}$ follows from the fact that two Abelian groups in duality are isomorphic, and the vector space pairing $\mathcal{V} \times \mathcal{V}^* \longrightarrow \mathbb{R}$ induces a perfect pairing $\tilde{\Lambda}/\Lambda \times \Lambda^\vee/\Lambda^\vee_{\text{root}} \longrightarrow \mathbb{R}/\mathbb{Z}$. Similarly $\tilde{\Lambda}^\vee/\Lambda^\vee \cong \Lambda/\Lambda_{\text{root}}$. The fact that $\pi_1(G) \cong \Lambda^\vee/\Lambda^\vee_{\text{coroot}}$ follows from Proposition 23.10. It remains to be shown that $Z(G) \cong \tilde{\Lambda}^\vee/\Lambda^\vee$. We know from Proposition 18.14 that $Z(G)$ is the intersection of the T_α. Thus $H \in \mathfrak{t}$ exponentiates into $Z(G)$ if it is in the kernel of all the root groups, that is, if $\alpha(H) \in \mathbb{Z}$ for all $\alpha \in \Phi$. This means that the exponential induces a surjection $\tilde{\Lambda}^\vee \longrightarrow Z(G)$. Since $\Lambda^\vee$ is the kernel of the exponential on $\mathfrak{t}$, the statement follows. $\square$

Proposition 23.12. *If G is semisimple and simply-connected, then $\tilde{\Lambda} = \Lambda$.*

Proof. This follows from (23.3) with $\pi_1(G) = 1$. □

Exercises

Exercise 23.1. If $\mathfrak{g}$ is a Lie algebra let $[\mathfrak{g}, \mathfrak{g}]$ be the vector space spanned by $[X, Y]$ with $X, Y \in \mathfrak{g}$. Show that $[\mathfrak{g}, \mathfrak{g}]$ is an ideal of $\mathfrak{g}$.

Exercise 23.2. Suppose that $\mathfrak{g}$ is a real or complex Lie algebra. Assume that there exists an invariant inner product $B : \mathfrak{g} \times \mathfrak{g} \longrightarrow \mathbb{C}$. Thus B is positive definite symmetric or Hermitian and satisfies the ad-invariance property (10.3). Let $\mathfrak{z}$ be the center of $\mathfrak{g}$. Show that the orthogonal complement of $\mathfrak{g}$ is $[\mathfrak{g}, \mathfrak{g}]$.

Exercise 23.3. Let G be a semisimple group of adjoint type, and let G' be its universal cover. Show that the fundamental group of G is isomorphic to the center of G'. (Both are finite Abelian groups.)

Exercise 23.4.

(i) Consider a simply-connected semisimple group G. Explain why $\Lambda = \tilde{\Lambda}$ in the notation of Theorem 23.2.
(ii) Using the description of the root lattices for each of the four classical Cartan types in Chap. 19, consider a simply-connected semisimple group G and compute the weight lattice Λ using Theorem 23.2. Confirm the following table:

Cartan type	Fundamental group
A_r	$\mathbb{Z}_{r+1}$
B_r	$\mathbb{Z}_2$
C_r	$\mathbb{Z}_2$
D_r, r odd	$\mathbb{Z}_4$
D_r, r even	$\mathbb{Z}_2 \times \mathbb{Z}_2$

Exercise 23.5. If $\mathfrak{g}$ is a Lie algebra, the *center* of $\mathfrak{g}$ is the set of all $Z \in \mathfrak{g}$ such that $[Z, X] = 0$ for all $X \in G$. Show that if G is a connected Lie group with Lie algebra $\mathfrak{g}$ then the center of $\mathfrak{g}$ is the Lie algebra of $Z(G)$.

Exercise 23.6. (i) Let $\mathfrak{g}$ be the Lie algebra of a compact Lie group G. If $\mathfrak{a}$ is an Abelian ideal, show that $\mathfrak{a}$ is contained in the center of $\mathfrak{g}$.
(ii) Show by example that this may fail without the assumption that $\mathfrak{g}$ is the Lie algebra of a compact Lie group. Thus give a Lie algebra and an Abelian ideal that is not central.

Exercise 23.7. Let G be a compact Lie group and $\mathfrak{g}$ its Lie algebra. Let T, $\mathfrak{t}$, and other notations be as in this chapter. Let $\mathfrak{t}'$ be the linear span of the coroots $\alpha^\vee = 2\pi i H_\alpha$. Let $\mathfrak{z}$ be the center of $\mathfrak{g}$. Show that $\mathfrak{t} = \mathfrak{t}' \oplus \mathfrak{z}$.

Part III

Noncompact Lie Groups

24

Complexification

Thus far, we have investigated the representations of compact connected Lie groups. In this chapter, we will see how the representation theory of compact connected Lie groups has implications for at least some noncompact Lie groups.

Let K be a connected Lie group. A *complexification* of K consists of a complex analytic group G with a Lie group homomorphism $i : K \longrightarrow G$ such that whenever $f : K \longrightarrow H$ is a Lie group homomorphism into a complex analytic group, there exists a unique analytic homomorphism $F : G \longrightarrow H$ such that $f = F \circ i$. This is a universal property, so it characterizes G up to isomorphism.

A consequence of this definition is that the finite-dimensional representations of K are in bijection with the finite-dimensional *analytic* representations of G. Indeed, we may take H to be $\mathrm{GL}(n, \mathbb{C})$. A finite-dimensional representation of K is a Lie group homomorphism $K \longrightarrow \mathrm{GL}(n, \mathbb{C})$, and so any finite-dimensional representation of K extends uniquely to an analytic representation of G.

Proposition 24.1. *The group* $\mathrm{SL}(n, \mathbb{C})$ *is the complexification of the Lie group* $\mathrm{SL}(n, \mathbb{R})$.

Proof. Given any complex analytic group H and any Lie group homomorphism $f : \mathrm{SL}(n, \mathbb{R}) \longrightarrow H$, the differential is a Lie algebra homomorphism $\mathfrak{sl}(n, \mathbb{R}) \longrightarrow \mathrm{Lie}(H)$. Since $\mathrm{Lie}(H)$ is a complex Lie algebra, this homomorphism extends uniquely to a complex Lie algebra homomorphism $\mathfrak{sl}(n, \mathbb{C}) \longrightarrow \mathrm{Lie}(H)$ by Proposition 11.3. By Theorems 13.5 and 13.6, $\mathrm{SL}(n, \mathbb{C})$ is simply connected, so by Theorem 14.2 this map is the differential of a Lie group homomorphism $F : \mathrm{SL}(n, \mathbb{C}) \longrightarrow H$. We need to show that F is analytic. Consider the commutative diagram

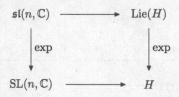

The top, left, and right arrows are all holomorphic maps, and exp : $\mathfrak{sl}(n, \mathbb{C}) \longrightarrow$ $SL(n, \mathbb{C})$ is a local homeomorphism in a neighborhood of the identity. Hence F is holomorphic near 1. If $g \in SL(n, \mathbb{C})$ and if $l(g) : SL(n, \mathbb{C}) \longrightarrow SL(n, \mathbb{C})$ and $l\big(F(g)\big) : H \longrightarrow H$ denote left translation with respect to g and $F(g)$, then $l(g)$ and $l\big(F(g)\big)$ are analytic, and $F = l\big(F(g)\big) \circ F \circ l(g)^{-1}$. Since F is analytic at 1, it follows that it is analytic at g.   □

We recall from Chap. 14, particularly the proof of Proposition 14.1, that if G is a Lie group and $\mathfrak{h}$ a Lie subalgebra of $\text{Lie}(G)$, then there is an involutory family of tangent vectors spanned by the left-invariant vector fields corresponding to the elements of $\mathfrak{h}$. Since these vector fields are left-invariant, this involutory family is invariant under left translation.

Proposition 24.2. *Let G be a Lie group and let $\mathfrak{h}$ be a Lie subalgebra of $\text{Lie}(G)$. Let H be a closed connected subset of G that is an integral submanifold of the involutory family associated with $\mathfrak{h}$, and suppose that $1 \in H$. Then H is a subgroup of G.*

One must not conclude from this that every Lie subalgebra of $\text{Lie}(G)$ is the Lie algebra of a closed Lie subgroup. For example, if $G = (\mathbb{R}/\mathbb{Z})^2$, then the one-dimensional subalgebra spanned by a vector $(x_1, x_2) \in \text{Lie}(G) = \mathbb{R}^2$ is the Lie algebra of a closed subgroup only if x_1/x_2 is rational or $x_2 = 0$.

Proof. Let $x \in H$ and let $U = \{y \in H \,|\, x^{-1}y \in H\}$.

We show that U is open in H. If $y \in U = H \cap xH$, both H and xH are integral submanifolds for the involutory family associated with $\mathfrak{h}$, since the vector fields corresponding to elements of $\mathfrak{h}$ are left-invariant. Hence by the uniqueness assertion of the local Frobenius theorem (Theorem 14.1) H and xH have the same intersection with a neighborhood of y in G, and it follows that U contains a neighborhood of y in H.

We next show that the complement of U is open in H. Suppose that y is an element $H - U$. Thus, $y \in H$ but $x^{-1}y \notin H$. By the local Frobenius theorem there exists an integral manifold V through $x^{-1}y$. Since H is closed, the intersection of V with a sufficiently small neighborhood of $x^{-1}y$ in G is disjoint from H. Replacing V by its intersection with this neighborhood, we may assume that the intersection $xV \cap H = \varnothing$. Since H and xV are both integral manifolds through y, they have the same intersection with a neighborhood of y in G, and so $xz \in V$ for z near y in H. Thus, $z \notin U$. It follows that $H - U$ is open.

We see that U is both open and closed in H and nonempty since $1 \in U$. Since H is connected, it follows that $U = H$. This proves that if $x, y \in H$, then $x^{-1}y \in H$. This implies that H is a subgroup of G. $\qquad\square$

Theorem 24.1. *Let K be a compact connected Lie group. Then K has a complexification $K \longrightarrow G$, where G is a complex analytic group. The induced map $\pi_1(K) \longrightarrow \pi_1(G)$ is an isomorphism. The Lie algebra of G is the complexification of the Lie algebra of K. Any faithful complex representation of K can be extended to a faithful analytic complex representation of G. Any analytic representation of G is completely reducible.*

Proof. By Theorem 4.2, K has a faithful complex representation, which is unitarizable, so we may assume that K is a closed subgroup of $U(n)$ for some n. The embedding $K \longrightarrow U(n)$ is the differential of a Lie algebra homomorphism $\mathfrak{k} \longrightarrow \mathfrak{gl}(n, \mathbb{C})$, where $\mathfrak{k}$ is the Lie algebra of K. This extends, by Proposition 11.3, to a homomorphism of complex Lie algebras $\mathfrak{k}_{\mathbb{C}} \longrightarrow \mathfrak{gl}(n, \mathbb{C})$, and we identify $\mathfrak{k}_{\mathbb{C}}$ with its image.

Let $P = \{e^{iX} | X \in \mathfrak{k}\} \subset GL(n, \mathbb{C})$, and let $G = PK$. Let $P' \subset GL(n, \mathbb{C})$ be the set of positive definite Hermitian matrices. By Theorem 13.4, the multiplication map $P' \times U(n) \longrightarrow GL(n, \mathbb{C})$ is a homeomorphism. Moreover, the exponentiation map from the vector space of Hermitian matrices to P' is a homeomorphism. Since $i\mathfrak{k}$ is a closed subspace of the real vector space of Hermitian matrices, P is a closed topological subspace of P', and $G = PK$ is a closed subset of $GL(n, \mathbb{C}) = P'U(n)$.

We associate with each element of $\mathfrak{k}_{\mathbb{C}}$ a left-invariant vector field on $GL(n, \mathbb{C})$ and consider the resulting involutory family on $GL(n, \mathbb{C})$. We will show that G is an integral submanifold of this involutory family. We must check that the left-invariant vector field associated with an element Z of $\mathfrak{k}_{\mathbb{C}}$ is everywhere tangent to G. It is easiest to check this separately in the cases $Z = Y$ and $Z = iY$ with $Y \in \mathfrak{k}$. Near the point $e^{iX}k \in G$, with $X \in \mathfrak{k}$ and $k \in K$, the path $t \longrightarrow e^{i(X + t\mathrm{Ad}(k)Y)}k$ is tangent to G when $t = 0$ and is also tangent to the path

$$t \mapsto e^{iX}e^{it\mathrm{Ad}(k)Y}k = e^{iX}ke^{itY}.$$

(The two paths are not identical if $[X, Y] \neq 0$, but this is not a problem.) The latter path is the left translate by $e^{iX}k$ of a path through the identity tangent to the left-invariant vector field corresponding to $iY \in \mathfrak{k}$. Since this vector field is left invariant, this shows that it is tangent to G at $e^{iX}k$. This settles the case $Z = iY$. The case where $Z = Y$ is similar and easier.

It follows from Proposition 24.2 that G is a closed subgroup of $GL(n, \mathbb{C})$. Since P is homeomorphic to a vector space, it is contractible, and since G is homeomorphic to $P \times K$, it follows that the inclusion $K \longrightarrow G$ induces an isomorphism of fundamental groups.

The Lie algebra of G is, by construction, $i\mathfrak{k} + \mathfrak{k} = \mathfrak{k}_{\mathbb{C}}$.

To show that G is the complexification of K, let H be a complex analytic group and $f : K \longrightarrow H$ be a Lie group homomorphism. We have an induced homomorphism $\mathfrak{k} \longrightarrow \mathrm{Lie}(H)$ of Lie algebras, which induces a homomorphism $\mathfrak{k}_\mathbb{C} = \mathrm{Lie}(G) \longrightarrow \mathrm{Lie}(H)$ of complex Lie algebras, by Proposition 11.3. If $\tilde{G}$ is the universal covering group of G, then by Proposition 14.2 we obtain a Lie group homomorphism $\tilde{G} \longrightarrow H$. To show that it factors through $G \cong \tilde{G}/\pi_1(G)$, we must show that the composite $\pi_1(G) \longrightarrow \tilde{G} \longrightarrow H$ is trivial. But this coincides with the composition $\pi_1(G) \cong \pi_1(K) \longrightarrow \tilde{K} \longrightarrow K \longrightarrow H$, where $\tilde{K}$ is the universal covering group of K, and the composition $\pi_1(K) \longrightarrow \tilde{K} \longrightarrow K$ is already trivial. Hence the map $\tilde{G} \longrightarrow H$ factors through G, proving that G has the universal property of the complexification.

We constructed G as an analytic subgroup of $\mathrm{GL}(n, \mathbb{C})$ starting with an arbitrary faithful complex representation of K. Looking at this another way, we have actually proved that *any* faithful complex representation of K can be extended to a faithful analytic complex representation of G. The reason is that if we started with another faithful complex representation and constructed the complexification using that one, we would have gotten a group isomorphic to G because the complexification is characterized up to isomorphism by its universal property.

It remains to be shown that analytic representations of G are completely reducible. If (π, V) is an analytic representation of G, then, since K is compact, by Proposition 2.1 there is a K-invariant inner product on V, and if U is an invariant subspace, then $V = U \oplus W$, where W is the orthogonal complement of U. Then we claim that W is G-invariant. Indeed, it is invariant under $\mathfrak{k}$ and hence under $\mathfrak{k}_\mathbb{C} = \mathfrak{k} \oplus i\mathfrak{k}$, which is the Lie algebra of G and, since G is connected, under G itself. $\square$

In addition to the analytic notion of complexification that we have already described, there is another notion, which we will call *algebraic complexification*. We will not need it, and the reader may skip the rest of this chapter with no loss of continuity. Still, it is instructive to consider complexification from the point of view of algebraic groups, so we digress to discuss it now. If $\mathcal{G}$ is an affine algebraic group defined over the real numbers, then $K = \mathcal{G}(\mathbb{R})$ is a Lie group and $G = \mathcal{G}(\mathbb{C})$ is a complex analytic group, and G is the *algebraic complexification* of K. We will assume that $\mathcal{G}(\mathbb{R})$ is Zariski-dense in $\mathcal{G}$ to exclude examples such as

$$\mathcal{G} = \{(x, y) \mid x^2 + y^2 = \pm 1\},$$

which is an algebraic group with group law $(x, y)(z, w) = (xz - yw, xw + yz)$, but which has one Zariski-connected component with no real points.

We see that the algebraic complexification is a functor *not* from the category of Lie groups but rather from the category of algebraic groups $\mathcal{G}$ defined over $\mathbb{R}$. So the algebraic complexification of a Lie group K depends on more than just the isomorphism class of K as a Lie group—it also depends on its realization as the group of real points of an algebraic group. We illustrate this point with an example.

Let G_a and G_m be the "additive group" and the "multiplicative group." These are algebraic groups such that for any field $G_a(F) \cong F$ (additive group) and $G_m(F) \cong F^\times$. The groups $\mathcal{G}_1 = G_a \times (\mathbb{Z}/2\mathbb{Z})$ and $\mathcal{G}_2 = G_m$ have isomorphic groups of real points since $\mathcal{G}_1(\mathbb{R}) \cong \mathbb{R} \times (\mathbb{Z}/2\mathbb{Z})$ and $\mathcal{G}_2(\mathbb{R}) \cong \mathbb{R}^\times$, and these are isomorphic as Lie groups. Their complexifications are $\mathcal{G}_1(\mathbb{C}) \cong \mathbb{C} \times (\mathbb{Z}/2\mathbb{Z})$ and $\mathcal{G}_2(\mathbb{C}) \cong \mathbb{C}^\times$. These groups are *not* isomorphic.

If $\mathcal{G}$ is an algebraic group defined over $F = \mathbb{R}$ or $\mathbb{C}$, and if $K = \mathcal{G}(F)$, then we call a complex representation $\pi : K \longrightarrow \mathrm{GL}(n, \mathbb{C})$ *algebraic* if there is a homomorphism of algebraic groups $\mathcal{G} \longrightarrow \mathrm{GL}(n)$ defined over $\mathbb{C}$ such that the induced map of rational points is π. (This amounts to assuming that the matrix coefficients of π are polynomial functions.) With this definition, the algebraic complexification has an interpretation in terms of representations like that of the analytic complexification.

Proposition 24.3. *If $G = \mathcal{G}(\mathbb{C})$ is the algebraic complexification of $K = \mathcal{G}(\mathbb{R})$, then any algebraic complex representation of K extends uniquely to an algebraic representation of G.*

Proof. This is clear since a polynomial function extends uniquely from $\mathcal{G}(\mathbb{R})$ to $\mathcal{G}(\mathbb{C})$. □

If K is a field and L is a Galois extension, we say that algebraic groups $\mathcal{G}_1$ and $\mathcal{G}_2$ defined over K are *L/K-Galois forms* of each other—or (more succinctly) *L/K-forms*—if there is an isomorphism $\mathcal{G}_1 \cong \mathcal{G}_2$ defined over L. If $K = \mathbb{R}$ and $L = \mathbb{C}$ this means that $K_1 = \mathcal{G}_1(\mathbb{R})$ and $K_2 = \mathcal{G}_2(\mathbb{R})$ have isomorphic algebraic complexifications. A $\mathbb{C}/\mathbb{R}$-Galois form is called a *real form*.

The example in Proposition 24.4 will help to clarify this concept.

Proposition 24.4. $\mathrm{U}(n)$ *is a real form of* $\mathrm{GL}(n, \mathbb{R})$.

Compare this with Proposition 11.4, which is the Lie algebra analog of this statement.

Proof. Let $\mathcal{G}_1$ be the algebraic group $\mathrm{GL}(n)$, and let

$$\mathcal{G}_2 = \{(A, B) \in \mathrm{Mat}_n \times \mathrm{Mat}_n \mid A \cdot {}^tA + B \cdot {}^tB = I, \ A \cdot {}^tB = B \cdot {}^tA\}.$$

The group law for G_2 is given by

$$(A, B)(C, D) = (AC - BD, AD + BC).$$

We leave it to the reader to check that this is a group. This definition is constructed so that $\mathcal{G}_2(\mathbb{R}) = \mathrm{U}(n)$ under the map $(A, B) \longrightarrow A + Bi$, when A and B are *real* matrices.

We show that $\mathcal{G}_2(\mathbb{C}) \cong \mathrm{GL}(n, \mathbb{C})$. Specifically, we show that if $g \in \mathrm{GL}(n, \mathbb{C})$ then there are unique matrices $(A, B) \in \mathrm{Mat}_n(\mathbb{C})$ such that $A \cdot {}^tA + B \cdot {}^tB = I$ and $A \cdot {}^tB = B \cdot {}^tA$ with $A + Bi = g$. We consider uniqueness first. We have

$$(A + Bi)({}^tA - {}^tBi) = (A\,{}^tA + B\,{}^tB) + (B^tA - A^tB)i = I,$$

so we must have $g^{-1} = {}^tA - {}^tBi$ and thus ${}^tg^{-1} = A - Bi$. We may now solve for A and B and obtain

$$A = \tfrac{1}{2}(g + {}^tg^{-1}), \qquad B = \tfrac{1}{2i}(g - {}^tg^{-1}). \tag{24.1}$$

This proves uniqueness. Moreover, if we define A and B by (24.1), then it is easy to see that $(A, B) \in G_2(\mathbb{C})$ and $A + Bi = g$.   □

It can be seen similarly that $\mathrm{SU}(n)$ and $\mathrm{SL}(n, \mathbb{R})$ are $\mathbb{C}/\mathbb{R}$ Galois forms of each other. One has only to impose in the definition of the second group G_2 an additional polynomial relation corresponding to the condition $\det(A+Bi) = 1$. (This condition, written out in terms of matrix entries, will not involve i, so the resulting algebraic group is defined over $\mathbb{R}$.)

Remark 24.1. Classification of Galois forms of a group is a problem in Galois cohomology. Indeed, the set of Galois forms of $\mathcal{G}$ is parametrized by $H^1(\mathrm{Gal}(L/K), \mathrm{Aut}(\mathcal{G}))$. See Springer [150], Satake [144] and III.1 of Serre [148]. Tits [162] contains the definitive classification over real, p-adic, finite, and number fields.

Galois forms are important because if G_1 and G_2 are Galois forms of each other, then we expect the representation theories of G_1 and G_2 to be related. We have already seen this principle applied (for example) in Theorem 14.3. Our next proposition gives a typical application.

Proposition 24.5. *Let* $\pi : \mathrm{GL}(n, \mathbb{R}) \longrightarrow \mathrm{GL}(m, \mathbb{C})$ *be an algebraic representation. Then* π *is completely reducible.*

This would *not* be true if we removed the assumption of algebraicity. For example, the representation $\pi : \mathrm{GL}(n, \mathbb{R}) \longrightarrow \mathrm{GL}(2, \mathbb{R})$ defined by

$$\pi(g) = \begin{pmatrix} 1 & \log|\det(g)| \\ & 1 \end{pmatrix}$$

is not completely reducible—and it is not algebraic.

Proof. Any irreducible algebraic representations of $\mathrm{GL}(n, \mathbb{R})$ can be extended to an algebraic representation of $\mathrm{GL}(n, \mathbb{C})$ and then restricted to $\mathrm{U}(n)$, where it is completely reducible because $\mathrm{U}(n)$ is compact.   □

The irreducible algebraic complex representations of $\mathrm{GL}(n, \mathbb{R})$ are the same as the irreducible algebraic complex representations of $\mathrm{GL}(n, \mathbb{C})$, which in turn are the same as the irreducible complex representations of $\mathrm{U}(n)$. (The latter are automatically algebraic, and indeed we will later construct them as algebraic representations.)

These finite-dimensional representations of $GL(n, \mathbb{R})$ may be parametrized by their highest weight vectors and classified as in the previous chapter. Their characters are given by the Weyl character formula.

Although the irreducible algebraic complex representations of $GL(n, \mathbb{R})$ are thus the same as the irreducible representations of the compact group $U(n)$, their significance is very different. These finite-dimensional representations of $GL(n, \mathbb{R})$ are *not* unitary (except for the one-dimensional ones). They therefore do not appear in the Fourier inversion formula (Plancherel theorem). Unlike $U(n)$, the noncompact group $GL(n, \mathbb{R})$ has unitary representations that are infinite-dimensional, and it is these infinite-dimensional representations that appear in the Plancherel theorem.

Exercises

Exercise 24.1. If F is a field, let

$$
SO_J(n, F) = \left\{ g \in SL(n, F) \mid g\, J\, {}^t g = J \right\}, \qquad J = \begin{pmatrix} & & 1 \\ & \cdot^{\cdot^{\cdot}} & \\ 1 & & \end{pmatrix}.
$$

Show that $SO_J(\mathbb{C})$ is the complexification of $SO(n)$. (Use Exercise 5.3.)

25

Coxeter Groups

As we will see in this chapter, Weyl groups and affine Weyl groups are examples of *Coxeter groups*, an important family of groups generated by "reflections."

Let G be a group, and let I be a set of generators of G, each of which has order 2. In practice, we will usually denote the elements of I by $\{s_1, s_2, \ldots, s_r\}$ or $\{s_0, \ldots, s_r\}$ with some definite indexing by integers. If s_i, $s_j \in I$, let $n(i,j) = n(s_i, s_j)$ be the order of $s_i s_j$. [Strictly speaking we should write $n(s_i, s_j)$ but prefer less uncluttered notation.] We assume $n(i,j)$ to be finite for all s_i, s_j. The pair (G, I) is called a *Coxeter group* if the relations

$$s_i^2 = 1, \qquad (s_i s_j)^{n(i,j)} = 1 \qquad (25.1)$$

are a presentation of G. This means that G is isomorphic to the quotient of the free group on a set of generators $\{\sigma_i\}$, one for each $s_i \in I$, by the smallest normal subgroup containing all elements

$$\sigma_i^2, \qquad (\sigma_i \sigma_j)^{n(i,j)},$$

and in this isomorphism each generator $\sigma_i \mapsto s_i$. Equivalently, G has the following universal property: if Γ is any other group having elements v_i (one for each generator s_i) satisfying the same relations (25.1), that is, if

$$v_i^2 = 1, \qquad (v_i v_j)^{n(i,j)} = 1,$$

then there exists a unique homomorphism $G \to \Gamma$ such that each $s_i \to v_i$.

A *word* representing an element w of a Coxeter group (W, I) is a sequence $(s_{i_1}, \ldots, s_{i_k})$ such that $w = s_{i_1} \cdots s_{i_k}$. The word is *reduced* if k is as small as possible. Thus, if the Coxeter group is a Weyl group and I the set of simple reflections, then k is the length of w. Less formally, we may abuse language by saying that $w = s_{i_1} \cdots s_{i_k}$ is a *reduced word* or *reduced decomposition* of w.

We return to the context of Chap. 20. Let $\mathcal{V}$ be a vector space, Φ a reduced root system in $\mathcal{V}$, and W the Weyl group. We partition Φ into positive and negative roots and denote by Σ the simple positive roots. Let $I = \{s_1, \ldots, s_r\}$ be of simple reflections. By definition, W is generated by the set I. Let $n(i, j)$ denote the order of $s_i s_j$. We will show that (W, I) is a Coxeter group. It is evident that the relations (25.1) are satisfied, but we need to see that they give a presentation of W.

Theorem 25.1. *Let W be the Weyl group of the root system Φ, and let $I = \{s_1, \ldots, s_r\}$ be the simple reflections. Then (W, I) is a Coxeter group.*

We will give a geometric proof of this fact, making use of the system of Weyl chambers. As it turns out, every Coxeter group has a geometric action on a simplicial complex, a *Coxeter complex*, which for Weyl groups is closely related to the action on Weyl chambers. This point of view leads to the theory of buildings. See Tits [163] and Abramenko and Brown [1] as well as Bourbaki [23].

Proof. Let (W', I') be the Coxeter group with generators $\{s_1', \ldots, s_j'\}$ and the relations (25.1) where $n(i, j)$ is the order of $s_i s_j$. Since the relations (25.1) are true in W we have a surjective homomorphism $W' \longrightarrow W$ sending $s_i' \mapsto s_i$. We must show that it is injective. Let $s_{i_1} \cdots s_{i_n} = s_{j_1} \cdots s_{j_m}$ be two decompositions of the same element w into products of simple reflections. We will show that we may go from one word $\mathfrak{w} = (s_{i_1}, \ldots, s_{i_n})$ to the other $\mathfrak{w} = (s_{j_1}, \ldots, s_{j_m})$, only making changes corresponding to relations in the Coxeter group presentation. That is, we may insert or remove a pair of adjacent equal s_i, or we may replace a segment $(s_i, s_j, s_i, \ldots)$ by $(s_j, s_i, s_j, \ldots)$ where the total number of s_i and s_j is each the order of $2n(s_i, s_j)$. This will show that $s_{i_1}' \cdots s_{i_n}' = s_{j_1}' \cdots s_{j_m}'$ so the homomorphism $W' \longrightarrow W$ is indeed injective.

Let $\mathcal{C}$ be the positive Weyl chamber. We have $s_{i_1} \cdots s_{i_n} \mathcal{C} = s_{j_1} \cdots s_{j_m} \mathcal{C}$. Let

$$w_1 = s_{i_1}, \ w_2 = s_{i_1} s_{i_2}, \ \cdots \ w_n = s_{i_1} \cdots s_{i_n}.$$

The sequence of chambers

$$\mathcal{C}, w_1 \mathcal{C}, w_2 \mathcal{C}, \ldots, w_n \mathcal{C} = w\mathcal{C} \tag{25.2}$$

are adjacent. We will say that $(\mathcal{C}, w_1 \mathcal{C}, w_2, \ldots, w\mathcal{C})$ is the *gallery* associated with the word $\mathfrak{w} = (s_{i_1}, \ldots, s_{i_k})$ representing w. We find a path p from a point in the interior of $\mathcal{C}$ to $w\mathcal{C}$ passing exactly through this sequence of chambers.

Similarly we have a gallery associated with the word $(s_{j_1}, \ldots, s_{j_m})$. We may similarly consider a path q from $\mathcal{C}$ to $w\mathcal{C}$ having the same endpoints as p passing through the chambers of this gallery. We will consider what happens when we deform p to q.

If $\alpha \in \Phi$ let $\mathfrak{H}_\alpha$ be the set of $v \in \mathcal{V}$ such that $\alpha^\vee(v) = 0$. This is the hyperplane perpendicular to the root α, and these hyperplanes are the walls of Weyl chambers. Let K_2 be the closed subset of $\mathcal{V}$ where two or more hyperplanes $\mathfrak{H}_\alpha$ intersect. It is a subset of codimension 2, that is, it is a (locally) finite union of codimension 2 linear subspaces of $\mathcal{V}$. Let K_3 be the closed subset of $\mathcal{V}$ consisting of points P such that three hyperplanes $\mathfrak{H}_{\alpha_1}, \mathfrak{H}_{\alpha_2}, \mathfrak{H}_{\alpha_3}$ pass through $\mathcal{V}$, with the roots α_1, α_2 and α_3 linearly independent. The subset K is of codimension 3 in $\mathcal{V}$. We have $K_2 \supset K_3$. The paths p and q do not pass through K_2.

Since K_3 has codimension 3 it is possible to deform p to q avoiding K_3. Let p_u with $u \in [0, 1]$ be such a deformation, with $p_0 = p$ and $p_1 = p'$. For each u the sequence of chambers through which p_u form the gallery associated to some word representing w. We consider what happens to this word when the gallery changes as u is varied.

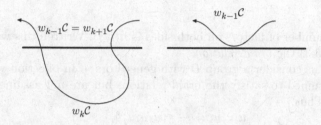

Fig. 25.1. Eliminating a crossing and recrossing of the same wall

There are two ways the word can change. If $i_k = i_{k+1}$ then $w_{k-1} = w_{k+1}$ and we have a crossing as in Fig. 25.1. The path may move to eliminate (or create) the crossing. This corresponds to eliminating (or inserting) a repeated $s_{i_k} = s_{i_{k+1}}$ from the word, and since s'_{i_k} has order 2 in the Coxeter group, the corresponding elements of the Coxeter group will be the same.

Since the deformation avoids K_3, the only other way that the word can change is if the deformation causes the path to cross K_2, that is, some point where two or more hyperplanes $\mathfrak{H}_{\alpha_1}, \mathfrak{H}_{\alpha_2}, \ldots, \mathfrak{H}_{\alpha_n}$ intersect, with the roots $\alpha_1, \ldots, \alpha_n$ in a two-dimensional subspace. In this case the transition looks like Fig. 25.2.

This can happen if $i_k = i_{k+2} = \cdots = i$ and $i_{k+1} = i_{k+3} = \cdots = j$, and the effect of the crossing is to replace a subword of the form $(s_i, s_j, s_i, \ldots)$ by an equivalent $(s_j, s_i, s_j, \ldots)$, where the total number of s_i and s_j is $2n(s_i, s_j)$. We have $s'_i s'_j s'_i \cdots = s'_j s'_i s'_j \cdots$ in W', so this type of transition also does not change the element of W'. We see that $s'_{i_1} \cdots s'_{i_n} = s'_{j_1} \cdots s'_{j_m}$, proving that $W \cong W'$. This concludes the proof that W is a Coxeter group.   □

The Coxeter group (W, I) has a close relative called the associated *braid group*. We note that in the Coxeter group (W, I) with generators s_i satisfying $s_i^2 = 1$, the relation $(s_i s_j)^{n(i,j)} = 1$ (true when $i \neq j$) can be written

$$s_i s_j s_i \cdots = s_j s_i s_j \cdots, \tag{25.3}$$

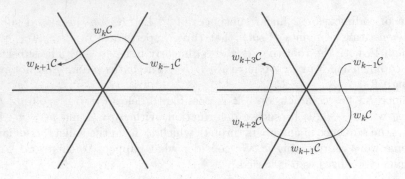

Fig. 25.2. Crossing K_2

where the number of factors on both sides is $n(i,j)$. Written this way, we call equation (25.3) the *braid relation*.

Now let us consider a group B with generators u_i in bijection with the s_i that are assumed to satisfy the braid relations but are *not* assumed to be of order two. Thus,

$$u_i u_j u_i u_j \cdots = u_j u_i u_j u_i \cdots ,\qquad (25.4)$$

where there are $n(i,j)$ terms on both sides. Note that since the relation $s_i^2 = 1$ is not true for the u_i, it is *not* true that $n(i,j)$ is the order of $u_i u_j$, and in fact $u_i u_j$ has infinite order. The group B is called the *braid group*.

The term *braid group* is used due to the fact that the braid group of type A_n is Artin's original braid group, which is a fundamental object in knot theory. Although Artin's braid group will not play any role in this book, abstract braid groups *will* play a role in our discussion of Hecke algebras in Chap. 46, and the relationship between Weyl groups and braid groups underlies many unexpected developments beginning with the use by Jones [91] of Hecke algebras in defining new knot invariants and continuing with the work of Reshetikhin and Turaev [135] based on the Yang–Baxter equation, with connections to quantum groups and applications to knot and ribbon invariants.

Consider a set of paths represented by a set of $n+1$ nonintersecting strings connected to two (infinite) parallel posts in $\mathbb{R}^3$ to be a *braid*. Braids are equivalent if they are homotopic. The "multiplication" in the braid group is concatenation: to multiply two braids, the endpoints of the first braid on the right post are tied to the endpoints of the second braid on the left post. In Fig. 25.3, we give generators u_1 and u_2 for the braid group of type A_2 and calculate their product. In Fig. 25.4, we consider $u_1 u_2 u_1$ and $u_2 u_1 u_2$; clearly these two braids are homotopic, so the braid relation $u_1 u_2 u_1 = u_2 u_1 u_2$ is satisfied.

We did not have to make the map n part of the defining data in the Coxeter group since $n(i,j)$ is just the order of $s_i s_j$. This is no longer true in the braid group. Coxeter groups are often finite, but the braid group (B, I) is infinite if $|I| > 1$.

$$u_1 \qquad \times \qquad u_2 \qquad = \qquad u_1 u_2$$

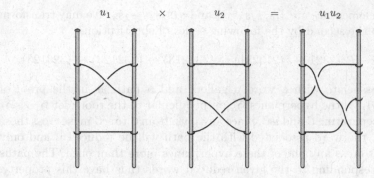

Fig. 25.3. Generators u_1 and u_2 of the braid group of type A_2 and $u_1 u_2$

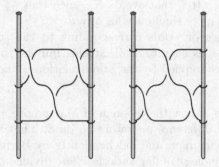

Fig. 25.4. The braid relation. *Left*: $u_1 u_2 u_1$. *Right*: $u_2 u_1 u_2$

Theorem 25.1 has an important complement due to Matsumoto [127] and (independently) by Tits. According to this result, if two reduced words represent the same element, then the corresponding elements represented by the same reduced words are equal in the braid group. (This is true for arbitrary Coxeter groups, but we will only prove it for Weyl groups and affine Weyl groups.) Both Theorem 25.1 and Matsumoto's theorem may be given proofs based on Proposition 20.7, and these may be found in Bourbaki [23]. We will give another geometric proof of Matsumoto's theorem based on ideas similar to those in the above proof of Theorem 25.1.

Theorem 25.2 (Matsumoto, Tits). *Let $w \in W$ have length $l(w) = r$. Let $s_{i_1} \cdots s_{i_r} = s_{j_1} \cdots s_{j_r}$ be two reduced decompositions of w into products of simple reflections. Then the corresponding words are equal in the braid group, that is, $u_{i_1} \cdots u_{i_r} = u_{j_1} \cdots u_{j_r}$.*

What we will actually prove is that if w of length k has two reduced decompositions $w = s_{i_1} \cdots s_{i_k} = s_{j_1} \cdots s_{j_k}$, then the word $(s_{i_1}, \ldots, s_{i_k})$ may be transformed into $(s_{j_1}, \ldots, s_{j_k})$ by a series of substitutions, in which a subword $(s_i, s_j, s_i, \ldots)$ is changed to $(s_j, s_i, s_j, \ldots)$, both subwords having $n(i, j)$ elements. For example, in the A_3 Weyl group, two words representing the long

Weyl group element w_0 are $s_1s_2s_1s_3s_2s_1$ and $s_3s_2s_3s_1s_2s_3$. We may transform the first into the second by the following series of substitutions:

$$(121321) \leftrightarrow (212321) \leftrightarrow (213231) \leftrightarrow (231213) \leftrightarrow (232123) \leftrightarrow (323123).$$

Proof. We associate with a word a gallery and a path as in the proof of Theorem 25.1. Of the hyperplanes $\mathfrak{H}_\alpha$ perpendicular to the roots, let $\mathfrak{H}_1 \cdots \mathfrak{H}_r$ be the ones separating $\mathcal{C}$ and $w\mathcal{C}$. Since any path from $\mathcal{C}$ to $w\mathcal{C}$ must cross these hyperplanes, the word associated with the path will be reduced if and only if it does not cross any one of these hyperplanes more than once. The paths p and q corresponding to the given reduced words thus have this property, and as in the proof of Theorem 25.1 it is easy to see that we may choose the deformation p_u from p to q that avoids K_3, such that p_u does not cross any of these hyperplanes more than once for any u.

Thus the sequence of words corresponding to the stages of p_u are all reduced words, and it is easy to see that this implies that the only transitions allowed are ones implied by the braid relations. Therefore $u_{i_1} \cdots u_{i_r} = u_{j_1} \cdots u_{j_r}$. $\qquad\square$

As a typical example of how the theorem of Matsumoto and Tits is used, let us define the *divided difference* operators D_i on $\mathcal{E}$. They were introduced by Lascoux and Schutzenberger, and independently by Bernstein, Gelfand, and Gelfand, in the cohomology of flag varieties. The divided difference operators are sometimes denoted ∂_i, but we will reserve that notation for the Demazure operators we will introduce below. D_i acts on the group algebra of the weight lattice Λ; this algebra was denoted $\mathcal{E}$ in Chap. 22. It has a basis e^λ indexed by weights $\lambda \in \Lambda$. We define

$$D_i f = (e^{\alpha_i} - 1)^{-1} (f - s_i f).$$

It is easy to check that $f - s_i f$ is divisible in $\mathcal{E}$ by $e^{\alpha_i} - 1$, so this operator maps $\mathcal{E}$ to itself.

More formally, let $\mathfrak{M}$ be the localization of $\mathcal{E}$ that is the subring of its field of fractions obtained by adjoining denominators of the form $H(\alpha) = (e^\alpha - 1)^{-1}$ with $\alpha \in \Phi$. It is convenient to think of the D_i as living in the ring $\mathfrak{D}$ of expressions of the form $\sum f_w \cdot w$ where $f_w \in \mathfrak{M}$, and the sum is over $w \in W$. We have $wfw^{-1} = w(f)$, that is, conjugation by a Weyl group element is the same as applying it to the element f of $\mathfrak{M}$. We have an obvious action of $\mathfrak{D}$ on $\mathfrak{M}$, and in this notation we write

$$D_i = (e^{\alpha_i} - 1)^{-1} (1 - s_i).$$

Because $e^{\alpha_i} - 1$ divides $f - s_i f$ for $f \in \mathcal{E}$, the operators D_i act on $\mathcal{E}$.

Proposition 25.1. *Let $n(i,j)$ be the order of $s_i s_j$ in W, where $i \neq j$. Then the D_i satisfy the braid relation*

$$D_i D_j D_i \cdots = D_j D_i D_j \cdots \qquad (25.5)$$

where the number of factors on both sides is $n(i,j)$. Moreover, this equals

$$\left[\prod_\alpha H(\alpha) \right] \sum_w (-1)^{l(w)} w$$

where the product is over roots α in the rank two root system spanned by α_i and α_j, and the sum is over the rank two Weyl group generated by s_i, s_j.

Proof. This calculation can be done separately for the four possible cases $n(i,j) = 2, 3, 4$ or 6. The case $n(i,j) = 2$ is trivial so let us assume $n(i,j) = 3$. We will show that

$$D_i D_j D_i = H(\alpha_i) H(\alpha_j) H(\alpha_i + \alpha_j) \sum_{w \in \langle s_i, s_j \rangle} (-1)^{l(w)} w, \qquad (25.6)$$

which implies that $D_i D_j D_i = D_j D_i D_j$. The left-hand side equals

$$H(\alpha_i)\,(1 - s_i)\,H(\alpha_j)\,(1 - s_j)\,H(\alpha_i)\,(1 - s_i).$$

This is the sum of eight terms $H(\alpha_i)\varepsilon_1 H(\alpha_j)\varepsilon_2 H(\alpha_i)\varepsilon_3$ where $\varepsilon_1 = 1$ or $-s_i$, etc. Expanding we get it in the form $\sum f_w \cdot w$ where each of the six coefficients f_w are easily evaluated. When $w = 1$ or s_1 there are two contributions, and for these we use the identity

$$H(\alpha + \beta)\,H(-\alpha) + H(\alpha)\,H(\beta) = H(\beta)\,H(\alpha + \beta).$$

The other four terms have only one contribution and are trivial to check. Each f_w turns out to be equal to $(-1)^{l(w)} H(\alpha_i) H(\alpha_j) H(\alpha_i + \alpha_j)$, proving (25.6). If $n(i,j) = 4$ or 6, the proof is similar (but more difficult). $\qquad \square$

Now we may give the first application of the theorem of Matsumoto and Tits. We may define, for $w \in W$ an operator D_w to be $D_{i_1} D_{i_2} \cdots D_{i_k}$ where $w = s_{i_1} \cdots s_{i_k}$ is a reduced expression for w. This is well-defined by the Matsumoto-Tits theorem. Indeed, given another reduced expression $w = s_{j_1} \cdots s_{j_k}$, then the content of the Matsumoto-Tits theorem is that the two deduced words are equal in the braid group, which means that we can go from $D_{i_1} D_{i_2} \cdots D_{i_k}$ to $D_{j_1} D_{j_2} \cdots D_{j_k}$ using only the braid relations, that is, by repeated applications of (25.5).

Similarly we may consider *Demazure operators* ∂_w indexed by $w \in W$. These were introduced by Demazure to describe the cohomology of line bundles over Schubert varieties, but they may also be used to give an efficient method of computing the characters of irreducible representations of compact Lie groups. Let

$$\partial_i f = \left(1 - e^{-\alpha_i}\right)^{-1} \left(f - e^{-\alpha_i}\,(s_i f)\right).$$

It is easy to see that $f - e^{-\alpha_i}\,(s_i f)$ is divisible by $1 - e^{-\alpha_i}$ so that this is in $\mathcal{E}$; in fact, this follows from the more precise formula in the following lemma.

Proposition 25.2. *We have*

$$\partial_i^2 = \partial_i, \qquad s_i \partial_i = \partial_i,$$

Let $f \in \mathcal{E}$. Then $\partial_i f$ is in $\mathcal{E}$ and is invariant under s_i, and if $s_i f = f$ then $\partial_i f = f$. If $f = e^\lambda$ with $\lambda \in \Lambda$, then we have

$$\partial_i e^\lambda = \begin{cases} e^\lambda + e^{\lambda - \alpha_i} + e^{\lambda - 2\alpha_i} + \cdots + e^{s_i \lambda} & \text{if } \alpha_i^\vee(\lambda) > 0; \\ -e^{\lambda + \alpha_i} - \cdots - e^{s_i \lambda - \alpha_i} & \text{if } \alpha_i^\vee(\lambda) \leqslant 0. \end{cases}$$

Proof. We have $s_i \partial_i = (1 - e^{\alpha_i})^{-1} (s - e^{\alpha_i})$ since $s_i e^\lambda s_i^{-1} = e^{s_i(\lambda)}$ and in particular $s_i e^{-\alpha_i} s_i^{-1} = e^{\alpha_i}$. Multiplying both the numerator and the denominator by $-e^{-\alpha_i}$ then shows that $s_i \partial_i = \partial_i$. This identity shows that for any $f \in \mathcal{E}$ the element $\partial_i f$ is s_i invariant. Moreover, if f is s_i-invariant, then $\partial_i f = f$ because $\partial_i f = (1 - e^{-\alpha_i})^{-1} (1 - e^{-\alpha_i}) f = f$. Since $\partial_i f$ is s_i invariant, we have $\partial_i^2 f = \partial_i f$. The action of $\mathfrak{D}$ on $\mathcal{E}$ is easily seen to be faithful so this proves $\partial_i^2 = \partial_i$ (or check this by direct computation. The last identity follows from the formula for a finite geometric series, $(1 - x)^{-1} (1 - x^{N+1}) = 1 + x + \cdots + x^N$ together with $s_i \lambda = \lambda - \alpha_i^\vee(\lambda) \alpha_i$. $\qquad\square$

It is easy to check that the Demazure and divided difference operators are related by $D_i = \partial_i - 1$.

Proposition 25.3. *The Demazure operators also satisfy the braid relations*

$$D_i D_j D_i \cdots = D_j D_i D_j \cdots \tag{25.7}$$

where the number of factors on both sides is $n(i, j)$.

Proof. Again there are different cases depending on whether $n(i, j) = 2, 3, 4$ or 6, but in each case this can be reduced to the corresponding relation (25.5) by use of $\partial_i^2 = \partial_i$. For example, if $n(i, j) = 3$, then expanding $0 = (\partial_i - 1)(\partial_j - 1)(\partial_i - 1) - (\partial_j - 1)(\partial_i - 1)(\partial_j - 1)$ and using $\partial_i^2 = \partial_i$ gives $\partial_i \partial_j \partial_i = \partial_j \partial_i \partial_j$. The other cases are similar. $\qquad\square$

Now, by the theorem of Matsumoto and Tits, we may define $\partial_w = \partial_{i_1} \cdots \partial_{i_k}$ where $w = s_{i_1} \cdots s_{i_k}$ is any reduced expression, and this is well defined. We return to the setting of Chap. 22. Thus, let λ be a dominant weight in $\Lambda = X^*(T)$, where T is a maximal torus in the compact Lie group G. Let χ_λ be the character of the corresponding highest weight module, which may be regarded as an element of $\mathcal{E}$.

Theorem 25.3 (Demazure). *Let w_0 be the long element in the Weyl group, and let λ be a dominant weight. Then*

$$\chi_\lambda = \partial_{w_0} e^\lambda.$$

This is an efficient method of computing χ_λ. Demazure also gave interpretations of ∂_w for other Weyl group elements as characters of T-modules of sections of line bundles over Schubert varieties.

Proof. Let $\partial_{w_0} = \sum f_w \cdot w$. We will prove that

$$f_w = \Delta^{-1} (-1)^{l(w)} e^{w(\rho)}. \tag{25.8}$$

where Δ is the Weyl denominator as in Chap. 22. This is sufficient, for then $\partial_{w_0} e^\lambda = \chi_\lambda$ when λ is dominant by the Weyl character formula.

Let $N = l(w_0)$. For each i, $l(s_i w_0) = N - 1$, so we may find a reduced word $s_i w_0 = s_{i_2} \cdots s_{i_N}$. Then $w_0 = s_i s_{i_2} \cdots s_{i_N}$ in which $i_1 = i$, so $\partial_{w_0} = \partial_i \partial_{s_i w_0}$. Since $\partial_i^2 = \partial_i$ and $s_i \partial_i = \partial_i$ this means that $\partial_i \partial_{w_0} = \partial_{w_0}$ and $s_i \partial_{w_0} = \partial_{w_0}$. A consequence is that $\partial_{w_0} e^\lambda$ is W-invariant for every weight λ (dominant or not). Therefore, if we write $\partial_{w_0} = \sum f_w \cdot w$ with $f_w \in \mathfrak{M}$ we have $w(f_{w'}) = f_{ww'}$. Since $w(\Delta^{-1}) = (-1)^{l(w)}$, we now have only to check (25.8) for one particular w. Fortunately when $w = w_0$ it is possible to do this without too much work. Choosing a reduced word, we have

$$\partial_{w_0} = \left(1 - e^{-\alpha_{i_1}}\right)^{-1} \left(1 - e^{-\alpha_{i_1}} s_{i_1}\right) \cdots \left(1 - e^{-\alpha_{i_N}}\right)^{-1} \left(1 - e^{-\alpha_{i_N}} s_{i_N}\right).$$

Expanding out the factors $1 - s_{i_k} e^{-\alpha_k}$ there is only one way to get w_0 in the decomposition $\sum f_w \cdot w$, namely we must take $-s_{i_k} e^{-\alpha_{i_k}}$ in every factor. Therefore,

$$f_{w_0} \cdot w_0 = \left(1 - e^{-\alpha_{i_1}}\right)^{-1} \left(-e^{-\alpha_1}\right) s_{i_1} \cdots \left(1 - e^{-\alpha_{i_N}}\right)^{-1} \left(-e^{-\alpha_N}\right) s_{i_N}$$

$$= (-1)^{l(w_0)} H(\alpha_1) s_{i_1} \cdots H(\alpha_{i_N}) s_{i_N}.$$

Moving the s_i to the right, this equals

$$(-1)^{l(w_0)} H(\alpha_{i_1}) H(s_{i_1} \alpha_{i_2}) H(s_{i_1} s_{i_2} \alpha_{i_3}) \cdots$$

Applying Proposition 20.10 to the reduced word $w_0 = s_{i_k} \cdots s_{i_1}$, this proves that

$$f_{w_0} = (-1)^{l(w_0)} \prod_{\alpha \in \Phi^+} H(\alpha) = \prod (-1)^{l(w_0)} \left(e^\alpha - 1\right)^{-1}.$$

Since $e^{w_0(\rho)} = e^{-\rho}$, this is equivalent to (25.8) in the case $w = w_0$. $\square$

Theorem 25.4. *The affine Weyl group is also a Coxeter group (generated by $s_0, \ldots, s_r$). Moreover, the analog of the Matsumoto-Tits theorem is true for the affine Weyl group: if w of length k has two reduced decompositions $w = s_{i_1} \cdots s_{i_k} = s_{j_1} \cdots s_{j_k}$, then the word $(s_{i_1}, \ldots, s_{i_k})$ may be transformed into $(s_{j_1}, \ldots, s_{j_k})$ by a series of substitutions, in which a subword $(s_i, s_j, s_i, \ldots)$ is changed to $(s_j, s_i, s_j, \ldots)$, both subwords having $n(i, j)$ elements.*

Proof. This may be proved by the same method as Theorem 25.1 and 25.2 (Exercise 25.3). □

As a last application of the theorem of Matsumoto and Tits, we discuss the *Bruhat order* on the Weyl group, which we will meet again in Chap. 27. This is a partial order, with the long Weyl group element maximal and the identity element minimal. If v and u are elements of the Weyl group W, then we write $u \leqslant v$ if, given a reduced decomposition $v = s_{i_1} \cdots s_{i_k}$ then there exists a subsequence $(j_1, \ldots, j_l)$ of $(i_1, \ldots, i_k)$ such that $u = s_{j_1} \cdots s_{j_l}$. By Proposition 20.4 we may assume that $u = s_{j_1} \cdots s_{j_l}$ is a reduced decomposition.

Proposition 25.4. *This definition does not depend on the reduced decomposition $v = s_{i_1} \cdots s_{i_k}$.*

Proof. By Theorem 25.2 it is sufficient to check that if $(i_1, \ldots, i_k)$ is changed by a braid relation, then we can still find a subsequence $(j_1, \ldots, j_l)$ representing u. We therefore find a subsequence of the form $(t, u, t, \ldots)$ where the number of elements is the order of $s_t s_u$, and we replace this by $(u, t, u, \ldots)$. We divide the subsequence $(j_1, \ldots, j_l)$ into three parts: the portion extracted from that part of $(i_1, \ldots, i_k)$ before the changed subsequence, the portion extracted from the changed subsequence, and the portion extracted from after the changed subsequence. The first and last part do not need to be altered. A subsequence can be extracted from the portion in the middle to represent any element of the dihedral group generated by s_t and s_u whether it is $(t, u, t, \ldots)$ or $(u, t, u, \ldots)$, so changing this portion has no effect. □

We now describe (without proof) the classification of the possible reduced root systems and their associated finite Coxeter groups. See Bourbaki [23] for proofs. If Φ_1 and Φ_2 are root systems in vector spaces $\mathcal{V}_1, \mathcal{V}_2$, then $\Phi_1 \cup \Phi_2$ is a root system in $\mathcal{V}_1 \oplus \mathcal{V}_2$. Such a root system is called *reducible*. Naturally, it is enough to classify the irreducible root systems.

The *Dynkin diagram* represents the Coxeter group in compact form. It is a graph whose vertices are in bijection with Σ. Let us label $\Sigma = \{\alpha_1, \ldots, \alpha_r\}$, and let $s_i = s_{\alpha_i}$. Let $\theta(\alpha_i, \alpha_j)$ be the angle between the roots α_i and α_j. Then

$$n(s_i, s_j) = \begin{cases} 2 \text{ if } \theta(\alpha_i, \alpha_j) = \frac{\pi}{2}, \\ 3 \text{ if } \theta(\alpha_i, \alpha_j) = \frac{2\pi}{3}, \\ 4 \text{ if } \theta(\alpha_i, \alpha_j) = \frac{3\pi}{4}, \\ 6 \text{ if } \theta(\alpha_i, \alpha_j) = \frac{5\pi}{6}. \end{cases}$$

These four cases arise in the rank 2 root systems $A_1 \times A_1$, A_2, B_2 and G_2, as the reader may confirm by consulting the figures in Chap. 19.

In the Dynkin diagram, we connect the vertices corresponding to α_i and α_j only if the roots are not orthogonal. If they make an angle of $2\pi/3$, we connect them with a single bond; if they make an angle of $6\pi/4$, we connect them with a double bond; and if they make an angle of $5\pi/6$, we connect them with a triple bond. The latter case only arises with the exceptional group G_2.

If α_i and α_j make an angle of $3\pi/4$ or $5\pi/6$, then these two roots have different lengths; see Figs. 19.4 and 19.6. In the Dynkin diagram, there will be a double or triple bond in these examples, and we draw an arrow from the long root to the short root. The triple bond (corresponding to an angle of $5\pi/6$) is rare—it is only found in the Dynkin diagram of a single group, the exceptional group G_2. If there are no double or triple bonds, the Dynkin diagram is called *simply laced*.

Fig. 25.5. The Dynkin diagram for the type A_5 root system

The root system of type A_n is associated with the Lie group $\mathrm{SU}(n+1)$. The corresponding abstract root system is described in Chap. 19. All roots have the same length, so the Dynkin diagram is simply laced. In Fig. 25.5 we illustrate the Dynkin diagram when $n = 5$. The case of general n is the same—exactly n nodes strung together in a line ($\bullet\!\!-\!\!\bullet\!\!-\!\cdots-\!\!\bullet$).

Fig. 25.6. The Dynkin diagram for the type B_5 root system

The root system of type B_n is associated with the odd orthogonal group $\mathrm{SO}(2n+1)$. The corresponding abstract root system is described in Chap. 19. There are both long and short roots, so the Dynkin diagram is not simply laced. See Fig. 25.6 for the Dynkin diagram of type B_5. The general case is the same ($\bullet\!\!-\!\!\bullet\!\!-\cdots-\!\!\bullet\!\!\Rrightarrow\!\!\bullet$), with the arrow pointing towards the α_n node corresponding to the unique short simple root.

Fig. 25.7. The Dynkin diagram for the type C_5 root system

The root system of type C_n is associated with the symplectic group $\mathrm{Sp}(2n)$. The corresponding abstract root system is described in Chap. 19. There are both long and short roots, so the Dynkin diagram is not simply laced. See Fig. 25.7 for the Dynkin diagram of type C_5. The general case is the same ($\bullet\!\!-\!\!\bullet\!\!-\cdots-\!\!\bullet\!\!\Lleftarrow\!\!\bullet$), with the arrow pointing from the α_n node corresponding to the unique long simple root, towards α_{n-1}.

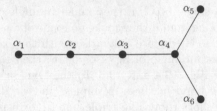

Fig. 25.8. The Dynkin diagram for the type D_6 root system

The root system of type D_n is associated with the even orthogonal group $O(2n)$. All roots have the same length, so the Dynkin diagram is simply-laced. See Fig. 25.8 for the Dynkin diagram of type D_6. The general case is similar, but the cases $n = 2$ or $n = 3$ are degenerate, and coincide with the root systems $A_1 \times A_1$ and A_3. For this reason, the family D_n is usually considered to begin with $n = 4$. See Fig. 30.2 and the discussion in Chap. 30 for further information about these degenerate cases.

These are the "classical" root systems, which come in infinite families. There are also five exceptional root systems, denoted E_6, E_7, E_8, F_4 and G_2. Their Dynkin diagrams are illustrated in Figs. 25.9–25.12.

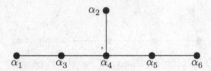

Fig. 25.9. The Dynkin diagram for the type E_6 root system

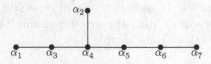

Fig. 25.10. The Dynkin diagram for the type E_7 root system

Fig. 25.11. The Dynkin diagram for the type E_8 root system

Fig. 25.12. The Dynkin diagrams of types F_4 (*left*) and G_2 (*right*)

Exercises

Exercise 25.1. For the root systems of types A_n, B_n, C_n, D_n and G_2 described in Chap. 19, identify the simple roots and the angles between them. Confirm that their Dynkin diagrams are as described in this chapter.

Exercise 25.2. Let Φ be a root system in a Euclidean space $\mathcal{V}$. Let W be the Weyl group, and let W' be the group of all linear transformations of $\mathcal{V}$ that preserve Φ. Show that W is a normal subgroup of W' and that W'/W is isomorphic to the group of all symmetries of the Dynkin diagram of the associated Coxeter group. (Use Proposition 20.13.)

Exercise 25.3. Prove Theorem 25.4 by imitating the proof of Theorem 25.1

Exercise 25.4. How many reduced expressions are there in the A_3 Weyl group representing the long Weyl group element?

Exercise 25.5. Let $\alpha_1,\ \ ,\alpha_r$ be the simple roots of a reduced irreducible root system Φ, and let α_0 be the affine root, so that $-\alpha_0$ is the highest root. By Proposition 20.1, the inner product $\langle \alpha_i, \alpha_j \rangle \leqslant 0$ when i, j are distinct with $1 \leqslant i, j \leqslant r$. Show that this statement remains true if $0 \leqslant i, j \leqslant r$.

The next exercise gives another interpretation of Proposition 20.4.

Exercise 25.6. Let W be a Weyl group. Let $w = s_{i_1} s_{i_2} \cdots s_{i_N}$ be a decomposition of w into a product of simple reflections. Construct a path through the sequence (25.2) of chambers as in the proof of Theorem 25.1. Observe that the word $(i_1, i_2, \ldots, i_N)$ representing w is reduced if and only if this path does not cross any of the hyperplanes $\mathcal{H}$ orthogonal to the roots twice. Suppose that the word is not reduced, and that it meets some hyperplane $\mathcal{H}$ in two points, P and Q. Then for some k, with notation as in (25.2), P lies between $w_{k-1}\mathcal{C}$ and $w_{k-1}s_{i_k}\mathcal{C}$. Similarly Q lies between $w_{l-1}\mathcal{C}$ and $w_{l-1}s_{i_l}\mathcal{C}$. Show that $i_k = i_l$, and that

$$w = s_{i_1} \cdots \hat{s}_{i_k} \cdots \hat{s}_{i_l} \cdots s_{i_N}$$

where the "hat" means that the two entries are omitted. (**Hint:** Reflect the segment of the path between P and Q in the hyperplane $\mathfrak{H}$.)

Exercise 25.7. Prove that the Bruhat order has the following properties.

(i) If $sv < v$ and $su < u$, then $u \leqslant v$ if and only if $su \leqslant sv$.
(ii) If $sv < v$ and $su > u$, then $u \leqslant v$ if and only if $u \leqslant sv$.
(iii) If $sv > v$ and $su > u$, then $u \geqslant v$ if and only if $su \geqslant sv$.
(iv) If $sv > v$ and $su < u$, then $u \geqslant v$ if and only if $u \geqslant sv$.

[**Hint:** Any one of these four properties implies the others. For example, to deduce (ii) from (i), replace u by su].

Observe that $su < u$ if and only if $l(su) < l(u)$, a condition that is easy to check. Therefore, (i) and (ii) give a convenient method of checking (recursively) whether $u \leqslant v$.

Exercise 25.8. Let w_0 be the long element in a Weyl group W. Show that if $u, v \in W$ then $u \leqslant v$ if and only if $uw_0 \geqslant vw_0$.

The Borel Subgroup

The Borel subgroup B of a (noncompact) Lie group G is a maximal closed and connected solvable subgroup. We will give several applications of the Borel subgroup in this chapter and the next. In this chapter, we will begin with the Iwasawa decomposition, an important decomposition involving the Borel subgroup. We will also show how invariant vectors with respect to the Borel subgroup give a convenient method of decomposing a representation into irreducibles. We will restrict ourselves here to complex analytic groups such as $GL(n, \mathbb{C})$ obtained by complexifying a compact Lie group. A more general Iwasawa decomposition will be found later in Chap. 29.

Let us begin with an example. Let $G = GL(n, \mathbb{C})$. It is the complexification of $K = U(n)$, which is a maximal compact subgroup. Let T be the maximal torus of K consisting of diagonal matrices with eigenvalues that have absolute value 1. The complexification $T_{\mathbb{C}}$ of T can be factored as TA, where A is the group of diagonal matrices with eigenvalues that are positive real numbers. Let B be the group of upper triangular matrices in G, and let B_0 be the subgroup of elements of B whose diagonal entries are positive real numbers. Finally, let N be the subgroup of unipotent elements of B. Recalling that a matrix is called *unipotent* if its only eigenvalue is 1, the elements of N are upper triangular matrices with diagonal entries that are all equal to 1. We may factor $B = TN$ and $B_0 = AN$. The subgroup N is normal in B and B_0, so these decompositions are semidirect products.

Proposition 26.1. *With $G = GL(n, \mathbb{C})$, $K = U(n)$, and B_0 as above, every element of $g \in G$ can be factored uniquely as bk where $b \in B_0$ and $k \in K$, or as $a\nu k$, where $a \in A$, $\nu \in N$, and $k \in K$. The multiplication maps $N \times A \times K \longrightarrow G$ and $A \times N \times K \longrightarrow G$ are diffeomorphisms.*

Proof. First let us consider $N \times A \times K \longrightarrow G$. Let $g \in G$. Let $v_1, \ldots, v_n$ be the rows of g. Then by the Gram–Schmidt orthogonalization algorithm, we find constants θ_{ij} $(i < j)$ such that v_n, $v_{n-1} + \theta_{n-1,n}v_n$, $v_{n-2} + \theta_{n-2,n-1}v_{n-1} + \theta_{n-2,n}v_n, \ldots$ are orthogonal. Call these vectors $u_n, \ldots, u_1$, and let

D. Bump, *Lie Groups*, Graduate Texts in Mathematics 225, DOI 10.1007/978-1-4614-8024-2_26, © Springer Science+Business Media New York 2013

$$\nu^{-1} = \begin{pmatrix} 1 & \theta_{12} & \cdots & \theta_{1n} \\ & 1 & & \theta_{2n} \\ & & \ddots & \vdots \\ & & & 1 \end{pmatrix},$$

so $u_1, \ldots, u_n$ are the rows of $\nu^{-1}g$. Let a be the diagonal matrix with diagonal entries $|u_1|, \ldots, |u_n|$. Then $k = a^{-1}\nu^{-1}g$ has orthonormal rows, and so $g = \nu a k = b_0 k$ is unitary with $b_0 = \nu a$. This proves that the multiplication map $N \times A \times K \longrightarrow G$ is surjective. It follows from the facts that $B_0 \cap K = \{1\}$ and that $A \cap N = \{1\}$ that it is injective. It is easy to see that the matrices a, ν, and k depend continuously on g, so the multiplication map $A \times N \times K \longrightarrow G$ has a continuous inverse and hence is a diffeomorphism.

As for the map $A \times N \times K \longrightarrow G$, this is the composition of the first map with a bijection $A \times N \times K \to N \times A \times K$, in which $(a, n, k) \mapsto (n', a, k)$ if $an = n'a$. The latter map is also a diffeomorphism, and the conclusion is proved.   □

The decomposition $G \cong A \times N \times K$ is called the *Iwasawa decomposition* of $\mathrm{GL}(n, \mathbb{C})$.

To give another example, if $G = \mathrm{GL}(n, \mathbb{R})$, one takes $K = \mathrm{O}(n)$ to be a maximal compact subgroup, A is the same group of diagonal real matrices with positive eigenvalues as in the complex case, and N is the group of upper triangular unipotent real matrices. Again there is an Iwasawa decomposition, and one may prove it by the Gram–Schmidt orthogonalization process.

In this section, we will prove an Iwasawa decomposition if G is a complex Lie group that is the complexification of a compact connected Lie group K. This result contains the first example of $G = \mathrm{GL}(n, \mathbb{C})$, though not the second example of $G = \mathrm{GL}(n, \mathbb{R})$. A more general Iwasawa decomposition containing both examples will be obtained in Theorem 29.2.

We say that a Lie algebra $\mathfrak{n}$ is *nilpotent* if there exists a finite chain of ideals

$$\mathfrak{n} = \mathfrak{n}_1 \supset \mathfrak{n}_2 \supset \cdots \supset \mathfrak{n}_N = \{0\}$$

such that $[\mathfrak{n}, \mathfrak{n}_k] \subseteq \mathfrak{n}_{k+1}$.

Example 26.1. Let F be a field, and let $\mathfrak{n}$ be the Lie algebra over F consisting of upper triangular nilpotent matrices in $\mathrm{GL}(n, F)$. Let

$$\mathfrak{n}_k = \{g \in \mathfrak{n} \mid g_{ij} = 0 \text{ if } j < i + k\}.$$

For example, if $n = 3$,

$$\mathfrak{n} = \mathfrak{n}_1 = \left\{ \begin{pmatrix} 0 & * & * \\ 0 & 0 & * \\ 0 & 0 & 0 \end{pmatrix} \right\}, \quad \mathfrak{n}_2 = \left\{ \begin{pmatrix} 0 & 0 & * \\ 0 & 0 & 0 \\ 0 & 0 & 0 \end{pmatrix} \right\}, \quad \mathfrak{n}_3 = \{0\}.$$

This Lie algebra is nilpotent.

We also say that a Lie algebra $\mathfrak{b}$ is *solvable* if there exists a finite chain of Lie subalgebras

$$\mathfrak{b} = \mathfrak{b}_1 \supset \mathfrak{b}_2 \supset \cdots \supset \mathfrak{b}_N = \{0\} \qquad (26.1)$$

such that $[\mathfrak{b}_i, \mathfrak{b}_i] \subseteq \mathfrak{b}_{i+1}$. It is not necessarily true that $\mathfrak{b}_i$ is an ideal in $\mathfrak{b}$. However, the assumption that $[\mathfrak{b}_i, \mathfrak{b}_i] \subseteq \mathfrak{b}_{i+1}$ obviously implies that $[\mathfrak{b}_i, \mathfrak{b}_{i+1}] \subseteq \mathfrak{b}_{i+1}$, so $\mathfrak{b}_{i+1}$ is an ideal in $\mathfrak{b}_i$.

Clearly, a nilpotent Lie algebra is solvable. The converse is not true, as the next example shows.

Example 26.2. Let F be a field, and let $\mathfrak{b}$ be the Lie algebra over F consisting of all upper triangular matrices in $\mathrm{GL}(n, F)$. Let

$$\mathfrak{b}_k = \{g \in \mathfrak{b} \mid g_{ij} = 0 \text{ if } j < i + k - 1\}.$$

Thus, if $n = 3$,

$$\mathfrak{b} = \mathfrak{b}_1 = \left\{ \begin{pmatrix} * & * & * \\ 0 & * & * \\ 0 & 0 & * \end{pmatrix} \right\}, \qquad \mathfrak{b}_2 = \left\{ \begin{pmatrix} 0 & * & * \\ 0 & 0 & * \\ 0 & 0 & 0 \end{pmatrix} \right\},$$

$$\mathfrak{b}_3 = \left\{ \begin{pmatrix} 0 & 0 & * \\ 0 & 0 & 0 \\ 0 & 0 & 0 \end{pmatrix} \right\}, \qquad \mathfrak{b}_4 = \{0\}.$$

This Lie algebra is solvable. It is *not* nilpotent.

Proposition 26.2. *Let $\mathfrak{b}$ be a Lie algebra, $\mathfrak{b}'$ an ideal of $\mathfrak{b}$, and $\mathfrak{b}'' = \mathfrak{b}/\mathfrak{b}'$. Then $\mathfrak{b}$ is solvable if and only if $\mathfrak{b}'$ and $\mathfrak{b}''$ are both solvable.*

Proof. Given a chain of Lie subalgebras (26.1) satisfying $[\mathfrak{b}_i, \mathfrak{b}_i] \subset \mathfrak{b}_{i+1}$, one may intersect them with $\mathfrak{b}'$ or consider their images in $\mathfrak{b}''$ and obtain corresponding chains in $\mathfrak{b}'$ and $\mathfrak{b}''$ showing that these are solvable.

Conversely, suppose that $\mathfrak{b}'$ and $\mathfrak{b}''$ are both solvable. Then there are chains

$$\mathfrak{b}' = \mathfrak{b}'_1 \supset \mathfrak{b}'_2 \supset \cdots \supset \mathfrak{b}'_M = \{0\}, \qquad \mathfrak{b}'' = \mathfrak{b}''_1 \supset \mathfrak{b}''_2 \supset \cdots \supset \mathfrak{b}''_N = \{0\}.$$

Let $\mathfrak{b}_i$ be the preimage of $\mathfrak{b}''_i$ in $\mathfrak{b}$. Splicing the two chains in $\mathfrak{b}$ as

$$\mathfrak{b} = \mathfrak{b}_1 \supset \mathfrak{b}_2 \supset \cdots \supset \mathfrak{b}_N = \mathfrak{b}' = \mathfrak{b}'_1 \supset \mathfrak{b}'_2 \supset \cdots \supset \mathfrak{b}'_M = \{0\}$$

shows that $\mathfrak{b}$ is solvable. $\qquad\square$

Proposition 26.3. (Dynkin) *Let $\mathfrak{g} \subset \mathfrak{gl}(V)$ be a Lie algebra of linear transformations over a field F of characteristic zero, and let $\mathfrak{h}$ be an ideal of $\mathfrak{g}$. Let $\lambda : \mathfrak{h} \longrightarrow F$ be a linear form. Then the space*

$$W = \{v \in V \mid Yv = \lambda(Y)v \text{ for all } Y \in \mathfrak{h}\}$$

is invariant under all of $\mathfrak{g}$.

Proof. If $W = 0$, there is nothing to prove, so assume $0 \neq v_0 \in W$. Fix an element $X \in \mathfrak{g}$. Let W_0 be the linear span of $v_0, Xv_0, X^2v_0, \ldots$, and let d be the dimension of W_0.

If $Z \in \mathfrak{h}$, then we will prove that

$$Z(W_0) \subseteq W_0 \text{ and the trace of } Z \text{ on } W_0 \text{ is } \dim(W_0) \cdot \lambda(Z). \qquad (26.2)$$

To prove this, note that

$$v_0, Xv_0, X^2v_0, \ldots, X^{d-1}v_0 \qquad (26.3)$$

is a basis of W_0. With respect to this basis, for suitable $c_{ij} \in F$, we have

$$ZX^iv_0 = \lambda(Z)X^iv_0 + \sum_{j<i} c_{ij}X^jv_0. \qquad (26.4)$$

This is proved by induction since

$$ZX^iv_0 = XZX^{i-1}v_0 - [X,Z]X^{i-1}v_0.$$

By the induction hypothesis, $XZX^{i-1}v_0$ is $X\lambda(Z)X^{i-1}v_0$ plus a linear combination of X^jv_0 with $j < i$, and $[X,Z]X^{i-1}v_0$ is $\lambda([X,Z])X^{i-1}v_0$ plus a linear combination of X^jv_0 with $j < i - 1$. The formula (26.4) follows. The invariance of W_0 under Z is now clear, and (26.2) also follows from (26.4) because with respect to the basis (26.3) the matrix of Z is upper triangular and the diagonal entries all equal $\lambda(Z)$.

Now let us show that $Xv_0 \in W$. Let $Y \in \mathfrak{h}$. What we must show is that $YXv_0 = \lambda(Y)Xv_0$. The space W_0 is invariant under both X (obviously) and Y (by (26.2) taking $Z = Y$). Thus, the trace of $[X,Y] = XY - YX$ on W_0 is zero. Since $Y \in \mathfrak{h}$ and $\mathfrak{h}$ is an ideal, $[X,Y] \in \mathfrak{h}$ and we may take $Z = [X,Y]$ in (26.2). Since the characteristic of F is 0, we see that $\lambda([X,Y]) = 0$. Now

$$YXv_0 = XYv_0 - [X,Y]v_0 = \lambda(Y)Xv_0 - \lambda([X,Y])v_0 = \lambda(Y)Xv_0,$$

as required. $\qquad \square$

Theorem 26.1. (Lie) *Let* $\mathfrak{b} \subseteq \mathfrak{gl}(V)$ *be a solvable Lie algebra of linear transformations over an algebraically closed field of characteristic zero. Assume that* $V \neq 0$.

(i) There exists a vector $v \in V$ *that is a simultaneous eigenvector for all of* $\mathfrak{b}$.
(ii) There exists a basis of V *with respect to which all elements of* $\mathfrak{b}$ *are represented by upper triangular matrices.*

Proof. To prove (i), we may clearly assume that $\mathfrak{b} \neq 0$. Let us first observe that $\mathfrak{b}$ has an ideal $\mathfrak{h}$ of codimension 1. Indeed, since $\mathfrak{b}$ is solvable, $[\mathfrak{b},\mathfrak{b}]$ is a proper ideal, and the quotient Lie algebra $\mathfrak{b}/[\mathfrak{b},\mathfrak{b}]$ is Abelian; hence any subspace at all of $\mathfrak{b}/[\mathfrak{b}, \mathfrak{b}]$ is an ideal. We choose a subspace of codimension 1, and let $\mathfrak{h}$ be its preimage in $\mathfrak{b}$.

Now $\mathfrak{h}$ is solvable and of strictly smaller dimension than $\mathfrak{b}$, so by induction there exists a simultaneous eigenvector v_0 for all of $\mathfrak{h}$. Let $\lambda : \mathfrak{h} \longrightarrow F$ be such that $Xv_0 = \lambda(X)v_0$. The space $W = \{v \in V \mid Xv = \lambda(X)v \text{ for all } X \in \mathfrak{h}\}$ is nonzero, and by Proposition 26.3 it is $\mathfrak{b}$-invariant. Let $Z \in \mathfrak{b} - \mathfrak{h}$. Since F is assumed to be algebraically closed, Z has an eigenvector on W, which will be an eigenvector v_1 for all of $\mathfrak{b}$ since it is already an eigenvector for $\mathfrak{h}$.

For (ii), the Lie algebra of linear transformations of V/Fv_1 induced by those of $\mathfrak{b}$ is solvable, so by induction this quotient space has a basis $\overline{v_2}, \ldots, \overline{v_d}$ with respect to which every $X \in \mathfrak{b}$ is upper triangular. This means that for suitable $a_{ij} \in F$, we have $X\overline{v_j} = \sum_{2 \leqslant i \leqslant j} a_{ij}\overline{v_i}$. Letting $v_2, \ldots, v_d$ be representatives of the cosets $\overline{v_i}$ in V, it follows that X is upper triangular with respect to the basis $v_1, \ldots, v_d$. $\qquad \square$

Let K be a compact connected Lie group and $\mathfrak{k}$ its Lie algebra. Let $\mathfrak{g} = \mathfrak{k}_{\mathbb{C}}$ be the analytic complexification of $\mathfrak{k}$, so that $\mathfrak{g}$ is the Lie algebra of the complex Lie group G that is the complexification of K. Let T be a maximal torus of K. We can embed its analytic complexification $T_{\mathbb{C}}$ into G by the universal property of the complexification.

Let Φ be the root system of K and let Φ^+ be the positive roots with respect to some ordering. If $\alpha \in \Phi$, let $\mathfrak{X}_\alpha \subset \mathfrak{g}$ be the α-eigenspace. By Proposition 18.6, $\mathfrak{X}_\alpha$ is one-dimensional, and we will denote by X_α a nonzero element. Define

$$\mathfrak{n} = \bigoplus_{\alpha \in \Phi^+} \mathfrak{X}_\alpha. \tag{26.5}$$

Then $\mathfrak{n}$ is a complex Lie subalgebra of $\mathfrak{g}$. Indeed, if α and β are positive roots, it is impossible that $\alpha = -\beta$, so by Proposition 18.4 (ii), $[X_\alpha, X_\beta] \subset \mathfrak{X}_{\alpha+\beta}$ if $\alpha + \beta$ is a positive root, and otherwise it is zero. In either case, it is in $\mathfrak{n}$.

Proposition 26.4. *The Lie algebra $\mathfrak{n}$ defined by (26.5) is nilpotent.*

Proof. Let Φ_k^+ be the set of positive roots α such that α is expressible as the sum of at least k simple positive roots. Thus, $\Phi_1^+ = \Phi$, $\Phi_1^+ \supset \Phi_2^+ \supset \Phi_3^+ \supset \cdots$, and eventually Φ_k^+ is empty. Define

$$\mathfrak{n}_k = \bigoplus_{\alpha \in \Phi_k^+} \mathfrak{X}_\alpha.$$

It follows from Proposition 18.4 (ii) that $[\mathfrak{n}, \mathfrak{n}_k] \subseteq \mathfrak{n}_{k+1}$, and eventually $\mathfrak{n}_k$ is zero, so $\mathfrak{n}$ is nilpotent. $\qquad \square$

Now let $\mathfrak{t}$ be the Lie algebra of T, and let $\mathfrak{b} = \mathfrak{t}_{\mathbb{C}} \oplus \mathfrak{n}$. Since $[\mathfrak{t}_{\mathbb{C}}, \mathfrak{X}_\alpha] \subseteq \mathfrak{X}_\alpha$, it is clear that $\mathfrak{b}$, like $\mathfrak{n}$, is closed under the Lie bracket and forms a complex Lie algebra. Moreover, since $\mathfrak{t}_{\mathbb{C}}$ is Abelian and normalizes $\mathfrak{n}$, we have $[\mathfrak{b}, \mathfrak{b}] \subset \mathfrak{n}$, and since $\mathfrak{n}$ is nilpotent and hence solvable, it follows that $\mathfrak{b}$ is solvable.

We aim to show that both $\mathfrak{n}$ and $\mathfrak{b}$ are the Lie algebras of closed complex Lie subgroups of G.

Proposition 26.5. *Let G be the complexification of a compact connected Lie group K, and let $\mathfrak{n}$ be as in (26.5). If $\pi : G \longrightarrow \mathrm{GL}(V)$ is any representation and $X \in \mathfrak{n}$, then $\pi(X)$ is nilpotent as a linear transformation; that is, $\pi(X)^N = 0$ for all sufficiently large N.*

We note that it is possible for a nilpotent Lie algebra of linear transformations to contain linear transformations that are not nilpotent. For example, an Abelian Lie algebra is nilpotent as a Lie algebra but might well contain linear transformations that are not nilpotent.

Proof. By Theorem 26.1, we may choose a basis of V such that all $\pi(X)$ are upper triangular for $X \in \mathfrak{b}$, where we are identifying $\pi(X)$ with its matrix with respect to the chosen basis. What we must show is that if $X \in \mathfrak{n}$, then the diagonal entries of this matrix are zero. It is sufficient to show this if $X \in \mathfrak{X}_\alpha$, where α is a positive root.

By the definition of a root, the character α of T is nonzero, and so its differential $d\alpha$ is nonzero. This means that there exists $H \in \mathfrak{t}$ such that $d\alpha(H) \neq 0$, and by (18.9) the commutator $[\pi(H), \pi(X_\alpha)]$ is a nonzero multiple of $\pi(X_\alpha)$. Because it is a nonzero multiple of the commutator of two upper triangular matrices, it follows that $\pi(X_\alpha)$ is an upper triangular matrix with zeros on the diagonal. Thus, it is nilpotent. $\square$

Theorem 26.2. *(i) Let G be the complexification of a compact connected Lie group K, let T be a maximal torus of K, let $\mathfrak{t}$ be the Lie algebra of T, and let $T_{\mathbb{C}}$ be its complexification. Let $\mathfrak{n}$ be as in (26.5), and let $\mathfrak{b} = \mathfrak{t}_{\mathbb{C}} \oplus \mathfrak{n}$. Let $N = \exp(\mathfrak{n})$ and $B = T_{\mathbb{C}}N$. Then N and B are closed Lie subalgebras of G and $\mathfrak{n}$ and $\mathfrak{b}$ are the Lie algebras of N and B.*

(ii) We may embed G in $\mathrm{GL}(n, \mathbb{C})$ for some n in such a way that K consists of unitary matrices, $T_{\mathbb{C}}$ consists of diagonal matrices, and B consists of upper triangular matrices.

(iii) If $\mathfrak{u}$ is a complex Lie subalgebra of $\mathfrak{n}$, and $U = \exp(\mathfrak{u})$, then U is a complex analytic subgroup of N and $\mathfrak{u}$ is its Lie algebra. If $\mathfrak{u}$ is a real Lie subalgebra of $\mathfrak{n}$, and $U = \exp(\mathfrak{u})$, then U is a Lie subgroup of N and $\mathfrak{u}$ is its Lie algebra.

(iv) Suppose that $\mathfrak{v}$ and $\mathfrak{w}$ are (complex) Lie subalgebras of $\mathfrak{n}$ such that $\mathfrak{n} = \mathfrak{v} \oplus \mathfrak{w}$. Let $V = \exp(\mathfrak{v})$ and $W = \exp(\mathfrak{w})$ so that by (iii) V and W are complex analytic subgroups of N. Then $V \cap W = \{1\}$ and $N = VW$.

The group B is called the *standard Borel subgroup* of G. A conjugate of B is called a *Borel subgroup*. A subgroup containing a Borel subgroup is called a *parabolic subgroup*. We will call a subgroup containing the standard Borel subgroup a *standard parabolic*.

Proof. We will prove parts (i) and (ii) simultaneously.

Let $\pi : K \longrightarrow \mathrm{GL}(V)$ be a faithful representation. We choose on V an inner product with respect to which $\pi(k)$ is unitary for $k \in K$. By Theorem 24.1, we may extend π to a faithful complex analytic representation of G. We have already noted that $\mathfrak{b}$ is a solvable Lie algebra, so by Theorem 26.1 we may find a basis $v_1, \ldots, v_n$ of V with respect to which the linear transformations $\pi(X)$ with $X \in \mathfrak{b}$ are upper triangular. This means that $\pi(X)v_i \in \sum_{j \leqslant i} F v_j$. We claim that we may assume that the v_i are orthonormal. This is accomplished by Gram–Schmidt orthonormalization. We first divide v_i by $|v_i|$ so v_i has length 1. Next we replace v_2 by $v_2 - \langle v_2, v_1 \rangle v_1$ and so forth so that the v_i are orthonormal. The matrices $\pi(X)$ with $X \in \mathfrak{b}$ remain upper triangular after these changes.

We identify G with its image in $\mathrm{GL}(n, \mathbb{C})$ and its Lie algebra with the corresponding Lie subalgebra of $\mathrm{Mat}_n(\mathbb{C}) = \mathfrak{gl}(n, \mathbb{C})$. Thus, we write X instead of $\pi(X)$ and regard it as a matrix.

Let

$$= \{ \exp(X) \mid X \in \mathfrak{n} \}. \tag{26.6}$$

We will show that N is a closed analytic subgroup of G with a Lie algebra that is N.

By Remark 8.1, if $X \in \mathfrak{n}$ and $Y = \exp(X)$, then

$$Y = I + X + \tfrac{1}{2}X^2 + \ldots + \tfrac{1}{n!}X^n.$$

This is now a series with only finitely many terms since X is nilpotent by Proposition 26.5. Moreover, $Y - I$ is a finite sum of upper triangular nilpotent matrices and hence is itself nilpotent, and reverting the exponential series, we have $X = \log(Y)$, where we define

$$\log(Y) = (Y - 1) - \tfrac{1}{2}(Y - 1)^2 + \tfrac{1}{3}(Y - 1)^3 - \cdots + (-1)^{n-1}\tfrac{1}{n}(Y - 1)^n$$

if Y is an upper triangular unipotent matrix. As with the exponential series, only finitely many terms are needed since $(Y - I)^n = 0$. This series defines a continuous map $\log : N \longrightarrow \mathfrak{n}$, which is the inverse of the exponential map. Therefore, $\mathfrak{n}$ is homeomorphic to N.

Next we show that N is a closed subset of $\mathrm{GL}(n, \mathbb{C})$ and in fact an affine subvariety. Let $\mathfrak{n}'$ be the Lie subalgebra of $\mathfrak{gl}(n, \mathbb{C})$ consisting of upper triangular nilpotent matrices, and let $\lambda_1, \ldots, \lambda_r$ be a set of linear functionals on $\mathfrak{n}'$ such that the intersection of the kernels of the λ_i is $\mathfrak{n}$. N may be characterized as follows. An element $g \in \mathrm{GL}(n, \mathbb{C})$ is in N if and only if it is upper triangular and unipotent, and each $\lambda_i (\log(g)) = 0$. These conditions comprise a set of polynomial equations characterizing N, showing that it is closed.

Next we show that N is a group. Indeed, its intersection with a neighborhood of the identity is a local group by Proposition 14.1. Thus, if g, h are near the identity in N, we have $gh \in N$, so $\phi_i(g, h) = 0$ where $\phi_i(g, h) =$

$\lambda_i\,(\exp(gh))$. Thus, the polynomial ϕ_i vanishes near the identity in $N \times N$, and since N is a connected affine subvariety of $\mathrm{GL}(n, \mathbb{C})$, this polynomial vanishes identically on all of N. Thus, N is closed under multiplication, and it is a group.

Since $[\mathfrak{t}_{\mathbb{C}}, \mathfrak{n}] \subset \mathfrak{n}$, the group $T_{\mathbb{C}}$ normalizes N, so $B = T_{\mathbb{C}}N$ is a subgroup of G. It is not hard to show that it is a closed Lie subgroup and its Lie algebra is $\mathfrak{b}$.

The same argument that proved that N is a Lie group proves (iii). In the case where $\mathfrak{u}$ is a complex Lie algebra, We simply take a larger set of linear functionals λ_i on $\mathfrak{n}'$ with a kernel that is the Lie subalgebra $\mathfrak{u}$ and argue identically to show first that $U = \exp(\mathfrak{u})$ is closed, and that it is a Lie subgroup. If $\mathfrak{u}$ is a real Lie algebra, we proceed in the same way but take the λ_i to be real linear.

We turn to the proof of (iv). We saw in the proof of (ii) that the map $\exp : \mathfrak{n} \longrightarrow N$ is surjective, and given by a polynomial expression, with a polynomial inverse $\log : N \longrightarrow \mathfrak{n}$. Moreover, $\exp$ takes $\mathfrak{v}$ to V and $\mathfrak{w}$ to W, while $\log$ takes V to $\mathfrak{v}$ and W to $\mathfrak{w}$. It follows that $V \cap W = \{1\}$ since if $g \in V \cap W$ then $\log(g) \in \mathfrak{v} \cap \mathfrak{w} = 0$, so $g = 1$.

To show that $N = VW$ we note that the multiplication map $V \times W \longrightarrow N$ has as its differential the inclusion $\mathfrak{v} \oplus \mathfrak{w} \longrightarrow \mathfrak{n}$. But this map is the identity map. Therefore, by the inverse function theorem multiplication $V \times W \longrightarrow N$ is onto a neighborhood of the identity and therefore has an analytic inverse. This means that there are analytic maps $\phi : \mathfrak{n} \longrightarrow V$ and $\psi : \mathfrak{n} \longrightarrow W$, defined by power series convergent near the identity, such that $\phi(X)\psi(X) = e^X$ for $X \in \mathfrak{n}$. We will argue that ϕ and ψ are polynomials. Let X_i be a basis of $\mathfrak{v}$ and let Y_j be a basis of $\mathfrak{w}$. Let $\lambda : \mathfrak{n} \longrightarrow \mathfrak{n}$ be the map $\lambda(X) = \log(\phi(X)) + \log(\psi(X))$. Then λ is the inverse map of the map μ that sends $X = \sum c_i X_i + \sum d_j Y_j$ to $\exp(\sum c_i X_i)\exp(\sum d_j Y_j)$. Regarding $\mathfrak{n}$ as a vector subspace of $\mathfrak{gl}(n, \mathbb{C}) = \mathrm{Mat}_n(\mathbb{C})$, we see that $\mu(X)$ is a finite linear combination of finite products of the $c_i X_i$ and $d_j Y_j$, where the products are taken in the sense of matrix multiplication. Inverting μ, we see that $\lambda(X)$ also is a linear combination of such finite products of the $c_i X_i$ and $d_j Y_j$. It is a *finite* such linear combination since only finitely many such products are nonzero: this is because the matrices X_i and Y_j are upper triangular and nilpotent. Projecting onto $\mathfrak{v}$ and $\mathfrak{w}$ and exponentiating, we see that ϕ and ψ are polynomials.

Since both sides are polynomials, the identity $\phi(X)\psi(X) = e^X$, already proved for X near 0, is true for all X and it follows that the multiplication map $V \times W \longrightarrow N$ is surjective. This proves that $N = VW$. $\qquad\square$

The Borel subgroup is a bit too big for the Iwasawa decomposition since it has a nontrivial intersection with K. Let $\mathfrak{a} = i\mathfrak{t}$. It is the Lie algebra of a closed connected Lie subgroup A of T. If we embed K and G into $\mathrm{GL}(n, \mathbb{C})$ as in Theorem 26.2, the elements of T are diagonal, and A consists of the subgroup of elements of T whose diagonal entries are positive real numbers. Let $B_0 = AN$.

Theorem 26.3. (Iwasawa decomposition) *With notations as in Theorem 26.2 and B_0 and A as above, each element of $g \in G$ can be factored uniquely as bk where $b \in B_0$ and $k \in K$, or as $a\nu k$ where $a \in A$, $\nu \in N$ and $k \in K$. The multiplication map $A \times N \times K \longrightarrow G$ is a diffeomorphism.*

Proof. Let $G' = \mathrm{GL}(n, \mathbb{C})$, $K' = U(n)$, A' be the subgroup of $\mathrm{GL}(n, \mathbb{C})$ consisting of diagonal matrices with positive real eigenvalues, and N' be the subgroup of upper triangular unipotent matrices in G'. By Theorem 26.2 (ii), we may embed G into G' for suitable n such that K ends up in K', N ends up in N', and A ends up in K_0'.

We have a commutative diagram

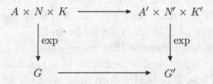

where the vertical arrows are multiplications and the horizontal arrows are inclusions. By Proposition 26.1, the composition

$$A \times N \times K \longrightarrow A' \times N' \times K' \longrightarrow G' \qquad (26.7)$$

is a diffeomorphism onto its image, and so the multiplication $A \times N \times K \longrightarrow G$ is a diffeomorphism onto its image. We must show that it is surjective.

Since A, N, and K are each closed in A', N', and K', respectively, the image of (26.7) is closed in G' and hence in G. We will show that this image is also open in G. We note that $\mathfrak{a} + \mathfrak{n} + \mathfrak{k} = \mathfrak{g}$ since $\mathfrak{t}_{\mathbb{C}} \subset \mathfrak{a} + \mathfrak{k}$, and each $\mathbb{C}\mathfrak{X}_\alpha \subset \mathfrak{n} + \mathfrak{k}$. It follows that the dimension of $A \times N \times K$ is greater than or equal to that of G. (These dimensions are actually equal, though we do not need this fact, since it is not hard to see that the sum $\mathfrak{a} + \mathfrak{n} + \mathfrak{k}$ is direct.) Since multiplication is a diffeomorphism onto its image, this image is open and closed in G. But G is connected, so this image is all of G, and the theorem is now clear. $\square$

As an application, we may now show why flag manifolds have a complex structure.

Theorem 26.4. *Let K be a compact connected Lie group and T a maximal torus. Then $X = K/T$ can be given the structure of a complex manifold in such a way that the translation maps $g : xT \longrightarrow gxT$ are holomorphic. This action of K can be extended to an action of the complexification G by holomorphic maps.*

Proof. By the Iwasawa decomposition, we may write $G = BK$. Since $B \cap K = T$, we have $G/B \cong K/T$, and this diffeomorphism is K-equivariant. Now G is a complex Lie group and B is a closed analytic subgroup, so the quotient G/B has the structure of a complex analytic manifold, and the action of G, *a fortiori* of K, consists of holomorphic maps.   □

We turn now to a different use of the Borel subgroup. If (π, V) is a finite-dimensional representation of K, then by Theorem 24.1, π can be extended to a complex analytic representation of G, and of course a complex analytic representation of G can be restricted back to K. So the categories of finite-dimensional representations of K, and the finite-dimensional analytic representations of G are equivalent.

Let λ be a weight. It is thus a character of $T_{\mathbb{C}}$. Now B is a semidirect product $T_{\mathbb{C}}N$ with N normal, so $T_{\mathbb{C}} \cong B/N$. This means that λ may be extended to a character of B with N in its kernel.

We will show that each irreducible representation of G has an N-fixed vector v that is unique up to scalar multiple. It is the highest weight vector of the representation. Thus, if λ is the dominant weight, we have $\pi(t)v = \lambda(t)v$ for $t \in T_{\mathbb{C}}$. Since v is N-fixed, we may also write $\pi(b)v = \lambda(b)v$ for $b \in B$. We will give some applications of this useful fact.

Let $\mathfrak{n}$ be the Lie algebra of N, defined by (26.5). We may similarly define $\mathfrak{n}_-$ to be the span of $\mathfrak{X}_\alpha$ with $\alpha \in \Phi_-$. It is also the Lie algebra of a Lie subgroup of G, which we will denote N_-. Let w_0 be a representative of the long Weyl group element. Then $\mathrm{Ad}(w_0)$ interchanges the positive and negative roots, so $\mathrm{Ad}(w_0)\mathfrak{n} = \mathfrak{n}_-$ and $w_0 N w_0^{-1} = N_-$.

Lemma 26.1. *The Lie algebra $\mathfrak{n}$ is generated by the X_α as α runs through the simple positive roots.*

Proof. Let $\mathfrak{n}'$ be the algebra generated by the X_α with α simple. Let us define the *height* of a positive root to be the number of simple roots into which it may be decomposed, counted with multiplicities. If $\alpha \in \Phi^+$ is not simple, we may write $\alpha = \beta + \gamma$ where β and γ are in Φ^+, and by induction on the height of α, we may assume that X_β and X_γ are in $\mathfrak{n}'$. By Corollary 18.1, $[X_\beta, X_\gamma]$ is a nonzero multiple of X_α and so X_α is in $\mathfrak{n}'$. Thus, $\mathfrak{n}' = \mathfrak{n}$.   □

Now $\mathfrak{g}$ is a complex Lie algebra and $\mathfrak{n}_-$, $\mathfrak{t}_{\mathbb{C}}$ and $\mathfrak{n}$ are Lie subalgebras, so we have homomorphisms $i_\mathfrak{n}$, $i_\mathfrak{t}$ and $i_{\mathfrak{n}_-}$ mapping $U(\mathfrak{n}_-)$, $\mathrm{U}(\mathfrak{t}_{\mathbb{C}})$ and $U(\mathfrak{n})$ into $U(\mathfrak{g})$. The Poincaré–Birkhoff–Witt theorem implies that these homomorphisms are injective, but we do not need that fact. If ξ is in $U(\mathfrak{n}_-)$, for example, we will use the same notation ξ for $i_\mathfrak{n}(\xi)$, which is an abuse of notation since we are omitting to prove that $i_\mathfrak{n}$ is injective. The multiplication map $U(\mathfrak{n}_-) \times U(\mathfrak{t}_{\mathbb{C}}) \times U(\mathfrak{n}) \longrightarrow U(\mathfrak{g})$ that sends (ξ, η, ζ) to $\xi\,\eta\zeta$ induces a linear map $\mu : U(\mathfrak{n}_-) \otimes U(\mathfrak{t}_{\mathbb{C}}) \otimes U(\mathfrak{n}) \longrightarrow U(\mathfrak{g})$. With the above abuse of notation, we denote the image of μ as $U(\mathfrak{n}_-)U(\mathfrak{t}_{\mathbb{C}})U(\mathfrak{n})$.

Proposition 26.6. (Triangular decomposition) *The linear map* μ :
$U(\mathfrak{n}_-) \otimes U(\mathfrak{t}_\mathbb{C}) \otimes U(\mathfrak{n}) \longrightarrow U(\mathfrak{g})$ *is surjective.*

Proof. Let $\mathfrak{R} = U(\mathfrak{n}_-)U(\mathfrak{t}_\mathbb{C})U(\mathfrak{n})$ be the image of μ. Since $\mathfrak{R}$ contains
generators of $U(\mathfrak{g})$, it is enough to show that it is closed under multiplication.
It is obvious that $U(\mathfrak{n}_-)\,\mathfrak{R} \subseteq \mathfrak{R}$. Moreover, since $[\mathfrak{t}_\mathbb{C}, \mathfrak{n}_-] \subseteq \mathfrak{n}_-$ we also have
$U(\mathfrak{t}_\mathbb{C})\,\mathfrak{R} \subseteq \mathfrak{R}$. It remains for us to show that $U(\mathfrak{n})\mathfrak{R} \subseteq \mathfrak{R}$. By Lemma 26.1,
$U(\mathfrak{n})$ is generated with X_α with α simple, so it is enough to show that $X_\alpha\mathfrak{R} \subseteq$
$\mathfrak{R}$ when α is simple. First, if $\beta \in \Phi_+$ then $X_\alpha X_{-\beta} - X_{-\beta}X_\alpha = [X_\alpha, X_{-\beta}] \in \mathfrak{n}$
unless $\beta = \alpha$, by Proposition 18.4, while if $\beta = \alpha$ we have $[X_\alpha, X_{-\alpha}] \in \mathfrak{t}_\mathbb{C}$.
On the other hand if $H \in \mathfrak{t}_\mathbb{C}$ then $[X_\alpha, H]$ is a constant multiple of X_α. As a
result of these relations, $X_\alpha\mathfrak{R} \subseteq \mathfrak{R}$. $\square$

Let (π, V) be an irreducible representation of K. We may extend (π, V)
to an irreducible analytic representation of G. Let λ be the highest weight.
Proposition 22.4 tells us that the weight space $V(\lambda)$ corresponding to the
highest weight λ is one-dimensional. Let v_λ be a nonzero element. If α is a
positive root, then $\pi(X_\alpha)v_\lambda = 0$ because it is in $V(\lambda + \alpha)$, which is zero.
(Otherwise, λ is not a highest weight.) On the other hand the triangular
decomposition gives some complementary information.

Theorem 26.5. *Let* (π, V) *be an irreducible representation of* K. *Extend*
(π, V) *to an irreducible analytic representation of* G. *Let* λ *be the highest*
weight. Then $V(\lambda) = V^N$ *is the space of* N-*invariants.*

Proof. Clearly $v \in V$ is N-invariant if and only if $\pi(X_\alpha) = 0$ for $\alpha \in \Phi^+$,
and as we have noted this is true if $v \in V(\lambda)$. We must show that N invari-
ance implies that $v \in V(\lambda)$. Since $T_\mathbb{C}$ normalizes N, V^N is $T_\mathbb{C}$-invariant and
can be decomposed into weight spaces. So we may assume that $v \in V(\mu)$
for some μ, and the problem is to prove that $\mu = \lambda$. We may write
$U(\mathfrak{g}) = U(\mathfrak{n}_-)U(\mathfrak{t}_\mathbb{C}) \oplus U(\mathfrak{n}_-)U(\mathfrak{t}_\mathbb{C})\mathfrak{J}$ where $\mathfrak{J}$ is the ideal of $U(\mathfrak{n})$ generated by
$\mathfrak{n}$, and $\mathfrak{J}v = 0$. Therefore, $U(\mathfrak{g})v = U(\mathfrak{n}_-)U(\mathfrak{t}_\mathbb{C})v = U(\mathfrak{n}_-)v$. But by Proposition 18.3 (iii) each weight in $U(\mathfrak{n}_-)v$ is $\preccurlyeq \mu$, so $U(\mathfrak{g})v$ does not contain $V(\lambda)$.
It is therefore a proper nonzero submodule. But V is irreducible, so this is a
contradiction. $\square$

Now suppose that we have a representation that we want to decompose into
irreducibles. Theorem 26.5 gives a strategy for obtaining this decomposition.
We remind the reader that if λ is a weight, that is, a character of $T_\mathbb{C}$, then we
may regard λ as a character of B in which N acts trivially, because $B/N \cong T_\mathbb{C}$.

Proposition 26.7. *Let* W *be a* G-*module that decomposes into a direct sum of*
finite-dimensional irreducible representations. Let λ *be a dominant weight, and*
let π_λ *be the irreducible* G-*module with highest weight* λ. *Then the multiplicity*
of π_λ *as a* G-*module in* W *equals the multiplicity of* λ *as a* B-*module in* W.

Proof. Since N acts trivially in λ as a B-module, every B-submodule of W isomorphic to λ is contained in W^N. By Theorem 26.5, each copy of π_λ contains a unique vector in W^N. The statement is therefore clear.   □

To give an example, let us decompose the symmetric algebra $\bigvee(\vee^2\mathbb{C}^n)$ for over the symmetric square of the standard module for $\mathrm{GL}(n,\mathbb{C})$.

Proposition 26.8. *Let Ω be the space of $n \times n$ symmetric complex matrices. Let $\mathcal{P}(\Omega)$ be the ring of polynomials on Ω with the $\mathrm{GL}(n,\mathbb{C})$ action $(gf)(X) = f({}^tg \cdot X \cdot g)$. Then $\mathcal{P}(\Omega)$ is isomorphic to the $\mathrm{GL}(n,\mathbb{C})$-module $\bigvee(\vee^2\mathbb{C}^n)$.*

Proof. For any module W the polynomial ring on W is isomorphic as a $\mathrm{GL}(n,\mathbb{C})$-module to the symmetric algebra on W^*. So it is sufficient to show that $\vee^2\mathbb{C}^n$ and Ω are dual modules. Indeed, if $V = \mathbb{C}^n$ and M is an $n \times n$ symmetric matrix then M induces a symmetric bilinear map $V \times V \longrightarrow \mathbb{C}$ by $(v_1, v_2) \longmapsto {}^t v_1 M v_2$. (We are thinking of v_1 and v_2 as column vectors.) The linear map $\vee^2 V \longrightarrow \mathbb{C}$ identifies Ω with the dual space of $\vee^2 V$.   □

We identify the weight lattice Λ of $U(n)$ or $\mathrm{GL}(n,\mathbb{C})$ with $\mathbb{Z}^n$ as follows: if $\lambda \in \mathbb{Z}^n$ then we identify λ with the character

$$\begin{pmatrix} t_1 & & \\ & \ddots & \\ & & t_n \end{pmatrix} \longmapsto \prod t_i^{\lambda_i}.$$

The weight λ is *dominant* if $\lambda_1 \geqslant \lambda_2 \geqslant \cdots \geqslant \lambda_n$. Assuming this, we say that λ is *even* if the λ_i are all even, and *effective* if $\lambda_n \geqslant 0$. It is not hard to see that λ is effective.

Theorem 26.6. *The $\mathrm{GL}(n,\mathbb{C})$-module $\bigvee(\vee^2\mathbb{C}^n)$ decomposes into a direct sum of irreducible representations, each with multiplicity one. Let λ be a dominant weight. The irreducible representation with highest weight λ occurs in this decomposition if and only if λ is even. Each occurs with multiplicity one.*

Proof. By Proposition 26.8 we may work with the representation $\mathcal{P}(\Omega)$. As we have explained, our task is to compute the N-invariants of the representation. If $X = (X_{ij}) \in \Omega$, let X_k $(1 \leqslant k \leqslant n)$ be the upper left $k \times k$ minor, that is

$$X_k = \det(X_{ij})_{1 \leqslant i,j \leqslant k}.$$

We consider X_k to be a polynomial function on Ω. It is simple to check that it is N-invariant, and we will show that the ring of N-invariants is generated by $X_1, \ldots, X_k$. Let Ω' be the subspace of Ω characterized by the nonvanishing of the X_k. It is a dense open set, so any element of $\mathcal{P}(\Omega)$ is determined by its restriction to Ω'. Now any double coset in $N \backslash \Omega' / N$ is equivalent to a diagonal element; indeed the element X is in the same double coset as

$$\begin{pmatrix} X_1 & & & \\ & X_2/X_1 & & \\ & & X_3/X_2 & \\ & & & \ddots \end{pmatrix}.$$

Clearly any polynomial of X restricts to the diagonal as a polynomial

$$\begin{pmatrix} x_1 & & & \\ & x_2 & & \\ & & x_3 & \\ & & & \ddots \end{pmatrix} = \sum_{\mu \in \mathbb{Z}^n} a(\mu) x^\mu, \qquad x^\mu = x_1^{\mu_1} \cdots x_n^{\mu_n},$$

and since the value of x^μ on X is $X_1^{\mu_1 - \mu_2} X_2^{\mu_2 - \mu_3} \cdots$ if this is a polynomial of X we must have $a(\mu) = 0$ unless $\mu_1 \geqslant \mu_2 \geqslant \cdots$. Thus the x^μ with μ dominant form a basis of the N-invariants. We have seen that there is one irreducible representation for each basis vector. Under the action in Proposition 26.8, the weight of the vector x^μ is 2μ. Hence the highest weights of the irreducible representations in $\bigvee(\vee^2 \mathbb{C}^n)$ are the even dominant weights. $\qquad\square$

The following proposition abstracts this situation. We will say that a representation of G that decomposes into a direct sum of finite-dimensional irreducible representations is *multiplicity-free* if no irreducible representation occurs in it with multiplicity greater than one. For example, we have just proved that $\bigvee(\vee^2 \mathbb{C}^n)$ is a multiplicity-free $GL(n, \mathbb{C})$-module. By Proposition 26.7 a method of proving that a module W is multiplicity-free is to show that W^N is multiplicity-free as a B-module. The next result exposes the underlying mechanism in the proof of Theorem 26.6.

In the following result we will assume that G is an affine algebraic group over $\mathbb{C}$. We note that all of the usual examples, $SL(n, \mathbb{C})$, $GL(n, \mathbb{C})$, $O(n, \mathbb{C})$, $Sp(2n, \mathbb{C})$, ... are affine algebraic groups. We continue to assume that G is the complexification of a compact Lie group, and this assumption is true for these examples as well.

Theorem 26.7. *Assume that G is an affine algebraic group over the complex numbers. Assume that it is also the complexification of a compact Lie group K. Let X be a complex affine algebraic variety on which the group G acts algebraically. Assume that the Borel subgroup B has a dense open orbit in X. Let W be the space of algebraic functions on X. Then W is a multiplicity-free G-module.*

The open orbit, if it exists, is always unique and dense, so the word "dense" could be eliminated from the statement. The theory of algebraic group actions is an important topic, and the standard monograph is Mumford, Fulton, and Kirwan [132]. Varieties (whether affine or not) with an open B-orbit are called *spherical*.

Proof. We need to prove that W decomposes into a direct sum of finite-dimensional modules.

We begin by showing that if $f \in W$ then the G-translates of f span a finite-dimensional vector space $W(f)$. Since the group action $G \times X \longrightarrow X$ is algebraic, if f is a polynomial function on X, then $(g, x) \longmapsto f(gx)$ is a polynomial function on $G \times X$ and so there exist polynomials ϕ_i on G and ψ_i on X such that $f(gx) = \sum_i \phi_i(g)\psi_i(x)$. Thus, the space $W(f)$ of left translates of f is spanned by the functions ψ_i and is finite-dimensional.

Now we embed X in an affine space, so that we may speak of the degree of a polynomial function. Let W_N be the direct sum of the $W(f)$ for f a polynomial of degree $\leqslant N$. Then W_N is finite-dimensional and G-invariant. Because K is compact, W_N decomposes into a direct sum of irreducible K-modules, which are also G-invariant subspaces since G is the complexification of K. Since $W_N \subset W_{N+1}$, may also choose these decompositions so that every irreducible that occurs in the decomposition of W_N is also in the decomposition of W_{N+1}. Taking the sum of all the irreducibles that occur in these decompositions shows that W is completely reducible.

Now let $x_0 \in X$ such that Bx_0 is open and dense. If $f \in W$ and $g \in G$, then the group action is by $(gf)(x) = f(g^{-1}x)$. So if f is in W^N and λ is a dominant weight such that $bf = \lambda(b)f$ for all $b \in B$, we have $f(bx_0) = \lambda(b)^{-1}f(x_0)$. Because Bx_0 is dense, this means that f is determined up to a scalar multiple by this condition. This shows that W^N is multiplicity-free as a B-module and therefore W is multiplicity-free as a G-module. $\square$

Exercises

Exercise 26.1. Let Ω be the vector space of $n \times n$ skew-symmetric matrices. $G = \mathrm{GL}(n, \mathbb{C})$ acts on Ω by the space of polynomial functions on Ω by $(gf)(X) = f({}^t g \cdot X \cdot g)$.

(i) Show that the symmetric algebra on the exterior square of the standard module of $\mathrm{GL}(n, \mathbb{C})$, that is, $\bigvee(\wedge^2 \mathbb{C}^n)$, is isomorphic as a G-module to the ring of polynomial functions on Ω.

(ii) Show that $\bigvee(\wedge^2 \mathbb{C}^n)$ decomposes as a direct sum of irreducible representations, each with multiplicity one, and that if $\lambda = (\lambda_1, \lambda_2, \ldots, \lambda_n)$ is a dominant weight, then λ occurs in this decomposition if and only if $\lambda_1 = \lambda_2$, $\lambda_3 = \lambda_4, \ldots$, and if n is odd, then $\lambda_n = 0$.

Exercise 26.2. Let $G = \mathrm{GL}(n, \mathbb{C})$, and let $H = \mathrm{O}(n, \mathbb{C})$. As in Proposition 26.8, let Ω be space of symmetric $n \times n$ complex matrices, and let Ω° be the open subset of invertible $n \times n$ matrices. Let $\mathcal{P}(\Omega)$ be the ring of polynomials on Ω, and let $\mathcal{P}(\Omega^\circ)$ be the space of polynomial functions on Ω°; it is generated by $\mathcal{P}(\Omega)$ together with $g \longmapsto \det(g)^{-1}$. The group G acts on both $\mathcal{P}(\Omega)$ and $\mathcal{P}(\Omega^\circ)$ as in Proposition 26.8. The stabilizer of $I \in \mathcal{P}(\Omega^\circ)$ is the group H, and the action on $\mathcal{P}(\Omega^\circ)$ is transitive, so $\mathcal{P}(\Omega^\circ)$ is in bijection with G/H. Let (π, V) be an irreducible representation of G.

(i) Show that (π, V) has a nonzero H-fixed vector if and only if its contragredient $(\hat{\pi}, V^*)$ does. [**Hint:** Show that $\hat{\pi}$ is equivalent to the representation $\pi' : G \longrightarrow V$ defined by $\pi'(g) = {}^t g^{-1}$ by comparing their characters.]

(ii) Assume that π has a nonzero H-fixed vector. By (i) there is an H-invariant linear functional $\phi : V \longrightarrow \mathbb{C}$. Define $\Phi : V \longrightarrow \mathcal{P}(\Omega^\circ)$ by letting $\Phi(v)$ be the function Φ_v defined by

$$\Phi_v(X) = \phi(^tgv), \qquad X = {}^tgg.$$

Show that this is well defined and that $\Phi_{gv}(X) = \Phi_v(^tgXg)$. Deduce that $v \longmapsto \Phi_v$ is an embedding of V into $\mathcal{P}(\Omega^\circ)$.

(iii) Show that π has a nonzero H-fixed vector if and only if π can be embedded in $\mathcal{P}(\Omega^\circ)$. [**Hint:** One direction is (ii). For the other, prove instead that π has an H-invariant linear functional.]

Remark: The argument in (ii) and (iii) is formally very similar to the proof of Frobenius reciprocity (Proposition 32.2) with $\mathcal{P}(\Omega^\circ)$ playing the role of the induced representation.)

(iv) *Show that an irreducible representation of $\mathcal{P}(\Omega^\circ)$ can be extended to $\mathcal{P}(\Omega)$ if and only if its highest weight λ is effective.*

(v) *Let π_λ be an irreducible representation of G with highest weight λ. Assume that λ is effective, so that λ is a partition. Show that π_λ has an $O(n)$-fixed vector if and only if λ is even.*

(vi) *Assume again that λ is effective, but only assume that π_λ has a fixed vector for $SO(n)$. What λ are possible?*

Exercise 26.3. The last exercise shows that if (π, V) is an irreducible representation of $GL(n, \mathbb{C})$, then the multiplicity of the trivial representation of $O(n, \mathbb{C})$ in its restriction to this subgroup is at most one. Show by example that there are other representations that can occur with higher multiplicity, for example when $n = 5$.

The next exercise is essentially a proof of the *Cauchy identity*, which is the subject of Chap. 38.

Exercise 26.4. Let $G = GL(n, \mathbb{C}) \times GL(n, \mathbb{C})$ acting on the ring $\mathcal{P}$ of polynomial functions on $\mathrm{Mat}_n(\mathbb{C})$ by

$$((g_1, g_2)f)(X) = f(^tg_1Xg_2),$$

$f \in \mathcal{P}$ and $X \in \mathrm{Mat}_n(\mathbb{C})$.

(i) Prove that $\mathcal{P}$ is isomorphic as a $GL(n, \mathbb{C}) \times GL(n, \mathbb{C})$ to the symmetric algebra on $V \otimes V$. (**Hint:** Adapt the proof of Proposition 26.8.)

(ii) Prove that $\mathcal{P}$ is isomorphic as a $GL(n, \mathbb{C}) \times GL(n, \mathbb{C})$ to the direct sum of all modules $\pi_\lambda \otimes \pi_\lambda$ as λ runs through the effective dominant weights. [**Hint:** Adapt the proof of Theorem 26.6. Note that if B is the standard Borel subgroup of $GL(n, \mathbb{C})$ then $B \times B$ is a Borel subgroup of $GL(n, \mathbb{C}) \times GL(n, \mathbb{C})$, so the problem is to find the $N \times N$ invariants in $\mathcal{P}$. These are, as in Theorem 26.6, again polynomials in certain minors of X.]

Exercise 26.5. Let G be a complex analytic Lie group and let H_1, H_2 be closed analytic subgroups. Then G acts on the homogeneous space G/H_1, as does its subgroup H_2. The quotient is the space of double cosets, $H_2 \backslash G / H_1$, which might also be obtained by letting H_1 act on the right on $H_2 \backslash G$.

(i) Show that if $\gamma \in H_1$ then the stabilizer in H_2 of the coset γH_1 is $H_\gamma = H_2 \cap \gamma H_1 \gamma^{-1}$. Deduce that the dimension of the orbit is $\dim(H_2) - \dim(H_\gamma)$

(ii) Show that H_2 has an open orbit on G/H_1 if and only if

$$\dim(H_\gamma) + \dim(G) = \dim(H_1) + \dim(H_2).$$

(iii) Show that H_2 has an open orbit on G/H_1 if and only if H_1 has an open orbit on $H_2 \backslash G$.

The Bruhat Decomposition

The Bruhat decomposition was discovered quite late in the history of Lie groups, which is surprising in view of its fundamental importance. It was preceded by Ehresmann's discovery of a closely related cell decomposition for flag manifolds. The Bruhat decomposition was axiomatized by Tits in the notion of a *Group with* (B, N) *pair* or *Tits' system*. This is a generalization of the notion of a Coxeter group, and indeed every (B, N) gives rise to a Coxeter group. We have remarked after Theorem 25.1 that Coxeter groups always act on simplicial complexes whose geometry is closely connected with their properties. As it turns out a group with (B, N) pair also acts on a simplicial complex, the *Tits' building*. We will not have space to discuss this important concept but see Tits [163] and Abramenko and Brown [1].

In this chapter, in order to be consistent with the notation in the literature on Tits' systems, particularly Bourbaki [23], we will modify our notation slightly. In other chapters such as the previous one, N denotes the subgroup (26.6) of the Borel subgroup. That group will appear in this Chapter also, but we will denote it as U, reserving the letter N for the normalizer of T. Similarly, in this chapter U_- will be the subgroup formerly denoted N_-.

Let $G = \mathrm{GL}(n, F)$, where F is a field, and let B be the Borel subgroup of upper triangular matrices in G. Taking $T \subset B$ to be the subgroup of diagonal matrices in G, the normalizer $N(T)$ consists of all monomial matrices. The Weyl group $W = N(T)/T \cong S_n$. If $w \in W$ is represented by $\omega \in N(T)$ then since $T \subset B$ the double coset $B\omega B$ is independent of the choice of representative ω, so by abuse of notation we write BwB for $B\omega B$. It is a remarkable and extremely important fact that $w \longrightarrow BwB$ is a bijection between the elements of W and the double cosets $B \backslash G / B$. We will prove the following *Bruhat decomposition*:

$$G = \bigcup BwB \quad \text{(disjoint)}. \tag{27.1}$$

The example of $\mathrm{GL}(2, F)$ is worth writing out explicitly. If $g = \begin{pmatrix} a & b \\ c & d \end{pmatrix}$, then $g \in B$ if $c = 0$. Therefore to prove the Bruhat decomposition, then for a representative ω of the long Weyl group element it will be convenient to take $\omega = \begin{pmatrix} 0 & -\Delta c^{-1} \\ c & 0 \end{pmatrix}$ where $\Delta = ad - bc$. Then this follows from the identity

$$\begin{pmatrix} a & b \\ c & d \end{pmatrix} = \begin{pmatrix} 1 & a/c \\ & 1 \end{pmatrix} \omega \begin{pmatrix} 1 & d/c \\ & 1 \end{pmatrix}$$

We will prove this and also obtain a similar statement in complex Lie groups. Specifically, if G is a complex Lie group obtained by complexification of a compact connected Lie group, we will prove a "Bruhat decomposition" analogous to (27.1) in G. A more general Bruhat decomposition will be found in Theorem 29.5.

We will prove the Bruhat decomposition for a group with a *Tits' system*, which consists of a pair of subgroups B and N satisfying certain axioms. The use of the notation N differs from that of Chap. 26, though the results of that chapter are very relevant here.

Let G be a group, and let B and N be subgroups such that $T = B \cap N$ is normal in N. Let W be the quotient group N/T. As with $\mathrm{GL}(n, F)$, we write wB instead of ωB when $\omega \in N$ represents the Weyl group element w, and similarly we will denote $Bw = B\omega$ and $BwB = B\omega B$.

Let G be a group with subgroups B and N satisfying the following conditions.

Axiom TS1. *The group $T = B \cap N$ is normal in N.*

Axiom TS2. *There is specified a set I of generators of the group $W = N/T$ such that if $s \in I$ then $s^2 = 1$.*

Axiom TS3. *Let $w \in W$ and $s \in I$. Then*

$$wBs \subset BwsB \cup BwB. \tag{27.2}$$

Axiom TS4. *Let $s \in I$. Then $sBs^{-1} \neq B$.*

Axiom TS5. *The group G is generated by N and B.*

Then we say that (B, N, I) is a *Tits' system*.

We will be particularly concerned with the double cosets $\mathcal{C}(w) = BwB$ with $w \in W$. Then Axiom TS3 can be rewritten

$$\mathcal{C}(w)\mathcal{C}(s) \subset \mathcal{C}(w) \cup \mathcal{C}(ws), \tag{27.3}$$

which is obviously equivalent to (27.2). Taking inverses, this is equivalent to

$$\mathcal{C}(s)\mathcal{C}(w) \subset \mathcal{C}(w) \cup \mathcal{C}(sw). \tag{27.4}$$

As a first example, let $G = \mathrm{GL}(n, F)$, where F is any field. Let B be the Borel subgroup of upper triangular matrices in G, let T be the standard "maximal torus" of all diagonal elements, and let N be the normalizer in G of T. Then B is the semidirect product of T with the normal subgroup U of upper triangular unipotent matrices. The group N consists of the monomial matrices, that is, matrices having exactly one nonzero entry in each row and column. Let $I = \{s_1, \ldots, s_{n-1}\}$ be the set of *simple reflections*, namely s_i is the image in $W = N/T$ of

$$\begin{pmatrix} I_{i-1} & & & \\ & 0 & 1 & \\ & 1 & 0 & \\ & & & I_{n-1-i} \end{pmatrix}.$$

We will prove in Theorem 27.1 below that this (B, N, I) is a Tits' system. The proof will require introducing a root system into $\mathrm{GL}(n, F)$. Of course, we have already done this if $F = \mathbb{C}$, but let us revisit the definitions in this new context.

Let $X^*(T)$ be the group of rational characters of T. In case F is a finite field, we don't want any torsion in $X^*(T)$; that is, we want $\chi \in X^*(T)$ to have infinite order so that $\mathbb{R} \otimes X^*(T)$ will be nonzero. So we define an element of $X^*(T)$ to be a character of $T(\overline{F})$, the group of diagonal matrices in $\mathrm{GL}(n, \overline{F})$, where $\overline{F}$ is the algebraic closure of F, of the form

$$\begin{pmatrix} t_1 & & \\ & \ddots & \\ & & t_n \end{pmatrix} \longmapsto t_1^{k_1} \ldots t_n^{k_n}, \tag{27.5}$$

where $k_i \in \mathbb{Z}$. Then $X^*(T) \cong \mathbb{Z}^n$, so $\mathcal{V} = \mathbb{R} \otimes X^*(T) \cong \mathbb{R}^n$.

As usual, we write the group law in $X^*(T)$ additively.

In this context, by a *root* of T in G we mean an element $\alpha \in X^*(T)$ such that there exists a group isomorphism x_α of F onto a subgroup X_α of G consisting of unipotent matrices such that

$$t\, x_\alpha(\lambda)\, t^{-1} = x_\alpha\big(\alpha(t)\,\lambda\big), \qquad t \in T, \lambda \in F. \tag{27.6}$$

(Strictly speaking, we should require that this identity be true as an equality of morphisms from the additive group into G.) There are $n^2 - n$ roots, which may be described explicitly as follows. If $1 \leqslant i, j \leqslant n$ and $i \neq j$, let

$$\alpha_{ij}(t) = t_i\, t_j^{-1} \tag{27.7}$$

when t is as in (27.5). Then $\alpha_{ij} \in X^*(T)$, and if E_{ij} is the matrix with 1 in the i, j position and 0's elsewhere, and if

$$x_\alpha(\lambda) = I + \lambda E_{ij},$$

then (27.6) is clearly valid. The set Φ consisting of α_{ij} is a root system; we leave the reader to check this but in fact it is identical to the root system of $GL(n, \mathbb{C})$ or its maximal compact subgroup $U(n)$ already introduced in Chap. 18 when $n = \mathbb{C}$. Let Φ^+ consist of the "positive roots" α_{ij} with $i < j$, and let Σ consist of the "simple roots" $\alpha_{i,i+1}$. We will sometimes denote the simple reflections $s_i = s_\alpha$, where $\alpha = \alpha_{i,i+1}$.

Suppose that α is a simple root. Let $T_\alpha \subset T$ be the kernel of α. Let M_α be the centralizer of T_α, and let P_α be the subgroup generated by B and M_α. By abuse of language, P_α is called a *minimal parabolic subgroup*. Observe that it is a parabolic subgroup since it contains the Borel subgroup. Strictly speaking it is not minimal amoung the parabolic subgroups, since the Borel itself is smaller. However it is minimal among non-Borel parabolic subgroups, and it is commonly called a minimal parabolic. in Chap. 30.) We have a semidirect product decomposition $P_\alpha = M_\alpha U_\alpha$, where U_α is the group generated by the $x_\beta(\lambda)$ with $\beta \in \Phi^+ - \{\alpha\}$. For example, if $n = 4$ and $\alpha = \alpha_{23}$, then

$$T_\alpha = \left\{ \begin{pmatrix} t_1 & & & \\ & t_2 & & \\ & & t_2 & \\ & & & t_4 \end{pmatrix} \right\}, \quad M_\alpha = \left\{ \begin{pmatrix} * & & & \\ & * & * & \\ & * & * & \\ & & & * \end{pmatrix} \right\},$$

$$P_\alpha = \left\{ \begin{pmatrix} * & * & * & * \\ & * & * & * \\ & * & * & * \\ & & & * \end{pmatrix} \right\}, \quad U_\alpha = \left\{ \begin{pmatrix} 1 & * & * & * \\ & 1 & & * \\ & & 1 & * \\ & & & 1 \end{pmatrix} \right\},$$

where $*$ indicates an arbitrary value.

Lemma 27.1. *Let $G = GL(n, F)$ for any field F, and let other notations be as above. If s is a simple reflection, then $B \cup \mathcal{C}(s)$ is a subgroup of G.*

Proof. First, let us check this when $n = 2$. In this case, there is only one simple root s_α where $\alpha = \alpha_{12}$. We check easily that

$$\mathcal{C}(s_\alpha) = B s_\alpha B = \left\{ \begin{pmatrix} a & b \\ c & d \end{pmatrix} \in GL(2, F) \,\Big|\, c \neq 0 \right\},$$

so $\mathcal{C}(s_\alpha) \cup B = G$.

In the general case, both $\mathcal{C}(s_\alpha)$ and B are subsets of P_α. We claim that their union is all of P_α. Both double cosets are right-invariant by U_α since $U_\alpha \subset B$, so it is sufficient to show that $\mathcal{C}(s_\alpha) \cup B \supset M_\alpha$. Passing to the quotient in $P_\alpha/U_\alpha \cong M_\alpha \cong GL(2) \times (F^\times)^{n-2}$, this reduces to the case $n = 2$ just considered. $\qquad \square$

We have an action of W on Φ as in Chap. 20. This action is such that if $\omega \in N$ represents the Weyl group element $w \in W$, we have

$$\omega x_\alpha(\lambda)\omega^{-1} \in x_{w(\alpha)}(F). \tag{27.8}$$

Other notations, such as the length function $l : W \longrightarrow \mathbb{Z}$, will be as in that chapter.

Lemma 27.2. *Let* $G = \mathrm{GL}(n, F)$ *for any field* F, *and let other notations be as above. If* α *is a simple root and* $w \in W$ *such that* $w(\alpha) \in \Phi^+$, *then* $\mathcal{C}(w)\mathcal{C}(s) = \mathcal{C}(ws)$.

Proof. We will show that

$$wBs \subseteq BwsB.$$

If this is known, then multiplying both left and right by B gives $\mathcal{C}(w)\mathcal{C}(s) = BwBsB \subseteq BwsB = \mathcal{C}(ws)$. The other inclusion is obvious, so this is sufficient. Let ω and σ be representatives of w and s as cosets in $N/T = W$, and let $b \in B$. We may write $b = tx_\alpha(\lambda)u$, where $t \in T$, $\lambda \in F$, and $u \in U_\alpha$. Then

$$\omega b\sigma = \omega t\omega^{-1} \cdot \omega x_\alpha(\lambda)\omega^{-1} \cdot \omega\sigma \cdot \sigma^{-1}u\sigma.$$

We have $\omega t\omega^{-1} \in T \subset B$ since $\omega \in N = N(T)$. We have $\omega x_\alpha(\lambda)\omega^{-1} \in x_{w(\alpha)}(F) \subset B$ using (27.8) and the fact that $w(\alpha) \in \Phi^+$. We have $\sigma^{-1}u\sigma \in U_\alpha \subset B$ since M_α normalizes U_α and $\sigma \in M_\alpha$. We see that $\omega b\sigma \in BwsB$ as required. $\qquad\square$

Proposition 27.1. *Let* $G = \mathrm{GL}(n, F)$ *for any field* F, *and let other notations be as above. If* $w, w' \in W$ *are such that* $l(ww') = l(w) + l(w')$, *then*

$$\mathcal{C}(ww') = \mathcal{C}(w) \cdot \mathcal{C}(w').$$

Proof. It is sufficient to show that if $l(w) = r$, and if $w = s_1 \ldots s_r$ is a decomposition into simple reflections, then

$$\mathcal{C}(w) = \mathcal{C}(s_1) \ldots \mathcal{C}(s_r). \tag{27.9}$$

Indeed, assuming we know this fact, let $w' = s'_1 \ldots s'_{r'}$ be a decomposition into simple reflections with $r' = l(r')$. Then $s_1 \ldots s_r s'_1 \ldots s'_{r'}$ is a decomposition of ww' into simple reflections with $l(ww') = r + r'$, so

$$\mathcal{C}(ww') = \mathcal{C}(s_1) \ldots \mathcal{C}(s_r)\mathcal{C}(s'_1) \ldots \mathcal{C}(s'_{r'}) = \mathcal{C}(w)\mathcal{C}(w').$$

To prove (27.9), let $s_r = s_\alpha$, and let $w_1 = s_1 \ldots s_{r-1}$. Then $l(w_1 s_\alpha) = l(w_1) + 1$, so by Propositions 20.2 and 20.5 we have $w'(\alpha) \in \Phi^+$. Thus, Lemma 27.2 is applicable and $\mathcal{C}(w) = \mathcal{C}(w_1)\mathcal{C}(s_r)$. By induction on r, we have $\mathcal{C}(w_1) = \mathcal{C}(s_1) \ldots \mathcal{C}(s_{r-1})$ and so we are done. $\qquad\square$

Theorem 27.1. *With* $G = \mathrm{GL}(n, F)$ *and* B, N, I *as above,* (B, N, I) *is a Tits' system in* G.

Proof. Only Axiom TS3 requires proof; the others can be safely left to the reader. Let $\alpha \in \Sigma$ such that $s = s_\alpha$.

First, suppose that $w(\alpha) \in \Phi^+$. In this case, it follows from Lemma 27.2 that $wBs \subset BwsB$.

Next suppose that $w(\alpha) \notin \Phi^+$. Then $ws_\alpha(\alpha) = w(-\alpha) = -w(\alpha) \in \Phi^+$, so we may apply the case just considered, with ws_α replacing w, to see that

$$wsBs \subset Bws^2B = BwB. \tag{27.10}$$

By Lemma 27.1, $B \cup BsB$ is a group containing a representative of the coset of $s \in N/T$, so $B \cup BsB = sB \cup sBsB$ and thus

$$Bs \subset sB \cup sBsB.$$

Using (27.10),

$$wBs \subset wsB \cup wsBsB \subset BwsB \cup BwB.$$

This proves Axiom TS3.   □

As a second example of a Tits' system, let K be a compact connected Lie group, and let G be its complexification. Let T be a maximal torus of K, let $T_{\mathbb{C}}$ be the complexification of T, and let B be the Borel subgroup of G as constructed in Chap. 26. Let N be the normalizer in G of $T_{\mathbb{C}}$, and let I be the set of simple reflections in $W = N/T$. We will prove that (B, N, I) is a Tits' system in G, closely paralleling the proof just given for $\mathrm{GL}(n, F)$. In fact, if $F = \mathbb{C}$ and $K = \mathrm{U}(n)$, so $G = \mathrm{GL}(n, \mathbb{C})$, the two examples, including the method of proof, exactly coincide.

The key to the proof is the construction of the minimal parabolic subgroup P_α corresponding to a simple root $\alpha \in \Sigma$. Chap. 30.) Let T_α be the kernel of α in T. The centralizer $C_K(T_\alpha)$ played a key role in Chap. 18, particularly in the proof of Theorem 18.1, where a homomorphism $i_\alpha : \mathrm{SU}(2) \longrightarrow C_K(T_\alpha)$ was constructed. This homomorphism extends to a homomorphism, which we will also denote as i_α, of the complexification $\mathrm{SL}(2, \mathbb{C})$ into the centralizer $C_G(T_\alpha)$ of T_α in G. Let P_α be the subgroup generated by $i_\alpha(\mathrm{SL}(2, \mathbb{C}))$ and B. Let M_α be the group generated by $i_\alpha(\mathrm{SL}(2, \mathbb{C}))$ and $T_{\mathbb{C}}$. Finally, let

$$\mathfrak{u}_\alpha = \bigoplus_{\substack{\beta \in \Phi^+ \\ \beta \neq \alpha}} \mathfrak{X}_\beta.$$

If $\beta_1, \beta_2 \in \{\beta \in \Phi^+ \mid \beta \neq \alpha\}$, then $\beta_1 + \beta_2 \neq 0$, and if $\beta_1 + \beta_2$ is a root, it is also in $\{\beta \in \Phi^+ \mid \beta \neq \alpha\}$. It follows from this observation and Proposition 18.4 that $\mathfrak{u}_\alpha$ is closed under the Lie bracket; that is, it is a complex Lie algebra of the Lie algebra denoted $\mathfrak{n}$ in Chap. 26. Theorem 26.2 (iii) shows that it is the Lie algebra of a complex Lie subgroup U_α of G.

Proposition 27.2. *Let G be the complexification of the compact connected Lie group K, let α be a simple positive root of G with respect to a fixed maximal torus T of K, and let other notations be as above. Then M_α normalizes U_α.*

Proof. It is clear that B normalizes U_α, so we need to show that $i_\alpha(\mathrm{SL}(2,\mathbb{C}))$ normalizes U_α. If $\gamma \in \{\beta \in \Phi^+ \,|\, \beta \neq \alpha\}$ and $\delta = \alpha$ or $-\alpha$, then $\gamma + \delta \neq 0$, and if $\gamma + \delta \in \Phi$, then $\gamma + \delta \in \{\beta \in \Phi^+ \,|\, \beta \neq \alpha\}$. Thus $[\mathfrak{X}_{\pm\alpha}, \mathfrak{X}_\gamma] \subseteq \mathfrak{u}_\alpha$, and since by Theorem 18.1 and Proposition 18.8 the Lie algebra of $i_\alpha(\mathrm{SL}(2,\mathbb{C}))$ is generated by $\mathfrak{X}_\alpha$ and $\mathfrak{X}_{-\alpha}$, it follows that the Lie algebra of $i_\alpha(\mathrm{SL}(2,\mathbb{C}))$ normalizes the Lie algebra of U_α. Since both groups are connected, it follows that $i_\alpha(\mathrm{SL}(2,\mathbb{C}))$ normalizes U_α. $\qquad\square$

Since M_α normalizes U_α, we may define P_α to be the semidirect product $M_\alpha U_\alpha$. An analog of Lemma 27.1 is true in this context.

Lemma 27.3. *Let G be the complexification of the compact connected Lie group K, and let other notations be as above. If s is a simple reflection, then $B \cup \mathcal{C}(s)$ is a subgroup of G.*

Proof. Indeed, if $s = s_\alpha$, then $B \cup \mathcal{C}(s) = P_\alpha$. From Theorem 18.1, the group M_α contains a representative of $s \in N/T$, so it is clear that $B \cup \mathcal{C}(s) \subset P_\alpha$. As for the other inclusion, both B and $\mathcal{C}(s)$ are invariant under right multiplication by U_α, so it is sufficient to show that $M_\alpha \in B \cup \mathcal{C}(s)$. Moreover, both B and $\mathcal{C}(s)$ are invariant under right multiplication by $T_\mathbb{C}$, so it is sufficient to show that $i_\alpha(\mathrm{SL}(2,\mathbb{C})) \subset B \cup \mathcal{C}(s)$. This is identical to Lemma 27.1 except that we work with $\mathrm{SL}(2,\mathbb{C})$ instead of $\mathrm{GL}(2,F)$. We have

$$i_\alpha \begin{pmatrix} a & b \\ c & d \end{pmatrix} \in \begin{cases} B & \text{if } c = 0, \\ \mathcal{C}(s) & \text{if } c \neq 0. \end{cases}$$

This completes the proof. $\qquad\square$

Theorem 27.2. *Let G be the complexification of the compact connected Lie group K. With B, N, I as above, (B, N, I) is a Tits' system in G.*

Proof. The proof of this is identical to Theorem 27.1. The analog of Lemma 27.2 is true, and the proof is the same except that we use Lemma 27.3 instead of Lemma 27.1. All other details are the same. $\qquad\square$

Now that we have two examples of Tits' systems, let us prove the Bruhat decomposition.

Theorem 27.3. *Let (B, N, I) be a Tits' system within a group G, and let W be the corresponding Weyl group. Then*

$$G = \bigcup_{w \in W} BwB, \tag{27.11}$$

and this union is disjoint.

Proof. Let us show that $\bigcup_{w \in W} \mathcal{C}(w)$ is a group. It is clearly closed under inverses. We must show that it is closed under multiplication.

Let us consider $\mathcal{C}(w_1) \cdot \mathcal{C}(w_2)$, where $w_1, w_2 \in W$. We show by induction on $l(w_2)$ that this is contained in a union of double cosets. If $l(w_2) = 0$, then $w_2 = 1$ and the assertion is obvious. If $l(w_2) > 0$, write $w_2 = sw_2'$, where $s \in I$ and $l(w_2') < l(w_2)$. Then, by Axiom TS3, we have

$$\mathcal{C}(w_1) \cdot \mathcal{C}(w_2) = Bw_1 Bsw_2'B \subset Bw_1 Bw_2'B \cup Bw_1 sBw_2'B,$$

and by induction this is contained in a union of double cosets.

We have shown that the right-hand side of (27.11) is a group, and since it clearly contains B and N, it must be all of G by Axiom TS5.

It remains to be shown that the union (27.11) is disjoint. Of course, two double cosets are either disjoint or equal, so assume that $\mathcal{C}(w) = \mathcal{C}(w')$, where $w, w' \in W$. We will show that $w = w'$.

Without loss of generality, we may assume that $l(w) \leqslant l(w')$, and we proceed by induction on $l(w)$. If $l(w) = 0$, then $w = 1$, and so $B = \mathcal{C}(w')$. Thus, in N/T, a representative for w' will lie in B. Since $B \cap N = T$, this means that $w' = 1$, and we are done in this case. Assume therefore that $l(w) > 0$ and that whenever $\mathcal{C}(w_1) = \mathcal{C}(w_1')$ with $l(w_1) < l(w)$ we have $w_1 = w_1'$.

Write $w = w''s$, where $s \in I$ and $l(w'') < l(w)$. Thus $w''s \in \mathcal{C}(w')$, and since s has order 2, we have

$$w'' \in \mathcal{C}(w')s \subset \mathcal{C}(w') \cup \mathcal{C}(w's)$$

by Axiom TS3. Since two double cosets are either disjoint or equal, this means that either

$$\mathcal{C}(w'') = \mathcal{C}(w') \qquad \text{or} \qquad \mathcal{C}(w'') = \mathcal{C}(w's).$$

Our induction hypothesis implies that either $w'' = w'$ or $w'' = w's$. The first case is impossible since $l(w'') < l(w) \leqslant l(w')$. Therefore $w'' = w's$. Hence $w = w''s = w'$, as required. □

We return to the second example of a Tits' system. Let K be a compact connected Lie group, G its complexification. Let B be the standard Borel subgroup, containing a maximal torus $T_{\mathbb{C}}$, with $T = T_{\mathbb{C}} \cap K$ the maximal torus of K. The group (26.6) which is usually denoted N will be denoted U (in this chapter only).

The flag manifold $X = K/T$ may be identified with $G/T_{\mathbb{C}}$ as in Theorem 26.4. We will use the Bruhat decomposition $G = \bigoplus BwB$ to look more closely at X.

By Theorem 26.4, X is a complex manifold. It is compact since it is a continuous image of K. We may decompose $X = \bigcup Y_w$ where w runs through the Weyl group and $Y_w = BwB/B$. Let us begin by looking more closely at Y_w. Let $U_+^w = U \cap wUw^{-1}$ and $U_-^w = U \cap wU_-w^{-1}$. The Lie algebra $\mathfrak{u}_+^w$ is the intersection of the Lie algebras of U and wUw^{-1}, so

$$u_+^w = \bigoplus_{\alpha \in \Phi^+ \cap w\Phi^+} \mathfrak{X}_\alpha,$$

and similarly

$$u_+^w = \bigoplus_{\alpha \in \Phi^+ \cap w\Phi^-} \mathfrak{X}_\alpha.$$

Proposition 27.3. *The map $u \mapsto uwB$ is a bijection of U_-^w onto Y_w.*

Proof. Clearly $BwB/B = UwB/B$. Moreover if $u, u' \in U$ then $uwB = u'wB$ if and only if $u^{-1}u' \in U_+^w$. We need to show that every coset in U/U_+^w has a unique representative from U_-^+. This follows from Theorem 26.2 (iv). □

The orbits of B under the left action of B on X are the Y_w. So the closure of Y_w is a union of other Y_u with $u \in W$. Which ones? We recall the *Bruhat order* that was introduced in Chap. 25. If $w = s_{i_1} \ldots s_{i_k}$ is a reduced decomposition, then $u \leqslant w$ if and only if u obtained by eliminating some of the factors. In other words, there is a subsequence $(j_1, \ldots, j_l)$ of $(i_1, \ldots, i_k)$ with $u = s_{j_1} \ldots s_{j_l}$. It was shown in Proposition 25.4 that this definition does not depend on the decomposition $w = s_{i_1} \ldots s_{i_k}$. Moreover, we may always arrange that $u = s_{j_1} \ldots s_{j_l}$ is a reduced decomposition.

Our goal is to prove that Y_u is contained in the closure of Y_w if and only if $u \leqslant v$ in the Bruhat order. To prove this, we introduce the *Bott-Samelson varieties*. If $1 \leqslant i \leqslant r$, where r is the semisimple rank of K, that is, the number of simple reflections, let P_i be minimal parabolic subgroup generated by s_i and B.

Proposition 27.4. *The minimal parabolic $P_i = C(1) \cup C(s_i)$. The quotient P_i/B is diffeomorphic to the projective line $\mathbb{P}^1(\mathbb{C})$.*

Proof. By Lemma 27.1, $C(1) \cup C(s_i)$ is a group, so $P_i = C(1) \cup C(s_i)$. Since $\mathrm{SL}(2, \mathbb{C})$ is simply-connected, the injection $i_{\alpha_k} : \mathfrak{sl}(2, \mathbb{C}) \longrightarrow \mathrm{Lie}(G)$ as in Proposition 18.8 induces a homomorphism $\mathrm{SL}(2, \mathbb{C}) \longrightarrow G$ whose image is in P_{i_k}. Since $i_{\alpha_i}(\mathrm{SL}(2, \mathbb{C}))$ contains s_i, we have $P_i = i_{\alpha_i}(\mathrm{SL}(2, \mathbb{C}))B$. Therefore P_i/B in bijection with $i_{\alpha_i}(\mathrm{SL}(2, \mathbb{C}))$ modulo its intersection with B. The quotient of $\mathrm{SL}(2, \mathbb{C})$ by its Borel subgroup is the projective line $\mathbb{P}^1(\mathbb{C})$. □

If $\mathfrak{w} = (i_1, \ldots, i_k)$, define a right action of B^k on $P_{i_1} \times \ldots \times P_{i_k}$ by

$$(p_1, \ldots, p_k) \cdot (b_1, \ldots, b_k) = (p_1 b_1, b_1^{-1} p_2 b_2, \ldots, b_{k-1}^{-1} p_k b_k), \qquad (27.12)$$

where $p_j \in P_{i_j}$ and $b_j \in B$. We are mainly interested in the case where $\mathfrak{w}$ is a reduced word. The quotient $Z_\mathfrak{w} = (P_{i_1} \times \ldots \times P_{i_k})/B^k$ is called a *Bott-Samelson variety*. We also have a map $Z_\mathfrak{w} \longrightarrow Z_{\mathfrak{w}'}$ where $\mathfrak{w}' = (i_1, \ldots, i_{k-1})$ in which the orbit of $(p_1, \ldots, p_k)$ goes to the orbit of $(p_1, \ldots, p_{k-1})$. This map is a fibration in which the typical fiber is $P_{i_k}/B \cong \mathbb{P}^1(\mathbb{C})$. Thus the Bott-Samelson variety is obtained by successive fiberings of $\mathbb{P}^1(\mathbb{C})$. In particular it is a compact manifold.

We have a map $\tau : Z_{\mathfrak{w}} \longrightarrow X$ induced by the map $(p_1, \ldots, p_k) \longmapsto$ $p_1 \ldots p_k B$. It is clearly well-defined. Let X_w be the closure of Y_w in X. It is called a *Schubert variety*. We will show that the image of $Z_{\mathfrak{w}}$ in X is precisely X_w. Although we will not discuss this point, both $Z_{\mathfrak{w}}$ and X_w are algebraic varieties. The variety $Z_{\mathfrak{w}}$ is less canonical, since it depends on the choice of a reduced word $\mathfrak{w}$. It is, however, easier to work with. For example $Z_{\mathfrak{w}}$ is smooth, whereas X_w can be singular.

Bott-Samelson varieties play a key role in many aspects of the theory. The map τ is a birational equivalence, so they resolve the singularities of the Schubert varieties. They are used in Demazure's calculation of the action of T on the spaces of sections of line bundles on X restricted to X_w as Demazure characters.

Theorem 27.4. *The image of τ is X_w. The Schubert variety X_w is the union of the Y_u for $u \leqslant w$ in the Bruhat order.*

Proof. Since $C(s_i)$ is dense in P_i, the set $C(s_{i_1}) \times \ldots \times C(s_{i_k})$ is dense in $P_{i_1} \times \ldots \times P_{i_k}$. Its image in X is $C(s_{i_1}) \ldots C(s_{i_k}) = C(w)$ by (27.9), and so the image of $C(s_{i_1}) \times \ldots \times C(s_{i_k})$ is Y_w, and it is dense in $\tau(Z_{\mathfrak{w}})$. On the other hand, the image of $\tau(Z_{\mathfrak{w}})$ is closed since $Z_{\mathfrak{w}}$ is compact. Thus $\tau(Z_{\mathfrak{w}})$ is the closure of Y_w, which by definition is X_w.

Now since $P_i = C(1) \cup C(s_i)$, it is clear that $\tau(Z_w)$ is the union of the $C(s_{j_1}) \ldots C(s_{j_l})/B$ as $(j_1, \ldots, j_l)$ runs through the subwords of $(i_1, \ldots, i_k)$. If $u = s_{j_1} \ldots s_{j_l}$ is a reduced decomposition, then by (27.9) this is $C(u)/B$, and so we obtain every Y_u for $u \leqslant w$. If the decomposition is not reduced, it is still a union of $C(v)/B$ for $v \leqslant u$, as follows easily from (27.3). $\square$

The *Borel-Weil theorem* realizes an irreducible representation of a compact Lie group or its complexification as an action on the space of sections of a holomorphic line bundle on the flag variety. This will be our next topic. The Bruhat decomposition will play a role in the discussion insofar as we will need to know that the big Bruhat cell $Bw_0 B$ is dense in G.

If (π, V) is a complex representation of the compact connected Lie group K, then it follows from the definition of the complexification G that π has a unique extension to an analytic representation $\pi : G \longrightarrow \mathrm{GL}(V)$. Similarly, the contragredient representation $\hat{\pi} : K \longrightarrow \mathrm{GL}(V^*)$ may be extended to an analytic representation of G. Let λ be the highest weight of π. By Proposition 22.8, the highest weight of $\hat{\pi}$ is $\hat{\lambda} = -w_0\lambda$, where w_0 is the long element of the Weyl group W.

Now let $X = G/B$ be the flag variety. We will construct a line bundle $\mathcal{L}_\lambda$ over X. This is a complex analytic manifold together with an analytic map $p : \mathcal{L}_\lambda \longrightarrow X$. The fibers of p are one-dimensional complex vector spaces. Moreover every point $x \in X$ has a neighborhood U such that the $p^{-1}(U)$ is a trivial bundle over U. This means that there is a complex analytic homeomorphism $\psi : p^{-1}(U) \longrightarrow U \times \mathbb{C}$ such that the composition of ψ with the projection $U \times \mathbb{C} \longrightarrow \mathbb{C}$ is p.

To construct $\mathcal{L}_\lambda$, define a right action of B on $G \times \mathbb{C}$, by $(g, \varepsilon)b = (gb, \hat{\lambda}(b)\varepsilon)$ for $b \in B$ and $(g, \varepsilon) \in G \times \mathbb{C}$. Then $\mathcal{L}_\lambda$ is the quotient $(G \times \mathbb{C})/B$. We will denote the orbit of (g, ε) by $[g, \varepsilon]$, so if $b \in B$ then $[g, \varepsilon] = [gb, \hat{\lambda}(b)\varepsilon]$. The map $p : \mathcal{L}_\lambda \longrightarrow X$ sends $[g, \varepsilon]$ to gB. We leave it to the reader to check that this is a line bundle.

A *section* of $\mathcal{L}_\lambda$ is a holomorphic map $s : X \longrightarrow \mathcal{L}_\lambda$ such that $p \circ s$ is the identity on X. It is well-known (and part of the Riemann-Roch theorem) that the space $\Gamma(\mathcal{L}_\lambda)$ of sections is finite-dimensional. We have compatible actions of G on X and on $\mathcal{L}_\lambda$ by left translation: if $\gamma \in G$ then $\gamma : X \longrightarrow X$ sends gB to γgB and $\gamma : \mathcal{L}_\lambda \longrightarrow \mathcal{L}_\lambda$ sends $[g, \varepsilon]$ to $[\gamma g, \varepsilon]$. Specifying a section is equivalent to giving a holomorphic map $\phi : G \longrightarrow \mathbb{C}$ such that $\phi(gb) = \hat{\lambda}(b)\phi(g)$; the last condition is needed so that $s(gB) = [g, \phi(g)]$ is well-defined. So $\Gamma(\mathcal{L}_\lambda)$ is isomorphic to the vector space $H(\lambda)$ of such holomorphic maps ϕ.

The compatibility is that these actions commute with p. One says that $\mathcal{L}_\lambda$ is an *equivariant* line bundle. Now we have an action of G on sections as follows. If $s \in \Gamma(\mathcal{L}_\lambda)$ and $\gamma \in G$ then γs is the section $\gamma s(gx) = \gamma(s(\gamma^{-1}x))$ for $x \in X$. This is equivalent to the action $\gamma\phi(g) = \phi(\gamma^{-1}g)$ on $H(\lambda)$.

Theorem 27.5. (Borel-Weil) *The space $\Gamma(\mathcal{L}_\lambda)$ is zero unless λ is dominant. If λ is dominant, then $\Gamma(\mathcal{L}_\lambda)$ is irreducible as a G-module, with highest weight λ.*

Proof. We will follow the now-familiar strategy of identifying the N-fixed vectors in the module. We will take for granted the well-known fact that the space of sections $\Gamma(\mathcal{L}_\lambda)$ is finite-dimensional. See Gunning and Rossi [60], Corollary 10 on page 241. Assume that $\Gamma(\mathcal{L}_\lambda)$ is nonzero. Let ϕ be the highest weight vector for some irreducible submodule. Then by Theorem 26.5 we have $\phi(ng) = \phi(g)$ for $\phi \in H(\lambda)$, so $\phi(nw_0b) = \hat{\lambda}(b)\,\phi(w_0)$. Thus ϕ is determined up to a constant on $Nw_0B = Bw_0B$. This is the big Bruhat cell, and it is open and dense in X. So ϕ is zero unless $\phi(w_0) \neq 0$ and we may normalize it so $\phi(nw_0b) = \hat{\lambda}(b)$. This shows that there can be (up to scalar multiple) at most one N-fixed vector, and therefore $\Gamma(\mathcal{L}_\lambda)$, which we are assuming to be nonzero, is irreducible. Also, since ϕ is the highest weight vector, we can use it to compute the highest weight. Let $t \in T_\mathbb{C}$. Then $t\phi(w_0) = \phi(t^{-1}w_0) = \phi(w_0 \cdot w_0^{-1}t^{-1}w_0) = \hat{\lambda}(w_0^{-1}t^{-1}w_0) = \lambda(t)$. So by Theorem 26.5 the highest weight in the unique irreducible submodule of $H(\lambda)$ is λ. In particular λ is dominant.

We have yet to show that if λ is dominant then $H(\lambda)$ is nonzero. The issue is whether the section whose existence on the big cell Bw_0B follows from the above considerations can be extended to the entire group. We will accomplish this by exhibiting a G-equivariant map $V \longrightarrow H(\lambda)$. y Proposition 22.8, the highest weight in $\hat{\pi} : G \longrightarrow \mathrm{GL}(V^*)$ is $\hat{\lambda}$, so if $\theta \in V^*$ is the highest weight vector then $\hat{\pi}(b)\theta = \hat{\lambda}(b)\theta$ for $b \in B$. Let us denote the dual pairing $V \times V^* \longrightarrow \mathbb{C}$ by $(v, v^*) \mapsto \langle v, v^* \rangle$. Define a map $v \mapsto \phi_v$ from v to the space of holomorphic functions on G by

$$\phi_v(g) = \langle \pi(g^{-1})v, \theta \rangle.$$

We have $\phi_v(gb) = \hat{\lambda}(b)\phi_v(g)$ since the left-hand side is

$$\phi_v(g) = \langle \pi(b^{-1})\pi(g^{-1})v, \theta \rangle = \langle \pi(g^{-1})v, \pi(b)\theta \rangle = \hat{\lambda}(b) \langle \pi(g^{-1})v, \theta \rangle.$$

So $\phi_v \in H(\lambda)$. It is clear that $\phi_{\gamma v}(g) = \phi_v(\gamma^{-1}g) = (\gamma\phi_v)(g)$, so the map $v \mapsto \phi_v$ is equivariant. $\qquad\square$

Exercises

Exercise 27.1. Explain why Y_w has complex dimension $l(w)$, or real dimension $2l(w)$. Also explain why Y_w is open in X_w. Since X_w is a union of Y_w and subsets of lower dimension, we may say that $l(w)$ is the dimension of the Schubert variety X_w.

If W is a finite Weyl group then W has a longest element w_0. By the last exercise, the Bruhat cell Bw_0B is the largest in the sense of dimension. It is therefore called the *big Bruhat cell*.

Exercise 27.2. Let $G = \mathrm{GL}(n, \mathbb{C})$. Show that $g = (g_{ij}) \in G$ is in the big Bruhat cell if and only if all the bottom left minors

$$g_{n,1}, \quad \begin{vmatrix} g_{n-1,1} & g_{n-1,2} \\ g_{n,1} & g_{n,2} \end{vmatrix}, \quad \begin{vmatrix} g_{n-2,1} & g_{n-2,2} & g_{n-2,3} \\ g_{n-1,1} & g_{n-1,2} & g_{n-1,3} \\ g_{n,1} & g_{n,2} & g_{n,3} \end{vmatrix}, \quad \cdots$$

are nonzero. If $n = 3$, give a similar interpretation of all the Bruhat cells.

Exercise 27.3. Show for an arbitrary reduced word that the fiber $\tau^{-1}(x)$ of the Bott-Samelson map τ is a single point for x in general position. ("General position" means that this is true on a dense open subset of X_w.)

Exercise 27.4. Let M be a manifold. Suppose that M has an open contractible subset Ω whose complement is a union of submaniolds of codimension $\geqslant 2$. Show that M is simply-connected. Use this to give another proof of 23.7, that the flag manifold is simply-connected.

Let $G = \mathrm{GL}_n(\mathbb{C})$. We give a concrete interpretation of Bott-Samelson variety as follows. The group G acts on the set $\mathfrak{X}$ of flags $\mathbf{U} = (U_0, \ldots, U_n)$ where $U_0 \subset U_1 \subset \ldots \subset U_n$ and each U_i is an i-dimensional vector subspace of $\mathbb{C}^n$. We fix a flag $\mathbf{V} = (V_0, \ldots, V_n)$, which we will call the *standard flag*. The Borel subgroup B may be taken to be the stabilizer of $\mathbf{V}$. The parabolic subgroup P_i is the set of $g \in G$ such that $gV_j = V_j$ for all $j \neq i$.

Thus let $\mathfrak{X}$ be the set of flags. We have a bijection between $X = G/B$ and $\mathfrak{X}$ in which the coset gB is in bijection with the flag $g\mathbf{V}$. In the same way we will describe $Z_{\mathfrak{w}}$ as the parameter space for a set of configurations of subspaces of $\mathbb{C}^n$ that are more complicated than simple flags but very similar in spirit. Let $\mathfrak{w} = (s_{h_1}, s_{h_2}, \ldots, s_{h_k})$ be a reduced word representing $w = s_{h_1} \ldots s_{h_k}$, and let $\mathfrak{Z}_{\mathfrak{w}}$ be the set of sequences $\mathcal{U} = (\mathbf{U}^0, \ldots, \mathbf{U}^k)$ of flags $\mathbf{U}^i = (U_0^i, \ldots, U_n^i)$ such $\mathbf{U}^0 = V$ is the standard flag and $U_j^i = U_{j-1}^i$ except when $j = h_i$.

Exercise 27.5. Let $p_1, \ldots, p_k$ in G. Let $\mathbf{U}^0 = \mathbf{V}$, and define a sequence $\mathcal{U} = (\mathbf{U}^0, \ldots, \mathbf{U}^k)$ of flags by

$$p_1 \ldots p_i \mathbf{U}^{i-1} = \mathbf{U}^i.$$

Show that $\mathcal{U} \in \mathfrak{Z}_{\mathfrak{w}}$ if and only if $(p_1, \ldots, p_k) \in P_{h_1} \times \ldots \times P_{h_k}$. Moreover if $(p_1', \ldots, p_k')$ is another element of $P_{h_1} \times \ldots \times P_{h_k}$ then we have $p_1' \ldots p_i' \, \mathbf{U}^{i-1} = \mathbf{U}^i$ if and only if $(p_1, \ldots, p_k)$ and $(p_1', \ldots, p_k')$ differ by an element of B^k under the right action (27.12). Conclude that the map $P_{h_1} \times \ldots \times P_{h_k} \longrightarrow \mathfrak{Z}_{\mathfrak{w}}$ induces a bijection $Z_w \longrightarrow \mathfrak{Z}_{\mathfrak{w}}$.

Exercise 27.6. Show that there is a commutative diagram

$$
\begin{array}{ccc}
\mathfrak{Z}_{\mathbf{w}} & \longrightarrow & Z_{\mathbf{w}} \\
\downarrow{\scriptstyle \Phi} & & \downarrow{\scriptstyle \phi} \\
\mathfrak{X}_w & \longrightarrow & X_w
\end{array}
$$

where the horizontal maps are the bijections described above, ϕ is the canonical map $Z_{\mathbf{w}} \to X_w$, and Φ is the map that sends the configuration $(\mathbf{U}^0, \ldots, \mathbf{U}^k)$ to its last flag $\mathbf{U}^k$.

For example, let $n = 3$. Then $\mathbf{U}^0 = (V_0, V_1, V_2, V_3)$ will be the standard flag. We have $U_j^i = U_{j-1}^i$ except when $(i, j) = (1, 1), (2, 2)$ or $(3, 1)$, which means that we may find subspaces W_1, U_2 and U_1 of dimensions $1, 2, 1$ such that $\mathbf{U}^1 = (V_0, W_1, V_2, V_3)$, $\mathbf{U}^2 = (V_0, W_1, U_2, V_3)$ and $\mathbf{U}^3 = (V_0, U_1, U_2, U_3)$. Thus we arrive at the following configuration:

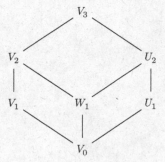

Vertical lines represent inclusions, subscripts dimensions. The Bott-Samelson space is a moduli space for such configurations, where (V_0, V_1, V_2, V_3) are fixed as the standard flag.

We may compute the fibers of the map ϕ by solving the equivalent problem of computing the fibers of Φ. In this case, the question is, given $\mathbf{U}^0$ (which is fixed) and $\mathbf{U}^3$ (representing a point in X_w), how many such configurations are there? The only unknown is W_1, but from the above inclusions, W_1 may be characterized as the intersection of V_2 and U_2. This will be a one-dimensional space (hence the only possibility for W_1) *except* in the case where $U_2 = V_2$. Thus if $x \in X_w$ is in general position the fiber $\phi^{-1}(x)$ consists of a single point. But if $U_2 = V_2$ then W_1 can be any one-dimensional subspace of U_2, so the fiber $\phi^{-1}(x)$ is $\mathbb{P}^1(\mathbb{C})$.

Exercise 27.7. (i) Show (for GL(3)) that if $\mathfrak{w} = (1,2)$ or $(2,1)$ ϕ is an isomorphism. (ii) Give a similar analysis when $\mathfrak{w} = (2,1,2)$.

Exercise 27.8. For GL(4), the Schubert variety X_w is singular if $\mathfrak{w} = (2,1,3,2)$ or $\mathfrak{w} = (1,3,2,1,3)$. Analyze the fibers of ϕ using this method. Here are the relevant configurations:

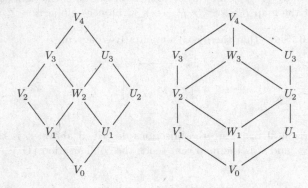

Exercise 27.9. Let $X_- = G/B_-$. Explain why X_- is diffeomorphic to the flag manifold X. Explore how the statement of the Borel-Weil theorem would change if instead of line bundles on X, we considered line bundles on X_-.

28

Symmetric Spaces

We have devoted some attention to an important class of homogeneous spaces of Lie groups, namely flag manifolds. Another important class is that of *symmetric spaces*. In differential geometry, a *symmetric space* is a Riemannian manifold in which around every point there is an isometry reversing the direction of every geodesic. Symmetric spaces generalize the non-Euclidean geometries of the sphere (compact with positive curvature) and the Poincaré upper half-plane (noncompact with negative curvature). Like these two examples, they tend to come in pairs, one compact and one noncompact. They were classified by E. Cartan.

Our approach to symmetric spaces will be to alternate the examination of examples with an explanation of general principles. In a few places (Remark 28.2, Theorem 28.2, Theorem 28.3, Proposition 28.3, and in the next chapter Theorem 29.5) we will make use of results from Helgason [66]. This should cause no problems for the reader. These are facts that need to be included to complete the picture, though we do not have space to prove them from scratch. They can be skipped without serious loss of continuity. In addition to Helgason [66], a second indispensable work on (mainly Hermitian) symmetric spaces is Satake [145].

It turns out that symmetric spaces (apart from Euclidean spaces) are constructed mainly as homogeneous spaces of Lie groups. In this chapter, an *involution* of a Lie group G is an automorphism of order 2.

Proposition 28.1. *Suppose that G is a connected Lie group with an involution θ. Assume that the group*

$$K = \{g \in G \mid \theta(g) = g\} \tag{28.1}$$

is a compact Lie subgroup. In this setting, $X = G/K$ is a symmetric space.

The involution θ is called a *Cartan involution* of G, and the involution it induces on the Lie algebra is called a *Cartan involution* of $\text{Lie}(G)$.

Proof. Clearly, G acts transitively on G/K, and K is the stabilizer of the base point x_0, that is, the coset $K \in G/K$. We put a positive definite inner product on the tangent space $T_{x_0}(X)$ that is invariant under the compact group K and also under θ. If $x \in X$, then we may find $g \in G$ such that $g(x_0) = x$, and g induces an isomorphism $T_{x_0}(X) \longrightarrow T_x(X)$ by which we may transfer this positive definite inner product to $T_x(X)$. Because the inner product on $T_{x_0}(X)$ is invariant under K, this inner product does not depend on the choice of g. Thus, X becomes a Riemannian manifold. The involution θ induces an automorphism of X that preserves geodesics through x_0, reversing their direction, so X is a symmetric space.   □

We now come to a striking algebraic fact that leads to the appearance of symmetric spaces in pairs. The involution θ induces an involution of $\mathfrak{g} = \mathrm{Lie}(G)$. The $+1$ eigenspace of θ is, of course, $\mathfrak{k} = \mathrm{Lie}(K)$. Let $\mathfrak{p}$ be the -1 eigenspace. Evidently,

$$[\mathfrak{k}, \mathfrak{k}] \subset \mathfrak{k}, \qquad [\mathfrak{k}, \mathfrak{p}] \subset \mathfrak{p}, \qquad [\mathfrak{p}, \mathfrak{p}] \subset \mathfrak{k}.$$

From this, it is clear that

$$\mathfrak{g}_c = \mathfrak{k} + i\mathfrak{p} \tag{28.2}$$

is a Lie subalgebra of $\mathfrak{g}_{\mathbb{C}} = \mathbb{C} \otimes \mathfrak{g}$. We observe that $\mathfrak{g}$ and $\mathfrak{g}_c$ have the same complexification; that is, $\mathfrak{g}_{\mathbb{C}} = \mathfrak{g} \oplus i\mathfrak{g} = \mathfrak{g}_c \oplus i\mathfrak{g}_c$.

The appearance of these two Lie algebras with a common complexification means that symmetric spaces come in pairs. To proceed further, we will make some assumptions, which we now explain.

Hypothesis 28.1. *Let G be a noncompact connected semisimple Lie group with Lie algebra $\mathfrak{g}$. Let θ be an involution of G such that the fixed subgroup K of θ is compact, as in Proposition 28.1. Let $\mathfrak{k}$ and $\mathfrak{p}$ be the $+1$ and -1 eigenspaces of θ on $\mathfrak{g}$, and let $\mathfrak{g}_c$ be the Lie algebra defined by (28.2). We will assume that $\mathfrak{g}_c$ is the Lie algebra of a second Lie group G_c that is compact and connected. Let $G_{\mathbb{C}}$ be the complexification of G_c (Theorem 24.1). We assume that the Lie algebra homomorphism $\mathfrak{g} \longrightarrow \mathfrak{g}_{\mathbb{C}}$ is the differential of a Lie group embedding $G \longrightarrow G_{\mathbb{C}}$ and that θ extends to an automorphism of $G_{\mathbb{C}}$, also denoted θ, which stabilizes G_c.*

This means G and G_c can be embedded compatibly in the complex analytic group $G_{\mathbb{C}}$. The involution θ extends to $\mathfrak{g}_{\mathbb{C}}$ and induces an involution on $\mathfrak{g}_c$ such that

$$X + iY \longmapsto X - iY, \qquad X \in \mathfrak{k}, Y \in \mathfrak{p}.$$

The last statement in Hypothesis 28.1 means that this θ is the differential of an automorphism of G_c. As a consequence the homogeneous space $X_c = G_c/K$ is also a symmetric space, again by Proposition 28.1. The symmetric spaces X and X_c, one noncompact and the other compact, are said to be in *duality* with each other.

Remark 28.1. We will see in Theorem 28.3 that every noncompact semisimple Lie group admits a Cartan involution θ such that this hypothesis is satisfied. Our proof of Theorem 28.3 will not be self-contained, but we do not really need to rely on it as motivation because we will give numerous examples in this chapter and the next where Hypothesis 28.1 is satisfied.

Remark 28.2. We do not specify G, K, and G_c up to isomorphism by this description since different K could correspond to the same pair G and θ. But K is always connected and contains the center of G (Helgason [66], Chap. VI, Theorem 1.1 on p. 252). If we replace G by a semisimple covering group, the center increases, so we must also enlarge K, and the quotient space G/K is unchanged. Hence, there is a unique symmetric space of noncompact type determined by the real semisimple Lie algebra $\mathfrak{g}$. By contrast, the symmetric space of compact type is *not* uniquely determined by $\mathfrak{g}_c$. There could be a finite number of different choices for G_c and K resulting in different compact symmetric spaces that have the same universal covering space. We will not distinguish a particular one as the dual of X but say that any one of these compact spaces is in duality with X. See Helgason [66], Chap. VII, for a discussion of this point and other subtleties in the compact case.

Example 28.1. Suppose that $G = \mathrm{SL}(n, \mathbb{R})$ and $K = \mathrm{SO}(n)$. Then $\mathfrak{g} = \mathfrak{sl}(n, \mathbb{R})$ and the involution $\theta : G \longrightarrow G$ is $\theta(g) = {}^t g^{-1}$. The induced involution on $\mathfrak{g}$ is $X \longrightarrow -{}^t X$. This $\mathfrak{p}$ consists of symmetric matrices, and $\mathfrak{g}_c$ consists of the skew-Hermitian matrices in $\mathfrak{sl}(n, \mathbb{C})$; that is, $\mathfrak{g}_c = \mathfrak{su}(n)$. The Lie groups $G = \mathrm{SL}(n, \mathbb{R})$ and $G_c = \mathrm{SU}(n)$ are subgroups of their common complexification $G_{\mathbb{C}} = \mathrm{SL}(n, \mathbb{C})$. The symmetric spaces $X = \mathrm{SL}(n, \mathbb{R})/\mathrm{SO}(n)$ and $X_c = \mathrm{SU}(n)/\mathrm{SO}(n)$ are in duality.

Let us obtain concrete realizations of the symmetric spaces G/K and G_c/K in Example 28.1. The group $\mathrm{GL}(n, \mathbb{R})$ acts on the cone $\mathcal{P}_n(\mathbb{R})$ of positive definite real symmetric matrices by the action

$$g : x \longmapsto g \, x \, {}^t g. \tag{28.3}$$

On the other hand, the group $\mathrm{U}(n)$ acts on the space $\mathcal{E}_n(\mathbb{R})$ of unitary symmetric matrices by the same formula (28.3). [The notation $\mathcal{E}_n(\mathbb{R})$ does not imply that the elements of this space are real matrices.]

Proposition 28.2. *Suppose that* $x \in \mathcal{P}_n(\mathbb{R})$ *or* $\mathcal{E}_n(\mathbb{R})$.

(i) There exists $g \in \mathrm{SO}(n)$ *such that* $g \, x \, {}^t g$ *is diagonal.*
(ii) The actions of $\mathrm{GL}(n, \mathbb{R})$ *and* $\mathrm{U}(n)$ *are transitive.*
(iii) Let $\mathfrak{p}$ *be the vector space of real symmetric matrices. We have*

$$\mathcal{P}_n(\mathbb{R}) = \{e^X \mid X \in \mathfrak{p}\}, \qquad \mathcal{E}_n(\mathbb{R}) = \{e^{iX} \mid X \in \mathfrak{p}\}.$$

See Theorem 45.6 in Chap. 45 for an application.

Proof. If $x \in \mathcal{P}_n(\mathbb{R})$, then (i) is, of course, just the spectral theorem. However, if $x \in \mathcal{E}_n(\mathbb{R})$, this statement may be less familiar. It is instructive to give a unified proof of the two cases. Give $\mathbb{C}^n$ its usual inner product, so $\langle u, v \rangle = \sum_i u_i \overline{v_i}$.

Let λ be an eigenvalue of x. We will show that the eigenspace $V_\lambda = \{v \in \mathbb{C}^n \mid xv = \lambda v\}$ is stable under complex conjugation. Suppose that $v \in V_\lambda$. If $x \in \mathcal{P}_n(\mathbb{R})$, then both x and λ are real, and simply conjugating the identity $xv = \lambda v$ gives $x\overline{v} = \lambda \overline{v}$. On the other hand, if $x \in \mathcal{E}_n(\mathbb{R})$, then $\overline{x} = {}^t x^{-1} = x^{-1}$ and $|\lambda| = 1$ so $\overline{\lambda} = \lambda^{-1}$. Thus, conjugating $xv = \lambda v$ gives $x^{-1}\overline{v} = \lambda^{-1}\overline{v}$, which implies that $x\overline{v} = \lambda\overline{v}$.

Now we can show that $\mathbb{C}^n$ has an orthonormal basis consisting of eigenvectors $v_1, \ldots, v_n$ such that $v_i \in \mathbb{R}^n$. The adjoint of x with respect to the standard inner product is x or x^{-1} depending on whether $x \in \mathcal{P}_n(\mathbb{R})$ or $\mathcal{E}_n(\mathbb{R})$. In either case, x is the matrix of a normal operator—one that commutes with its adjoint—and $\mathbb{C}^n$ is the orthogonal direct sum of the eigenspaces of x. Each eigenspace has an orthonormal basis consisting of real vectors. Indeed, if $v_1, \ldots, v_k$ is a basis of V_λ, then since we have proved that $\overline{v_i} \in V_\lambda$, the space is spanned by $\frac{1}{2}(v_i + \overline{v_i})$ and $\frac{1}{2i}(v_i - \overline{v_i})$; selecting a basis from this spanning set and applying the usual Gram–Schmidt orthogonalization process gives an orthonormal basis of real vectors.

In either case, we see that $\mathbb{C}^n$ has an orthonormal basis consisting of eigenvectors $v_1, \ldots, v_n$ such that $v_i \in \mathbb{R}^n$. Let $xv_i = \lambda_i v_i$. Then, if $k \in O(n)$ is the matrix with columns x_i and d is the diagonal matrix with diagonal entries λ_i, we have $xk = kd$ so $k^{-1}xk = \delta$. As $k^{-1} = {}^t k$ we may take the matrix $g = k^{-1}$. If the determinant of k is -1, we can switch the sign of the first entry without harm, so we may assume $k \in SO(n)$ and (i) is proved.

For (i), we have shown that each orbit in $\mathcal{P}_n(\mathbb{R})$ or $\mathcal{E}_n(\mathbb{R})$ contains a diagonal matrix. The eigenvalues are positive real if $x \in \mathcal{P}_n(\mathbb{R})$ or of absolute value 1 if $x \in \mathcal{E}_n(\mathbb{R})$. In either case, applying the action (28.3) with $g \in GL(n, \mathbb{R})$ or $U(n)$ diagonal will reduce to the identity, proving (ii). For (iii), we use (ii) to write an arbitrary element x of $\mathcal{P}_n(\mathbb{R})$ or $\mathcal{E}_n(\mathbb{R})$ as kdk^{-1}, where k is orthogonal and d diagonal. The eigenvalues of d are either positive real if $x \in \mathcal{P}_n(\mathbb{R})$ or of absolute value 1 if $x \in \mathcal{E}_n(\mathbb{R})$. Thus, $d = e^Y$, where Y is real or purely imaginary, and $x = e^X$ or e^{iX}, where $X = kYk^{-1}$ or $-ikYk^{-1}$ is real. □

In the action (28.3) of $GL(n, \mathbb{R})$ or $U(n)$ on $\mathcal{P}_n(\mathbb{R})$ or $\mathcal{E}_n(\mathbb{R})$, the stabilizer of I is $O(n)$, so we may identify the coset spaces $GL(n, \mathbb{R})/O(n)$ and $U(n)/O(n)$ with $\mathcal{P}_n(\mathbb{R})$ and $\mathcal{E}_n(\mathbb{R})$, respectively. The actions of $SL(n, \mathbb{R})$ and $SU(n)$ on $\mathcal{P}_n(\mathbb{R})$ and $\mathcal{E}_n(\mathbb{R})$ are not transitive. Let $\mathcal{P}_n^\circ(\mathbb{R})$ and $\mathcal{E}_n^\circ(\mathbb{R})$ be the subspaces of matrices of determinant 1. Then the actions of $SL(n, \mathbb{R})$ and $SU(n)$ on $\mathcal{P}_n^\circ(\mathbb{R})$ and $\mathcal{E}_n^\circ(\mathbb{R})$ are transitive, so we may identify $\mathcal{P}_n^\circ(\mathbb{R}) = SL(n, \mathbb{R})/SO(n)$ and $\mathcal{E}_n^\circ(\mathbb{R}) = SU(n)/SO(n)$. Thus, we obtain concrete models of the dual symmetric spaces $\mathcal{P}_n^\circ(\mathbb{R})$ and $\mathcal{E}_n^\circ(\mathbb{R})$.

We say that a symmetric space X is *reducible* if its universal cover decomposes into a product of two lower-dimensional symmetric spaces. If X is *irreducible* (i.e., not reducible) and not a Euclidean space, then it is classified into one of four types, called I, II, III, and IV. We next explain this classification.

Example 28.2. If K_0 is a compact Lie group, then K_0 is itself a compact symmetric space, the geodesic reversing involution being $k \longmapsto k^{-1}$. A symmetric space of this type is called *Type II*.

Example 28.3. Suppose that G is itself obtained by complexification of a compact connected Lie group K_0 and that the involution θ of G is the automorphism of G as a real Lie group induced by complex conjugation. This means that on the Lie algebra $\mathfrak{g} = \mathfrak{k}_0 \oplus i\mathfrak{k}_0$ of G, where $\mathfrak{k}_0 = \mathrm{Lie}(K)$, the involution θ sends $X + iY \longmapsto X - iY$, $Y \in \mathfrak{k}_0$. The fixed subgroup of θ is K_0, and the symmetric space is G/K_0. A symmetric space of this type is called *Type IV*. It is noncompact.

We will show that the Type II and Type IV symmetric spaces are in duality. For this, we need a couple of lemmas. If R is a ring and $e, f \in R$ we call e and f *orthogonal central idempotents* if $ex = xe$ and $fx = xf$ for all $x \in R$, $e^2 = e$, $f^2 = f$, and $ef = fe$.

Lemma 28.1. (Peirce decomposition) *Let R be a ring, and let e and f be orthogonal central idempotents. Assume that $1 = e + f$. Then Re and Rf are (two-sided) ideals of R, and each is a ring with identity elements e and f, respectively. The ring R decomposes as $Re \oplus Rf$.*

Proof. It is straightforward to see that Re is closed under multiplication and is a ring with identity element e and similarly for Rf. Since $1 = e + f$, we have $R = Re + Rf$, and $Re \cap Rf = 0$ because if $x \in Re \cap Rf$ we can write $x = re = r'f$, so $x = r'f^2 = ref = 0$. □

Lemma 28.2. *Regard $\mathbb{C} \otimes \mathbb{C} = \mathbb{C} \otimes_{\mathbb{R}} \mathbb{C}$ as a $\mathbb{C}$-algebra with scalar multiplication $a(x \otimes y) = ax \otimes y$, $a \in \mathbb{C}$. Then $\mathbb{C} \otimes \mathbb{C}$ and $\mathbb{C} \oplus \mathbb{C}$ are isomorphic as $\mathbb{C}$-algebras.*

Proof. Let

$$e = \tfrac{1}{2}(1 \otimes 1 + i \otimes i), \qquad f = \tfrac{1}{2}(1 \otimes 1 - i \otimes i). \qquad (28.4)$$

It is easily checked that e and f are orthogonal central idempotents whose sum is the identity element $1 \otimes 1$, and so we obtain a Peirce decomposition by Lemma 28.1. The ideals generated by e and f are both isomorphic to $\mathbb{C}$. □

Theorem 28.1. *Let K_0 be a compact connected Lie group. Then the compact and noncompact symmetric spaces of Examples 28.2 and 28.3 are in duality.*

Proof. Let $\mathfrak{g}$ and $\mathfrak{k}_0$ be the Lie algebras of G and K_0, respectively. We have $\mathfrak{g} = \mathbb{C} \otimes \mathfrak{k}_0$. The involution $\theta : \mathfrak{g} \longrightarrow \mathfrak{g}$ takes $a \otimes X \longrightarrow \bar{a} \otimes X$. By Lemma 28.2, we have $\mathfrak{g}_{\mathbb{C}} = \mathbb{C} \otimes \mathbb{C} \otimes \mathfrak{k}_0 \cong \mathbb{C} \otimes \mathfrak{k}_0 \oplus \mathbb{C} \otimes \mathfrak{k}_0$. Now θ induces the automorphism

$$\theta : a \otimes b \otimes X \longrightarrow a \otimes \bar{b} \otimes X, \qquad a, b \in \mathbb{C}, \, X \in \mathfrak{k}_0.$$

The $+1$ and -1 eigenspaces are spanned by vectors of the form $1 \otimes 1 \otimes X$ and $1 \otimes i \otimes X$ ($X \in \mathfrak{k}_0$), so the Lie algebra $\mathfrak{g}_c$ as in (28.2) will be spanned by vectors of the form $1 \otimes 1 \otimes X$ and $i \otimes i \circ X$, and the Lie algebra $\mathfrak{k}$ is $1 \otimes 1 \otimes \mathfrak{k}_0$.

Thus, with e and f as in (28.4), $\mathfrak{g}_c$ is the $\mathbb{R}$-linear span of $e \otimes \mathfrak{k}_0$ and $f \otimes \mathfrak{k}_0$. We can identify

$$\mathfrak{g}_c = e \otimes \mathfrak{k}_0 \oplus f \otimes \mathfrak{k}_0 \cong \mathfrak{k}_0 \oplus \mathfrak{k}_0.$$

The involution θ interchanges these two components, and since $1 \otimes 1 = e + f$, $\mathfrak{k} = 1 \otimes \mathfrak{k}_0 \cong \mathfrak{k}_0$ embedded diagonally in $\mathfrak{k}_0 \otimes \mathfrak{k}_0$.

From this description, we see that $\mathfrak{g}_c$ is the Lie algebra of $K \times K$, which we take to be the group G_c. The involution $\theta : K \times K \longrightarrow K \times K$ is $\theta(x, y) = (y, x)$, and K is embedded diagonally. This differs from the description of the compact symmetric space of Type II in Example 28.2, but it is equivalent. We may see this as follows. We can map $K \longrightarrow G_c/K$ by $x \longrightarrow (x, 1)K$. The involution sends this to $(1, x)K = (x^{-1}, 1)K$ since $(x, x) \in K$ embedded diagonally. Thus, if we represent the cosets of G_c/K this way, the symmetric space is parameterized by K, and the involution corresponds to $x \longrightarrow x^{-1}$. $\square$

If G/K and G_c/K are noncompact and compact symmetric spaces in duality, and if G/K and G_c/K are not of types IV and II, they are said to be of types III and I, respectively.

Theorem 28.2. *Let G be a noncompact, connected semisimple Lie group with an involution θ satisfying Hypothesis 28.1. Then K is a maximal compact subgroup of G. Indeed, if K' is any compact subgroup of G, then K' is conjugate to a subgroup of K.*

Proof. This follows from Helgason [66], Theorem 2.1 of Chap. VI on page 246. (Note the hypothesis that K be compact in our Proposition 28.1.) The proof in [66] depends on showing that G/K is a space of constant negative curvature. A compact group of isometries of such a space has a fixed point ([66], Theorem 13.1 of Chap. I on page 75). Now if K' fixes $xK \in G/K$, then $x^{-1}K'x \subseteq K$. $\square$

A semisimple real Lie algebra $\mathfrak{g}$ is *compact* if and only if the Killing form is negative definite. If this is the case, then $\mathrm{ad}(\mathfrak{g})$ is contained in the Lie algebra of the compact orthogonal group with respect to this negative definite quadratic form, and it follows that $\mathfrak{g}$ is the Lie algebra of a compact Lie group. A semisimple Lie algebra is *simple* if it has no proper nontrivial ideals.

Theorem 28.3. *If $\mathfrak{g}$ is a noncompact Lie algebra, then there exists a noncompact Lie group G with Lie algebra $\mathfrak{g}$ and a Cartan involution θ of G with fixed*

*points that are a maximal subgroup K of G so that G/K is a symmetric space
of noncompact type. In particular, Hypothesis 28.1 is satisfied. If $\mathfrak{g}$ is simple,
then G/K is irreducible, and this construction gives a one-to-one correspondence between the simple real Lie algebras and the irreducible noncompact
symmetric spaces of noncompact type.*

Although we will not need this fact, it is very striking that the classification of irreducible symmetric spaces of noncompact type is the same as the
classification of noncompact real forms of the semisimple Lie algebras.

Proof. It follows from Helgason [66], Chap. III, Theorem 6.4 on p. 181, that
$\mathfrak{g}$ has a compact form; that is, a compact Lie algebra $\mathfrak{g}_c$ with an isomorphic
complexification. It follows from Theorems 7.1 and 7.2 in Chap. III of [66]
that we may arrange things so that $\mathfrak{g}_c = \mathfrak{k} + i\mathfrak{p}$ and $\mathfrak{g} = \mathfrak{k} + \mathfrak{p}$, where $\mathfrak{k}$ and $\mathfrak{p}$
are the $+1$ and -1 eigenspaces of a Cartan involution θ, and that this Cartan
involution is essentially unique. Let G_c be the adjoint group of $\mathfrak{g}_c$; that is, the
group generated by exponentials of endomorphisms $\mathrm{ad}(X)$ with $X \in \mathfrak{g}_c$. It is
a compact Lie group with Lie algebra $\mathfrak{g}_c$—see Helgason [66], Chap. II, Section
5. Thus, G_c is a group of linear transformations of $\mathfrak{g}_c$, but we extend them to
complex linear transformations of $\mathfrak{g}_\mathbb{C}$, and so G_c and the other groups $G, G_\mathbb{C}$,
and K that we will construct will all be subgroups of $\mathrm{GL}(\mathfrak{g}_\mathbb{C})$. Let $G_\mathbb{C}$ be the
complexification of G_c. The conjugation of $\mathfrak{g}_\mathbb{C}$ with respect to $\mathfrak{g}$ induces an
automorphism of $G_\mathbb{C}$ as a real Lie group with a fixed-point set that can be
taken to be G. The Cartan involution θ induces an involution of G with a
fixed-point set K that is a subgroup with Lie algebra $\mathfrak{k}$.   □

In Table 28.1, we give the classification of Cartan [31] of the Type I and
Type III symmetric spaces. (The symmetric spaces of Type II and Type IV,
as we have already seen, correspond to complex semisimple Lie algebras.)

In Table 28.1, the group $\mathrm{SO}^*(2n)$ consists of all elements of $\mathrm{SO}(2n, \mathbb{C})$ that
stabilize the skew-Hermitian form

$$x_1\overline{x_{n+1}} + x_2\overline{x_{n+2}} + \ldots + x_n\overline{x_{2n}} - x_{n+1}\overline{x_1} - x_{n+2}\overline{x_2} - \cdots - x_{2n}\overline{x_n}.$$

The subgroups $\mathrm{S}(\mathrm{O}(p) \times \mathrm{O}(q))$ and $\mathrm{S}(\mathrm{U}(p) \times \mathrm{U}(q))$ are the subgroups of
$\mathrm{O}(p) \times \mathrm{O}(q)$ and $\mathrm{U}(p) \times \mathrm{U}(q)$ consisting of elements of determinant 1. Cartan
considered the special cases $q = 1$ significant enough to warrant independent
classifications. The group $\mathrm{S}(\mathrm{O}(p) \times \mathrm{O}(1)) \cong \mathrm{O}(p)$, and we have written K this
way for types *BII* and *DII*.

For the exceptional groups, we have only described the Lie algebra of the
maximal compact subgroup. We have given the real form from the classifica-
tion of Tits [162]. In this classification, $^2E_{6,2}^{16} = {}^iE_{6,r}^d$, for example, where i,
d, and r are numbers whose significance we will briefly discuss. They will all
reappear in the next chapter.

The number $i = 1$ if the group is an *inner form* and 2 if it is an *outer
form*. As we mentioned in Remark 24.1, real forms of G_c are parameterized by
elements of $H^1(\mathrm{Gal}(\mathbb{C}/\mathbb{R}), \mathrm{Aut}(G_\mathbb{C}))$. If the defining co-cycle is in the image of

Table 28.1. Real forms and Type I and Type III symmetric spaces

Cartan's class	G	G_c	K° or $\mathfrak{k}$	Dimension rank	Absolute/rel. root systems
AI	$\mathrm{SL}(n,\mathbb{R})$	$\mathrm{SU}(n)$	$\mathrm{SO}(n)$	$\frac{1}{2}(n-1)(n+2)$ $n-1$	A_{n-1} A_{n-1}
AII	$\mathrm{SL}(n,\mathbb{H})$	$\mathrm{SU}(2n)$	$\mathrm{Sp}(2n)$	$(n-1)(2n+1)$ $n-1$	A_{2n-1} A_{n-1}
$AIII$	$\mathrm{SU}(p,q)$ $p,q>1$	$\mathrm{SU}(p+q)$	$S(\mathrm{U}(p)\times\mathrm{U}(q))$	$2pq$ $\min(p,q)$	A_{p+q-1} $\begin{cases} C_p & (p=q) \\ BC_p & (p>q) \end{cases}$
AIV	$\mathrm{SU}(p,1)$	$\mathrm{SU}(p+1)$	$S(\mathrm{U}(p)\times\mathrm{U}(q))$	$2p$ 1	A_p BC_1
BI	$\mathrm{SO}(p,q)$ $p,q>1$ $p+q$ odd	$\mathrm{SO}(p+q)$	$S(\mathrm{O}(p)\times\mathrm{O}(q))$	pq $\min(p,q)$	$B_{(p+q-1)/2}$ $\begin{cases} B_q & (p>q) \\ D_p & (p=q) \end{cases}$
BII	$\mathrm{SO}(p,1)$ $p+1$ odd	$\mathrm{SO}(p+1)$	$\mathrm{O}(p)$	$2p$ 1	$B_{p/2}$ B_1
DI	$\mathrm{SO}(p,q)$ $p,q>1$ $p+q$ even	$\mathrm{SO}(p+q)$	$S(\mathrm{O}(p)\times\mathrm{O}(q))$	pq $\min(p,q)$	$D_{(p+q)/2}$ $\begin{cases} B_q & (p>q) \\ D_p & (p=q) \end{cases}$
DII	$\mathrm{SO}(p,1)$ $p+1$ even	$\mathrm{SO}(p+1)$	$\mathrm{O}(p)$	$2p$	$D_{(p+1)/2}$ A_1
$DIII$	$\mathrm{SO}^*(2n)$	$\mathrm{SO}(2n)$	$\mathrm{U}(n)$	$n-1$ $m=[n/2]$	D_n $\begin{cases} C_m & n=2m \\ BC_m & n=2m+1 \end{cases}$
CI	$\mathrm{Sp}(2n,\mathbb{R})$	$\mathrm{Sp}(2n)$	$\mathrm{U}(n)$	$n(n+1)$ n	C_n C_n
CII	$\mathrm{Sp}(2p,2q)$	$\mathrm{Sp}(2p+2q)$	$\mathrm{Sp}(2p)\times\mathrm{Sp}(2q)$	$4pq$ $\min(p,q)$	C_{p+q} $\begin{cases} BC_q & (p>q) \\ C_p & (p=q) \end{cases}$
EI	$^1E_{6,6}^0$	E_6	$\mathfrak{sp}(8)$	42	$E_6 \quad E_6$
EII	$^2E_{6,4}^2$	E_6	$\mathfrak{su}(6)\times\mathfrak{su}(2)$	40	$E_6 \quad F_4$
$EIII$	$^2E_{6,2}^{16}$	E_6	$\mathfrak{so}(10)\times\mathfrak{u}(1)$	32	$E_6 \quad G_2$
EIV	$^1E_{6,2}^{28}$	E_6	$\mathfrak{f}_4$	26	$E_6 \quad A_2$
EV	$E_{7,7}^0$	E_7	$\mathfrak{so}(10)\times\mathfrak{u}(1)$	70	$E_7 \quad E_7$
EVI	$E_{7,4}^9$	E_7	$\mathfrak{so}(12)\times\mathfrak{su}(2)$	64	$E_7 \quad F_4$
$EVII$	$E_{7,3}^{28}$	E_7	$\mathfrak{e}_6\times\mathfrak{u}(1)$	54	$E_7 \quad C_3$
$EVIII$	$E_{8,8}^0$	E_8	$\mathfrak{so}(16)$	128	$E_8 \quad E_8$
EIX	$E_{8,4}^{28}$	E_8	$\mathfrak{e}_7\times\mathfrak{su}(2)$	112	$E_8 \quad F_4$
FI	$F_{4,4}^0$	F_4	$\mathfrak{sp}(6)\times\mathfrak{su}(2)$	28	$F_4 \quad F_4$
FII	$F_{4,1}^{21}$	F_4	$\mathfrak{so}(9)$	16	$F_4 \quad A_1$
G	$G_{2,2}^0$	G_2	$\mathfrak{su}(2)\times\mathfrak{su}(2)$	8	$G_2 \quad G_2$

$$H^1\big(\mathrm{Gal}(\mathbb{C}/\mathbb{R}),\mathrm{Inn}(G_c)\big) \longrightarrow H^1\big(\mathrm{Gal}(\mathbb{C}/\mathbb{R}),\mathrm{Aut}(G_c)\big),$$

where $\mathrm{Inn}(G_c)$ is the group of inner automorphisms, then the group is an inner form. Looking ahead to the next chapter, where we introduce the Satake diagrams, G is an inner form if and only if the symmetry of the Satake diagram, corresponding to the permutation $\alpha \longmapsto -\theta(\alpha)$ of the relative root system, is trivial. Thus, from Fig. 29.3, we see that $SO(6,6)$ is an inner form, but the quasisplit group $SO(7,5)$ is an outer form. For the exceptional groups, only E_6 admits an outer automorphism (corresponding to the nontrivial automorphism of its Dynkin diagram). Thus, for the other exceptional groups, this parameter is omitted from the notation.

The number r is the *(real) rank*—the dimension of the group $A = \exp(\mathfrak{a})$, where $\mathfrak{a}$ is a maximal Abelian subspace of $\mathfrak{p}$. The number d is the dimension of the *anisotropic kernel*, which is the maximal compact subgroup of the centralizer of A. Both A and M will play an extensive role in the next chapter.

We have listed the rank for the groups of classical type but not the exceptional ones since for those the rank is contained in the Tits' classification.

For classification matters we recommend Tits [162] supplemented by Borel [20]. The definitive classification in this paper, from the point of view of algebraic groups, includes not only real groups but also groups over p-adic fields, number fields, and finite fields. Knapp [106], Helgason [66], Onishchik and Vinberg [166], and Satake [144] are also very useful.

Example 28.4. Consider $SL(2,\mathbb{R})/SO(2)$ and $SU(2)/SO(2)$. Unlike the general case of $SL(n,\mathbb{R})/SO(n)$ and $SU(n)/SO(n)$, these two symmetric spaces have complex structures. Specifically, $SL(2,\mathbb{R})$ acts transitively on the Poincaré upper half-plane $\mathfrak{H} = \{z = x + iy \,|\, x, y \in \mathbb{R}, y > 0\}$ by linear fractional transformations:

$$SL(2,\mathbb{R}) \ni \begin{pmatrix} a & b \\ c & d \end{pmatrix} : z \longmapsto \frac{az+b}{cz+d}.$$

The stabilizer of the point $i \in \mathfrak{H}$ is $SO(2)$, so we may identify $\mathfrak{H}$ with $SL(2,\mathbb{R})/SO(2)$. Equally, let $\mathfrak{R}$ be the Riemann sphere $\mathbb{C} \cup \{\infty\}$, which is the same as the complex projective line $\mathbb{P}^1(\mathbb{C})$. The group $SU(2)$ acts transitively on $\mathfrak{R}$, also by linear fractional transformations:

$$SU(2) \ni \begin{pmatrix} a & b \\ -\bar{b} & \bar{a} \end{pmatrix} : z \longmapsto \frac{az+b}{-\bar{b}z+\bar{a}}, \qquad |a|^2 + |b|^2 = 1.$$

Again, the stabilizer of i is $SO(2)$, so we may identify $SU(2)/SO(2)$ with $\mathfrak{R}$.

Both $\mathfrak{H}$ and $\mathfrak{R}$ are naturally complex manifolds, and the action of $SL(2,\mathbb{R})$ or $SU(2)$ consists of holomorphic mappings. They are examples of *Hermitian symmetric spaces*, which we now define. A *Hermitian manifold* is the complex analog of a Riemannian manifold. A Hermitian manifold is a complex manifold on which there is given a (smoothly varying) positive definite Hermitian inner product on each tangent space (which has a complex structure because the space is a complex manifold). The real part of this positive definite Hermitian inner product is a positive definite symmetric bilinear form, so a Hermitian manifold becomes a Riemannian manifold. A real-valued symmetric bilinear form B on a complex vector space V is the real part of a positive definite Hermitian form H if and only if it satisfies

$$B(iX, iY) = B(X, Y),$$

for if this is true it is easy to check that

$$H(X, Y) = \tfrac{1}{2}\big(B(X, Y) - iB(iX, Y)\big)$$

is the unique positive definite Hermitian form with real part H. From this remark, a complex manifold is Hermitian by our definition if and only if it is Hermitian by the definition in Helgason [66].

A symmetric space X is called *Hermitian* if it is a Hermitian manifold that is homogeneous with respect to a group of holomorphic Hermitian isometries that is connected and contains the geodesic-reversing reflection around each point. Thus, if $X = G/K$, the group G consists of holomorphic mappings, and if $g(x) = y$ for $x, y \in X$, $g \in X$, then g induces an isometry between the tangent spaces at x and y.

The irreducible Hermitian symmetric spaces can easily be recognized by the following criterion.

Proposition 28.3. *Let $X = G/K$ and $X_c = G_c/K$ be a pair of irreducible symmetric spaces in duality. If one is a Hermitian symmetric space, then they both are. This will be true if and only if the center of K is a one-dimensional central torus Z. In this case, the rank of G_c equals the rank of K.*

In a nutshell, if K has a one-dimensional central torus, then there exists a homomorphism of $\mathbb{T}$ into the center of K. The image of $\mathbb{T}$ induces a group of isometries of X fixing the base point $x_0 \in X$ corresponding to the coset of K. The content of the proposition is that X may be given the structure of a complex manifold in such a way that the maps on the tangent space at x_0 induced by this family of isometries correspond to multiplication by $\mathbb{T}$, which is regarded as a subgroup of $\mathbb{C}^\times$.

Proof. See Helgason [66], Theorem 6.1 and Proposition 6.2, or Wolf [176], Corollary 8.7.10, for the first statement. The latter reference has two other very interesting conditions for the space to be symmetric. The fact that G_c and K are of equal rank is contained in Helgason [66] in the first paragraph of "Proof of Theorem 7.1 (ii), algebraic part" on p. 383. □

A particularly important family of Hermitian symmetric spaces are the *Siegel upper half-spaces* $\mathfrak{H}_n$, also known as *Siegel spaces* which generalize the Poincaré upper half-plane $\mathfrak{H} = \mathfrak{H}_1$. We will discuss this family of examples in considerable detail since many features of the general case are already present in this example and are perhaps best understood with an example in mind.

In this chapter, if F is a field (always $\mathbb{R}$ or $\mathbb{C}$), the symplectic group is

$$\mathrm{Sp}(2n, F) = \{g \in \mathrm{GL}(2n, F) \,|\, g \, J^t g = J\}, \qquad J = \begin{pmatrix} & -I_n \\ I_n & \end{pmatrix}.$$

Write $g = \begin{pmatrix} A & B \\ C & D \end{pmatrix}$, where A, B, C, and D are $n \times n$ blocks. Multiplying out the condition $g \, J^t g = J$ gives the conditions

$$\begin{array}{ll} A \cdot {}^t B = B \cdot {}^t A, & C \cdot {}^t D = D \cdot {}^t C, \\ A \cdot {}^t D - B \cdot {}^t C = I, & D \cdot {}^t A - C \cdot {}^t B = I. \end{array} \tag{28.5}$$

The condition $g \, J^t g = J$ may be expressed as $(gJ)({}^t gJ) = -I$, so gJ and ${}^t gJ$ are negative inverses of each other. From this, we see that ${}^t g$ is also symplectic, and so (28.5) applied to ${}^t g$ gives the further relations

$$\begin{array}{ll} {}^t A \cdot C = {}^t C \cdot A, & {}^t B \cdot D = {}^t D \cdot B, \\ {}^t A \cdot D - {}^t C \cdot B = I; & {}^t D \cdot A - {}^t B \cdot C = I. \end{array} \tag{28.6}$$

If $A + iB \in \mathrm{U}(n)$, where the matrices A and B are real, then $A \cdot {}^t A + B \cdot {}^t B = I$ and $A \cdot {}^t B = B \cdot {}^t A$. Thus, if we take $D = A$ and $C = -B$, then (28.5) is satisfied. Thus,

$$A + iB \longmapsto \begin{pmatrix} A & B \\ -B & A \end{pmatrix} \tag{28.7}$$

maps $\mathrm{U}(n)$ into $\mathrm{Sp}(2n, \mathbb{R})$ and is easily checked to be a homomorphism.

If W is a Hermitian matrix, we write $W > 0$ if W is positive definite.

If $\Omega \subset \mathbb{R}^r$ is any connected open set, we can form the *tube domain over* Ω. This is the set of all elements of $\mathbb{C}^r$ that has imaginary parts in Ω. Let $\mathfrak{H}_n$ be the tube domain over $\mathcal{P}_n(\mathbb{R})$. Thus, $\mathfrak{H}_n$ is the space of all symmetric complex matrices $Z = X + iY$ where X and Y are real symmetric matrices such that $Y > 0$.

Proposition 28.4. *If* $Z \in \mathfrak{H}_n$ *and* $g = \begin{pmatrix} A & B \\ C & D \end{pmatrix} \in \mathrm{Sp}(2n, \mathbb{R})$, *then* $CZ + D$ *is invertible. Define*

$$g(Z) = (AZ + B)(CZ + D)^{-1}. \tag{28.8}$$

Then $g(Z) \in \mathfrak{H}_n$, *and (28.8) defines an action of* $\mathrm{Sp}(2n, \mathbb{R})$ *on* $\mathfrak{H}_n$. *The action is transitive, and the stabilizer of* $iI_n \in \mathfrak{H}_n$ *is the image of* $\mathrm{U}(n)$ *under the embedding (28.7). If* W *is the imaginary part of* $g(Z)$ *then*

$$W = (\overline{Z}{}^t C + {}^t D)^{-1} Y (CZ + D)^{-1}. \tag{28.9}$$

Proof. Using (28.6), one easily checks that

$$\tfrac{1}{2i}\left((\overline{Z}{}^t C + {}^t D)(AZ + B) - (\overline{Z}{}^t A + {}^t B)(CZ + D)\right) = \tfrac{1}{2i}(Z - \overline{Z}) = Y, \tag{28.10}$$

where Y is the imaginary part of Z. From this it follows that $CZ + D$ is invertible since if it had a nonzero nullvector v, then we would have ${}^t\overline{v}Yv = 0$, which is impossible since $Y > 0$.

Therefore, we may make the definition (28.8). To check that $g(Z)$ is symmetric, the identity $g(Z) = {}^tg(Z)$ is equivalent to

$$(AZ + B)(Z^tC + {}^tD) = (Z^tA + {}^tB)(CZ + D),$$

which is easily confirmed using (28.5) and (28.6).

Next we show that the imaginary part W of $g(Z)$ is positive definite. Indeed W equals $\frac{1}{2i}\big(g(Z) - \overline{g(Z)}\big)$. Using the fact that $\overline{g(Z)}$ is symmetric and (28.10), this is

$$\tfrac{1}{2i}\big((AZ + B)(CZ + D)^{-1} - (\overline{Z}{}^tC + {}^tD)^{-1}(\overline{Z}{}^tA + {}^tB)\big).$$

Simplifying this gives (28.9). From this it is clear that W is Hermitian and that $W > 0$. It is real, of course, though that is less clear from this expression. Since it is real and positive definite Hermitian, it is a positive definite symmetric matrix.

It is easy to check that $g(g'(Z)) = (gg')(Z)$, so this is a group action. To show that this action is transitive, we note that if $Z = X + iY \in \mathfrak{H}_n$, then

$$\begin{pmatrix} I & -X \\ & I \end{pmatrix} \in \mathrm{Sp}(2n, \mathbb{R}),$$

and this matrix takes Z to iY. Now if $h \in \mathrm{GL}(n, \mathbb{R})$, then

$$\begin{pmatrix} g & \\ & {}^tg^{-1} \end{pmatrix} \in \mathrm{Sp}(2n, \mathbb{R}),$$

and this matrix takes iY to iY', where $Y' = gY^tg$. Since $Y > 0$, we may choose g so that $Y' = I$. This shows that any element in $\mathfrak{H}_n$ may be moved to iI_n, and the action is transitive.

To check that $\mathrm{U}(n)$ is the stabilizer of iI_n is quite easy, and we leave it to the reader. $\square$

Example 28.5. By Proposition 28.4, we can identify $\mathfrak{H}_n$ with $\mathrm{Sp}(2n, \mathbb{R})/\mathrm{U}(n)$. The fact that it is a Hermitian symmetric space is in accord with Proposition 28.3, since $\mathrm{U}(n)$ has a central torus. In the notation of Proposition 28.1, if $G = \mathrm{Sp}(2n, \mathbb{R})$ and $K = \mathrm{U}(n)$ are embedded via (28.7), then the compact group G_c is $\mathrm{Sp}(2n)$, where as usual $\mathrm{Sp}(2n)$ denotes $\mathrm{Sp}(2n, \mathbb{C}) \cap \mathrm{U}(2n)$.

We will investigate the relationship between Examples 28.1 and 28.5 using a fundamental map, the *Cayley transform*. For clarity, we introduce this first in the more familiar context of the Poincaré upper half-plane (Example 28.4), which is a special case of Example 28.5.

We observe that the action of $G_c = \mathrm{SU}(2)$ on the compact dual $X_c = \mathrm{SU}(2)/\mathrm{SO}(2)$ can be extended to an action of $G_{\mathbb{C}} = \mathrm{SL}(2, \mathbb{C})$. Indeed, if we identify X_c with the Riemann sphere $\mathfrak{R}$, then the action of SU(2) was by linear fractional transformations and so is the extension to SL(2, $\mathbb{C}$).

As a consequence, we have an action of $G = $ SL(2, $\mathbb{R}$) on X_c since $G \subset G_{\mathbb{C}}$ and $G_{\mathbb{C}}$ acts on X_c. This is just the action by linear fractional transformations on $\mathfrak{R} = \mathbb{C} \cup \{\infty\}$. There are three orbits: $\mathfrak{H}$, the projective real line $\mathbb{P}^1(\mathbb{R}) = \mathbb{R} \cup \{\infty\}$, and the lower half-plane $\overline{\mathfrak{H}}$.

The *Cayley transform* is the element $c \in$ SU(2) given by

$$c = \tfrac{1}{\sqrt{2i}} \begin{pmatrix} 1 & -i \\ 1 & i \end{pmatrix}, \qquad \text{so} \qquad c^{-1} = \tfrac{1}{\sqrt{2i}} \begin{pmatrix} i & i \\ -1 & 1 \end{pmatrix}. \qquad (28.11)$$

Interpreted as a transformation of $\mathfrak{R}$, the Cayley transform takes $\mathfrak{H}$ to the unit disk

$$\mathfrak{D} = \{w \in \mathbb{C} \,|\, |w| < 1\}.$$

Indeed, if $z \in \mathfrak{H}$, then

$$c(z) = \frac{z - i}{z + i},$$

and since z is closer to i than to $-i$, this lies in $\mathfrak{D}$. The effect of the Cayley transform is shown in Fig. 28.1.

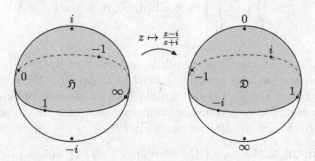

Fig. 28.1. The Cayley transform

The significance of the Cayley transform is that it relates a bounded symmetric domain $\mathfrak{D}$ to an unbounded one, $\mathfrak{H}$. We will use both $\mathfrak{H}$ and $\mathfrak{D}$ together when thinking about the *boundary* of the noncompact symmetric space embedded in its compact dual.

Since SL(2, $\mathbb{R}$) acts on $\mathfrak{H}$, the group $c\,$SL(2, $\mathbb{R})\,c^{-1}$ acts on $c(\mathfrak{H}) = \mathfrak{D}$. This group is

$$\text{SU}(1,1) = \left\{ \begin{pmatrix} a & b \\ \bar{b} & \bar{a} \end{pmatrix} \,\Big|\, |a|^2 - |b|^2 = 1 \right\}.$$

The Cayley transform is generally applicable to Hermitian symmetric spaces. It was shown by Cartan and Harish-Chandra that Hermitian symmetric spaces could be realized on bounded domains in $\mathbb{C}^n$. Piatetski-Shapiro [133] gave unbounded realizations. Korányi and Wolf [111, 112] gave a completely general theory relating bounded symmetric domains to unbounded ones by means of the Cayley transform.

Now let us consider the Cayley transform for $\mathfrak{H}_n$. Let $G = \mathrm{Sp}(2n, \mathbb{R})$, $K = \mathrm{U}(n)$, $G_c = \mathrm{Sp}(2n)$, and $G_{\mathbb{C}} = \mathrm{Sp}(2n, \mathbb{C})$. Let

$$c = \frac{1}{\sqrt{2i}} \begin{pmatrix} I_n & -iI_n \\ I_n & iI_n \end{pmatrix}, \qquad c^{-1} = \frac{1}{\sqrt{2i}} \begin{pmatrix} iI_n & iI_n \\ -I_n & I_n \end{pmatrix}.$$

They are elements of $\mathrm{Sp}(2n)$. We will see that the scenario uncovered for $\mathrm{SL}(2, \mathbb{R})$ extends to the symplectic group.

Our first goal is to show that $\mathfrak{H}_n$ can be embedded in its compact dual, a fact already noted when $n = 1$. The first step is to interpret G_c/K as an analog of the Riemann sphere $\mathfrak{R}$, a space on which the actions of both groups G and G_c may be realized as linear fractional transformations. Specifically, we will construct a space $\mathfrak{R}_n$ that contains a dense open subspace $\mathfrak{R}_n^\circ$ that can be naturally identified with the vector space of all complex symmetric matrices. What we want is for $G_{\mathbb{C}}$ to act on $\mathfrak{R}_n$, and if $g \in G_{\mathbb{C}}$, with both $Z, g(Z) \in \mathfrak{R}_n^\circ$, then $g(Z)$ is expressed in terms of Z by (28.8).

Toward the goal of constructing $\mathfrak{R}_n$, let

$$P = \left\{ \begin{pmatrix} h & \\ & {}^t h^{-1} \end{pmatrix} \begin{pmatrix} I & X \\ & I \end{pmatrix} \,\middle|\, h \in \mathrm{GL}(n, \mathbb{C}), X \in \mathrm{Mat}_n(\mathbb{C}), X = {}^t X \right\}. \tag{28.12}$$

This group is called the *Siegel parabolic subgroup* of $\mathrm{Sp}(2n, \mathbb{C})$. (The term *parabolic subgroup* will be formally defined in Chap. 30.) We will define $\mathfrak{R}_n$ to be the quotient $G_{\mathbb{C}}/P$. Let us consider how an element of this space can (usually) be regarded as a complex symmetric matrix, and the action of $G_{\mathbb{C}}$ is by linear fractional transformations as in (28.8).

Proposition 28.5. *We have $PG_c = \mathrm{Sp}(2n, \mathbb{C})$ and $P \cap G_c = cKc^{-1}$.*

Proof. Indeed, P contains a Borel subgroup, the group B of matrices (28.12) with g upper triangular, so $PG_c = \mathrm{Sp}(2n, \mathbb{C})$ follows from the Iwasawa decomposition (Theorem 26.3). The group K is $\mathrm{U}(n)$ embedded via (28.7), and it is easy to check that

$$cKc^{-1} = \left\{ \begin{pmatrix} g & \\ & {}^t g^{-1} \end{pmatrix} \,\middle|\, g \in \mathrm{U}(n) \right\}. \tag{28.13}$$

It is clear that $cKc^{-1} \subseteq P \cap \mathrm{Sp}(2n)$. To prove the converse inclusion, it is straightforward to check that any unitary matrix in P is actually of the form (28.13), and so $P \cap G_c \subseteq cKc^{-1}$. $\square$

We define $\mathfrak{R}_n = G_{\mathbb{C}}/P$. We define $\mathfrak{R}_n^\circ$ to be the set of cosets gP, where $g = \begin{pmatrix} A & B \\ C & D \end{pmatrix} \in \mathrm{Sp}(2n, \mathbb{C})$ and $\det(C)$ is nonsingular.

Lemma 28.3. *Suppose that*

$$g = \begin{pmatrix} A & B \\ C & D \end{pmatrix}, \qquad g' = \begin{pmatrix} A' & B' \\ C' & D' \end{pmatrix},$$

are elements of $G_{\mathbb{C}}$. Then $gP = g'P$ if and only if there exists a matrix $h \in GL(n, \mathbb{C})$ such that $Ah = A'$ and $Ch = C'$. If C is invertible, then AC^{-1} is a complex symmetric matrix. If this is true, a necessary and sufficient condition for $gP = g'P$ is that C' is also invertible and that $AC^{-1} = A'(C')^{-1}$.

Proof. Most of this is safely left to the reader. We only point out the reason that AC^{-1} is symmetric. By (28.6), the matrix ${}^tCA$ is symmetric, so ${}^tC^{-1} \cdot {}^tCA \cdot C^{-1} = AC^{-1}$ is also. $\qquad\square$

Let $\mathcal{R}_n$ be the vector space of $n \times n$ complex symmetric matrices. By the Lemma 28.3, the map $\sigma : \mathcal{R}_n \longrightarrow \mathfrak{R}_n^{\circ}$ defined by

$$\sigma(Z) = \begin{pmatrix} Z & -I \\ I & \end{pmatrix} P$$

is a bijection, and we can write

$$\sigma(Z) = \begin{pmatrix} A & B \\ C & D \end{pmatrix} P$$

if and only if $AC^{-1} = Z$.

Proposition 28.6. *If $\sigma(Z)$ and $g(\sigma(Z))$ are both in $\mathfrak{R}_n^{\circ}$, where $g = \begin{pmatrix} A & B \\ C & D \end{pmatrix}$ is an element of $\mathrm{Sp}(2n, \mathbb{C})$, then $CZ + D$ is invertible and*

$$g(\sigma(Z)) = \sigma((AZ + B)(CZ + D)^{-1}).$$

Proof. We have

$$g(\sigma(Z)) = \begin{pmatrix} A & B \\ C & D \end{pmatrix} \begin{pmatrix} Z & -I \\ I & \end{pmatrix} P = \begin{pmatrix} AZ + B & -A \\ CZ + D & -C \end{pmatrix} P.$$

Since we are assuming this is in $\mathfrak{R}_n^{\circ}$, the matrix $CZ + D$ is invertible by Lemma 28.3, and this equals $\sigma((AZ + B)(CZ + D)^{-1})$. $\qquad\square$

In view of Proposition 28.6 we will identify $\mathcal{R}_n$ with its image in $\mathfrak{R}_n^{\circ}$. Thus, elements of $\mathfrak{R}_n^{\circ}$ become for us complex symmetric matrices, and the action of $\mathrm{Sp}(2n, \mathbb{C})$ is by linear fractional transformations.

We can also identify $\mathfrak{R}_n$ with the compact symmetric space G_c/K by means of the composition of bijections

$$G_c/K \longrightarrow G_c/cKc^{-1} \longrightarrow G_{\mathbb{C}}/P = \mathfrak{R}_n.$$

The first map is induced by conjugation by $c \in G_c$. The second map is induced by the inclusion $G_c \longrightarrow G_{\mathbb{C}}$ and is bijective by Proposition 28.5, so we may regard the embedding of $\mathfrak{H}_n$ into $\mathfrak{R}_n$ as an embedding of a noncompact symmetric space into its compact dual.

So far, the picture is extremely similar to the case where $n = 1$. We now come to an important difference. In the case of $\mathrm{SL}(2, \mathbb{R})$, the topological boundary of $\mathfrak{H}$ (or $\mathfrak{D}$) in $\mathfrak{R}$ was just a circle consisting of a single orbit of $\mathrm{SL}(2, \mathbb{R})$ or even its maximal compact subgroup $\mathrm{SO}(2)$.

When $n \geqslant 2$, however, the boundary consists of a finite number of orbits, each of which is the union of smaller pieces called *boundary components*. Each boundary component is a copy of a Siegel space of lower dimension. The boundary components are infinite in number, but each is a copy of one of a finite number of standard ones. Since the structure of the boundary is suddenly interesting when $n \geqslant 2$, we will take a closer look at it. For more information about boundary components, which are important in the theory of automorphic forms and the algebraic geometry of arithmetic quotients such as $\mathrm{Sp}(2n, \mathbb{Z}) \backslash \mathfrak{H}_n$, see Ash, Mumford, Rapoport, and Tai [11], Baily [13], Baily and Borel [14], and Satake [142, 144].

The first step is to map $\mathfrak{H}_n$ into a bounded region. Writing $Z = X + iY$, where X and Y are real symmetric matrices, $Z \in \mathfrak{H}_n$ if and only if $Y > 0$. So Z is on the boundary if Y is positive *semidefinite* yet has 0 as an eigenvalue. The multiplicity of 0 as an eigenvalue separates the boundary into several pieces that are separate orbits of G. (These are *not* the boundary components, which we will meet presently.)

If we embed $\mathfrak{H}_n$ into $\mathfrak{R}_n$, a portion of the border is at "infinity"; that is, it is in $\mathfrak{R}_n - \mathfrak{R}_n^\circ$. We propose to examine the border by applying c, which maps $\mathfrak{H}_n$ into a region with a closure that is wholly contained in $\mathfrak{R}_n$.

Proposition 28.7. *The image of $\mathfrak{H}_n$ under c is*

$$\mathfrak{D}_n = \{ W \in \mathfrak{R}_n^\circ \mid I - \overline{W}W > 0 \}.$$

The group $c\, \mathrm{Sp}(2n, \mathbb{R})\, c^{-1}$, acting on $\mathfrak{D}_n$ by linear fractional transformations, consists of all symplectic matrices of the form

$$\begin{pmatrix} A & B \\ \overline{B} & \overline{A} \end{pmatrix}. \tag{28.14}$$

(Note that, since W is symmetric, $I - \overline{W}W$ is Hermitian.)

Proof. The condition on W to be in $c(\mathfrak{H})$ is that the imaginary part of

$$c^{-1}(W) = -i(W - I)(W + I)^{-1}$$

be positive definite. This imaginary part is

$$Y = -\tfrac{1}{2}\big((W - I)(W + I)^{-1} + (\overline{W} - I)(\overline{W} + I)^{-1}\big) =$$
$$-\tfrac{1}{2}\big((W - I)(W + I)^{-1} + (\overline{W} + I)^{-1}(\overline{W} - I)\big),$$

where we have used the fact that both $\overline{W}$ and $(\overline{W}-I)(\overline{W}+I)^{-1}$ are symmetric to rewrite the second term. This will be positive definite if and only if $(\overline{W}+I)Y(W+I)$ is positive definite. This equals

$$-\tfrac{1}{2}((\overline{W}+I)(W-I) + (\overline{W}-I)(W+I)) = I - \overline{W}W.$$

Since $\mathrm{Sp}(2n,\mathbb{R})$ maps $\mathfrak{H}_n$ into itself, $c\,\mathrm{Sp}(2n,\mathbb{R})\,c^{-1}$ maps $\mathfrak{D}_n = c(\mathfrak{H}_n)$ into itself. We have only to justify the claim that this group consists of the matrices of form (28.14). For $g = \begin{pmatrix} A\ B \\ C\ D \end{pmatrix} \in \mathrm{Sp}(2n,\mathbb{C})$ to have the property that $c^{-1}g\,c$ is real, we need $c^{-1}g\,c = \overline{c^{-1}g\,c}$, so

$$c\,\overline{c}^{-1}\overline{\begin{pmatrix} A\ B \\ C\ D \end{pmatrix}} = \begin{pmatrix} A\ B \\ C\ D \end{pmatrix} c\,\overline{c}^{-1}, \qquad c\,\overline{c}^{-1} = \sqrt{-i}\begin{pmatrix} 0\ I_n \\ I_n\ 0 \end{pmatrix}.$$

This implies that $C = \overline{B}$ and $D = \overline{A}$. □

Proposition 28.8. (i) *The closure of $\mathfrak{D}_n$ is contained within $\mathfrak{R}_n^\circ$. The boundary of $\mathfrak{D}_n$ consists of all complex symmetric matrices W such that $I - \overline{W}W$ is positive semidefinite but such that $\det(I - \overline{W}W) = 0$.*

(ii) *If W and W' are points of the closure of $\mathfrak{D}_n$ in $\mathfrak{R}_n$ that are congruent modulo cGc^{-1}, then the ranks of $I - \overline{W}W$ and $I - \overline{W'}W'$ are equal.*

(iii) *Let W be in the closure of $\mathfrak{D}_n$, and let r be the rank of $I - \overline{W}W$. Then there exists $g \in cGc^{-1}$ such that $g(W)$ has the form*

$$\begin{pmatrix} W_r \\ & I_{n-r} \end{pmatrix}, \qquad W_r \in \mathfrak{D}_{n-r}. \tag{28.15}$$

Proof. The diagonal entries in $\overline{W}W$ are the squares of the lengths of the rows of the symmetric matrix W. If $I - \overline{W}W$ is positive definite, these must be less than 1. So $\mathfrak{D}_n$ is a bounded domain within the set $\mathfrak{R}_n^\circ$ of symmetric complex matrices. The rest of (i) is also clear.

For (ii), if $g \in cGc^{-1}$, then by Proposition 28.7 the matrix g has the form (28.14). Using the fact that both W and W' are symmetric, we have

$$I - \bar{W}'W = I - (\bar{W}^{\,t}B + {}^tA)^{-1}(\bar{W}^{\,t}\bar{A} + \overline{{}^tB})(AW + B)(\bar{B}W + \bar{A})^{-1}.$$

Both W and W' are in $\mathfrak{R}_n^\circ$, so by Proposition 28.6 the matrix $\overline{B}W + \bar{A}$ is invertible. Therefore, the rank of $I - \bar{W}'W'$ is the same as the rank of

$$(\bar{W}^{\,t}B + {}^tA)(I - \bar{W}'W')(\bar{B}W + \bar{A}) =$$
$$(\bar{W}^{\,t}B + {}^tA)(\bar{B}W + \bar{A}) - (\bar{W}^{\,t}\bar{A} + \overline{{}^tB})(AW + B).$$

Using the relations (28.6), this equals $I - \bar{W}W$.

To prove (iii), note first that if $u \in \mathrm{U}(n) \subset cGc^{-1}$, then

$$cGc^{-1} \ni \begin{pmatrix} u \\ & \bar{u} \end{pmatrix} : W \longmapsto u\,W\,{}^tu.$$

Taking u to be a scalar, we may assume that -1 is not an eigenvalue of W. Then $W + I$ is nonsingular so $Z = c^{-1}W = -i(W-I)(W+I)^{-1} \in \mathfrak{R}_n^\circ$. Since Z is in the closure of $\mathfrak{H}$, it follows that $Z = X + iY$, where X and Y are real symmetric and Y is positive semidefinite. There exists an orthogonal matrix k such that $D = kYk^{-1}$ is diagonal with nonnegative eigenvalues. Now

$$\gamma(Z) = iD, \qquad \gamma = \begin{pmatrix} k & \\ & k \end{pmatrix}\begin{pmatrix} I & -X \\ & I \end{pmatrix} \in G.$$

Thus, denoting $W' = c\gamma c^{-1}W$,

$$W' = c(iD) = (D-I)(D+I)^{-1}.$$

Like D, the matrix W' is diagonal, and its diagonal entries equal to 1 correspond to the diagonal entries of D equal to 0. These correspond to diagonal entries of $I - \overline{W'}W'$ equal to 0, so the diagonal matrices D and $I - \overline{W'}W'$ have the same rank. But by (ii), the ranks of $I - \overline{W}W$ and $I - \overline{W'}W'$ are equal, so the rank of D is r. Clearly, W' has the special form (28.15).   □

Now let us fix $r < n$ and consider

$$\mathfrak{B}_r = \left\{ \begin{pmatrix} W_r & \\ & I_{n-r} \end{pmatrix} \,\Big|\, W_r \in \mathfrak{D}_{n-r} \right\}.$$

By Proposition 28.7, the subgroup of cGc^{-1} of the form

$$\begin{pmatrix} A_r & 0 & B_r & 0 \\ 0 & I_{n-r} & 0 & 0 \\ \overline{B_r} & 0 & \overline{A_r} & 0 \\ 0 & 0 & 0 & I_{n-r} \end{pmatrix}$$

is isomorphic to $\mathrm{Sp}(2r, \mathbb{R})$, and $\mathfrak{B}_r$ is homogeneous with respect to this subgroup. Thus, $\mathfrak{B}_r$ is a copy of the lower-dimensional Siegel space $\mathfrak{D}_r$ embedded into the boundary of $\mathfrak{D}_n$.

Any subset of the boundary that is congruent to a $\mathfrak{B}_r$ by an element of cGc^{-1} is called a *boundary component*. There are infinitely many boundary components, but each of them resembles one of these standard ones. The closure of a boundary component is a union of lower-dimensional boundary components. L Now let us consider the union of the zero-dimensional boundary components, that is, the set of all elements equivalent to $\mathfrak{B}_0 = \{I_n\}$. By Proposition 28.8, it is clear that this set is characterized by the *vanishing* of $I - \overline{W}W$. In other words, this is the set $\mathcal{E}_n(\mathbb{R})$.

If D is a bounded convex domain in $\mathbb{C}^r$, homogeneous under a group G of holomorphic mappings, the *Bergman–Shilov boundary* of D is the unique minimal closed subset B of the topological boundary ∂D such that a function holomorphic on D and continuous on its closure will take its maximum (in absolute value). See Korányi and Wolf [112] for further information, including the fact that a bounded symmetric domain must have such a boundary.

Theorem 28.4. *The domain $\mathfrak{D}_n$ has $\mathcal{E}_n(\mathbb{R})$ as its Bergman–Shilov boundary.*

Proof. Let f be a holomorphic function on $\mathfrak{D}_n$ that is continuous on its closure. We will show that f takes its maximum on the set $\mathcal{E}_n(\mathbb{R})$. This is sufficient because G acts transitively on $\mathcal{E}_n(\mathbb{R})$, so the set $\mathcal{E}_n(\mathbb{R})$ cannot be replaced by any strictly smaller subspace with the same maximizing property.

Suppose $x \in \overline{\mathfrak{D}_n}$ maximizes $|f|$. Let $\mathfrak{B}$ be the boundary component containing x, so $\mathfrak{B}$ is congruent to some $\mathfrak{B}_r$. If $r > 0$, then noting that the restriction of f to $\mathfrak{B}$ is a holomorphic function there, the maximum modulus principle implies that f is constant on $\mathfrak{B}$ and hence $|f|$ takes the same maximum value on the boundary of $\mathfrak{B}$, which intersects $\mathcal{E}_n(\mathbb{R})$. $\square$

We now see that *both* the dual symmetric spaces $\mathcal{P}_n(\mathbb{R})$ and $\mathcal{E}_n(\mathbb{R})$ appear in connection with $\mathfrak{H}_n$. The construction of $\mathfrak{H}_n$ involved building a tube domain over the cone $\mathcal{P}_n(\mathbb{R})$, while the dual $\mathcal{E}_n(\mathbb{R})$ appeared as the Bergman–Shilov boundary. (Since $\mathcal{P}_n^\circ(\mathbb{R})$ and $\mathcal{E}_n(\mathbb{R})^\circ$ are in duality, it is natural to extend the notion of duality to the reducible symmetric spaces $\mathcal{P}_n(\mathbb{R})$ and $\mathcal{E}_n(\mathbb{R})$ and to say that these are in duality.)

This scenario repeats itself: there are four infinite families and one isolated example of Hermitian symmetric spaces that appear as tube domains over cones. In each case, the space can be mapped to a bounded symmetric domain by a Cayley transform, and the compact dual of the cone appears as the Bergman–Shilov boundary of the cone. These statements follow from the work of Koecher, Vinberg, and Piatetski-Shapiro [133], culminating in Korányi and Wolf [111, 112].

Let us describe this setup in a bit more detail. Let V be a real vector space with an inner product $\langle\ ,\ \rangle$. A *cone* $\mathcal{C} \subset V$ is a convex open set consisting of a union of rays through the origin but not containing any line. The *dual cone* to $\mathcal{C}$ is $\{x \in V \,|\, \langle x, y \rangle > 0 \text{ for all } y \in \mathcal{C}\}$. If $\mathcal{C}$ is its own dual, it is naturally called *self-dual*. It is called *homogeneous* if it admits a transitive automorphism group.

A homogeneous self-dual cone is a symmetric space. It is not irreducible since it is invariant under similitudes (i.e, transformations $x \longmapsto \lambda x$ where $\lambda \in \mathbb{R}^\times$). The orbit of a typical point under the commutator subgroup of the group of automorphisms of the cone sits inside the cone, inscribed like a hyperboloid, though this description is a little misleading since it may be the constant locus of an invariant of degree > 2. For example, $\mathcal{P}_n^\circ(\mathbb{R})$ is the locus of $\det(x) = 1$, and $\det$ is a homogeneous polynomial of degree n.

Homogeneous self-dual cones were investigated and applied to symmetric domains by Koecher, Vinberg, and others. A *Jordan algebra* over a field F is a nonassociative algebra over F with multiplication that is commutative and satisfies the weakened associative law $(ab)a^2 = a(ba^2)$. The basic principle is that if $\mathcal{C} \subset V$ is a self-dual convex cone, then V can be given the structure of a Jordan algebra in such a way that $\mathcal{C}$ becomes the set of squares in V.

In addition to Satake [145] Chap. I, Sect. 8, see Ash, Mumford, Rapoport, and Tai [11], Chap. II, for good explanations, including a discussion of the boundary components of a self-dual cone.

Example 28.6. Let $D = \mathbb{R}$, $\mathbb{C}$, or $\mathbb{H}$. Let $d = 1, 2$ or 4 be the real dimension of D. Let $\mathcal{J}_n(D)$ be the set of Hermitian matrices in $\mathrm{Mat}_n(D)$, which is a Jordan algebra. Let $\mathcal{P}_n(D)$ be the set of positive definite elements. It is a self-dual cone of dimension $n + (d/2)n(n-1)$. It is a reducible symmetric space, but the elements of $g \in \mathcal{P}_n(D)$ such that multiplication by g as an $\mathbb{R}$-linear transformation of $\mathrm{Mat}_n(D)$ has determinant 1 is an irreducible symmetric space $\mathcal{P}_n^\circ(D)$ of dimension $n + (d/2)n(n-1) - 1$. The dual $\mathcal{E}_n^\circ(D)$ is a compact Hermitian symmetric space.

Example 28.7. The set defined by the inequality $x_0 > \sqrt{x_1^2 + \cdots + x_n^2}$ in $\mathbb{R}^{n+1}$ is a self-dual cone, which we will denote $\mathcal{P}(n, 1)$. The group of automorphisms is the group of similitudes for the quadratic form $x_0^2 - x_1^2 - \cdots - x_n^2$. The derived group is $\mathrm{SO}(n, 1)$, and its homogeneous space $\mathcal{P}^\circ(n, 1)$ can be identified with the orbit of $(1, 0, \ldots, 0)$, which is the locus of the hyperboloid $x_0^2 - x_1^2 - \cdots - x_n^2 = 1$. The following special cases are worth noting: $\mathcal{P}(2, 1) \cong \mathcal{P}_2(\mathbb{R})$ can be identified with the Poincaré upper half-plane, $\mathcal{P}^\circ(3, 1)$ can be identified with $\mathcal{P}_2(\mathbb{C})$, and $\mathcal{P}^\circ(5, 1)$ can be identified with $\mathcal{P}_2(\mathcal{H})$.

Example 28.8. The *octonions* or *Cayley numbers* are a nonassociative algebra $\mathbb{O}$ over $\mathbb{R}$ of degree 8. The automorphism group of $\mathbb{O}$ is the exceptional group G_2. The construction of Example 28.6 applied to $D = \mathbb{O}$ does not produce a Jordan algebra if $n > 3$. If $n \leqslant 3$, then $\mathcal{J}_n(\mathbb{O})$ is a Jordan algebra containing a self-dual cone $\mathcal{P}_n(\mathbb{O})$. But $\mathcal{P}_2(\mathbb{O})$ is the same as $\mathcal{P}(9, 1)$. Only the 27-dimensional *exceptional* Jordan algebra $\mathcal{J}_3(\mathbb{O})$, discovered in 1947 by A. A. Albert, produces a new cone $\mathcal{P}_3(\mathbb{O})$. It contains an irreducible symmetric space of codimension 1, $\mathcal{P}_3^\circ(\mathbb{O})$, which is the locus of a cubic invariant. Let $\mathcal{E}_3^\circ(\mathbb{O})$ denote the compact dual. The Cartan classification of these 26-dimensional symmetric spaces is *EIV*.

The tube domain $\mathfrak{H}(\mathcal{C})$ over a self-dual cone $\mathcal{C}$, consisting of all $X + iY \in \mathbb{C} \otimes V$, is a Hermitian symmetric space. These examples are extremely similar to the case of the Siegel space. For example, we can embed $\mathfrak{H}(\mathcal{C})$ in its compact dual $\mathfrak{R}(\mathcal{C})$, which contains $\mathfrak{R}^\circ(\mathcal{C}) = \mathbb{C} \otimes V$ as a dense open set. A Cayley transform $c : \mathfrak{R}(\mathcal{C}) \longrightarrow \mathfrak{R}(\mathcal{C})$ takes $\mathfrak{H}(\mathcal{C})$ into a bounded symmetric domain $\mathfrak{D}(\mathcal{C})$, the closure of which is contained in $\mathfrak{R}^\circ(\mathcal{C})$. The Bergman–Shilov boundary can be identified with the compact dual of the (reducible) symmetric space $\mathcal{C}$, and its preimage under c consists of $X + iY \in \mathbb{C} \otimes V$ with $Y = 0$, that is, with the vector space V.

The nonassociative algebras $\mathbb{O}$ and $\mathcal{J}_3(\mathbb{O})$ are crucial in the construction of the exceptional groups and Lie algebras. See Adams [3], Jacobson [88], Onishchik and Vinberg [166] and Schafer [146] for constructions. Freudenthal [50] observed a phenomenon involving some symmetric spaces known as the *magic square*. Freudenthal envisioned a series of geometries over the algebras $\mathbb{R}, \mathbb{C}, \mathbb{H}$, and $\mathbb{O}$, and found a remarkable symmetry, which we will present momentarily. Tits [161] gave a uniform construction of the exceptional Lie algebras that sheds light on the magic square. See also Allison [6]. The paper of Baez [12] gives a useful survey of matters related to the exceptional groups, including the magic square and much more. A recent paper on the magic square, in the geometric spirit of Freudenthal's original approach, is Landsberg and Manivel [115]. And Tits' theory of buildings ([163], [1]) had its roots in his investigations of the geometry of the exceptional groups ([134]).

We will now take a look at the magic square Let us denote $\mathfrak{R}(\mathcal{C})$ as $\mathfrak{R}_n(D)$ if $\mathcal{C} = \mathcal{P}_n(D)$. We associate with this $\mathcal{C}$ three groups $G_n(D)$, $G'_n(D)$, and $G''_n(D)$ such that $G''_n(D) \supset G'_n(D) \supset G_n(D)$ and such that $G''(D)/G'_n(D) = \mathfrak{R}_n(D)$, while $G'_n(D)/G_n(D) = \mathcal{E}_n(D)$. Thus $G_n(\mathbb{R}) = \mathrm{SO}(n)$ and $G'_n(\mathbb{R}) = \mathrm{GL}(n, \mathbb{R})$, while $G''_n(\mathbb{R}) = \mathrm{Sp}(2n, \mathbb{R})$.

These groups are tabulated in Fig. 28.2 together with the noncompact duals that produce tube domains. Note that the symmetric spaces $\mathrm{U}(n) \times \mathrm{U}(n)/\mathrm{U}(n) = \mathrm{U}(n)$ and $\mathrm{GL}(2n, \mathbb{C})/\mathrm{U}(n) = \mathcal{P}_3(\mathbb{C})$ of the center column are of Types II and IV, respectively. The "magic" consists of the fact that the square is symmetric.

D	$\mathbb{R}$	$\mathbb{C}$	$\mathbb{H}$		$\mathbb{R}$	$\mathbb{C}$	$\mathbb{H}$
$G_n(D)$	$\mathrm{SO}(n)$	$\mathrm{U}(n)$	$\mathrm{Sp}(2n)$		–	–	–
$G'_n(D)$	$\mathrm{U}(n)$	$\mathrm{U}(n) \times \mathrm{U}(n)$	$\mathrm{U}(2n)$		$\mathrm{GL}(n, \mathbb{R})$	$\mathrm{GL}(n, \mathbb{C})$	$\mathrm{GL}(n, \mathbb{H})$
$G''_n(D)$	$\mathrm{Sp}(2n)$	$\mathrm{U}(2n)$	$\mathrm{SO}(4n)$		$\mathrm{Sp}(2n, \mathbb{R})$	$\mathrm{GU}(n, n)$	$\mathrm{SO}(4n)^*$

Fig. 28.2. The 3×3 square. *Left*: compact forms. *Right*: noncompact forms

We have the following numerology:

$$\dim G''_n(D) + 2 \dim G(D) = 3 \dim G'_n(D). \qquad (28.16)$$

Indeed, $\dim G''_n(D) - \dim G'_n(D)$ is the dimension of the tube domain, and this is twice the dimension $\dim G'(D) - \dim G_n(D)$ of the cone.

Although in presenting the 3×3 square—valid for all n—in Fig. 28.2 it seems best to take the full unitary groups in the second rows and columns, this does not work so well for the 4×4 magic square. Let us therefore note that we can also use modified groups that we call $H_n(D) \subset H'_n(D) \subset H''_n$, which are the derived groups of the $G_n(D)$. We must modify (28.16) accordingly:

$$\dim H''(D) + 2 \dim H(D) = 3 \dim H'_n(D) + 3. \qquad (28.17)$$

See Fig. 28.3 for the resulting "reduced" 3×3 magic square.

If $n = 3$, the reduced 3×3 square can be extended, resulting in Freudenthal's magic square, which we consider next. It will be noted that in Cartan's list (Table 28.1) some of the symmetric spaces have an $\mathrm{SU}(2)$ factor in K. Since

D	$\mathbb{R}$	$\mathbb{C}$	$\mathbb{H}$
$Hn(D)$	$SO(n)$	$SU(n)$	$Sp(2n)$
$H_n'(D)$	$SU(n)$	$SU(n) \times SU(n)$	$SU(2n)$
$H_n''(D)$	$Sp(2n)$	$SU(2n)$	$SO(4n)$

$\mathbb{R}$	$\mathbb{C}$	$\mathbb{H}$
$\frac{1}{2}n(n-1)$	n^2-1	$n(2n+1)$
n^2-1	$2n^2-2$	$4n^2-1$
$n(2n+1)$	$4n^2-1$	$2n(4n-1)$

Fig. 28.3. *Left*: the reduced 3×3 square. *Right*: dimensions

$SU(2)$ is the multiplicative group of quaternions of norm 1, these spaces have a quaternionic structure analogous to the complex structure shown by Hermitian symmetric spaces, where K contains a U(1) factor (Proposition 28.3). See Wolf [175]. Of the exceptional types, EII, EIV, EIX, FI, and G are quaternionic. Observe that in each case the dimension is a multiple of 4. Using some of these quaternionic symmetric spaces it is possible to extend the magic square in the special case where $n = 3$ by a fourth group $H_n'''(D)$ such that $H_n''(D) \times SU(2)$ is the maximal compact subgroup of the relevant noncompact form. It is also possible to add a fourth column when $n = 3$ due to existence of the exceptional Jordan algebra and $\mathcal{P}_3(\mathbb{O})$.

The magic square then looks like Fig. 28.4. In addition to (28.17), there is a similar relation,

$$\dim\ H'''(D) + 2\dim\ H'(D) = 3\dim H_n''(D) + 5, \qquad (28.18)$$

which suggests that the quaternionic symmetric spaces—they are FI, EII, EVI, and EIX in Cartan's classification—should be thought of as "quaternionic tube domains" over the corresponding Hermitian symmetric spaces.

Exercises

In the exercises, we look at the complex unit ball, which is a Hermitian symmetric space that is *not* a tube domain. For these spaces, Piatetski-Shapiro [133] gave unbounded realizations that are called *Siegel domains of Type II*. (Siegel domains of Type I are tube domains over self-dual cones.)

D	$\mathbb{R}$	$\mathbb{C}$	$\mathbb{H}$	$\mathbb{O}$
$H_3(D)$	$SO(3)$	$SU(3)$	$Sp(6)$	F_4
$H_3'(D)$	$SU(3)$	$SU(3) \times SU(3)$	$SU(6)$	E_6
$H_3''(D)$	$Sp(6)$	$SU(6)$	$SO(12)$	E_7
$H_3'''(D)$	F_4	E_6	E_7	E_8

$\mathbb{R}$	$\mathbb{C}$	$\mathbb{H}$	$\mathbb{O}$
3	8	21	52
8	16	35	78
21	35	66	133
52	78	133	248

Fig. 28.4. *Left*: the magic square. *Right*: dimensions

Exercise 28.1. The group $G = \mathrm{SU}(n, 1)$ consists of solutions to

$$t\overline{g}\begin{pmatrix} I_n & \\ & -1 \end{pmatrix} g = \begin{pmatrix} I_n & \\ & -1 \end{pmatrix}, \qquad g \in \mathrm{GL}(n+1, \mathbb{C}).$$

Let

$$\mathcal{B}_n = \left\{ w = \begin{pmatrix} w_1 \\ \vdots \\ w_n \end{pmatrix} \,\middle|\, |w_1|^2 + \cdots + |w_n|^2 < 1 \right\}$$

be the complex unit ball. Write

$$g = \begin{pmatrix} A & b \\ c & d \end{pmatrix}, \qquad A \in \mathrm{Mat}_n(\mathbb{C}), b \in \mathrm{Mat}_{n \times 1}(\mathbb{C}), c \in \mathrm{Mat}_{1,n}(\mathbb{C}), d \in \mathbb{C}.$$

If $w \in \mathcal{B}_n$, show that $cw + d$ is invertible. (This is a 1×1 matrix, so it can be regarded as a complex number.) Define

$$g(w) = (Aw + b)(cw + d)^{-1}. \tag{28.19}$$

Show that $g(w) \in \mathcal{B}_n$ and that this defines an action of $\mathrm{SU}(n, 1)$ on $\mathcal{B}_n$.

Exercise 28.2. Let $\mathcal{H}_n \in \mathbb{C}^n$ be the bounded domain

$$\mathcal{H}_n = \left\{ z = \begin{pmatrix} z_1 \\ \vdots \\ z_n \end{pmatrix} \,\middle|\, 2 \operatorname{Im}(z_1) > |z_2|^2 + \cdots + |z_n|^2 \right\}.$$

Show that there are holomorphic maps $c : \mathcal{H}_n \longrightarrow \mathcal{B}_n$ and $c^{-1} : \mathcal{B}_n \longrightarrow \mathcal{H}_n$ that are inverses of each other and are given by

$$c(z) = \begin{pmatrix} (z_1 - i)(z_1 + i)^{-1} \\ \sqrt{2i} z_2(z_1 + i)^{-1} \\ \vdots \\ \sqrt{2i} z_n(z_1 + i)^{-1} \end{pmatrix}, \qquad c^{-1}(w) = \begin{pmatrix} i(1 + w_1)(1 - w_1)^{-1} \\ \sqrt{2i} w_2(1 - w_1)^{-1} \\ \vdots \\ \sqrt{2i} w_n(1 - w_1)^{-1} \end{pmatrix}.$$

Note: If we extend the action (28.19) to allow $g \in \mathrm{GL}(n+1, \mathbb{C})$, these "Cayley transforms" are represented by the matrices

$$c = \begin{pmatrix} 1/\sqrt{2i} & & -i/\sqrt{2i} \\ & I_{n-1} & \\ 1/\sqrt{2i} & & i/\sqrt{2i} \end{pmatrix}, \qquad c^{-1} = \begin{pmatrix} i/\sqrt{2i} & & i/\sqrt{2i} \\ & I_{n-1} & \\ -1/\sqrt{2i} & & 1/\sqrt{2i} \end{pmatrix}.$$

Exercise 28.3. Show that $c^{-1}\mathrm{SU}(n, 1)c = \mathrm{SU}_\xi$, where SU_ξ is the group of all $g \in \mathrm{GL}(n, \mathbb{C})$ satisfying $g \xi \, {}^t\overline{g}^{-1} = \xi$, where

$$\xi = \begin{pmatrix} & & -i \\ & I_{n-1} & \\ i & & \end{pmatrix}.$$

Show that SU_ξ contains the noncompact "Heisenberg" unipotent subgroup

$$H = \left\{ \begin{pmatrix} 1 & i\bar{b} & \frac{i}{2}|b|^2 + ia \\ & I_{n-1} & b \\ & & 1 \end{pmatrix} \,\middle|\, b \in \mathrm{Mat}_{n,1}(\mathbb{C}), a \in \mathbb{R} \right\}.$$

Let us write

$$z = \begin{pmatrix} z_1 \\ z_2 \\ \vdots \\ z_n \end{pmatrix} = \begin{pmatrix} z_1 \\ \zeta \end{pmatrix}, \qquad \zeta = \begin{pmatrix} z_2 \\ \vdots \\ z_n \end{pmatrix}.$$

According to (28.19), a typical element of H should act by

$$z_1 \longmapsto z_1 + i\bar{b}\zeta + \frac{i}{2}|b|^2 + ia,$$
$$\zeta \longmapsto \zeta + b.$$

Check directly that H is invariant under such a transformation. Also show that SU_ξ contains the group

$$M = \left\{ \begin{pmatrix} u & & \\ & h & \\ & & \bar{u}^{-1} \end{pmatrix} \,\middle|\, u, v \in \mathbb{C}^\times, h \in \mathrm{U}(n-1) \right\}.$$

Describe the action of this group. Show that the subgroup of SU_ξ generated by H and M is transitive on $\mathcal{H}_n$, and deduce that the action of $SU(n,1)$ on $\mathcal{B}_n$ is also transitive.

Exercise 28.4. Observe that the subgroup $K = S(\mathrm{U}(n) \times \mathrm{U}(1))$ of $SU(n,1)$ acts transitively on the topological boundary of $\mathcal{B}_n$, and explain why this shows that the Bergman–Shilov boundary is the *whole* topological boundary. Contrast this with the case of $\mathfrak{D}_n$.

Exercise 28.5. Emulate the construction of $\mathfrak{R}_n$ and $\mathfrak{R}_n^\circ$ to show that the compact dual of $\mathcal{B}_n$ has a dense open subset that can be identified with $\mathbb{C}^n$ in such a way that $G_\mathbb{C} = \mathrm{GL}(n+1, \mathbb{C})$ acts by (28.19). Show that $\mathcal{B}_n$ can be embedded in its compact dual, just as $\mathfrak{D}_n$ is in the case of the symplectic group.

29

Relative Root Systems

In this chapter, we will consider root systems and Weyl groups associated with a Lie group G. We will assume that G satisfies the assumptions in Hypothesis 28.1 of the last chapter. Thus, G is semisimple and comes with a compact dual G_c. In Chap. 18, we associated with G_c a root system and Weyl group. That root system and Weyl group we will call the *absolute* root system Φ and Weyl group W. We will introduce another root system Φ_{rel}, called the *relative* or *restricted* root system, and a Weyl group W_{rel} called the *relative Weyl group*. The relation between the two root systems will be discussed. The structures that we will find give Iwasawa and Bruhat decompositions in this context. This chapter may be skipped with no loss of continuity.

As we saw in Theorem 28.3, *every* semisimple Lie group admits a Cartan decomposition, and Hypothesis 28.1 will be satisfied. The assumption of semisimplicity can be relaxed—it is sufficient for G to be *reductive*, though in this book we only define the term "reductive" when G is a complex analytic group. A more significant generalization of the results of this chapter is that relative and absolute root systems and Weyl groups can be defined whenever G is a reductive algebraic group defined over a field F. If F is algebraically closed, these coincide. If $F = \mathbb{R}$, they coincide with the structures defined in this chapter. But reductive groups over p-adic fields, number fields, or finite fields have many applications, and this reason alone is enough to prefer an approach based on algebraic groups. For this, see Borel [20] as well as Borel and Tits [21], Tits [162] (and other papers in the same volume), and Satake [144].

Consider, for example, the group $G = \mathrm{SL}(r, \mathbb{H})$, the construction of which we recall. The group $\mathrm{GL}(r, \mathbb{H})$ is the group of units of the central simple algebra $\mathrm{Mat}_r(\mathbb{H})$ over $\mathbb{R}$. We have $\mathbb{C} \otimes \mathbb{H} \cong \mathrm{Mat}_2(\mathbb{C})$ as $\mathbb{C}$-algebras. Consequently, $\mathbb{C} \otimes \mathrm{Mat}_r(\mathbb{H}) \cong \mathrm{Mat}_{2r}(\mathbb{C})$. The *reduced norm* $\nu : \mathrm{Mat}_r(\mathbb{H}) \longrightarrow \mathbb{R}$ is a map determined by the commutativity of the diagram

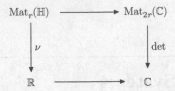

(See Exercise 29.1.) The restriction of the reduced norm to $\mathrm{GL}(r, \mathbb{H})$ is a homomorphism $\nu : \mathrm{GL}(r, \mathbb{H}) \longrightarrow \mathbb{R}^{\times}$ with a kernel that is the group $\mathrm{SL}(r, \mathbb{H})$. It is a real form of $\mathrm{SL}(2r, \mathbb{R})$, or of the compact group $G_c = \mathrm{SU}(2r)$, and we may associate with it the Weyl group and root system W and Φ of $\mathrm{SU}(2r)$ of type A_{2r-1}. This is the *absolute root system*. On the other hand, there is also a *relative* or *restricted* root system and Weyl group, which we now describe.

Let K be the group of $g \in \mathrm{SL}(r, \mathbb{H})$ such that $g^{t}\bar{g} = I$, where the bar denotes the conjugation map of $\mathbb{H}$. By Exercise 5.7, K is a compact group isomorphic to $\mathrm{Sp}(2r)$. The Cartan involution θ of Hypothesis 28.1 is the map $g \longmapsto {}^{t}\bar{g}^{-1}$.

We will denote by $\mathbb{R}_{+}^{\times}$ the multiplicative group of the positive real numbers. Let $A \cong (\mathbb{R}_{+}^{\times})^r$ be the subgroup

$$\begin{pmatrix} t_1 & & \\ & \ddots & \\ & & t_r \end{pmatrix}, \qquad t_i \in \mathbb{R}_{+}^{\times}, \qquad \prod t_i = 1.$$

The centralizer of A consists of the group

$$C_G(A) = \left\{ \begin{pmatrix} t_1 & & \\ & \ddots & \\ & & t_r \end{pmatrix} \middle| t_i \in \mathbb{H}^{\times} \right\}.$$

The group $M = C_G(A) \cap K$ consists of all elements with $|t_i| = 1$. The group of norm 1 elements in $\mathbb{H}^{\times}$ is isomorphic to $\mathrm{SU}(2)$ by Exercise 5.7 with $n = 1$. Thus M is isomorphic to $\mathrm{SU}(2)^r$.

On the other hand, the normalizer of $N_G(A)$ consists of all monomial quaternion matrices. The quotient $W_{\mathrm{rel}} = N_G(A)/C_G(A)$ is of type A_{r-1}. The "restricted roots" are "rational characters" of the group A, of the form $\alpha_{ij} = t_i t_j^{-1}$, with $i \neq j$. We can identify $\mathfrak{g} = \mathrm{Lie}(G)$ with $\mathrm{Mat}_n(\mathbb{H})$, in which case the subspace of $\mathfrak{g}$ that transforms by α_{ij} consists of all elements of $\mathfrak{g}$ having zeros everywhere except in the i, j position. In contrast with the absolute root system, where the eigenspace $\mathfrak{X}_\alpha$ of a root is always one-dimensional (see Proposition 18.6), these eigenspaces are all four-dimensional.

We see from this example that the group $\mathrm{SL}(n, \mathbb{H})$ looks like $\mathrm{SL}(n, \mathbb{R})$, but the root eigenspaces are "fattened up." The role of the torus T in Chap. 18 will be played by the group $C_G(A)$, which may be thought of as a "fattened up" and non-Abelian replacement for the torus.

We turn to the general case and to the proofs.

Proposition 29.1. *Assume that the assumptions of Hypothesis 28.1 are satisfied. Then the map*

$$(Z, k) \longmapsto \exp(Z)k \tag{29.1}$$

is a diffeomorphism $\mathfrak{p} \times K \longrightarrow G$.

Proof. Choosing a faithful representation (π, V) of the compact group G_c, we may embed G_c into $\mathrm{GL}(V)$. We may find a positive definite invariant inner product on V and, on choosing an orthonormal basis, we may embed G_c into $\mathrm{U}(n)$, where $n = \dim(V)$. The Lie algebra $\mathfrak{g}_{\mathbb{C}}$ is then embedded into $\mathfrak{gl}(n, \mathbb{C})$ in such a way that $\mathfrak{k} \subseteq \mathfrak{u}(n)$ and $\mathfrak{p}$ is contained in the space $\mathfrak{P}$ of $n \times n$ Hermitian matrices. We now recall from Theorem 13.4 and Proposition 13.7 that the formula (29.1) defines a diffeomorphism $\mathfrak{P} \times \mathrm{U}(n) \longrightarrow \mathrm{GL}(n, \mathbb{C})$. It follows that it gives a diffeomorphism of $\mathfrak{p} \times K$ onto its image. It also follows that (29.1) has nonzero differential everywhere, and as $\mathfrak{p} \times K$ and G have the same dimension, we get an open mapping $\mathfrak{p} \times K \longrightarrow G$. On the other hand, $\mathfrak{p} \times K$ is closed in $\mathfrak{P} \times \mathrm{U}(n)$, so the image of (29.1) is closed as well as open in G. Since G is connected, it follows that (29.1) is surjective. $\qquad\square$

If $\mathfrak{a}$ is an Abelian Lie subalgebra of $\mathfrak{g}$ such that $\mathfrak{a} \subset \mathfrak{p}$, we say $\mathfrak{a}$ is an *Abelian subspace* of $\mathfrak{p}$. This expression is used instead of "Abelian subalgebra" since $\mathfrak{p}$ itself is not a Lie subalgebra of $\mathfrak{g}$. We will see later in Theorem 29.3 that a maximal Abelian subspace $\mathfrak{a}$ of $\mathfrak{p}$ is unique up to conjugation.

Proposition 29.2. *Assume that the assumptions of Hypothesis 28.1 are satisfied. Let $\mathfrak{a}$ be a maximal Abelian subspace of $\mathfrak{p}$. Then $A = \exp(\mathfrak{a})$ is a closed Lie subgroup of G, and $\mathfrak{a}$ is its Lie algebra. There exists a θ-stable maximal torus T of G_c such that A is contained in the complexification $T_{\mathbb{C}}$ regarded as a subgroup of $G_{\mathbb{C}}$. If $r = \dim(\mathfrak{a})$, then $A \cong (\mathbb{R}_+^{\times})^r$. Moreover, $A_c = \exp(i\mathfrak{a})$ is a compact torus contained in T. We have $T = A_c T_M$, where $T_M = (T \cap K)^{\circ}$.*

Proof. By Proposition 15.2, A is an Abelian group. By Proposition 29.1, the restriction of $\exp$ to $\mathfrak{p}$ is a diffeomorphism onto its image, which is closed in G, and since $\mathfrak{a}$ is closed in $\mathfrak{p}$ it follows that $\exp(\mathfrak{a})$ is closed and isomorphic as a Lie group to the vector space $\mathfrak{a} \cong \mathbb{R}^r$. Exponentiating, the group $A \cong (\mathbb{R}_+^{\times})^r$.

Let $A_c = \exp(i\mathfrak{a}) \subset G_c$. By Proposition 15.2, it is an Abelian subgroup. We will show that it is closed. If it is not, consider its topological closure $\overline{A}_c$. This is a closed connected Abelian subgroup of the compact group G_c and hence a torus by Theorem 15.2. Since θ induces -1 on $\mathfrak{p}$, it induces the automorphism $g \longmapsto g^{-1}$ on A_c and hence on $\overline{A}_c$. Therefore, the Lie algebra of $\overline{A}_c$ is contained in the -1 eigenspace $i\mathfrak{p}$ of θ in $\mathrm{Lie}(G_c)$. Since $i\mathfrak{a}$ is a maximal Abelian subspace of $i\mathfrak{p}$, it follows that $i\mathfrak{a}$ is the Lie algebra of $\overline{A}_c$, and therefore $\overline{A}_c = \exp(i\mathfrak{a}) = A_c$.

Now let T be a maximal θ-stable torus of G_c containing A_c. We will show that T is a maximal torus of G_c. Let $T' \supseteq T$ be a maximal torus. Let $\mathfrak{t}'$ and $\mathfrak{t}$ be the respective Lie algebras of T' and T. Suppose that $H \in \mathfrak{t}'$. If $Y \in \mathfrak{t}$, then $[Y, {}^{\theta}H] = [{}^{\theta}Y, H] = -[Y, H] = 0$ since $\mathfrak{t}$ is θ-stable and $Y, H \in \mathfrak{t}'$, which is Abelian. Thus, both H and $^\theta H$ are in the centralizer of $\mathfrak{t}$. Now we can write $H = H_1 + H_2$, where $H_1 = \frac{1}{2}(H + {}^\theta H)$ and $H_2 = \frac{1}{2}(H - {}^\theta H)$. Note that the torus S_i, which is the closure of $\{\exp(tH_i) \,|\, t \in \mathbb{R}\}$, is θ stable – indeed θ is trivial on S_1 and induces the automorphism $x \longmapsto x^{-1}$ on S_2. Also $S_i \subseteq T'$ centralizes T. Consequently, TS_i is a θ-stable torus and, by maximality of T, $S_i \subseteq T$. It follows that $H_i \in \mathfrak{t}$, and so $H \in \mathfrak{t}$. We have proven that $\mathfrak{t}' = \mathfrak{t}$ and so $T = T'$ is a maximal torus.

It remains to be shown that $T = A_c T_M$. It is sufficient to show that the Lie algebra of T decomposes as $i\mathfrak{a} \oplus \mathfrak{t}_M$, where $\mathfrak{t}_M$ is the Lie algebra of T_M. Since θ stabilizes T, it induces an endomorphism of order 2 of $\mathfrak{t} = \mathrm{Lie}(T)$. The $+1$ eigenspace is $\mathfrak{t}_M = \mathfrak{t} \cap \mathfrak{k}$ since the $+1$ eigenspace of θ on $\mathfrak{g}_c$ is $\mathfrak{k}$. On the other hand, the -1 eigenspace of θ on $\mathfrak{t}$ contains $i\mathfrak{a}$ and is contained in $i\mathfrak{p}$, which is the -1 eigenspace of θ on $\mathfrak{g}_c$. Since $\mathfrak{a}$ is a maximal Abelian subspace of $\mathfrak{p}$, it follows that the -1 eigenspace of θ on $\mathfrak{t}$ is exactly $i\mathfrak{a}$, so $\mathfrak{t} = i\mathfrak{a} \oplus \mathfrak{t}_M$. □

Lemma 29.1. *Let $Z \in \mathrm{GL}(n, \mathbb{C})$ be a Hermitian matrix. If $g \in \mathrm{GL}(n, \mathbb{C})$ commutes with $\exp(Z)$, then g commutes with Z.*

Proof. Let $\lambda_1, \ldots, \lambda_h$ be the distinct eigenvalues of Z. Let us choose a basis with respect to which Z has the matrix

$$\begin{pmatrix} \lambda_1 I_{r_1} & & \\ & \ddots & \\ & & \lambda_h I_{r_h} \end{pmatrix}.$$

Then $\exp(Z)$ has the same form with λ_i replaced by $\exp(\lambda_i)$. Since the λ_i are distinct real numbers, the $\exp(\lambda_i)$ are also distinct, and it follows that g has the form

$$\begin{pmatrix} g_1 & & \\ & \ddots & \\ & & g_h \end{pmatrix},$$

where g_i is an $r_i \times r_i$ block. Thus g commutes with Z. □

Proposition 29.3. *In the context of Proposition 29.2, let $M = C_G(A) \cap K$. Then $C_G(A) = MA$ and $M \cap A = \{1\}$, so $C_G(A)$ is the direct product of M and A. The group T_M is a maximal torus of M.*

The compact group M is called the *anisotropic kernel*.

Proof. Since $M \subseteq K$ and $A \subseteq \exp(\mathfrak{p})$, and since by Proposition 29.1 $K \cap \exp(\mathfrak{p}) = \{1\}$, we have $M \cap A = \{1\}$. We will show that $C_G(A) = MA$. Let $g \in M$. By Proposition 29.1, we may write $g = \exp(Z)k$, where $Z \in \mathfrak{p}$ and $k \in K$. If $a \in A$, then a commutes with $\exp(Z)k$. We will show that any $a \in A$ commutes with $\exp(Z)$ and with k individually. From this we will deduce that $\exp(Z) \in A$ and $k \in M$.

By Theorem 4.2, G_c has a faithful complex representation $G_c \longrightarrow GL(V)$. We extend this to a representation of $G_{\mathbb{C}}$ and $\mathfrak{g}_{\mathbb{C}}$. Giving V a G_c-invariant inner product and choosing an orthonormal basis, G_c is realized as a group of unitary matrices. Therefore $\mathfrak{g}_c$ is realized as a Lie algebra of skew-Hermitian matrices, and since $i\mathfrak{p} \subseteq \mathfrak{g}_c$, the vector space $\mathfrak{p}$ consists of Hermitian matrices.

We note that $\theta(Z) = -Z$, $\theta(a) = a^{-1}$, and $\theta(k) = k$. Thus if we apply the automorphism θ to the identity $a\exp(Z)k = \exp(Z)ka$, we get $a^{-1}\exp(-Z)k = \exp(-Z)ka^{-1}$. Since this is true for all $a \in A$, both $\exp(-Z)k$ and $\exp(Z)k$ are in $C_G(A)$. It follows that $\exp(2Z) = \big(\exp(Z)k\big)\big(\exp(-Z)k\big)^{-1}$ is in $C_G(A)$. Since $\exp(2Z)$ commutes with A, by Lemma 29.1, Z commutes with the elements of A (in our matrix realization) and hence $\mathrm{ad}(Z)\mathfrak{a} = 0$. Because $\mathfrak{a}$ is a maximal Abelian subspace of $\mathfrak{p}$, it follows that $Z \in \mathfrak{a}$. Also, k centralizes A since $\exp(Z)k$ and $\exp(Z)$ both do, and so $\exp(Z) \in A$ and $k \in M$.

It is clear that $T_M = (T \cap K)^\circ$ is contained in $C_G(A)$ and K, so T_M is a torus in M. Let T'_M be a maximal torus of M containing T_M. Then $A_c T'_M$ is a connected Abelian subgroup of $C_G(A)$ containing $T = A_c T_M$, and since T is a maximal torus of G_c we have $A_c T'_M = T$. Therefore, $T'_M \subset T$. It is also contained in K and connected. This proves that $T_M = T'_M$ is a maximal torus of M. $\qquad\square$

We say that a quasicharacter of $A \cong (\mathbb{R}_+^\times)^r$ is a *rational character* if it can be extended to a complex analytic character of $A_{\mathbb{C}} = \exp(\mathfrak{a}_{\mathbb{C}})$. We will denote by $X^*(A)$ the group of rational characters of A. We recall from Chap. 15 that $X^*(A_c)$ is the group of all characters of the compact torus A_c.

Proposition 29.4. *Every rational character of A has the form*

$$(t_1,\ldots,t_r) \longmapsto t_1^{k_1} \cdots t_r^{k_r}, \qquad k_i \in \mathbb{Z}. \tag{29.2}$$

The groups $X^*(A)$ and $X^*(A_c)$ are isomorphic: extending a rational character of A to a complex analytic character of $A_{\mathbb{C}}$ and then restricting it to A_c gives every character of A_c exactly once.

Proof. Obviously (29.2) is a rational character. Extending any rational character of A to an analytic character of $A_{\mathbb{C}}$ and then restricting it to A_c gives a homomorphism $X^*(A) \longrightarrow X^*(A_c)$, and since the characters of $X^*(A_c)$ are classified by Proposition 15.4, we see that every rational character has the form (29.2) and that the homomorphism $X^*(A) \longrightarrow X^*(A_c)$ is an isomorphism. $\qquad\square$

Since the compact tori T and A_c satisfy $T \supset A_c$, we may restrict characters of T to A_c. Some characters may restrict trivially. In any case, if $\alpha \in X^*(T)$ restricts to $\beta \in X^*(A) = X^*(A_c)$, we write $\alpha|\beta$. Assuming that α and hence β are not the trivial character, as in Chap. 18 we will denote by $\mathfrak{X}_\beta$ the β-eigenspace of T on $\mathfrak{g}_{\mathbb{C}}$. We will also denote by $\mathfrak{X}_\alpha^{\mathrm{rel}}$ the α-eigenspace of A_c on $\mathfrak{g}_{\mathbb{C}}$. Since $X^*(A_c) = X^*(A)$, we may write

$$\mathfrak{X}_{\alpha}^{\mathrm{rel}} = \{X \in \mathfrak{g}_{\mathbb{C}} \mid \mathrm{Ad}(a)X = \alpha(a)X \text{ for all } a \in A\}.$$

We will see by examples that $\mathfrak{X}_{\alpha}^{\mathrm{rel}}$ may be more than one-dimensional. However, $\mathfrak{X}_{\beta}$ is one-dimensional by Proposition 18.6, and we may obviously write

$$\mathfrak{X}_{\alpha}^{\mathrm{rel}} = \bigoplus_{\substack{\beta \in X^*(T) \\ \beta \mid \alpha}} \mathfrak{X}_{\beta}.$$

Let Φ be the set of $\beta \in X^*(T)$ such that $\mathfrak{X}_{\beta} \neq 0$, and let Φ_{rel} be the set of $\alpha \in X^*(A)$ such that $\mathfrak{X}_{\alpha}^{\mathrm{rel}} \neq 0$.

If $\beta \in X^*(T)$, let $d\beta : \mathfrak{t} \longrightarrow \mathbb{C}$ be the differential of β. Thus

$$d\beta(H) = \frac{d}{dt}\beta(e^{tH})\Big|_{t=0}, \qquad H \in \mathfrak{t}.$$

As in Chap. 18, the linear form $d\beta$ is pure imaginary on the Lie algebra $\mathfrak{t}_M \oplus i\mathfrak{a}$ of the compact torus T. This means that $d\beta$ is real on $\mathfrak{a}$ and purely imaginary on $\mathfrak{t}_M$.

If $\alpha \in \Phi_{\mathrm{rel}}$, the α-eigenspace $\mathfrak{X}_{\alpha}^{\mathrm{rel}}$ can be characterized by either the condition (for $X \in \mathfrak{X}_{\alpha}^{\mathrm{rel}}$)

$$\mathrm{Ad}(a)X = \alpha(a)X, \qquad a \in A,$$

or

$$[H, X] = d\alpha(H)\,X, \qquad H \in \mathfrak{a}. \tag{29.3}$$

Let $c : \mathfrak{g}_{\mathbb{C}} \longrightarrow \mathfrak{g}_{\mathbb{C}}$ denote the conjugation with respect to $\mathfrak{g}$. Thus, if $Z \in \mathfrak{g}_{\mathbb{C}}$ is written as $X + iY$, where $X, Y \in \mathfrak{g}$, then $c(Z) = X - iY$ so $\mathfrak{g} = \{Z \in \mathfrak{g}_{\mathbb{C}} \mid c(Z) = Z\}$. Let $\mathfrak{m}$ be the Lie algebra of M. Thus, the Lie algebra of $C_G(A) = MA$ is $\mathfrak{m} \oplus \mathfrak{a}$. It is the 0-eigenspace of A on $\mathfrak{g}$, so

$$\mathfrak{g}_{\mathbb{C}} = \mathbb{C}(\mathfrak{m} \oplus \mathfrak{a}) \oplus \bigoplus_{\alpha \in \Phi_{\mathrm{rel}}} \mathfrak{X}_{\alpha} \tag{29.4}$$

is the decomposition into eigenspaces.

Proposition 29.5. *(i) In the context of Proposition 29.2, if $\alpha \in \Phi_{\mathrm{rel}}$, then $\mathfrak{X}_{\alpha}^{\mathrm{rel}} \cap \mathfrak{g}$ spans $\mathfrak{X}_{\alpha}^{\mathrm{rel}}$.*
(ii) If $0 \neq X \in \mathfrak{X}_{\alpha}^{\mathrm{rel}} \cap \mathfrak{g}$, then $\theta(X) \in \mathfrak{X}_{-\alpha}^{\mathrm{rel}} \cap \mathfrak{g}$ and $[X, \theta(X)] \neq 0$.
(iii) The space $\mathfrak{X}_{\alpha}^{\mathrm{rel}} \cap \mathfrak{g}$ is invariant under $\mathrm{Ad}(MA)$.
(iv) If $\alpha, \alpha' \in \Phi_{\mathrm{rel}}$, and if $X_{\alpha} \in \mathfrak{X}_{\alpha}^{\mathrm{rel}}$, $X_{\alpha'} \in \mathfrak{X}_{\alpha'}^{\mathrm{rel}}$, then

$$[X_{\alpha}, X_{\alpha'}] \in \begin{cases} \mathbb{C}(\mathfrak{m} \oplus \mathfrak{a}) & \text{if } \alpha' = -\alpha, \\ \mathfrak{X}_{\alpha+\alpha'} & \text{if } \alpha + \alpha' \in \Phi_{\mathrm{rel}}, \end{cases}$$

while $[X_{\alpha}, X_{\alpha'}] = 0$ if $\alpha' \neq -\alpha$ and $\alpha + \alpha' \notin \Phi$.

This is in contrast with the situation in Chap. 18, where the spaces $\mathfrak{X}_{\alpha}$ did *not* intersect the Lie algebra of the compact Lie group.

Proof. We show that we may find a basis $X_1, \ldots, X_h$ of the complex vector space $\mathfrak{X}_\alpha^{\mathrm{rel}}$ such that $X_i \in \mathfrak{g}$. Suppose that $X_1, \ldots, X_h$ are a maximal linearly independent subset of $\mathfrak{X}_\alpha^{\mathrm{rel}}$ such that $X_i \in \mathfrak{g}$. If they do not span $\mathfrak{X}_\alpha^{\mathrm{rel}}$, let $0 \neq Z \in \mathfrak{X}_\alpha^{\mathrm{rel}}$ be found that is not in their span. Then $c(Z) \in \mathfrak{X}_\alpha^{\mathrm{rel}}$ since applying c to (29.3) gives the same condition, with Z replaced by $c(Z)$. Now

$$\tfrac{1}{2}(Z + c(Z)), \qquad \tfrac{1}{2i}(Z - c(Z)),$$

are in $\mathfrak{g}$, and at least one of them is not in the span of $X_1, \ldots, X_i$ since Z is not. We may add this to the linearly independent set $X_1, \ldots, X_h$, contradicting the assumed maximality. This proves (i).

For (ii), let us show that θ maps $\mathfrak{X}_\alpha^{\mathrm{rel}}$ to $\mathfrak{X}_{-\alpha}^{\mathrm{rel}}$. Indeed, if $X \in X_\alpha^{\mathrm{rel}}$, then for $a \in A$ we have $\mathrm{Ad}(a)X = \alpha(a)X_\alpha$. Since $\theta(a) = a^{-1}$, replacing a by its inverse and applying θ, it follows that $\mathrm{Ad}(a)\theta(X) = \alpha(a^{-1})\,\theta(X)$. Since the group law in $X^*(A)$ is written additively, $(-\alpha)(a) = \alpha(a^{-1})$. Therefore $\theta(X) \in \mathfrak{X}_{-\alpha}$.

Since θ and c commute, if $X \in \mathfrak{g}$, then $\theta(X) \in \mathfrak{g}$.

The last point we need to check for (ii) is that if $0 \neq X \in \mathfrak{X}_\alpha^{\mathrm{rel}} \cap \mathfrak{g}$, then $[X, \theta(X)] \neq 0$. Since $\mathrm{Ad} : G_c \longrightarrow \mathrm{GL}(\mathfrak{g}_c)$ is a real representation of a compact group, there exists a positive definite symmetric bilinear form B on $\mathfrak{g}_c$ that is G_c-invariant. We extend B to a symmetric $\mathbb{C}$-bilinear form $B : \mathfrak{g}_\mathbb{C} \times \mathfrak{g}_\mathbb{C} \longrightarrow \mathbb{C}$ by linearity. We note that $Z = X + \theta(X) \in \mathfrak{k}$ since $\theta(Z) = Z$ and $Z \in \mathfrak{q}$. In particular $Z \in \mathfrak{g}_c$. It cannot vanish since X and $\theta(X)$ lie in $\mathfrak{X}_\alpha$ and $\mathfrak{X}_{-\alpha}$, which have a trivial intersection. Therefore, $B(Z, Z) \neq 0$. Choose $H \in \mathfrak{a}$ such that $d\alpha(H) \neq 0$. We have

$$B\big(X + \theta(X), [H, X - \theta(X)]\big) = B\big(Z, d\alpha(H)Z\big) \neq 0.$$

On the other hand, by (10.3) this equals

$$-B\big([X + \theta(X), X - \theta(X)], H\big) = 2B\big([X, \theta(X)], H\big).$$

Therefore, $[X, \theta(X)] \neq 0$.

For (ii), we will prove that $\mathfrak{X}_\alpha^{\mathrm{rel}}$ is invariant under $C_G(A)$, which contains M. Since $\mathfrak{g}$ is obviously an Ad-invariant real subspace of $\mathfrak{g}_\mathbb{C}$ it will follow that $\mathfrak{X}_\alpha^{\mathrm{rel}} \cap \mathfrak{g}$ is $\mathrm{Ad}(M)$-invariant. Since $C_G(A)$ is connected by Theorem 16.6, it is sufficient to show that $\mathfrak{X}_\alpha^{\mathrm{rel}}$ is invariant under $\mathrm{ad}(Z)$ when Z is in the Lie algebra centralizer of $\mathfrak{a}$. Thus, if $H \in \mathfrak{a}$ we have $[H, Z] = 0$. Now if $X \in \mathfrak{X}_\alpha^{\mathrm{rel}}$ we have

$$[H, [Z, X]] = [[H, Z], X] + [Z, [H, X]] = [Z, d\alpha(H)X] = d\alpha(H)[Z, X].$$

Therefore, $\mathrm{Ad}(Z)X = [Z, X] \in \mathfrak{X}_\alpha^{\mathrm{rel}}$.

Part (iv) is entirely similar to Proposition 18.4 (ii), and we leave it to the reader. □

The roots in Φ can now be divided into two classes. First, there are those that restrict nontrivially to A and hence correspond to roots in Φ_{rel}. On the other hand, some roots do restrict trivially, and we will show that these correspond to roots of the compact Lie group M. Let $\mathfrak{m} = \mathrm{Lie}(M)$.

Proposition 29.6. *Suppose that $\beta \in \Phi$. If the restriction of β to A is trivial, then $\mathfrak{X}_\beta$ is contained in the complexification of $\mathfrak{m}$ and β is a root of the compact group M with respect to T_M.*

Proof. We show that $\mathfrak{X}_\beta$ is θ-stable. Let $X \in \mathfrak{X}_\beta$. Then

$$[H, X] = d\beta(H)X, \qquad H \in \mathfrak{t}. \tag{29.5}$$

We must show that $\theta(X)$ has the same property. Applying θ to (29.5) gives

$$[\theta(H), \theta(X)] = d\beta(H)\,\theta(X), \qquad H \in \mathfrak{t}.$$

If $H \in \mathfrak{t}_M$, then $\theta(H) = H$ and we have (29.5) with $\theta(X)$ replacing X. On the other hand, if $H \in i\mathfrak{a}$ we have $\theta(H) = -H$, but by assumption $d\beta(H) = 0$, so we have (29.5) with $\theta(X)$ replacing X in this case, too. Since $\mathfrak{t} = \mathfrak{t}_M \oplus i\mathfrak{a}$, we have proved that $\mathfrak{X}_\beta$ is θ-stable.

If $a \in A$ and $X \in \mathfrak{X}_\beta$, then $\mathrm{Ad}(a)X$ is trivial, so a commutes with the one-parameter subgroup $t \longmapsto \exp(tX)$, and therefore $\exp(tX)$ is contained in the centralizer of A in $G_{\mathbb{C}}$. This means that $\exp(tX)$ is contained in the complexification of the Lie algebra of $C_G(A)$, which by Proposition 29.3 is $\mathbb{C}(\mathfrak{m} \oplus \mathfrak{a})$. Since θ is $+1$ on $\mathfrak{m}$ and -1 on $\mathfrak{a}$, and since we have proved that $\mathfrak{X}_\beta$ is θ-stable, we have $X \in \mathbb{C}\mathfrak{m}$. $\qquad\square$

Now let $\mathcal{V} = \mathbb{R} \otimes X^*(T)$, $\mathcal{V}_M = \mathbb{R} \otimes X^*(T_M)$, and $\mathcal{V}_{\mathrm{rel}} = \mathbb{R} \otimes X^*(A) = \mathbb{R} \otimes X^*(A_c)$. Since $T = T_M A_c$ by Proposition 29.2, we have $\mathcal{V} = \mathcal{V}_M \oplus \mathcal{V}_{\mathrm{rel}}$. In particular, we have a short exact sequence

$$0 \longrightarrow \mathcal{V}_M \longrightarrow \mathcal{V} \longrightarrow \mathcal{V}_{\mathrm{rel}} \longrightarrow 0. \tag{29.6}$$

Let Φ_M be the root system of M with respect to T_M. The content of Proposition 29.6 is that the roots of G_c with respect to T that restrict trivially to A are roots of M with respect to T_M.

We choose on $\mathcal{V}$ an inner product that is invariant under the absolute Weyl group $N_{G_c}(T)/T$. This induces an inner product on $\mathcal{V}_{\mathrm{rel}}$ and, if α is a root, there is a reflection $s_\alpha : \mathcal{V}_{\mathrm{rel}} \longrightarrow \mathcal{V}_{\mathrm{rel}}$ given by (18.1).

Proposition 29.7. *In the context of Proposition 29.2, let $\alpha \in \Phi_{\mathrm{rel}}$. Let $A_\alpha \subset A$ be the kernel of α, let $G_\alpha \subset G$ be its centralizer, and let $\mathfrak{g}_\alpha \subset \mathfrak{g}$ be the Lie algebra of G_α. There exist $X_\alpha \in \mathfrak{X}_\alpha \cap \mathfrak{g}$ such that if $X_{-\alpha} = -\theta(X_\alpha)$ and $H_\alpha = [X_\alpha, X_{-\alpha}]$, then $d\alpha(H_\alpha) = 2$. We have*

$$[H_\alpha, X_\alpha] = 2X_\alpha, \qquad [H_\alpha, X_{-\alpha}] = -2X_{-\alpha}. \tag{29.7}$$

There exists a Lie group homomorphism $i_\alpha : \mathrm{SL}(2, \mathbb{R}) \longrightarrow G_\alpha$ such that the differential $di_\alpha : \mathfrak{sl}(2, \mathbb{R}) \longrightarrow \mathfrak{g}_\alpha$ maps

$$\begin{pmatrix} 1 & \\ & -1 \end{pmatrix} \longmapsto H_\alpha, \qquad \begin{pmatrix} 0 & 1 \\ 0 & 0 \end{pmatrix} \longmapsto X_\alpha, \qquad \begin{pmatrix} 0 & 0 \\ 1 & 0 \end{pmatrix} \longmapsto X_{-\alpha}. \tag{29.8}$$

The Lie group homomorphism i_α extends to a complex analytic homomorphism $\mathrm{SL}(2, \mathbb{C}) \longrightarrow G_{\mathbb{C}}$.

Proof. Choose $0 \neq X_\alpha \in \mathfrak{X}_\alpha$. By Proposition 29.5, we may choose $X_\alpha \in \mathfrak{g}$, and denoting $X_{-\alpha} = -\theta(X_a)$ we have $X_{-\alpha} \in \mathfrak{X}_{-\alpha} \cap \mathfrak{g}$ and $H_\alpha = [X_\alpha, X_{-\alpha}] \neq 0$. We claim that $H_\alpha \in \mathfrak{a}$. Observe that $H_\alpha \in \mathfrak{g}$ since X_α and $X_{-\alpha}$ are in $\mathfrak{g}$, and applying θ to H_α gives $[X_{-\alpha}, X_\alpha] = -H_\alpha$. Therefore, $H_\alpha \in \mathfrak{p}$. Now if $H \in \mathfrak{a}$ we have

$$[H, H_\alpha] = [[H, X_\alpha], X_{-\alpha}] + [X_\alpha, [H, X_{-\alpha}]] =$$
$$[\mathrm{d}\alpha(H)X_\alpha, X_{-\alpha}] + [X_\alpha, -\mathrm{d}\alpha(H)X_{-\alpha}] = 0.$$

Since $\mathfrak{a}$ is a maximal Abelian subspace of $\mathfrak{p}$, this means that $H_\alpha \in \mathfrak{a}$.

Now $iH_\alpha \in i\mathfrak{p}$, $Z = X_\alpha - X_{-\alpha} \in \mathfrak{k}$, and $Y = i(X_\alpha + X_{-\alpha}) \in i\mathfrak{p}$ are all elements of $\mathfrak{g}_c = \mathfrak{k} \oplus i\mathfrak{p}$. We have

$$[iH_\alpha, Z] = \mathrm{d}\alpha(H_\alpha)Y, \qquad [iH_\alpha, Y] = -\mathrm{d}\alpha(H_\alpha)Z,$$

and

$$[Y, Z] = 2iH_\alpha.$$

Now $\mathrm{d}\alpha(H_\alpha) \neq 0$. Indeed, if $\mathrm{d}\alpha(H_\alpha) = 0$, then $\mathrm{ad}(Z)^2 Y = 0$ while $\mathrm{ad}(Z)Y \neq 0$, contradicting Lemma 18.1, since $Z \in \mathfrak{k}$. After replacing X_α by a positive multiple, we may assume that $\mathrm{d}\alpha(H) = 2$.

Now at least we have a Lie algebra homomorphism $\mathfrak{sl}(2, \mathbb{R}) \longrightarrow \mathfrak{g}$ with the effect (29.8), and we have to show that it is the differential of a Lie group homomorphism $\mathrm{SL}(2, \mathbb{R}) \longrightarrow G$. We begin by constructing the corresponding map $\mathrm{SU}(2) \longrightarrow G_c$. Note that iH_α, Y, and Z are all elements of $\mathfrak{g}_c$, and so we have a homomorphism $\mathfrak{su}(2) \longrightarrow \mathfrak{k}$ that maps

$$\begin{pmatrix} i & \\ & -i \end{pmatrix} \longmapsto iH_\alpha, \qquad \begin{pmatrix} & i \\ i & \end{pmatrix} \longmapsto Y, \qquad \begin{pmatrix} & 1 \\ -1 & \end{pmatrix} \longmapsto Z.$$

By Theorem 14.2, there exists a homomorphism $\mathrm{SU}(2) \longrightarrow G_c$. Since $\mathrm{SL}(2, \mathbb{C})$ is the analytic complexification of $\mathrm{SU}(2)$, and $G_\mathbb{C}$ is the analytic complexification of G_c, this extends to a complex analytic homomorphism $\mathrm{SL}(2, \mathbb{C}) \longrightarrow G_\mathbb{C}$. The restriction to $\mathrm{SL}(2, \mathbb{R})$ is the sought-after embedding.

Lastly, we note that X_α and $X_{-\alpha}$ centralize A_α since $[H, X_{\pm\alpha}] = 0$ for H in the kernel $\mathfrak{a}_\alpha$ of $\mathrm{d}\alpha : \mathfrak{a} \longrightarrow \mathbb{R}$, which is the Lie algebra of A_α. Thus, the Lie algebra they generate is contained in $\mathfrak{g}_\alpha$, and its exponential is contained in G_α. $\qquad \square$

Theorem 29.1. *In the context of Proposition 29.7, the set Φ_rel of restricted roots is a root system. If $\alpha \in \Phi_\mathrm{rel}$, there exists $w_\alpha \in K$ that normalizes A and that induces on $X^*(A)$ the reflection s_α.*

Proof. Let

$$w_\alpha = i_\alpha \begin{pmatrix} & 1 \\ -1 & \end{pmatrix}.$$

We note $w_\alpha \in K$. Indeed, it is the exponential of

$$di_\alpha \left(\frac{\pi}{2} \begin{pmatrix} & 1 \\ -1 & \end{pmatrix} \right) = \frac{\pi}{2} (X_\alpha - X_{-\alpha}) \in \mathfrak{k}$$

since

$$\exp \left(t \begin{pmatrix} & 1 \\ -1 & \end{pmatrix} \right) = \begin{pmatrix} \cos(t) & \sin(t) \\ -\sin(t) & \cos(t) \end{pmatrix}.$$

Now w_α centralizes A_α by Proposition 29.7. Also

$$\mathrm{ad} \left(\frac{\pi}{2} \begin{pmatrix} & 1 \\ -1 & \end{pmatrix} \right) : \begin{pmatrix} 1 & \\ & -1 \end{pmatrix} \longmapsto - \begin{pmatrix} 1 & \\ & -1 \end{pmatrix}$$

in $\mathrm{SL}(2,\mathbb{R})$, and applying i_α gives $\mathrm{Ad}(w_\alpha)H_\alpha = -H_\alpha$. Since $\mathfrak{a}$ is spanned by the codimension 1 subspace $\mathfrak{a}_\alpha$ and the vector H_α, it follows that (in its action on $\mathcal{V}_{\mathrm{rel}}$) w_α has order 2 and eigenvalue -1 with multiplicity 1. It therefore induces the reflection s_α in its action on $\mathcal{V}_{\mathrm{rel}}$.

Now the proof that Φ_{rel} is a root system follows the structure of the proof of Theorem 18.2. The existence of the simple reflection w_α in the Weyl group implies that s_α preserves the set Φ.

For the proof that if α and β are in Φ then $2\langle \alpha, \beta \rangle / \langle \alpha, \alpha \rangle \in \mathbb{Z}$, we adapt the proof of Proposition 18.10. If $\lambda \in X^*(A_c)$, we will denote (in this proof only) by $\mathfrak{X}_\lambda$ the λ-eigenspace of A_c in the adjoint representation. We normally use this notation only if $\lambda \neq 0$ is a root. If $\lambda = 0$, then $\mathfrak{X}_\lambda$ is the complexified Lie algebra of $C_G(A)$; that is, $\mathbb{C}(\mathfrak{m} \oplus \mathfrak{a})$. Let

$$W = \bigoplus_{k \in \mathbb{Z}} \mathfrak{X}_{\beta + k\alpha} \subseteq \mathfrak{X}_\mathbb{C}.$$

We claim that W is invariant under $i_\alpha(\mathrm{SL}(2,\mathbb{C}))$. To prove this, it is sufficient to show that it is invariant under $di_\alpha(\mathfrak{sl}(2,\mathbb{C}))$, which is generated by X_α and $X_{-\alpha}$, since these are the images under i_α of a pair of generators of $\mathfrak{sl}(2,\mathbb{C})$ by (29.8). These are the images of di_α and i_α, respectively. From (29.7), we see that $\mathrm{ad}(X_\alpha)\mathfrak{X}_\gamma \in \mathfrak{X}_{\gamma+2\alpha}$ and $\mathrm{ad}(X_{-\alpha})\mathfrak{X}_\gamma \in \mathfrak{X}_{\gamma-2\alpha}$, proving that $i_\alpha(\mathrm{SL}(2,\mathbb{C}))$ is invariant. In particular, W is invariant under $w_\alpha \in \mathrm{SL}(2,\mathbb{C})$. Since $\mathrm{ad}(w_\alpha)$ induces s_α on $\mathcal{V}_{\mathrm{rel}}$, it follows that the set $\{\beta + k\alpha | k \in \mathbb{Z}\}$ is invariant under s_α and, by (18.1), this implies that $2\langle \alpha, \beta \rangle / \langle \alpha, \alpha \rangle \in \mathbb{Z}$. $\square$

The group $W_{\mathrm{rel}} = N_G(A)/C_G(A)$ is the *relative Weyl group*. In Theorem 29.1 we constructed simple reflections showing that W_{rel} contains the abstract Weyl group associated with the root system Φ_{rel}. An analog of Theorem 18.3 is true – W_{rel} is generated by the reflections and hence coincides with the abstract Weyl group. We note that by Theorem 29.1 the generators of W_{rel} can be taken in K, so we may write $W_{\mathrm{rel}} = N_K(A)/C_K(A)$.

Although we have proved that Φ_{rel} is a root system, we have *not* proved that it is reduced. In fact, it may not be—we will give examples where the type of Φ_{rel} is BC_q and is *not* reduced! In Chap. 20, except for Proposition 20.18, it was assumed that the root system was reduced. Proposition 20.18 contains all we need about nonreduced root systems.

The relationship between the three root systems Φ, Φ_M, and Φ_{rel} can be expressed in a "short exact sequence of root systems,"

$$0 \longrightarrow \Phi_M \longrightarrow \Phi \longrightarrow \Phi_{\text{rel}} \longrightarrow 0, \qquad (29.9)$$

embedded in the short exact sequence (29.6) of Euclidean spaces. Of course, this is intended symbolically rather than literally. What we mean by this "short exact sequence" is that, in accord with Proposition 29.6, each root of M can be extended to a unique root of G_c; that the roots in Φ that are not thus extended from M are precisely those that restrict to a nonzero root in Φ_{rel}; and that every root in Φ_{rel} is a restricted root.

Proposition 29.8. *If $\alpha \in \Phi_{\text{rel}}^+$ is a simple positive root, then there exists a $\beta \in \Phi^+$ such that β is a simple positive root and $\beta | \alpha$. Moreover, if $\beta \in \Phi^+$ is a simple positive root with a restriction to A that is nonzero, then its restriction is a simple root of Φ_{rel}^+.*

Proof. Find a root $\gamma \in \Phi$ whose restriction to A is α. Since we have chosen the root systems compatibly, γ is a positive root. We write it as a sum of positive roots: $\gamma = \sum \beta_i$. Each of these restricts either trivially or to a relative root in Φ_{rel}^+, and we can write α as the sum of the nonzero restrictions of β_i, which are positive roots. Because α is simple, exactly one restricted β_i can be nonzero, and taking β to be this β_i, we have $\beta | \alpha$.

The last statement is clear. □

We see that the restriction map induces a surjective mapping from the set of simple roots in Φ that have nonzero restrictions to the simple roots in Φ_{rel}. The last question that needs to be answered is when two simple roots of Φ can have the same nonzero restriction to Φ_{rel}.

Proposition 29.9. *Let $\beta \in \Phi^+$. Then $-\theta(\beta) \in \Phi^+$. The roots β and $-\theta(\beta)$ have the same restriction to A. If β is a simple positive root, then so is $-\theta(\beta)$, and if α is a simple root of Φ_{rel} and β, β' are simple roots of Φ_{rel} both restricting to α, then either $\beta' = \beta$ or $\beta' = -\theta(\beta)$.*

Proof. The fact that β and $-\theta(\beta)$ have the same restriction follows from Proposition 29.5 (ii). It follows immediately that $-\theta(\beta)$ is a positive root in Φ. The map $\beta \longmapsto -\theta(\beta)$ permutes the positive roots, is additive, and therefore preserves the simple positive roots.

Suppose that α is a simple root of Φ_{rel} and β, β' are simple roots of Φ_{rel} both restricting to α. Since $\beta - \beta'$ has trivial restriction to A_c, it is θ-invariant. Rewrite $\beta - \beta' = \theta(\beta - \beta')$ as $\beta + \big(-\theta(\beta)\big) = \beta' + \big(\theta(-\beta')\big)$. This expresses the sum of two simple positive roots as the sum of another two simple positive roots. Since the simple positive roots are linearly independent by Proposition 20.18, it follows that either $\beta' = \beta$ or $\beta' = -\theta(\beta)$. □

The symmetry $\beta \longmapsto -\theta(\beta)$ of the Weyl group is reflected by a symmetry of the Dynkin diagram. It may be shown that if G_c is simply connected, this symmetry corresponds to an outer automorphism of $G_{\mathbb{C}}$. Only the Dynkin diagrams of types A_n, D_n, and E_6 admit nontrivial symmetries, so unless the absolute root system is one of these types, $\beta = -\theta(\beta)$.

The relationship between the three root systems in the "short exact sequence" (29.9) may be elucidated by the "Satake diagram," which we will now discuss. Tables of Satake diagrams may be found in Table VI on p. 532 of Helgason [66], p. 124 of Satake [144], or in Table 4 on p. 229 of Onishchik and Vinberg [166]. The diagrams in Tits [162] look a little different from the Satake diagram but contain the same information.

In addition to the Satake diagrams we will work out, a few different examples are explained in Goodman and Wallach [56].

Knapp [106] contains a different classification based on tori (Cartan subgroups) that (in contrast with our "maximally split" torus T), are maximally anisotropic, that is, are split as little as possible. Knapp also discusses the relationships between different tori by Cayley transforms. In this classification the Satake diagrams are replaced by "Vogan diagrams."

In the Satake diagram, one starts with the Dynkin diagram of Φ. We recall that the nodes of the Dynkin diagram correspond to simple roots of G_c. Those corresponding to roots that restrict trivially to A are colored dark. By Proposition 29.6, these correspond to the simple roots of the anisotropic kernel M, and indeed one may read the Dynkin diagram of M from the Satake diagram simply by taking the colored roots.

In addition to coloring some of the roots, the Satake diagram records the effect of the symmetry $\beta \longmapsto -\theta(\beta)$ of the Dynkin diagram. In the "exact sequence" (29.9), corresponding nodes are mapped to the same node in the Dynkin diagram of Φ_{rel}. We will discuss this point later, but for examples of diagrams with nontrivial symmetries see Figs. 29.3b and 29.5.

As a first example of a Satake diagram, consider $\mathrm{SL}(3, \mathbb{H})$. The Satake diagram is $\bullet\!-\!\circ\!-\!\bullet\!-\!\circ\!-\!\bullet$. The symmetry $\beta \longmapsto -\theta(\beta)$ is trivial. From this Satake diagram, we can read off the Dynkin diagram of $M \cong \mathrm{SU}(2) \times \mathrm{SU}(2) \times \mathrm{SU}(2)$ by erasing the uncolored dots to obtain the disconnected diagram $\bullet \quad \bullet \quad \bullet$ of type $A_1 \times A_1 \times A_1$.

On the other hand, in this example, the relative root system is of type A_2. We can visualize the "short exact sequence of root systems" as in Fig. 29.1, where we have indicated the destination of each simple root in the inclusion $\Phi_M \longrightarrow \Phi$ and the destinations of those simple roots in Φ that restrict nontrivially in the relative root system.

As a second example, let $F = \mathbb{R}$, and let us consider the group $G = \mathrm{SO}(n,1)$. In this example, we will see that G has real rank 1 and that the relative root system of G is of type A_1. Groups of real rank 1 are in many ways the simplest groups. Their symmetric spaces are direct generalizations of the Poincaré upper half-plane, and the symmetric space of $\mathrm{SO}(n,1)$ is of-

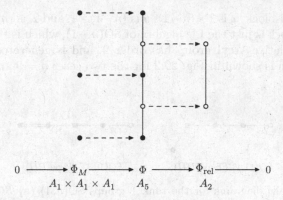

$$
\begin{array}{ccccccccc}
0 & \longrightarrow & \Phi_M & \longrightarrow & \Phi & \longrightarrow & \Phi_{\text{rel}} & \longrightarrow & 0 \\
& & A_1 \times A_1 \times A_1 & & A_5 & & A_2 & &
\end{array}
$$

Fig. 29.1. The "short exact sequence of root systems" for $\mathrm{SL}(3, \mathbb{H})$

ten referred to as *hyperbolic n-space*. (It is *n*-dimensional.) We have seen in Example 28.7 that this symmetric space can be realized as a hyperboloid.

We will see, consistent with our description of $\mathrm{SL}(n, \mathbb{H})$ as a "fattened up" version of $\mathrm{SL}(n, \mathbb{R})$, that $\mathrm{SO}(n, 1)$ can be seen as a "fattened up" version of $\mathrm{SO}(2, 1)$.

We originally defined $G = \mathrm{SO}(n, 1)$ to be the set of $g \in \mathrm{GL}(n+1, \mathbb{R})$ such that $g\, J\,{}^t g = J$, where $J = J_1$ and

$$
J_1 = \begin{pmatrix} I_n & \\ & -1 \end{pmatrix}.
$$

However, we could just as easily take $J = J_2$ and

$$
J_2 = \begin{pmatrix} & & 1 \\ & I_{n-1} & \\ 1 & & \end{pmatrix}
$$

since this symmetric matrix also has eigenvalues 1 with multiplicity n and -1 with multiplicity -1. Thus, if

$$
u = \begin{pmatrix} 1/\sqrt{2} & & -1/\sqrt{2} \\ & I_{n-1} & \\ 1/\sqrt{2} & & 1/\sqrt{2} \end{pmatrix},
$$

then $u \in \mathrm{O}(n+1)$ and $u\, J_1\,{}^t u = J_2$. It follows that if $g\, J_1\,{}^t g = J_1$, then $h = ugu^{-1}$ satisfies $h\, J_2\,{}^t h = J_2$. The two orthogonal groups are thus equivalent, and we will take $J = J_2$ in the definition of $\mathrm{O}(n, 1)$. Then we see that the Lie algebra of G is

$$
\left\{ \begin{pmatrix} a & x & 0 \\ y & T & -{}^t x \\ 0 & -{}^t y & -a \end{pmatrix} \,\middle|\, T = -{}^t T \right\}.
$$

Here a is a 1×1 block, x is $1 \times (n-1)$, y is $(n-1) \times 1$, and T is $(n-1) \times (n-1)$. The middle block is just the Lie algebra of $SO(n-1)$, which is the anisotropic kernel. The relative Weyl group has order 2, and is generated by J_2. The Satake diagram is shown in Fig. 29.2 for the two cases $n = 9$ and $n = 10$.

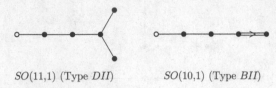

$SO(11,1)$ (Type DII) $SO(10,1)$ (Type BII)

Fig. 29.2. Satake diagrams for the rank 1 groups $SO(n,1)$ (a) $SO(11, 1)$ (Type DII) (b) $SO(10, 1)$ (Type BII)

A number of rank 1 groups, such as $SO(n,1)$ can be found in Cartan's list. Notably, among the exceptional groups, we find Type FII. Most of these can be thought of as "fattened up" versions of $SL(2, \mathbb{R})$ or $SO(2,1)$, as in the two cases above. Some rank 1 groups have relative root system of type BC_1.

At the other extreme, let us consider the groups $SO(n,n)$ and $SO(n + 1, n - 1)$. The group $SO(n,n)$ is *split*. This means that the anisotropic kernel is trivial and that the absolute and relative root systems Φ and Φ_{rel} coincide. We can take $G = \{g \in \mathrm{GL}(2n, \mathbb{R}) \mid g\, J\, {}^t g = J\}$, where

$$ J = \begin{pmatrix} & & 1 \\ & \cdot^{\cdot^{\cdot}} & \\ 1 & & \end{pmatrix}. $$

We leave the details of this case to the reader. The Satake diagram is shown in Fig. 29.3 when $n = 6$.

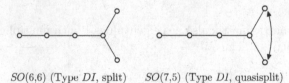

$SO(6,6)$ (Type DI, split) $SO(7,5)$ (Type DI, quasisplit)

Fig. 29.3. Split and quasisplit even orthogonal groups (a) $SO(6, 6)$ (Type DI, split) (b) SO (7, 5) (Type DI, quasisplit)

A more interesting case is $SO(n + 1, n - 1)$. This group is *quasisplit*. This means that the anisotropic kernel M is Abelian. Since M contains no roots, there are no colored roots in the Dynkin diagram of a quasisplit group. A split group is quasisplit, but not conversely, as this example shows. This group is not split since the relative root systems Φ and Φ_{rel} differ. We can take $G = \{g \in \mathrm{GL}(2n, \mathbb{R}) \mid gJ^t g = J\}$ where now

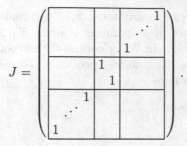

We can take A to be the group of matrices of the form

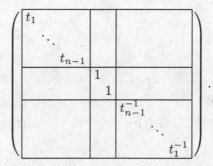

For $n = 5$, the Lie algebra of SO$(6, 4)$ is shown in Fig. 29.4. For $n = 6$, the Satake diagram of SO$(7, 5)$ is shown in Fig. 29.3.

$$\begin{pmatrix}
t_1 & \boxed{x_{12}} & x_{13} & x_{14} & x_{15} & x_{16} & x_{17} & x_{18} & x_{19} & 0 \\
x_{21} & t_2 & \boxed{x_{23}} & x_{24} & x_{25} & x_{26} & x_{27} & x_{28} & 0 & -x_{19} \\
x_{31} & x_{32} & t_3 & \boxed{x_{34}} & x_{35} & x_{36} & x_{37} & 0 & -x_{28} & -x_{18} \\
x_{41} & x_{42} & x_{43} & t_4 & \boxed{x_{45}} & \boxed{x_{46}} & 0 & -x_{37} & -x_{27} & -x_{17} \\
x_{51} & x_{52} & x_{53} & x_{54} & 0 & t_5 & -x_{45} & -x_{35} & -x_{25} & -x_{15} \\
x_{61} & x_{62} & x_{63} & x_{64} & -t_5 & 0 & -x_{46} & -x_{36} & -x_{26} & -x_{16} \\
x_{71} & x_{72} & x_{73} & 0 & -x_{54} & -x_{64} & -t_4 & -x_{34} & -x_{24} & -x_{14} \\
x_{81} & x_{82} & 0 & -x_{73} & -x_{53} & -x_{63} & -x_{43} & -t_3 & -x_{23} & -x_{13} \\
x_{91} & 0 & -x_{82} & -x_{72} & -x_{52} & -x_{62} & -x_{42} & -x_{32} & -t_2 & -x_{12} \\
0 & -x_{91} & -x_{81} & -x_{71} & -x_{51} & -x_{61} & -x_{41} & -x_{31} & -x_{21} & -t_1
\end{pmatrix}$$

Fig. 29.4. The Lie algebra of quasisplit SO$(6, 4)$

The circling of the x_{45} and x_{46} positions in Fig. 29.4 is slightly misleading because, as we will now explain, these do not correspond exactly to roots.

Indeed, each of the circled coordinates x_{12}, x_{23}, and x_{34} corresponds to a one-dimensional subspace of $\mathfrak{g}$ spanning a space $\mathfrak{X}_{\alpha_i}$, where $i = 1, 2, 3$ are the first three simple roots in Φ. In contrast, the root spaces $\mathfrak{X}_{\alpha_4}$ and $\mathfrak{X}_{\alpha_5}$ are *divided* between the x_{45} and x_{46} positions. To see this, the torus T in $G_c \subset G_{\mathbb{C}}$ consists of matrices

$$
t = \begin{pmatrix}
e^{it_1} & & & & & & & \\
& e^{it_2} & & & & & & \\
& & e^{it_3} & & & & & \\
& & & e^{it_4} & & & & \\
& & & & \cos(t_5) & \sin(t_5) & & \\
& & & & -\sin(t_5) & \cos(t_5) & & \\
& & & & & & e^{-it_4} & \\
& & & & & & & e^{-it_3} \\
& & & & & & & & e^{-it_2} \\
& & & & & & & & & e^{-it_1}
\end{pmatrix}
$$

with $t_i \in \mathbb{R}$. The simple roots are

$$
\alpha_1(t) = e^{i(t_1 - t_2)}, \quad \alpha_2(t) = e^{i(t_2 - t_3)}, \quad \alpha_3(t) = e^{i(t_3 - t_4)},
$$

and

$$
\alpha_4(t) = e^{i(t_4 - t_5)}, \quad \alpha_5(t) = e^{i(t_4 + t_5)}.
$$

The eigenspaces $\mathfrak{X}_{\alpha_4}$ and $\mathfrak{X}_{\alpha_5}$ are spanned by X_{α_4} and X_{α_5}, where

$$
X_{\alpha_4} = \begin{pmatrix}
0 & 0 & 0 & 0 & 0 & 0 & 0 & 0 \\
0 & 0 & 0 & 0 & 0 & 0 & 0 & 0 \\
0 & 0 & 0 & 1 & i & 0 & 0 & 0 \\
0 & 0 & 0 & 0 & 0 & -1 & 0 & 0 \\
0 & 0 & 0 & 0 & 0 & -i & 0 & 0 \\
0 & 0 & 0 & 0 & 0 & 0 & 0 & 0 \\
0 & 0 & 0 & 0 & 0 & 0 & 0 & 0 \\
0 & 0 & 0 & 0 & 0 & 0 & 0 & 0
\end{pmatrix},
$$

and its conjugate is X_{α_5}.

The involution θ is transpose-inverse. In its effect on the torus T, $\theta(t^{-1})$ does not change t_1, t_2, t_3, or t_4 but sends $t_5 \longmapsto -t_5$. Therefore, $-\theta$ interchanges the simple roots α_4 and α_5, as indicated in the Satake diagram in Figs. 29.3 and 29.4.

As a last example, we look next at the Lie group $\mathrm{SU}(p, q)$, where $p > q$. We will see that this has type BC_q. Recall from Chap. 19 that the root system of type BC_q can be realized as all elements of the form

$$
\pm \mathbf{e}_i \pm \mathbf{e}_j (i < j), \qquad \pm \mathbf{e}_i, \qquad \pm 2\mathbf{e}_i,
$$

where $\mathbf{e}_i$ are standard basis vectors of $\mathbb{R}^n$. See Fig. 19.5 for the case $q = 2$. We defined $\mathrm{U}(p, q)$ to be

$$
\{ g \in \mathrm{GL}(p + q, \mathbb{C}) \mid g \, J \, {}^t\bar{g} = J \},
$$

where $J = J_1$, but (as with the group $O(n, 1)$ discussed above) we could just as well take $J = J_2$, where now

$$J_1 = \begin{pmatrix} I_p & \\ & -I_q \end{pmatrix}, \qquad J_2 = \begin{pmatrix} & & I_q \\ & I_{p-q} & \\ I_q & & \end{pmatrix}.$$

This has the advantage of making the group A diagonal. We can take A to be the group of matrices of the form

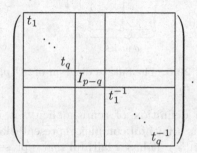

Now the Lie algebra of $SU(p, q)$ consists of

$$\left\{ \begin{pmatrix} a & x & b \\ y & u & -{}^t\bar{x} \\ c & -{}^t\bar{y} & -{}^t\bar{a} \end{pmatrix} \middle| b, c, u \text{ skew-Hermitian} \right\}.$$

Considering the action of the adjoint representation, the roots $t_i t_j^{-1}$ appear in a, the roots $t_i t_j$ and t_i^2 appear in b, the roots $t_i^{-1} t_j^{-1}$ and t_i^{-2} appear in c, the roots t_i appear in x, and the roots t_i^{-1} appear in y. Identifying $\mathbb{R} \otimes X^*(A) = \mathbb{R}^n$ in such a way that the rational character t_i corresponds to the standard basis vector $\mathbf{e}_i$, we see that Φ_{rel} is a root system of type BC_q. The Satake diagram is illustrated in Fig. 29.5.

We turn now to the Iwasawa decomposition for G admitting a Cartan decomposition as in Hypothesis 28.1. The construction is rather similar to what we have already done in Chap. 26.

Proposition 29.10. *Let G, G_c, K, $\mathfrak{g}$, and θ satisfy Hypothesis 28.1. Let M and A be as in Propositions 29.2 and 29.3. Let Φ and Φ_{rel} be the absolute and relative root systems, and let Φ^+ and Φ_{rel}^+ be the positive roots with respect to compatible orders. Let*

$$\mathfrak{n} = \bigoplus_{\alpha \in \Phi_{\mathrm{rel}}^+} (\mathfrak{X}_\alpha \cap \mathfrak{g}).$$

Then $\mathfrak{n}$ is a nilpotent Lie algebra. It is the Lie algebra of a closed subgroup N of G. The group N is normalized by M and by A. We may embed the complexification $G_\mathbb{C}$ of G into $GL(n, \mathbb{C})$ for some n in such a way that $G \subseteq GL(n, \mathbb{R})$, $G_c \subseteq U(n)$, $K \subseteq O(n)$, N is upper triangular, and $\theta(g) = {}^t g^{-1}$.

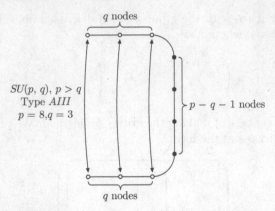

Fig. 29.5. The Satake diagram of $SU(p,q)$

Proof. As part of the definition of semisimplicity, it is assumed that the semisimple group G has a faithful complex representation. Since we may embed $GL(n,\mathbb{C})$ in $GL(2n,\mathbb{R})$, it has a faithful real representation. We may assume that $G \subseteq GL(V)$, where V is a real vector space. We may then assume that the complexification $G_\mathbb{C} \subseteq GL(V_\mathbb{C})$, where $V_\mathbb{C} = \mathbb{C} \otimes V$ is the complexified vector space.

The proof that $\mathfrak{n}$ is nilpotent is identical to Proposition 26.4 but uses Proposition 29.5 (iv) instead of Proposition 18.4 (ii). By Lie's Theorem 26.1, we can find an $\mathbb{R}$-basis $v_1, \ldots, v_n$ of V such that each $X \in \mathfrak{n}$ is upper triangular with respect to this basis. It is nilpotent as a matrix by Proposition 26.5.

Choose a G_c-invariant inner product on $V_\mathbb{C}$ (i.e., a positive definite Hermitian form $\langle \ , \ \rangle$). It induces an inner product on V; that is, its restriction to V is a positive definite $\mathbb{R}$-bilinear form. Now applying Gram–Schmidt orthogonalization to the basis $v_1, \ldots, v_n$, we may assume that they are orthonormal. This does not alter the fact that $\mathfrak{n}$ consists of upper triangular matrices. It follows by imitating the argument of Theorem 26.2 that $N = \exp(\mathfrak{n})$ is a Lie group with Lie algebra $\mathfrak{n}$. The group M normalizes N because its Lie algebra normalizes the Lie algebra of N by Proposition 18.4 (iii), so the Lie algebra of N is invariant under $\mathrm{Ad}(MA)$.

We have $G \subseteq GL(n,\mathbb{R})$ since G stabilizes V. It is also clear that $G_c \subseteq U(n)$ since v_i are an orthonormal basis and the inner product $\langle \ , \ \rangle$ was chosen to be G_c-invariant. Since $K \subseteq G \cap G_c$, we have $K \subseteq O(n)$.

It remains to be shown that $\theta(g) = {}^t g^{-1}$ for $g \in G$. Since G is assumed to be connected in Hypothesis 28.1, it is sufficient to show that $\theta(X) = -{}^t X$ for $X \in \mathfrak{g}$, and we may treat the cases $X \in \mathfrak{k}$ and $X \in \mathfrak{p}$ separately. If $X \in \mathfrak{k}$, then X is skew-symmetric since $K \subseteq O(n)$. Thus, $\theta(X) = X = -{}^t X$. On the other hand, if $X \in \mathfrak{p}$, then $iX \in \mathfrak{g}_c$, and iX is skew-Hermitian because $G_c \subseteq U(n)$. Thus, X is symmetric, and $\theta(X) = -X = -{}^t X$.   □

Since M normalizes N, we have a Lie subgroup $B = MAN$ of G. We may call it the (standard) $\mathbb{R}$-Borel subgroup of G. (If G is split or quasisplit, one may omit the "$\mathbb{R}$-" from this designation.) Let $B_0 = AN$.

Theorem 29.2. (Iwasawa decomposition) *With notations as above, each element of $g \in G$ can be factored uniquely as bk, where $b \in B_0$ and $k \in K$, or as $a\nu k$ where $a \in A$, $\nu \in N$, and $k \in K$. The multiplication map $A \times N \times K \longrightarrow G$ is a diffeomorphism.*

Proof. This is nearly identical to Theorem 26.3, and we mostly leave the proof to the reader. We consider only the key point that $\mathfrak{g} = \mathfrak{a} + \mathfrak{n} + \mathfrak{k}$. It is sufficient to show that $\mathfrak{g}_{\mathbb{C}} = \mathbb{C}\,\mathfrak{a} + \mathbb{C}\,\mathfrak{n} + \mathbb{C}\,\mathfrak{k}$. We have $\mathfrak{t}_{\mathbb{C}} \subseteq \mathbb{C}\,\mathfrak{a} + \mathbb{C}\,\mathfrak{m} \subseteq \mathbb{C}\,\mathfrak{a} + \mathbb{C}\,\mathfrak{k}$, so it is sufficient to show that $\mathbb{C}\,\mathfrak{n} + \mathbb{C}\,\mathfrak{k}$ contains $\mathfrak{X}_\beta$ for each $\beta \in \Phi$. If β restricts trivially to A, then $\mathfrak{X}_\beta \subseteq \mathbb{C}\,\mathfrak{m}$ by Proposition 29.6, so we may assume that β restricts nontrivially. Let α be the restriction of β. If $\beta \in \Phi^+$, then $\mathfrak{X}_\beta \subseteq \mathfrak{X}_\alpha \subset \mathbb{C}\,\mathfrak{n}$. On the other hand, if $\beta \in \Phi^-$ and $X \in \mathfrak{X}_\beta$, then $X + \theta(X) \in \mathbb{C}\,\mathfrak{k}$ and $\theta(X) \in \mathfrak{X}_{-\beta} \subseteq \mathfrak{X}_{-\alpha} \subset \mathbb{C}\,\mathfrak{n}$. In either case, $\mathfrak{X}_\beta \subset \mathbb{C}\,\mathfrak{k} + \mathbb{C}\,\mathfrak{n}$. $\qquad\square$

Our next goal is to show that the maximal Abelian subspace $\mathfrak{a}$ is unique up to conjugacy. First, we need an analog of Proposition 18.14 (ii). Let us say that $H \in \mathfrak{p}$ is *regular* if it is contained in a unique maximal Abelian subspace of $\mathfrak{p}$ and *singular* if it is not regular.

Proposition 29.11. *(i) If H is regular and $Z \in \mathfrak{p}$ satisfies $[H, Z] = 0$, then $Z \in \mathfrak{a}$.*

(ii) An element $H \in \mathfrak{a}$ is singular if and only if $d\alpha(H) = 0$ for some $\alpha \in \Phi_{\mathrm{rel}}$.

Proof. The element H is singular if and only if there is some $Z \in \mathfrak{p} - \mathfrak{a}$ such that $[Z, H] = 0$, for if this is the case, then H is contained in at least two distinct maximal Abelian subspaces, namely $\mathfrak{a}$ and any maximal Abelian subspace containing the Abelian subspace $\mathbb{R}Z + \mathbb{R}H$. Conversely, if no such Z exists, then any maximal Abelian subgroup containing H must obviously coincide with $\mathfrak{a}$.

Now (i) is clear.

We also use this criterion to prove (ii). Consider the decomposition of $Z \in \mathfrak{p}$ in the eigenspace decomposition (29.4):

$$Z = Z_0 + \sum_{\alpha \in \Phi_{\mathrm{rel}}} Z_\alpha, \qquad Z_0 \in \mathbb{C}(\mathfrak{m} \oplus \mathfrak{a}),\; Z_\alpha \in \mathfrak{X}_\alpha^{\mathrm{rel}}.$$

We have

$$0 = [H, Z] = [H, Z_0] + \sum_{\alpha \in \Phi_{\mathrm{rel}}} [H, Z_\alpha] = \sum_{\alpha \in \Phi_{\mathrm{rel}}} d\alpha(H) Z_\alpha.$$

Thus, for all $\alpha \in \Phi_{\mathrm{rel}}$, we have either $d\alpha(H) = 0$ or $Z_\alpha = 0$. So if $d\alpha(H) \neq 0$ for all H then all $Z_\alpha = 0$ and $Z = Z_0 \in \mathbb{C}(\mathfrak{m} \oplus \mathfrak{a})$. Since $Z \in \mathfrak{p}$, this implies that $Z \in \mathfrak{a}$, and so H is regular. On the other hand, if $d\alpha = 0$ for some α, then we can take $Z = Z_\alpha - \theta(Z_\alpha)$ for nonzero $Z_\alpha \in \mathfrak{X}_\alpha^{\mathrm{rel}} \cap \mathfrak{g}$ and $[Z, H] = 0$, $Z \in \mathfrak{p} - \mathfrak{a}$. $\qquad\square$

Theorem 29.3. *Let $\mathfrak{a}_1$ and $\mathfrak{a}_2$ be two maximal Abelian subspaces of $\mathfrak{p}$. Then there exists a $k \in \mathfrak{k}$ such that $\mathrm{Ad}(k)\mathfrak{a}_1 = \mathfrak{a}_2$.*

Thus, the relative root system does not dependent in any essential way on the choice of $\mathfrak{a}$. The argument is similar to the proof of Theorem 16.4.

Proof. By Proposition 29.11 (ii), $\mathfrak{a}_1$ and $\mathfrak{a}_2$ contain regular elements H_1 and H_2. We will show that $[\mathrm{Ad}(k)H_1, H_2] = 0$ for some $k \in \mathfrak{k}$. Choose an Ad-invariant inner product $\langle \, , \, \rangle$ on $\mathfrak{g}$, and choose $k \in K$ to maximize $\langle \mathrm{Ad}(k)H_1, H_2 \rangle$. If $Z \in \mathfrak{k}$, then since $\langle \mathrm{Ad}(e^{tZ})H_1, H_2 \rangle$ is maximal when $t = 0$, we have

$$0 = \frac{\mathrm{d}}{\mathrm{d}t} \left\langle \mathrm{Ad}(e^{tZ})\mathrm{Ad}(k)H_1, H_2 \right\rangle = - \left\langle [\mathrm{Ad}(k)H_1, Z], H_2 \right\rangle.$$

By Proposition 10.3, this equals $\langle Z, [\mathrm{Ad}(k)H_1, H_2] \rangle$. Since both $\mathrm{Ad}(k)H_1$ and H_2 are in $\mathfrak{p}$, their bracket is in $\mathfrak{k}$, and the vanishing of this inner product for all $Z \in \mathfrak{k}$ implies that $[\mathrm{Ad}(k)H_1, H_2] = 0$.

Now take $Z = \mathrm{Ad}(k)H_1$ in Proposition 29.11 (i). We see that $\mathrm{Ad}(k)H_1 \in \mathfrak{a}_2$, and since both $\mathrm{Ad}(k)H_1$ and H_2 are regular, it follows that $\mathrm{Ad}(k)\mathfrak{a}_1 = \mathfrak{a}_2$. $\square$

Theorem 29.4. *With notations as above, $G = KAK$.*

Proof. Let $g \in G$. Let $p = g\theta(g)^{-1} = g\,{}^tg$. We will show that $p \in \exp(\mathfrak{p})$. By Proposition 29.1, we can write $p = \exp(Z)\,k_0$, where $Z \in \mathfrak{p}$ and $k_0 \in K$, and we want to show that $k_0 = 1$. By Proposition 29.10, we may embed $G_\mathbb{C}$ into $\mathrm{GL}(n, \mathbb{C})$ in such a way that $G \subseteq \mathrm{GL}(n, \mathbb{R})$, $G_c \subseteq \mathrm{U}(n)$, $K \subseteq \mathrm{O}(n)$, and $\theta(g) = {}^tg^{-1}$. In the matrix realization, p is a positive definite symmetric matrix. By the uniqueness assertion in Theorem 13.4, it follows that $k_0 = 1$ and $p = \exp(Z)$.

Now, by Theorem 29.3, we can find $k \in K$ such that $\mathrm{Ad}(k)Z = H \in \mathfrak{a}$. It follows that $kpk^{-1} = a^2$, where $a = \exp(\mathrm{Ad}(k)H/2) \in A$. Now

$$(a^{-1}kg)\theta(a^{-1}kg)^{-1} = a^{-1}kg\theta(g)^{-1}k^{-1}a = a^{-1}kpk^{-1}a^{-1} = 1.$$

Therefore, $a^{-1}kg \in K$, and it follows that $g \in KaK$. $\square$

Finally, there is the *Bruhat decomposition*. Let B be the $\mathbb{R}$-Borel subgroup of G. If $w \in W$, let $\omega \in N_G(A)$ represent W. Clearly, the double coset $B\omega B$ does not depend on the choice of representative ω, and we denote it BwB.

Theorem 29.5. (Bruhat decomposition) *We have*

$$G = \bigcup_{w \in W_{\mathrm{rel}}} BwB.$$

Proof. Omitted. See Helgason [66], p. 403. $\square$

Exercises

Exercise 29.1. Show that $\mathbb{C} \otimes \mathrm{Mat}_n(\mathbb{H}) \cong \mathrm{Mat}_{2n}(\mathbb{C})$ as $\mathbb{C}$-algebras and that the composition

$$\mathrm{Mat}_n(\mathbb{H}) \longrightarrow \mathbb{C} \otimes \mathrm{Mat}_n(\mathbb{H}) \cong \mathrm{Mat}_{2n}(\mathbb{C}) \xrightarrow{\det} \mathbb{C}$$

takes values in $\mathbb{R}$.

Exercise 29.2. Compute the Satake diagrams for $\mathrm{SO}(p,q)$ with $p \geqslant q$ for all p and q.

Exercise 29.3. Prove an analog of Theorem 18.3 showing that W_{rel} is generated by the reflections constructed in Theorem 29.1.

30

Embeddings of Lie Groups

In this chapter, we will contemplate how Lie groups embed in one another. Our aim is not to be systematic or even completely precise but to give the reader some tools for thinking about the relationships between different Lie groups.

If G is a Lie group and H a subgroup, then there exists a chain of Lie subgroups of G,

$$G = G_0 \supset G_1 \supset \cdots \supset G_n = H$$

such that each G_i is maximal in G_{i-1}. Dynkin [45–47] classified the maximal subgroups of semisimple complex analytic groups. Thus, the lattice of semisimple complex analytic subgroups of such a group is known.

Let K_1 and K_2 be compact connected Lie groups, and let G_1 and G_2 be their complexifications. Given an embedding $K_1 \longrightarrow K_2$, there is a unique analytic embedding $G_1 \longrightarrow G_2$. The converse is also true: given an analytic embedding $G_1 \longrightarrow G_2$, then K_1 embeds as a compact subgroup of G_2. However, any compact subgroup of G_2 is conjugate to a subgroup of K_2 (Theorem 28.2), so K_1 is conjugate to a subgroup of K_2. Thus, embeddings of compact connected Lie groups and analytic embeddings of their complexifications are essentially the same thing. To be definite, let us specify that in this chapter we are talking about analytic embeddings of complex analytic groups, with the understanding that the ideas will be applicable in other contexts. By a "torus," we therefore mean a group analytically isomorphic to $(\mathbb{C})^n$ for some n. We will allow ourselves to be a bit sloppy in this chapter, and we will sometimes write $O(n)$ when we should really write $O(n, \mathbb{C})$.

So let us start with embeddings of complex analytic Lie groups. A useful class of complex analytic groups that is slightly larger than the semisimple ones is the class of *reductive* complex analytic groups. A complex analytic group G (connected, let us assume) is called *reductive* if its linear analytic representations are completely reducible. For example, $\mathrm{GL}(n, \mathbb{C})$ is reductive, though it is not semisimple.

D. Bump, *Lie Groups*, Graduate Texts in Mathematics 225,
DOI 10.1007/978-1-4614-8024-2_30, © Springer Science+Business Media New York 2013

Examples of groups that are *not* reductive are *parabolic* subgroups. Let G be the complexification of the compact connected Lie group K, and let B be the Borel subgroup described in Theorem 26.2. A subgroup of G containing B is called a *standard parabolic subgroup*. (Any conjugate of a standard parabolic subgroup is called *parabolic*.)

As an example of a group that is not reductive, let $P \subset \mathrm{GL}(n, \mathbb{C})$ be the maximal parabolic subgroup consisting of matrices

$$\begin{pmatrix} g_1 & * \\ & g_2 \end{pmatrix}, \qquad g_1 \in \mathrm{GL}(r, \mathbb{C}), g_2 \in \mathrm{GL}(s, \mathbb{C}), \quad r + s = n.$$

In the standard representation corresponding to the inclusion $P \longrightarrow \mathrm{GL}(n, \mathbb{C})$, the set of matrices which have last s entries that are zero is a P-invariant subspace of $\mathbb{C}^n$ that has no invariant complement. Therefore, this representation is not completely reducible, and so P is not reductive.

If G is the complexification of a connected compact group, then analytic representations of G are completely reducible by Theorem 24.1. It turns out that the converse is true—a connected complex analytic reductive group is the complexification of a compact Lie group. We will not prove this, but it is useful to bear in mind that whatever we prove for complexifications of connected compact groups is applicable to the class of reductive complex analytic Lie groups.

Even if we restrict ourselves to finding reductive subgroups of reductive Lie groups, the problem is very difficult. After all, any faithful representation gives an embedding of a Lie group in another. There is an important class of embeddings for which it is possible to give a systematic discussion. Following Dynkin, we call an embedding of Lie groups or Lie algebras *regular* if it takes a maximal torus into a maximal torus and roots into roots. Our first aim is to show how regular embeddings can be recognized using *extended Dynkin diagrams*.

We will use orthogonal groups to illustrate some points. It is convenient to take the orthogonal group in the form

$$\mathrm{O}_J(n, F) = \left\{ g \in \mathrm{GL}(n, F) \,|\, g\, J^t g = J \right\}, \qquad J = \begin{pmatrix} & & 1 \\ & \cdot^{\cdot^{\cdot}} & \\ 1 & & \end{pmatrix}.$$

We will take the realization $\mathrm{O}_J(n, \mathbb{C}) \cap \mathrm{U}(n) \cong \mathrm{O}(n)$ of the usual orthogonal group in Exercise 5.3 with the maximal torus T consisting of diagonal elements of $\mathrm{O}_J(n, \mathbb{C}) \cap \mathrm{U}(n)$. Then, as in Exercise 24.1, $\mathrm{O}_J(n, \mathbb{C})$ is the analytic complexification of the usual orthogonal group $\mathrm{O}(n)$. We can take the ordering of the roots so that the root eigenspaces $\mathfrak{X}_\alpha$ with $\alpha \in \Phi^+$ are upper triangular.

We recall that the root system of type D_n is the root system for $\mathrm{SO}(2n)$. Normally, one only considers D_n when $n \geqslant 4$. The reason for this is that the Lie groups $\mathrm{SO}(4)$ and $\mathrm{SO}(6)$ have root systems of types $A_1 \times A_1$ and A_3, respectively. To see this, consider the Lie algebra of type $\mathrm{SO}(8)$. This consists of the set of all matrices of the form in Fig. 30.1.

$$\begin{pmatrix} t_1 & x_{12} & x_{13} & x_{14} & x_{15} & x_{16} & x_{17} & 0 \\ x_{21} & t_2 & x_{23} & x_{24} & x_{25} & x_{26} & 0 & -x_{17} \\ x_{31} & x_{32} & t_3 & x_{34} & x_{35} & 0 & -x_{26} & -x_{16} \\ x_{41} & x_{42} & x_{43} & t_4 & 0 & -x_{35} & -x_{25} & -x_{15} \\ x_{51} & x_{52} & x_{53} & 0 & -t_4 & -x_{34} & -x_{24} & -x_{14} \\ x_{61} & x_{62} & 0 & -x_{53} & -x_{43} & -t_3 & -x_{23} & -x_{13} \\ x_{71} & 0 & -x_{62} & -x_{52} & -x_{42} & -x_{32} & -t_2 & -x_{12} \\ 0 & -x_{71} & -x_{61} & -x_{51} & -x_{41} & -x_{31} & -x_{21} & -t_1 \end{pmatrix}$$

Fig. 30.1. The Lie algebra of SO(8)

The Lie algebra t of T consists of the subalgebra of diagonal matrices, where all $x_{ij} = 0$. The 24 roots α are such that each $\mathfrak{X}_\alpha$ is characterized by the nonvanishing of exactly one x_{ij}. We have circled the $\mathfrak{X}_\alpha$ corresponding to the four simple roots and drawn lines to indicate the graph of the Dynkin diagram. (Note that each x_{ij} occurs in two places. We have only circled the x_{ij} in the upper half of the diagram.)

The middle 6×6 block, shaded in Fig. 30.1, is the Lie algebra of SO(6), and the very middle 4×4 block, shaded dark, is the Lie algebra of SO(4). Looking at the simple roots, we can see the inclusions of Dynkin diagrams in Fig. 30.2. The shadings of the nodes correspond to the shadings in Fig. 30.1.

The coincidences of root systems $D_2 = A_1 \times A_1$ and $D_3 = A_3$ are worth explaining from another point of view. We may realize the group SO(4) concretely as follows. Let $V = \text{Mat}_2(\mathbb{C})$. The determinant is a nondegenerate quadratic form on the four-dimensional vector space V. Since all nondegenerate quadratic forms are equivalent, the group of linear transformations of V preserving the determinant may thus be identified with SO(4). We consider the group

$$G = \{(g_1, g_2) \in \text{GL}(2, \mathbb{C}) \times \text{GL}(2, \mathbb{C}) \mid \det(g_1) = \det(g_2)\}.$$

This group acts on V by

$$(g_1, g_2) : X \longmapsto g_1 X g_2^{-1}.$$

This action preserves the determinant, so we have a homomorphism $G \longrightarrow$ O(4). There is a kernel Z^Δ consisting of the scalar matrices in $\text{GL}(2, \mathbb{C})$

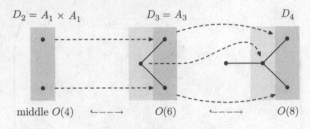

$$D_2 = A_1 \times A_1 \qquad D_3 = A_3 \qquad D_4$$

middle $O(4)$ $\longleftarrow\,-\longrightarrow$ $O(6)$ $\longleftarrow\,-\longrightarrow$ $O(8)$

Fig. 30.2. The inclusions $SO(4) \to SO(6) \to SO(8)$

embedded diagonally. We therefore have an injective homomorphism $G/Z^\Delta \longrightarrow O(4)$. Both groups have dimension 6, so this homomorphism is a surjection onto the connected component $SO(4)$ of the identity.

Using the fact that $\mathbb{C}$ is algebraically closed, the subgroup $SL(2,\mathbb{C}) \times SL(2,\mathbb{C})$ of G maps surjectively onto $SO(4)$. The kernel of the map

$$SL(2,\mathbb{C}) \times SL(2,\mathbb{C}) \longrightarrow SO(4)$$

has order 2, and we may identify the simply-connected group $SL(2,\mathbb{C}) \times SL(2,\mathbb{C})$ as the double cover $\mathrm{Spin}(4,\mathbb{C})$. Since $SO(4)$ is a quotient of $SL(2,\mathbb{C}) \times SL(2,\mathbb{C})$, we see why its root system is of type $A_1 \times A_1$.

Remark 30.1. Although we could have worked with $SL(2,\mathbb{C}) \times SL(2,\mathbb{C})$ at the outset, over a field F that was not algebraically closed, it is often better to use the realization $G/Z^\Delta \cong SO(4)$. The reason is that if F is *not* algebraically closed, the image of the homomorphism $SL(2,F) \times SL(2,F) \longrightarrow SO(4,F)$ may not be all of $SO(4)$. Identifying $SL(2) \times SL(2)$ with the algebraic group $\mathrm{Spin}(4)$, this is a special instance of the fact that the covering map $\mathrm{Spin}(n) \longrightarrow SO(n)$, though surjective over an algebraically closed field, is not generally surjective on rational points over a field that is not algebraically closed. A surjective map may instead be obtained by working with the group of similitudes $\mathrm{GSpin}(n)$, which when $n = 4$ is the group G. This is analogous to the fact that the homomorphism $SL(2,F) \longrightarrow PGL(2,F)$ is not surjective if F is algebraically closed, which is why the adjoint group $PGL(2,F)$ of $SL(2)$ is constructed as $GL(2,F)$ modulo the center, not $SL(2)$ modulo the center.

We turn next to $SO(6)$. Let W be a four-dimensional complex vector space. There is a homomorphism $GL(W) \longrightarrow GL(\wedge^2 W) \cong GL(6,\mathbb{C})$, namely the exterior square map, and there is a homomorphism

$$GL(\wedge^2 W) \xrightarrow{\wedge^2} GL(\wedge^4 W) \cong \mathbb{C}^\times.$$

The latter map is symmetric since in the exterior algebra

$$(v_1 \wedge \ldots \wedge v_r) \wedge (w_1 \wedge \ldots \wedge w_s) = (-1)^{rs}(w_1 \wedge \ldots \wedge w_s) \wedge (v_1 \wedge \ldots \wedge v_r).$$

(Each v_i has to move past each w_j producing rs sign changes.) Hence we may regard $\wedge^2$ as a quadratic form on $\mathrm{GL}(\wedge^2 W)$. The subspace preserving the determinant is therefore isomorphic to $\mathrm{SO}(6)$. The composite

$$\mathrm{GL}(W) \xrightarrow{\wedge^2} \mathrm{GL}(\wedge^2 W) \xrightarrow{\wedge^2} \mathrm{GL}(\wedge^4 W) \cong \mathbb{C}^\times$$

is the determinant, so the image of $\mathrm{SL}(W) = \mathrm{SL}(4,\mathbb{C})$ in $\mathrm{GL}(\wedge^2 W)$ is therefore contained in $\mathrm{SO}(6)$. Both $\mathrm{SL}(4,\mathbb{C})$ and $\mathrm{SO}(6)$ are 15-dimensional and connected, so we have constructed a homomorphism onto $\mathrm{SO}(6)$. The kernel consists of $\{\pm 1\}$, so we see that $\mathrm{SO}(6) \cong \mathrm{SL}(4,\mathbb{C})/\{\pm I\}$. Since $\mathrm{SO}(6)$ is a quotient of $\mathrm{SL}(4,\mathbb{C})$, we see why its root system is of type A_3.

The maps discussed so far, involving $\mathrm{SO}(2n)$ with $n = 2, 3$, and 4, are regular. Sometimes (as in these examples) regular embeddings can be recognized by inclusions of ordinary Dynkin diagrams, but a fuller picture will emerge if we introduce the *extended Dynkin diagram*.

Let K be a compact connected Lie group with maximal torus T. Let G be its complexification. Let Φ, Φ^+, Σ, and other notations be as in Chap. 18.

Proposition 30.1. *Suppose in this setting that S is any set of roots such that if $\alpha, \beta \in S$ and if $\alpha + \beta \subset \Phi$, then $\alpha + \beta \in S$. Then*

$$\mathfrak{h} = \mathfrak{t}_\mathbb{C} \oplus \bigoplus_{\alpha \in S} \mathfrak{X}_\alpha$$

is a Lie subalgebra of $\mathrm{Lie}(G)$.

Proof. It is immediate from Proposition 18.4 (ii) and Proposition 18.3 (ii) that this vector space is closed under the bracket. $\qquad \square$

We will not worry too much about verifying that $\mathfrak{h}$ is the Lie algebra of a closed Lie subgroup of G except to remark that we have some tools for this, such as Theorem 14.3.

We have already introduced the Dynkin diagram in Chap. 25. We recall that the Dynkin diagram is obtained as a graph whose vertices are in bijection with Σ. Let us label $\Sigma = \{\alpha_1, \ldots, \alpha_r\}$, and let $s_i = s_{\alpha_i}$. Let $\theta(\alpha_i, \alpha_j)$ be the angle between the roots α_i and α_j. Then

$$n(s_i, s_j) = \begin{cases} 2 & \text{if } \theta(\alpha_i, \alpha_j) = \frac{\pi}{2}, \\[2mm] 3 & \text{if } \theta(\alpha_i, \alpha_j) = \frac{2\pi}{3}, \\[2mm] 4 & \text{if } \theta(\alpha_i, \alpha_j) = \frac{3\pi}{4}, \\[2mm] 6 & \text{if } \theta(\alpha_i, \alpha_j) = \frac{5\pi}{6}. \end{cases}$$

The *extended Dynkin diagram* adjoins to the graph of the Dynkin diagram one more node, which corresponds to the negative root α_0 such that $-\alpha_0$

is the highest weight vector in the adjoint representation. The negative root α_0 is sometimes called the *affine root*, because of its role in the affine root system (Chap. 23). As in the usual Dynkin diagram, we connect the vertices corresponding to α_i and α_j only if the roots are not orthogonal. If they make an angle of $2\pi/3$, we connect them with a single bond; if they make an angle of $6\pi/4$, we connect them with a double bond; and if they make an angle of $5\pi/6$, we connect them with a triple bond.

The basic paradigm is that if we remove a node from the extended Dynkin diagram, what remains will be the Dynkin diagram of a subgroup of G. To get some feeling for why this is true, let us consider an example in the exceptional group G_2. We may take S in Proposition 30.1 to be the set of six long roots. These form a root system of type A_2, and $\mathfrak{h}$ is the Lie algebra of a Lie subgroup isomorphic to $\mathrm{SL}(3, \mathbb{C})$. Since $\mathrm{SL}(3, \mathbb{C})$ is the complexification of the simply-connected compact Lie group $\mathrm{SU}(2)$, it follows from Theorem 14.3 that there is a homomorphism $\mathrm{SL}(3, \mathbb{C}) \longrightarrow G$.

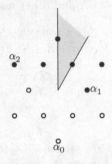

Fig. 30.3. The exceptional root α_0 of G_2 ($\bullet$ = positive roots)

The ordinary Dynkin diagram of G_2 does not reflect the existence of this embedding. However, from Fig. 30.3, we see that the roots α_2 and α_0 can be taken as the simple roots of $\mathrm{SL}(3, \mathbb{C})$. The embedding $\mathrm{SL}(3, \mathbb{C})$ can be understood as an inclusion of the A_2 (ordinary) Dynkin diagram in the extended G_2 Dynkin diagram (Fig. 30.4).

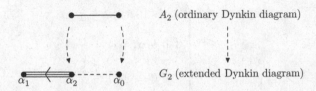

Fig. 30.4. The inclusion of $\mathrm{SL}(3)$ in G_2

Let us consider some more extended Dynkin diagrams. If $n > 2$, and if G is the odd orthogonal group $SO(2n + 1)$, its root system is of type B_n, and its extended Dynkin diagram is as in Fig. 30.5. We confirm this in Fig. 30.6 for $SO(9)$ – that is, when $n = 4$ – by explicitly marking the simple roots $\alpha_1, \ldots, \alpha_n$ and the largest root α_0.

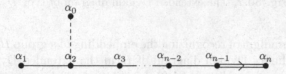

Fig. 30.5. The extended Dynkin diagram of type B_n

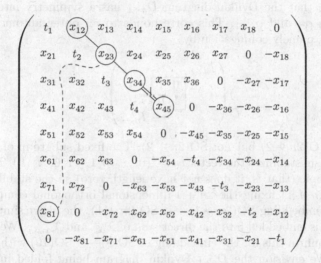

Fig. 30.6. The Lie algebra of $SO(9)$

Next, if $n \geqslant 5$ and $G = SO(2n)$, the root system of G is D_n, and the extended Dynkin diagram is as in Fig. 30.7. For example if $n = 5$, the configuration of roots is as in Fig. 30.8.

We leave it to the reader to check the extended Dynkin diagrams of the symplectic group $Sp(2n)$, which is of type C_n (Fig. 30.9).

The extended Dynkin diagram of type A_n ($n \geqslant 2$) is shown in Fig. 30.10. It has the feature that removing a node leaves the diagram connected. Because of this, the paradigm of finding subgroups of a Lie group by examining the extended Dynkin diagram does *not* produce any interesting examples for $SL(n + 1)$ or $GL(n + 1)$.

We already encountered the extended Dynkin diagram of G_2 is in Fig. 30.4. The extended Dynkin diagrams of all the exceptional groups are listed in Fig. 30.11.

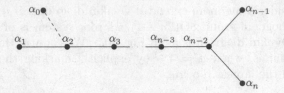

Fig. 30.7. The extended Dynkin diagram of type D_n

Our first paradigm of recognizing the embedding of a group H in G by embedding the ordinary Dynkin diagram of H in the extended Dynkin diagram of G predicts the embedding of $\mathrm{SO}(2n)$ in $\mathrm{SO}(2n+1)$ but not the embedding of $\mathrm{SO}(2n+1)$ in $\mathrm{SO}(2n+2)$. For this we need another paradigm, which we call *root folding*.

We note that the Dynkin diagram D_{n+1} has a symmetry interchanging the vertices α_n and α_{n+1}. This corresponds to an outer automorphism of $\mathrm{SO}(2n+2)$, namely conjugation by

$$\begin{pmatrix} I_{n-1} & & \\ & \begin{array}{cc} 0 & 1 \\ 1 & 0 \end{array} & \\ & & I_{n-1} \end{pmatrix},$$

which is in $\mathrm{O}(2n+2)$ but not $\mathrm{SO}(2n+2)$. The fixed subgroup of this outer automorphism stabilizes the vector $v_0 = {}^t(0,\ldots,0,1,-1,0,\ldots,0)$. This vector is not isotropic (that is, it does not have length zero) so the stabilizer is the group $\mathrm{SO}(2n+1)$ fixing the $2n+1$-dimensional orthogonal complement of v_0. In this embedding $\mathrm{SO}(2n+1) \longrightarrow \mathrm{SO}(2n+1)$, the short simple root of $\mathrm{SO}(2n+1)$ is embedded into the direct sum of $\mathfrak{X}_{\alpha_n}$ and $\mathfrak{X}_{\alpha_{n+1}}$. We invite the reader to confirm this for the embedding of $\mathrm{SO}(9) \longrightarrow \mathrm{SO}(10)$ with the above matrices. We envision the D_{n+1} Dynkin diagram being folded into the B_n diagram, as in Fig. 30.12.

The Dynkin diagram of type D_4 admits a rare symmetry of order 3 (Fig. 30.13). This is associated with a phenomenon known as *triality*, which we now discuss.

Referring to Fig. 30.1, the groups $\mathfrak{X}_{\alpha_i}$ ($i = 1, 2, 3, 4$) correspond to x_{12}, x_{23}, x_{34} and x_{35}, respectively. The Lie algebra will thus have an automorphism τ that sends $x_{12} \longrightarrow x_{34} \longrightarrow x_{35} \longrightarrow x_{12}$ and fixes x_{23}. Let us consider the effect on $\mathfrak{t}_{\mathbb{C}}$, which is the subalgebra of elements t with all $x_{ij} = 0$. Noting that $d\alpha_1(t) = t_1 - t_2$, $d\alpha_2(t) = t_2 - t_3$, $d\alpha_3(t) = t_3 - t_4$, and $d\alpha_4(t) = t_3 + t_4$, we must have

$$\tau : \begin{cases} t_1 - t_2 \longmapsto t_3 - t_4 \\ t_2 - t_3 \longmapsto t_2 - t_3 \\ t_3 - t_4 \longmapsto t_3 + t_4 \\ t_3 + t_4 \longmapsto t_1 - t_2 \end{cases},$$

from which we deduce that

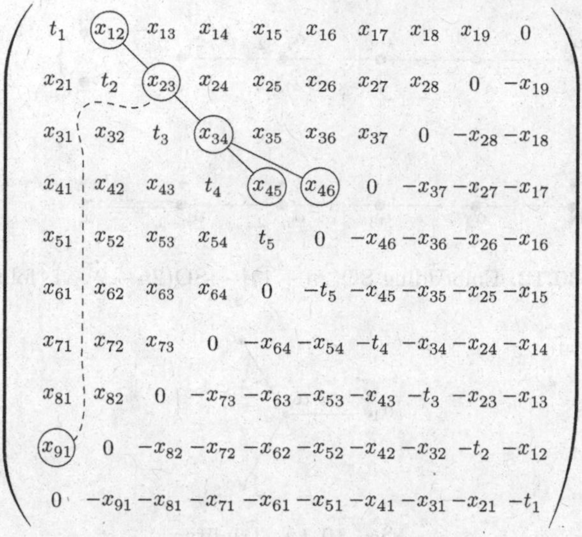

$$\begin{pmatrix}
t_1 & \boxed{x_{12}} & x_{13} & x_{14} & x_{15} & x_{16} & x_{17} & x_{18} & x_{19} & 0 \\
x_{21} & t_2 & \boxed{x_{23}} & x_{24} & x_{25} & x_{26} & x_{27} & x_{28} & 0 & -x_{19} \\
x_{31} & x_{32} & t_3 & \boxed{x_{34}} & x_{35} & x_{36} & x_{37} & 0 & -x_{28} & -x_{18} \\
x_{41} & x_{42} & x_{43} & t_4 & \boxed{x_{45}} & \boxed{x_{46}} & 0 & -x_{37} & -x_{27} & -x_{17} \\
x_{51} & x_{52} & x_{53} & x_{54} & t_5 & 0 & -x_{46} & -x_{36} & -x_{26} & -x_{16} \\
x_{61} & x_{62} & x_{63} & x_{64} & 0 & -t_5 & -x_{45} & -x_{35} & -x_{25} & -x_{15} \\
x_{71} & x_{72} & x_{73} & 0 & -x_{64} & -x_{54} & -t_4 & -x_{34} & -x_{24} & -x_{14} \\
x_{81} & x_{82} & 0 & -x_{73} & -x_{63} & -x_{53} & -x_{43} & -t_3 & -x_{23} & -x_{13} \\
\boxed{x_{91}} & 0 & -x_{82} & -x_{72} & -x_{62} & -x_{52} & -x_{42} & -x_{32} & -t_2 & -x_{12} \\
0 & -x_{91} & -x_{81} & -x_{71} & -x_{61} & -x_{51} & -x_{41} & -x_{31} & -x_{21} & -t_1
\end{pmatrix}$$

Fig. 30.8. The Lie algebra of SO(10)

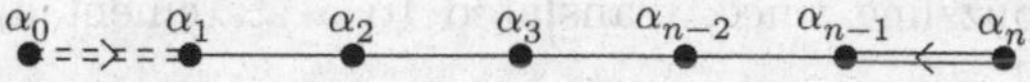

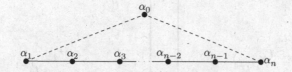

Fig. 30.9. The extended Dynkin diagram of type C_n

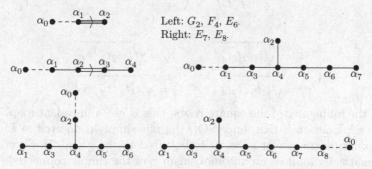

Fig. 30.10. The extended Dynkin diagram of type A_n

Left: G_2, F_4, E_6.
Right: E_7, E_8.

Fig. 30.11. Extended Dynkin diagram of the exceptional groups

$$\tau(t_1) = \tfrac{1}{2}(t_1 + t_2 + t_3 - t_4),$$
$$\tau(t_2) = \tfrac{1}{2}(t_1 + t_2 - t_3 + t_4),$$
$$\tau(t_3) = \tfrac{1}{2}(t_1 - t_2 + t_3 + t_4),$$
$$\tau(t_4) = \tfrac{1}{2}(t_1 - t_2 - t_3 - t_4).$$

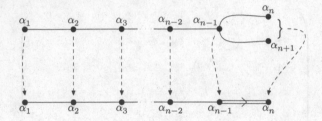

Fig. 30.12. Embedding $SO(2n+1) \hookrightarrow SO(2n+2)$ as "folding"

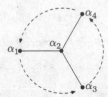

Fig. 30.13. Triality

At first this is puzzling since, translated to a statement about the group, we have

$$\tau \begin{pmatrix} t_1 & & & & & & & \\ & t_2 & & & & & & \\ & & t_3 & & & & & \\ & & & t_4 & & & & \\ & & & & t_4^{-1} & & & \\ & & & & & t_3^{-1} & & \\ & & & & & & t_2^{-1} & \\ & & & & & & & t_1^{-1} \end{pmatrix} = \begin{pmatrix} t_1' & & & & & & & \\ & t_2' & & & & & & \\ & & t_3' & & & & & \\ & & & t_4' & & & & \\ & & & & t_4'^{-1} & & & \\ & & & & & t_3'^{-1} & & \\ & & & & & & t_2'^{-1} & \\ & & & & & & & t_1'^{-1} \end{pmatrix},$$

where

$$t_1' = \sqrt{t_1 t_2 t_3 t_4^{-1}}, \qquad t_2' = \sqrt{t_1 t_2 t_3^{-1} t_4},$$

$$t_3' = \sqrt{t_1 t_2^{-1} t_3 t_4}, \qquad t_4' = \sqrt{t_1 t_2^{-1} t_3^{-1} t_4^{-1}}.$$

Due to the ambiguity of the square roots, this is not a univalent map.

The explanation is that since $SO(8)$ is not simply-connected, a Lie algebra automorphism cannot necessarily be lifted to the group. However, there is automatically induced an automorphism τ of the simply-connected double cover $\mathrm{Spin}(8)$. The center of $\mathrm{Spin}(8)$ is $(\mathbb{Z}/2\mathbb{Z}) \times (\mathbb{Z}/2\mathbb{Z})$, which has an automorphism of order 3 that does not preserve the kernel (of order 2) of τ. If we divide $\mathrm{Spin}(8)$ by its entire center $(\mathbb{Z}/2\mathbb{Z}) \times (\mathbb{Z}/2\mathbb{Z})$, we obtain the adjoint group $PGO(8)$, and the triality automorphism of $\mathrm{Spin}(8)$ induces an automorphism of order 3 of $PGO(8)$. To summarize, triality is an automorphism of order 3 of either $\mathrm{Spin}(8)$ or $PGO(8)$ but *not* of $SO(8)$.

The fixed subgroup of τ in either Spin(8) or PGO(8) is the exceptional group G_2, and the inclusion of G_2 in Spin(8) can be understood as a folding of roots. The unipotent subgroup corresponding to a short simple root of G_2 is included diagonally in the three root groups $\exp(\mathfrak{X}_{\alpha_i})$, $(i = 1, 3, 4)$ of Spin(8) as in Fig. 30.14 (left).

Triality has the following interpretation. The quadratic space V of dimension 8 on which SO(8) acts can be given the structure of a nonassociative algebra known as the *octonions* or *Cayley numbers*.

If $f_1 : V \longrightarrow V$ is any nonsingular orthogonal linear transformation, there exist linear transformations f_2 and f_3 such that

$$f_1(xy) = f_2(x)f_3(y).$$

The linear transformations f_2 and f_3 are only determined up to sign. The maps $f_1 \longmapsto f_2$ and $f_1 \longmapsto f_3$, though thus not well-defined as an automorphisms of SO(8), *do* lift to well-defined automorphisms of Spin(8), and the resulting automorphism $f_1 \longmapsto f_2$ is the triality automorphism. Triality permutes the three orthogonal maps f_1, f_2, and f_3 cyclicly. Note that if $f_1 = f_2 = f_3$, then f_1 is an automorphism of the octonion ring, so the fixed group G_2 is the automorphism group of the octonions. See Chevalley [36], p.188. As an alternative to Chevalley's approach, one may first prove a local form of triality as in Jacobson [88] and then deduce the global form. See also Schafer [146]. Over an algebraically closed field, the octonion algebra is unique. Over the real numbers there are two forms, which correspond to the compact group O(8) and the split form O(4, 4).

So far, the examples we have given of folding correspond to automorphisms of the group G. For an example that does not, consider the embedding of G_2 into Spin(7) (Fig. 30.14, right).

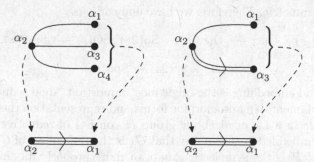

Fig. 30.14. The group G_2 embedded in Spin(8) and Spin(7)

A frequent way in which large subgroups of a Lie group arise is as fixed points of automorphisms, usually involutions. Many of these subgroups can be understood by the the paradigms explained above. A list of such subgroups can be found in Table 28.1, for in this list, the compact subgroup K is the fixed point of an involution in the compact group G_c, and this relationship is also true for the complexifications. For example, the first entry, corresponding to Cartan's classification AI, is the symmetric space with $G_c = \mathrm{SU}(n)$ and the subgroup $K = \mathrm{SO}(n)$. Assuming that we use the version of the orthogonal group in Exercise 5.3, the involution θ is $g \mapsto J\,{}^t g^{-1} J$, where J is given by (5.3). This involution extends to the complexification $\mathrm{SL}(n,\mathbb{C})$, and the fixed point set is the subgroup $\mathrm{SO}(n,\mathbb{C})$. If n is odd, then every simple root eigenspace of $\mathrm{SO}(n,\mathbb{C})$ embeds in the direct sum of one or two simple root eigenspaces of $\mathrm{SL}(2,\mathbb{C})$, and the embedding may be understood as an example of the root folding paradigm. But if $n = 2r$ is even, then one of the roots of $\mathrm{SO}(2r)$, namely the simple root $\mathbf{e}_{r-1} + \mathbf{e}_r$, involves non-simple roots of $\mathrm{SL}(n)$.

Suppose that V_1 and V_2 are quadratic spaces (that is, vector spaces equipped with nondegenerate symmetric bilinear forms). Then $V_1 \oplus V_2$ is naturally a quadratic space, so we have an embedding $\mathrm{O}(V_1) \times \mathrm{O}(V_2) \longrightarrow \mathrm{O}(V_1 \oplus V_2)$. The same is true if V_1 and V_2 are symplectic (that is, equipped with nondegenerate skew-symmetric bilinear forms). It follows that we have embeddings

$$\mathrm{O}(r) \times \mathrm{O}(s) \longrightarrow \mathrm{O}(r+s), \qquad \mathrm{Sp}(2r) \times \mathrm{Sp}(2s) \longrightarrow \mathrm{Sp}\big(2(r+s)\big).$$

These embeddings can be understood as embeddings of extended Dynkin diagrams *except* in the orthogonal case where r and s are both odd (Exercise 30.2.

Also, if V_1 and V_2 are vector spaces with bilinear forms $\beta_i : V_i \times V_i \longrightarrow \mathbb{C}$, then there is a bilinear form B on $V_1 \otimes V_2$ such that

$$B(v_1 \otimes v_2, v_1' \otimes v_2') = \beta_1(v_1, v_1')\,\beta_2(v_2, v_2').$$

If both β_1 and β_2 are either symmetric or skew-symmetric, then B is symmetric. If one of β_1 and β_2 is symmetric and the other skew-symmetric, then B is skew-symmetric. Therefore, we have embeddings

$$\mathrm{O}(r) \times \mathrm{O}(s) \longrightarrow \mathrm{O}(rs), \qquad \mathrm{Sp}(2r) \times \mathrm{O}(s) \longrightarrow \mathrm{Sp}(2rs),$$

$$\mathrm{Sp}(2r) \times \mathrm{Sp}(2s) \to \mathrm{Sp}(4rs). \tag{30.1}$$

The second embedding is the single most important "dual reductive pair," which is fundamental in automorphic forms and representation theory. A *dual reductive pair* in a Lie or algebraic group H consists of reductive subgroups G_1 and G_2 embedded in such a way that G_1 is the centralizer of G_2 in H and conversely. If H is the symplectic group, or more properly its "metaplectic" double cover, then H has an important infinite-dimensional representation ω introduced by Weil [172]. Weil showed in [173] that in many cases the restriction of the Weil representation to a dual reductive pair can be used to understand classical correspondences of automorphic forms due to Siegel. The importance of this phenomenon cannot be overstated. From Weil's point of view this phenomenon is a global one, but Howe [73] gave better foundations, including a local theory. This is a topic that transcends Lie theory since in much of the literature one will consider $O(s)$ or $Sp(2r)$ as algebraic groups defined over a p-adic field or a number field (and its adele ring). Expositions of pure Lie group applications may be found in Howe and Tan [78] and Goodman and Wallach [56].

The classification of dual reductive pairs in $Sp(2n)$, described in Weil [173] and Howe [73], has its origins in the theory of algebras with involutions, due to Albert [5]. The connection between algebras with involutions and the theory of algebraic groups was emphasized earlier by Weil [171]. A modern and immensely valuable treatise on algebras with involutions and their relations with the theory of algebraic groups may be found in Knus, Merkurjev, Rost, and Tignol [107].

A classification of dual reductive pairs in exceptional groups is in Rubenthaler [138]. These examples have proved interesting in the theory of automorphic forms since an analog of the Weil representation is available.

So far, our point of view has been to start with a group G and understand its large subgroups H, and we have a set of examples sufficient for understanding most, but not all such pairs. Let us consider the alternative question: given H, how can we embed it in a larger group G?

Suppose, therefore that $\pi : H \to \mathrm{GL}(V)$ is a representation. We assume that it is faithful and irreducible. Then we get an embedding of H into $\mathrm{GL}(V)$. However sometimes there is a smaller subgroup $G \subset \mathrm{GL}(V)$ such that the image of π is contained in G. A frequent case is that G is an orthogonal or symplectic group. These cases may be classified by considering the theory of the Frobenius-Schur indicator, which is discussed in the exercises to this chapter and again in Chap. 43. The Frobenius-Schur indicator $\epsilon(\pi)$ is the multiplicity of the trivial character in the generalized character $g \mapsto \chi(g^2)$, where χ is the character of π. It equals 0 unless $\pi = \hat{\pi}$ is self-contragredient, in which case either it equals 1 and π is orthogonal, or -1 and π is symplectic. This means that if $\epsilon(\pi) = 1$, then we may take $G = O(n)$ where $n = \dim(V)$, while if $\epsilon(\pi) = -1$, then $\dim(V)$ is even, and we may take $G = Sp(n)$.

The examples (30.1) can be understood this way. Here's a couple more. Let $\mathfrak{H} = \mathrm{SL}(2)$, and let π be the symmetric k-th power representation. The vector space V is $k + 1$-dimensional. Exercise 22.15 computes the Frobenius-Schur indicator, and we see that H embeds in $\mathrm{SO}(k + 1)$ if k is even, and $Sp(k + 1)$ if k is odd. For another example, if H is any simple Lie group, then the adjoint representation is orthogonal since the Killing form on the Lie algebra is a nondegenerate symmetric bilinear form. Thus for example we get an embedding of $\mathrm{SL}(3)$ into $\mathrm{SO}(8)$.

As a final topic, we discuss parabolic subgroups. Just as regular subgroups of G can be read off from the extended Dynkin diagram, the parabolic subgroups can be read off from the regular Dynkin diagram. Let $\Sigma' \subset \Sigma$ be any proper subset of the set of simple roots. Then Σ' is the set of vertices of a (possibly disconnected) Dynkin diagram $\mathcal{D}'$ contained in that of G. There will be a unique parabolic subgroup P such that, for a simple root $\alpha \in \Sigma$, the space $\mathfrak{X}_{-\alpha}$ is contained in the Lie algebra of P if and only if $\alpha \in S$.

The roots $\mathfrak{X}_{-\alpha}$ and $\mathfrak{X}_{\alpha}$ with $\alpha \in S$ together with $\mathfrak{t}_{\mathbb{C}}$ generate a Lie algebra $\mathfrak{m}$, which is the Lie algebra of a reductive Lie group M, and

$$\mathfrak{u} = \bigoplus_{\substack{\alpha \in \Phi^+ \\ \mathfrak{X}_\alpha \not\subseteq \mathfrak{m}}} \mathfrak{X}_\alpha$$

is the Lie algebra of a unipotent subgroup U of P. (By *unipotent* we mean here that its image in any analytic representation of G consists of unipotent matrices.) The group $P = MU$. This factorization is called the *Levi decomposition*. The subgroup U of P is normal, so this decomposition is a semidirect product. The group M is called the *Levi factor*, and the group U is called the *unipotent radical* of P.

We illustrate all this with an example from the symplectic group. We take $G = \mathrm{Sp}(2n)$ to be $\{g \mid {}^t g J g = J\}$, where

$$J = \begin{pmatrix} & & & & -1 \\ & & & \cdot^{\cdot^{\cdot}} & \\ & & -1 & & \\ & 1 & & & \\ & \cdot^{\cdot^{\cdot}} & & & \\ 1 & & & & \end{pmatrix}.$$

This realization of the symplectic group has the advantage that the $\mathfrak{X}_\alpha$ corresponding to positive roots $\alpha \in \Phi^+$ all correspond to upper triangular matrices. We see from Fig. 30.9 that removing a node from the Dynkin diagram of type C_n gives a smaller diagram, disconnected unless we take an end vertex, of type $A_{r-1} \times C_{n-r}$. This is the Dynkin diagram of a maximal parabolic subgroup with Levi factor $M = \mathrm{GL}(r) \times \mathrm{Sp}(2(n-r))$. The subgroup looks like this:

$$M = \left\{ \left(\begin{array}{c} \boxed{g} \\ \boxed{h} \\ \boxed{g'} \end{array} \right) \,\middle|\, g \in \mathrm{GL}(r), h \in \mathrm{Sp}(2m) \right\}, \quad U = \left\{ \left(\begin{array}{ccc} \boxed{I_r} & * & * \\ & \boxed{I_{2m}} & * \\ & & \boxed{I_r} \end{array} \right) \right\}.$$

Here $m = n - r$. In the matrix M, the matrix g' depends on g; it is determined by the requirement that the given matrix be symplectic. Figure 30.15 shows the parabolic subgroup with Levi factor $\mathrm{GL}(3) \times \mathrm{Sp}(4)$ in $\mathrm{GL}(10)$. Its Lie algebra is shaded here: the Lie algebra of M shaded dark and the Lie algebra of U is shaded light.

The Levi factor $M = \mathrm{GL}(3) \times \mathrm{Sp}(4)$ is a proper subgroup of the larger group $\mathrm{Sp}(6) \times \mathrm{Sp}(4)$, which can be read off from the extended Dynkin diagram. The Lie algebra of $\mathrm{Sp}(6) \times \mathrm{Sp}(4)$ is shaded dark in Fig. 30.16.

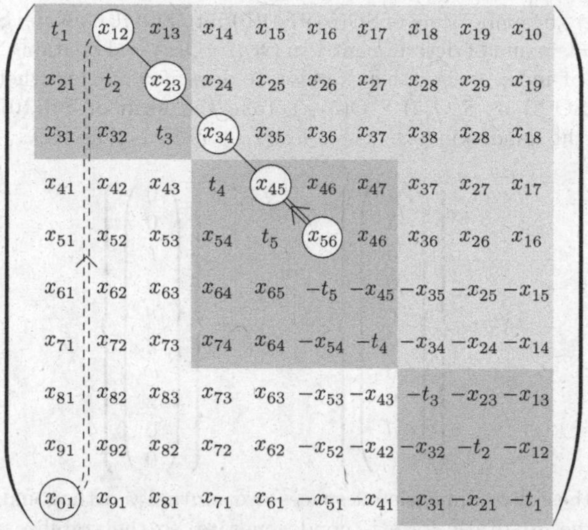

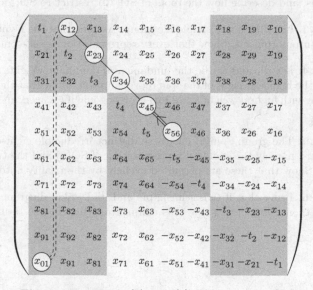

Fig. 30.15. A parabolic subgroup of Sp(10)

Fig. 30.16. The Sp(6) × Sp(4) subgroup of Sp(10)

Exercises

Exercise 30.1. Discuss as many as possible of the embeddings $K \hookrightarrow G_c$ in Table 28.1 of Chap. 28 using the extended Dynkin diagram of G_c.

Exercise 30.2. In doing the last exercise, one case you may have trouble with is the embedding of $S(O(p) \times O(q))$ into $SO(p+q)$ when p and q are both odd. To get some insight, consider the embedding of $SO(5) \times SO(5)$ into $SO(10)$. (*Note:* $S(O(p) \times O(q))$ is the group of elements of determinant 1 in $O(p) \times O(q)$ and contains $SO(p) \times SO(q)$ as a subgroup of index 2. For this exercise, it does not matter whether you work with $SO(5) \times SO(5)$ or $S(O(5) \times O(5))$.) Take the form of $SO(10)$ in Fig. 30.8. This stabilizes the quadratic form $x_1 x_{10} + x_2 x_9 + x_3 x_8 + x_4 x_7 + x_5 x_6$. Consider the subspaces

$$
V_1 = \left\{ \begin{pmatrix} a \\ b \\ 0 \\ 0 \\ c \\ -c \\ 0 \\ 0 \\ d \\ e \end{pmatrix} \right\}, \qquad
V_2 = \left\{ \begin{pmatrix} 0 \\ 0 \\ t \\ u \\ v \\ v \\ w \\ x \\ 0 \\ 0 \end{pmatrix} \right\}.
$$

Observe that these five-dimensional spaces are mutually orthogonal and that the restriction of the quadratic form is nondegenerate, so the stabilizers of these two spaces are mutually centralizing copies of $SO(5)$. Compute the Lie algebras of these two subgroups, and describe how the roots of $SO(10)$ restrict to $SO(5) \times SO(5)$.

Exercise 30.3. Let G be a semisimple Lie group. Assume that the Dynkin diagram of G has no automorphisms. Show that every representation is self-contragredient.

Exercise 30.4. Let ϖ_1 and ϖ_2 be the fundamental dominant weights for $\mathrm{Spin}(5)$, so that ϖ_2 is the highest weight of the spin representation. Show that the irreducible representation with highest weight $k\varpi_1 + l\varpi_2$ is orthogonal if l is even, and symplectic of l is odd.

Exercise 30.5. The group $\mathrm{Spin}(8)$ has three distinct irreducible eight-dimensional representations, namely the standard representation of $SO(8)$ and the two spin representations. Show that these are permuted cyclicly by the triality automorphism.

Exercise 30.6. Prove that if G is semisimple and its Dynkin diagram has no automorphisms, then every element in G is conjugate to its inverse. Is the converse true?

31

Spin

This chapter does not depend on the last few chapters, and may be read at any point after Chap. 23, or even earlier. The results of Chap. 23 are not used here, but are illustrated by the results of this chapter.

We will take a closer look at the groups $\mathrm{SO}(N)$ and their double covers, $\mathrm{Spin}(N)$. We assume that $N \geqslant 3$ and that $N = 2n + 1$ or $2n$. In this Chapter, we will take a closer look at the groups $\mathrm{SO}(N)$ and their double covers, $\mathrm{Spin}(N)$. These groups have remarkable "spin" representations of dimension 2^n, where $N = 2n$ or $2n + 1$. We will first show that this follows from the Weyl theorem of Chap. 22. We will then take a different point of view and give a different construction, using Clifford algebras and a uniqueness principle.

The group $\mathrm{Spin}(N)$ was constructed at the end of Chap. 13 as the universal cover of $\mathrm{SO}(N)$. Since we proved that $\pi_1(\mathrm{SO}(N)) \cong \mathbb{Z}/2\mathbb{Z}$, it is a double cover. In this chapter, we will construct and study the interesting and important *spin representations* of the group $\mathrm{Spin}(N)$. We will also show how to compute the center of $\mathrm{Spin}(N)$.

Let $G = \mathrm{SO}(N)$ and let $\tilde{G} = \mathrm{Spin}(N)$. We will take G in the realization of Exercise 5.3; that is, as the group of unitary matrices satisfying $g\, J\, {}^t g = J$, where J is (5.3). Let $p : \tilde{G} \longrightarrow G$ be the covering map. Let T be the diagonal torus in G, and let $\tilde{T} = p^{-1}(T)$. Thus $\ker(p) \cong \pi_1(\mathrm{SO}(N)) \cong \mathbb{Z}/2\mathbb{Z}$.

Proposition 31.1. *The group $\tilde{T}$ is connected and is a maximal torus of $\tilde{G}$.*

Proof. Let $\Pi \subset \tilde{G}$ be the kernel of p. The connected component $\tilde{T}^\circ$ of the identity in $\tilde{T}$ is a torus of the same dimension as T, so it is a maximal torus in $\tilde{G}$. Its image in G is isomorphic to $\tilde{T}^\circ/(\tilde{T}^\circ \cap \Pi) \cong \tilde{T}^\circ \Pi/\Pi$. This is a torus of G contained in T, and of the same dimension as T, so it is all of T. Thus, the composition $\tilde{T}^\circ \longrightarrow \tilde{T} \xrightarrow{p} T$ is surjective. We see that $\tilde{T}/\Pi \cong T \cong \tilde{T}^\circ \Pi/\Pi$ canonically and therefore $\tilde{T} = \tilde{T}^\circ \Pi$.

We may identify Π with the fundamental group $\pi_1(G)$ by Theorem 13.2. It is a discrete normal subgroup of $\tilde{G}$ and hence central in $\tilde{G}$ by Proposition 23.1.

Thus it is contained in every maximal torus by Proposition 18.14, particularly in $\tilde{T}^\circ$. Thus $\tilde{T}^\circ = \tilde{T}^\circ \Pi = \tilde{T}$ and so $\tilde{T}$ is connected and a maximal torus.   □

Composition with p is a homomorphism $X^*(T) \longrightarrow X^*(\tilde{T})$, which induces an isomorphism $\mathbb{R} \otimes X^*(T) \longrightarrow \mathbb{R} \otimes X^*(\tilde{T})$. We will identify these two vector spaces, which we denote by $\mathcal{V}$. From the short exact sequence

$$1 \longrightarrow \pi_1(G) \longrightarrow \tilde{T} \longrightarrow T \longrightarrow 1,$$

we have a short exact sequence

$$0 \longrightarrow X^*(T) \longrightarrow X^*(\tilde{T}) \longrightarrow X^*\big(\pi_1(G)\big) \longrightarrow 0. \qquad (31.1)$$

(Surjectivity of the last map uses Exercise 4.2.) We recall that $\Lambda_{\text{root}} \subseteq X^*(T) \subseteq \Lambda$, where Λ and Λ_{root} are the root and weight lattices.

A typical element of T has the form

$$t = \begin{cases} \begin{pmatrix} t_1 & & & & & & \\ & \ddots & & & & & \\ & & t_n & & & & \\ & & & 1 & & & \\ & & & & t_n^{-1} & & \\ & & & & & \ddots & \\ & & & & & & t_1^{-1} \end{pmatrix} & \text{if } N = 2n+1 \text{ is odd,} \\[4em] \begin{pmatrix} t_1 & & & & & \\ & \ddots & & & & \\ & & t_n & & & \\ & & & t_n^{-1} & & \\ & & & & \ddots & \\ & & & & & t_1^{-1} \end{pmatrix} & \text{if } N = 2n \text{ is even.} \end{cases} \qquad (31.2)$$

In either case, $\mathcal{V}$ is spanned by $\mathbf{e}_1, \ldots, \mathbf{e}_n$, where $\mathbf{e}_i(t) = t_i$. The root system, as we have already seen in Chap. 19, consists of all $\pm \mathbf{e}_i \pm \mathbf{e}_j$ $(i \neq j)$, with the additional roots $\pm \mathbf{e}_i$ included only if $N = 2n+1$ is odd. Order the roots so that the positive roots are $\mathbf{e}_i \pm \mathbf{e}_j$ $(i < j)$ and (if N is odd) $\mathbf{e}_i$. This is the ordering that makes the root eigenspaces $\mathfrak{X}_\alpha$ upper triangular. See Fig. 30.1 and Fig. 19.3 for the groups SO(8) and SO(9).

It is easy to check that the simple roots are

$$\alpha_1 = \mathbf{e}_1 - \mathbf{e}_2,$$
$$\alpha_2 = \mathbf{e}_2 - \mathbf{e}_3,$$
$$\vdots$$
$$\alpha_{n-1} = \mathbf{e}_{n-1} - \mathbf{e}_n$$
$$\alpha_n = \begin{cases} \mathbf{e}_{n-1} + \mathbf{e}_n & \text{if } N = 2n, \\ \mathbf{e}_n & \text{if } N = 2n+1. \end{cases} \qquad (31.3)$$

The Weyl group may now be described.

Theorem 31.1. *The Weyl group W of $\mathrm{O}(N)$ has order $2^n \cdot n!$ if $N = 2n + 1$ and order $2^{n-1} \cdot n!$ if $N = 2n$. It has as a subgroup the symmetric group S_n, which simply permutes the t_i in the action on T, or dually the $\mathbf{e}_i$ in its action on $\mathcal{V}$. It also has a subgroup H consisting of transformations of the form*

$$t_i \longmapsto t_i^{\pm 1} \qquad or \qquad \mathbf{e}_i \longmapsto \pm \mathbf{e}_i.$$

If $N = 2n + 1$, then H consists of all such transformations, and its order is 2^n. If $N = 2n$, then H only contains transformations that change an even number of signs. In either case, H is a normal subgroup of W and $W = H \cdot S_n$ is a semidirect product.

Proof. Regarding S_n and H as groups of linear transformations of $\mathcal{V}$, the group H is normalized by S_n, and $H \cap S_n = \{1\}$, so the semidirect product $H \cdot S_n$ exists and has order $2^n n!$ or $2^{n-1} n!$ depending on whether $|H| = 2^n$ or 2^{n-1}. We must show that this is exactly the group generated by the simple reflections.

The simple reflections with respect to $\alpha_1, \dots, \alpha_{n-1}$ are identical with the simple reflections in the Weyl group S_n of $\mathrm{U}(n)$, which is clear since we may embed $\mathrm{U}(n) \longrightarrow \mathrm{O}(2n)$ or $\mathrm{O}(2n + 1)$ by

$$g \longmapsto \begin{pmatrix} g & \\ & g^* \end{pmatrix} \qquad or \qquad \begin{pmatrix} g & & \\ & 1 & \\ & & g^* \end{pmatrix},$$

where

$$g^* = \begin{pmatrix} & & 1 \\ & \iddots & \\ 1 & & \end{pmatrix} {}^t g^{-1} \begin{pmatrix} & & 1 \\ & \iddots & \\ 1 & & \end{pmatrix}.$$

Under this embedding, the Weyl group S_n of $\mathrm{U}(n)$ gets embedded in the Weyl group of $\mathrm{O}(N)$. In its action on the torus, the t_i are simply permuted, and in the action on $X^*(T)$, the $\mathbf{e}_i$ are permuted. The simple i-th simple reflection in S_n has the sends α_i to its negative ($1 \leqslant i \leqslant n - 1$) while permuting the other positive roots, so it coincides with the i-th simple reflection in $\mathrm{SO}(N)$.

Now let us consider the simple reflection with respect to α_n. If $N = 2n + 1$, then since $\alpha_n = \mathbf{e}_n$ this just has the effect $\mathbf{e}_n \longmapsto -\mathbf{e}_n$, and all other $\mathbf{e}_i \longmapsto \mathbf{e}_i$. A representative in $N(T)$ can be taken to be

$$w_n = \begin{pmatrix} I_{n-1} & & \\ & \begin{matrix} 0 & 0 & 1 \\ 0 & -1 & 0 \\ 1 & 0 & 0 \end{matrix} & \\ & & I_{n-1} \end{pmatrix}.$$

It is clear that all elements of the group H described in the statement of the theorem that change the sign of exactly one $\mathbf{e}_i$ can be generated by conjugating

w_n by elements of S_n and that these generate H. Thus W contains HS_n. On the other hand, all simple reflections are contained in HS_n, so $W = HS_n$ in this case.

If $N = 2n$, then since $\alpha_n = \mathbf{e}_{n-1} + \mathbf{e}_n$, the simple reflection in α_n has the effect $\mathbf{e}_{n-1} \longmapsto -\mathbf{e}_n$, $\mathbf{e}_n \longmapsto -\mathbf{e}_{n-1}$. A representative in $N(T)$ can be taken to be

$$
w_n = \begin{pmatrix}
I_{n-2} & & & & & & \\
 & 0 & 0 & 1 & 0 & & \\
 & 0 & 0 & 0 & 1 & & \\
 & 1 & 0 & 0 & 0 & & \\
 & 0 & 1 & 0 & 0 & & \\
 & & & & & I_{n-2}
\end{pmatrix}.
$$

If we multiply this by the simple reflection in α_{n-1}, which just interchanges $\mathbf{e}_{n-1}$ and $\mathbf{e}_n$, we get the element of the group H that changes the signs of $\mathbf{e}_{n-1}$ and $\mathbf{e}_n$ and leaves everything else fixed. It is clear that all elements of the group H described in the statement of the theorem that change the sign of exactly two $\mathbf{e}_i$ can be generated by conjugating this element of W by elements of S_n and that these generate H. Again W contains HS_n, and again all simple reflections are contained in HS_n, so $W = HS_n$ in this case.   □

Proposition 31.2. *The weight lattice* $\Lambda = X^*(\tilde{T})$ *consists of all elements of* $\mathcal{V}$ *of the form*

$$
\frac{1}{2}\left(\sum_{i=1}^{n} c_i \mathbf{e}_i \right), \tag{31.4}
$$

where $c_i \in \mathbb{Z}$ *are either all even or all odd.*

Proof. From our determination of the simple reflections, which generate W, the W-invariant inner product on $\mathcal{V} = \mathbb{R} \otimes \Lambda$ may be chosen so that the $\mathbf{e}_i$ are orthonormal. By Proposition 18.10 every weight λ is in the lattice $\tilde{\Lambda}$ of $\lambda \in \mathcal{V}$ such that $2\langle \lambda, \alpha \rangle / \langle \alpha, \alpha \rangle \in \mathbb{Z}$ for α in the root lattice. Since we know the root system, it is easy to see that (31.4) consists of the weights 31.4.

We could now invoke Proposition 23.12. But since Proposition 23.12 is somewhat deep, let us give a simple alternative argument that avoids it. We know that $\mathbb{Z}^n \subset \Lambda$ since $\mathbb{Z}^n = X^*(T)$ is the weight lattice of $SO(N)$, contained in $\Lambda = X^*(\tilde{T})$ by means of the homomorphism $\tilde{T} \to T$. By (31.1) $\mathbb{Z}^n$ is a subgroup of index two in Λ. Since $\mathbb{Z}^n$ is of index two in $\tilde{\Lambda}$ and since $\Lambda \subseteq \tilde{\Lambda}$ by Proposition 18.10, we see that $\Lambda = \tilde{\Lambda}$.   □

From (31.3), we can compute the fundamental dominant weights ϖ_i. If $N = 2n + 1$ is odd, these are

$$\varpi_1 = \mathbf{e}_1,$$
$$\varpi_2 = \mathbf{e}_1 + \mathbf{e}_2,$$
$$\vdots$$
$$\varpi_{n-1} = \mathbf{e}_1 + \mathbf{e}_2 + \ldots + \mathbf{e}_{n-1},$$
$$\varpi_n = \tfrac{1}{2}(\mathbf{e}_1 + \mathbf{e}_2 + \ldots + \mathbf{e}_{n-1} + \mathbf{e}_n).$$

On the other hand, if $N = 2n$ is even, the last two are a little changed. In this case, the fundamental weights are

$$\varpi_1 = \mathbf{e}_1,$$
$$\varpi_2 = \mathbf{e}_1 + \mathbf{e}_2,$$
$$\vdots$$
$$\varpi_{n-1} = \tfrac{1}{2}(\mathbf{e}_1 + \mathbf{e}_2 + \ldots + \mathbf{e}_{n-1} - \mathbf{e}_n),$$
$$\varpi_n = \tfrac{1}{2}(\mathbf{e}_1 + \mathbf{e}_2 + \ldots + \mathbf{e}_{n-1} + \mathbf{e}_n).$$

Of course, to check the correctness of these weights, what one must check is that $2\langle \varpi_i, \alpha_j \rangle / \langle \alpha_j, \alpha_j \rangle = 1$ if $i = j$, and 0 if $i \neq j$, and this is easily done.

We say that a weight is *integral* if it is in $X^*(T)$ and *half-integral* if it is not. Thus a weight is integral if it is of the form (31.4) with the c_i even, and half-integral if they are odd. Dominant integral weights, of course, are highest weight vectors of representations of $\mathrm{SO}(N)$. By Proposition 31.2, the dominant half-integral weights are highest weight vectors of representations of $\mathrm{Spin}(N)$. They are *not* highest weight vectors of representations of $\mathrm{SO}(N)$.

If $N = 2n+1$, we see that just the last fundamental weight is half-integral, but if $N = 2n$, the last two fundamental weights are half-integral. The representations with highest weight vectors ϖ_n (when $N = 2n + 1$) or ϖ_{n-1} and ϖ_n (when $N = 2n$) are called the *spin representations*.

Theorem 31.2. *(i) If* $N = 2n + 1$, *the dimension of the spin representation* $\pi(\varpi_n)$ *is* 2^n. *The weights that occur with nonzero multiplicity in this representation all occur with multiplicity one; they are*

$$\tfrac{1}{2}(\pm \mathbf{e}_1 \pm \mathbf{e}_2 \pm \ldots \pm \mathbf{e}_n).$$

(ii) If $N = 2n$, *the dimensions of the spin representations* $\pi(\varpi_{n-1})$ *and* $\pi(\varpi_n)$ *are each* 2^{n-1}. *The weights that occur with nonzero multiplicity in this representation all occur with multiplicity one; they are*

$$\tfrac{1}{2}(\pm \mathbf{e}_1 \pm \mathbf{e}_2 \pm \ldots \pm \mathbf{e}_n),$$

where the number of minus signs is odd for $\pi(\varpi_{n-1})$ *and even for* $\pi(\varpi_n)$.

Proof. There is enough information in Proposition 22.4 to determine the weights in the spin representations.

Specifically, let $\lambda = \varpi_n$ and $N = 2n+1$ or $2n$, or $\lambda = \varpi_{n-1}$ if $N = 2n$. Let $S(\lambda)$ be as in Exercise 22.1. Then it is not hard to check that $S(\lambda)$ is exactly the set of weights stated in the theorem. By Proposition 22.4, $S(\lambda) \supseteq \operatorname{supp} \chi_\lambda$. On the other hand, it is easy to check that $S(\lambda)$ consists of a single Weyl group orbit, namely the orbit of the highest weight vector λ, so $S(\lambda) \subseteq \operatorname{supp} \chi_\lambda$, and, for this orbit, Proposition 22.4 also tells us that each weight appears in χ_λ with multiplicity exactly one. $\qquad\qquad\qquad\qquad\qquad\qquad\square$

The center of $\mathrm{SO}(N)$ consists of $\{\pm I_N\}$ if N is even but is trivial if N is odd. The center of $\mathrm{Spin}(N)$ is more subtle, but we now have the tools to compute it.

Theorem 31.3. *If $N = 2n + 1$, then $Z(G) \cong \mathbb{Z}/2\mathbb{Z}$. If $N = 2n$, then $Z(G) \cong \mathbb{Z}/4\mathbb{Z}$ if n is odd, while $Z(G) \cong (\mathbb{Z}/2\mathbb{Z}) \times (\mathbb{Z}/2\mathbb{Z})$ if n is even.*

Proof. $X^*(\tilde{T})$ is described explicitly by Proposition 31.2, and we have also described the simple roots, which generate Λ_{root}. We leave the verification that $X^*(\tilde{T})/\Lambda_{\mathrm{root}}$ is as described to the reader. The result follows from Theorem 23.2. $\qquad\qquad\qquad\qquad\qquad\qquad\qquad\qquad\qquad\qquad\square$

Now let us consider the spin representations from a different point of view. If V is a complex vector space of dimension N with a nondegenerate quadratic form q, and if $W \subset V$ is a maximal subspace on which q restricts to zero, we will call W *Lagrangian*. We will see that the dimension of such a Lagrangian subspace W is n where $N = 2n$ or $2n + 1$, so the exterior algebra $\bigwedge W$ has dimension 2^n, and we will construct the spin representation on this vector space.

To construct the spin representation, we will make use of the properties of Clifford algebras. We digress to develop what we need. For more about Clifford algebras, see Artin [9], Chevalley [36], Goodman and Wallach [56], and Lawson and Michelsohn [118].

By a $\mathbb{Z}/2\mathbb{Z}$-*graded algebra* we mean an F-algebra A that decomposes into a direct sum $A_0 \oplus A_1$, where A_0 is a subalgebra, with $A_i \cdot A_j \subseteq A_{i+j}$ where $i+j$ is modulo 2. We require that F be contained in A_0 and that F is central in A, but A_0 may be strictly larger than F. An element a of A is called *homogeneous* if it is in A_i with $i \in \{0, 1\}$ and then we call $\deg(a) = i$ the degree of a.

If A and B are $\mathbb{Z}/2\mathbb{Z}$-graded algebras, then we may define a $\mathbb{Z}/2\mathbb{Z}$-graded algebra $A \otimes B$. As a vector space, this is the usual tensor product of A and B, with the following $\mathbb{Z}/2\mathbb{Z}$ grading:

$$(A \otimes B)_0 = A_0 \otimes B_0 \oplus A_1 \otimes B_1, \qquad (A \otimes B)_1 = A_1 \otimes B_0 \oplus A_0 \otimes B_1.$$

The multiplication involves a sign as follows. It is sufficient to define the product of two homogeneous elements, and then we define

$$(a \otimes b)(a' \otimes b') = (-1)^{\deg(b) \cdot \deg(a')} aa' \otimes bb'. \tag{31.5}$$

Every $\mathbb{Z}/2\mathbb{Z}$-graded algebra A has an automorphism of order 2 that is 1 on A_0 and -1 on A_1. We will denote this operation by $a \longmapsto \bar{a}$. We will encounter the algebra $M = \text{Mat}_2(F)$ with the following $\mathbb{Z}/2\mathbb{Z}$-grading: M_0 consists of matrices $\begin{pmatrix} a & b \\ c & d \end{pmatrix}$ with $b = c = 0$, and M_1 consists of matrices with $a = d = 0$. Now if A is a $\mathbb{Z}/2\mathbb{Z}$-graded algebra, then we may identify elements of $A \otimes M$ with 2×2 matrices with coefficients in A by mapping

$$a \otimes \begin{pmatrix} 1 & 0 \\ 0 & 0 \end{pmatrix} + b \otimes \begin{pmatrix} 0 & 1 \\ 0 & 0 \end{pmatrix} + c \otimes \begin{pmatrix} 0 & 0 \\ 1 & 0 \end{pmatrix} + d \otimes \begin{pmatrix} 0 & 0 \\ 0 & 1 \end{pmatrix}$$

to the matrix $\begin{pmatrix} a & b \\ c & d \end{pmatrix}$. However we do not use ordinary matrix multiplication in this ring. Indeed, by the sign rule (31.5) the multiplication in this "matrix ring" is twisted by conjugation:

$$\begin{pmatrix} a & b \\ c & d \end{pmatrix} \begin{pmatrix} a' & b' \\ c' & d' \end{pmatrix} = \begin{pmatrix} aa' + b\overline{c'} & ac' + b\overline{d'} \\ ca' + dc' & c\overline{b'} + dd' \end{pmatrix}.$$

Let us denote this $\mathbb{Z}/2\mathbb{Z}$-graded algebra $M \otimes A$ as $M(A)$.

We recall that a ring is *simple* if it has no proper nontrivial ideals.

Proposition 31.3. *If A is a simple $\mathbb{Z}/2\mathbb{Z}$-graded algebra then so is $M(A)$.*

Proof. Let I be a nonzero ideal. If $m = \begin{pmatrix} a & b \\ c & d \end{pmatrix}$ is a nonzero element of I, then one of a, b, c, d is nonzero. Left and/or right multiplying by $\begin{pmatrix} 0 & 1 \\ 1 & 0 \end{pmatrix}$ we may assume that $a \neq 0$. Then left and right multiplying by $\begin{pmatrix} 1 & 0 \\ 0 & 0 \end{pmatrix}$ we may assume $m = \begin{pmatrix} a & 0 \\ 0 & 0 \end{pmatrix}$. Since A is simple, I contains $\begin{pmatrix} 1 & 0 \\ 0 & 0 \end{pmatrix}$. Similarly it contains $\begin{pmatrix} 0 & 0 \\ 0 & 1 \end{pmatrix}$. Adding these two elements $\begin{pmatrix} 1 & 0 \\ 0 & 1 \end{pmatrix}$ is in the ideal I, which is thus not proper. $\square$

We will also encounter the $\mathbb{Z}/2\mathbb{Z}$-graded algebra $D(F)$ which is a two-dimensional algebra over F generated by an element ζ of $\mathbb{Z}/2\mathbb{Z}$-degree 1 that satisfies $\zeta^2 = -1$. Then $A \otimes D(F)$ will be denoted $D(A)$.

Proposition 31.4. *In the graded ring $D(A)$ we have $D(A)_0 \cong A$ as a ring.*

Proof. We may identify $D(A)$ with $A \oplus A$ as a vector space in which $a \otimes 1 + b \otimes \zeta$ is identified with the ordered pair (a, b). In view of (31.5) the multiplication is

$$(a, b)(c, d) = (ac - b\bar{d}, ad + b\bar{c}).$$

Now $D(A)_0$ consists of pairs (a, b) with a of degree zero and b of degree 1, and for this subring, the multiplication is

$$(a, b)(c, d) = (ac + bd, ad + bc).$$

Now every element of A can be written uniquely as $a + b$ with $a \in A_0$ and $b \in A_1$. Then $a + b \longmapsto (a, b)$ is clearly an isomorphism $A \longrightarrow D(A)_0$.   □

Let F be a field, which for simplicity we assume has characteristic not equal to 2. By a *quadratic space* we mean a vector space V (over F) together with a symmetric bilinear form $B = B_V : V \times V \longrightarrow F$. We say that the quadratic space V is *nondegenerate* if the symmetric bilinear form B is non-degenerate. Let $q(x) = q_V(x) = B(x, x)$. This is a quadratic form, and giving B is equivalent to giving q since $B(x, y) = \frac{1}{2}(q(x + y) - q(x) - q(y))$.

The *Clifford algebra* $C(V)$ will be an F-algebra characterized by a universal property: it comes with a map $\iota : V \longrightarrow C(V)$ and if $x, y \in V$ then $\iota(x)^2 = q(x)$. The universal property is that if A is any F-algebra with a linear map $j : V \longrightarrow A$ satisfying $j(x)^2 = q(x)$ in A, then there exists a unique algebra homomorphism $J : C(V) \longrightarrow A$ such that $j = J \circ \iota$.

Instead of verifying $j(x)^2 = q(x)$ it may be more convenient to verify the equivalent condition $j(x)j(y) + j(y)j(x) = 2B(x, y)$, for $x, y \in V$ since the latter condition is linear, so it is sufficient to verify it on a subset of V that spans it. The bilinear condition is equivalent to $j(x)^2 = q(x)$ since

$$xy + yx = (x + y)^2 - x^2 - y^2, \qquad 2B(x, y) = q(x + y) - q(x) - q(y).$$

In order to construct the Clifford algebra, we may take the tensor algebra $T(V)$ modulo the ideal $\mathfrak{I} = \mathfrak{I}_V$ generated by elements of the form $x \otimes y + y \otimes x - 2B(x, y)$ with $x, y \in V$.

Proposition 31.5. (i) *The Clifford algebra is a $\mathbb{Z}/2\mathbb{Z}$-graded algebra.*
(ii) *Suppose that V is the orthogonal direct sum of two subspaces U and W. Then $C(U \oplus W) \cong C(U) \otimes C(W)$.*
(iii) *The dimension of $C(V)$ is $2^{\dim(V)}$.*
(iv) *The map $i : V \longrightarrow C(V)$ is injective.*
(v) *If $v_1, v_2, \ldots, v_d$ is a basis of V, then the set of products*

$$v_{i_1} v_{i_2} \ldots v_{i_k} \qquad (1 \leqslant i_1 < i_2 < \ldots < i_k \leqslant d)$$

is a basis of $C(V)$. Here we are using (iii) to identify $v \in V$ with $i(v) \in C(V)$.

Proof. The tensor algebra $T = T(V)$ is a graded algebra in which the homogeneous part of degree k is $\otimes^k V$. Let

$$T_i = \bigoplus_{k \equiv i \bmod 2} \otimes^k V, \qquad (i = 0, 1).$$

Let $\mathfrak{R}$ be the vector space in T spanned by the relations $x \otimes y + y \otimes x - 2B(x, y)$ so that $T\mathfrak{R}T = \mathfrak{I}$. Clearly $\mathfrak{R} \subset T_0$ and so the ideal $\mathfrak{I}$ is homogeneous in the sense that $\mathfrak{I} = (\mathfrak{I} \cap T_0) \oplus (\mathfrak{I} \cap T_1)$. This implies that the quotient $C(V) = T(V)/\mathfrak{I}$ inherits the $\mathbb{Z}/2\mathbb{Z}$-grading from $T(V)$.

For (ii), we have linear maps $i_U : U \longrightarrow C(U)$ and $i_W : W \longrightarrow C(W)$. Define $j : U \oplus W \longrightarrow C(U) \otimes C(W)$ by $j(u) = i_U(u) \otimes 1$ on U and $j(w) = 1 \otimes i_W(w)$ on W. Using the fact that U and W are orthogonal, we have $uw = -wu$ in $C(U \oplus W)$, from which it follows that $j(x)^2 = q(x)$ for $x \in U \oplus W$. (Indeed $j(x)^2$ and $q(x)$ equal $q(u) + q(w)$ if $x = u + w$ with $u \in U$ and $w \in W$.) Therefore there exists a ring homomorphism $J : C(U \oplus W) \longrightarrow C(U) \otimes C(W)$ such that $J \circ i_{U \oplus W} = j$. The map is surjective since its image contains the generators $i_U U \otimes 1$ and $1 \otimes i_W W$. To see that it is injective, we compose it with the canonical map $T(U \oplus W) \longrightarrow C(U \oplus W)$. The kernel $\mathfrak{I}_{U \oplus W}$ is generated by the relations $x \otimes y + y \otimes x - 2B(x, y)$ with x and y in either U or W, and in each of the four cases, these are mapped to zero by j. Hence the induced map $C(U \oplus W) \longrightarrow C(U) \otimes C(W)$ is injective, and indeed is an isomorphism.

Next let us show that $\dim C(V) = 2^{\dim(V)}$. We will argue by induction on $\dim(V)$. If $\dim(V) = 1$, then let v be a basis vector and $a = q(v)$. The ring $C(V)$ is easily seen to be spanned as an F-vector space by 1 and v with one relation $v^2 = a$, and clearly v is not zero. Thus $\dim C(V) = 2 = 2^{\dim(V)}$.

So by induction we may assume that $\dim(V) > 1$. We may always find a nonzero vector u and a vector subspace W of codimension 1 that is orthogonal to u. Indeed, if the bilinear form B is degenerate, we may take u to be a nonzero element of the kernel, and W to be any vector space of codimension 1 not containing u. On the other hand, if V is nondegenerate, we may find a vector u with $q(u) \neq 0$; in this case we take W to be the orthogonal complement of $U = Fu$. Now $C(V) \cong C(U) \otimes C(W)$. By induction, $C(U)$ has dimension $2^{\dim(U)}$ and $C(W)$ has dimension $2^{\dim(W)}$ and the statement follows.

If $v_1, \ldots, v_d$ are a basis of V, it is easy to see using the generating relations that the vector space spanned by

$$i_V(v_{i_1}) \ldots i_V(v_{i_d}), \qquad i_1 < \ldots < i_d$$

is closed under multiplication, so these span $C(V)$. Since they are $2^{\dim(V)}$ in number, they are linearly independent. This proves (v) and (vi). $\square$

A vector v is called *isotropic* if $q(v) = 0$. Similarly, a subspace W of the quadratic space V is *isotropic* if $B(x, y) = 0$ for $x, y \in W$. If $F = \mathbb{R}$ there may be no nonzero isotropic subspaces (if the quadratic form is positive definite), but if F is $\mathbb{C}$ and V is nondegenerate, we will see that the dimension of a maximal isotropic subspace W of V will be n if $\dim(V) = 2n$ or $2n + 1$. If $\dim(W) = n$ and $\dim(V) = 2n$ or $2n + 1$ we call the isotropic subspace W *Lagrangian*. It follows from Witt's theorem (see Lang [116]) that maximal isotropic subspaces are conjugated transitively by $O(N)$, and if V is nondegenerate these are the Lagrangian subspaces, provided Lagrangian subspaces exist. This is always true if F is algebraically closed.

Let V be a two-dimensional quadratic space. Then V is called a *hyperbolic plane* if it is nondegenerate, and if V has a basis x and y of linearly independent isotropic vectors. We may multiply x by a nonzero constant and also assume that $B(x, y) = \frac{1}{2}$.

Proposition 31.6. *If V is a hyperbolic plane then $C(V) \cong M(F)$ as $\mathbb{Z}/2\mathbb{Z}$-graded algebras.*

Proof. Let $X = \begin{pmatrix} 0 & 1 \\ 0 & 0 \end{pmatrix}$ and $Y = \begin{pmatrix} 0 & 0 \\ 1 & 0 \end{pmatrix}$. With x, y such that $q(x) = q(y) = 0$ and $B(x, y) = \frac{1}{2}$, we have $x^2 = y^2 = 0$ and $xy + yx = 1$ in $C(V)$. Since X and Y satisfy the same relations, the universal property of the Clifford algebra implies that there is a homomorphism $C(V) \longrightarrow \mathrm{Mat}_2(F)$. Since $\mathrm{Mat}_2(F)$ is generated by X and Y, the map is surjective, and since both algebras have dimension four, it is an isomorphism. The $\mathbb{Z}/2\mathbb{Z}$-gradings are compatible. $\square$

Lemma 31.1. *Assume that F is algebraically closed and V is nondegenerate. If $\dim(V) \geqslant 2$ then V may be decomposed as $V_0 \oplus V'$ where V_0 is a hyperbolic plane and V' is its orthogonal complement.*

Proof. Let v be any vector with $q(v) \neq 0$, and let w be any nonzero vector in the orthogonal complement of v. Then $q(w) \neq 0$ also since V otherwise it is in the kernel of the associated symmetric bilinear form B, but B is nondegenerate. Let $q(v) = a^2$ and $q(w) = b^2$. Then $x = bv - aw$ and $y = bv + aw$ are linearly independent isotropic vectors since $a, b \neq 0$. Clearly the space V_0 spanned by x and y is a hyperbolic plane, and we may take V' to be its orthogonal complement. $\square$

Proposition 31.7. *If F is algebraically closed and V is a nondegenerate quadratic space of dimension $2n$ or $2n + 1$, then V contains Lagrangian subspaces W and W' such that $W \cap W' = 0$, and B induces a nondegenerate pairing $W \times W' \longrightarrow F$. If $\dim(V) = 2n + 1$ then the one-dimensional orthogonal complement of $W + W'$ is spanned by a vector z such that $q(z) = 1$.*

Proof. Using the Lemma 31.1 repeatedly, we may decompose $V = V_1 \oplus V_2 \oplus \ldots \oplus V_n \oplus V'$ where V_i are hyperbolic planes and V' is either zero or one-dimensional. Each V_i is spanned by two isotropic vectors x_i and y_i such that $B(x_i, y_i) = \frac{1}{2}$. Let W be the space spanned by the x_i and W' be the space spanned by the y_i. We have $x_i y_j + y_j x_i = \delta_{ij}$ (Kronecker delta). If $\dim(V) = 2n + 1$ then the orthogonal complement V_0 of $W + W'$ is one-dimensional, and since V is nondegenerate, if q is a basis vector then $q(z) \neq 0$ for nonzero $z \in V_0$. Since F is algebraically closed, we may scale z so that $q(z) = 1$. $\square$

Let us assume that $\dim(V) = 2n$ or $2n + 1$. Let us also assume that V has a decomposition $V = W \oplus W' \oplus V_0$ where W and W' are Lagrangian subspaces (i.e. isotropic subspaces of dimension n) that are dually paired by B. Of course W and W' are not orthogonal, but the space V_0 is assumed to be the orthogonal complement of $W \oplus W'$. It is zero if $\dim(V) = 2n$ but it is one-dimensional if $\dim(V) = 2n + 1$, and in this case we assume that it is spanned by a vector z with $q(z) = 1$. We call such a decomposition $V = W \oplus W' \oplus V_0$ a *Lagrangian decomposition*. By Proposition 31.7 there is always a Lagrangian decomposition if F is algebraically closed.

Given a Lagrangian decomposition of V we will describe a representation of $C(V)$ in the exterior algebra $\bigwedge W$ on W. This module is known as the *Fermionic Fock space* and we will denote it as Ω.

Proposition 31.8. *Given a Lagrangian decomposition $V = W \oplus W' \oplus V_0$ of a nondegenerate quadratic space, let $\Omega = \bigwedge W$. There exists an algebra homomorphism $\omega : C(V) \longrightarrow \mathrm{End}(W)$ in which, for $\xi \in \Omega$ we have*

$$\omega(x)\xi = x \wedge \xi, \qquad x \in W, \tag{31.6}$$

$$\omega(y)\xi = \sum_{i=1}^{k} 2B(y, w_i)(-1)^{i+1} w_1 \wedge \ldots \wedge \widehat{w_i} \wedge \ldots \wedge w_k, \qquad y \in W'. \tag{31.7}$$

if $\xi = w_1 \wedge \ldots \wedge w_k$, where the "hat" over $\widehat{w_i}$ means that this factor is omitted. Also if $\dim(V) = 2n + 1$, let z be the chosen element of V_0 with $q(z) = 1$. If $\xi \in \Omega$ is homogeneous of degree i, then $\omega(z)\xi = (\ 1)^i \xi$.

Proof. We can define ω by the (31.6) and (31.7) and (if N is odd) the requirement that $\omega(z)\xi = (-1)^i \xi$. Regarding (31.7) this is well-defined by the universal property of the exterior power because it is easy to check that the right-hand side is multiplied by -1 if w_i and w_{i+1} are interchanged, so it is alternating. We have to check that ω is an algebra homomorphism.

We will show that if $x, y \in V$ then

$$\omega(x)\omega(y) + \omega(y)\omega(x) = 2B(x, y). \tag{31.8}$$

If $x, y \in W$ or $x, y \in W'$, both sides are zero. If $x \in W$ and $y = w'$, and $\xi = w_1 \wedge \ldots \wedge w_k$ then $\omega(y)\omega(x)\xi$ consists of $k + 1$ terms. All but one of these are the k terms in $\omega(x)\omega(y)\xi$ but with opposite sign, and the one term that is not cancelled equals $2B(y, x)\xi$, as required. This proves (31.8) in the case $\dim(V) = 2n$. If $\dim(V) = 2n + 1$ we have also to check this if $y = z$ and either $x = z$ or $x \in W$ or $x \in W'$. If $x \in W$ or W', both sides of (31.8) vanish by definition of $\omega(z)$ since $\omega(x)$ has graded degree ± 1. If $x = y = z$, then both sides of (31.8) are multiplication by 2, so (31.8) is proved.

By the universal property of the Clifford algebra, (31.8) implies that there is a homomorphism $C(V) \longrightarrow \mathrm{End}(\Omega)$ as required.   □

We will denote by $C_i(V)$ the homogeneous part of degree i in the $\mathbb{Z}/2\mathbb{Z}$-grading. In other words, with $i = 0$ or 1, $C_i(V) = A_i$ if $A = C(V)$.

Theorem 31.4. *Let V be a quadratic space with a nondegenerate symmetric bilinear form and a Lagrangian decomposition $V = W \oplus W' \oplus V_0$, and let*

$\Omega = \bigwedge W$ as in Proposition 31.8. Assume that the ground field contains an element i such that $i^2 = -1$. Let $R = C(V)$ if $\dim(V) = 2n$ and $R = C_0(V)$ if $\dim(V) = 2n + 1$. Then R is a simple ring with Ω its irreducible module, and in fact the representation $\pi : R \longrightarrow \mathrm{End}(\Omega)$ in Proposition 31.8 is an isomorphism.

Proof. Since by Lemma 31.1 the even-dimensional subspace $W \oplus W'$ is an orthogonal direct sum of hyperbolic planes, it follows from Proposition 31.6 that $A = C(W \oplus W')$ is a simple algebra. If $\dim(V) = 2n$ then $R = A$.

On the other hand, suppose that $\dim(V) = 2n + 1$. Then taking $\zeta = iz$, where $q(z) = 1$ as in Proposition 31.8, we see that $C(V_0) \cong D(F)$. Therefore $C(V) \cong D(A)$ where $A = C(W \oplus W')$ and by Proposition 31.4 we have $R = D(A)_0 \cong A$.

In either case, it is a simple algebra by Proposition 31.3. The homomorphism $\pi : R \longrightarrow \mathrm{End}(\Omega)$ must be an isomorphism since it cannot have a kernel (by the simplicity of R) and both rings have the same dimension 2^{2n}. $\square$

We will construct a representation of the $\mathrm{Spin}(N)$ on Ω (with Ω as in Theorem 31.4) by first constructing a *projective representation* of $\mathrm{O}(N)$. We therefore digress to review projective representations and their relations to true representations.

If V is a complex vector space, $\mathrm{PGL}(V)$ is $\mathrm{GL}(V)/Z$ where Z is the center of $\mathrm{GL}(V)$, that is, the group of scalar linear transformations of V. Let $P : \mathrm{GL}(V) \longrightarrow \mathrm{PGL}(V)$ be the projection map. A *projective representation* of a group G a homomorphism $\pi : G \longrightarrow \mathrm{PGL}(V)$. Equivalently, we may describe the projective representation by giving a map $\pi' : G \longrightarrow \mathrm{GL}(V)$ such that $P \circ \pi' = \pi$.

We review the connection between projective representations and central extensions. By a *central extension* of G by an Abelian group A we mean a group $\hat{G}$ with an subgroup isomorphic to A contained in its center, such that (identifying this subgroup with A) we have $\hat{G}/A \cong G$. In other words we have a short exact sequence

$$1 \longrightarrow A \longrightarrow \hat{G} \xrightarrow{p} G \longrightarrow 1$$

with the image of A contained in the center of $\hat{G}$. We are interested in the case where $A \subseteq \mathbb{C}^\times$, the group of complex numbers of absolute value 1.

Suppose $(\hat{\pi}, V)$ is a representation of $\hat{G}$. Assume that $\hat{\pi}(a)$ is a scalar linear transformation for all $a \in A$. For example, by Schur's Lemma, this is true if $\hat{\pi}$ is irreducible. Then the map $P \circ \hat{\pi}$ from $\hat{\pi}$ to $\mathrm{PGL}(V)$ is constant on the cosets of A. It thus gives a projective representation of G, the *projective representation associated with* $\hat{\pi}$.

Proposition 31.9. *Suppose that $\pi : G \longrightarrow \mathrm{PGL}(V)$ is a projective representation of G. Then there exists a central extension $\hat{G}$ of G by $\mathbb{C}^\times$ and a representation of $\hat{G}$ such that π is the projective representation associated with $\hat{\pi}$.*

Proof. Choose a map $\pi' : G \longrightarrow \mathrm{GL}(V)$ such that $P \circ \pi' = \pi$. Then $\pi'(g_1 g_2)$ differs from $\pi'(g_1)\pi'(g_2)$ by a scalar linear transformation, since these have the same image under P. Thus there is a map $\phi : G \times G \longrightarrow \mathbb{C}^\times$ such that

$$\pi'(g_1)\pi'(g_2) = \phi(g_1, g_2)\pi'(g_1 g_2) \tag{31.9}$$

Applying π' to $(g_1 g_2)g_3 = g_1(g_2 g_3)$ gives the "cocycle relation"

$$\phi(g_1, g_2)\phi(g_1 g_2, g_3) = \phi(g_1, g_2 g_3)\phi(g_2, g_3).$$

Let $\hat{G}$ be (as a set) the Cartesian product $G \times \mathbb{C}^\times$, and we make it a group by defining

$$(g_1, \varepsilon_1)(g_2, \varepsilon_2) = (g_1 g_2, \phi(g_1, g_2)\varepsilon_1 \varepsilon_2). \tag{31.10}$$

The cocycle relation implies that this group law is associative.

Now define $\hat{\pi} : \hat{G} \longrightarrow \mathrm{GL}(V)$ by

$$\hat{\pi}(g, \varepsilon) = \varepsilon \pi'(g). \tag{31.11}$$

Then it is easy to see that (31.9) and (31.10) imply that $\hat{\pi}$ is a representation, and it is clear that the associated projective representation is π. $\qquad \square$

Remark 31.1. If the map ϕ in (31.9) takes values in a subgroup A of $\mathbb{C}^\times$, then we may obtain a true representation of a central extension of G by A by exactly the same construction: as a set, the extension is $G \times A$, and the multiplication is defined by the same formula (31.10). For example if G is a simply-connected Lie group the next Proposition shows that we may we may even take $A = 1$ and obtain a true representation of G.

Proposition 31.10. *Suppose that G is a simply-connected Lie group, and let $\pi : G \longrightarrow \mathrm{PGL}(V)$ be a projective representation. Then there exists a representation $\hat{\pi} : G \longrightarrow \mathrm{GL}(V)$ such that π is the projective representation associated with $\hat{\pi}$. Moreover we may assume that $\hat{\pi}(G) \subseteq \mathrm{SL}(V)$.*

Proof. Let $d = \dim(V)$. The natural map $\mathrm{SL}(V) \longrightarrow \mathrm{PGL}(V)$ has kernel of order d, consisting of scalar linear transformations εI_V where ε is an d-th root of unity. Hence this is a covering map. Since G is simply-connected, by Proposition 13.4 we may find a continuous map $\pi' : G \longrightarrow \mathrm{SL}(V)$ such that $P \circ \pi' = \pi$. Now as in the proof of Proposition 31.9 we may define $\phi : G \times G \longrightarrow \mathrm{SL}(V)$ such that (31.9) is true, and ϕ is continuous since π' is. Taking determinants on both sides, $\phi(g)$ is an d-th root of unity. Proceeding as in the proof of Proposition 31.9 we then obtain a true representation $\hat{\pi}$ of a central extension

$$1 \longrightarrow \mu_d \longrightarrow \hat{G} \xrightarrow{p} G \longrightarrow 1$$

where μ_d is the group of d-th roots of unity in $\mathbb{C}$. Since μ_d is discrete, the restriction of p to the connected component $\hat{G}^\circ$ of the identity in $\hat{G}$ is a covering map, but since G is simply-connected, this restriction is an isomorphism. Let $s : G \longrightarrow \hat{G}^\circ$ be the inverse map. Then $\hat{\pi} = \hat{\pi}' \circ s$ is a true representation of G whose associated projective representation is π. By (31.11) this construction gives it values in $\mathrm{SL}(V)$. $\qquad \square$

Now the method by which we will construct the spin representations of orthogonal groups, or more precisely their double covers, may be revealed. The Clifford algebra has a crucial property: it has only one or two classes of simple modules, and if Ω is such a module, then $\mathrm{O}(V)$ gets a projective representation on Ω by the following Proposition. It follows that we have a representation of a central extension of $\mathrm{O}(V)$, and these are the spin representations.

Proposition 31.11. *Let G be a group, and let R be a $\mathbb{C}$-algebra that has a unique isomorphism class of simple modules. Let Ω be such a module, and let $\omega : R \longrightarrow \mathrm{End}_{\mathbb{C}}(\Omega)$ be the $\mathbb{C}$-algebra homomorphism defined by $\omega(r)v = r \cdot v$. Let $\rho : G \longrightarrow \mathrm{Aut}(R)$ be a group homomorphism. Then there exists a projective representation $\pi : G \longrightarrow \mathrm{GL}(\Omega)$ such that for $r \in R, g \in G$ and $v \in \Omega$ we have*

$$\pi(g)\,\omega(r) = \omega\big(\rho(g)r\big)\pi(g). \tag{31.12}$$

Proof. Given $g \in G$, define another R-module structure on Ω by means of the homomorphism ${}^g\omega : R \longrightarrow \mathrm{End}_{\mathbb{C}}(\Omega)$ given by ${}^g\omega\big(\rho(g)r\big) = \omega(r)$, $r \in R$. Denote by ${}^g\Omega$ the vector space Ω with this R-module structure. Since R has a unique isomorphism class of simple modules, we may find a $\pi(g) : \Omega \longrightarrow {}^g\Omega$ that is an R-module homomorphism. By Schur's Lemma, it is determined up to isomorphism. The fact that it is an R-module homomorphism amounts to the identity (31.12). To show that π is a projective representation, we need to show that $\pi(g_1)\pi(g_2)$ and $\pi(g_1 g_2)$ are the same, up to a constant multiple. Indeed, both satisfy (31.12) with $g = g_1 g_2$, and so these two endomorphisms are proportional. □

To apply this, we may take $R = C(V)$ if $\dim(V) = 2n$ or $C(V)_0$ if $\dim(V) = 2n + 1$ as in Theorem 31.4. Since $\mathrm{O}(V)$ acts by automorphisms on V, hence on R, we obtain a projective representation of $\mathrm{O}(V)$ on $\Omega = \bigwedge W$ for W a Lagrangian subspace. In order to apply Proposition 31.10, we restrict to the connected subgroup $\mathrm{SO}(V)$, whose universal cover we denote $\mathrm{Spin}(V)$. Except in the case where $\dim(V) = 2$, we have already shown that this is a central extension such that the cover map $\mathrm{Spin}(V) \longrightarrow \mathrm{SO}(V)$ has degree 2. We see that there exists a true representation $\pi : \mathrm{Spin}(V) \longrightarrow \mathrm{GL}(\Omega)$, and that the image of this may be taken inside of $\mathrm{SL}(\Omega)$.

To compare this with our previous computation of the spin representations, let $w_1, \ldots, w_n$ be a basis of W, and let $w'_1, \ldots, w'_n$ be the dual basis of W', characterized by $2B(w_i, w'_j) = \delta_{ij}$. If $\dim(V)$ is even, then $w_1, \ldots, w_n, w'_n, \ldots, w'_1$ form a basis $\mathcal{B}$ of V; if $\dim(V)$ is odd, we supplement these by a basis vector v_0 of V_0, and let $\mathcal{B}$ be $w_1, \ldots, w_n, v_0, w'_n, \ldots, w'_1$. Let T be the maximal torus of $\mathrm{SO}(V)$ that is diagonal with respect to the basis $\mathcal{B}$. If we identify these with the standard basis of $\mathbb{C}^N$ then T consists of the elements (31.2). Let $\tilde{T}$ be the preimage of T in $\mathrm{Spin}(V)$.

Proposition 31.12. *Let $\tilde{t} \in \tilde{T}$. Assume with the above identifications that the image of t in T is (31.2). Then the eigenvalues of $\sigma(t)$ are the 2^n values $\prod t_i^{\pm 1/2}$ for an appropriate choice of the square roots.*

Proof. Let $t \in T$ be the element corresponding to $\tilde{t} \in \tilde{T}$. By (31.12) we have, for $r \in R$

$$\sigma(\tilde{t})\omega(r) = \omega(\rho(t)r)\sigma(\tilde{t}). \tag{31.13}$$

Here $\rho : \mathrm{SO}(V) \longrightarrow \mathrm{Aut}(R)$ is obtained by extending the action of $\mathrm{SO}(V)$ on V to automorphisms of the Clifford algebra, and $\omega : R \longrightarrow \mathrm{End}(\Omega)$ is the representation of Proposition 31.8. By (31.7) the vector $1 \in \Omega$ is characterized by being the unique (up to constant multiple) nonzero vector annihilated by $\omega(W')$, and since $\rho(t)W' = W'$ it follows from this characterization that $1 \in W$ is an eigenvector of $\sigma(\tilde{t})$. Let $\sigma(\tilde{t})1 = \lambda 1$, where $\lambda \in \mathbb{C}^\times$ is to be determined.

Let $r = w_{i_1} \ldots w_{i_k}$ with $i_1 < \ldots < i_k$, where the multiplication is in the Clifford algebra. We have $\rho(t)r = t_{i_1} \ldots t_{i_k} r$ so by (31.13) we see that $\omega(r)1$ is also an eigenvector of $\sigma(\tilde{t})$, with eigenvalue $t_{i_1} \ldots t_{i_k} \lambda$. By (31.6) we have $\omega(r)1 = w_{i_1} \wedge \ldots \wedge w_{i_k}$, so a basis of Ω consisting of eigenvectors of $\tilde{T}$ consist of the elements $w_{i_1} \wedge \ldots \wedge w_{i_k}$ and the eigenvalues are $\lambda t_{i_1} \ldots t_{i_k}$. Now since $\sigma(\tilde{T}) \subset \mathrm{SL}(\Omega)$, the product of these eigenvalues is 1, that is

$$\left(\lambda^2 \prod_{i=1}^{n} t_i \right)^{2^{n-1}} = 1.$$

Now $\lambda^2 \prod_{i=1}^{n} t_i$ must depend continuously on $\tilde{t}$, and since it is a 2^{n-1}-st root of unity, it is constant. Clearly $\lambda = 1$ when $\tilde{t} = 1$, so $\lambda^2 \prod_{i=1}^{n} t_i = 1$. Therefore we may write $\lambda = \prod_{i=1}^{n} t_i^{-1/2}$ for some choice of the square root, and the statement follows. $\square$

Comparing Proposition 31.12 with Theorem 31.2, we see from this computation of the character that the representation we have constructed is the same spin representation $\pi(\varpi_n)$ described in that theorem when $\dim(V)$ is odd; when $\dim(V)$ is even, it is the direct sum of the two spin representations $\pi(\varpi_{n-1})$ and $\pi(\varpi_n)$.

This approach, based on Proposition 31.12 is a variant the construction of the *Weil representation*, or *oscillator representation*, a projective representation of the symplectic group of a local field which was introduced in the great paper [172] in order to explain Siegel's work on the theory of quadratic forms. The analogy between the Weil representation and the spin representation was emphasized and applied in the very interesting papers Howe [75, 77]. Let F be a field, and consider the action of $\mathrm{Sp}(2n, F)$ on a vector space V of dimension $2n$ with a nondegenerate bilinear form $B : V \times V \longrightarrow F$ which satisfies $B(x, y) = -B(y, x)$. There is again a *symplectic Clifford algebra* more commonly called the *Weyl algebra* whose definition is similar to the orthogonal case, except for a sign: in this case the relations to be satisfied are $xy - yx = B(x, y)$ for $x, y \in V$. Thus if W is a maximal isotropic subspace (of dimension n) then elements of W commute, rather than anticommute. Therefore the module Ω would not be the exterior algebra on W but the symmetric algebra, and it should be infinite-dimensional. As in the orthogonal case it is indeed true that if F is a locally compact field (e.g. $\mathbb{R}$, $\mathbb{C}$, a finite or p-adic field) then the Clifford algebra has a unique irreducible representation though one takes not (as this reasoning might suggest) the symmetric algebra but rather the Schwartz space on W. This uniqueness produces a projective representation of $\mathrm{Sp}(2n, F)$ by the same principle based on Proposition 31.11.

The symplectic Clifford algebra (Weyl algebra) is a quotient of the universal enveloping algebra of the Heisenberg Lie algebra $\mathfrak{h}$, which has generators X_i, Y_i, $1 \leqslant i \leqslant n$ and the central element Z, with relations $[X_i, Y_i] = Z$. More precisely, if we divide $U(\mathfrak{h})$ by the ideal generated by $Z - \lambda$, where λ is a nonzero complex number, we obtain the symplectic Clifford algebra. So the fact that the Weyl algebra has a unique module is equivalent to the *Stone-von Neumann Theorem* which asserts that the Heisenberg group, or its Lie algebra has a unique irreducible module with a given nontrivial central character. See Lion and Vergne [119] Sect. 1.6 for an account of the Weil representation featuring the Stone von Neumann theorem.

Exercises

Exercise 31.1. Check the details in the proof of Theorem 31.2. That is, verify that $S(\lambda)$ is exactly the set of characters stated in the theorem and that it consists of just the W orbit of λ.

Exercise 31.2. Prove that the restriction of the spin representation of $\mathrm{Spin}(2n+1)$ to $\mathrm{Spin}(2n)$ is the sum of the two spin representations of $\mathrm{Spin}(2n)$.

Exercise 31.3. Prove that the restriction of either spin representation of $\mathrm{Spin}(2n)$ to $\mathrm{Spin}(2n - 1)$ is the spin representation of $\mathrm{Spin}(2n)$.

Exercise 31.4. Show that one of the spin representations of $\mathrm{Spin}(6)$ gives an isomorphism $\mathrm{Spin}(6) \cong \mathrm{SU}(4)$. What is the significance of the fact that there are two spin representations?

For another spin exercise, see Exercise 30.5.

Exercise 31.5. Verify the description of $X^*(\tilde{T})/\Lambda_{\mathrm{root}}$ in Theorem 31.3.

Exercise 31.6. Let G be a compact connected Lie group whose root system is of type G_2. (See Fig. 19.6.) Prove that G is simply-connected.

Duality and Other Topics

32

Mackey Theory

Given a subgroup H of a finite group G, and a representation π of H, there is an induced representation π^G of G. Mackey theory is concerned with intertwining operators between a pair of induced representations. It is based on a very simple idea: if the two representations are induced from subgroups H_1 and H_2, then every such intertwining operator is convolution with suitable function Δ, which has left and right translation properties by H_1 and H_2. This leads to a method of calculating the space of intertwining operators, based on the double cosets $H_2 \backslash G / H_1$.

If H is a subgroup of the finite group G, and if (π, V) is a representation of H, then we define the *induced representation* (π^G, V^G) as follows. The vector space V^G consists of all maps $f : G \longrightarrow V$ that satisfy $f(hg) = \pi(h) f(g)$ when $h \in H$. The representation $\pi^G : G \longrightarrow \mathrm{GL}(V^G)$ is by right translation

$$(\pi^G(g)f)(x) = f(xg).$$

It is easy to see that if $f \in V^G$, then so is $\pi^G(g) f$, and that π^G is a representation. We will sometimes denote the representation (π^G, V^G) as $\mathrm{Ind}_H^G(\pi)$. If V happens to be one-dimensional, we may identify $V = \mathbb{C}$. Also in Theorem 32.1 the vector space $\mathrm{Hom}(V_1, V_2)$ plays a role; if V_1 and V_2 are one-dimensional we may identify $\mathrm{Hom}(V_1, V_2)$ with $\mathbb{C}$.

We begin with an instructive example of how Mackey theory is used in practice. Let $G = \mathrm{GL}(2, F)$ where $F = \mathbb{F}_q$ is a finite field, and let B be the Borel subgroup of upper triangular matrices. Let χ_1 and χ_2 be characters of $F^\times$. Let χ be the character

$$\chi \begin{pmatrix} y_1 & * \\ & y_2 \end{pmatrix} = \chi_1(y_1)\chi_2(y_2) \tag{32.1}$$

of B. Similarly, let μ_1 and μ_2 be two other characters of $F^\times$, and let μ be the corresponding character of B.

Proposition 32.1. *The representation* $\mathrm{Ind}_B^G(\chi)$ *is of degree* $q + 1$. *It is irreducible unless* $\chi_1 = \chi_2$. *Moreover, it is isomorphic to* $\mathrm{Ind}_B^G(\mu)$ *if and only if either* $\chi_1 = \mu_1$ *and* $\chi_2 = \mu_2$ *or* $\chi_1 = \mu_2$ *and* $\chi_2 = \mu_1$.

This completely classifies the *principal series representations* of $\mathrm{GL}(2, F)$. The irreducibles of this type are about half the irreducible representation of $\mathrm{GL}(2, F)$. The proof will be complete, assuming one fact that will be proved later in the chapter. The reason for this deviation from linear ordering of the material is heuristic—we assume that the reader will be more cheerful while reading the proof of Theorem 32.1 given an example of how the theorem is used.

Proof. The index of B in G is easily seen to be $q + 1$, so this is the dimension of the induced representation. We recall that the vector space for the representation $\mathrm{Ind}_B^G(\chi)$ consists of the space of functions $f : G \longrightarrow \mathbb{C}$ such that $f(bg) = \chi(b) f(g)$. Let us call this space V_χ. The key calculation is to compute $\mathrm{Hom}_G(V_\chi, V_\mu)$. We will show that

$$\dim \mathrm{Hom}_G(V_\chi, V_\mu) =$$

$$\left\{ \begin{array}{l} 1 \text{ if } \chi_1 = \mu_1, \, \chi_2 = \mu_2 \\ 0 \text{ otherwise} \end{array} \right\} + \left\{ \begin{array}{l} 1 \text{ if } \chi_1 = \mu_2, \, \chi_2 = \mu_1 \\ 0 \text{ otherwise} \end{array} \right\}. \tag{32.2}$$

Before showing how Mackey theory can be used to prove (32.2), let us observe that this implies the proposition. First, if $\chi_1 \neq \chi_2$, then it shows that $\mathrm{Hom}_G(V_\chi, V_\chi)$ is one-dimensional, so V_χ is irreducible. Moreover, (32.2) shows exactly when there is a nonzero intertwining operator $V_\chi \longrightarrow V_\mu$, and the second statement is easily deduced.

To prove (32.2), we make use of Mackey's theorem, which we will prove later in the chapter. We recall that if f_1 and f_2 are functions on G, their *convolution* is the function

$$(f_1 * f_2)(g) = \sum_{h \in G} f_1(gh) \, f_2(h^{-1}) = \sum_{h \in G} f_1(h) \, f_2(h^{-1} g).$$

Mackey's theorem (Theorem 32.1 below) asserts that any intertwining operator $T : V_\chi \longrightarrow V_\mu$ is of the form $Tf = \Delta * f$ where $\Delta : G \longrightarrow \mathbb{C}$ is a function satisfying

$$\Delta(b_2 g \, b_1) = \mu(b_2) \Delta(g) \chi(b_1).$$

Such a function is determined by its values on a set of representatives for the double cosets $B \backslash G / B$. By the Bruhat decomposition, there are just two double cosets:

$$G = B \, 1 B \cup B w_0 B, \qquad w_0 = \begin{pmatrix} & 1 \\ 1 & \end{pmatrix},$$

where 1 is, of course, the identity matrix. [A quick proof is given below (27.1).]

So what we will prove is that $\Delta(1) = 0$ unless $\chi_1 = \mu_1$ and $\chi_2 = \mu_2$, and that $\Delta(w_0) = 1$ unless $\chi_1 = \mu_2$ and $\chi_2 = \mu_1$. Indeed,

$$\Delta(1) = \Delta\left(\begin{pmatrix} t_1 & \\ & t_2 \end{pmatrix} \begin{pmatrix} 1 & \\ & 1 \end{pmatrix} \begin{pmatrix} t_1^{-1} & \\ & t_2^{-1} \end{pmatrix} \right) = \mu\begin{pmatrix} t_1 & \\ & t_2 \end{pmatrix} \Delta(1) \chi \begin{pmatrix} t_1 & \\ & t_2 \end{pmatrix}^{-1},$$

that is,

$$\Delta(1) = \mu_1(t_1)\mu_2(t_2)\chi_1(t_1)^{-1}\chi_2(t_2)^{-1}\Delta(1).$$

Unless $\chi_1 = \mu_1$ and $\chi_2 = \mu_2$, we may choose t_1 and t_2 so that

$$\mu_1(t_1)\mu_2(t_2)\chi_1(t_1)^{-1}\chi_2(t_2)^{-1} \neq 1,$$

proving $\Delta(1) = 0$. The proof that $\Delta(w_0) = 0$ unless $\chi_1 = \mu_2$ and $\chi_2 = \mu_1$ is similar.   □

Now let us treat Mackey theory more systematically. We will work with finite groups and with representations over an arbitrary ground field F. In this generality, representations may not be completely reducible. Before considering Mackey theory in general, we will give two *functorial* interpretations of Frobenius reciprocity that correspond to the two special cases where $H_1 = G$ and $H_2 = G$.

Let G be a finite group, F a field, and $F[G]$ the group algebra. If $\pi : G \longrightarrow GL(V)$ is an representation in an F-vector space V, then V becomes an $F[G]$ module by

$$\left(\sum_{g \in G} c_g \cdot g \right) v = \sum_{g \in G} c_g \pi(g) v, \qquad \sum_{g \in G} c_g \cdot g \in F[G],$$

and, conversely, if V is an $F[G]$-module, then $\pi : G \longrightarrow GL(V)$ defined by $\pi(g)v = gv$ is a representation. Thus, the categories of complex representations of G and $F[G]$-modules are equivalent. In either case, we may refer to V as a *G-module*. An intertwining operator for two representations is the same as an $F[G]$-module homomorphism for the corresponding $F[G]$-modules, and we call such a map a *G-module homomorphism*.

Also, if (σ, U) is a representation of G, then we can restrict σ to H to obtain a representation of H. We call U_H the corresponding H-module. Thus, as sets, U and U_H are equal.

Proposition 32.2. (Frobenius reciprocity, first version) *Let H be a subgroup of G and let (π, V) be a representation of H. Let (σ, U) be a representation of G. Then*

$$\mathrm{Hom}_G(U, V^G) \cong \mathrm{Hom}_H(U_H, V). \tag{32.3}$$

In this isomorphism, $J \in \mathrm{Hom}_G(U, V^G)$ and $j \in \mathrm{Hom}_H(U_H, V)$ correspond if and only if $j(u) = J(u)(1)$ and $J(u)(g) = j(\sigma(g)u)$.

Proof. Given $J \in \operatorname{Hom}_G(U, V^G)$, define $j(u) = J(u)(1)$. We show that j is in $\operatorname{Hom}_H(U_H, V)$. Indeed, if $h \in H$, we have

$$j\big(\sigma(h)u\big) = J\big(\sigma(h)u\big)(1) = \big(\pi^G(h)\, J(u)\big)(1)$$

because $J : U \longrightarrow V^G$ is G-equivariant. This equals $J(u)(1.h) = J(u)(h.1) = \pi(h)\, J(u)(1) = \pi(h)j(u)$ because $h \in H$ and $J(u) \in V^G$. Therefore, $j \in \operatorname{Hom}_H(U_H, V)$.

Conversely, if $j \in \operatorname{Hom}_H(U_H, V)$ and $u \in U$, we define $J(u) : G \longrightarrow V$ by $J(u)(g) = j\big(\sigma(g)u\big)$. We leave it to the reader to check that $J(u) \in V^G$ and that $J : U \to V^G$ is G-equivariant. We also leave it to the reader to check that $J \mapsto j$ and $j \mapsto J$ are inverse maps and so $\operatorname{Hom}_G(U, V^G) \cong \operatorname{Hom}_H(U_H, V)$.
□

If the ground field $F = \mathbb{C}$, then we may reinterpret this statement in terms of characters. If η and χ are the characters of U and V, respectively, and if χ^G is the character of the representation of G on V^G, then by Theorem 2.5 we may express Proposition 32.2 by the well-known character identity

$$\langle \chi, \eta \rangle_H = \langle \chi^G, \eta \rangle_G. \tag{32.4}$$

Dual to (32.3) there is also a natural isomorphism

$$\operatorname{Hom}_G(V^G, U) \cong \operatorname{Hom}_H(V, U). \tag{32.5}$$

This is slightly more difficult than Proposition 32.2, and it also involves ideas that we will need in our discussion of Mackey theory. We will approach this by means of a universal property.

Proposition 32.3. *Let H be a subgroup of G and let (π, V) be a representation of H. If $v \in V$, define $\epsilon(v) : G \longrightarrow V$ by*

$$\epsilon(v)(g) = \begin{cases} \pi(g)v & \text{if } g \in H, \\ 0 & \text{otherwise.} \end{cases}$$

Then $\epsilon(v) \in V^G$, and $\epsilon : V \longrightarrow V^G$ is H-equivariant. Let (σ, U) be a representation of G. If $j : V \longrightarrow U$ is any H-module homomorphism, then there exists a unique G-module homomorphism $J : V^G \longrightarrow U$ such that $j = J \circ \epsilon$. We have

$$J(f) = \sum_{\gamma \in G/H} \sigma(\gamma)j\big(f(\gamma^{-1})\big). \tag{32.6}$$

Proof. It is easy to check that $\epsilon(v) \in V^G$ and that if $h \in H$, then

$$\epsilon\big(\pi(h)v\big) = \pi^G(h)\epsilon(v). \tag{32.7}$$

Thus ϵ is H-equivariant.

We prove that if $f \in V^G$, then

$$f = \sum_{G/H} \pi^G(\gamma)\,\epsilon\big(f(\gamma^{-1})\big). \tag{32.8}$$

Using (32.7), each term on the right-hand side of (32.8) is independent of the choice of representatives γ of the cosets in G/H. Let us apply the right-hand side to $g \in G$. We get

$$\sum_{G/H} \epsilon\big(f(\gamma^{-1})\big)(g\gamma).$$

Only one coset representative γ of G/H contributes since, by the definition of ϵ, the contribution is zero unless $g\gamma \in H$. Since we have already noted that each term on the right-hand side of (32.8) is independent of the choice of γ modulo right multiplication by an element of H, we may as well choose $\gamma = g^{-1}$. We obtain $\epsilon\big(f(g)\big)(1) = f(g)$. This proves (32.8).

Suppose now that $J : V^G \longrightarrow U$ is G-equivariant and that $j = J \circ \epsilon$. Then, using (32.8),

$$J(f) = \sum_{G/H} J\big(\pi^G(\gamma)\epsilon(f(\gamma^{-1}))\big) = \sum_{G/H} \sigma(\gamma)(J \circ \epsilon)\big(f(\gamma^{-1})\big)$$

so J must satisfy (32.6). We leave it to the reader to check that J defined by (32.6) is independent of the choice of representatives γ for G/H. We check that it is G-equivariant. If $g \in G$, we have

$$J\big(\pi^G(g)f\big) = \sum_{\gamma \in G/H} \sigma(\gamma)j\big(f(\gamma^{-1}g)\big).$$

The variable change $\gamma \longrightarrow g\gamma$ permutes the cosets in G/H and shows that

$$J\big(\pi^G(g)f\big) = \sum_{\gamma \in G/H} \sigma(g\gamma)j\big(f(\gamma^{-1})\big) = \sigma(g)J(f),$$

as required. □

Corollary 32.1. (Frobenius reciprocity, second version) *If H is a subgroup of the finite group G, and if (σ, U) and (π, V) are representations of G and H, respectively, then $\mathrm{Hom}_G(V^G, U) \cong \mathrm{Hom}_H(V, U)$, and in this isomorphism $j \in \mathrm{Hom}_H(V, U)$ corresponds to $J \in \mathrm{Hom}_G(V^G, U)$ if and only if they are related by (32.6).*

Proof. This is a direct restatement of Proposition 32.3. □

We turn next to Mackey theory. In the following statement, $\mathrm{Hom}(V_1, V_2)$ means $\mathrm{Hom}_F(V_1, V_2)$, the space of all linear maps.

Theorem 32.1. (Mackey's theorem, geometric version) *Suppose that* G *is a finite group,* H_1 *and* H_2 *subgroups, and* (π_1, V_1) *and* (π_2, V_2) *represen-* *tations of* H_1 *and* H_2, *respectively. Then* $\mathrm{Hom}_G(V_1^G, V_2^G)$ *is isomorphic as a* *vector space to the space of all functions* $\Delta : G \longrightarrow \mathrm{Hom}(V_1, V_2)$ *that satisfy*

$$\Delta(h_2 g h_1) = \pi_2(h_2) \circ \Delta(g) \circ \pi_1(h_1), \qquad h_i \in H_i. \tag{32.9}$$

In this isomorphism an intertwining operator $\Lambda : V_1^G \longrightarrow V_2^G$ *corresponds to* Δ *if* $\Lambda(f) = \Delta * f$ $(f \in V_1^G)$, *where the "convolution"* $\Delta * f$ *is defined by*

$$(\Delta * f)(g) = \sum_{\gamma \in G/H_1} \Delta(\gamma) f(\gamma^{-1} g). \tag{32.10}$$

Proof. Let Δ satisfying (32.9) be given. It is easy to check, using (32.9) and the fact that $f \in V_1^G$, that (32.10) is independent of the choice of coset representatives γ for G/H_1. Moreover, if $h_2 \in H_2$, then the variable change $\gamma \longrightarrow h_2 \gamma$ permutes the cosets of G/H_1, and again using (32.9), this variable change shows that $\Delta * f \in V_2^G$. Thus $f \longrightarrow \Delta * f$ is a well-defined map $V_1^G \longrightarrow V_2^G$, and using the fact that G acts on both these spaces by right translation, it is straightforward to see that $\Lambda(f) = \Delta * f$ defines an intertwining operator $V_1^G \longrightarrow V_2^G$.

To show that this map $\Delta \mapsto \Lambda$ is an isomorphism of the space of Δ satisfy- ing (32.9) to $\mathrm{Hom}_G(V_1^G, V_2^G)$, we make use of Corollary 32.1. We must relate the space of Δ satisfying (32.9) to $\mathrm{Hom}_{H_1}(V_1, V_2^G)$. Given $\lambda \in \mathrm{Hom}_{H_1}(V_1, V_2^G)$ corresponding to $\Lambda \in \mathrm{Hom}_G(V_1^G, V_2^G)$ as in that corollary, define $\Delta : G \longrightarrow \mathrm{Hom}(V_1, V_2)$ by $\Delta(g)v_1 = \lambda(v_1)(g)$. The condition that $\lambda(v_1) \in V_2^G$ for all $v_1 \in V_1$ is equivalent to

$$\Delta(h_2 g) = \pi_2(h_2) \circ \Delta(g), \qquad h_2 \in H_2,$$

and the condition that $\lambda : V_1 \longrightarrow V_2^G$ is H_1-equivariant is equivalent to

$$\Delta(g h_1) = \Delta(g) \circ \pi_1(h_1), \qquad h_1 \in H_1.$$

Of course, these two properties together are equivalent to (32.9). We see that Corollary 32.1 implies a linear isomorphism between the space of functions Δ satisfying (32.9) and the elements of $\mathrm{Hom}_G(V_1^G, V_2^G)$. We have only to show that this correspondence is given by (32.10). In (32.6), we take $H = H_1$, $(\sigma, U) = (\pi_2^G, V_2^G)$, and $j = \lambda$. Then $J = \Lambda$ and (32.6) gives us, for $f \in V_1^G$,

$$\Lambda(f) = \sum_{\gamma \in G/H_1} \pi_2^G(\gamma) \lambda(f(\gamma^{-1})).$$

Applying this to $g \in G$,

$$\Lambda(f)(g) = \sum_{\gamma \in G/H_1} \lambda(f(\gamma^{-1}))(g\gamma) = \sum_{\gamma \in G/H_1} \Delta(g\gamma) f(\gamma^{-1}).$$

Making the variable change $\gamma \longrightarrow g^{-1}\gamma$, this equals (32.10). $\square$

Remark 32.1. Although we are working here with finite groups, Mackey's theorem is (since Bruhat [26]) a standard tool in representation theory of Lie groups also. The function Δ becomes a distribution.

Remark 32.2. Suppose that H_1, H_2, and (π_i, V_i) are as in Theorem 32.1. The function $\Delta : G \longrightarrow \mathrm{Hom}(V_1, V_2)$ associated with an intertwining operator $\Lambda : V_1^G \longrightarrow V_2^G$ is clearly determined by its values on a set of representatives for the double cosets in $H_2 \backslash G / H_1$. The simplest case is when Δ is supported on a single double coset $H_2 \gamma H_1$. In this case, we say that the intertwining operator Λ is *supported on $H_2 \gamma H_1$*.

Proposition 32.4. *In the setting of Theorem 32.1, let $\gamma \in G$. Let $H_\gamma = H_2 \cap \gamma H_1 \gamma^{-1}$. Define two representations (π_1^γ, V_1) and (π_2^γ, V_2) of H_γ as follows. The representation π_2^γ is just the restriction of π_2 to H_γ. On the other hand, we define $\pi_1^\gamma(h) = \pi_1(\gamma^{-1} h \gamma)$ for $h \in H_\gamma$. The space of intertwining operators $\Lambda : V_1^G \longrightarrow V_2^G$ supported on $H_2 \gamma H_1$ is isomorphic to $\mathrm{Hom}_{H_\gamma}(\pi_1^\gamma, \pi_2^\gamma)$, the space of all $\delta : V_1 \longrightarrow V_2$ such that*

$$\delta \circ \pi_1^\gamma(h) = \pi_2^\gamma(h) \circ \delta, \qquad h \in H_\gamma. \tag{32.11}$$

Proof. If $\Delta : G \longrightarrow \mathrm{Hom}(V_1, V_2)$ is associated with Λ as in Theorem 32.1, then Δ is by assumption supported on $H_2 \gamma H_1$, and (32.9) implies that Δ is determined by $\delta = \Delta(\gamma)$. This is subject to a consistency condition derived from (32.9). If $h \in H_\gamma$, then $\gamma h' = h \gamma$, where $h' = \gamma^{-1} h \gamma$. We have $h \in H_2$ and $h' \in H_1$, so by (32.9) the map $\delta : V_1 \longrightarrow V_2$ must satisfy (32.11). Conversely, if (32.11) is assumed, it is not hard to see that

$$\Delta(g) = \begin{cases} \pi_2(h_2) \delta \pi_1(h_1) & \text{if } g = h_2 \gamma h_1 \in H_2 \gamma H_1, h_i \in H_i, \\ 0 & \text{if } g \notin H_2 \gamma H_1, \end{cases}$$

is a well-defined function $G \longrightarrow \mathrm{Hom}(V_1, V_2)$ satisfying (32.9), and the corresponding intertwining operator Λ is supported on $H_2 \gamma H_1$. $\square$

Theorem 32.2. (Mackey's theorem, algebraic version) *In the setting of Theorem 32.1, let $\gamma_1, \ldots, \gamma_h$ be a complete set of representatives of the double cosets in $H_2 \backslash G / H_1$. With $\gamma = \gamma_i$, let π_i^γ be as in Proposition 32.4. We have*

$$\dim \mathrm{Hom}_G(V_1^G, V_2^G) = \sum_{i=1}^h \dim \mathrm{Hom}_{H_{\gamma_i}}(\pi_1^{\gamma_i}, \pi_2^{\gamma_i}). \tag{32.12}$$

Proof. If Δ is as in Theorem 32.1, write $\Delta = \sum_i \Delta_i$, where

$$\Delta_i(g) = \begin{cases} \Delta(g) & \text{if } g \in H_2 \gamma_i H_1, \\ 0 & \text{otherwise.} \end{cases}$$

Then Δ_i satisfy (32.9). Let Λ_i be the intertwining operator. Then Λ_i is supported on a single double coset, and the dimension of the space of such intertwining operators is computed in Proposition 32.4. $\square$

Corollary 32.2. *Assume that the ground field F is of characteristic zero. Let H_1 and H_2 be subgroups of G and let (π, V) be an irreducible representation of H_1. Let $\gamma_1, \ldots, \gamma_h$ be a complete set of representatives of the double cosets in $H_2 \backslash G / H_1$. If $\gamma \in G$, let $H_\gamma = H_2 \cap \gamma H_1 \gamma^{-1}$, and let $\pi^\gamma : H_\gamma \longrightarrow \mathrm{GL}(V)$ be the representation $\pi^\gamma(g) = \pi(\gamma^{-1} g \gamma)$. Then the restriction of π^G to H_2 is isomorphic to*

$$\bigoplus_{i=1}^{h} \mathrm{Ind}_{H_{\gamma_i}}^{H_2}(\pi^{\gamma_i}). \tag{32.13}$$

In a word, first inducing and then restricting gives the same result as restricting, then inducing. This way of explaining the result is a pithy oversimplification that has to be correctly understood. More precisely, there are different ways we can restrict, namely given γ we may restrict to H_γ, then induce; we have to sum over all these different ways. And the different ways depend only on the double coset $H_2 \gamma H_1$.

Proof. Since we are assuming that the characteristic of F is zero, representations are completely reducible and it is enough to show that the multiplicity of an irreducible representation (π_2, V_2) in π^G is the same as the multiplicity of π_2 in the direct sum (32.13). The multiplicity of π_2 in π^G is

$$\dim \mathrm{Hom}_{H_2}(V^G, V_2) = \dim \mathrm{Hom}_G(V^G, V_2^G) = \sum_{i=1}^{h} \dim \mathrm{Hom}_{H_{\gamma_i}}(\pi^{\gamma_i}, \pi_2^{\gamma_i})$$

by Frobenius reciprocity and Theorem 32.2. One more application of Frobenius reciprocity shows that this equals

$$\sum_{i=1}^{h} \dim \mathrm{Hom}_{H_2}\left(\mathrm{Ind}_{H_{\gamma_i}}^{H_2}(\pi^{\gamma_i}), \pi_2\right).$$

$\square$

Next we will reinterpret induced representations as obtained by "extension of scalars" as explained in Chap. 11. We must extend the setup there to noncommutative rings. In particular, we recall the basics of tensor products over noncommutative rings. Let R be a ring, not necessarily commutative, and let W be a right R-module and V a left R-module. If C is an Abelian group (written additively), a map $f : W \times V \longrightarrow C$ is called *balanced* if (for $w, w_1, w_2 \in W$ and $v, v_1, v_2 \in V$)

$$f(w_1 + w_2, v) = f(w_1, v) + f(w_2, v),$$

$$f(w, v_1 + v_2) = f(w, v_1) + f(w, v_2),$$

and if $r \in R$,

$$f(wr, v) = f(w, rv).$$

The tensor product $W \otimes_R V$ is an Abelian group with a balanced map $T :$ $W \times V \longrightarrow W \otimes_R V$ such that if $f : W \times V \longrightarrow C$ is any balanced map into an Abelian group C, then there exists a unique homomorphism $F : W \otimes_R V \longrightarrow C$ of Abelian groups such that $f = F \circ T$. The balanced map T is usually denoted $T(w, v) = w \otimes v$.

Remark 32.3. The tensor product always exists and is characterized up to isomorphism by this universal property. If R is noncommutative, then $W \otimes_R V$ does not generally have an R-module structure. However, in special cases it is a module. If A is another ring, we call W an (A, R)-*bimodule* if it is a left A-module and a right R-module, and if these module structures are compatible in the sense that if $w \in W$, $a \in A$, and $r \in R$, then $a(wr) = (aw)r$. If W is an (A, R)-bimodule, then $W \otimes_R V$ has the structure of a left A-module with multiplication satisfying

$$a(w \otimes v) = aw \otimes v, \qquad a \in A.$$

If R is a subring of A, then A is itself an (A, R)-bimodule. Therefore, if V is a left R-module, we can consider $A \otimes_R V$ and this is a left A-module.

Proposition 32.5. *If R is a subring of A and V is a left R-module, let V' be the left A-module $A \otimes_R V$. We have a homomorphism $i : V \longrightarrow V'$ of R-modules defined by $i(v) = 1 \otimes v$. If U is any left A-module and $j : V \longrightarrow U$ is an R-module homomorphism, then there exists a unique A-module homomorphism $J : V' \longrightarrow U$ such that $j = J \circ i$.*

Proof. Suppose that $J : V' \longrightarrow U$ is A-linear and satisfies $j = J \circ i$. Then

$$J(a \otimes v) = J\big(a(1 \otimes v)\big) = aJ(1 \otimes v) = aJ\big(i(v)\big) = aj(v).$$

Since V' is spanned by elements of the form $a \otimes v$, this proves that J, if it exists, is unique.

To show that J exists, note that we have a balanced map $A \times V \longrightarrow U$ given by $(a, v) \longrightarrow aj(v)$. Hence, there exists a unique homomorphism $J :$ $V' = A \otimes_R V \longrightarrow U$ of Abelian groups such that $J(a \otimes v) = aj(v)$. It is straightforward to see that this J is A-linear and that $J \circ i = j$. $\square$

Proposition 32.6. *If R is a subring of A, U is a left A-module, and V is a left R-module, we have a natural isomorphism*

$$\mathrm{Hom}_R(V, U) \cong \mathrm{Hom}_A(A \otimes_R V, U). \tag{32.14}$$

Proof. This is a direct generalization of Proposition 11.1 (ii). It is also essentially equivalent to Proposition 32.5. Indeed, composition with $i : V \longrightarrow$ $V' = A \otimes_R V$ is a map $\mathrm{Hom}_A(V', U) \longrightarrow \mathrm{Hom}_R(V, U)$, and the content of Proposition 32.5 is that this map is bijective. $\square$

Proposition 32.7. *Suppose that H is a subgroup of G and V is an H-module. Then V is a module for the group ring $F[H]$, which is a subring of $F[G]$. We have an isomorphism*

$$V^G \cong F[G] \otimes_{F[H]} V$$

as G-modules.

Proof. Comparing Proposition 32.3 and Proposition 32.5, the G-modules V^G and $F[G] \otimes_{F[H]} V$ satisfy the same universal property, so they are isomorphic. □

Finally, if $F = \mathbb{C}$, let us recall the formula for the character of the induced representation. If χ is a class function of the subgroup H of G, let $\dot{\chi} : G \longrightarrow \mathbb{C}$ be the function

$$\dot{\chi}(g) = \begin{cases} \chi(g) & \text{if } g \in H, \\ 0 & \text{otherwise}, \end{cases}$$

and let $\chi^G : G \longrightarrow \mathbb{C}$ be the function

$$\chi^G(g) = \sum_{x \in H \backslash G} \dot{\chi}(xgx^{-1}). \tag{32.15}$$

We note that since χ is assumed to be a class function, each term depends only on the coset of x in $H \backslash G$. We may, of course, also write

$$\chi^G(g) = \frac{1}{|H|} \sum_{x \in G} \dot{\chi}(xgx^{-1}). \tag{32.16}$$

Clearly, χ^G is a class function on G.

Proposition 32.8. *Let (π, V) be a complex representation of the subgroup H of the finite group G with character χ. Then the character of the induced representation π^G is χ^G.*

Proof. Let η be the character of a representation (σ, U) of G. We will prove that the class function χ^G satisfies Frobenius reciprocity in its classical form (32.4). This suffices because χ^G is determined by the inner product values $\langle \chi^G, \eta \rangle$. We have

$$\langle \chi^G, \eta \rangle_G = \frac{1}{|G|} \sum_{g \in G} \frac{1}{|H|} \sum_{x \in G} \dot{\chi}(xgx^{-1}) \, \eta(g) =$$

$$\frac{1}{|G|} \sum_{g \in G} \frac{1}{|H|} \sum_{h \in H} \sum_{\substack{x \in G \\ xgx^{-1} = h}} \chi(h) \, \eta(g).$$

Given $h \in H$, we can enumerate the pairs $(g, x) \in G \times G$ that satisfy $xgx^{-1} = h$ by noting that they are the pairs $(x^{-1}hx, x)$ with $x \in G$. So the sum equals

$$\frac{1}{|G|}\frac{1}{|H|}\sum_{h\in H}\sum_{x\in G}\chi(h)\,\eta(x^{-1}hx) = \frac{1}{|H|}\sum_{h\in H}\chi(h)\,\eta(h) = \langle\chi,\eta\rangle_H$$

since $\eta(x^{-1}hx) = \eta(h)$. $\qquad\qquad\qquad\qquad\qquad\qquad\qquad\qquad\qquad\square$

Exercises

Exercise 32.1. Some points in the proof of Proposition 32.2 were left to the reader. Write out a complete proof.

Exercise 32.2. Let H_1, H_2, and H_3 be subgroups of G, with (π_i, V_i) a representation of H_i. Let there be given intertwining operators $\Lambda_1 : V_1^G \to V_2^G$ and $\Lambda_2 : V_2^G \to V_3^G$. Let $\Delta_1 : G \to \mathrm{Hom}(V_1, V_2)$ and $\Delta_2 : G \to \mathrm{Hom}(V_2, V_3)$ being the corresponding functions as in Theorem 32.1. Express the $\Delta : G \to \mathrm{Hom}(V_1, V_3)$ corresponding to the composition $\Lambda_2 \circ \Lambda_1$ in terms of Δ_1 and Δ_2.

Exercise 32.3. Let H be a subgroup of G, and $\psi : H \to \mathbb{C}^\times$ a linear character. Prove that the ring of G-module endomorphisms of the induced representation ψ^G is isomorphic to the convolution ring of functions $\Delta : G \to \mathbb{C}^\times$ such that

$$\Delta(h_2 g h_1) = \psi(h_2)\,\Delta(g)\,\psi(h_1), \qquad h_1, h_2 \in H.$$

What can you say about ψ^G if this ring is commutative?

Exercise 32.4. Let $G = \mathrm{GL}(2, F)$, where F is a finite field. Let B be the Borel subgroup of upper triangular matrices, and let N be its subgroup of unipotent matrices. Let $\psi_F : F \to \mathbb{C}$ be any nontrivial character. Define a character of N as follows:

$$\psi\begin{pmatrix} 1 & x \\ & 1 \end{pmatrix} = \psi_F(x).$$

Let χ be a linear character of B as in (32.1). Show that up to scalar multiple there is a unique intertwining operator $\mathrm{Ind}_B^G(\chi) \to \mathrm{Ind}_N^G(\psi)$.

Exercise 32.5. Let H be the non-Abelian group of order q^3 consisting of all matrices

$$\begin{pmatrix} 1 & x & z \\ & 1 & y \\ & & 1 \end{pmatrix}.$$

The center Z of matrices with $x = y = 0$. The subgroup A of matrices with $x = 0$ is Abelian but not central. Let χ and ψ be two linear characters of A.

(i) Assume that χ and ψ have nontrivial restrictions to Z. Let χ^H and ψ^H be the induced representations. Use Mackey theory to prove that

$$\dim\,\mathrm{Hom}_H(\chi^H, \psi^H) = \begin{cases} 1 & \text{if } \chi, \psi \text{ have the same restriction to } Z; \\ 0 & \text{otherwise.} \end{cases}$$

(ii) Prove that χ^H is irreducible, and that χ^H, ψ^H are isomorphic if and only if χ and ψ have the same restriction to Z.

(iii) Prove that given a nontrivial central character θ of Z, H has a unique irreducible representation with central character θ. This is the *Stone–von Neumann theorem* for finite fields.

33

Characters of $\mathrm{GL}(n, \mathbb{C})$

In the next few chapters, we will construct the irreducible representations of the symmetric group in parallel with the irreducible algebraic representations of $\mathrm{GL}(n, \mathbb{C})$. In this chapter, we will construct some generalized characters of $\mathrm{GL}(n, \mathbb{C})$. The connection with the representation theory of S_k will become clear later.

A complex representation (π, V) of $\mathrm{GL}(n, \mathbb{C})$ is *algebraic* if the matrix coefficients of $\pi(g)$ are polynomial functions in the matrix coefficients g_{ij} of $g = (g_{ij}) \in \mathrm{GL}(n, \mathbb{C})$ and of $\det(g)^{-1}$. Thus, if we choose a basis of V, then $\pi(g)$ becomes a matrix $(\pi(g)_{kl})$ with $1 \leqslant k, l \leqslant \dim(V)$, and for each k, l we require that there be a polynomial P_{kl} with $n^2 + 1$ entries such that

$$\pi(g)_{kl} = P_{kl}\big(g_{11}, \ldots, g_{nn}, \det(g)^{-1}\big).$$

The assumption that a representation is algebraic is similar to the assumption that it is analytic—it rules out representations such as complex conjugation $\mathrm{GL}(n, \mathbb{C}) \to \mathrm{GL}(n, \mathbb{C})$. It is not hard to show (using the Weyl character formula) that every analytic representation of $\mathrm{GL}(n, \mathbb{C})$ is algebraic, and of course the converse is also true.

A character χ is *algebraic* if it is the character of an algebraic representation. A *generalized character*, also called a *virtual character*, is the difference between two characters. If $G = \mathrm{GL}(n, \mathbb{C})$, or more generally any algebraic group, we will say a generalized character is *algebraic* if it is $\chi_1 - \chi_2$, where χ_1 and χ_2 are algebraic.

If R is a commutative ring, we will denote by $R_{\mathrm{sym}}[x_1, \ldots, x_n]$ the ring of symmetric polynomials in $x_1, \ldots, x_n$ having coefficients in R. Let e_k and $h_k \in \mathbb{Z}_{\mathrm{sym}}[x_1, \ldots, x_n]$ be the kth *elementary* and *complete* symmetric polynomials in n variables. Specifically,

$$e_k(x_1, \ldots, x_n) = \sum_{1 \leqslant i_1 < i_2 < \cdots < i_k \leqslant n} x_{i_1} x_{i_2} \ldots x_{i_k},$$

D. Bump, *Lie Groups*, Graduate Texts in Mathematics 225,
DOI 10.1007/978-1-4614-8024-2_33, © Springer Science+Business Media New York 2013

$$h_k(x_1, \ldots, x_n) = \sum_{1 \leqslant i_1 \leqslant i_2 \leqslant \cdots \leqslant i_k \leqslant n} x_{i_1} x_{i_2} \cdots x_{i_k}.$$

If $k > n$, then $e_k = 0$, although this is not true for h_k. Our convention is that $e_0 = h_0 = 1$.

Let $E(t)$ be the generating function for the elementary symmetric polynomials:

$$E(t) = \sum_{k=0}^{n} e_k t^k.$$

Then

$$E(t) = (1 + x_1 t)(1 + x_2 t) \cdots (1 + x_n t) \tag{33.1}$$

since expanding the right-hand side and collecting the coefficients of t^k will give each monomial in the definition of e_k exactly once. Similarly, if

$$H(t) = \sum_{k=0}^{\infty} h_k t^k,$$

then

$$H(t) = \prod_{i=0}^{n} (1 + x_i t + x_i^2 t^2 + \cdots) = (1 - x_1 t)^{-1} \cdots (1 - x_n t)^{-1}. \tag{33.2}$$

We see that

$$H(t)E(-t) = 1.$$

Equating the coefficients in this identity gives us recursive relations

$$h_k - e_1 h_{k-1} + e_2 h_{k-2} - \cdots + (-1)^k e_k = 0, \qquad k > 0. \tag{33.3}$$

These can be used to express the h's in terms of the e's or vice versa.

Proposition 33.1. *The ring $\mathbb{Z}_{\mathrm{sym}}[x_1, \ldots, x_n]$ is generated as a $\mathbb{Z}$-algebra by $e_1, \ldots, e_n$, and they are algebraically independent. Thus, $\mathbb{Z}_{\mathrm{sym}}[x_1, \ldots, x_n] = \mathbb{Z}[e_1, \ldots, e_n]$ is a polynomial ring. It is also generated by $h_1, \ldots, h_n$, which are algebraically independent, and $\mathbb{Z}_{\mathrm{sym}}[x_1, \ldots, x_n] = \mathbb{Z}[h_1, \ldots, h_n]$.*

Proof. The fact that the e_i generate $\mathbb{Z}_{\mathrm{sym}}[x_1, \ldots, x_n]$ is Theorem 6.1 on p. 191 of Lang [116], and their algebraic independence is proved on p. 192 of that reference. The fact that $h_1, \ldots, h_n$ also generate follows since (33.3) can be solved recursively to express the e_i in terms of the h_i. The h_i must be algebraically independent since if they were dependent the transcendence degree of the field of fractions of $\mathbb{Z}_{\mathrm{sym}}[x_1, \ldots, x_n]$ would be less than n, so the e_i would also be algebraically dependent, which is a contradiction. $\square$

If V is a vector space, let $\wedge^k V$ and $\vee^k V$ denote the kth exterior and symmetric powers. If $T : V \longrightarrow W$ is a linear transformation, then there are induced linear transformations $\wedge^k T : \wedge^k V \longrightarrow \wedge^k W$ and $\vee^k T : \vee^k V \longrightarrow \vee^k W$.

Proposition 33.2. *If V is an n-dimensional vector space and $T : V \longrightarrow V$ an endomorphism, and if $t_1, \ldots, t_n$ are its eigenvalues with multiplicities (that is, each eigenvalue is listed with its multiplicity as a root of the characteristic polynomial), then*

$$\operatorname{tr} \wedge^k T = e_k(t_1, \ldots, t_n) \tag{33.4}$$

and

$$\operatorname{tr} \vee^k T = h_k(t_1, \ldots, t_n). \tag{33.5}$$

Proof. First, assume that T is diagonalizable and that $v_1, \ldots, v_n$ are its eigenvectors, so $Tv_i = t_i v_i$. Then a basis of $\wedge^k V$ consists of the vectors

$$v_{i_1} \wedge \cdots \wedge v_{i_k}, \qquad 1 \leqslant i_1 < i_2 < \cdots < i_k \leqslant n,$$

and this is an eigenvector of $\wedge^k T$ with eigenvalue $t_{i_1} \cdots t_{i_k}$. Summing these eigenvalues gives $e_k(t_1, \ldots, t_n)$. Thus, (33.4) is true if T is diagonalizable. Similarly, a basis of $\vee^k V$ consists of the vectors

$$v_{i_1} \vee \cdots \vee v_{i_k}, \qquad 1 \leqslant i_1 \leqslant i_2 \leqslant \cdots \leqslant i_k \leqslant n,$$

so (33.5) is also true if T is diagonalizable.

In the general case, both sides of (33.4) or (33.5) are continuous functions of the matrix entries of T. The left-hand side of (33.4) is continuous because if we refer T to a fixed basis, then $\operatorname{tr} \wedge^k T$ is the sum of the $\binom{n}{k}$ principal minors of its matrix with respect to this basis, and the right-hand side is continuous because it is a coefficient in the characteristic polynomial of T. Since the diagonalizable matrices are dense in GL(n, $\mathbb{C}$), it follows that (33.4) is true for all T. As for (33.5), the h's are polynomial functions in the e's, as we see by solving (33.3) recursively, so the right-hand side of (33.5) is also continuous, and (33.5) is also proved. $\square$

Theorem 33.1. *Let $f(x_1, \ldots, x_n)$ be a symmetric polynomial with integer coefficients. Define a function ψ_f on GL(n, $\mathbb{C}$) as follows. If $t_1, \ldots, t_n$ are the eigenvalues of g, let*

$$\psi_f(g) = f(t_1, \ldots, t_n). \tag{33.6}$$

Then ψ_f is an algebraic generalized character of GL(n, $\mathbb{C}$).

As in Proposition 33.2, there may be repeated eigenvalues. If this is the case, we count each eigenvalue with the multiplicity with which it occurs as a root of the characteristic polynomial.

Proof. Let us call a symmetric polynomial f *constructible* if ψ_f is a generalized character of GL(n, $\mathbb{C}$). The generalized characters of GL(n, $\mathbb{C}$) form a ring since the direct sum and tensor product operations on GL(n, $\mathbb{C}$)-modules correspond to addition and multiplication of characters. Since

$$\psi_{f_1 \pm f_2} = \psi_{f_1} \pm \psi_{f_2}, \qquad \psi_{f_1 f_2} = \psi_{f_1} \psi_{f_2},$$

it follows that the constructible polynomials also form a ring. The e_k are constructible by Proposition 33.2 and generate $\mathbb{Z}_{\text{sym}}[x_1, \ldots, x_n]$ by Proposition 33.1. Thus, the ring of constructible polynomials is all of $\mathbb{Z}_{\text{sym}}[x_1, \ldots, x_n]$.

□

In addition to the elementary and complete symmetric polynomials, we have the *power sum* symmetric polynomials

$$p_k(x_1, \ldots, x_n) = x_1^k + \cdots + x_n^k. \tag{33.7}$$

Theorem 33.2. *Let G be a group, let χ be a character of G, and let k be a nonnegative integer. Then $g \mapsto \chi(g^k)$ is a virtual character of G.*

Proof. Let χ be the character corresponding to the representation $\pi : G \to GL(n, \mathbb{C})$. If ψ is any generalized character of $GL(n, \mathbb{C})$, then $\psi \circ \pi$ is a generalized character of G. We take $\psi = \psi_{p_k}$, which is a generalized character by Theorem 33.1. If $t_1, \ldots, t_n$ are the eigenvalues of $\pi(g)$, then $t_1^k, \ldots, t_n^k$ are the eigenvalues of $\pi(g^k)$. Hence

$$(\psi_{p_k} \circ \pi)(g) = \chi(g^k), \tag{33.8}$$

proving that $\chi(g^k)$ is a generalized character.   □

Proposition 33.3. (Newton) *The polynomials p_k generate $\mathbb{Q}_{\text{sym}}[x_1, \ldots, x_n]$ as a $\mathbb{Q}$-algebra.*

Proof. We will make use of the identity

$$\log(1 + t) = \sum_{k=1}^{\infty} \frac{(-1)^{k-1}}{k} t^k.$$

Replacing t by tx_i in this identity, summing over the x_i, and using (33.1), we see that

$$\log E(t) = \sum_{k=1}^{\infty} \frac{(-1)^{k-1}}{k} p_k t^k.$$

Exponentiating this identity,

$$\sum_{k=0}^{\infty} e_k t^k = \exp\left(\sum_{k=1}^{\infty} \frac{(-1)^{k-1}}{k} p_k t^k \right).$$

Expanding and collecting the coefficients of t^k expresses e_k as a polynomial in the p's, with rational coefficients.   □

Let us return to the context of Theorem 33.2. Let G be a group and χ the character of a representation $\pi : G \longrightarrow \mathrm{GL}(n, \mathbb{C})$. As we saw in that theorem, the functions $g \longrightarrow \chi_k(g) = \chi(g^k)$ are generalized characters; indeed they are the functions $\psi_{p_k} \circ \pi$. They are conveniently computable and therefore useful. The operations $\chi \longrightarrow \chi_k$ on the ring of generalized characters of G are called the *Adams operations*. See also the exercises in Chap. 22 for more about the Adams operations.

Let us consider an example. Consider the polynomial

$$s(x_1, \ldots, x_n) = \sum_{i \neq j} x_i^2 x_j + 2 \sum_{i < j < k} x_i x_j x_k. \tag{33.9}$$

We find that

$$p_1^3 = \sum_i x_i^3 + 3 \sum_{i \neq j} x_i^2 x_j + 6 \sum_{i < j < k} x_i x_j x_k,$$

so

$$s = \tfrac{1}{3}(p_1^3 - p_3). \tag{33.10}$$

Hence, if $\pi : G \longrightarrow \mathrm{GL}(n, \mathbb{C})$ is a representation affording the character χ, then we have

$$(\psi_s \circ \pi)(g) = \tfrac{1}{3}\left(\chi(g)^3 - \chi(g^3)\right). \tag{33.11}$$

Such a composition of a representation with a ψ_f is called a *plethysm*. The expression on the right-hand side is useful for calculating the values of this function, which we have proved is a virtual character of $\mathrm{GL}(n, \mathbb{C})$, provided we know the values of the character χ. We will show in the next chapter that (for this particular s) this plethysm is actually a proper character. Indeed, we will actually prove that ψ_s is a character of $\mathrm{GL}(n, \mathbb{C})$, not just a virtual character. This will require ideas different from those than used in this chapter.

EXERCISES

Exercise 33.1. Express each of the sets of polynomials $\{e_k \mid k \leqslant 5\}$ and $\{p_k \mid k \leqslant 5\}$ in terms of the other.

Exercise 33.2. Here is the character table of S_4.

	1	(123)	(12)(34)	(12)	(1234)
χ_1	1	1	1	1	1
χ_2	1	1	1	-1	-1
χ_3	3	0	-1	1	-1
χ_4	3	0	-1	-1	1
χ_5	2	-1	2	0	0

Let s be as in (33.9). Using (33.11), compute $\psi_s \circ \pi$ when (π, V) is an irreducible representation with character χ_i for each i, and decompose the resulting class function into irreducible characters, confirming that it is a generalized character.

34

Duality Between S_k and $\mathrm{GL}(n, \mathbb{C})$

Let V be a complex vector space, and let $\bigotimes^k V = V \otimes \cdots \otimes V$ be the k-fold tensor of V. (Unadorned $\otimes$ means $\otimes_{\mathbb{C}}$.) We consider this to be a right module over the group ring $\mathbb{C}[S_k]$, where $\sigma \in S_k$ acts by permuting the factors:

$$(v_1 \otimes \cdots \otimes v_k)\sigma = v_{\sigma(1)} \otimes \cdots \otimes v_{\sigma(k)}. \tag{34.1}$$

It may be checked that with this definition

$$((v_1 \otimes \cdots \otimes v_k)\sigma)\,\tau = (v_1 \otimes \cdots \otimes v_k)(\sigma\tau).$$

If A is $\mathbb{C}$-algebra and V is an A-module, then $\bigotimes^k V$ has an A-module structure; namely, $a \in A$ acts diagonally:

$$a(v_1 \otimes \cdots \otimes v_k) = av_1 \otimes \cdots \otimes av_k.$$

This action commutes with the action (34.1) of the symmetric group, so it makes $\bigotimes^k V$ an $(A, \mathbb{C}[S_k])$-bimodule. Suppose that $\rho : S_k \longrightarrow \mathrm{GL}(N_\rho)$ is a representation. Then N_ρ is an S_k-module, so by Remark 32.3

$$V_\rho = \left(\bigotimes^k V \right) \otimes_{\mathbb{C}[S_k]} N_\rho \tag{34.2}$$

is a left A-module.

We can take $A = \mathrm{End}(V)$. Embedding $\mathrm{GL}(V) \longrightarrow A$, we obtain a representation of $\mathrm{GL}(V)$ parametrized by a module N_ρ of S_k. Thus, V_ρ is a $\mathrm{GL}(V)$-module. This is the basic construction of Frobenius–Schur duality.

We now give a reinterpretation of the symmetric and exterior powers, which were used in the proof of Theorem 33.1. Let $\mathbb{C}_{\mathrm{sym}}$ be a left $\mathbb{C}[S_k]$-module for the trivial representation, and let $\mathbb{C}_{\mathrm{alt}}$ be a $\mathbb{C}[S_k]$-module for the alternating character. Thus, $\mathbb{C}_{\mathrm{alt}}$ is $\mathbb{C}$ with the S_k-module structure

$$\sigma x = \varepsilon(\sigma)\, x,$$

for $\sigma \in S_k$, $x \in \mathbb{C}_{\mathrm{alt}}$, where $\varepsilon : S_k \to \{\pm 1\}$ is the alternating character.

D. Bump, *Lie Groups*, Graduate Texts in Mathematics 225,
DOI 10.1007/978-1-4614-8024-2_34, © Springer Science+Business Media New York 2013

Proposition 34.1. *Let V be a vector space over $\mathbb{C}$. We have functorial isomorphisms*

$$\wedge^k V \cong \left(\bigotimes^k V \right) \otimes_{\mathbb{C}[S_k]} \mathbb{C}_{\mathrm{alt}}, \qquad \vee^k V \cong \left(\bigotimes^k V \right) \otimes_{\mathbb{C}[S_k]} \mathbb{C}_{\mathrm{sym}}.$$

Here "functorial" means that if $T : V \longrightarrow W$ is a linear transformation, then we have a commutative diagram

$$
\begin{array}{ccc}
\wedge^k V & \xrightarrow{\cong} & \left(\bigotimes^k V \right) \otimes_{\mathbb{C}[S_k]} \mathbb{C}_{\mathrm{alt}} \\
\downarrow & & \downarrow \\
\wedge^k W & \xrightarrow{\cong} & \left(\bigotimes^k W \right) \otimes_{\mathbb{C}[S_k]} \mathbb{C}_{\mathrm{alt}}
\end{array}
$$

and in particular if $V = W$, this implies that $\wedge^k V \cong \left(\bigotimes^k V \right) \otimes_{\mathbb{C}[S_k]} \mathbb{C}_{\mathrm{alt}}$ as $\mathrm{GL}(V)$-modules.

Proof. The proofs of these isomorphisms are similar. We will prove the first. It is sufficient to show that the right-hand side satisfies the universal property of the exterior kth power. We recall that this is the following property of $\wedge^k V$. Given a vector space W, a k-linear map $f : V \times \cdots \times V \longrightarrow W$ is *alternating* if

$$f\left(v_{\sigma(1)}, \ldots, v_{\sigma(k)} \right) = \varepsilon(\sigma) \, f(v_1, \ldots, v_k).$$

The universal property is that any such alternating map factors uniquely through $\wedge^k V$. That is, the map $(v_1, \ldots, v_k) \mapsto v_1 \wedge \cdots \wedge v_k$ is itself alternating, and given any alternating map $f : V \times \cdots \times V \longrightarrow W$ there exists a unique linear map $F : \wedge^k V \longrightarrow W$ such that $f(v_1, \ldots, v_k) = F(v_1 \wedge \cdots \wedge v_k)$. We will show that $\left(\bigotimes^k V \right) \otimes_{\mathbb{C}[S_k]} \mathbb{C}_{\mathrm{alt}}$ has the same universal property.

We are identifying the underlying space of $\mathbb{C}_{\mathrm{alt}}$ with $\mathbb{C}$, so $1 \in \mathbb{C}_{\mathrm{alt}}$. There exists a map $i : V \times \cdots \times V \to \left(\bigotimes^k V \right) \otimes_{\mathbb{C}[S_k]} \mathbb{C}_{\mathrm{alt}}$ given by

$$i(v_1, \ldots, v_k) = (v_1 \otimes \cdots \otimes v_k) \otimes_{\mathbb{C}[S_k]} 1.$$

Let $f : V \times \cdots \times V \to W$ be an alternating k-linear map into a vector space W. We must show that there exists a unique linear map

$$F : \left(\bigotimes^k V \right) \otimes_{\mathbb{C}[S_k]} \mathbb{C}_{\mathrm{alt}} \to W$$

such that $f = F \circ i$. Uniqueness is clear since the image of i spans the space $\left(\bigotimes^k V \right) \otimes_{\mathbb{C}[S_k]} \mathbb{C}_{\mathrm{alt}}$. To prove existence, we observe first that by the universal property of the tensor product there exists a linear map $f' : \bigotimes^k V \to W$ such that $f(v_1, \ldots, v_k) = f'(v_1 \otimes \cdots \otimes v_k)$. Now consider the map

$$\left(\bigotimes^k V \right) \times \mathbb{C}_{\mathrm{alt}} \to W$$

defined by $(\xi, t) \longmapsto t f'(\xi)$. It follows from the fact that f is alternating that this map is $\mathbb{C}[S_k]$-balanced and consequently induces a map

$$F : \left(\bigotimes^k V \right) \otimes_{\mathbb{C}[S_k]} \mathbb{C}_{\mathrm{alt}} \to W.$$

This is the map we are seeking. We see that $\left(\bigotimes^k V \right) \otimes_{\mathbb{C}[S_k]} \mathbb{C}_{\mathrm{alt}}$ satisfies the same universal property as the exterior power, so it is naturally isomorphic to $\wedge^k V$.   □

For the rest of this chapter, fix n and let $V = \mathbb{C}^n$. If $\rho : S_k \longrightarrow \mathrm{GL}(N_\rho)$ is any representation, then (34.2) defines a module V_ρ for $\mathrm{GL}(n, \mathbb{C})$.

Theorem 34.1. *Let $\rho : S_k \longrightarrow \mathrm{GL}(N_\rho)$ be a representation. Let V_ρ be as in (34.2). There exists a homogeneous symmetric polynomial s_ρ of degree k in n variables such that if $\psi_\rho(g)$ is the trace of $g \in \mathrm{GL}(n, \mathbb{C})$ on V_ρ, and if $t_1, \ldots, t_n$ are the eigenvalues of g, then*

$$\psi_\rho(g) = s_\rho(t_1, \ldots, t_n). \tag{34.3}$$

Proof. First let us prove this for g restricted to the subgroup of diagonal matrices. Let $\xi_1, \ldots, \xi_n$ be the standard basis of V. In other words, identifying V with $\mathbb{C}^n$, let $\xi_i = (0, \ldots, 1, \ldots, 0)$, where the 1 is in the ith position. The vectors $(\xi_{i_1} \otimes \cdots \otimes \xi_{i_k}) \otimes \nu$, where ν runs through a basis of N_ρ, and $1 \leqslant i_1 \leqslant \cdots \leqslant i_k \leqslant n$ span V_ρ. They will generally not be linearly independent, but there will be a linearly independent subset that forms a basis of V_ρ. For g diagonal, if $g(\xi_i) = t_i \xi_i$, then $(\xi_{i_1} \otimes \cdots \otimes \xi_{i_k}) \otimes \nu$ will be an eigenvector for g in V_ρ with eigenvalue $t_{i_1} \cdots t_{i_k}$. Thus, we see that there exists a homogeneous polynomial s_ρ of degree k such that (34.3) is true for diagonal matrices g.

To see that s_ρ is symmetric, we have pointed out that the action of S_k on $\otimes^k V$ commutes with the action of $\mathrm{GL}(n, \mathbb{C})$. In particular, it commutes with the action of the permutation matrices in $\mathrm{GL}(n, \mathbb{C})$, which form a subgroup isomorphic to S_n. These permute the eigenvectors $(\xi_{i_1} \otimes \cdots \otimes \xi_{i_k}) \otimes \nu$ of g and hence their eigenvalues. Thus, the polynomial s_ρ must be symmetric.

Since the eigenvalues of a matrix are equal to the eigenvalues of any conjugate, we see that (34.3) must be true for any matrix that is conjugate to a diagonal matrix. Since these are dense in $\mathrm{GL}(n, \mathbb{C})$, (34.3) follows for all g by continuity.   □

Proposition 34.2. *Let $\rho_i : S_k \longrightarrow \mathrm{GL}(N_{\rho_i})$ $(i = 1, \ldots, h)$ be the irreducible representations of S_k and let $d_1, \ldots, d_h$ be their respective degrees. Then*

$$p_1^k = \sum_i d_i s_{\rho_i}. \tag{34.4}$$

Proof. If R is a ring and M a right R-module, then

$$M \otimes_R R \cong M. \tag{34.5}$$

(To prove this standard isomorphism, observe that $m \otimes r \mapsto mr$ and $m \mapsto m \otimes 1$ are inverse maps between the two Abelian groups.) If M is an (S, R)-bimodule, then this is an isomorphism of S-modules. Consequently,

$$\bigotimes^k V \cong \left(\bigotimes^k V \right) \otimes_{\mathbb{C}[S_k]} \mathbb{C}[S_k].$$

The multiplicity of ρ_i in the regular representation is d_i, that is, $\mathbb{C}[S_k] \cong \bigoplus d_i N_{\rho_i}$, and hence

$$\bigotimes^k V \cong \bigoplus_i d_i \left(\bigotimes^k V \right) \otimes_{\mathbb{C}[S_k]} N_{\rho_i} = \bigoplus_i d_i V_{\rho_i}. \tag{34.6}$$

Taking characters, we obtain (34.4). $\square$

Recall that we ended the last chapter by asserting that ψ_s is a proper character of $\mathrm{GL}(n, \mathbb{C})$, where s is the polynomial in (33.10). We now have the tools to prove this.

Let $k = 3$, and let ρ_i be the irreducible representations of degree 2 of S_3. We will take ρ_1 to be the trivial representation, $\rho_2 = \varepsilon$ to be the alternating representation, and ρ_3 to be the irreducible two-dimensional representation. If $g \in \mathrm{GL}(n, \mathbb{C})$ has eigenvalues $t_1, \dots, t_n$, then the value at g of the character of the representation of $\mathrm{GL}(n, \mathbb{C})$ on the module $\bigotimes^3 V$ is

$$p_1^3(t_1, \dots, t_n) = \left(\sum t_i \right)^3 = \sum t_i^3 + 3 \sum_{i \neq j} t_i^2 t_j + 6 \sum_{i < j < k} t_i t_j t_k.$$

The right-hand side of (34.4) consists of three terms. First, corresponding to ρ_1 and the symmetric cube $\vee^3 V \cong V_{\rho_1}$ representation of $\mathrm{GL}(n, \mathbb{C})$ is

$$h_3 = \sum t_i^3 + \sum_{i \neq j} t_i^2 t_j + \sum_{i < j < k} t_i t_j t_k.$$

Second, corresponding to ρ_2 and the exterior cube $\wedge^3 V \cong V_{\rho_2}$ representation of $\mathrm{GL}(n, \mathbb{C})$ is

$$e_3 = \sum_{i < j < k} t_i t_j t_k.$$

Finally, corresponding to ρ_3, the associated module V_{ρ_3} of $\mathrm{GL}(n, \mathbb{C})$ affords the character ψ_{ρ_3}, and the associated symmetric polynomial s_{ρ_3} occurs with coefficient $d_3 = 2$. This satisfies the equation

$$p_1^3 = h_3 + e_3 + 2s_{\rho_3},$$

from which we easily calculate that s_{ρ_3} is the polynomial in (33.10).

The conjugacy classes of S_k are parametrized by the *partitions* of k. A *partition* of k is a decomposition of k into a sum of positive integers. Thus, the partitions of 5 are

$$5, \quad 4+1, \quad 3+2, \quad 3+1+1, \quad 2+2+1, \quad 2+1+1+1, \quad 1+1+1+1+1.$$

Note that the partitions $3+2$ and $2+3$ are considered equal. We may arrange the terms in a partition into descending order. Hence, a partition λ of k may be more formally defined to be a sequence of nonnegative integers $(\lambda_1, \ldots, \lambda_l)$ such that $\lambda_1 \geqslant \lambda_2 \geqslant \cdots \geqslant \lambda_l \geqslant 0$ and $\sum_i \lambda_i = k$. It is sometimes convenient to allow some of the parts λ_i to be zero, in which case we identify two sequences if they differ only by trailing zeros. Thus, $(3, 2, 0, 0)$ is considered to be the same partition as $(3, 2)$. The *length* or *number of parts* $l(\lambda)$ of the partition λ is the largest i such that $\lambda_i \neq 0$, so the length of the partition $(3, 2)$ is two. We will denote by $p(k)$ the number of partitions of k, so that $p(5) = 7$.

If λ is a partition of k, there is another partition, called the *conjugate* partition and denoted λ^t, which may be constructed as follows. We construct from λ a diagram in which the ith row is a series of λ_i boxes. Thus, the diagram corresponding to the partition $\lambda = (3, 2)$ is

Having constructed the diagram, we transpose it, and the corresponding partition is the conjugate partition, denoted λ^t. Hence, the transpose of the preceding diagram is

and so the partition of 5 conjugate to $\lambda = (3, 2)$ is $\lambda^t = (2, 1, 1)$. These types of diagrams are called *Young diagrams* or *Ferrers' diagrams*.

More formally, the *diagram* $D(\lambda)$ of a partition λ is the set of $(i, j) \in \mathbb{Z}^2$ such that $0 \leqslant i$ and $0 \leqslant j \leqslant \lambda_i$. We associate with each pair (i, j) the box in the ith row and the jth column, where the convention is that the row index i increases as one moves downward and the column index j increases as one moves to the right, so that the boxes lie in the fourth quadrant.

Suppose that $\mu = \lambda^t$. Then $(i, j) \in D(\lambda)$ if and only if $(j, i) \in D(\mu)$. Therefore,

$$j \leqslant \lambda_i \iff i \leqslant \mu_j. \tag{34.7}$$

If G is a finite group, let $X(G)$ be the additive group of generalized characters of G. It is isomorphic to the free Abelian group generated by the isomorphism classes of irreducible representations. Because $X(G)$ has a well-known ring structure, it is usually called the character ring of G, but we will not use the multiplication in $X(G)$ at all. To us it is simply an additive Abelian group, the group of generalized characters.

Let $\mathcal{R}_k = X(S_k)$. Its rank, as a free $\mathbb{Z}$-module is equal to the number $p(k)$ of partitions of k. Our convention is $\mathcal{R}_0 = \mathbb{Z}$.

Although we do not need the ring structure on $\mathcal{R}_k$ itself, we will introduce a multiplication $\mathcal{R}_k \times \mathcal{R}_l \to \mathcal{R}_{k+l}$, which makes $\mathcal{R} = \bigoplus_k \mathcal{R}_k$ into a graded ring. The multiplication in $\mathcal{R}$ is as follows. If θ, ρ are representations of S_k and S_l, respectively, then $\theta \otimes \rho$ is a representation of $S_k \times S_l$, which is a subgroup of S_{k+l}. We will always use the unadorned symbol $\otimes$ to denote $\otimes_{\mathbb{C}}$.

We let $\theta \circ \rho$ be the representation obtained by inducing $\theta \otimes \rho$ from $S_k \times S_l$ to S_{k+l}. This multiplication, at first defined only for genuine representations, extends to virtual representations by additivity, and so we get a multiplication $\mathcal{R}_k \times \mathcal{R}_l \to \mathcal{R}_{k+l}$. It follows from the principle of transitivity of induction that this multiplication is associative, and since the subgroups $S_k \times S_l$ and $S_l \times S_k$ are conjugate in S_{k+l}, it is also commutative.

Now let us introduce another graded ring. Let n be a fixed integer, and let $x_1, \ldots, x_n$ be indeterminates. We consider the ring

$$\Lambda^{(n)} = \mathbb{Z}_{\mathrm{sym}}[x_1, \ldots, x_n]$$

of symmetric polynomials with integer coefficients in $x_1, \ldots, x_n$, graded by degree. By Proposition 33.1, $\Lambda^{(n)}$ is a polynomial ring in the symmetric polynomials $e_1, \ldots, e_n$,

$$\Lambda^{(n)} \cong \mathbb{Z}[e_1, \ldots, e_n] \tag{34.8}$$

or equally, in terms of the symmetric polynomials h_i,

$$\Lambda^{(n)} \cong \mathbb{Z}[h_1, \ldots, h_n].$$

$\Lambda^{(n)}$ is a graded ring. We have $\Lambda^{(n)} = \bigoplus \Lambda_k^{(n)}$, where $\Lambda_k^{(n)}$ consists of all homogeneous polynomials of degree k in $\Lambda^{(n)}$.

Proposition 34.3. *The homogeneous part $\Lambda_k^{(n)}$ is a free Abelian group of rank equal to the number of partitions of k into no more than n parts.*

Proof. Let $\lambda^{(n)}$ be such a partition. Thus, $\lambda^{(n)} = (\lambda_1, \ldots, \lambda_n)$, where $\lambda_1 \geqslant \lambda_2 \geqslant \cdots \geqslant \lambda_n \geqslant 0$ and $\sum_i \lambda_i = k$. Let

$$m_\lambda(x_1, \ldots, x_n) = \sum x_1^{\alpha_1} \cdots x_n^{\alpha_n},$$

where $(\alpha_1, \ldots, \alpha_n)$ runs over all distinct permutations of $(\lambda_1, \ldots, \lambda_n)$. Clearly, the m_λ form a $\mathbb{Z}$-basis of $\Lambda_k^{(n)}$, and therefore $\Lambda_k^{(n)}$ is a free Abelian group of rank equal to the number of partitions of k into no more than n parts.   □

In Theorem 34.1, we associated with each irreducible representation ρ of S_k an element s_ρ of $\Lambda_k^{(n)}$. Thus, there exists a homomorphism of Abelian groups $\mathrm{ch}_k^{(n)} : \mathcal{R}_k \to \Lambda_k^{(n)}$ such that $\mathrm{ch}_k^{(n)}(\rho) = s_\rho$. Let $\mathrm{ch}^{(n)} : \mathcal{R} \longrightarrow \Lambda^{(n)}$ be the homomorphism of graded rings that is $\mathrm{ch}_k^{(n)}$ on the homogeneous part $\mathcal{R}_k$ of degree k.

Proposition 34.4. *The map* $\mathrm{ch}^{(n)}$ *is a surjective homomorphism of graded rings. The map* $\mathrm{ch}_k^{(n)}$ *in degree* k *is an isomorphism if* $n \geqslant k$.

Proof. The main thing to check is that the group law $\circ$ that was introduced in the ring $\mathcal{R}$ corresponds to multiplication of polynomials. Indeed, let θ and ρ be representations of S_k and S_l, respectively. Then $\theta \otimes \rho$ is an $S_k \times S_l$-module, and by Proposition 32.7, $\theta \circ \rho$ is the representation of S_{k+l} attached to $\mathbb{C}[S_{k+l}] \otimes_{\mathbb{C}[S_k \times S_l]} (N_\theta \otimes N_\rho)$. Therefore,

$$V_{\theta \circ \rho} = (\otimes^{k+l} V) \otimes_{\mathbb{C}[S_{k+l}]} \mathbb{C}[S_{k+l}] \otimes (N_\theta \otimes N_\rho),$$

which by (34.5) is isomorphic to

$$
\begin{aligned}
(\otimes^{k+l} V) &\otimes_{\mathbb{C}[S_k \times S_l]} (N_\theta \otimes N_\rho) \\
&\cong \left((\otimes^k V) \otimes (\otimes^l V) \right) \otimes_{\mathbb{C}[S_k] \otimes \mathbb{C}[S_l]} (N_\theta \otimes N_\rho) \\
&\cong (\otimes^k V \otimes_{\mathbb{C}[S_k]} N_\theta) \otimes (\otimes^l V \otimes_{\mathbb{C}[S_l]} N_\rho) = V_\theta \otimes V_\rho.
\end{aligned}
$$

Consequently the trace of $g \in \mathrm{GL}(n, \mathbb{C})$ on $V_{\theta \circ \rho}$ is the product of the traces on V_θ and V_ρ. It follows that for representations θ and ρ of S_k and S_l, we have $s_{\theta \circ \rho} = s_\theta \, s_\rho$. Hence, $\mathrm{ch}^{(n)}$ is multiplicative and therefore is a homomorphism of graded rings. It is surjective because a set of generators—the elementary symmetric polynomials e_i—are in the image. If $n \geqslant k$, then the ranks of $\mathcal{R}_k$ and $\Lambda_k^{(n)}$ both equal $p(k)$, so surjectivity implies that it is an isomorphism.   □

We will denote by $\mathbf{e}_k, \mathbf{h}_k \in \mathcal{R}_k$ the classes of the alternating representation and the trivial representation, respectively. It follows from Proposition 34.1 that $\mathrm{ch}^{(n)}(\mathbf{e}_k) = e_k$ and $\mathrm{ch}^{(n)}(\mathbf{h}_k) = h_k$.

Proposition 34.5. $\mathcal{R}$ *is a polynomial ring in an infinite number of generators,* $\mathcal{R} = \mathbb{Z}[\mathbf{e}_1, \mathbf{e}_2, \mathbf{e}_3, \ldots] = \mathbb{Z}[\mathbf{h}_1, \mathbf{h}_2, \mathbf{h}_3, \ldots]$.

Proof. To show that the $\mathbf{e}_i$ generate $\mathcal{R}$, it is sufficient to show that the ring they generate contains an arbitrary element u of $\mathcal{R}_k$ for any fixed k. Take $n \geqslant k$. Since $e_1, \ldots, e_n$ generate the ring $\Lambda^{(n)}$, there exists a polynomial f with integer coefficients such that $f(e_1, \ldots, e_n) = \mathrm{ch}(u)$. Then $\mathrm{ch}^{(n)}$ applied to $f(\mathbf{e}_1, \ldots, \mathbf{e}_n)$ gives $\mathrm{ch}(u)$, and it follows from the injectivity assertion in Proposition 34.4 that $f(\mathbf{e}_1, \ldots, \mathbf{e}_n) = u$.

To see that the $\mathbf{e}_i$ are algebraically independent, if f is a polynomial with integer coefficients such that $f(\mathbf{e}_1, \ldots, \mathbf{e}_n) = 0$, then since applying $\mathrm{ch}^{(n)}$ we have $f(e_1, \ldots, e_n) = 0$, by Proposition 33.1 it follows that $f = 0$.

Identical arguments work for the h's using Proposition 33.1.   □

The rings $\Lambda^{(n)}$ may be combined as follows. We have a homomorphism

$$r_n : \Lambda^{(n+1)} \longrightarrow \Lambda^{(n)}, \qquad x_{n+1} \longrightarrow 0. \tag{34.9}$$

It is easy to see that in this homomorphism $e_i \mapsto e_i$ if $i \leqslant n$ while $e_{n+1} \mapsto 0$, and so in the inverse limit

$$\Lambda = \varprojlim \Lambda^{(n)} \tag{34.10}$$

there exists a unique element whose image under the projection $\Lambda \to \Lambda^{(n)}$ is e_k for all $n \geqslant k$; we naturally denote this element e_k, and (34.8) implies that

$$\Lambda \cong \mathbb{Z}[e_1, e_2, e_3, \ldots]$$

is a polynomial ring in an infinite number of variables, and similarly

$$\Lambda \cong \mathbb{Z}[h_1, h_2, h_3, \ldots].$$

In the natural grading on Λ, e_i and h_i are homogeneous of degree i. Since the rank of $\Lambda_k^{(n)}$ equals the number of partitions of k into no more than n parts, the rank of Λ equals the number of partitions of k.

Proposition 34.6. *We have* $r_n \circ \mathrm{ch}^{(n+1)} = \mathrm{ch}^{(n)}$ *as maps* $\mathcal{R} \longrightarrow \Lambda^{(n)}$.

Proof. It is enough to check this on $\mathbf{e}_1, \mathbf{e}_2, \ldots$ since they generate $\mathcal{R}$ by Proposition 33.1. Both maps send $\mathbf{e}_k \overset{\bullet}{\longrightarrow} e_k$ if $k \leqslant n$, and $\mathbf{e}_k \longrightarrow 0$ if $k > n$.   □

Now turning to the inverse limit (34.10), the homomorphisms $\mathrm{ch}^{(n)} : \mathcal{R} \to \Lambda^{(n)}$ are compatible with the homomorphisms $\Lambda^{(n+1)} \to \Lambda^{(n)}$, and so there is induced a ring homomorphism $\mathrm{ch} : \mathcal{R} \to \Lambda$.

Theorem 34.2. *The map* $\mathrm{ch} : \mathcal{R} \longrightarrow \Lambda$ *is a ring isomorphism.*

Proof. This is clear from Proposition 34.4.   □

Theorem 34.3. *The rings* $\mathcal{R}$ *and* Λ *admit automorphisms of order* 2 *that interchange* $\mathbf{e}_i \longleftrightarrow \mathbf{h}_i$ *and* $e_i \longleftrightarrow h_i$.

Proof. Of course, it does not matter which ring we work in. Since $\Lambda \cong \mathbb{Z}[e_1, e_2, e_3, \ldots]$, and since the e_i are algebraically independent, if $u_1, u_2, \ldots$ are arbitrarily elements of Λ, there exists a unique ring homomorphism $\Lambda \longrightarrow \Lambda$ such that $e_i \longrightarrow u_i$. What we must show is that if we take the $u_i = h_i$, then this same homomorphism maps $h_i \longrightarrow u_i$. This follows from the fact that the recursive identity (33.3), from which we may solve for the e's in terms of the h's or conversely, is unchanged if we interchange $e_i \longleftrightarrow h_i$.   □

We will usually denote the involution of Theorem 34.3 as ι.

EXERCISES

Exercise 34.1. Let $\mathbf{s} = \mathbf{h}_1 \mathbf{h}_2 - \mathbf{h}_3$. Show that $'\mathbf{s} = \mathbf{s}$.

The Jacobi–Trudi Identity

For another account that derives the Jacobi-Trudi identity as a determinantal identity for characters of S_n using Mackey theory see Kerber [100]. The point of view in Zelevinsky [178] is slightly different but also similar in spirit. We take up his Hopf algebra approach in the exercises. For us, the details were worked out some years ago in the Stanford senior thesis of Karl Rumelhart.

An important question is to characterize the symmetric polynomials that correspond to irreducible characters of S_k. These are called *Schur polynomials*.

If $A = (a_{ij})$ and $B = (b_{ij})$ are square $N \times N$ matrices, and if $I, J \subset \{1, 2, 3, \ldots, n\}$ are two subsets of cardinality r, where $1 \leqslant r \leqslant n$, the minors

$$\det(a_{ij} \,|\, i \in I, j \in J), \qquad \det(b_{ij} \,|\, i \notin I, j \notin J),$$

are called *complementary*.

Proposition 35.1. *Let A be a matrix of determinant 1, and let $B = {}^t A^{-1}$. Each minor of A equals $\pm$ the complementary minor of B.*

This is a standard fact from linear algebra. For example, if

$$A = \begin{pmatrix} a_{11} & a_{12} & a_{13} & a_{14} \\ a_{21} & a_{22} & a_{23} & a_{24} \\ a_{31} & a_{32} & a_{33} & a_{34} \\ a_{41} & a_{42} & a_{43} & a_{44} \end{pmatrix}, \qquad B = \begin{pmatrix} b_{11} & b_{12} & b_{13} & b_{14} \\ b_{21} & b_{22} & b_{23} & b_{24} \\ b_{31} & b_{32} & b_{33} & b_{34} \\ b_{41} & b_{42} & b_{43} & b_{44} \end{pmatrix},$$

then

$$a_{23} = - \begin{vmatrix} b_{11} & b_{12} & b_{14} \\ b_{31} & b_{32} & b_{34} \\ b_{41} & b_{42} & b_{44} \end{vmatrix}, \qquad \begin{vmatrix} a_{12} & a_{13} \\ a_{32} & a_{33} \end{vmatrix} = - \begin{vmatrix} b_{21} & b_{24} \\ b_{41} & b_{44} \end{vmatrix}.$$

It is not hard to give a rule for the sign in general, but we will not need it.

Proof. Let us show how to prove this fact using exterior algebra. Suppose that A is an $N \times N$ matrix. Let $V = \mathbb{C}^N$. Then $\wedge^N V$ is one-dimensional, and we

$$
\begin{array}{ccc}
(\wedge^k V) \times (\wedge^{N-k} V) & \xrightarrow{\ (\wedge^k A,\ \wedge^{N-k} A)\ } & (\wedge^k V) \times (\wedge^{N-k} V) \\
\wedge \big\downarrow & & \big\downarrow \wedge \\
\wedge^N V & \xrightarrow{\ \wedge^N A\ } & \wedge^N V \\
\eta \big\downarrow & & \big\downarrow \eta \\
\mathbb{C} & \xrightarrow{\ \det A\ } & \mathbb{C}
\end{array}
$$

fix an isomorphism $\eta : \wedge^N V \longrightarrow \mathbb{C}$. If $1 \leqslant k \leqslant N$, and if $A : V \longrightarrow V$ is any linear transformation, we have a commutative diagram: The vertical arrows marked $\wedge$ are multiplications in the exterior algebra. The vertical map $\eta \circ \wedge :$ $(\wedge^k V) \times (\wedge^{N-k} V) \longrightarrow \mathbb{C}$ is a nondegenerate bilinear Indeed, let $v_1, \ldots, v_N$ be a basis of V chosen so that

$$
\eta(v_1 \wedge \cdots \wedge v_N) = 1.
$$

Then a pair of dual bases of $\wedge^k V$ and $\wedge^{N-k} V$ with respect to this pairing are

$$
v_{i_1} \wedge \cdots \wedge v_{i_k}, \qquad \pm v_{j_1} \wedge \cdots \wedge v_{j_{N-k}},
$$

where $i_1 < \cdots < i_k$, $j_1 < \cdots < j_{N-k}$, and the two subsets

$$
\{i_1, \ldots, i_k\}, \qquad \{j_1, \ldots, j_{N-k}\},
$$

of $\{1, \ldots, N\}$ are complementary. [The sign of the second basis vector will be $(-1)^d$, where $d = (i_1 - 1) + (i_2 - 2) + \cdots + (i_k - k)$.] If $\det(A) = 1$, then the bottom arrow is the identity map, and therefore we see that the map $\wedge^{N-k} A : \wedge^{N-k} V \to \wedge^{N-k} V$ is the inverse of the adjoint of $\wedge^k A : \wedge^k V \to \wedge^k V$ with respect to this dual pairing. Hence, if we use the above dual bases to compute matrices for these two maps, the matrix of $\wedge^{N-k} A$ is the transpose of the inverse of the matrix of $\wedge^k A$. Thus, if B is the inverse of the adjoint of A with respect to the inner product on V for which $v_1, \ldots, v_N$ are an orthonormal basis, then the matrix of $\wedge^{N-k} B$ is the same as the matrix of $\wedge^k A$. Now, with respect to the chosen dual bases, the coefficients in the matrix of $\wedge^k A$ are the $k \times k$ minors of A, while the matrix coefficients of $\wedge^{N-k} B$ are (up to sign) the complementary $(N-k) \times (N-k)$ minors of B. Hence, these are equal. $\qquad\square$

Proposition 35.2. *Suppose that* $\lambda = (\lambda_1, \ldots, \lambda_r)$ *and* $\mu = (\mu_1, \ldots, \mu_s)$ *are conjugate partitions of* k. *Then the* $r + s$ *numbers*

$$
\begin{array}{ll}
s + i - \lambda_i, & (i = 1, \ldots, r), \\
s - j + \mu_j + 1, & (j = 1, \ldots, s),
\end{array}
$$

are $1, 2, 3, \ldots, r + s$ *rearranged.*

Another proof of this combinatorial lemma may be found in Macdonald [124], I.1.7.

Proof. First note that the $r + s$ integers all lie between 0 and $r + s$. Indeed, if $1 \leqslant i \leqslant r$, then

$$0 \leqslant s + i - \lambda_i \leqslant s + r$$

because s is greater than or equal to the length $l(\mu) = \lambda_1 \geqslant \lambda_i$, so $s + i - \lambda_i \geqslant s - \lambda_i \geqslant 0$, and $s + i - \lambda_i \leqslant s + i \leqslant s + r$; and if $1 \leqslant j \leqslant s$, then

$$0 \leqslant s - j + \mu_j + 1 \leqslant s + r$$

since $s - j + \mu_j + 1 \geqslant s - j \geqslant 0$, and $\mu_j \leqslant \mu_1 = l(\lambda) \leqslant r$, so $s - j + \mu_j + 1 \leqslant s + \mu_j \leqslant s + r$.

Thus, it is sufficient to show that there are no duplications between these $s + r$ numbers. The sequence $s + i - \lambda_i$ is strictly increasing, so there can be no duplications in it, and similarly there can be no duplications among the $s - j + \mu_j + 1$. We need to show that $s + i - \lambda_i \neq s - j + \mu_j + 1$ for all $1 \leqslant i \leqslant r$, $1 \leqslant j \leqslant s$, that is,

$$\lambda_i + \mu_j + 1 \neq i + j. \tag{35.1}$$

There are two cases. If $j \leqslant \lambda_i$, then by (34.7) we have also $i \leqslant \mu_j$, so $\lambda_i + \mu_j + 1 > \lambda_i + \mu_j \geqslant i + j$. On the other hand, if $j > \lambda_i$, then by (34.7), $i > \lambda_j$, so

$$i + j \geqslant \lambda_i + \mu_j + 2 > \lambda_i + \mu_j + 1.$$

In both cases, we have (35.1). $\qquad\square$

We will henceforth denote the multiplication in $\mathcal{R}$, which was denoted in Chap. 34 with the symbol $\circ$, by the usual notations for multiplication. Thus, what was formerly denoted $\theta \circ \rho$ will be denoted $\theta\rho$, etc. Observe that the ring $\mathcal{R}$ is commutative.

We recall that $\mathbf{e}_k$ and $\mathbf{h}_k \in \mathcal{R}_k$ denote the sign character and the trivial character of S_k, respectively.

Proposition 35.3. *We have*

$$\mathbf{h}_k - \mathbf{e}_1\mathbf{h}_{k-1} + \mathbf{e}_2\mathbf{h}_{k-2} - \cdots + (-1)^k \mathbf{e}_k = 0 \tag{35.2}$$

if $k \geqslant 1$.

Proof. Choose $n \geqslant k$ so that the characteristic map $\mathrm{ch}^{(n)} : \mathcal{R}_k \to \Lambda_k^{(n)}$ is injective. It is then sufficient to prove that $\mathrm{ch}^{(n)}$ annihilates the left-hand side. Since $\mathrm{ch}^{(n)}(\mathbf{e}_i) = e_i$ and $\mathrm{ch}^{(n)}(\mathbf{h}_i) = h_i$, this follows from (33.3). $\qquad\square$

Proposition 35.4. *Let $\lambda = (\lambda_1, \ldots, \lambda_r)$ and $\mu = (\mu_1, \ldots, \mu_s)$ be conjugate partitions of k. Then*

$$\det(h_{\lambda_i - i + j})_{1 \leqslant i, j \leqslant r} = \pm \det(e_{\mu_i - i + j}). \tag{35.3}$$

Our convention is that if $r < 0$, then $h_r = e_r = 0$. (Also, remember that $h_0 = e_0 = 1$.) As an example, if $\lambda = (3,3,1)$, then $\mu = \lambda^t = (3,2,2)$, and we have

$$\begin{vmatrix} h_3 & h_4 & h_5 \\ h_2 & h_3 & h_4 \\ 0 & h_0 & h_1 \end{vmatrix} = \begin{vmatrix} e_3 & e_4 & e_5 \\ e_1 & e_2 & e_3 \\ e_0 & e_1 & e_2 \end{vmatrix}.$$

Later, in Proposition 35.1 we will see that the sign in (35.3) is always $+$. This could be proved now by carefully keeping track of the sign, but this is more trouble than it is worth because we will determine the sign in a different way.

Proof. We may interpret (33.3) as saying that the Toeplitz matrix

$$\begin{pmatrix} h_0 & h_1 & \cdots & h_{r+s-1} \\ & h_0 & \cdots & h_{r+s-2} \\ & & \ddots & \vdots \\ & & & h_0 \end{pmatrix} \tag{35.4}$$

is the transpose inverse of

$$\begin{pmatrix} e_0 & & \\ e_1 & e_0 & \\ \vdots & & \ddots \\ e_{r+s-1} & e_{r+s-2} & e_0 \end{pmatrix} \tag{35.5}$$

conjugated by

$$\begin{pmatrix} 1 & & & \\ & -1 & & \\ & & \ddots & \\ & & & (-1)^{r+s-1} \end{pmatrix}.$$

We only need to compute the minors up to sign, and conjugation by the latter matrix only changes the signs of these minors. Hence, it follows from Proposition 35.1 that each minor of (35.4) is, up to sign, the same as the complementary minor of (35.5). Let us choose the minor of (35.4) with columns $s+1,\ldots,s+r$ and rows $s+i-\lambda_i$ $(i = 1,\ldots,r)$. This minor is the left-hand side of (35.3). By Proposition 35.2, the complementary minor of (35.5) is formed with columns $1,\ldots,s$ and rows $s-j+\mu_j+1$ $(j = 1,\ldots,s)$. After conjugating this matrix by

$$\begin{pmatrix} & & 1 \\ & \cdot^{\cdot^{\cdot}} & \\ 1 & & \end{pmatrix},$$

we obtain the right-hand side of (35.3). $\qquad\square$

Suppose that $\lambda = (\lambda_1, \ldots, \lambda_r)$ is a partition of k. Then we will denote

$$\mathbf{e}_\lambda = \mathbf{e}_{\lambda_1} \cdots \mathbf{e}_{\lambda_r}, \qquad \mathbf{h}_\lambda = \mathbf{h}_{\lambda_1} \cdots \mathbf{h}_{\lambda_r}.$$

Referring to the definition of multiplication in the ring $\mathcal{R}_k$, we see that $\mathbf{e}_\lambda$ and $\mathbf{h}_\lambda$ are the characters of S_k induced from the sign and trivial characters, respectively, of the subgroup $S_{\lambda_1} \times \cdots \times S_{\lambda_r}$. We will denote this group by S_λ.

There is a partial ordering on partitions. We write $\lambda \succcurlyeq \mu$ if

$$\lambda_1 + \cdots + \lambda_i \geqslant \mu_1 + \cdots + \mu_i, \qquad (i = 1, 2, 3, \dots).$$

Since $\mathcal{R}_k$ is the character ring of S_k, it has a natural inner product, which we will denote $\langle \, , \, \rangle$. Our objective is to compute the inner product $\langle \mathbf{e}_\lambda, \mathbf{h}_\mu \rangle$.

Proposition 35.5. *Let $\lambda = (\lambda_1, \dots, \lambda_r)$ and $\mu = (\mu_1, \dots, \mu_s)$ be partitions of k. Then*

$$\langle \mathbf{h}_\lambda, \mathbf{e}_\mu \rangle = \langle \mathbf{e}_\lambda, \mathbf{h}_\mu \rangle. \tag{35.6}$$

This inner product is equal to the number of $r \times s$ matrices with each coefficient equal to either 0 or 1 such that the sum of the ith row is equal to λ_i and the sum of the jth column is equal to μ_j. This inner product is nonzero if and only if $\mu^t \succcurlyeq \lambda$. If $\mu^t = \lambda$, then the inner product is 1.

Proof. Computing the right- and left-hand sides of (35.6) both lead to the same calculation, as we shall see. For definiteness, we will compute the left-hand side of (35.6). Note that

$$\langle \mathbf{h}_\lambda, \mathbf{e}_\mu \rangle = \dim \operatorname{Hom}_{S_k} \left(\operatorname{Ind}_{S_\lambda}^{S_k}(1), \operatorname{Ind}_{S_\mu}^{S_k}(\varepsilon) \right),$$

where ε is the alternating character of S_λ, and $\operatorname{Ind}_{S_\mu}^{S_k}(\varepsilon)$ denotes the corresponding induced representation of S_k. This is because $\mathbf{e}_{\mu_i} \in \mathcal{R}_{\mu_i}$ is the alternating character of S_{λ_i}, and the multiplication in $\mathcal{R}$ is defined so that the product $\mathbf{e}_\mu = \mathbf{e}_{\mu_1} \cdots \mathbf{e}_{\mu_s}$ is obtained by induction from S_μ.

By Mackey's theorem, we must count the number of double cosets in $S_\mu \backslash S_k / S_\lambda$ that support intertwining operators. (See Remark 32.2.) Simply counting these double cosets is sufficient because the representations that we are inducing are both one-dimensional, so each space on the right-hand side of (32.12) is either one-dimensional (if the coset supports an intertwining operator) or zero-dimensional (if it doesn't).

First, we will show that the double cosets in $S_\mu \backslash S_k / S_\lambda$ may be parametrized by $s \times r$ matrices with nonnegative integer coefficients such that the sum of the ith row is equal to μ_i and the sum of the jth column is equal to λ_j. Then we will show that the double cosets that support intertwining operators are precisely those that have no entry > 1. This will prove the first assertion.

We will identify S_k with the group of $k \times k$ permutation matrices. (A *permutation matrix* is one that has only zeros and ones as entries, with exactly one nonzero entry in each row and column.) Then S_λ is the subgroup consisting of elements of the form

$$\begin{pmatrix} D_1 & 0 & \cdots & 0 \\ 0 & D_2 & \cdots & 0 \\ \vdots & \vdots & \ddots & \vdots \\ 0 & 0 & \cdots & D_r \end{pmatrix},$$

where D_i is a $\lambda_i \times \lambda_i$ permutation matrix. Let $g \in S_k$ represent a double coset in $S_\mu \backslash S_k / S_\lambda$. Let us write g in block form,

$$\begin{pmatrix} G_{11} & G_{12} & \cdots & G_{1r} \\ G_{21} & G_{22} & \cdots & G_{2r} \\ \vdots & \vdots & \ddots & \vdots \\ G_{s1} & G_{s2} & \cdots & G_{sr} \end{pmatrix}, \tag{35.7}$$

where G_{ij} is a $\mu_i \times \lambda_j$ block. Let γ_{ij} be the rank of G_{ij}, which is the number of nonzero entries. Then the matrix $r \times s$ matrix (γ_{ij}) is independent of the choice of representative of the double coset. It has the property that the sum of the ith row is equal to μ_i and the sum of the jth column is equal to λ_j. Moreover, it is easy to see that any such matrix arises from a double coset in this manner and determines the double coset uniquely. This establishes the correspondence between the matrices (γ_{ij}) and the double cosets.

Next we show that a double coset supports an intertwining operator if and only if each $\gamma_{ij} \leqslant 1$. A double coset $S_\mu g S_\lambda$ supports an intertwining operator if and only if there exists a nonzero function $\Delta : S_k \to \mathbb{C}$ with support in $S_\mu g S_\lambda$ such that

$$\Delta(\tau h \sigma) = \varepsilon(\tau)\Delta(h) \tag{35.8}$$

for $\tau \in S_\mu$, $\sigma \in S_\lambda$.

First, suppose the matrix (γ_{ij}) is given such that for some particular i, j, we have $\gamma = \gamma_{ij} > 1$. Then we may take as our representative of the double coset a matrix g such that

$$G_{ij} = \begin{pmatrix} I_\gamma & 0 \\ 0 & 0 \end{pmatrix}.$$

Now there exists a transposition $\sigma \in S_\lambda$ and a transposition $\tau \in S_\mu$ such that $g = \tau g \sigma$. Indeed, we may take τ to be the transposition $(12) \in S_{\lambda_j} \subset S_\lambda$ and σ to be the transposition $(12) \in S_{\mu_i} \subset S_\mu$. Now, by (35.8),

$$\Delta(g) = \Delta(\tau g \sigma) = -\Delta(g),$$

so $\Delta(g) = 0$ and therefore Δ is identically zero. We see that if any $\gamma_{ij} > 1$, then the corresponding double coset does not support an intertwining operator.

On the other hand, if each $\gamma_{ij} \leqslant 1$, then we will show that for g a representative of the corresponding double coset, $g^{-1} S_\mu g \cap S_\lambda = \{1\}$, or

$$S_\mu g \cap g S_\lambda = \{g\}. \tag{35.9}$$

Indeed, suppose that $\tau \in S_\mu$ and $\sigma \in S_\lambda$ such that $\tau g = g\sigma$. Writing

$$\tau = \begin{pmatrix} \tau_{\mu_1} & & \\ & \tau_{\mu_2} & \\ & & \ddots \end{pmatrix}, \qquad \sigma = \begin{pmatrix} \sigma_{\lambda_1} & & \\ & \sigma_{\lambda_2} & \\ & & \ddots \end{pmatrix},$$

with $\tau_{\mu_i} \in S_{\mu_i}$ and $\sigma_{\lambda_i} \in S_{\lambda_i}$ and letting g be as in (35.7), we have $\tau_{\mu_i} G_{ij} = G_{ij} \sigma_{\lambda_j}$. If $\tau_{\mu_i} \neq I$, then

$$\tau_{\mu_i} \begin{pmatrix} G_{i1} \cdots G_{ir} \end{pmatrix} \neq \begin{pmatrix} G_{i1} \cdots G_{ir} \end{pmatrix}$$

since the rows of the second matrix are distinct. Thus $\tau_{\mu_i} G_{ij} \neq G_{ij}$ for some i. Since G_{ij} has at most one nonzero entry, it is impossible that after reordering the rows (which is the effect of left multiplication by τ_{μ_i}) this nonzero entry could be restored to its original position by reordering the columns (which is the effect of right multiplication by $\sigma_{\lambda_j}^{-1}$). Thus, $\tau_{\mu_i} G_{ij} \neq G_{ij}$ implies that $\tau_{\mu_i} G_{ij} \neq G_{ij} \sigma_{\lambda_j}$. This contradiction proves (35.9).

Now (35.9) shows that each element of the double coset has a unique representation as $\tau g \sigma$ with $\tau \in S_\mu$ and $\sigma \in S_\lambda$. Hence, we may define

$$\Delta(h) = \begin{cases} \varepsilon(\tau) & \text{if } h = \tau g \sigma \text{ with } \tau \in S_\mu \text{ and } \sigma \in S_\lambda, \\ 0 & \text{otherwise,} \end{cases}$$

and this is well-defined. Hence, such a double coset does support an intertwining operator.

Now we have asserted further that (35.6) is nonzero if and only if $\mu^t \succcurlyeq \lambda$ and that if $\mu^t = \lambda$, then the inner product is 1. Let us ask, therefore, for given λ and μ, whether we can construct a matrix (γ_{ij}) with each $\gamma_{ij} = 0$ or 1 such that the sum of the ith row is μ_i and the sum of the jth column is λ_j. Let $\nu = \mu^t$. Then

$$\nu_i = \text{card}\,\{j \mid \mu_j \geqslant i\}.$$

That is, ν_i is the number of rows that will accommodate up to i 1's. Now $\nu_1 + \nu_2 + \cdots + \nu_t$ is equal to the number of rows that will take a 1, plus the number of rows that will take two 1's, and so forth. Let us ask how many 1's we may put in the first t columns. *Each nonzero entry must lie in a different row*, so to put as many 1's as possible in the first t columns, we should put ν_t of them in those rows that will accommodate t nonzero entries, ν_{t-1} of them in those rows that will accommodate $t-1$ entries, and so forth. Thus, $\nu_1 + \cdots + \nu_t$ is the maximum number of 1's we can put in the first t columns. We need to place $\lambda_1 + \cdots + \lambda_t$ ones in these rows, so in order for the construction to be possible, what we need is

$$\lambda_1 + \cdots + \lambda_t \leqslant \nu_1 + \cdots + \nu_t$$

for each t, that is, for $\nu \succcurlyeq \lambda$. It is easy to see that if $\nu = \lambda$, then the location of the ones in the matrix (γ_{ij}) is forced so that in this case there exists a unique intertwining operator. $\qquad\square$

Corollary 35.1. *If λ and μ are partitions of k, then we have $\mu^t \succcurlyeq \lambda^t$ if and only if $\lambda \succcurlyeq \mu$.*

Proof. This is equivalent to the statement that $\mu^t \succcurlyeq \lambda$ if and only if $\lambda^t \succcurlyeq \mu$. In this form, this is contained in the preceding proposition from the identity (35.6) together with the characterization of the nonvanishing of that inner product. Of course, one may also give a direct combinatorial argument. □

Theorem 35.1. (Jacobi–Trudi identity) *Let $\lambda = (\lambda_1, \ldots, \lambda_r)$ and $\mu = (\mu_1, \ldots, \mu_s)$ be conjugate partitions of k. We have the identity*

$$\det(\mathbf{h}_{\lambda_i - i + j})_{1 \leqslant i, j \leqslant r} = \det(\mathbf{e}_{\mu_i - i + j})_{1 \leqslant i, j \leqslant s} \tag{35.10}$$

in $\mathcal{R}_k$. We denote this element (35.10) as $\mathbf{s}_\lambda$. It is an irreducible character of S_k and may be characterized as the unique irreducible character that occurs with positive multiplicity in both $\mathrm{Ind}_{S_\mu}^{S_k}(\varepsilon)$ and $\mathrm{Ind}_{S_\lambda}^{S_k}(1)$; it occurs with multiplicity one in each of them. The $p(k)$ characters $\mathbf{s}_\lambda$ are all distinct, and are all the irreducible characters of S_k.

Proof. Let $n \geqslant k$, so that $\mathrm{ch}^{(n)} : \mathcal{R}_k \to \Lambda_k^{(n)}$ is injective. Applying ch to (35.10) and using (35.3), we see that the left- and right-hand sides are either equal or negatives of each other. We will show that the inner product of the left-hand side with the right-hand side of (35.10) equals 1. Since the inner product is positive definite, this will show that the left- and right-hand sides are actually equal. Moreover, if $\sum d_i \chi_i$ is the decomposition of (35.10) into irreducibles, this inner product is $\sum_i d_i^2$, so knowing that the inner product is 1 will imply that $\mathbf{s}_\lambda$ is either an irreducible character, or the negative of an irreducible character.

We claim that expanding the determinant on the left-hand side of (35.10) gives a sum of terms of the form $\pm \mathbf{h}_{\lambda'}$ where each $\lambda' \succcurlyeq \lambda$ and the term $\mathbf{h}_\lambda$ occurs exactly once. Indeed, the terms in the expansion of the determinant are of the form

$$\mathbf{h}_{\lambda_1 - 1 + j_1} \mathbf{h}_{\lambda_2 - 2 + j_2} \cdots \mathbf{h}_{\lambda_r - r + j_r},$$

where $(j_1, \ldots, j_r)$ is a permutation of $(1, 2, \ldots, r)$. If we arrange the indices $\lambda_i - i + j_i$ into descending order as $\lambda_1', \lambda_2', \ldots$, then λ_1' is greater than or equal to $\lambda_1 - 1 + j_1$. Moreover, $j_1 \geqslant 1$ so

$$\lambda_1' \geqslant \lambda_1 - 1 + j_1 \geqslant \lambda_1,$$

and similarly $j_1 + j_2 \geqslant 3$ so

$$\lambda_1' + \lambda_2' \geqslant (\lambda_1 - 1 + j_1) + (\lambda_2 - 2 + j_2) \geqslant \lambda_1 + \lambda_2,$$

and so forth.

Similarly, expanding the right-hand side gives a sum of terms of the form $\pm \mathbf{e}_{\mu'}$, where $\mu' \succcurlyeq \mu$, and the term $\mathbf{e}_\mu$ also occurs exactly once.

Now let us consider $\langle \mathbf{h}_{\lambda'}, \mathbf{e}_{\mu'} \rangle$. By Proposition 35.5, if this is nonzero we have $(\mu')^t \succcurlyeq \lambda'$. Since $\lambda' \succcurlyeq \lambda$ and $\mu' \succcurlyeq \mu$, which implies $\mu^t \succcurlyeq (\mu')^t$ by Corollary 35.1, we have

$$\lambda = \mu^t \succcurlyeq (\mu')^t \succcurlyeq \lambda' \succcurlyeq \lambda.$$

Thus, we must have $\lambda' = \lambda$. It is easy to see that this implies that $(j_1, \ldots, j_r) = (1, 2, \ldots, r)$, so the monomial $\mathbf{e}_\lambda$ occurs exactly once in the expansion of $\det(\mathbf{h}_{\lambda_i - i + j})$. A similar analysis applies to $\det(\mathbf{e}_{\mu_i - i + j})$.

We see that the inner product of the left- and right-hand sides of (35.10) equals 1, which implies everything except that $\mathbf{s}_\lambda$ and not $-\mathbf{s}_\lambda$ is an irreducible character of S_k. To see this, we form the inner product $\langle \mathbf{s}_\lambda, \mathbf{h}_\mu \rangle$. The same considerations show that this inner product is 1. Since $\mathbf{h}_\mu$ is a proper character [it is the character of $\mathrm{Ind}_{S_\mu}^{S_k}(1)$] this implies that it is $\mathbf{s}_\lambda$, and not $-\mathbf{s}_\lambda$, is an irreducible character.

We have just noted that $\mathbf{s}_\lambda$ occurs with positive multiplicity in $\mathbf{h}_\lambda$, which is the character of the representation $\mathrm{Ind}_{S_\lambda}^{S_k}(1)$. Similar considerations show that $\langle \mathbf{s}_\lambda, \mathbf{e}_\mu \rangle = 1$ and $\mathbf{e}_\lambda$ is the character of the representation $\mathrm{Ind}_{S_\mu}^{S_k}(\epsilon)$. By Proposition 35.5, $\langle \mathbf{e}_\mu, \mathbf{h}_\lambda \rangle = 1$, so there cannot be any other representation that occurs with positive multiplicity in both.

This characterization of $\mathbf{s}_\lambda$ shows that it cannot equal $\mathbf{s}_\mu$ for any $\mu \neq \lambda$, so the irreducible characters $\mathbf{s}_\lambda$ are all distinct. Their number is $p(k)$, which is also the number of conjugacy classes in S_k (i.e., the total number of irreducible representations). We have therefore constructed all of them. $\square$

Theorem 35.2. *If λ and μ are conjugate partitions, and if ι is the involution of Theorem 34.3, then ${}^\iota \mathbf{s}_\lambda = \mathbf{s}_\mu$ and ${}^\iota s_\lambda = s_\mu$.*

Proof. Since ${}^\iota \mathbf{h}_\lambda = \mathbf{e}_\lambda$ and ${}^\iota \mathbf{e}_\lambda = \mathbf{h}_\lambda$, this follows from the Jacobi-Trudi identity. $\square$

EXERCISES

Exercise 35.1. Let λ and μ be partitions of k. Show that

$$\langle \mathbf{h}_\lambda, \mathbf{h}_\mu \rangle = \langle \mathbf{e}_\lambda, \mathbf{e}_\mu \rangle$$

and that this inner product is equal to the number of $r \times s$ matrices with each coefficient a nonnegative integer such that the sum of the ith row is equal to λ_i, and the sum of the jth column is equal to μ_j.

Exercise 35.2. Give a combinatorial proof of Corollary 35.1.

Exercise 35.3. If λ, μ are a partitions of k, let $T_{\mathrm{sh}}(\lambda\mu)$ be the coefficient of h_μ when s_λ is expressed in terms of the h_μ, that is,

$$s_\lambda = T_{\mathrm{sh}}(\lambda, \mu) h_\mu.$$

Similarly we will define T_{xy} when x, y are s, e or h to denote the transition matrices between the bases s_λ, e_λ and h_λ of Λ_k.

(i) Show that $T_{sh}(\lambda, \mu) = 0$ unless $\mu \geqslant \lambda$.
(ii) Show that $T_{hs}(\lambda, \mu) = 0$ unless $\mu \geqslant \lambda$.
(iii) Show that $T_{se}(\lambda, \mu) = 0$ unless $\mu \geqslant \lambda^t$.
(iv) Show that $T_{es}(\lambda, \mu) = 0$ unless $\mu^t \geqslant \lambda$.
(v) Show that $T_{he}(\lambda, \mu) = 0$ unless $\mu^t \geqslant \lambda$.
(v) Show that $T_{eh}(\lambda, \mu) = 0$ unless $\mu^t \geqslant \lambda$.

Zelevinsky [178] shows how the ring $\mathcal{R}$ may be given the structure of a graded Hopf algebra. This extra algebraic structure (actually introduced earlier by Geissinger) encodes all the information about the representations of S_k that comes from Mackey theory. Moreover, a similar structure exists in a ring $\mathcal{R}(q)$ analogous to $\mathcal{R}$, constructed from the representations of $\mathrm{GL}(k, \mathbb{F}_q)$, which we will consider in Chap. 47. Thus, Zelevinsky is able to give a unified discussion of important aspects of the two theories. In the next exercises, we will establish the basic fact that $\mathcal{R}$ is a Hopf algebra.

We begin by reviewing the notion of a Hopf algebra. We recommend Majid [125] for further insight. (Apart from its use as an introduction to quantum groups, this is good for gaining facility with Hopf algebra methods such as the Sweedler notation.) Let A be a commutative ring. An A-algebra is normally defined to be a ring R with a homomorphism u into the center of R. The homomorphism u (called the *unit*) then makes R into an A-module. The multiplication map $R \times R \to R$ is A-bilinear hence induces a linear map $m : R \otimes R \to R$. The associative law for multiplication may be interpreted as the commutativity of the diagram:

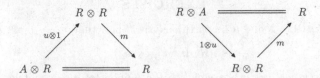

We also have commutative diagrams
Here we are identifying R with $A \otimes R$ by the canonical isomorphism $x \mapsto 1 \otimes x$.

As an alternative viewpoint, given an A-module R with linear maps $u : A \to R$ and $m : R \otimes R \to R$ subject to these commutative diagrams, R is an A-algebra. The change of viewpoint in replacing the bilinear multiplication map $R \times R \to R$ with the linear map $R \otimes R \to R$ is a simple but useful one, since it allows us to transport the notion to other contexts. For example, now we can dualize it.

The dual notion to an algebra is that of a *coalgebra*. The definition and axioms are obtained by reversing all the arrows. That is, we require an A-module R together with linear maps $\Delta : R \to R \otimes R$ and $\epsilon : R \to A$ such that we have commutative diagrams

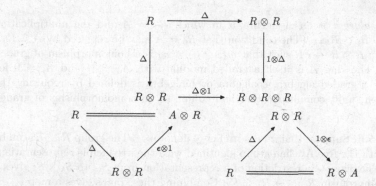

Exercise 35.4. Let R be a algebra that is also a coalgebra. Show that the three statements are equivalent.

(i) The comultiplication $\Delta : R \longrightarrow R \otimes R$ and counit $R \to A$ are homomorphisms of algebras.

(ii) The multiplication $m : R \otimes R \longrightarrow R$ and unit $A \to R$ are homomorphisms of coalgebras.

(iii) The following diagram is commutative:
Here τ is the "transposition" map $R \otimes R \to R \otimes R$ that sends $x \otimes y$ to $y \otimes x$. We will refer to this property as the *Hopf axiom*.

$$
\begin{array}{ccccc}
R \otimes R & \xrightarrow{\;\Delta \otimes \Delta\;} & R \otimes R \otimes R \otimes R & \xrightarrow{\;1 \otimes \tau \otimes 1\;} & R \otimes R \otimes R \otimes R \\
\downarrow{\scriptstyle m} & & & & \downarrow{\scriptstyle m \otimes m} \\
R & \xrightarrow{\hspace{4cm}\Delta\hspace{4cm}} & & & R \otimes R
\end{array}
$$

If these three equivalent conditions are satisfied, then R is called a *bialgebra*. Note that this definition is self-dual. For example, if A is a field and R is a finite-dimensional bialgebra, then the dual space R^* is also a bialgebra, with comultiplication being the adjoint of multiplication, etc.

Exercise 35.5. Let G be a finite group, $A = \mathbb{C}$ and let R be the group algebra. Define a map $\Delta : R \to R \otimes R$ by extending the diagonal map $G \to G \times G$ to a linear map $R \to R \otimes R$, and let $\epsilon : R \to A$ be the augmentation map that sends every element of G to 1. Show that R is a bialgebra.

As a variant, all these notions have graded versions. Let A be a commutative ring. A *graded A-module* R is an A-module R with a sequence $\{R_0, R_1, R_2, \ldots\}$ of submodules such that $R = \bigoplus R_i$, and a homomorphism $R \longrightarrow S$ of graded A-modules is a homomorphism that takes R_i into S_i. The tensor product $R \otimes S = R \otimes_A S$ of two graded A-modules is a graded A-module with

$$(R \otimes S)_m = \bigoplus_{k+l=m} R_k \otimes S_l.$$

A *graded A-algebra* is an A-algebra R in which $R_0 = A$ and the multiplication satisfies $R_k \cdot R_l \subset R_{k+l}$. (The condition that $R_0 = A$ may be replaced by $A \subseteq R_0$.) The map $m : R \otimes R \longrightarrow R$ such that $m(x \otimes y) = xy$ is a homomorphism of graded A-modules. The ring A is itself a graded module with $A_0 = A$ and $A_i = 0$ for $i > 0$. Now a graded algebra, coalgebra or bialgebra is defined by requiring the multiplication, unit, comultiplication and counit to be homomorphisms of graded modules.

Exercise 35.6. Suppose that $k+l = m$. Let $\otimes$ denote $\otimes_{\mathbb{Z}}$. The group $\mathcal{R}_k \otimes \mathcal{R}_l$ can be identified with the free Abelian group identified with the irreducible representations of $S_k \times S_l$. (Explain.) So restriction of a representation from S_m to $S_k \times S_l$ gives a group homomorphism $\mathcal{R}_m \longrightarrow \mathcal{R}_k \otimes \mathcal{R}_l$. Combining these maps gives a map

$$\Delta : \mathcal{R}_m \longrightarrow \bigoplus_{k+l=m} \mathcal{R}_k \otimes \mathcal{R}_l = (\mathcal{R} \otimes \mathcal{R})_m.$$

Show that this homomorphism of graded $\mathbb{Z}$-algebras makes $\mathcal{R}$ into a graded coalgebra.

Exercise 35.7. (Zelevinsky [178])

(i) Let $k+l = p+q = m$. Representing elements of the symmetric group as matrices, show that a complete set of double coset representatives for $(S_p \times S_q) \backslash S_m / (S_k \times S_l)$ consists of the matrices

$$\begin{pmatrix} I_a & 0 & 0 & 0 \\ 0 & 0 & 0 & I_c \\ 0 & 0 & I_d & 0 \\ 0 & I_b & 0 & 0 \end{pmatrix},$$

where $a + b = k$, $c + d = l$, $a + c = p$, and $b + d = q$.

(ii) Use (i) and Mackey theory to prove that $\mathcal{R}$ is a graded bialgebra over $\mathbb{Z}$.

Hint: Both parts are similar to parts of the proof of Proposition 35.5.

A bialgebra R is called a *Hopf algebra* if it satisfies the following additional condition. There must be a map $S : R \to R$ such that the following diagram is commutative:

$$
\begin{array}{ccccc}
R \otimes R & \xleftarrow{\;\Delta\;} & R & \xrightarrow{\;\Delta\;} & R \otimes R \\
\Big\downarrow{\scriptstyle 1 \otimes S} & & \Big\downarrow{\scriptstyle \epsilon} & & \Big\downarrow{\scriptstyle S \otimes 1} \\
& & A & & \\
& & \Big\downarrow{\scriptstyle u} & & \\
R \otimes R & \xrightarrow{\;m\;} & R & \xleftarrow{\;m\;} & R \otimes R
\end{array}
$$

Exercise 35.8. Show that a group algebra is a Hopf algebra. The antipode is the map $S(g) = g^{-1}$.

Exercise 35.9. Show that $\mathcal{R}$ is a Hopf algebra. We have $S(\mathbf{h}_k) = (-1)^k \mathbf{e}_k$ and $S(\mathbf{e}_k) = (-1)^k \mathbf{h}_k$.

Exercise 35.10. Let H be a Hopf algebra. The *Hopf square* map $\sigma : H \to H$ is $m \circ \Delta$. Prove that if H is commutative as a ring, then σ is a ring homomorphism.

The next exercise is from the 2013 senior thesis of Seth Shelley-Abrahamson. Similar statements relate the higher Hopf power maps to other wreath products. Interest in the Hopf square map and higher-power maps has been stimulated by recent investigations of Diaconis, Pang, and Ram.

Exercise 35.11. Let H_k be the *hyperoctahedral group* of $k \times k$ matrices g such that g has one nonzero entry in every row and column, and every nonzero entry is ± 1. The order of H_k is $k!2^k$ and H_k is isomorphic to the Weyl group of Cartan type B_k (or C_k). Given a character χ of S_k, we may induce χ to H_k, then restrict it back to S_k. Thus, we get a self-map of $\mathcal{R}_k$.

 (i) Use Mackey theory to show that this map is the Hopf square map.
 (ii) Let θ_k be the function on S_k that has a value on a permutation σ is 2^n where n is the number of cycles in σ. Show that θ_k is a character of S_k and that the map of (i) multiplies every character of S_k by σ_k.

Schur Polynomials and $\mathrm{GL}(n, \mathbb{C})$

Now let $s_\mu(x_1, \ldots, x_n)$ be the symmetric polynomial $\mathrm{ch}^{(n)}(\mathbf{s}_\mu)$; we will use the same notation s_μ for the element $\mathrm{ch}(\mathbf{s}_\mu)$ of the inverse limit ring Λ defined by (34.10). These are the *Schur polynomials*.

Theorem 36.1. *Assume that $n \geqslant l(\lambda)$. We have*

$$
s_\lambda(x_1, \ldots, x_n) = \frac{\begin{vmatrix} x_1^{\lambda_1+n-1} & x_2^{\lambda_1+n-1} & \cdots & x_n^{\lambda_1+n-1} \\ x_1^{\lambda_2+n-2} & x_2^{\lambda_2+n-2} & \cdots & x_n^{\lambda_2+n-2} \\ \vdots & & & \vdots \\ x_1^{\lambda_n} & x_2^{\lambda_n} & \cdots & x_n^{\lambda_n} \end{vmatrix}}{\begin{vmatrix} x_1^{n-1} & x_2^{n-1} & \cdots & x_n^{n-1} \\ x_1^{n-2} & x_2^{n-2} & \cdots & x_n^{n-2} \\ \vdots & & & \vdots \\ x_1 & x_2 & \cdots & x_n \\ 1 & 1 & \cdots & 1 \end{vmatrix}},
\tag{36.1}
$$

provided that n is greater than or equal to the length of the partition k, so that we may denote $\lambda = (\lambda_1, \ldots, \lambda_n)$ (possibly with trailing zeros). In this case $s_\lambda \neq 0$.

It is worth recalling that the Vandermonde determinant in the denominator can be factored:

$$
\begin{vmatrix} x_1^{n-1} & x_2^{n-1} & \cdots & x_n^{n-1} \\ x_1^{n-2} & x_2^{n-2} & \cdots & x_n^{n-2} \\ \vdots & & & \vdots \\ x_1 & x_2 & \cdots & x_n \\ 1 & 1 & \cdots & 1 \end{vmatrix} = \prod_{i<j}(x_i - x_j).
$$

It is also worth noting, since it is not immediately obvious from the expression (36.1), that the Schur polynomial s_λ in $n+1$ variables restricts to the Schur

D. Bump, *Lie Groups*, Graduate Texts in Mathematics 225,
DOI 10.1007/978-1-4614-8024-2_36, © Springer Science+Business Media New York 2013

polynomial also denoted s_λ under the map (34.9). This is of course clear from Proposition 34.6 and the fact that $\mathrm{ch}(\mathbf{s}_\lambda) = s_\lambda$.

Proof. Let $e_k^{(i)}$ be the kth elementary symmetric matrix in $n-1$ variables

$$x_1, \ldots, x_{i-i}, x_{i+1}, \ldots, x_n,$$

omitting x_i. We have, using (33.1) and (33.2) and omitting one variable in (33.1),

$$\sum_{k=0}^{\infty} (-1)^k e_k^{(i)} t^k = \prod_{\substack{j \neq i}}^{n} (1 - x_j t),$$

$$\sum_{k=0}^{\infty} h_k t^k = \prod_{i=1}^{n} (1 - x_i t)^{-1},$$

and therefore

$$\left[\sum_{k=0}^{\infty} (-1)^k e_k^{(i)} t^k \right] \left[\sum_{k=0}^{\infty} h_k t^k \right] = (1 - t x_i)^{-1} = 1 + t x_i + t^2 x_i^2 + \cdots.$$

Comparing the coefficients of t^r in this identity, we have

$$\sum_{k=0}^{\infty} (-1)^k e_k^{(i)} h_{r-k} = x_i^r.$$

(Our convention is that $e_k^{(i)} = h_k = 0$ if $k < 0$, and also note that $e_k^{(i)} = 0$ if $k \geqslant n$.) Therefore, we have

$$\begin{pmatrix} h_{\lambda_1} & h_{\lambda_1+1} & \cdots & h_{\lambda_1+n-1} \\ h_{\lambda_2-1} & h_{\lambda_2} & \cdots & h_{\lambda_2+n-2} \\ \vdots & \vdots & & \vdots \\ h_{\lambda_n-n+1} & h_{\lambda_n-n+2} & \cdots & h_{\lambda_n} \end{pmatrix} \begin{pmatrix} \pm e_{n-1}^{(1)} & \pm e_{n-1}^{(2)} & \cdots & \pm e_{n-1}^{(n)} \\ \mp e_{n-2}^{(1)} & \mp e_{n-2}^{(2)} & \cdots & \mp e_{n-2}^{(n)} \\ \vdots & \vdots & & \vdots \\ e_0^{(1)} & e_0^{(2)} & \cdots & e_0^{(n)} \end{pmatrix} =$$

$$\begin{pmatrix} x_1^{\lambda_1+n-1} & x_2^{\lambda_1+n-1} & \cdots & x_n^{\lambda_1+n-1} \\ x_1^{\lambda_2+n-2} & x_2^{\lambda_2+n-2} & \cdots & x_2^{\lambda_2+n-2} \\ \vdots & & & \vdots \\ x_1^{\lambda_n} & x_2^{\lambda_n} & \cdots & x_n^{\lambda_n} \end{pmatrix}.$$

Denote the determinant of the second factor on the left-hand side by D. Taking determinants,

$$s_\lambda D = \begin{vmatrix} x_1^{\lambda_1+n-1} & x_2^{\lambda_1+n-1} & \cdots & x_n^{\lambda_1+n-1} \\ x_1^{\lambda_2+n-2} & x_2^{\lambda_2+n-2} & \cdots & x_n^{\lambda_2+n-2} \\ \vdots & & & \vdots \\ x_1^{\lambda_n} & x_2^{\lambda_n} & \cdots & x_n^{\lambda_n} \end{vmatrix}. \tag{36.2}$$

Hence, we have only to prove that D is equal to the denominator in (36.1), and this follows from (36.2) by taking $\lambda = (0, \ldots, 0)$ since $s_{(0,\ldots,0)} = 1$. $\quad\square$

Suppose that V and W are vector spaces over a field of characteristic zero and $B : V \times \cdots \times V \longrightarrow W$ is a symmetric k-linear map. Let $Q : V \longrightarrow W$ be the function $Q(v) = B(v, \ldots, v)$. The function B can be reconstructed from Q, and this process is called *polarization*. For example, if $k = 2$ we have

$$B(v, w) = \frac{1}{2} \left(Q(v + w) - Q(v) - Q(w) \right),$$

as we may see by expanding the right-hand side and using $B(v, w) = B(w, v)$.

Proposition 36.1. *Let U and W be vector spaces over a field of characteristic zero and let $B : U \times \cdots \times U \longrightarrow W$ be a symmetric k-linear map. Let $Q : U \longrightarrow W$ be the function $Q(u) = B(u, \ldots, u)$. If $u_1, \ldots, u_k \in U$, and if $S \subset I = \{1, 2, \ldots, k\}$, let $u_S = \sum_{i \in S} u_i$. We have*

$$B(u_1, \ldots, u_k) = \frac{1}{k!} \left[\sum_{S \subseteq I} (-1)^{k - |S|} Q(u_S) \right].$$

Proof. Expanding $Q(u_S) = B(u_S, \ldots, u_S)$ and using the k-linearity of B, we have

$$Q(u_S) = \sum_{i_1, \ldots, i_k \in S} B(u_{i_1}, u_{i_2}, \ldots, u_{i_k}).$$

Therefore,

$$\sum_{S \subseteq I} (-1)^{k - |S|} Q(u_S) = \sum_{\substack{1 \leqslant i_1 \leqslant k \\ \vdots \\ 1 \leqslant i_k \leqslant k}} B(u_{i_1}, \ldots, u_{i_k}) \sum_{S \supseteq \{i_1, \ldots, i_k\}} (-1)^{k - |S|}.$$

Suppose that there are repetitions among the list $i_1, \ldots, i_k$. Then there will be some $j \in I$ such that $j \notin \{i_1, \ldots, i_k\}$, and pairing those subsets containing j with those not containing j, we see that the sum $\sum_{S \supseteq \{i_1, \ldots, i_k\}} (-1)^{k - |S|} = 0$. Hence, we need only consider those terms where $\{i_1, \ldots, i_k\}$ is a permutation of $\{1, \ldots, k\}$. Remembering that B is symmetric, these terms all contribute equally and the result follows. $\square$

Theorem 36.2. *Let λ be a partition of k, and let $n \geqslant l(\lambda)$. Then there exists an irreducible representation $\pi_\lambda = \pi_\lambda^{\mathrm{GL}(n)}$ of $\mathrm{GL}(n, \mathbb{C})$ with character χ_λ such that if $g \in \mathrm{GL}(n, \mathbb{C})$ has eigenvalues $t_1, \ldots, t_n$, then*

$$\chi_\lambda(g) = s_\lambda(t_1, \ldots, t_n). \tag{36.3}$$

The restriction of π_λ to $\mathrm{U}(n)$ is an irreducible representation of $\mathrm{U}(n)$. If $\mu \neq \lambda$ is another partition of k with $n \geqslant l(\mu)$, then χ_λ and χ_μ are distinct.

Proof. We know that the representation exists by applying Theorem 34.1 to the irreducible representation (ρ, N_ρ) of S_k with character $\mathbf{s}_\lambda$. The problem is to prove the irreducibility of the module $V_\rho = \left(\bigotimes^k V \right) \otimes_{\mathbb{C}[S_k]} N_\rho$, which as the character χ_λ by Theorem 34.1. (As in Theorem 34.1, we are taking $V = \mathbb{C}^n$.)

Let B be the ring of endomorphisms of $\bigotimes^k V$ that commute with the action of S_k. We will show that B is spanned by the linear transformations

$$v_1 \otimes \cdots \otimes v_k \longrightarrow g v_1 \otimes \cdots \otimes g v_k, \qquad g \in \mathrm{GL}(n, \mathbb{C}). \tag{36.4}$$

We have an isomorphism $\bigotimes^k \mathrm{End}(V) \cong \mathrm{End} \left(\bigotimes^k V \right)$. In this isomorphism, $f_1 \otimes \cdots \otimes f_k \in \bigotimes^k \mathrm{End}(V)$ corresponds to the endomorphism $v_1 \otimes \cdots \otimes v_k \longrightarrow f_1(v_1) \otimes \cdots \otimes f_k(v_k)$. Conjugation in $\mathrm{End} \left(\bigotimes^k V \right)$ by an element of $\sigma \in S_k$ in the action (34.1) on $\bigotimes^k V$ corresponds to the transformation

$$f_1 \otimes \cdots \otimes f_k \longrightarrow f_{\sigma(1)} \otimes \cdots \otimes f_{\sigma(k)}$$

of $\bigotimes^k \mathrm{End}(V)$. If $\xi \in \bigotimes^k \mathrm{End}(V)$ commutes with this action, then ξ is a linear combination of elements of the form $B(f_1, \ldots, f_k)$, where $B : \mathrm{End}(V) \times \cdots \times \mathrm{End}(V) \longrightarrow \bigotimes^k \mathrm{End}(V)$ is the symmetric k-linear map

$$B(f_1, \ldots, f_k) = \sum_{\sigma \in S_k} f_{\sigma(1)} \otimes \cdots \otimes f_{\sigma(k)}.$$

It follows from Proposition 36.1 that the vector space of such elements of $\bigotimes^k \mathrm{End}(V)$ is spanned by those of the form $Q(f) = B(f, \ldots, f)$ with $f \in \mathrm{End}(V)$. Since $\mathrm{GL}(n, \mathbb{C})$ is dense in $\mathrm{End}(V)$, the elements $Q(f)$ with f invertible span the same vector space. This proves that the transformations of the form (36.4) span the space of transformations of $\bigotimes^k V$ commuting with the action of S_k.

We temporarily restrict the action of $\mathrm{GL}(n, \mathbb{C}) \times S_k$ on $\bigotimes^k V$ to the compact subgroup $\mathrm{U}(n) \times S_k$. Representations of a compact group are completely reducible, and the irreducible representations of $\mathrm{U}(n) \times S_k$ are of the form $\pi \otimes \rho$, where π is an irreducible representation of $\mathrm{U}(n)$ and ρ is an irreducible representation of S_k. Thus, we write

$$\bigotimes^k V \cong \sum_i \pi_i \otimes \rho_i, \tag{36.5}$$

where the π_i and ρ_i are irreducible representations of $\mathrm{U}(n)$ and S_k, respectively. We take the π_i to be left $\mathrm{U}(n)$-modules and the ρ_i to be right S_k-modules. This is because the commuting actions we have defined on $\bigotimes^k V$ have $\mathrm{U}(n)$ acting on the left and S_k acting on the right.

The subspace of $\bigotimes^k V$ corresponding to $\pi_i \otimes \rho_i$ is actually $\mathrm{GL}(n, \mathbb{C})$-invariant. This is because it is a complex subspace invariant under the

Lie algebra action of $\mathfrak{u}(n)$ and hence is invariant under the action of the complexified Lie algebra $\mathfrak{u}(n) + i\mathfrak{u}(n) = \mathfrak{gl}(n, \mathbb{C})$ and therefore under its exponential, GL($n, \mathbb{C}$). So we may regard the decomposition (36.5) as a decomposition with respect to GL($n, \mathbb{C}$) $\times S_k$.

We claim that there are no repetitions among the isomorphism classes of the representations ρ_i of S_k that occur. This is because if $\rho_i \cong \rho_j$, then if we denote by f an intertwining map $\rho_i \longrightarrow \rho_j$ and by τ an arbitrary nonzero linear transformation from the space of π_i to the space of π_j, then $\tau \otimes f$ is a map from the space of $\pi_i \otimes \rho_i$ to the space of $\pi_j \otimes \rho_j$ that commutes with the action of S_k. Extending it by zero on direct summands in (36.5) beside $\pi_i \otimes \rho_i$ gives an endomorphism of $\bigotimes^k V$ that commutes with the action of S_k. It therefore is in the span of the endomorphisms (36.4). But this is impossible because those endomorphisms leave $\pi_i \otimes \rho_i$ invariant and this one does not. This contradiction shows that the ρ_i all have distinct isomorphism classes.

It follows from this that at most one ρ_i can be isomorphic to the contragredient representation of ρ_λ. Thus, in $V_\rho = \left(\bigotimes^k V \right) \otimes_{\mathbb{C}[S_k]} N_\rho$ at most one term can survive, and that term will be isomorphic to π_i as a GL($n, \mathbb{C}$) module for this unique i. We know that V_ρ is nonzero since by Theorem 36.1 the polynomial $s_\lambda \neq 0$ under our hypothesis that $l(\lambda) \leqslant n$. Thus, such a π_i does exist, and it is irreducible as a U(n)-module a $fortiori$ as a GL($n, \mathbb{C}$)-module.

It remains to be shown that if $\mu \neq \lambda$, then $\chi_\mu \neq \chi_\lambda$. Indeed, the Schur polynomials s_μ and s_λ are distinct since the partition λ can be read off from the numerator in (36.1). $\qquad \square$

We have constructed an irreducible representation of GL($n, \mathbb{C}$) for every partition $\lambda = (\lambda_1, \ldots, \lambda_n)$ of length $\leqslant n$.

Proposition 36.2. *Suppose that $n \geqslant l(\lambda)$. Let*

$$\lambda' = (\lambda_1 - \lambda_n, \lambda_2 - \lambda_n, \ldots, \lambda_{n-1} - \lambda_n, 0).$$

In the ring $\Lambda^{(n)}$ of symmetric polynomials in n variables, we have

$$s_\lambda(x_1, \ldots, x_n) = e_n(x_1, \ldots, x_n)^{\lambda_n} s_{\lambda'}(x_1, \ldots, x_n). \qquad (36.6)$$

In terms of the characters of GL($n, \mathbb{C}$), we have

$$\chi_\lambda(g) = \det(g)^{\lambda_n} \chi_{\lambda'}(g). \qquad (36.7)$$

Note that $e_n(x_1, \ldots, x_n) = x_1 \cdots x_n$. Caution: This identity is special to $\Lambda^{(n)}$. The corresponding statement is not true in Λ.

Proof. It follows from (36.1) that $s_\lambda(x_1, \ldots, x_n)$ is divisible by $(x_1 \cdots x_n)^{\lambda_n}$. Indeed, each entry of the first column of the matrix in the numerator is divisible by $x_1^{\lambda_n}$, so we may pull $x_1^{\lambda_n}$ out of the first column, $x_2^{\lambda_n}$ out of the second column, and so forth, obtaining (36.6).

If the eigenvalues of g are $t_1, \ldots, t_n$, then $e_n(t_1, \ldots, t_n) = t_1 \cdots t_n = \det(g)$ and (36.7) follows from (36.6) and (36.3). $\qquad \square$

Although we have constructed many irreducible characters of GL($n, \mathbb{C}$), it is not true that every character is a χ_λ for some partition λ. What we are missing are those of the form $\det(g)^{-m}\chi_\lambda(g)$, where $m > 0$ and χ_λ is not divisible by $\det(g)^m$. We may slightly expand the parametrization of the irreducible characters of GL($n, \mathbb{C}$) as follows. Let λ be a sequence of n integers, $\lambda_1 \geqslant \lambda_2 \geqslant \cdots \geqslant \lambda_n$. (We no longer assume that the λ are nonnegative; if $\lambda_n < 0$, such a λ is not a partition.) Then we can define a character of GL($n, \mathbb{C}$) by (36.7) since even if λ is not a partition, λ' is still a partition. We will denote this representation by $\pi_\lambda^{\mathrm{GL}(n)}$, and its character by χ_λ.

We now have a representation $\pi_\lambda^{\mathrm{GL}(n)}$ for each $\lambda \in \mathbb{Z}^n$ such that $\lambda_1 \geqslant \lambda_2 \geqslant \cdots \geqslant \lambda_n$. We will show that we have all the irreducible finite-dimensional analytic representations. We will call such a λ a *dominant weight*. Thus, the dominant weight λ is a partition if and only if $\lambda_n \geqslant 0$. We call λ the *highest weight* of the representation $\pi_\lambda^{\mathrm{GL}(n)}$. This terminology is consistent with that introduced in Chap. 21.

Proposition 36.3. *Let π be a finite-dimensional irreducible representation of* U(n). *Then π is isomorphic to the restriction of $\pi_\lambda^{\mathrm{GL}(n)}$ for some λ.*

Proof. Let $G = $ U(n). By Schur orthogonality, it is enough to show that the characters of the $\pi_\lambda = \pi_\lambda^{\mathrm{GL}(n)}$ are dense in the space of class functions in $L^2(G)$. We refer to a symmetric polynomial in $\alpha_1, \ldots, \alpha_n$ and their inverses as a *symmetric Laurent polynomial*. We regard a symmetric Laurent functions as class functions on U(n) by applying it to the eigenvalues of $g \in$ U(n). Every symmetric polynomial is a linear combination of the characters of the π_λ with λ a partition, so expanding the set of λ to dominant weights gives us all symmetric Laurent polynomials. Remembering that the eigenvalues α_i of g satisfy $|\alpha_i| = 1$, we may approximate an arbitrary L^2 function by a symmetric Laurent polynomial by symmetrically truncating its Fourier expansion.   □

Lemma 36.1. *If f is an analytic function on* GL($n, \mathbb{C}$), *then f is determined by its restriction to* U(n).

Proof. We show that if $f|_{\mathrm{U}(n)} = 0$ then $f = 0$. Let $\mathfrak{g}$ be the Lie algebra of U(n) of consisting of skew-Hermitian matrices. Then the exponential map $\exp : \mathfrak{g} \longrightarrow$ U(n) is surjective, so $f \circ \exp$ is zero on $\mathfrak{g}$. Since f is analytic, so is $f \circ \exp$ and it follows that $f \circ \exp$ is zero on $\mathfrak{g} \oplus i\mathfrak{g}$ which is all Mat$_n(\mathbb{C})$. So $f = 0$ in a neighborhood of the identity in GL($n, \mathbb{C}$), so it vanishes identically.   □

Proposition 36.4. *Let π_1 and π_2 be analytic representations of* GL($n, \mathbb{C}$). *If π_1 and π_2 have isomorphic restrictions to* U(n), *they are isomorphic.*

Proof. We may assume that π_1 and π_2 act on the same complex vector space V, and that $\pi_1(g) = \pi_2(g)$ when $g \in$ U(n). Applying Lemma 36.1 to the matrix coefficients of π_1 and π_2 it follows that $\pi_1(g) = \pi_2(g)$ for all $g \in$ GL($n, \mathbb{C}$).   □

Theorem 36.3. *Every finite-dimensional representation of the group* U(n) *extends uniquely to an analytic representation of* GL($n, \mathbb{C}$). *The irreducible complex representations of* U(n), *or equivalently the irreducible analytic complex representations of* GL($n, \mathbb{C}$), *are precisely the* $\pi_\lambda^{\mathrm{GL}(n)}$ *parametrized by the dominant weights* λ.

Proof. The fact that irreducible representations of U(n) extend to analytic representations follows from the fact that such a representation is a $\pi_\lambda^{\mathrm{GL}(n)}$, proved in Proposition 36.3. Since U(n) is compact, each representation is a direct sum of irreducibles, and it follows that each representation of U(n) extends to an analytic representation. The uniqueness of the extension follows from Proposition 36.4. The last statement now follows from Proposition 36.3. $\qquad\square$

Proposition 36.5. *Suppose that* λ *is a partition and* $l(\lambda) > n$. *Then we have* $s_\lambda(x_1, \ldots, x_n) = 0$ *in the ring* $\Lambda^{(n)}$.

Proof. If $N = l(\lambda)$, then $\lambda = (\lambda_1, \ldots, \lambda_N)$, where $\lambda_N > 0$ and $N > n$. Apply the homomorphism r_{N-1} defined by (34.9), noting that $r_{N-1}(e_N) = 0$, since e_N is divisible by x_N, and r_{N-1} consists of setting $x_N = 0$. It follows from (36.6) that r_{N-1} annihilates s_λ. We may apply r_{N-2}, etc., until we reach $\Lambda^{(n)}$ and so $s_\lambda = 0$ in $\Lambda^{(n)}$. $\qquad\square$

Theorem 36.4. *If* λ *is a partition of* k *let* ρ_λ *denote the irreducible representation of* S_k *affording the character* $\mathbf{s}_\lambda$ *constructed in Theorem 35.1. If, moreover,* $l(\lambda) \leqslant n$, *let* π_λ *denote the irreducible representation of* GL($n, \mathbb{C}$) *constructed in Theorem 36.2. Let* $V = \mathbb{C}^n$ *denote the standard module of* GL($n, \mathbb{C}$). *The* GL($n, \mathbb{C}$) $\times S_k$ *module* $\bigotimes^k V$ *is isomorphic to* $\bigoplus_\lambda \pi_\lambda \otimes \rho_\lambda$, *where the sum is over partitions of* k *of length* $\leqslant n$.

Proof. Most of this was proved in the proof of Theorem 36.2. Particularly, we saw there that each irreducible representation of S_k occurring in (36.5) occurs at most once and is paired with an irreducible representation of GL($n, \mathbb{C}$). If $l(\lambda) \leqslant n$, we saw in the proof of Theorem 36.2 that ρ_λ does occur and is paired with π_λ. The one fact that was not proved there is that ρ_λ with $l(\lambda) > n$ do not occur, and this follows from Proposition 36.5. $\qquad\square$

Schur Polynomials and S_k

Frobenius [51] discovered that the characters of the symmetric group can be computed using symmetric functions. We will explain this from our point of view. We highly recommend Curtis [39] as an account, both historical and mathematical, of the work of Frobenius and Schur on representation theory.

We remind the reader that the elements of $\mathcal{R}_k$, as generalized characters, are class functions on S_k. The conjugacy classes of S_k are parametrized by the partitions as follows. Let $\lambda = (\lambda_1, \ldots, \lambda_r)$ be a partition of k. Let $\mathcal{C}_\lambda$ be the conjugacy class consisting of products of disjoint cycles of lengths $\lambda_1, \lambda_2, \ldots$. Thus, if $k = 7$ and $\lambda = (3, 3, 1)$, then $\mathcal{C}_\lambda$ consists of the conjugates of $(123)(456)(7) = (123)(456)$. We say that the partition λ is the *cycle type* of the permutations in the conjugacy class $\mathcal{C}_\lambda$. Let $z_\lambda = |S_k|/|\mathcal{C}_\lambda|$.

The *support* of $\sigma \in S_k$ is the set of $x \in \{1, 2, 3, \ldots, k\}$ such that $\sigma(x) \neq x$.

Proposition 37.1. *Let m_r be the number of i such that $\lambda_i = r$. Then*

$$z_\lambda = \prod_{r=1}^{k} r^{m_r} m_r! \,. \tag{37.1}$$

Proof. z_λ is the order of the centralizer of a representative element $g \in \mathcal{C}_\lambda$. This centralizer is easily described.

First, we consider the case where g contains only cycles of length r in its decomposition into disjoint cycles. In this case (denoting $m_r = m$), $k = rm$ and we may write $g = c_1 \cdots c_m$, where c_m is a cycle of length r. The centralizer $C_{S_k}(g)$ contains a normal subgroup N of order r^m generated by $c_1, \ldots, c_m$. The quotient $C_{S_k}(g)/N$ can be identified with S_m since it acts by conjugation on the m cyclic subgroups $\langle c_1 \rangle, \ldots, \langle c_m \rangle$. Thus, $|C_{S_k}(g)| = r^m m! \,.$

In the general case where g has cycles of different lengths, its centralizer is a direct product of groups such as the one just described. $\square$

We showed in the previous chapter that the irreducible characters of S_k are also parametrized by the partitions of k—namely to a partition μ there corresponds an irreducible representation $\mathbf{s}_\mu$. Our aim is to compute $\mathbf{s}_\mu(g)$ when $g \in \mathcal{C}_\lambda$ using symmetric functions.

Proposition 37.2. *The character values of the irreducible representations of S_k are rational integers.*

Proof. Using the Jacobi–Trudi identity (Theorem 35.1), $\mathbf{s}_\lambda$ is a sum of terms of the form $\pm \mathbf{h}_\mu$ for various partitions μ. Each $\mathbf{h}_\mu$ is the character induced from the trivial character of S_μ, so it has integer values.   □

Let $\mathbf{p}_\lambda$ $(k \geqslant 1)$ be the *conjugacy class indicator*, which we define to be the function

$$\mathbf{p}_\lambda(g) = \begin{cases} z_\lambda & \text{if } g \in \mathcal{C}_\lambda, \\ 0 & \text{otherwise.} \end{cases}$$

As a special case, $\mathbf{p}_k$ will denote the indicator of the conjugacy class of the k-cycle, corresponding to the partition $\lambda = (k)$. The term '*conjugacy class indicator*' is justified by the following result.

Proposition 37.3. *If $g \in \mathcal{C}_\lambda$, then $\langle \mathbf{s}_\mu, \mathbf{p}_\lambda \rangle = \mathbf{s}_\mu(g)$.*

Proof. We have

$$\langle \mathbf{s}_\mu, \mathbf{p}_\lambda \rangle = \frac{1}{|S_k|} \sum_{x \in \mathcal{C}_\lambda} z_\lambda \mathbf{s}_\mu(x).$$

The summand is constant on $\mathcal{C}_\lambda$ and equals $z_\lambda \mathbf{s}_\mu(g)$ for any fixed representative g. The cardinality of $\mathcal{C}_\lambda$ is $|S_k|/z_\lambda$ and the result follows.   □

It is clear that the $\mathbf{p}_\lambda$ are orthogonal. More precisely, we have

$$\langle \mathbf{p}_\lambda, \mathbf{p}_\mu \rangle = \begin{cases} z_\lambda & \text{if } \lambda = \mu, \\ 0 & \text{otherwise.} \end{cases} \tag{37.2}$$

This is clear since $\mathbf{p}_\lambda$ is supported on the conjugacy class $\mathcal{C}_\lambda$, which has cardinality $|S_k|/z_\lambda$.

We defined $\mathbf{p}_\lambda$ as a class function. We now show it is a generalized character.

Proposition 37.4. *If λ is a partition of k, then $\mathbf{p}_\lambda \in \mathcal{R}_k$.*

Proof. The inner products $\langle \mathbf{p}_\lambda, \mathbf{s}_\mu \rangle$ are rational integers by Propositions 37.2 and 37.3. By Schur orthogonality, we have $\mathbf{p}_\lambda = \sum_\mu \langle \mathbf{p}_\lambda, \mathbf{s}_\mu \rangle \mathbf{s}_\mu$, so $\mathbf{p}_\lambda \in \mathcal{R}_k$.   □

Proposition 37.5. *If $h = l(\lambda)$, so $\lambda = (\lambda_1, \ldots, \lambda_h)$ and $\lambda_h > 0$, then*

$$\mathbf{p}_\lambda = \mathbf{p}_{\lambda_1} \mathbf{p}_{\lambda_2} \cdots \mathbf{p}_{\lambda_h}.$$

Proof. From the definitions, $\mathbf{p}_{\lambda_1} \cdots \mathbf{p}_{\lambda_h}$ is induced from the class function f on the subgroup S_λ of S_k which has a value on $(\sigma_1, \ldots, \sigma_h)$ that is

$$\begin{cases} \lambda_1 \cdots \lambda_h & \text{if each } \sigma_i \text{ is a } \lambda_i\text{-cycle}, \\ 0 & \text{otherwise}. \end{cases}$$

The formula (32.15) may be used to compute this induced class function. It is clear that $\mathbf{p}_{\lambda_1} \cdots \mathbf{p}_{\lambda_h}$ is supported on the conjugacy class of cycle type λ, and so it is a constant multiple of $\mathbf{p}_\lambda$. We write $\mathbf{p}_{\lambda_1} \cdots \mathbf{p}_{\lambda_h} = c\mathbf{p}_\lambda$ and use a trick to show that $c = 1$. By Proposition 37.3, since $\mathbf{h}_k = \mathbf{s}_{(k)}$ is the trivial character of S_k, we have $\langle \mathbf{h}_k, \mathbf{p}_\lambda \rangle_{S_k} = 1$. On the other hand, by Frobenius reciprocity, $\langle \mathbf{h}_k, \mathbf{p}_{\lambda_1} \cdots \mathbf{p}_{\lambda_h} \rangle_{S_k} = \langle \mathbf{h}_k, f \rangle_{S_\lambda}$. As a class function, $\mathbf{h}_k$ is just the constant function on S_k equal to 1, so this inner product is

$$\prod_i \langle \mathbf{h}_{\lambda_i}, \mathbf{p}_{\lambda_i} \rangle_{S_{\lambda_i}} = 1.$$

Therefore, $c = 1$. $\square$

Proposition 37.6. *We have*

$$k\mathbf{h}_k = \sum_{r=1}^{k} \mathbf{p}_r \, \mathbf{h}_{k-r}. \tag{37.3}$$

Proof. Let λ be a partition of k. Let m_s be the number of λ_i equal to s. We will prove

$$\langle \mathbf{p}_r \mathbf{h}_{k-r}, \mathbf{p}_\lambda \rangle = r m_r. \tag{37.4}$$

By Frobenius reciprocity, this inner product is $\langle f, \mathbf{p}_\lambda \rangle_{S_r \times S_{k-r}}$, where f is the function on $S_r \times S_{k-r}$ which as a value on (σ, τ), with $\sigma \in S_r$ and $\tau \in S_{k-r}$ that is

$$\begin{cases} r & \text{if } \sigma \text{ is an } r\text{-cycle}, \\ 0 & \text{otherwise}. \end{cases}$$

The value of $f\mathbf{p}_\lambda$ restricted to $S_r \times S_{k-r}$ will be zero on (σ, τ) unless σ is an r-cycle [since $f(\sigma, \tau)$ must be nonzero] and τ has cycle type λ', where λ' is the partition obtained from λ by removing one part of length r [since $\mathbf{p}_\lambda(\sigma, \tau)$ must be nonzero]. The number of such pairs (σ, τ) is $|S_r| \cdot |S_{k-r}|$ divided by the product of the orders of the centralizers in S_r and S_{k-r}, respectively, of an r-cycle and of a permutation of cycle type λ'. That is,

$$\frac{|S_r| \cdot |S_{k-r}|}{r \cdot r^{m_r - 1}(m_r - 1)! \prod_{s \neq r} s^{m_s} m_s!}.$$

The value of $f\mathbf{p}_\lambda$ on these conjugacy classes is $r z_\lambda$. Therefore,

$$\langle f, \mathbf{p}_\lambda \rangle_{S_r \times S_{k-r}} = \frac{1}{|S_r| \cdot |S_{k-r}|} \left[\frac{|S_r| \cdot |S_{k-r}|}{r \cdot r^{m_r - 1}(m_r - 1)! \prod_{s \neq r} s^{m_s} m_s!} \right] r z_\lambda,$$

which equals $r m_r$. This proves (37.4).

We note that since λ is a partition of k, and since λ has m_r cycles of length r, we have $k = \sum_{r=1}^{k} r m_r$. Therefore,

$$\left\langle \sum_{r=1}^{k} \mathbf{p}_r \mathbf{h}_{k-r}, \mathbf{p}_\lambda \right\rangle = \sum_r r m_r = k = \langle k\mathbf{h}_k, \mathbf{p}_\lambda \rangle.$$

Because this is true for every λ, we obtain (37.3). $\square$

Let $p_\lambda = p_{\lambda_1} p_{\lambda_2} \cdots \in \Lambda_k$, where p_k is defined by (33.7).

Proposition 37.7. *We have*

$$kh_k = \sum_{r=1}^{k} p_r h_{k-r}. \tag{37.5}$$

Proof. We recall from (33.2) that

$$\sum_{k=0}^{\infty} h_k t^k = \prod_{i=1}^{n} (1 - x_i t)^{-1},$$

which we differentiate logarithmically to obtain

$$\frac{\sum_{k=0}^{\infty} k h_k t^{k-1}}{\sum_{k=0}^{\infty} h_k t^k} = \sum_{i=1}^{n} \frac{d}{dt} \log(1 - x_i t)^{-1}.$$

Since

$$\frac{d}{dt} \log(1 - x_i t)^{-1} = \sum_{r=1}^{\infty} x_i^r t^{r-1},$$

we obtain

$$\sum_{k=1}^{\infty} k h_k t^{k-1} = \left[\sum_{k=0}^{\infty} h_k t^k \right] \sum_{r=1}^{\infty} p_r\, t^{r-1}.$$

Equating the coefficients of t^{k-1}, the result follows. $\square$

Theorem 37.1. *We have* $\mathrm{ch}(\mathbf{p}_\lambda) = p_\lambda$.

Proof. We have $\mathbf{p}_\lambda = \mathbf{p}_{\lambda_1} \mathbf{p}_{\lambda_2} \cdots$. Hence, it is sufficient to show that $\mathrm{ch}(\mathbf{p}_k) = p_k$. This follows from the fact that they satisfy the same recursion formula—compare (37.5) with (37.3)—and that $\mathrm{ch}(\mathbf{h}_k) = h_k$. $\square$

Now we may determine the irreducible characters of S_k.

Theorem 37.2. *Express each symmetric polynomial p_λ as a linear combination of the s_μ:*

$$p_\lambda = \sum_\mu c_{\lambda\mu} s_\mu.$$

Then the coefficient $c_{\lambda\mu}$ is the value of the irreducible character $\mathbf{s}_\mu$ on elements of the conjugacy class $\mathcal{C}_\lambda$.

Proof. Since $n \geqslant k$, $\mathrm{ch} : \mathcal{R}_k \to \Lambda_k$ is injective, and it follows that

$$\mathbf{p}_\lambda = \sum_\mu c_{\lambda\mu} \mathbf{s}_\mu.$$

Taking the inner product of this relation with $\mathbf{s}_\mu$, we see that

$$c_{\lambda\mu} = \langle \mathbf{p}_\lambda, \mathbf{s}_\mu \rangle .$$

The result follows from Proposition 37.3.   □

Here is a variant of Theorem 37.2. Let

$$\Delta = \prod_{i<j}(x_i - x_j) = \det(x_j^{n-i})$$

be the Vandermonde determinant. which is the denominator in (36.1).

Theorem 37.3. (Frobenius) *Let μ be a partition of k of length $\leqslant n$, and let λ be another partition of k. Let $c_{\lambda\mu}$ be the value of the character $\mathbf{s}_\mu$ on elements of the conjugacy class $\mathcal{C}_\lambda$. Then $c_{\lambda\mu}$ is the coefficient of*

$$x_1^{\mu_1 + n - 1} x_2^{\mu_2 + n - 2} \cdots x_n^{\mu_n} \tag{37.6}$$

in the polynomal $p_\lambda \Delta$.

Proof. By Theorem 37.2, we have $p_\lambda = \sum_\mu c_{\lambda\mu} s_\mu$, and by (36.1) this means that

$$p_\lambda \Delta = \sum_\mu c_{\lambda\mu} \det(x_j^{\mu_i + n - i}),$$

the determinant being the determinant in the numerator in (36.1). The monomial (37.6) appears only in the μ term and the statement follows.   □

As an example of Theorem 37.2, let us verify the irreducible characters of S_3. We have

$$\begin{aligned}
s_{(3)} &= h_3 = \sum x_i^3 + \sum_{i \neq j} x_i^2 x_j + \sum_{i<j<k} x_i x_j x_k, \\
s_{(21)} &= \quad\quad\quad \sum_{i \neq j} x_i^2 x_j + 2\sum_{i<j<k} x_i x_j x_k, \\
s_{(111)} &= e_{(3)} = \quad\quad\quad\quad\quad \sum_{i<j<k} x_i x_j x_k.
\end{aligned}$$

and

$$\begin{aligned}
p_{(3)} &= \sum x_i^3, \\
p_{(21)} &= \sum x_i^3 + \sum_{i \neq j} x_i^2 x_j, \\
p_{(111)} &= \sum x_i^3 + 3\sum_{i \neq j} x_i^2 x_j + 6\sum_{i<j<k} x_i x_j x_k,
\end{aligned}$$

so

$$\begin{aligned}
p_{(111)} &= s_{(3)} + s_{(111)} + 2s_{(21)} , \\
p_{(3)} &= s_{(3)} + s_{(111)} - s_{(21)} , \\
p_{(21)} &= s_{(3)} - s_{(111)}.
\end{aligned}$$

These coefficients are precisely the coefficients in the character table of S_3:

	1	(123)	(12)
$\mathbf{s}_{(3)}$	1	1	1
$\mathbf{s}_{(111)}$	1	1	−1
$\mathbf{s}_{(21)}$	2	−1	0

Before we leave the representation theory of the symmetric group, let us recall the involution ι of Proposition 34.3 and Theorem 35.2, which interchanges $\mathbf{s}_\lambda$ with $\mathbf{s}_\mu$, where $\mu = \lambda^t$ is the conjugate partition. It has a concrete interpretation in this context.

Lemma 37.1. *Let H be a subgroup of the finite group G. Let χ be a character of H, and let ρ be a one-dimensional character of G, which we may restrict to H. The induced character $(\rho\chi)^G$ equals $\rho\chi^G$.*

Thus, it does not matter whether we multiply by ρ before or after inducing to G.

Proof. This may be proved either directly from the definition of the induced representation or by using (32.15).   □

Theorem 37.4. *If $\mathbf{f}$ is a class function on S_k, its involute ${}^\iota\mathbf{f}$ is the result of multiplying $\mathbf{f}$ by the alternating character ε of $S_k^{\bullet}$.*

We refrain from denoting ${}^\iota\mathbf{f}$ as $\varepsilon\mathbf{f}$ because the graded ring $\mathcal{R}$ has a different multiplication.

Proof. Let us denote by $\tau : \mathcal{R}_k \longrightarrow \mathcal{R}_k$ the linear map that takes a class function $\mathbf{f}$ on S_k and multiplies it by ε, and assemble the τ in different degrees to a linear map of $\mathcal{R}$ to itself. We want to prove that τ and ι are the same. By the definition of the $\mathbf{e}_k$ and $\mathbf{h}_k$, they are interchanged by τ, and by Theorem 35.2 they are interchanged by ι. Since the $\mathbf{e}_k$ generate $\mathcal{R}$ as a ring, the result will follow if we check that τ is a ring homomorphism.

Applying Lemma 37.1 with $G = S_{k+l}$, $H = S_k \times S_l$, and $\rho = \varepsilon$ shows that multiplying the characters χ and η of S_k and S_l each by ε to obtain the characters ${}^\tau\chi$ and ${}^\tau\eta$ and then inducing the character ${}^\tau\chi \otimes {}^\tau\eta$ of $S_k \times S_l$ to S_{k+l} gives the same result as inducing $\chi \otimes \eta$ and multiplying it by ε. This shows that τ is a ring homomorphism.   □

EXERCISES

Exercise 37.1. Compute the character table of S_4 using symmetric polynomials by the method of this chapter.

Exercise 37.2. Prove the identity

$$ke_k = \sum_{r=1}^{k}(-1)^r p_r e_{k-r}.$$

Let us say that a partition λ is a *ribbon partition* if its Young diagram only has entries in the first row and column. The ribbon partitions of k are of the form $(k-r, 1^r)$ with $0 \leqslant r \leqslant k$, where the notation means the partition with one part of length $k-r$ and r parts of length 1.

Exercise 37.3. Show that

$$p_k = \sum_{r=0}^{k}(-1)^r s_{(k-r,1^r)}.$$

[**Hint:** This may be proved by multiplying the denominator in (36.1) by p^r and manipulating the result.]

See Exercise 40.1 for a generalization.

Exercise 37.4. Let s_λ be an irreducible character of S_k, where λ is a partition of k. Let σ be a k-sycle. Show that $s_\lambda(\sigma)$ is 0, 1 or -1. For which partitions is it nonzero?

The Cauchy Identity

Suppose that $\alpha_1, \ldots, \alpha_n$ and $\beta_1, \ldots, \beta_m$ are two sets of variables. The *Cauchy identity* asserts that

$$\prod_{i=1}^{n} \prod_{j=1}^{m} (1 - \alpha_i \beta_j)^{-1} = \sum_\lambda s_\lambda(\alpha_1, \ldots, \alpha_n) \, s_\lambda(\beta_1, \ldots, \beta_m), \qquad (38.1)$$

where the sum is over all partitions λ (of all k). The series is absolutely convergent if all $|\alpha_i|, |\beta_i| < 1$. It can also be regarded as an equality of formal power series.

The general context for our discussion of the Cauchy identity will be the Frobenius–Schur duality. For other approaches, see Exercises 26.4 and 38.4.

We recall from Chap. 34 that the characteristic map $\mathrm{ch} : \mathcal{R} \longrightarrow \Lambda^{(N)}$ allows us to interpret a character (or class function) on the symmetric group S_k as a symmetric polynomial in N variables that is homogeneous of degree k.

Here is a simple fact we will need. Notations are as in Chap. 37.

Proposition 38.1. *Let k be a nonnegative integer. Then we have the following identity in the ring $\Lambda^{(N)}$ of symmetric polynomials.*

$$\sum_{\lambda \ a \ partition \ of \ k} z_\lambda^{-1} p_\lambda = h_k.$$

Proof. In view of Theorem 37.1 it is sufficient to show in $\mathcal{R}$ that

$$\sum_{\lambda \ a \ partition \ of \ k} z_\lambda^{-1} \mathbf{p}_\lambda = \mathbf{h}_k.$$

We consider both sides as functions on S_k. By definition, $\mathbf{p}_\lambda$ is the function supported on the single conjugacy class $\mathcal{C}_\lambda$ with value z_λ on that class; summing over all conjugacy classes, $\sum_\lambda z_\lambda^{-1} \mathbf{p}_\lambda$ is the constant function equal to 1 on S_k, that is, $\mathbf{h}_k$. $\qquad \square$

Next we will consider symmetric polynomials in two sets of variables, $\alpha_1, \ldots, \alpha_n$ and $\beta_1, \ldots, \beta_m$. Consider a polynomial f in $\alpha_1, \ldots, \alpha_n$ and $\beta_1, \ldots, \beta_m$ that is (for fixed β) symmetric in $\alpha_1, \ldots, \alpha_n$ and homogeneous of degree k, and also (for fixed α) symmetric in $\beta_1, \ldots, \beta_m$ and homogeneous of degree l. Then we may transfer this by the Frobenius–Schur duality to $\mathcal{R}_k \otimes \mathcal{R}_l$. In other words, we may find an element ξ in $\mathcal{R}_k \otimes \mathcal{R}_l$ such that $\mathrm{ch}^{(n)} \otimes \mathrm{ch}^{(m)}(\xi)$ is the given symmetric polynomial in two sets of variables.

Proposition 38.2. *Let k be a nonnegative integer. Then*

$$\sum_{\lambda \text{ a partition of } k} s_\lambda(\alpha)\, s_\lambda(\beta) = \sum_{\lambda \text{ a partition of } k} z_\lambda^{-1} p_\lambda(\alpha)\, p_\lambda(\beta). \qquad (38.2)$$

Proof. Both sides polynomials in the α_i and β_i that are symmetric and homogeneous of degree k in either set of variables. Use the Frobenius–Schur duality to transfer the function on the right-hand side to a function on $S_k \times S_k$. In view of Theorem 37.1 this is the function

$$\Delta(\sigma, \tau) = \sum_{\lambda \text{ a partition of } k} z_\lambda^{-1} \mathbf{p}_\lambda(\sigma)\, \mathbf{p}_\lambda(\tau)$$

that maps $(\sigma, \tau) \in S_k \times S_k$ to the function that has the value z_λ if σ and τ are in the conjugacy class $\mathcal{C}_\lambda$, and is zero if σ and τ are not conjugate. This function may be characterized as follows: if f is a class function, then

$$\frac{1}{k!} \sum_{\tau \in S_k} \Delta(\sigma, \tau)\, f(\tau) = f(\sigma).$$

Indeed, if σ is in the conjugacy class $\mathcal{C}_\lambda$, there are $|\mathcal{C}_\lambda|$ values of τ, namely the conjugates of σ, for which there is a contribution of $z_\lambda f(\tau) = z_\lambda f(\sigma)$, and since $|\mathcal{C}_\lambda| z_\lambda = k!$, the statement follows. Thus, Δ is the *reproducing kernel for class functions*. It is characterized by this property, together with the fact (with τ fixed) $\Delta(\sigma, \tau)$ is constant on conjugacy classes of σ, and similarly for τ with σ fixed. Now

$$\sum_{\lambda \text{ a partition of } k} \mathbf{s}_\lambda(\sigma)\, \mathbf{s}_\lambda(\tau)$$

is also a class function in σ and τ separately, and it has the same reproducing property, as a consequence of Schur orthogonality. Hence these are equal. We see that

$$\sum_{\lambda \text{ a partition of } k} \mathbf{s}_\lambda(\sigma)\, \mathbf{s}_\lambda(\tau) = \sum_{\lambda \text{ a partition of } k} z_\lambda^{-1} \mathbf{p}_\lambda(\sigma)\, \mathbf{p}_\lambda(\tau),$$

and applying $\mathrm{ch} \otimes \mathrm{ch}$ we obtain (38.2). $\qquad \square$

Theorem 38.1. (Cauchy) *Suppose $\alpha_1, \ldots, \alpha_n$ and $\beta_1, \ldots, \beta_m$ are complex numbers of absolute value < 1. Then*

$$\prod_{i=1}^{n}\prod_{j=1}^{m}(1 - \alpha_i\beta_j)^{-1} = \sum_{\lambda} s_\lambda(\alpha_1, \ldots, \alpha_n)\, s_\lambda(\beta_1, \ldots, \beta_m). \qquad (38.3)$$

The sum is over all partitions λ.

Proof. Using (33.2) in the nm variables $\alpha_i\beta_j$, the left-hand side equals

$$\sum_{k=0} h_k(\alpha_i\beta_j),$$

so it is sufficient to show

$$\sum_{\lambda \text{ a partition of } k} s_\lambda(\alpha_1, \ldots, \alpha_n)\, s_\lambda(\beta_1, \ldots, \beta_m) = h_k(\alpha_i\beta_j). \qquad (38.4)$$

By (38.2) this equals

$$\sum_{\lambda \text{ a partition of } k} z_\lambda^{-1} p_\lambda(\alpha)\, p_\lambda(\beta).$$

We now make the observation that $p_k(\alpha_i\beta_j)$, which is the kth power sum symmetric polynomial in nm variables $\alpha_i\beta_j$, equals $p_k(\alpha)p_k(\beta)$, and so $p_\lambda(\alpha_i\beta_j) = p_\lambda(\alpha)p_\lambda(\beta)$. The statement now follows from Proposition 38.1. $\qquad\square$

The Cauchy identity may be interpreted as describing the decomposition of the symmetric algebra over the tensor product representation of $\mathrm{GL}_n \times \mathrm{GL}_m$, as we will now explain. Let G be a group, and let $\pi : G \longrightarrow \mathrm{GL}(\Omega)$ be a representation on some vector space Ω. Let $g \in G$, and let $\alpha_1, \ldots, \alpha_N$ be the eigenvalues of $\pi(g)$. Then $h_k(\alpha) = h_k(\alpha_1, \ldots, \alpha_N)$ and $e_k(\alpha) = e_k(\alpha_1, \ldots, \alpha_k)$ are the traces of $\pi(g)$ on the kth symmetric and exterior powers $\vee^k \Omega$ and $\wedge^k \Omega$, respectively. Therefore,

$$\sum_{k=0}^{\infty} h_k(\alpha) = \prod_{i=1}^{N}(1 - \alpha_i)^{-1} \quad \text{and} \quad \sum_{k=0}^{\infty} e_k(\alpha) = \prod_{i=1}^{N}(1 + \alpha_i)$$

may be regarded as the characters of g on the symmetric and exterior algebras.

The symmetric algebra is infinite-dimensional, so strictly speaking the trace of an endomorphism only has a provisional meaning. Indeed the first series is only convergent if $|\alpha_i| < 1$, but there are several ways of handling this. One may try to choose g so that its eigenvalues are < 1, or one may simply regard the series as formal. Or, assuming no $\alpha_i = 1$, one may regard the series as obtained from

$$\sum_{k=0}^{\infty} h_k(\alpha)t^k = \prod_{i=1}^{N}(1 - t\alpha_i)^{-1}$$

by analytic continuation in t.

Proposition 38.3. *Let $G = \mathrm{GL}_n(\mathbb{C}) \times \mathrm{GL}_m(\mathbb{C})$ acting on the tensor product $\Omega = \mathbb{C}^n \otimes \mathbb{C}^m$ of the standard modules of $\mathrm{GL}_n(\mathbb{C})$ and $\mathrm{GL}_m(\mathbb{C})$. Then the symmetric algebra*

$$\bigvee \Omega \cong \bigoplus_\lambda \pi_\lambda^{\mathrm{GL}_n} \otimes \pi_\lambda^{\mathrm{GL}_m} \tag{38.5}$$

as G-modules, where the summation is over all partitions λ of length $\leqslant \min(m, n)$.

Proof. If g has eigenvalues α_i and h has eigenvalues β_j, then (g, h) has eigenvalues $\alpha_i \beta_j$ on Ω, hence has trace $h_k(\alpha_i \beta_j)$ on $\vee^k \Omega$. By the Cauchy identity in the form (38.4), this equals $\sum s_\lambda(\alpha) \, s_\lambda(\beta)$ where the sum is over partitions of k. [If the length of the partition λ is $> n$, we interpret $s_\lambda(\alpha)$ as zero.] Combining the contributions over all k, the statement follows.   □

There is a *dual Cauchy identity*.

Theorem 38.2. *Suppose $\alpha_1, \ldots, \alpha_n$ and $\beta_1, \ldots, \beta_m$ are complex numbers of absolute value < 1. Then*

$$\prod_{i=1}^{n} \prod_{j=1}^{m} (1 + \alpha_i \beta_j) = \sum_\lambda s_\lambda(\alpha_1, \ldots, \alpha_n) \, s_{\lambda^t}(\beta_1, \ldots, \beta_m). \tag{38.6}$$

Note that now each partition λ is paired with its conjugate partition λ^t. This may be regarded as a decomposition of the exterior algebra on $\mathrm{Mat}_n(\mathbb{C})^*$.

Proof. Let $\alpha_1, \ldots, \alpha_n$ be fixed complex numbers, and let $\Lambda^{(m)}$ be the ring of symmetric polynomials in $\beta_1, \ldots, \beta_m$ with integer coefficients. We recall from Theorems 34.3 and 35.2 that Λ has an involution ι that interchanges s_λ and $s_{\lambda'}$. We have to be careful how we use ι because it does *not* induce an involution of $\Lambda^{(m)}$. Indeed, it is possible that in $\Lambda^{(m)}$ one of s_λ and $s_{\lambda'}$ is zero and the other is not, so no involution exists that simply interchanges them.

We write the Cauchy identity in the form

$$\prod_{i=1}^{n} \left[\sum_{k=0}^{\infty} \alpha_i^k h_k(\beta_1, \ldots, \beta_m) \right] = \sum_\lambda s_\lambda(\alpha_1, \ldots, \alpha_n) \, s_\lambda(\beta_1, \ldots, b_m).$$

This is true for all m, and therefore we may write

$$\prod_{i=1}^{n} \left[\sum_{k=0}^{\infty} \alpha_i^k h_k \right] = \sum_\lambda s_\lambda(\alpha_1, \ldots, \alpha_n) \, s_\lambda,$$

where the h_k on the left and the second occurrence of s_λ on the right are regarded as elements of the ring Λ, which is the inverse limit (34.10) of the rings $\Lambda^{(m)}$, while α_i and $s_\lambda(\alpha_1, \ldots, \alpha_n)$ are regarded as complex numbers. To this identity we may apply ι and obtain

$$\prod_{i=1}^{n} \left[\sum_{k=0}^{\infty} \alpha_i^k e_k \right] = \sum_{\lambda} s_\lambda(\alpha_1, \ldots, \alpha_n)\, s_{\lambda'},$$

and now we specialize from Λ to $\Lambda^{(m)}$ and obtain (38.6). □

In this chapter and the next, we will give some applications of the Cauchy identity. First some preliminaries. If λ, μ, ν are partitions, there is defined a nonnegative integer $c_{\mu\nu}^\lambda$ called the *Littlewood–Richardson* coefficient. It is (by definition) zero unless $|\lambda| = |\mu| + |\nu|$, where we recall that $|\lambda| = \sum \lambda_i$ is the sum of the parts, that is, λ is a partition of $|\lambda|$. There is a combinatorial description of $c_{\mu\nu}^\lambda$, but we will not describe it (except in special cases). For this *Littlewood–Richardson rule* see Macdonald [124] or Stanley [153].

The next theorem, asserting the equivalence of three definitions of $c_{\mu\nu}^\lambda$, shows that the Littlewood–Richardson coefficients have three distinct representation theoretic interpretations. They have other interpretations too. For example, they describe the structure constants in the cohomology ring of Grassmannians with respect to the basis of cohomology classes corresponding to Schubert cycles. If $\lambda = (\lambda_1, \lambda_2, \ldots, \lambda_n)$ is a partition of length $\leqslant n$, we will denote by $\pi_\lambda^{\mathrm{GL}(n)}$ the irreducible representation of $\mathrm{GL}(n)$ parametrized by λ.

Let G be a group and H a subgroup. A rule describing how irreducible representations of G decompose into irreducibles when restricted to H is called a *branching rule*. The *tensor product rule* describing how the tensor product $\pi \otimes \pi'$ of irreducibles π, π' of H decomposes into irreducibles of H may be thought of a branching rule. Indeed, $\pi \otimes \pi'$ extends to an irreducible representation of $H \times H$, so the tensor product rule is really a branching rule for H embedded in $G = H \times H$ diagonally.

All three definitions of the Littlewood–Richardson coefficients may be characterized as branching rules. To specify a branching rule, we need to specify an embedding of a group H in a larger group G. The embeddings $H \to G$ in the three branching rules as follows. The first is the embedding of $S_k \times S_l \longrightarrow S_{k+l}$ that we worked with in Chap. 34. The second is the diagonal embedding $\mathrm{GL}(n, \mathbb{C}) \longrightarrow \mathrm{GL}(n, \mathbb{C}) \times \mathrm{GL}(n, \mathbb{C})$. The third is the *Levi* embedding of $\mathrm{GL}(p, \mathbb{C}) \times \mathrm{GL}(q, \mathbb{C}) \longrightarrow \mathrm{GL}(p + q, \mathbb{C})$ as follows:

$$(g, h) \longmapsto \begin{pmatrix} g & \\ & h \end{pmatrix}, \qquad g \in \mathrm{GL}_p(\mathbb{C}), h \in \mathrm{GL}_q(\mathbb{C}). \tag{38.7}$$

As usual, we are only interested in analytic representations of $\mathrm{GL}(n, \mathbb{C})$, which are the same as representations of $\mathrm{U}(n)$, so we could equally well work with the embeddings $\mathrm{U}(n) \longrightarrow \mathrm{U}(n) \times \mathrm{U}(n)$ and the Levi embedding $\mathrm{U}(p) \times \mathrm{U}(q) \longrightarrow \mathrm{U}(p + q)$.

Remarkably, these three branching rules involve the same coefficients $c_{\mu\nu}^\lambda$. This is the content of the next result.

Theorem 38.3. *Let λ, μ, ν be partitions such that $|\lambda| = |\mu| + |\nu|$. Then the following three definitions of $c_{\mu\nu}^\lambda$ are equivalent. We will denote $k = |\mu|$ and $l = |\nu|$.*

(i) Let ρ_λ be the irreducible representation of S_{k+l} with character $\mathbf{s}_\lambda$. Then $c_{\mu\nu}^\lambda$ is the multiplicity of $\rho_\mu \otimes \rho_\nu$ in the restriction of ρ_λ to $S_k \times S_l$.

(ii) Let $n \geqslant |\lambda|$. Then $c_{\mu\nu}^\lambda$ is the multiplicity of $\pi_\lambda^{\mathrm{GL}(n)}$ in the decomposition of the representation $\pi_\mu^{\mathrm{GL}(n)} \otimes \pi_\nu^{\mathrm{GL}(n)}$ of $\mathrm{GL}(n)$ into irreducibles.

(ii) Let $p \geqslant k$ and $q \geqslant l$. Then $c_{\mu\nu}^\lambda$ is the multiplicity of $\pi_\mu^{\mathrm{GL}(p)} \otimes \pi_\nu^{\mathrm{GL}(q)}$ in the restriction of $\pi_\mu^{\mathrm{GL}(p+q)}$ to $\mathrm{GL}(p,\mathbb{C}) \times \mathrm{GL}(q,\mathbb{C})$ into irreducibles.

Proof. We note that (i) can be expressed as the identity

$$\mathbf{s}_\mu \mathbf{s}_\nu = \sum_\mu c_{\mu\nu}^\lambda \mathbf{s}_\lambda,$$

since taking the inner product of the left-hand side with $\mathbf{s}_\lambda$ and using Frobenius reciprocity gives the coefficient of $\rho_\mu \otimes \rho_\nu$ in the restriction of ρ_λ from S_{k+l} to $S_k \times S_l$. On the other hand (ii) can be expressed as the identity

$$s_\mu(x_1,\ldots,x_n)\, s_\nu(x_1,\ldots,x_n) = \sum_\lambda c_{\mu\nu}^\lambda s_\lambda(x_1,\ldots,x_n)$$

in the ring $\Lambda^{(n)}$ of symmetric polynomials. Indeed, substituting for x_i the eigenvalues of $g \in \mathrm{GL}(n,\mathbb{C})$, the Schur polynomial s_λ becomes the character of $\pi_\lambda^{\mathrm{GL}(n)}$, and the left-hand side becomes the character of $\pi_\mu^{\mathrm{GL}(n)} \otimes \pi_\nu^{\mathrm{GL}(n)}$. Thus the equivalence of (i) and (ii) follows from Proposition 34.4.

As for the equivalence of (ii) and (iii), we give an argument based on the Cauchy identity. Comparing the characters of $\pi_\lambda^{\mathrm{GL}(p+q)}$ and

$$\bigoplus_{\mu,\nu} c_{\mu\nu}^\lambda \pi_\mu^{\mathrm{GL}(p)} \otimes \pi_\nu^{\mathrm{GL}(q)}$$

on the matrix (38.7), we see that (iii) is equivalent to the identity

$$s_\lambda(\alpha_1,\ldots,\alpha_p,\beta_1,\ldots,\beta_q) = \sum_{\mu,\nu} c_{\mu\nu}^\lambda s_\mu(\alpha_1,\ldots,\alpha_p) s_\nu(\beta_1,\ldots,\beta_q),$$

where $\alpha_1,\ldots,\alpha_p$ are the eigenvalues of g and $\beta_1,\ldots,\beta_q$ are the eigenvalues of h. In a more succinct notation, we write the left-hand side $s_\lambda(\alpha,\beta)$, so what we need to prove is

$$s_\lambda(\alpha,\beta) = \sum_{\mu,\nu} c_{\mu\nu}^\lambda s_\mu(\alpha) s_\nu(\beta). \tag{38.8}$$

Let $\gamma_1,\ldots,\gamma_n$ be arbitrary complex numbers. By the Cauchy identity for

$$\sum_\lambda s_\lambda(\alpha,\beta) s_\lambda(\gamma) = \prod_{i,k} (1-\alpha_i\gamma_k)^{-1} \prod_{j,k} (1-\beta_j\gamma_k)^{-1}.$$

Also by the Cauchy identity this equals

$$\left(\sum_{\mu} s_{\mu}(\alpha)s_{\mu}(\gamma)\right)\left(\sum_{\nu} s_{\nu}(\beta)s_{\nu}(\gamma)\right) = \sum_{\mu,\nu} s_{\mu}(\alpha)s_{\nu}(b)\sum_{\lambda} c^{\lambda}_{\mu\nu}s_{\lambda}(\gamma).$$

Since the functions $s_{\lambda}(\gamma)$ are linearly independent as λ varies, we may compare the coefficients of $s_{\lambda}(\gamma)$ and obtain (38.8). $\qquad\square$

It is worth pondering the mechanism behind the proof that (ii) is equivalent to (iii). We will reconsider it after some preliminaries.

We begin with the notion of a *correspondence* in the sense of Howe, who wrote many papers on this subject: see Howe [75, 77]. A correspondence is a bijection between a set of irreducible representations of a group G and another group H. The relevant examples arise in the following manner.

Let $\mathfrak{G}$ be a group with a representation Θ, and let G and H be subgroups that centralize each other. Thus, we have a homomorphism $G \times H \longrightarrow \mathfrak{G}$. (Often this homomorphism is injective so $G \times H$ is a subgroup of $\mathfrak{G}$, but we do not require this.) We assume given a representation Θ of $\mathfrak{G}$ with the following property: when Θ is restricted to $G \times H$, it becomes a direct sum $\pi_i \otimes \pi'_i$, where π_i are irreducible representations of G, and π'_i are irreducible representations of H. We assume that each $\pi_i \otimes \pi'_i$ occurs with multiplicity at most one, and moreover, there are no repetitions between the representations π_i and no repetitions among the π'_i. (This definition is adequate if G and H are compact but might need to be generalized slightly if they are not.) If this condition is satisfied, we say the representation Θ *induces a correspondence* for G and H. The correspondence is the bijection $\pi_i \longleftrightarrow \pi'_i$. Here are some examples.

- Let $G = S_k$, $H = \mathrm{GL}(n, \mathbb{C})$, and $\mathfrak{G} = G \times H$. The representation Θ is the action on $\otimes^k \mathbb{C}^n$ in Theorem 36.4. That theorem implies that Θ induces a correspondence. Indeed, by Theorem 36.4 the correspondence is the bijection $\rho_{\lambda} \longleftrightarrow \pi^{\mathrm{GL}(n)}_{\lambda}$, as λ runs through partitions of k that have length $\leqslant n$. Thus, the Frobenius–Schur duality is an example of a correspondence.

- Consider $G = \mathrm{GL}(n, \mathbb{C})$, $H = \mathrm{GL}(m, \mathbb{C})$ acting on $\Omega = \mathbb{C}^n \otimes \mathbb{C}^m$ as above. Since $(g \otimes I_m)(I_n \otimes h) = (g \otimes h) = (I_n \otimes h)(g \otimes I_m)$, the actions of G and H commute. Let Θ be the action on the symmetric algebra of Ω. It is actually a representation of $\mathfrak{G} = \mathrm{GL}(\Omega)$. As we have already explained, when restricted to $G \times H$, the Cauchy identity implies the decomposition (38.5), so Θ induces a correspondence. This is the bijection $\pi^{\mathrm{GL}(n)}_{\lambda} \longleftrightarrow \pi^{\mathrm{GL}(m)}_{\lambda}$ as λ runs through all partitions of length $\leqslant \min(m, n)$. This equivalence is sometimes referred to as $\mathrm{GL}(n) \times \mathrm{GL}(m)$-*duality*.

- Howe conjectured [73], and it was eventually proved, that if G and H are reductive subgroups of $\mathrm{Sp}(2N, F)$, where F is a local field (including $\mathbb{R}$ or $\mathbb{C}$), then the Weil (oscillator) representation induces a correspondence. In some cases one of the groups of the correspondence must be replaced by a covering group. In one most important case, $G = \mathrm{Sp}(2n)$ and $H = \mathrm{O}(m)$, where $nm = N$, so the correspondence relates representations of symplectic groups (or their double covers) to representations of an orthogonal group. This phenomenon is known as *Howe duality*. It is closely related to the theta liftings in the theory of automorphic forms. Here the Weil representation is a projective representation of $\mathrm{Sp}(2N, F)$ with a construction that is similar to the construction of a projective representation of the orthogonal groups in Chap. 31. In place of the Clifford algebra one uses a Heisenberg group or the symplectic Clifford algebra, often called the Weyl algebra.

Now let us consider the following abstract situation. Let G and G' be groups. Let H and H' be subgroups of G and H, respectively. We will assume that G and G' are subgroups of a larger group $\mathfrak{G}$ such that H' is the centralizer of G, and H is the centralizer of G'. Now let us assume that $\mathfrak{G}$ has a representation Θ that induces correspondences between G and H', and between G' and H.

We summarize this situation by a "see-saw" diagram:

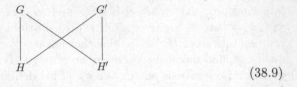

$$(38.9)$$

The vertical lines are inclusions, and the diagonal lines are correspondences. Now we can show that the pairs G, H and G', H' *have the same branching rule* (except inverted with respect to inclusion).

Proposition 38.4. *Let there be given a see-saw (38.9). Let π_i^G and $\pi_i^{H'}$ be corresponding representations of G and H', and let $\pi_j^{G'}$ and π_j^H be corresponding representations of G' and H. Then the multiplicity of π_j^H in π_i^G equals the multiplicity of $\pi_i^{H'}$ in $\pi_j^{G'}$.*

Proof. We may express the correspondences as follows:

$$\Theta|_{G \times H'} = \bigoplus_{i \in I} \pi_i^G \otimes \pi_i^{H'}, \qquad \Theta|_{H \times G'} = \bigoplus_{j \in J} \pi_j^H \otimes \pi_j^{G'}, \qquad (38.10)$$

for suitable indexing sets I and J. We first observe that if σ is an irreducible of H that occurs in any $\pi_i^G|_H$, then $\sigma = \pi_j^H$ for some j. Indeed, it follows from the second decomposition that the π_j^H are precisely the irreducibles of H that occur in the restriction of Θ to H, from which this statement is clear. Therefore, we may find integers $c(i, j)$ such that

$$\pi_i^G|_H = \bigoplus_{j \in J} c(i, j) \pi_j^H \qquad (38.11)$$

and similarly

$$\pi_j^{G'}|_{H'} = \bigoplus_{i \in I} d(i,j)\pi_i^{H'}.$$

What we must prove is that $c(i,j) = d(i,j)$. Now combining the first equation in (38.10) with (38.11) we get

$$\Theta_{H \times H'} = \bigoplus_{i,j} c(i,j)\pi_j^H \otimes \pi_i^{H'}$$

and similarly

$$\Theta_{H \times H'} = \bigoplus_{i,j} d(i,j)\pi_j^H \otimes \pi_i^{H'}.$$

Comparing, the statement follows. □

Now let us reconsider the equivalence of (ii) and (iii) in Theorem 38.3. This may be understood as a reflection of the following see-saw:

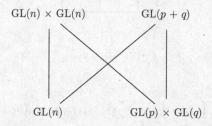

The left vertical line is the diagonal embedding $GL(n) \longrightarrow GL(n) \times GL(n)$, and the right vertical line is the Levi embedding $GL(p) \times GL(q) \longrightarrow GL(p+q)$. The ambient group is $GL(\Omega)$ where $\Omega = \mathbb{C}^n \otimes \mathbb{C}^{p+q}$ acting on the symmetric algebra $\bigvee \Omega$. More specifically, with $H = GL(n)$ and $G' = GL(p+q)$, $H \times G'$ acts on $\mathbb{C}^n \otimes \mathbb{C}^{p+q}$ in the obvious way; with $G = GL(n) \times GL(n)$ and $H' = GL(p) \times GL(q)$ we use the isomorphism

$$\mathbb{C}^n \otimes \mathbb{C}^{p+q} \cong (\mathbb{C}^n \otimes \mathbb{C}^p) \oplus (\mathbb{C}^n \otimes \mathbb{C}^q)$$

with the first $GL(n)$ and $GL(p)$ acting on the first component, and the second $GL(n)$ and $GL(q)$ on the second component. Proposition 38.4 asserts that the two branching rules are the same, which is the equivalence of (ii) and (iii) in Theorem 38.3.

The paper of Howe, Tan, and Willenbring [79] gives many more examples of see-saws applied to branching rules. Kudla [113] showed that many constructions in the theory of automorphic forms could be explained by see-saws.

Branching rules are important for many problems and are the subject of considerable literature. Branching rules for the orthogonal and symplectic groups are discussed in Goodman and Wallach [56], Chap. 8. King [101] is a useful survey of branching rules for classical groups. Many branching rules are programmed into Sage.

Exercises

Exercise 38.1. Let n and m be integers. Define a bijection between partitions $\lambda = (\lambda_1, \ldots, \lambda_n)$ with $\lambda_1 \leqslant m$ and partitions $\mu = (\mu_1, \ldots, \mu_m)$ with $\mu_1 \leqslant n$ as follows. The shapes of the partitions λ and μ must sit as complementary pieces in an $n \times m$ box, with the λ_i being the lengths of the rows of one piece, and the μ_j being the lengths of the columns of the other. For example, suppose $n = 3$ and $\mu = 5$ we could have $\lambda = (4, 2, 1)$ and $\mu = (3, 2, 2, 1)$. As usual, a partition may be padded with zeros, so we identify this μ with $(3, 2, 2, 1, 0)$, and the diagram is as follows:

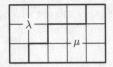

(i) Show that

$$(y_1, \ldots, y_m)^n s_{\lambda'}\left(y_1^{-1}, \ldots, y_m^{-1}\right) = s_\mu(y_1, \ldots, y_m).$$

(ii) Prove that

$$\prod_{i=1}^n \prod_{j=1}^m (x_i + y_j) = \sum s_\lambda(x_1, \ldots, x_n)\, s_\mu(y_1, \ldots, y_m),$$

where the sum is over λ and μ related as explained above.

Exercise 38.2. Give another proof of Proposition 38.1 as follows. Show that

$$\prod_i (1 - \alpha_i t)^{-1} = \sum_\lambda z_\lambda p_\lambda(\alpha_1, \ldots, \alpha_n) t^{|\lambda|} \tag{38.12}$$

by writing the left-hand side as

$$\prod_i \exp\left(\sum_k \frac{\alpha_i^k}{k} t^k\right) = \prod_k \exp\left(\frac{p_k(\alpha_1, \ldots, \alpha_k)}{k} t^k\right),$$

expanding and making use of (37.1).

The next two exercises lead to another proof of the Cauchy identity.

Exercise 38.3. Let G be any compact group. Let τ be an antiautomorphism of G, that is, a continuous map that satisfies $\tau(gh) = \tau(h)\tau(g)$. Assume that $\tau(g)$ is conjugate to g for all $g \in G$. For example, we could take $G = U(n)$ and τ to be the transpose map.

(i) Let (π, V) be an irreducible representation of G. Let $\pi' : G \longrightarrow \mathrm{GL}(V)$ be the map $\pi'(g) = \pi(\tau(g)^{-1})$. Show that (π', V) is isomorphic to the contragredient representation $(\hat{\pi}, V^*)$.

(ii) Let $G \times G$ act on the ring of matrix coefficients $\mathfrak{M}_\pi$ of π by

$$(g, h)f(x) = f(\tau(g)xh).$$

Show that this representation is isomorphic to $\pi \otimes \pi$ as $(G \times G)$-modules (**Hint:** Use Exercise 2.4.)

Let us call a function on $\mathrm{GL}(n, \mathbb{C})$ *regular* if, as a function of $g = (g_{ij})$ it is a polynomial in the g_{ij} and $\det(g)^{-1}$. A *regular function* on $\mathrm{Mat}_n(\mathbb{C})$ is a polynomial. Thus, it is a regular function on $\mathrm{GL}(n, \mathbb{C})$, but one that is a polynomial in just the coordinate functions g_{ij} and not involving the inverse determinant. The rings $\mathcal{O}(\mathrm{Mat}_n(\mathbb{C}))$ and $\mathcal{O}(\mathrm{GL}((n, \mathbb{C}))$ of regular functions on $\mathrm{Mat}_n(\mathbb{C})$ and $\mathrm{GL}(n, \mathbb{C})$ are just the affine rings of algebraic geometry. The fact that the regular functions on Mat_n are a subring of the regular functions on $\mathrm{GL}(n)$ reflects the fact that Mat_n contains $\mathrm{GL}(n)$ as an open subset.

Exercise 38.4. (i) Show that every matrix coefficient of $\mathrm{U}(n)$ extends uniquely to a regular function on $\mathrm{GL}(n, \mathbb{C})$, so the ring of matrix coefficients on $\mathrm{U}(n)$ may be identified with $\mathcal{O}(\mathrm{GL}(n, \mathbb{C})$. Deduce that the ring of matrix coefficients of $\mathrm{U}(n)$ may be identified with the $\mathcal{O}(\mathrm{GL}(n, \mathbb{C}))$. Let $\mathrm{GL}(n, \mathbb{C}) \times \mathrm{GL}(n, \mathbb{C})$ act on functions on either $\mathrm{GL}(n, \mathbb{C})$ or $\mathrm{Mat}_n(\mathbb{C})$ by

$$(g_1, g_2)f(h) = f(^t g_1 h g_2).$$

(ii) Show that

$$\mathcal{O}(\mathrm{GL}(n, \mathbb{C})) \cong \bigoplus_{\lambda \text{ a dominant weight}} \pi \otimes \pi$$

as $\mathrm{GL}(n, \mathbb{C}) \times \mathrm{GL}(n, \mathbb{C})$-modules.

(iii) Show that the component $\pi \otimes \pi$ in this decomposition extends to a space of regular functions on Mat_n if and only if λ is a partition, and deduce that

$$\mathcal{O}(\mathrm{Mat}_n(\mathbb{C})) \cong \bigoplus_{\lambda \text{ a partition}} \pi \otimes \pi.$$

Explain why this proves (38.5) when $m = n$.

(iv) Explain why the Cauchy identity when $m = n$ implies the general case, and deduce the Cauchy identity from (iii).

Exercise 38.5. (i) Let $\alpha = (\alpha_1, \alpha_2, \ldots)$, $\beta = (\beta_1, \beta_2, \ldots)$, $\gamma = (\gamma_1, \gamma_2, \ldots)$, $\delta = (\delta_1, \delta_2, \ldots)$ be three sets of variables. Using (38.8), evaluate $\sum_\nu s_\nu(\alpha, \beta) s_\nu(\gamma, \delta)$ in two different ways and obtain the identity

$$\sum_\nu c^\nu_{\lambda\mu} c^\nu_{\theta\tau} = \sum_{\phi, \psi, \xi, \eta} c^\lambda_{\phi\psi} c^\mu_{\xi\eta} c^\theta_{\phi\xi} c^\tau_{\psi\eta}. \tag{38.13}$$

(ii) Show that (38.13) implies the Hopf axiom, that is, the commutativity of the diagram in Exercise 35.4.

Let $\alpha = (\alpha_1, \alpha_2, \ldots)$, $\beta = (\beta_1, \beta_2, \ldots)$ be two sets of variables. Define the *supersymmetric Schur polynomial* (Littlewood [120] (pages 66–70), Berele and Remmel [15], Macdonald [123], Bump and Gamburd [29]) by the formula

$$s_\lambda(\alpha/\beta) = \sum_{\mu,\nu} c_{\mu\nu}^\lambda s_\mu(\alpha) s_{\nu^t}(\beta)$$

where ν^t is the conjugate partition.

Exercise 38.6. Prove the *supersymmetric Cauchy identity*

$$\sum_\nu s_\nu(\alpha/\beta) s_\nu(\gamma/\delta) = \prod_{i,j}(1 - \alpha_i\gamma_j)^{-1} \prod_{i,j}(1 + \alpha_i\delta_j) \prod_{i,j}(1 + \beta_i\delta_j) \prod_{i,j}(1 - \beta_i\delta_j)^{-1}.$$

(Hint: Use the involution.)

39

Random Matrix Theory

In this chapter, we will work not with $GL(n, \mathbb{C})$ but with its compact subgroup $U(n)$. As in the previous chapters, we will consider elements of $\mathcal{R}_k$ as generalized characters on S_k. If $\mathbf{f} \in \mathcal{R}_k$, then $f = \mathrm{ch}^{(n)}(\mathbf{f}) \in \Lambda_k^{(n)}$ is a symmetric polynomial in n variables, homogeneous of weight k. Then $\psi_f : U(n) \longrightarrow \mathbb{C}$, defined by (33.6), is the function on $U(n)$ obtained by applying f to the eigenvalues of $g \in U(n)$. We will denote $\psi_f = \mathrm{Ch}^{(n)}(\mathbf{f})$. Thus, $\mathrm{Ch}^{(n)}$ maps the additive group of generalized characters on S_k to the additive group of generalized characters on $U(n)$. It extends by linearity to a map from the Hilbert space of class functions on S_k to the Hilbert space of class functions on $U(n)$.

Proposition 39.1. *Let $\mathbf{f}$ be a class function on S_k. Write $\mathbf{f} = \sum_\lambda c_\lambda \mathbf{s}_\lambda$, where the sum is over the partitions of k. Then*

$$|\mathbf{f}|^2 = \sum_\lambda |c_\lambda|^2, \qquad |\mathrm{Ch}^{(n)}(\mathbf{f})|^2 = \sum_{l(\lambda) \leqslant n} |c_\lambda|^2.$$

Proof. The $\mathbf{s}_\lambda$ are orthonormal by Schur orthogonality, so $|\mathbf{f}|^2 = \sum |c_\lambda|^2$. By Theorem 36.2, $\mathrm{Ch}^{(n)}(\mathbf{s}_\lambda)$ are distinct irreducible characters when λ runs through the partitions of k with length $\leqslant n$, while, by Proposition 36.5, $\mathrm{Ch}^{(n)}(\mathbf{s}_\lambda) = 0$ if $l(\lambda) > n$. Therefore, we may write

$$\mathrm{Ch}^{(n)}(\mathbf{f}) = \sum_{l(\lambda) \leqslant n} c_\lambda \, \mathrm{Ch}^{(n)}(\mathbf{s}_\lambda),$$

and the $\mathrm{Ch}^{(n)}(\mathbf{s}_\lambda)$ in this decomposition are orthonormal by Schur orthogonality on $U(n)$. Thus, $|\mathrm{Ch}^{(n)}(\mathbf{f})|^2 = \sum_{l(\lambda) \leqslant n} |c_\lambda|^2$. $\square$

Theorem 39.1. *The map $\mathrm{Ch}^{(n)}$ is a contraction if $n < k$ and an isometry if $n \geqslant k$. In other words, if $\mathbf{f}$ is a class function on S_k;*

$$|\mathrm{Ch}^{(n)}(\mathbf{f})| \leqslant |\mathbf{f}|$$

with equality when $n \geqslant k$.

Proof. This follows immediately from Proposition 39.1 since if $n \geqslant k$ every partition of k has length $\leqslant n$.   □

Theorem 39.1 is a powerful tool for transferring computations from one group to another, in this case from the unitary group to the symmetric group. The underlying principle is that of a correspondence introduced in the last chapter. This is not unlike Proposition 38.4, where we showed how correspondences may be used to transfer a branching rule from one pair of groups to another.

We will illustrate Theorem 39.1 with a striking result of Diaconis and Shahshahani [42], who showed by this method that the traces of large random unitary matrices are normally distributed. We will give a second example of using a correspondence to transfer a calculation from one group to another below in the theorem of Keating and Snaith, were we will employ $\mathrm{GL}_n \times \mathrm{GL}_m$ duality in a similar way.

A measure is called a *probability measure* if its total volume is 1. Suppose that X and Y are topological spaces and that X is endowed with a Borel probability measure $d\mu_X$. Let $f : X \longrightarrow Y$ be a continuous function. We can push the measure $d\mu_X$ forward to probability measure $d\mu_Y$ on Y, defined by

$$\int_Y \phi(y)\, d\mu_Y(y) = \int_X \phi\big(f(x)\big)\, d\mu_X(x)$$

for measurable functions on Y. Concretely, this measure gives the distribution of the values $f(x)$ when $x \in X$ is a random variable.

For example, the trace of a Haar random unitary matrix $g \in \mathrm{U}(n)$ is distributed with a measure $d\mu_n$ on $\mathbb{C}$ satisfying

$$\int_{\mathrm{U}(n)} \phi\big(\mathrm{tr}(g)\big)\, dg = \int_{\mathbb{C}} \phi(z)\, d\mu_n(z). \tag{39.1}$$

We say that a sequence ν_n of Borel probability measures on a space X converges *weakly* to a measure ν if $\int_X \phi(x)\, d\nu_n(x) \longrightarrow \int_X \phi(x)\, d\nu(x)$ for all bounded continuous functions ϕ on X. We will see that the measures μ_n converge weakly as $n \longrightarrow \infty$ to a fixed Gaussian measure

$$d\mu(z) = \frac{1}{\pi} e^{-(x^2 + y^2)}\, dx \wedge dy, \qquad z = x + iy. \tag{39.2}$$

Let us consider how surprising this is! As n varies, the number of eigenvalues increases and one might expect the standard deviation of the traces to increase with n. This is what would happen were the eigenvalues of a random symmetric matrix uncorrelated. *That it converges to a fixed Gaussian measure means that the eigenvalues of a random unitary matrix are quite evenly distributed around the circle.*

Intuitively, the eigenvalues "repel" and tend not to lie too close together. This is reflected in the property of the trace—that its distribution does not spread out as n is increased. This can be regarded as a reflection of (17.3). Because of the factor $|t_i - t_j|^2$, matrices with close eigenvalues have small Haar measure in $U(n)$. Dyson [48] gave the following analogy. Consider the eigenvalues of a Haar random matrix distributed on the unit circle to be like the distribution of charged particles in a Coulomb gas. At a certain temperature ($T = \frac{1}{2}$), this model gives the right distribution. The exercises introduce Dyson's "pair correlation" function that quantifies the tendency of the eigenvalues to repel at close ranges. Figure 39.1 shows the probability density

$$R_2(1, \theta) = n^2 - \frac{\sin^2(n\theta/2)}{\sin^2(\theta/2)} \tag{39.3}$$

that there are eigenvalues at both e^{it} and $e^{i(t+\theta)}$ as a function of θ (for $n = 10$). (Consult the exercises for the definition of R_m and a proof that R_2 is given by (39.3).) We can see from this figure that the probability is small when θ is small, but is essentially independent of θ if θ is moderate.

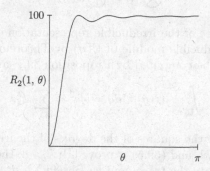

Fig. 39.1. The pair correlation $R_2(1, \theta)$ when $n = 10$

Weak convergence requires that for any continuous bounded function ϕ

$$\lim_{n \to \infty} \int_C \phi(z) \, d\mu_n(z) = \int_C \phi(z) \, d\mu(z),$$

or in other words

$$\lim_{n \to \infty} \int_{U(n)} \phi(\operatorname{tr}(g)) \, dg = \int_C \phi(x + iy) \, d\mu(z). \tag{39.4}$$

Remarkably, if $\phi(z)$ is a polynomial in z and $\overline{z}$, this identity is *exactly* true for sufficiently large n, depending only on the degree of the polynomial! Of course, a polynomial is not a bounded continuous function, but we will deduce weak convergence from this fact about polynomial functions.

Proposition 39.2. Let $k, l \geqslant 0$. Then

$$\int_{U(n)} \operatorname{tr}(g)^k \, \overline{\operatorname{tr}(g)^l} \, dg = 0 \quad \text{if } k \neq l,$$

while

$$\int_{U(n)} |\mathrm{tr}(g)|^{2k}\, dg \leqslant k!\,,$$

with equality when $n \geqslant k$.

Proof. If $k \neq l$, then the variable change $g \longrightarrow e^{i\theta} g$ multiplies the left-hand side by $e^{i(k-l)\theta} \neq 1$ for θ in general position, so the integral vanishes.

Assume that $k = l$. We show that

$$\int_{U(n)} |\mathrm{tr}(g)|^{2k}\, dg = k! \tag{39.5}$$

provided $k \leqslant n$. Note that if $V = \mathbb{C}^n$ is the standard module for $U(n)$, then $\mathrm{tr}(g)^k$ is the trace of g acting on $\bigotimes^k V$ as in (36.4). As in (34.6), we may decompose

$$\bigotimes^k V = \bigoplus_\lambda d_\lambda V_\lambda,$$

where d_λ is the degree of the irreducible representation of S_k with character s_λ, and V_λ is an irreducible module of $U(n)$ by Theorem 36.2. The L_2-norm of $f(g) = \mathrm{tr}(g)^k$ can be computed by Proposition 39.1, and we have

$$\int_{U(n)} |\mathrm{tr}(g)|^{2k} dg = |f|^2 = \sum_\lambda d_\lambda^2.$$

Of course, the sum of the squares of the degrees of the irreducible representations of S_k is $|S_k| = k!$, and (39.5) is proved. If $k > n$, then the same method can be used to evaluate the trace, and we obtain $\sum_\lambda d_\lambda^2$, where now the sum is restricted to partitions of length $\leqslant n$. This is $< k!$. $\square$

Theorem 39.2. *Suppose that $\phi(z)$ is a polynomial in z and $\overline{z}$ of degree $\leqslant 2n$. Then*

$$\int_{U(n)} \phi(\mathrm{tr}(g))\, dg = \int_{\mathbb{C}} \phi(z)\, d\mu(z), \tag{39.6}$$

where $d\mu$ is the measure (39.2).

Proof. It is sufficient to prove this if $\phi(z) = z^k \overline{z}^l$. If $\deg(\phi) \leqslant 2n$, then $k + l \leqslant 2n$ so either $k \neq l$ or both $k, l \leqslant n$, and in either case Proposition 39.2 implies that the left-hand side equals 0 if $k \neq l$ and $k!$ if $k = l$. What we must therefore show is

$$\int_{\mathbb{C}} z^k \overline{z}^l\, d\mu(z) = \begin{cases} k! \text{ if } k = l, \\ 0 \text{ if } k \neq l. \end{cases}$$

The measure $d\mu(z)$ is rotationally symmetric, and if $k \neq l$, then replacing z by $e^{i\theta} z$ multiplies the left-hand side by $e^{i\theta(k-l)}$, so the integral is zero in that case. Assume therefore that $\phi(x + iy) = |z|^{2k}$. Then using polar coordinates (so $z = x + iy = re^{i\theta}$) the integral equals

$$\int_{\mathbb{C}} |z|^{2k} d\mu(z) = \tfrac{1}{\pi} \int_{-\infty}^{\infty} \int_{-\infty}^{\infty} (x^2 + y^2)^k e^{-(x^2+y^2)} \, dx \, dy =$$

$$\tfrac{1}{\pi} \int_0^{2\pi} \int_0^{\infty} r^{2k} e^{-2r} r \, dr \, d\theta = 2 \int_0^{\infty} r^{2k+1} e^{-2r} \, dr = \Gamma(k+1) = k!$$

and the theorem is proved. □

This establishes (39.4) when ϕ is a polynomial—indeed the sequence becomes stationary for large n. However, it does not establish weak convergence. To this end, we will study the Fourier transforms of the measures μ_n and μ.

The Fourier transform of a probability measure ν on $\mathbb{R}^N$ is called its *characteristic function*. Concretely,

$$\hat{\nu}(y_1, \ldots, y_N) = \int_{\mathbb{R}^N} e^{i(x_1 y_1 + \cdots + x_N y_N)} d\nu(x), \qquad x = (x_1, \ldots, x_N).$$

Theorem 39.3. *Let $\nu_1, \nu_2, \nu_3, \ldots$ and ν be probability measures on $\mathbb{R}^N$. Suppose that the characteristic functions $\hat{\nu_i}(y_1, \ldots, y_N) \longrightarrow \hat{\nu}(y_1, \ldots, y_N)$ pointwise for all $(y_1, \ldots, y_N) \in \mathbb{R}^N$. Then the measures ν_i converge weakly to ν.*

Proof omitted. A proof may be found in Billingsley [18], Theorem 26.3 (when $N = 1$) and Sect. 28 (for general N). The precise statement we need is on p. 383 before Theorem 29.4. □

In the case at hand, we wish to compare probability measures on $\mathbb{C} = \mathbb{R}^2$, and it will be most convenient to define the Fourier transform as a function of $w = u + iv \in \mathbb{C}$. Let

$$\hat{\mu}(w) = \int_{\mathbb{C}} e^{i(zw + \overline{z}\overline{w})} d\mu(z)$$

and similarly for the $\hat{\mu}_n$.

Proposition 39.3. *The functions $\hat{\mu}_n$ converge uniformly on compact subsets of $\mathbb{C}$ to $\hat{\mu}$.*

Proof. The function $\hat{\mu}$ is easily computed. As the Fourier transform of a Gaussian distribution, $\hat{\mu}$ is also Gaussian and in fact $\hat{\mu}(w) = e^{-|w|^2}$. We write this as a power series:

$$\hat{\mu}(w) = F(|w|), \qquad F(r) = \sum_{k=0}^{\infty} \frac{1}{k!} (-1)^k r^{2k}.$$

The radius of convergence of this power series is ∞.

We have

$$\hat{\mu}_n(w) = \int_{\mathbb{C}} \left[\sum_{k=0}^{\infty} \sum_{l=0}^{\infty} \frac{i^{k+l} z^k \, w^k \, \overline{z}^l \, \overline{w}^l}{k! \, l!} \right] d\mu_n(z)$$

$$= \sum_{k=0}^{\infty} \sum_{l=0}^{\infty} \frac{i^{k+l}}{k! \, l!} \left[\int_{\mathbb{C}} z^k \, \overline{z}^l \, d\mu_n(z) \right] w^k \, \overline{w}^l.$$

The interchange of the summation and the integration is justified since the measure $d\mu_n$ is compactly supported, and the series is uniformly convergent when z is restricted to a compact set. By Proposition 39.2 and the definition (39.1) of μ_n, the integral inside brackets vanishes unless $k = l$, so

$$\hat{\mu}_n(w) = F_n(|w|), \qquad F_n(r) = \sum_{k=0}^{\infty} a_{k,n} \frac{(-1)^k}{k!} r^{2k},$$

$$a_{k,n} = \frac{1}{k!} \int_{\mathbb{C}} |z|^{2k} d\mu_n(z).$$

By Proposition 39.2 the coefficients $a_{k,n}$ satisfy $0 \leqslant a_{k,n} \leqslant 1$ with equality when $k > n$. We have

$$|F(r) - F_n(r)| = \left| \sum_{k=n}^{\infty} (1 - a_{k,n}) \frac{(-1)^k}{k!} r^{2k} \right| \leqslant \sum_{k=n}^{\infty} \frac{r^{2k}}{k!},$$

which converges to 0 uniformly as $n \longrightarrow \infty$ when r is restricted to a compact set. $\qquad\square$

Corollary 39.1. *The measures μ_n converge weakly to μ.*

Proof. This follows immediately from the criterion of Theorem 39.3. $\qquad\square$

Since we have not proved Theorem 39.3, let us point out that we can immediately prove (39.4) for a fairly big set of test functions ϕ. For example, if ϕ is the Fourier transform of an integrable function ψ with compact support, we can write

$$\int_{\mathrm{U}(n)} \phi(\mathrm{tr}(g)) \, dg = \int_{\mathbb{C}} \phi(z) \, d\mu_n(z) = \int_{\mathbb{C}} \psi(w) \, \hat{\mu}_n(w) \, du \wedge dv, \qquad w = u + iv,$$

by the Plancherel formula and, since we have proved that $\hat{\mu}_n \longrightarrow \mu$ uniformly on compact sets (39.4) is clear for such ϕ.

Diaconis and Shahshahani [42] proved a much stronger statement to the effect that the quantities

$$\mathrm{tr}(g), \mathrm{tr}(g^2), \ldots, \mathrm{tr}(g^r),$$

where g is a Haar random element of $\mathrm{U}(n)$, are distributed like the moments of r independent Gaussian random variables. Strikingly, what the proof requires is the full representation theory of the symmetric group in the form of Theorem 37.1!

Proposition 39.4. *We have*

$$\int_{U(n)} |\mathrm{tr}(g)|^{2k_1} |\mathrm{tr}(g^2)|^{2k_2} \cdots |\mathrm{tr}(g^r)|^{2k_r} \, dg \leqslant \prod_{j=1}^{r} j^{k_j} k_j! \qquad (39.7)$$

with equality provided $k_1 + 2k_2 + \cdots + rk_r \leqslant n$.

Proof. Let $k = k_1 + 2k_2 + \cdots + rk_r$, and let λ be the partition of k containing k_1 entries equal to 1, k_2 entries equal to 2, and so forth. By Theorem 37.1, we have $\mathrm{Ch}^{(n)}(\mathbf{p}_\lambda) = \psi_{p_\lambda}$. This is the function

$$g \mapsto \mathrm{tr}(g)^{k_1} \mathrm{tr}(g^2)^{k_2} \cdots \mathrm{tr}(g^r)^{k_r}$$

since $p_\lambda = p_{\lambda_1} \cdots p_{\lambda_r}$, and applying p_{λ_i} to the eigenvalues of g gives $\mathrm{tr}(g^{\lambda_i})$.

The left-hand side of (39.7) is thus the L^2 norm of $\mathrm{Ch}^{(n)}$, and if $k \leqslant n$, then by Theorem 39.1 we may compute this L^2 norm in S_k. It equals

$$\frac{1}{|S_k|} \sum_{\sigma \in S_k} |\mathbf{p}_\lambda(\sigma)|^2 = z_\lambda$$

by (37.2). This is the right-hand side of (39.7). If $k > n$, the proof is identical except that Theorem 39.1 only gives an inequality in (39.7). $\qquad\square$

Theorem 39.4. (Diaconis and Shahshahani) *The joint probability distribution of the* $(\mathrm{tr}(g), \mathrm{tr}(g^2), \ldots, \mathrm{tr}(g^r))$ *near* $(z_1, \ldots, z_r) \in \mathbb{C}^r$ *is a measure weakly converging to*

$$\prod_{j=1}^{r} \frac{1}{j} \pi e^{-\pi|z_j|^2/j} \, dx_j \wedge dy_j. \qquad (39.8)$$

Thus, the distributions of $\mathrm{tr}(g), \mathrm{tr}(g^2), \ldots, \mathrm{tr}(g^r)$ are as a sequence of independent random variables in Gaussian distributions.

Proof. Indeed, this follows along the lines of Corollary 39.1 using the fact that the moments of the measure (39.8)

$$\int_{\mathbb{C}} |z_1|^{2k_1} |z_2|^{2k_2} \cdots |z_r|^{2k_r} \prod_{j=1}^{r} \frac{1}{j} \pi e^{-\pi|z_j|^2/j} \, dx_j \wedge dy_j = \prod_{j=1}^{r} j^{k_j} k_j!,$$

agree with (39.7). $\qquad\square$

By an *ensemble* we mean a topological space with elements that are matrices, given a probability measure. Random matrix theory is concerned with the statistical distribution of the eigenvalues of the matrices in the ensemble, particularly *local* statistical facts such as the spacing of these eigenvalues.

The original focus of random matrix theory was not on unitary matrices but on random *Hermitian* matrices. The reason for this had to do with the origin of the theory in nuclear physics. In quantum mechanics, an *observable* quantity such as energy or angular momentum is associated with a Hermitian operator acting on a Hilbert space with elements that correspond to possible states of a physical system. An eigenvector corresponds to a state in which the observable has a definite value, which equals the eigenvalue of the operator on that eigenvector. The Hermitian operator corresponding to the energy level of the physical system (a typical observable) is called the *Hamiltonian*. A Hamiltonian operator is typically positive definite.

It was observed by Wigner and his collaborators that although the spectra of atomic nuclei (emitting or absorbing neutrons) were hopeless to calculate from first principles, the spacing of the eigenvalues still obeyed statistical laws that could be studied. To this end, random Hermitian operators were studied, first by Wigner, Gaudin, Mehta, and Dyson. The book of Mehta [128] is the standard treatise on the subject from the point of view taken by this physics-inspired literature. The papers of Dyson [49] also greatly repay study. The more recent books of Anderson, Guionnet, and Zeitouni [7], Deift [40], Katz and Sarnak [95] and the handbook [4] are all strongly recommended.

Although the Hilbert space on which the Hermitian operator corresponding to an observable acts is infinite-dimensional, one may truncate the operator, replacing the Hilbert space with a finite-dimensional invariant subspace. The operator is then realized as a Hermitian matrix.

To study the local properties of the eigenvalues, one seeks to give the real vector space of Hermitian matrices a probability measure which is invariant under the action of the unitary group by conjugation, since one is interested in the eigenvalues, and these are preserved under conjugation. The usual way is to assume that the matrix entries are independent random variables with normal (i.e., Gaussian) distributions. This probability space is called the *Gaussian unitary ensemble* (GUE). Two other ensembles model physical systems with time reversal symmetry. For these, the type of symmetry depends on whether reversing the direction of time multiplies the operator by ± 1. The ensemble that models systems with a Hamilton that is unchanged under time-reversal consists of real symmetric matrices and is called the *Gaussian orthogonal ensemble* (GOE). The ensemble modeling systems with a Hamiltonian that is antisymmetric under time-reversal can be represented by quaternionic Hermitian matrices and is called the *Gaussian symplectic ensemble* (GSE). See Dyson [48] and Mehta [128] for further information about this point.

The space of positive definite Hermitian matrices is an open subset of the space of all Hermitian matrices, and this space is isomorphic to the Type IV symmetric space $GL(n, \mathbb{C})/U(n)$, under the map which associates with the coset $gU(n)$ in the symmetric space the Hermitian matrix $g\,^t\bar{g}$. Similarly the positive-definite parts of the GOE and GSE are $GL(n, \mathbb{R})/O(n)$ and $GL(n, \mathbb{H})/Sp(2n)$ with associated probability measures.

Dyson [48] shifted focus from the Gaussian ensembles to the *circular ensembles* that are the compact duals of the symmetric spaces $GL(n, \mathbb{C})/U(n)$, $GL(n, \mathbb{R})/O(n)$ and $GL(n, \mathbb{H})/Sp(2n)$. For example, by Theorem 28.1, the dual of $GL(n, \mathbb{C})/U(n)$ is just $U(n)$. Haar measure makes this symmetric space into the *circular unitary ensemble* (CUE). The ensemble is called *circular* because the eigenvalues of a unitary matrix lie on the unit circle instead of the real line. It is the CUE that we have studied in this chapter. Note that in the GUE, we cannot use Haar measure to make $GL(n, \mathbb{C})/U(n)$ into a measure space, since we want a probability measure on each ensemble, but the noncompact group $GL(n, \mathbb{C})$ has infinite volume. This is an important advantage of the CUE over the GUE. And as Dyson observed, as far as the *local* statistics of random matrices are concerned—for examples, with matters of spacing of eigenvalues—the circular ensembles are faithful mirrors of the Gaussian ones. The circular orthogonal and symplectic ensembles (COE and CSE) are similarly the measure spaces $U(n)/O(n)$ and $U(2n)/Sp(2n)$ with their unique invariant probability measures.

In recent years, random matrix theory has found a new application in the study of the zeros of the Riemann zeta function and similar arithmetic data. The observation that the distribution of the zeros of the Riemann zeta function should have a local distribution similar to that of the eigenvalues of a random Hermitian matrix in the GUE originated in a conversation between Dyson and Montgomery, and was confirmed numerically by Odlyzko and others; see Rubinstein [139]. See Katz and Sarnak [94] and Conrey [38] for surveys of this field, Keating and Snaith [99] for a typical paper from the extensive literature. The paper of Keating and Snaith is important because it marked a paradigm shift away from the study of the *spacing* of the zeros of $\zeta(s)$ to the distribution of the *values* of $\zeta(\frac{1}{2}+it)$, which are, in the new paradigm, related to the values of the characteristic polynomial of a random matrix.

Theorem 39.5. (Keating and Snaith) *Let k be a nonnegative integer. Then*

$$\int_{U(n)} |\det(g - I)|^{2k}\, dg = \prod_{j=0}^{n-1} \frac{j!(j+2k)!}{(j+k)!^2} \tag{39.9}$$

This was proved by Keating and Snaith using the Selberg integral. However an alternative proof was found by Alex Gamburd (see Bump and Gamburd [29]) which we will give here. This proof is similar to that of Theorem 39.4 in that we will transfer the computation from $U(n)$ to another group. Whereas in Theorem 39.4 we used the Frobenius–Schur duality to transfer the computation to the symmetric group S_k, here we will use the Cauchy identity to transfer the computation from $U(n)$ to $U(2k)$. The two procedures are extremely analogous and closely related to each other.

Proof. Let $t_1, \ldots, t_k$ and $u_1, \ldots, u_k$ be complex numbers. We will show that

$$\int_{U(n)} \prod_{i=1}^{k} (1 + t_i \det(g))(u_i + \det(g)^{-1})\, dg = s_{(n^k)}(t_1, \ldots, t_k, u_1, \ldots, u_k). \tag{39.10}$$

Here $(n^k) = (n, \ldots, n, 0, \ldots, 0)$ is the partition with k nonzero parts, each equal to n. Taking $t_i = u_i = 1$ gives $\int_{U(n)} |1 + \det(g)|^{2k} \, dg$, because $|\det(g)| = 1$ so $\det(g)^{-1} = \overline{\det(g)}$. This equals the left-hand side of (39.9) because if g is a unitary matrix so is $-g$. Now $s_{(n^k)}(1, \ldots, 1)$ with $2k$ entries equal to 1 is the dimension of an irreducible representation of $U(2k)$, which may be evaluated using the Weyl dimension formula (Theorem 22.10). We leave it to the reader to check that this dimension equals the right-hand side of (39.9).

Thus, consider the left-hand side of (39.10). If the eigenvalues of g are $\alpha_1, \ldots, \alpha_n$, by the dual Cauchy identity the integrand equals

$$\prod_{i=1}^{k} \prod_{j=1}^{n} (1 + t_i \alpha_j)(1 + u_i \alpha_j) \det(g)^{-k}$$

$$= \sum_{\lambda} s_\lambda(\alpha_1, \ldots, \alpha_n) \, s_{\lambda^t}(t_1, \ldots, t_n, u_1, \ldots, u_n) \, \overline{\det(g)}^k.$$

Now each $s_\lambda(\alpha_1, \ldots, \alpha_n)$ is the character of an irreducible representation of $U(n)$ if it is nonzero, that is, if the length of λ is $\leqslant n$. In particular $\det(g)^n = s_{(k^n)}(\alpha_1, \ldots, \alpha_n)$. So by Schur orthogonality, integrating over g picks off the contribution of a single term, with $\lambda = (k^n)$ and $\lambda^t = (n^k)$. This proves (39.10).   □

Exercises

Let $m \leqslant n$. The m-*level* correlation function of Dyson [48] for unitary statistics is a function R_m on $\mathbb{T}^m$ defined by the requirement that if f is a test function on $\mathbb{T}^m$ (piecewise continuous, let us say) then

$$\int_{\mathbb{T}^m} R_m(t_1, \ldots, t_m) f(t_1, \ldots, t_m) \, dt_1 \cdots dt_m = \int_{U(n)} \sum{}^{*} f(t_{i_1}, \ldots, t_{i_m}) \, dg, \quad (39.11)$$

where the sum is over all distinct m-tuples $(i_1, \ldots, i_m)$ of distinct integers between 1 and n, and $t_1, \ldots, t_n$ are the eigenvalues of g. Intuitively, this function gives the probability density that $t_1, \ldots, t_n$ are the eigenvalues of $g \in U(n)$.

The purpose of the exercises is to prove (and generalize) Dyson's formula

$$R_m(t_1, \ldots, t_m) = \det \left(s_n(\theta_j - \theta_k) \right)_{j,k}, \qquad t_i = e^{i\theta_j}, \quad (39.12)$$

where

$$s_n(\theta) = \begin{cases} \dfrac{\sin(n\theta/2)}{\sin(\theta/2)} & \text{if } \theta \neq 0, \\ n & \text{if } \theta = 0. \end{cases}$$

As a special case, when $m = 2$, the graph of the "pair correlation" $R_2(1, \theta)$ may be found in Fig. 39.1. This shows graphically the repulsion of the zeros – as we can see, the probability of two zeros being close together is small, but for moderate distances there is no correlation.

Exercise 39.1. If $m = n$, prove that

$$R_n(t_1, \ldots, t_n) = \det(A \cdot {}^t\bar{A}), \qquad A = \begin{pmatrix} 1 & t_1 & \cdots & t_1^{n-1} \\ 1 & t_2 & \cdots & t_2^{n-1} \\ \vdots & & & \vdots \\ 1 & t_n & \cdots & t_n^{n-1} \end{pmatrix}.$$

[Since $n = m$, the matrix A is square and we have $\det(A \cdot {}^t\bar{A}) = |\det(A)|^2$. Reduce to the case where the test function f is symmetric. Then use the Weyl integration formula.]

Exercise 39.2. Show that

$$R_m(x_1, \ldots, x_m) = \frac{1}{(n-m)!} \int_{\mathbb{T}^{n-m}} R_n(x_1, \ldots, x_n) \, dx_1 \cdots dx_n.$$

Exercise 39.3. Prove that when $m \leqslant n$ we have

$$R_m(t_1, \ldots, t_m) = \det(A \cdot {}^t\bar{A}), \qquad A = \begin{pmatrix} 1 & t_1 & \cdots & t_1^{n-1} \\ 1 & t_2 & \cdots & t_2^{n-1} \\ \vdots & & & \vdots \\ 1 & t_m & \cdots & t_m^{n-1} \end{pmatrix}.$$

Observe that if $m < n$, then A is not square, so we may no longer factor the determinant. Deduce Dyson's formula (39.12).

Exercise 39.4. (Bump, Diaconis and Keller [30]) Generalize Dyson's formula as follows. Let λ be a partition of length $\leqslant n$. The measure $|\chi_\lambda(g)|^2 \, dg$ is a probability measure, and we may define an m-level correlation function for it exactly as in (39.11). Denote this as $R_{m,\lambda}$. Prove that

$$R_{m,\lambda}(t_1, \ldots, t_n) = \det(A \cdot {}^t\bar{A}), \qquad A = \begin{pmatrix} t_1^{-\lambda_1} & t_1^{1-\lambda_2} & \cdots & t_1^{-\lambda_n+n-1} \\ t_2^{-\lambda_1} & t_2^{1-\lambda_2} & \cdots & t_2^{-\lambda_n+n-1} \\ \vdots & & & \vdots \\ t_m^{-\lambda_1} & t_m^{1-\lambda_2} & \cdots & t_m^{-\lambda_n+n-1} \end{pmatrix}.$$

Exercise 39.5. Let us consider the distribution of the traces of $g \in \mathrm{SU}(2)$. In this cases the traces are real valued so we must modify (39.1) to read

$$\int_{\mathrm{SU}(2)} \phi(\mathrm{tr}(g)) \, dg = \int_{\mathbb{R}} \phi(x) \, d\mu(x).$$

Since $|\mathrm{tr}(g)| \leqslant 2$, and since the map $g \mapsto -g$ takes $\mathrm{SU}(2)$ to itself, the measure $d\mu$ will be even and supported between -2 and 2. Show that

$$\frac{1}{2\pi} \int_{-\infty}^{\infty} \sqrt{4 - x^2} \, x^{2k} \, dx = \frac{1}{k+1} \binom{2k}{k}$$

and deduce that

$$d\mu(x) = \frac{1}{2\pi} \sqrt{4 - x^2} \, dx.$$

Symmetric Group Branching Rules and Tableaux

If $G \supset H$ are groups, a *branching rule* is an explicit description of how representations of G decompose into irreducibles when restricted to H. By Frobenius reciprocity, this is equivalent to asking how representations of H decompose into irreducibles on induction to G. In this chapter, we will obtain the branching rule for the symmetric groups.

Suppose that λ is a partition of k and that μ is a partition of l with $k \leqslant l$. We write $\lambda \subseteq \mu$ or $\lambda \supseteq \mu$ if the Young diagram of λ is contained in the Young diagram of μ. Concretely this means that $\lambda_i \leqslant \mu_i$ for all i. If $\lambda \neq \mu$, we write $\lambda \subset \mu$ or $\mu \supset \lambda$.

We will denote by ρ_λ the irreducible representation of S_k parametrized by λ. We follow the notation of the last chapter in regarding elements of $\mathcal{R}_k$ as generalized characters of S_k. Thus, $\mathbf{s}_\lambda$ is the character of the representation ρ_λ.

Proposition 40.1. *Let λ be a partition of k, and let μ be a partition of $k-1$. Then*

$$\langle \mathbf{s}_\lambda, \mathbf{s}_\mu \mathbf{e}_1 \rangle = \begin{cases} 1 \ \textit{if } \lambda \supset \mu, \\ 0 \ \textit{otherwise.} \end{cases}$$

Proof. Applying ch, it is sufficient to show that

$$e_1 s_\mu = \sum_{\lambda \supset \mu} s_\lambda.$$

We work in $\Lambda^{(n)}$ for any sufficiently large n; of course $n = k$ is sufficient. Let Δ denote the denominator in (36.1), and let

$$M = \begin{vmatrix} x_1^{\mu_n} & x_2^{\mu_n} & \cdots & x_n^{\mu_n} \\ x_1^{\mu_{n-1}+1} & x_2^{\mu_{n-1}+1} & \cdots & x_n^{\mu_{n-1}+1} \\ \vdots & \vdots & & \vdots \\ x_1^{\mu_1+n-1} & x_2^{\mu_1+n-1} & \cdots & x_n^{\mu_1+n-1} \end{vmatrix}. \tag{40.1}$$

By (36.1), we have $s_\mu = M/\Delta$ and $e_1 = \sum x_i$, so

$$\Delta e_1 s_\mu = \sum_{i=1}^n x_i M = \sum_{i=1}^n \begin{vmatrix} x_1^{\mu_n} & \cdots & x_i^{\mu_n+1} & \cdots & x_n^{\mu_n} \\ \vdots & & \vdots & & \vdots \\ x_1^{\mu_{n-j}+j} & \cdots & x_i^{\mu_{n-j}+j+1} & \cdots & x_n^{\mu_{n-j}+j} \\ \vdots & & \vdots & & \vdots \\ x_1^{\mu_1+n-1} & \cdots & x_i^{\mu_1+n} & \cdots & x_n^{\mu_1+n-1} \end{vmatrix}. \tag{40.2}$$

We claim that this equals

$$\sum_{j=1}^n \begin{vmatrix} x_1^{\mu_n} & \cdots & x_i^{\mu_n} & \cdots & x_n^{\mu_n} \\ \vdots & & \vdots & & \vdots \\ x_1^{\mu_{n-j}+j+1} & \cdots & x_i^{\mu_{n-j}+j+1} & \cdots & x_n^{\mu_{n-j}+j+1} \\ \vdots & & \vdots & & \vdots \\ x_1^{\mu_1+n-1} & \cdots & x_i^{\mu_1+n-1} & \cdots & x_n^{\mu_1+n-1} \end{vmatrix}. \tag{40.3}$$

In (40.2), we have increased the exponent in exactly one column of M by one and then summed over columns; in (40.3), we have increased the exponent in exactly one row of M by one and then summed over rows. In either case, expanding the determinants and summing over i or j gives the result of first expanding M and then in each resulting monomial increasing the exponent of exactly one x_i by one. These are the same set of terms, so (40.2) and (40.3) are equal.

In (40.3), not all terms may be nonzero. Two consecutive rows will be the same if $\mu_{n-j} + j + 1 = \mu_{n-j+1} + j + 1$, that is, if $\mu_{n-j} = \mu_{n-j+1}$. In this case, the determinant is zero. Discarding these terms, (40.3) is the sum of all s_λ as λ runs through those partitions of k that contain μ. □

Theorem 40.1. *Let λ be a partition of k and let μ be a partition of $k - 1$. The following are equivalent.*

(i) *The representation ρ_λ occurs in the representation of S_k induced from the representation S_μ of $S_{k-1} \subset S_k$; in this case it occurs with multiplicity one.*

(ii) *The representation ρ_μ occurs in the representation of S_k restricted from the representation S_λ of $S_k \supset S_{k-1}$; in this case it occurs with multiplicity one.*

(iii) *The partition $\mu \subset \lambda$.*

Proof. Statements (i) and (ii) are equivalent by Frobenius reciprocity. Noting that S_1 is the trivial group, we have $S_{k-1} = S_{k-1} \times S_1$. By definition, $\mathbf{s}_\mu \mathbf{e}_1$ is the character of S_k induced from the character $\mathbf{s}_\mu \otimes \mathbf{e}_1$ of $S_{k-1} \times S_1$. With this in mind, this theorem is just a paraphrase of Proposition 40.1. □

A representation is *multiplicity-free* if in its decomposition into irreducibles, no irreducible occurs with multiplicity greater than 1.

Corollary 40.1. *If ρ is an irreducible representation of S_{k-1}, then the representation of S_k induced from ρ is multiplicity-free; and if τ is an irreducible representation of S_k then the representation of S_{k-1} restricted from τ is multiplicity-free.*

Proof. This is an immediate consequence of the theorem.   □

Let λ be a partition of k. By a *standard (Young) tableau of shape λ*, we mean a labeling of the diagram of λ by the integers 1 through k in such a way that entries increase in each row and column. As we explained earlier, we represent the diagram of a partition by a series of boxes. This is more convenient than a set of dots since we can then represent a tableau by putting numbers in the boxes to indicate the labeling.

For example, the standard tableaux of shape $(3, 2)$ are:

The following theorem makes use of the following chain of groups:

$$S_k \supset S_{k-1} \supset \cdots \supset S_1.$$

These have the remarkable property that the restriction of each irreducible representation of S_i to S_{i-1} is multiplicity-free and the branching rule is explicitly known. Although this is a rare phenomenon, there are a couple of other important cases:

$$\mathrm{U}(n) \supset \mathrm{U}(n-1) \supset \cdots \supset \mathrm{U}(1),$$

and

$$\mathrm{O}(n) \supset \mathrm{O}(n-1) \supset \cdots \supset \mathrm{O}(2).$$

Theorem 40.2. *If λ is a partition of k, the degree of the irreducible representation ρ_λ of S_k associated with λ is equal to the number of standard tableaux of shape λ.*

Proof. Removing the top box (labeled k) from a tableau of shape λ results in another tableau, of shape μ (say), where $\mu \subset \lambda$. Thus, the set of tableaux of shape λ is in bijection with the set of tableaux of shape μ, where μ runs through the partitions of $k-1$ contained in λ.

The restriction of ρ_λ to S_{k-1} is the direct sum of the irreducible representations ρ_μ, where μ runs through the partitions of $k-1$ contained in λ, and by induction the degree of each such ρ_μ equals the number of tableaux of shape μ. The result follows.   □

Tableaux are an important topic in combinatorics. Fulton [53] and Stanley [153] have extensive discussions of tableaux, and there is a very good discussion of standard tableaux in Knuth [109].

A famous formula, due to Frame, Robinson, and Thrall, for the number of tableaux of shape λ—that is, the degree of ρ_λ—is the *hook length formula*. There are many proofs in the literature. For a variety of proofs see Fulton [53], Knuth [109], Macdonald [124], Manivel [126], Sagan [141] (with anecdote), and Stanley [153]. The hook length formula is equivalent to an older formula of Frobenius and (independently) Young, which is treated in Exercise 40.4.

For each box B in the diagram of λ, the *hook* at B consists of B, all boxes to the right and below. The hook length is the length of the hook. For example, Fig. 40.1 shows a hook for the partition $\lambda = (5, 5, 4, 3, 3)$ of 20. This hook has length 5.

Theorem 40.3. (Hook length formula) *Let λ be a partition of k. The number of standard tableaux of shape λ equals $k!$ divided by the product of the lengths of the hooks.*

For the example, we have indicated the lengths of the hooks in Fig. 40.1. By the hook length formula, we see that the number of tableaux of shape λ is

$$\frac{20!}{9 \cdot 8 \cdot 7 \cdot 4 \cdot 2 \cdot 8 \cdot 7 \cdot 6 \cdot 3 \cdot 1 \cdot 6 \cdot 5 \cdot 4 \cdot 1 \cdot 4 \cdot 3 \cdot 2 \cdot 3 \cdot 2 \cdot 1} = 34,641,750,$$

and this is the degree of the irreducible representation ρ_λ of S_{20}.

Proof. See Exercise 40.5.   □

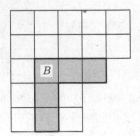

Fig. 40.1. The hook length formula for $\lambda = (5, 5, 4, 3, 3)$

Proposition 40.1 is a special case of *Pieri's formula*, which we explain and prove. First, we give a bit of background on the Littlewood–Richardson rule, of which Pieri's formula is itself a special case.

The multiplicative structure of the ring $\mathcal{R} \cong \Lambda$ is of intense interest. If λ and μ are partitions of r and k, respectively, then we can decompose

$$\mathbf{s}_\lambda \mathbf{s}_\mu = \sum_\lambda c^\nu_{\lambda\mu} \mathbf{s}_\nu,$$

where the sum is over partitions ν of $r + k$. The coefficients $c_{\lambda\mu}^{\nu}$ are called the *Littlewood–Richardson* coefficients. They are integers since the $\mathbf{s}_{\nu}$ are a $\mathbb{Z}$-basis of the free Abelian group $\mathcal{R}_{r+k}$.

Applying $\mathrm{ch}^{(n)}$, we may also write

$$s_{\lambda} s_{\mu} = \sum_{\lambda} c_{\lambda\mu}^{\nu} s_{\nu}$$

as a decomposition of Schur polynomials, or $\chi_{\lambda} \chi_{\mu} = \sum c_{\lambda\mu}^{\nu} \chi_{\nu}$ in terms of the irreducible characters of $U(n)$ parametrized by λ, μ, and ν. Using the fact that the $\mathbf{s}_{\lambda}$ are orthonormal, we have also

$$c_{\lambda\mu}^{\nu} = \langle \mathbf{s}_{\lambda} \mathbf{s}_{\mu}, \mathbf{s}_{\nu} \rangle.$$

Proposition 40.2. *The coefficients $c_{\lambda\mu}^{\nu}$ are nonnegative integers.*

Proof. This is clear from any one of the characterizations in Theorem 38.3.

$\square$

Given that the Littlewood–Richardson coefficients are nonnegative integers, a natural question is to ask for a combinatorial interpretation. Can $c_{\lambda\mu}^{\nu}$ be realized as the cardinality of some set? The answer is yes, and this interpretation is known as the *Littlewood–Richardson rule*. We refer to Fulton [53], Stanley [153], or Macdonald [124] for a full discussion of the Littlewood–Richardson rule.

Even just to state the Littlewood–Richardson rule in full generality is slightly complex, and we will content ourselves with a particularly important special case. This is where $\lambda = (r)$ or $\lambda = (1, \ldots, 1)$, so $\mathbf{s}_{\lambda} = \mathbf{h}_r$ or $\mathbf{e}_r$. This simple and useful case of the Littlewood–Richardson rule is called *Pieri's formula*. We will now state and prove it.

If $\mu \subset \lambda$ are partitions, we call the pair (μ, λ) a *skew partition* and denote it $\lambda \backslash \mu$. Its diagram is the set-theoretic difference between the diagrams of λ and μ. We call the skew partition $\lambda \backslash \mu$ a *vertical strip* if its diagram does not contain more than one box in any given row. It is called a *horizontal strip* if its diagram does not contain more than one box in any given column.

For example, if $\mu = (3, 3)$, then the partitions λ of 8 such that $\lambda \backslash \mu$ is a vertical strip are $(4, 4)$, $(4, 3, 1)$, and $(3, 3, 1, 1)$. The diagrams of these skew partitions are the shaded regions in Fig. 40.2.

Theorem 40.4. (Pieri's formula) *Let μ be a partition of k, and let $r \geqslant 0$. Then $\mathbf{s}_{\mu} \mathbf{e}_r$ is the sum of the $\mathbf{s}_{\lambda}$ as λ runs through the partitions of $k + r$ containing μ such that $\lambda \backslash \mu$ is a vertical strip. Also, $\mathbf{s}_{\mu} \mathbf{h}_r$ is the sum of the $\mathbf{s}_{\lambda}$ as λ runs through the partitions of $k + r$ such that $\lambda \backslash \mu$ is a horizontal strip.*

Proof. Since by Theorems 34.3 and 35.2 applying the involution ι interchanges $\mathbf{e}_r$ and $\mathbf{h}_r$ and also interchanges $\mathbf{s}_{\mu}$ and $\mathbf{s}_{\lambda}$, the second statement follows from the first, which we prove.

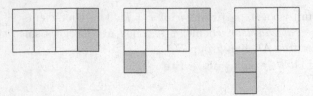

Fig. 40.2. Vertical strips

The proof that $\mathbf{s}_\mu \mathbf{e}_r$ is the sum of the $\mathbf{s}_\lambda$ as λ runs through the partitions of $k + r$ containing μ such that $\lambda \backslash \mu$ is a vertical strip is actually identical to the proof of Proposition 40.1. Choose $n \geqslant k + r$ and, applying ch, it is sufficient to prove the corresponding result for Schur polynomials.

With notations as in that proof, we see that $\Delta e_r s_\mu$ equals the sum of $\binom{k}{r}$ terms, each of which is obtained by multiplying M, defined by (40.1), by a monomial $x_{i_1} \cdots x_{i_r}$, where $i_1 < \cdots < i_r$. Multiplying M by $x_{i_1} \cdots x_{i_r}$ amounts to increasing the exponent of x_{i_r} in the i_rth column by one. Thus, we get $\Delta e_r s_\mu$ if we take M, increase the exponents in r columns each by one, and then add the resulting $\binom{k}{r}$ determinants.

We claim that this gives the same result as taking M, increasing the exponents in r rows each by one, and then adding the resulting $\binom{k}{r}$ determinants. Indeed, either way, we get the result of taking each monomial occurring in the expansion of the determinant M, increasing the exponents of exactly r of the x_i each by one, and adding all resulting terms.

Thus, $e_r s_\mu$ equals the sum of all terms (36.1) where $(\lambda_1, \ldots, \lambda_n)$ is obtained from $(\mu_1, \ldots, \mu_n)$ by increasing exactly r of the μ_i by one. Some of these terms may not be partitions, in which case the determinant in the numerator of (36.1) will be zero since it will have repeated rows. The terms that remain will be the partitions of $k + r$ length such that $\lambda \backslash \mu$ is a vertical strip. These partitions all have length $\leqslant n$ because we chose n large enough. Thus, $e_r s_\mu$ is the sum of s_λ for these λ, as required. □

Exercises

The next problem generalizes Exercise 37.3. If λ and μ are partitions such that the Young diagram of λ contains that of μ, then the pair (λ, μ) is called a *skew shape* and is denoted $\lambda \backslash \mu$. Its Young diagram is the set-theoretic difference between the Young diagrams of λ and μ. The skew shape is called a *ribbon shape* if the diagram is connected and contains no 2×2 squares. For example, if $\lambda = (5, 4, 4, 3, 2)$ and $\mu = (5, 3, 2, 2, 2)$ then the skew shape $\lambda \backslash \mu$ is a ribbon shape. Its diagram is the shaded region in the following figure.

If $\lambda \backslash \mu$ is a ribbon shape, we call its *height*, denoted $\mathrm{ht}(\lambda \backslash \mu)$ one less than the number of rows involved in its Young diagram. In the example, the height is 2.

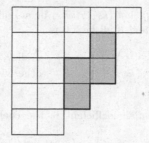

The following result is called the *Murnaghan–Nakayama rule* (Stanley [153]). It is the combinatorial basis of the Boson–Fermion correspondence in the theory of infinite-dimensional Lie algbras.

Exercise 40.1. Let μ be a partition of k and r a positive integer. Show that

$$p_r s_\mu = \sum_\lambda (-1)^{\text{ht}(\lambda \setminus \mu)},$$

where the sum is over all partitions λ of $k + r$ such that $\lambda \setminus \mu$ is a ribbon shape.

[**Hint**: If $\lambda \in \mathbb{Z}^n$, let

$$F(\lambda) = \det(x_i^{\lambda_j})/\Delta,$$

where Δ is the denominator in (36.1). Thus if $\rho = (n-1, n-2, \ldots, 0)$, then (36.1) can be written $F(\lambda + \rho) = s_\lambda$. Show that

$$p_r s_\lambda = \sum_{k=1}^n F(\lambda + \rho + r\mathbf{e}_k),$$

where $\mathbf{e}_k = (0, \ldots, 1, \ldots, 0)$ is the kth standard basis vector of $\mathbb{Z}^n$. Show that each term in this sum is either zero or $\pm s_\mu$ where $\lambda \setminus \mu$ is a ribbon shape.]

Exercise 40.2. Since the h_k generate the ring $\mathcal{R}$, knowing how to multiply them gives complete information about the multiplication in $\mathcal{R}$. Thus, Pieri's formula contains full information about the Littlewood–Richardson coefficients. This exercise gives a concrete illustration. Using Pieri's formula (or the Jacobi–Trudi identity), check that

$$\mathbf{h}_2 \mathbf{h}_1 - \mathbf{h}_3 = \mathbf{s}_{(21)}.$$

Use this to show that

$$\mathbf{s}_{(21)} \mathbf{s}_{(21)} = \mathbf{s}_{(42)} + \mathbf{s}_{(411)} + \mathbf{s}_{(33)} + 2\mathbf{s}_{(321)} + \mathbf{s}_{(3111)} + \mathbf{s}_{(222)} + \mathbf{s}_{(2211)}.$$

Exercise 40.3. Let λ be a partition of k into at most n parts. Prove that the number of standard tableaux of shape λ is

$$\int_{U(n)} \text{tr}(g)^k \, \overline{\chi_\lambda(g)} \, dg.$$

(**Hint:** Use Theorems 40.2 and 36.4.)

Exercise 40.4. (Frobenius) Let $(k_1, \ldots, k_r)$ be a sequence of integers whose sum is k. The *multinomial coefficient* if all $k_i \geqslant 0$ is

$$\binom{k}{k_1, \ldots, k_r} = \begin{cases} \frac{k!}{k_1! \cdots k_r!} & \text{if all } k_i \geqslant 0, \\ 0 & \text{otherwise.} \end{cases}$$

(i) Show that this multinomial coefficient is the coefficient of $t_1^{k_1} \cdots t_r^{k_r}$ in the expansion of $(\sum_{i=1}^r t_i)^k$.

(ii) Prove that if λ is a partition of k into at most n parts, then the number of standard tableaux of shape λ is

$$\sum_{w \in S_n} (-1)^{l(w)} \binom{k}{\lambda_1 - 1 + w(1), \lambda_2 - 2 + w(2), \ldots, \lambda_n - n + w(n)}.$$

For example, let $\lambda = (3, 2) = (3, 2)$ and $k = 5$. The sum is

$$\binom{5}{3, 2} - \binom{5}{4, 1} = 10 - 5 = 5,$$

the number of standard tableaux with shape λ. (**Hint:** Use Theorems 37.3 and 40.2.)

(iii) Let λ be a partition of k into at most n parts. Let $\mu = \lambda + \delta$, where $\delta = (n-1, n-2, \ldots, 1, 0)$. Show that the number of standard tableaux of shape λ is

$$\frac{k!}{\prod_i \mu_i!} \left(\prod_{i<j} (\mu_i - \mu_j) \right).$$

[**Hint:** Show that

$$\frac{\prod_i \mu_i!}{k!} \sum_{w \in S_n} (-1)^{l(w)} \binom{k}{\mu_1 - n + w(1), \mu_2 - n + w(2), \ldots, \mu_n - n + w(n)}$$

is a polynomial of degree $\frac{1}{2}n(n-1)$ in $\mu_1, \ldots, \mu_n$, and that it vanishes when $\mu_i = \mu_j$.]

Continuing from the previous exercise:

Exercise 40.5. (i) Show that the product of the hooks in the ith row is

$$\frac{\mu_i!}{\prod_{j>i} \mu_i - \mu_j}.$$

(ii) Prove the hook length formula.

41

Unitary Branching Rules and Tableaux

In this chapter, representations of both $\mathrm{GL}(n, \mathbb{C})$ and $\mathrm{GL}(n-1, \mathbb{C})$ occur. To distinguish the two, we will modify the notation introduced before Theorem 36.3 as follows. If λ is a partition (of any k) of length $\leqslant n$, or more generally an integer sequence $\lambda = (\lambda_1, \ldots, \lambda_n)$ with $\lambda_1 \geqslant \lambda_2 \geqslant \cdots$, we will denote by $\pi_\lambda^{\mathrm{GL}_n}$ or more simply as π_λ the representation of $\mathrm{GL}(n, \mathbb{C})$ parametrized by λ. On the other hand, if μ is a partition of length $\leqslant n-1$, or more generally an integer sequence $\mu = (\mu_1, \ldots, \mu_{n-1})$ with $\mu_1 \geqslant \mu_2 \geqslant \cdots$, we will denote by $\pi_\mu^{\mathrm{GL}_{n-1}}$ or (more simply) as π_μ' the representation of $\mathrm{GL}(n-1, \mathbb{C})$ parametrized by μ.

We embed $\mathrm{GL}(n-1, \mathbb{C}) \longrightarrow \mathrm{GL}(n, \mathbb{C})$ by

$$g \longmapsto \begin{pmatrix} g & \\ & 1 \end{pmatrix}. \tag{41.1}$$

It is natural to ask when the restriction of π_λ to $\mathrm{GL}(n-1, \mathbb{C})$ contains π_μ'. Since algebraic representations of $\mathrm{GL}(n, \mathbb{C})$ correspond precisely to representations of its maximal compact subgroup, this is equivalent to asking for the branching rule from $\mathrm{U}(n)$ to $\mathrm{U}(n-1)$.

This question has a simple and beautiful answer in Theorem 41.1 below. We say that the integer sequences $\lambda = (\lambda_1, \ldots, \lambda_n)$ and $\mu = (\mu_1, \ldots, \mu_{n-1})$ *interlace* if

$$\lambda_1 \geqslant \mu_1 \geqslant \lambda_2 \geqslant \mu_2 \geqslant \cdots \geqslant \mu_{n-1} \geqslant \lambda_n.$$

Proposition 41.1. *Suppose that λ_n and μ_{n-1} are nonnegative, so the integer sequences λ and μ are partitions. Then λ and μ interlace if and only if $\lambda \supset \mu$ and the skew partition $\lambda \backslash \mu$ is a horizontal strip.*

This is obvious if one draws a diagram.

Proof. Assume that $\lambda \supset \mu$ and $\lambda \backslash \mu$ is a horizontal strip. Then $\lambda_j \geqslant \mu_j$ because $\lambda \supset \mu$. We must show that $\mu_j \geqslant \lambda_{j+1}$. If it is not, $\lambda_j \geqslant \lambda_{j+1} > \mu_j$, which

D. Bump, *Lie Groups*, Graduate Texts in Mathematics 225,
DOI 10.1007/978-1-4614-8024-2_41, © Springer Science+Business Media New York 2013

implies that the diagram of $\lambda\backslash\mu$ contains two entries in the $\mu_j + 1$ column, namely in the j and $j+1$ rows, which is a contradiction since $\lambda\backslash\mu$ was assumed to be a horizontal strip. We have proved that λ and μ interlace. The converse is similar.   □

Theorem 41.1. *Let* $\lambda = (\lambda_1, \ldots, \lambda_n)$ *and* $\mu = (\mu_1, \ldots, \mu_{n-1})$ *be integer sequences with* $\lambda_1 \geqslant \lambda_2 \geqslant \cdots$ *and* $\mu_1 \geqslant \mu_2 \geqslant \cdots$. *Then the restriction of* π_λ *to* $\mathrm{GL}(n-1, \mathbb{C})$ *contains a copy of* π'_μ *if and only if* λ *and* μ *interlace. The restriction of* π_λ *is multiplicity-free.*

Proof. We restriction the representation π_λ of $\mathrm{GL}_n(\mathbb{C})$ to $\mathrm{GL}_{n-1}(\mathbb{C})$ in two stages. First, we restrict it to $\mathrm{GL}_{n-1}(\mathbb{C}) \times \mathrm{GL}_1(\mathbb{C})$, and then we restrict it from $\mathrm{GL}_{n-1}(\mathbb{C}) \times \mathrm{GL}_1(\mathbb{C})$ to $\mathrm{GL}_{n-1}(\mathbb{C})$. Here $(g, h) \in \mathrm{GL}_{n-1}(\mathbb{C}) \times \mathrm{GL}_1(\mathbb{C})$ is embedded in $\mathrm{GL}_n(\mathbb{C})$ as in (38.7). Every irreducible character of $\mathrm{GL}_1(\mathbb{C}) = \mathbb{C}^\times$ is of the form $\alpha \longmapsto \alpha^k$ for some $k \in \mathbb{Z}$, and we will denote this character as π''_k.

We may order the eigenvalues of (g, h) so that $\alpha_1, \ldots, \alpha_{n-1}$ are the eigenvalues of g, and α_n is the eigenvalue of h. Since $s_\lambda(\alpha_1, \ldots, \alpha_n)$ is a homogeneous polynomial of degree $|\lambda| = \sum \lambda_i$, and since $s_\mu(\alpha_1, \ldots, \alpha_{n-1})\alpha_n^k$ is homogeneous of degree $|\mu| + k$, $\pi'_\mu \otimes \pi''_k$ can occur in the restriction of π_λ to $\mathrm{GL}_{n-1}(\mathbb{C}) \times \mathrm{GL}_1(\mathbb{C})$ if and only if $k = |\lambda| - |\mu|$. In other words, the fact that Schur polynomials are homogeneous implies that the multiplicity of π'_μ in π_λ restricted to $\mathrm{GL}_{n-1}(\mathbb{C})$ equals the multiplicity of $\pi'_\mu \otimes \pi''_k$ to $\mathrm{GL}_{n-1}(\mathbb{C}) \times \mathrm{GL}_1(\mathbb{C})$ where $k = |\lambda| - |\mu|$. By Theorem 38.3 this equals the Littlewood–Richardson coefficient $c^\lambda_{\mu\nu}$ where $\nu = (k)$, and by Pieri's formula (Theorem 40.4) this equals 1 if $\lambda\backslash\mu$ is a horizontal strip, 0 otherwise. By Proposition 41.1 this means that the partitions λ and μ must interlace.   □

We can now give a combinatorial formula for the degree of the irreducible representation π_λ of $\mathrm{GL}(n, \mathbb{C})$, where $\lambda = (\lambda_1, \ldots, \lambda_n)$ and $\lambda_1 \geqslant \cdots \geqslant \lambda_n$. A *Gelfand–Tsetlin pattern* of degree n consists of n decreasing integer sequences of lengths $n, n-1, \ldots, 1$ such that each adjacent pair interlaces. For example, if the top row is $3, 2, 1$, there are eight possible Gelfand–Tsetlin patterns:

$$
\begin{array}{ccc}
3\ \ 2\ \ 1 & 3\ \ 2\ \ 1 & 3\ \ 2\ \ 1 \\
\ \ 3\ \ 2 & \ \ 3\ \ 2 & \ \ 3\ \ 1 \\
\ \ \ \ 3 & \ \ \ \ 2 & \ \ \ \ 3
\end{array}
$$

$$
\begin{array}{ccc}
3\ \ 2\ \ 1 & 3\ \ 2\ \ 1 & 3\ \ 2\ \ 1 \\
\ \ 3\ \ 1 & \ \ 3\ \ 1 & \ \ 2\ \ 2 \\
\ \ \ \ 2 & \ \ \ \ 1 & \ \ \ \ 2
\end{array}
$$

$$
\begin{array}{ccc}
3\ \ 2\ \ 1 & & 3\ \ 2\ \ 1 \\
\ \ 2\ \ 1 & \text{and} & \ \ 2\ \ 1 \\
\ \ \ \ 2 & & \ \ \ \ 1
\end{array}
$$

Theorem 41.2. *The degree of the irreducible representation* π_λ *of* $\mathrm{GL}(n, \mathbb{C})$ *equals the number of Gelfand–Tsetlin patterns whose top row is* λ.

Thus, $\dim\left(\pi_{(3,2,1)}\right) = 8$.

Proof. The proof is identical in structure to Theorem 40.2. The Gelfand–Tsetlin patterns of shape λ can be counted by noting that striking the top row gives a Gelfand–Tsetlin pattern with a top row that is a partition μ of length $n-1$ that interlaces with λ. By induction, the number of such patterns is equal to the dimension of π'_μ, and the result now follows from the branching rule of Theorem 41.1. □

Just as with the symmetric group, the dimension of an irreducible representation of $\mathrm{U}(n)$ can be expressed as the number of tableaux of a certain type. By a *semistandard Young tableau* of shape λ we mean a filling of the boxes in the Young diagram of shape λ in which the columns are strictly increasing but the rows are only weakly increasing.

Proposition 41.2. *Let λ be a partition of length $\leqslant n$. The degree of the irreducible representation π_λ of $\mathrm{GL}(n,\mathbb{C})$ equals the number semistandard Young tableaux of shape λ with entries in $\{1, 2, \ldots, n\}$.*

Proof. In view of Theorem 41.2 it is sufficient to exhibit a bijection between these tableaux and the Gelfand–Tsetlin patterns with top row λ. We will explain how to go from the tableau to the Gelfand–Tsetlin pattern. Given a tableau, the top row of the Gelfand–Tsetlin pattern is the shape: of the tableau:

$$
\begin{array}{|c|c|c|c|c|}
\hline
1 & 1 & 1 & 2 & 3 \\
\hline
\end{array}
\quad\quad
\left\{ \begin{array}{ccc} 5 & 2 & 1 \end{array} \right\}
$$

(with rows $\boxed{2\,|\,2}$ and $\boxed{3}$ below)

Removing all boxes labeled n gives a second tableau, with entries in $1, 2, \ldots, n-1$. Its shape is the second row of the Gelfand–Tsetlin pattern:

$$
\left\{ \begin{array}{ccc} 5 & 2 & 1 \\ & 4 & 3 \end{array} \right\}
$$

We continue removing the boxes labeled $n-1$, and the resulting shape is the third row:

$$
\begin{array}{|c|c|c|}
\hline
1 & 1 & 1 \\
\hline
\end{array}
\quad\quad
\left\{ \begin{array}{ccc} 5 & 2 & 1 \\ & 4 & 3 \\ & & 3 \end{array} \right\}
$$

Continuing in this way we obtain a Gelfand–Tsetlin pattern. We leave it to the reader to convince themselves that this is a bijection. □

The relationship between representation theory and the combinatorics of tableaux is subtle and interesting. It can be understood as just an analogy, but at a deeper level, it can be understood as a reflection of the theory of quantum groups. We start by explaining the analogy.

There is an algorithm, the *Robinson–Schensted–Knuth (RSK) algorithm*, which describes bijections between pairs of tableaux of the same shape (or of conjugate shapes) and various combinatorial objects. Historically, the RSK algorithm first occurred in Robinson's work on the representation theory of the symmetric group [136]. It was rediscovered in the early 1960s by Schensted [147], who was motivated the question of the longest increasing subsequence of an integer sequence, and substantially generalized by Knuth [108]. It has applications in various fields from linguistics to algebraic geometry. We will comment mainly on its connections with representation theory, so we begin by pointing out how it gives combinatorial analogs of the correspondences that we are familiar with, Frobenius–Schur duality and $GL_n \times GL_m$ duality. We will focus on Frobenius–Schur duality.

Let us recapitulate two facts. Let λ be a partition of k with length $\leqslant n$.

- Let $SYT(\lambda)$ be the set of standard tableaux of shape λ having entries in $\{1, \ldots, k\}$. Then by Proposition 40.2 the cardinality of $ST(k)$ equals the degree of the irreducible representation ρ_λ of S_k corresponding to λ.
- Let $SSYT(\lambda, n)$ be the set of semistandard tableau of shape λ having entries in $\{1, \ldots, n\}$. Then by Proposition 41.2 the cardinality of SSYT (λ, n) equals the degree of the irreducible representation $\pi_\lambda^{GL(n)}$ with highest weight λ.

For more about the RSK algorithm, see Fulton [53], Knuth [109] Sect. 5.1.4, or Stanley [153], van Leeuwen [164] and the original papers. Our interest here in the RSK algorithm comes from the fact that it is the basis of the combinatorial side of a series of analogies between results in representation theory and combinatorics. There are three main versions of RSK, each analogous to a fact in representation theory. Briefly, the three representation-theoretic facts in question are:

- The decomposition of $\mathbb{C}[S_k]$ under the action of $S_k \times S_k$ by left and right translation;
- Frobenius–Schur duality;
- $GL(n) \times GL(m)$ duality.

We will take these one at a time, focussing on the second.

The first version of RSK (Robinson) gives a bijection between S_k and the set of pairs of standard tableaux of the same shape λ. That is, between S_k and the disjoint union

$$\bigsqcup_{\lambda \text{ a partition of } k} SYT(\lambda) \times SYT(\lambda).$$

So $k!$, the cardinality of S_k, equals the number of pairs of standard tableaux of size k with the same shape. Beyond this combinatorial reason, let us observe another representation-theoretic reason that these two sets have the same cardinality. Indeed, the cardinality of any finite group equals the sum of the squares of the degrees of its irreducible representations.

The second version of RSK (Schensted) gives a bijection between the set of sequences $\{m_1, \ldots, m_k\}$ with $m_i \in \{1, \ldots, n\}$ (called *words*) and

$$\bigsqcup_{\substack{\lambda \text{ a partition of } k \\ l(\lambda) \leqslant n}} \text{SYT}(\lambda) \times \text{SSYT}(\lambda, n). \tag{41.2}$$

Let us again observe that these sets have the same cardinality, which we may prove using Frobenius–Schur duality in the form

$$\otimes^k \mathbb{C}^n \cong \bigoplus_{\substack{\lambda \text{ a partition of } k \\ l(\lambda) \leqslant n}} \rho_\lambda \otimes \pi_\lambda^{\text{GL}(n)}.$$

Indeed, the dimension of the left-hand side is n^k, which is the cardinality of the set of integer sequences; the right-hand side has the same cardinality as the above disjoint union. So the second RSK bijection is a combinatorial analog of Frobenius–Schur duality.

The third RSK bijection (Knuth) is between the set of $n \times m$ matrices whose entries are nonnegative integers is in bijection with

$$\bigsqcup_\lambda \text{SSYT}(\lambda, m) \times \text{SSYT}(\lambda, n).$$

This may be thought of as a combinatorial analog of $\text{GL}_n \times \text{GL}_m$ duality. In particular there is a combinatorial proof of the Cauchy identity (see Stanley [153]). Knuth also found a variant for matrices whose entries are 0 and 1 and

$$\bigsqcup_\lambda \text{SSYT}(\lambda, m) \times \text{SSYT}(\lambda^t, n),$$

where λ^t is the conjugate partition. This is related to the dual Cauchy identity, and the version of $\text{GL}_n \times \text{GL}_m$ duality for the exterior algebra on $\mathbb{C}^n \times \mathbb{C}^m$.

All of these combinatorial bijections are based on one process, called *Schensted insertion*, which we will explain. Given a semistandard Young tableaux Q of shape λ and an integer m, there is a tableau $\boxed{m} \to Q$ whose shape μ is obtained from λ by adding one box (somewhere).

To compute $\boxed{m} \to Q$ we insert the m into the first row at its "best" location. It can go at the end if it is larger or equal to all of the entries in the row, and if this is true, the algorithm terminates. Otherwise, it will have to displace or "bump" one of the entries in the row. It displaces the first entry that its greater than it. The displaced entry is then inserted into the second row. If it is greater than or equal to all the entries in the row, then we add the entry at the end of the row and the algorithm terminates. Otherwise, we continue by inserting the bumped entry into the third row, and so forth.

Let us do an example. We will calculate $\boxed{1} \longrightarrow Q$ where Q is the following tableau:

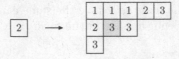

We've shaded the box where the inserted $\boxed{1}$ will go. The 1 bumps the 2 which is then inserted into the second row:

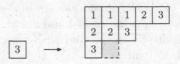

again we've shaded the location where the inserted 2 will go. The 3 that is bumped will then go in the third row. This time it will go at the end:

$$\boxed{3} \longrightarrow$$

The algorithm therefore terminates and we see that $\boxed{1} \longrightarrow Q$ is the tableau:

Now let us explain the second RSK algorithm mentioned above, which is the bijection of $\{1, \ldots, n\}^k$ with (41.2). We begin with a sequence $(m_1, \ldots, m_k)$ and the empty tableau, which we will denote by $\varnothing$. We insert the m_i one by one, finally ending up with a tableau

$$Q = \boxed{m_1} \to \boxed{m_2} \to \cdots \to \boxed{m_k} \to \varnothing$$

Actually $\boxed{m_k} = \boxed{m_k} \to \varnothing$, so this is the same as $\boxed{m_1} \to \cdots \to \boxed{m_k}$.

Clearly Q is in $\mathrm{SSYT}(\lambda, n)$ for some partition λ of k. We obtain another tableau P, called the *recording tableau* of the same shape which has the entries $\{1, 2, 3, \ldots, k\}$ by putting 1 in the first box that was created (necessarily the upper left-hand corner), then 2 in the second box that was created, and so forth. The recording tableau is a clearly a standard tableau. The bijection maps the sequence $(m_1, \ldots, m_k)$ to the pair (P, Q) in (41.2).

During the years prior to the late 1980s, one could say that tableau combinatorics and representation theory existed in parallel. The RSK algorithm existed as a combinatorial analog of the Frobenius–Schur duality correspondence (and $\mathrm{GL}_n \times \mathrm{GL}_m$ duality), but a direct connection between these topics was missing until Kashiwara described *crystals* as an aspect of the developing theory of quantum groups. The book [72] of Hong and Kang and the paper Kashiwara [93] are good introductions to this topic.

We will not give a complete definition of crystals here. Our goal is to describe them sufficiently well to explain their connection with the RSK algorithm. Crystals are purely combinatorial analogs of Lie group representations, but now the connection is more than an analogy: crystals are derived from Lie group representations by a process of deformation.

Let us begin with a Lie group G (say a compact Lie group or its complexification) with weight lattice Λ. Let $\lambda \in \Lambda$ be a dominant weight. The crystal $\mathcal{B}_\lambda$ is a combinatorial analog of the irreducible representation π_λ^G having highest weight λ.

The crystal $\mathcal{B}_\lambda$ is a directed graph with vertices that may be identified with some type of tableaux (at least for the classical Cartan types). Its cardinality (i.e., the number of vertices in this graph) is equal to the dimension of π_λ^G. The edges of the graph are labeled by integer $1 \leqslant i \leqslant r$ where r is the semisimple rank, that is, the number of simple roots. There is a weight function $\mathrm{wt} : \mathcal{B}_\lambda \to \Lambda$ such that if P and Q are vertices with an edge $P \xleftarrow{i} Q$, then

$$\mathrm{wt}(Q) = \mathrm{wt}(P) + \alpha_i.$$

If χ_λ is the character of π_λ^G, then

$$\chi_\lambda = \sum_{P \in \mathcal{B}_\lambda} \mathrm{e}^{\mathrm{wt}(P)}. \tag{41.3}$$

(The notation e^λ is as in Chap. 22.)

Let us consider the case $G = \mathrm{GL}(n, \mathbb{C})$. The semisimple rank r is $n - 1$. Assuming that the dominant weight λ is a partition, the vertices of $\mathcal{B}_\lambda$ are the semistandard Young tableaux in $\mathrm{SSYT}(\lambda, n)$, and the weight function is easy to describe: the weight of a tableau is $\mu = (\mu_1, \ldots, \mu_n)$ where μ_i is the number of boxes labeled i in the tableau. Figure 41.1 shows the crystal with $n = 3$ and $\lambda = (3, 1)$: The edges are labeled 1 or 2. These correspond to the simple roots α_1 and α_2.

In the case $G = GL(n, \mathbb{C})$ where we have identified the vertices with tableaux, the edges have the following meaning: if there is an edge labeled i from P to Q, then Q is obtained from P by changing an entry labeled i to $i + 1$. (But if there is more than one box labeled i, deciding which is to be changed is not entirely straightforward.)

For $\mathrm{GL}(n)$, the weight function may be described as follows: Λ, we recall, is identified with $\mathbb{Z}^n$, in which $\mu = (\mu_1, \ldots, \mu_n)$ is identified with the character $\mu \in \Lambda X^*(T)$, where T is the diagonal torus and the character μ maps

$$\mathbf{z} = \begin{pmatrix} t_1 & & \\ & \ddots & \\ & & t_n \end{pmatrix} \longmapsto \prod t_i^{\mu_i}.$$

Then if $P \in \mathrm{SSYT}(\lambda, n)$ we define $\mu = \mathrm{wt}(P)$ by letting μ_i be the number of entries in the tableau equal to i. So (41.3) becomes the combinatorial formula for Schur polynomial:

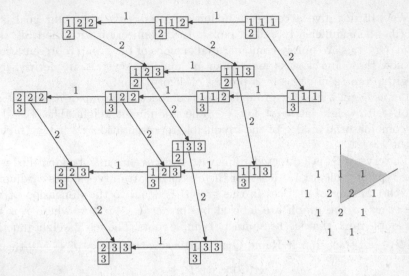

Fig. 41.1. The crystal with highest weight $\lambda = (3,1,0)$. The weight diagram (see Chap. 21) is supplied to the right, to orient the reader

$$s_\lambda(z_1, \ldots, z_n) = \sum_P \mathbf{z}^{\mathrm{wt}(P)} \tag{41.4}$$

where $\mathbf{z}^{\mathrm{wt}(P)}$ is the product of z_i as i runs through the entries in the tableau P. We will not prove (41.4), which is due to Littlewood, but proofs may be found in Fulton [53] or Stanley [153].

Crystals have a purely combinatorial tensor product rule that exactly parallels the decomposition rule for tensor products of Lie group representations. That is, if $\mathcal{C}$ and $\mathcal{C}'$ are crystals, a crystal $\mathcal{C} \otimes \mathcal{C}'$ is defined which is the disjoint union of "irreducible" crystals, each isomorphic to a crystal of the type $\mathcal{B}_\lambda$. If λ, μ and ν are dominant weights, then the number of copies of $\mathcal{B}_\lambda$ in the decomposition of $\mathcal{B}_\mu \otimes \mathcal{B}_\nu$ equals the multiplicity of π_λ^G in $\pi_\mu^G \otimes \pi_\nu^G$.

Crystals give an explanation of the RSK algorithm. The point is that the tensor product operation is closely related to Schensted insertion. Let $\mathcal{B} = \mathcal{B}_{(1)}$. This is Kashiwara's *standard crystal*. It looks like this:

$$\boxed{1} \xrightarrow{1} \boxed{2} \xrightarrow{2} \cdots \xrightarrow{n-1} \boxed{n}.$$

Now we have an isomorphism

$$\mathcal{B} \otimes \mathcal{B}_\lambda \cong \bigsqcup_\mu \mathcal{B}_\mu$$

as μ runs through all the partitions with Young diagrams that are obtained from λ by adding one box. In this isomorphism, if P is a tableau of shape λ, the element $\boxed{i} \otimes P$ in $\mathcal{B} \otimes \mathcal{B}_\lambda$ corresponds to the tableaux $\boxed{i} \to P$ obtained by Schensted insertion in one of the $\mathcal{B}_\lambda$. Which $\mathcal{B}_\mu$ it lives in depends on the row in which the Schensted insertion terminates. The crystal analog of Frobenius–Schur duality expresses $\otimes^k \mathcal{B}$ as a disjoint union of copies of $\mathcal{B}_\lambda$ as λ runs through the partitions of k; the number of times $\mathcal{B}_\lambda$ occurs equals the number of standard tableaux of shape λ.

Crystals are a concrete link between the combinatorics of tableaux and representation theory. Let V_λ be the module for the representation π_λ^G. We would like to deform the group G to obtain a "quantum group" depending on a parameter q. The quantum group q should be G in the limit. This scenario does not quite work, but instead, one may replace G by its (slightly modified) universal enveloping algebra which is a Hopf algebra (Exercise 41.3). This has a deformation $U_q(\mathfrak{g})$. This Hopf algebra is the quantized enveloping algebra. It was introduced by Drinfeld and (independently) Jimbo in 1986, in response to developments in mathematical physics. If the parameter $q \longrightarrow 1$ we recover $U(\mathfrak{g})$. If $q \longrightarrow 0$, the Hopf algebra $U_q(\mathfrak{g})$ does not have a limit but its modules do. They "crystalize" and the crystal is a basis of the resulting module. We refer to Hong and Kang [72] for an account following Kashiwara.

Exercises

Exercise 41.1. Illustrate the bijection described in the proof of Proposition 41.2 by translating the eight Gelfand–Tsetlin patterns with top row $\{3, 2, 1\}$ into tableaux.

Exercise 41.2. Illustrate the second RSK bijection by showing that the word $(1, 2, 3, 2, 3, 1)$ corresponds to the tableau Q with recording tableau P where

$$Q = \begin{array}{|c|c|c|} \hline 1 & 1 & 2 \\ \hline 2 & 3 \\ \cline{1-2} 3 \\ \cline{1-1} \end{array} \qquad P = \begin{array}{|c|c|c|} \hline 1 & 2 & 4 \\ \hline 3 & 5 \\ \cline{1-2} 3 \\ \cline{1-1} \end{array}$$

Exercise 41.3. Let $\mathfrak{g}$ be a Lie algebra. Let $U = U(\mathfrak{g})$ be its universal enveloping algebra.

(i) Show that the map $\Delta : \mathfrak{g} \to U \otimes U$ defined by $\Delta(X) = X \otimes 1 + 1 \otimes X$ satisfies $\Delta(X)\Delta(Y) - \Delta(Y)\Delta(X) = \Delta([X, Y])$ and conclude that Δ extends to a ring homomorphism $U \to U \otimes U$.
(ii) Prove that U is a bialgebra with comultiplication Δ. You will have to define a co-unit.
(iii) Let $S : \mathfrak{g} \to U$ be the map $S(X) = -X$. Show that S extends to a linear map such that is, antimultiplicative, that is $S(ab) = S(b)\,S(a)$.
(iv) Show that U is a Hopf algebra with antipode S.

42

Minors of Toeplitz Matrices

This chapter can be read immediately after Chap. 39. It may also be skipped without loss of continuity. It gives further examples of how Frobenius–Schur duality can be used to give information about problems related to random matrix theory.

Let $f(t) = \sum_{n=-\infty}^{\infty} d_n t^n$ be a Laurent series representing a function $f : \mathbb{T} \longrightarrow \mathbb{C}$ on the unit circle. We consider the *Toeplitz matrix*

$$
T_{n-1}(f) = \begin{pmatrix} d_0 & d_1 & \cdots & d_{n-1} \\ d_{-1} & d_0 & \cdots & d_{n-2} \\ \vdots & \vdots & & \vdots \\ d_{1-n} & d_{2-n} & \cdots & d_0 \end{pmatrix}.
$$

Szegö [157] considered the asymptotics of $D_{n-1}(f) = \det\left(T_{n-1}(f)\right)$ as $n \longrightarrow \infty$. He proved, under certain assumptions, that if

$$
f(t) = \exp\left(\sum_{-\infty}^{\infty} c_n t^n\right),
$$

then

$$
D_{n-1}(f) \sim \exp\left(nc_0 + \sum_{k=1}^{\infty} k c_k c_{-k}\right). \tag{42.1}
$$

In other words, the ratio is asymptotically 1 as $n \longrightarrow \infty$. See Böttcher and Silbermann [22] for the history of this problem and applications of Szegö's theorem.

A generalization of Szegö's theorem was given by Bump and Diaconis [28], who found that the asymptotics of *minors* of Toeplitz matrices had a similar formula. Very strikingly, the irreducible characters of the symmetric group appear in the formula.

D. Bump, *Lie Groups*, Graduate Texts in Mathematics 225,
DOI 10.1007/978-1-4614-8024-2_42, © Springer Science+Business Media New York 2013

One may form a minor of a Toeplitz matrix by either *striking* some rows and columns or by *shifting* some rows and columns. For example, if we strike the second row and first column of $T_4(f)$, we get

$$\begin{pmatrix} d_1 & d_2 & d_3 & d_4 \\ d_{-1} & d_0 & d_1 & d_2 \\ d_{-2} & d_{-1} & d_0 & d_1 \\ d_{-3} & d_{-2} & d_{-1} & d_0 \end{pmatrix}.$$

This is the same result as we would get by simply *shifting* the indices in $T_3(f)$; that is, it is the determinant $\det(d_{\lambda_i - i + j})_{1 \leqslant i, j \leqslant 4}$ where λ is the partition (1). The most general Toeplitz minor has the form $\det(d_{\lambda_i - \mu_j - i + j})$, where λ and μ are partitions. The asymptotic formula of Bump and Diaconis holds λ and μ fixed and lets $n \longrightarrow \infty$.

The formula with μ omitted [i.e., for $\det(d_{\lambda_i - i + j})$] is somewhat simpler to state than the formula, involving Laguerre polynomials, with both λ and μ. We will content ourselves with the special case where μ is trivial.

We will take the opportunity in the proof of Theorem 42.1 to correct a minor error in [28]. The statement before (3.4) of [28] that "... the only terms that survive have $\alpha_k = \beta_k$" is only correct for terms of degree $\leqslant n$. (We thank Barry Simon for pointing this out.)

If λ is a partition, let χ_λ denote the character of $\mathrm{U}(n)$ defined in Chap. 36.

We will use the notation like that at the end of Chap. 22, which we review next. Although we hark back to Chap. 22 for our notation, the only "deep" fact that we need from Part II of this book is the Weyl integration formula. For example, the Weyl character formula in the form that we need it is identical to the combination of (36.1) and (36.3). The proof of Theorem 42.1 in [28], based on the Jacobi–Trudi and Cauchy identities, did not make use of the Weyl integration formula, so even this aspect of the proof can be made independent of Part II.

Let T be the diagonal torus in $\mathrm{U}(n)$. We will identify $X^*(T) \cong \mathbb{Z}^n$ by mapping the character (22.15) to $(k_1, \ldots, k_n)$. If $\chi \in X^*(T)$ we will use the "multiplicative" notation e^χ for χ so as to be able to form linear combinations of characters yet still write $X^*(T)$ additively. The Weyl group W can be identified with the symmetric group S_n acting on $X^*(T) = \mathbb{Z}^n$ by permuting the characters. Let $\mathcal{E}$ be the free Abelian group on $X^*(T)$. (This differs slightly from the use of $\mathcal{E}$ at the end of Chap. 22.)

Elements of $\mathcal{E}$ are naturally functions on T. Since each conjugacy class of $\mathrm{U}(n)$ has a representative in T, and two elements of T are conjugate in G if and only if they are equivalent by W, class functions on G are the same as W-invariant functions on W. In particular, a W-invariant element of $\mathcal{E}$ may be regarded as a function on the group. We write the Weyl character formula in the form (22.17) with $\delta = (n-1, n-2, \ldots, 1, 0)$ as in (22.16).

If λ and μ are partitions of length $\leqslant n$, let

$$D_{n-1}^{\lambda, \mu}(f) = \det(d_{\lambda_i - \mu_j - i + j}).$$

It is easy to see that this is a minor in a larger Toeplitz matrix.

Theorem 42.1 (Heine, Szegö, Bump, Diaconis). *Let* $f \in L^1(\mathbb{T})$ *be given, with* $f(t) = \sum_{n=-\infty}^{\infty} d_n t^n$. *Let* λ *and* μ *be partitions of length* $\leqslant n$. *Define a function* $\Phi_{n,f}$ *on* $\mathrm{U}(n)$ *by* $\Phi_{n,f}(g) = \prod_{i=1}^{n} f(t_i)$, *where* t_i *are the eigenvalues of* $g \in \mathrm{U}(n)$. *Then*

$$D_{n-1}^{\lambda,\mu}(f) = \int_{\mathrm{U}(n)} \Phi_{n,f}(g)\, \overline{\chi_\lambda(g)}\, \chi_\mu(g)\, \mathrm{d}g\,.$$

If λ and μ are trivial, this is the classical *Heine–Szegö* identity. Historically, a "Hermitian" precursor of this formula may be found in Heine's 1878 treatise on spherical functions, but the "unitary" version seems due to Szegö. The following proof of the general case is different from that given by Bump and Diaconis, who deduced this formula from the Jacobi–Trudi identity.

Proof. By the Weyl integration formula in the form (22.18), and the Weyl character formula in the form (22.17), we have

$$\int_{\mathrm{U}(n)} \Phi_{n,f}(g)\overline{\chi_\lambda(g)}\chi_\mu(g)\, \mathrm{d}g$$

$$= \frac{1}{n!} \int_{T} \Phi_{n,f}(t) \left(\sum_{w \in W} (-1)^{l(w)} \mathrm{e}^{w(\mu+\delta)} \right) \left(\sum_{w' \in W} (-1)^{l(w')} \mathrm{e}^{-w'(\lambda+\delta)} \right) \mathrm{d}t$$

$$= \frac{1}{n!} \int_{T} \Phi_{n,f}(t) \left(\sum_{w,w' \in W} (-1)^{l(w)+l(w')} \mathrm{e}^{w(\mu+\delta)-w'(\lambda+\delta)} \right) \mathrm{d}t.$$

Interchanging the order of summation and integration, replacing w by $w'w$, and then making the variable change $t \longmapsto w't$, we get

$$\frac{1}{n!} \sum_{w' \in W} \left[\sum_{w \in W} \int_{T} \Phi_{n,f}(t) \left((-1)^{l(w)} \mathrm{e}^{w(\mu+\delta)-\lambda-\delta} \right) \mathrm{d}t \right].$$

Each w' contributes equally, and we may simply drop the summation over w' and the $1/n!$ to get

$$\sum_{w \in W} \int_{T} \Phi_{n,f}(t) \left((-1)^{l(w)} \mathrm{e}^{w(\mu+\delta)-\lambda-\delta} \right) \mathrm{d}t.$$

Now, as a function on T, the weight $\mathrm{e}^{w(\mu+\delta)-\lambda-\delta}$ has the effect

$$\begin{pmatrix} t_1 & & \\ & \ddots & \\ & & t_n \end{pmatrix} \longmapsto \prod_{i=1}^{n} t_i^{\mu_{w(i)}+(n-w(i))-\lambda_i-(n-i)} = \prod_{i=1}^{n} t_i^{\mu_{w(i)}-w(i)-\lambda_i+i}.$$

Thus, the integral is

$$\sum_{w \in W} (-1)^{l(w)} \prod_{i=1}^{n} \int_{\mathbb{T}} \left(\sum_{-\infty}^{\infty} d_k t_i^k \right) t_i^{\mu_{w(i)} - w(i) - \lambda_i + i} dt$$

$$= \sum_{w \in W} (-1)^{l(w)} \prod_{i=1}^{n} d_{-\mu_{w(i)} + w(i) + \lambda_i - i}.$$

Since the Weyl group is S_n and $(-1)^{l(w)}$ is the sign character, by the definition of the determinant, this is the determinant $D_{n-1}^{\lambda,\mu}(f)$. $\square$

As we already mentioned, we will only consider here the special case where μ is $(0, \ldots, 0)$. We refer to [28] for the general case. If μ is trivial, then Theorem 42.1 reduces to the formula

$$D_{n-1}^{\lambda}(f) = \int_{U(n)} \Phi_{n,f}(g) \overline{\chi_\lambda(g)} \, dg, \tag{42.2}$$

where

$$D_{n-1}^{\lambda}(f) = \det(d_{\lambda_i - i + j}).$$

Theorem 42.2 (Szegö, Bump, Diaconis). *Let*

$$f(t) = \exp \left(\sum_{-\infty}^{\infty} c_k t^k \right),$$

where we assume that

$$\sum_k |c_k| < \infty, \qquad \text{and} \qquad \sum_k |k| |c_k|^2 < \infty.$$

Let λ be a partition of m. Let $\mathbf{s}_\lambda : S_k \longrightarrow \mathbb{Z}$ be the irreducible character associated with λ. If $\xi \in S_m$, let $\gamma_k(\xi)$ denote the number of k-cycles in the decomposition of ξ into a product of disjoint cycles, and define

$$\Delta(f, \xi) = \prod_{k=1}^{\infty} (k c_k)^{\gamma_k(\xi)}.$$

(The product is actually finite.) Then

$$D_{n-1}^{\lambda}(f) \sim \frac{1}{m!} \sum_{\xi \in S_m} \mathbf{s}_\lambda(\xi) \, \Delta(f, \xi) \exp \left(n c_0 + \sum_{k=1}^{\infty} k c_k c_{-k} \right).$$

Proof. Our assumption that $\sum |c_k| < \infty$ implies that

$$\int_{U(n)} \exp\left(\sum |c_k|\,|\mathrm{tr}(g^k)|\right) \mathrm{d}g < \infty,$$

which is enough to justify all of the following manipulations. (We will use the assumption that $\sum |kc_k|^2 < \infty$ later.)

First, take λ to be trivial, so that $m = 0$. This special case is Szegö's original theorem. By (42.2),

$$D_{n-1}(f) = \int_{U(n)} \exp\left(\sum_k c_k\,\mathrm{tr}(g^k)\right) \mathrm{d}g = \int_{U(n)} \prod_k \exp\left(c_k\,\mathrm{tr}(g^k)\right) \mathrm{d}g.$$

We can pull out the factor $\exp(nc_0)$ since $\mathrm{tr}(1) = n$, substitute the series expansion for the exponential function, and group together the contributions for k and $-k$. We get

$$e^{nc_0} \int_{U(n)} \prod_k \left[\sum_{\alpha_k=0}^{\infty} \frac{c_k^{\alpha_k}}{\alpha_k!}\mathrm{tr}(g^k)^{\alpha_k}\right] \left[\sum_{\alpha_k=0}^{\infty} \frac{c_{-k}^{\beta_k}}{\beta_k!}\overline{\mathrm{tr}(g^k)}^{\beta_k}\right] \mathrm{d}g$$

$$= e^{nc_0} \sum_{(\alpha_k)} \sum_{(\beta_k)} \int_{U(n)} \left(\prod_k \frac{c_k^{\alpha_k}}{\alpha_k!}\mathrm{tr}(g^k)^{\alpha_k}\right) \left(\prod_k \frac{c_{-k}^{\beta_k}}{\beta_k!}\overline{\mathrm{tr}(g^k)}^{\beta_k}\right) \mathrm{d}g,$$

where the sum is now over all sequences (α_k) and (β_k) of nonnegative integers. The integrand is multiplied by $e^{i\theta(\sum k\alpha_k - \sum k\beta_k)}$ when we multiply g by $e^{i\theta}$. This means that the integral is zero unless $\sum k\alpha_k = \sum k\beta_k$. Assuming this, we look more closely at these terms. By Theorem 37.1, in notation introduced in Chap. 39, the function $g \longmapsto \prod_k \mathrm{tr}(g^k)^{\alpha_k}$ is $\mathrm{Ch}^{(n)}(\mathbf{p}_\nu)$, where ν is a partition of $r = \sum k\alpha_k = \sum k\beta_k$ with $\alpha_k = \alpha_k(\nu)$ parts of size k, and similarly we will denote by σ the partition of r with β_k parts of size k. This point was discussed in the last chapter in connection with (39.7). We therefore obtain

$$D_{n-1}^{\lambda}(f) = e^{nc_0} \sum_{r=0}^{\infty} C(r,n),$$

where

$$C(r,n) = \sum_{\nu,\,\sigma \text{ partitions of } r} \left(\prod_k \frac{c_k^{\alpha_k} c_{-k}^{\beta_k}}{\alpha_k!\beta_k!}\right) \left\langle \mathrm{Ch}^{(n)}(\mathbf{p}_\nu), \mathrm{Ch}^{(n)}(\mathbf{p}_\sigma)\right\rangle.$$

Now consider the terms with $r \leqslant n$. When $r \leqslant n$, by Theorem 39.1, the characteristic map from $\mathcal{R}_r$ to the space of class functions in $L^2(G)$ is an isometry, and if $\nu = \nu'$, then by (37.2) we have

$$\left\langle \mathrm{Ch}^{(n)}(\mathbf{p}_\nu), \mathrm{Ch}^{(n)}(\mathbf{p}_\sigma)\right\rangle_{U(n)} = \langle \mathbf{p}_\nu, \mathbf{p}_\sigma\rangle_{S_r} = \begin{cases} z_\nu & \text{if } \nu = \sigma, \\ 0 & \text{otherwise.} \end{cases}$$

(This is the same fact we used in the proof of Proposition 39.4.) Thus, when $r \leqslant n$, we have $C(r, n) = C(r)$ where, using the explicit form (37.1) of z_ν, we have

$$C(r) = e^{nc_0} \sum_{\nu \text{ a partition of } r} z_\nu \left(\prod_k \frac{(c_k c_{-k})^{\alpha_k}}{(\alpha_k!)^2} \right)$$

$$= e^{nc_0} \sum_{\nu \text{ a partition of } r} \left(\prod_k \frac{(k c_k c_{-k})^{\alpha_k}}{\alpha_k!} \right).$$

Now

$$\sum_r C(r) = e^{nc_0} \prod_k \sum_{\alpha_k = 0}^{\infty} \left(\frac{(k c_k c_{-k})^{\alpha_k}}{\alpha_k!} \right) = e^{nc_0} \prod_k \exp(k c_k c_{-k}),$$

so as $n \longrightarrow \infty$, the series $\sum_r C(r, n)$ stabilizes to the series $\sum_r C(r)$ that converges to the right-hand side of (42.1).

To prove (42.1), we must bound the tails of the series $\sum_r C(r, n)$. It is enough to show that there exists an absolutely convergent series $\sum_r |D(r)| < \infty$ such that $|C(r, n)| \leqslant |D(r)|$. First, let us consider the case where $c_k = \overline{c_{-k}}$. In this case, we may take $D(r) = C(r)$. The absolute convergence of the series $\sum |D(r)|$ follows from our assumption that $\sum |k| \, |c_k|^2 < \infty$ and the Cauchy–Schwarz inequality. In this case,

$$C(r, n) = \left\| \sum_{\nu \text{ a partition of } r} \left(\prod_k \frac{c_k^{\alpha_k}}{\alpha_k!} \right) \mathrm{Ch}^{(n)}(\mathbf{p}_\nu) \right\|^2,$$

where, as before, $\alpha_k = \alpha_k(\nu)$ is the number of parts of size k of the partition ν and the inner product is taken in $U(n)$. Invoking the fact from Theorem 39.1 that the $\mathrm{Ch}^{(n)}$ is a contraction, this is bounded by

$$C(r, n) = \left\| \sum_{\nu \text{ a partition of } r} \left(\prod_k \frac{c_k^{\alpha_k}}{\alpha_k!} \right) \mathbf{p}_\nu \right\|^2,$$

where now the inner product is taken in S_r, and of course this is $C(r)$. If we do not assume $c_k = \overline{c_{-k}}$, we may use the Cauchy–Schwarz inequality and bound $C(r, n)$ by

$$\left\| \sum_{\nu \text{ a partition of } r} \left(\prod_k \frac{c_k^{\alpha_k}}{\alpha_k!} \right) \mathrm{Ch}^{(n)}(\mathbf{p}_\nu) \right\| \cdot \left\| \sum_{\sigma \text{ a partition of } r} \left(\prod_k \frac{c_{-k}^{\beta_k}}{\beta_k!} \right) \mathrm{Ch}^{(n)}(\mathbf{p}_\sigma) \right\|.$$

Each norm is dominated by the corresponding norm in R_k and, proceeding as before, we obtain the same bound with c_k replaced by $\max(|c_k|, |c_{-k}|)$.

Now (42.1) is proved, which is the special case with λ trivial. We turn now to the general case.

We will make use of the identity

$$\mathbf{s}_\lambda = \sum_{\mu \text{ a partition of } m} z_\mu^{-1} \mathbf{s}_\lambda(\xi_\mu)\, \mathbf{p}_\mu$$

in the ring of class functions on S_m, where for each μ, ξ_μ is a representative of the conjugacy class $\mathcal{C}_\mu$ of cycle type μ. This is clear since $z_\mu^{-1}\mathbf{p}_\mu$ is the characteristic function of $\mathcal{C}_\mu$, so this function has the correct value at every group element. Applying the characteristic map in the ring of class functions on $U(n)$, we have

$$\chi_\lambda = \sum_{\mu \text{ a partition of } m} z_\mu^{-1} \mathbf{s}_\lambda(\xi_\mu)\, \mathrm{Ch}^{(n)}(\mathbf{p}_\mu).$$

For each μ, let $\gamma_k(\xi_\mu)$ be the number of cycles of length k in the decomposition of ξ_μ into a product of disjoint cycles. By Theorem 37.1, we may write this identity

$$\chi_\lambda = \sum_{\mu \text{ a partition of } m} z_\mu^{-1} \mathbf{s}_\lambda(\xi_\mu) \prod_k \mathrm{tr}(g^k)^{\gamma_k(\xi_\mu)}.$$

Now, proceeding as before from (42.2), we see that $D_{n-1}^\lambda(f)$ equals

$$e^{nc_0} \sum_{\mu \text{ a partition of } m} z_\mu^{-1} \mathbf{s}_\lambda(\xi_\mu)$$

$$\times \int_{U(n)} \prod_k \left(\sum_{\alpha_k=0}^\infty \frac{c_k^{\alpha_k}}{\alpha_k!} \mathrm{tr}(g^k)^{\alpha_k} \right) \left(\sum_{\beta_k=0}^\infty \frac{c_{-k}^{\beta_k}}{\beta_k!} \overline{\mathrm{tr}(g^k)}^{\beta_k+\gamma_k(\xi_\mu)} \right) dg.$$

Since S_m contains $m!/z_\lambda$ elements of cycle type μ and $\mathbf{s}_\lambda$ has the same value $\mathbf{s}_\lambda(\xi_\mu)$ on all of them, we may write this as

$$e^{nc_0} \frac{1}{m!} \sum_{\xi \in S_m} \mathbf{s}_\lambda(\xi)$$

$$\times \sum_{(\alpha_k)} \sum_{(\beta_k)} \int_{U(n)} \left(\prod_k \frac{c_k^{\alpha_k}}{\alpha_k!} \mathrm{tr}(g^k)^{\alpha_k} \right) \left(\prod_k \frac{c_{-k}^{\beta_k}}{\beta_k!} \overline{\mathrm{tr}(g^k)}^{\beta_k+\gamma_k(\xi)} \right) dg.$$

As in the previous case, the contribution vanishes unless $\sum k\alpha_k = \sum k\beta_k + m$, and we assume this. We get

$$D_{n-1}^\lambda(f) = e^{nc_0} \frac{1}{m!} \sum_{\xi \in S_m} \mathbf{s}_\lambda(\xi) \sum_{r=0}^\infty C(r,n,\xi),$$

where now

$$C(r,n,\xi)$$

$$= \sum_{\substack{(\alpha_k) \\ \sum k\alpha_k = r+m}} \sum_{\substack{(\beta_k) \\ \sum k\beta_k = r}} \int_{U(n)} \left(\prod_k \frac{c_k^{\alpha_k}}{\alpha_k!} \mathrm{tr}(g^k)^{\alpha_k} \right) \left(\prod_k \frac{c_{-k}^{\beta_k}}{\beta_k!} \overline{\mathrm{tr}(g^k)}^{\beta_k+\gamma_k(\xi)} \right) dg.$$

If $r \leqslant n$, then (as before) the contribution is zero unless $\alpha_k = \beta_k + \gamma_k$. In this case,

$$\int_{U(n)} \prod_k \mathrm{tr}(g^k)^{\alpha_k} \overline{\mathrm{tr}(g^k)}^{\beta_k + \gamma_k} \, dg = \prod_k (\beta_k + \gamma_k)! \, k^{\beta_k + \gamma_k},$$

and using this value, we see that when $r \leqslant n$ we have $C(r, n, \xi) = C(r, \xi)$, where

$$C(r, \xi) = \Delta(f, \xi) \sum_{\substack{(\beta_k) \\ \sum k\beta_k = r}} \frac{(kc_k c_{-k})^{\beta_k}}{\beta_k!}.$$

The series is

$$\sum_r C(r, \xi) = \Delta(f, \xi) \exp\left(nc_0 + \sum_{k=1}^{\infty} kc_k c_{-k} \right),$$

so the result will follow as before if we can show that $|C(r, n, \xi)| < |D(r, \xi)|$ where $\sum |D(r, \xi)| < \infty$. The method is the same as before, based on the fact that the characteristic map is a contraction, and we leave it to the reader.   □

Exercises

Exercise 42.1 (Bump et al. [30]).

(i) If f is a continuous function on $\mathbb{T}$, show that there is a well-defined continuous function $u_f : U(n) \longrightarrow U(n)$ such that if $t_i \in \mathbb{T}$ and $h \in U(n)$, we have

$$u_f\left(h \begin{pmatrix} t_1 & \\ & \ddots \\ & & t_n \end{pmatrix} h^{-1} \right) = h \begin{pmatrix} f(t_1) & \\ & \ddots \\ & & f(t_n) \end{pmatrix} h^{-1}.$$

(ii) If g is an $n \times n$ matrix, with $n \geqslant m$, let $E_m(g)$ denote the sum of the $\binom{n}{m}$ principal $m \times m$ minors of g. Thus, if $n = 4$, then $E_2(g)$ is

$$\begin{vmatrix} g_{11} & g_{12} \\ g_{21} & g_{22} \end{vmatrix} + \begin{vmatrix} g_{11} & g_{13} \\ g_{31} & g_{33} \end{vmatrix} + \begin{vmatrix} g_{11} & g_{14} \\ g_{41} & g_{44} \end{vmatrix} + \begin{vmatrix} g_{22} & g_{23} \\ g_{32} & g_{33} \end{vmatrix} + \begin{vmatrix} g_{22} & g_{24} \\ g_{42} & g_{44} \end{vmatrix} + \begin{vmatrix} g_{33} & g_{34} \\ g_{43} & g_{44} \end{vmatrix}.$$

Prove that if $f(t) = \sum d_k t^k$, then

$$\int_{U(n)} E_m(u_f(g)) \, \overline{\chi_\lambda(g)} \, \chi_\mu(g) \, dg = E_m(T_{n-1}^{\mu, \lambda}),$$

where $T_{n-1}^{\mu, \lambda}$ is the $n \times n$ matrix whose i, jth entry is $d_{\lambda_i - \mu_j - i + j}$. (**Hint:** Deduce this from the special case $m = n$.)

43

The Involution Model for S_k

Let $\sigma_1 = 1$, $\sigma_2 = (12)$, $\sigma_3 = (12)(34),\dots$ be the conjugacy classes of involutions in S_k. It was shown by Klyachko and by Inglis et al. [82] that it is possible to specify a set of characters $\psi_1, \psi_2, \psi_3, \dots$ of degree 1 of the centralizers of $\sigma_1, \sigma_2, \sigma_3, \dots$ such that the direct sum of the induced representations of the ψ_i contains every irreducible representation exactly once. In the next chapter, we will see that translating this fact and related ones to the unitary group gives classical facts about symmetric and exterior algebra decompositions due to Littlewood [120].

If (π, V) is a self-contragredient irreducible complex representation of a compact group G, we may classify π as *orthogonal (real)* or *symplectic (quaternionic)*. We will now explain this classification due to Frobenius and Schur [52]. We recall that the *contragredient representation* to (π, V) is the representation $\hat{\pi} : G \longrightarrow \mathrm{GL}(V^*)$ on the dual space V^* of V defined by $\hat{\pi}(g) = \pi(g^{-1})^*$, which is the adjoint of $\pi(g^{-1})$. Its character is the complex conjugate of the character of π.

Proposition 43.1. *The irreducible complex representation π is self-contragredient if and only if there exists a nondegenerate bilinear form $B : V \times V \longrightarrow \mathbb{C}$ such that*

$$B\big(\pi(g)v, \pi(g)w\big) = B(v, w). \tag{43.1}$$

The form B is unique up to a scalar multiple. We have $B(w, v) = \epsilon B(v, w)$, where $\epsilon = \pm 1$.

Proof. To emphasize the symmetry between V and V^*, let us write the dual pairing $V \times V^* \longrightarrow \mathbb{C}$ in the symmetrical form $L(v) = [\![v, L]\!]$. The contragredient representation thus satisfies $[\![\pi(g)v, L]\!] = [\![v, \hat{\pi}(g^{-1})L]\!]$, or $[\![\pi(g)v, \hat{\pi}(g)L]\!] = [\![v, L]\!]$. Any bilinear form $B : V \times V \longrightarrow \mathbb{C}$ is of the form $B(v, w) = [\![v, \lambda(w)]\!]$, where $\lambda : V \longrightarrow V^*$ is a linear isomorphism. It is clear that (43.1) is satisfied if and only if λ intertwines π and $\hat{\pi}$.

Since π and $\hat{\pi}$ are irreducible, Schur's lemma implies that λ, if it exists, is unique up to a scalar multiple, and the same conclusion follows for B.

D. Bump, *Lie Groups*, Graduate Texts in Mathematics 225,
DOI 10.1007/978-1-4614-8024-2_43, © Springer Science+Business Media New York 2013

Now $(v, w) \mapsto B(w, v)$ has the same property as B, and so $B(w, v) = \epsilon B(v, w)$ for some constant ϵ. Applying this identity twice, $\epsilon^2 B(v, w) = B(v, w)$ so $\epsilon = \pm 1$.   □

If (π, V) is self-contragredient, let ϵ_π be the constant ϵ in Proposition 43.1; otherwise let $\epsilon_\pi = 0$. If $\epsilon_\pi = 1$, then we say that π is *orthogonal* or *real*; if $\epsilon_\pi = -1$, we say that π is *symplectic* or *quaternionic*. We call ϵ_π the *Frobenius-Schur indicator* of π.

Theorem 43.1 (Frobenius and Schur). *Let* (π, V) *be an irreducible representation of the compact group* G. *Then*

$$\epsilon_\pi = \int_G \chi(g^2)\, dg.$$

Proof. We have $p_2 = h_2 - e_2$ in $\Lambda^{(n)}$. Indeed, $p_2(x_1, \ldots, x_n)$ equals

$$\sum_i x_i^2 = \left(\sum_i x_i^2 + \sum_{i<j} x_i x_j \right) - \left(\sum_{i<j} x_i x_j \right)$$
$$= h_2(x_1, \ldots, x_n) - e_2(x_1, \ldots, x_n).$$

By (33.8) and Proposition 33.2, this means that

$$\chi(g^2) = \mathrm{tr}\big(\vee^2 \pi(g) \big) - \mathrm{tr}\big(\wedge^2 \pi(g) \big).$$

We see that ϵ_π is

$$\int_G \mathrm{tr}\big(\vee^2 \pi(g) \big)\, dg - \int_G \mathrm{tr}\big(\wedge^2 \pi(g) \big)\, dg.$$

Thus, what we need to know is that $\vee^2 \pi(g)$ contains the trivial representation if and only if $\epsilon_\pi = 1$, while $\wedge^2 \pi(g)$ contains the trivial representation if and only if $\epsilon_\pi = -1$.

If $\vee^2 \pi(g)$ contains the trivial representation, let $\xi \in \vee^2 V$ be a $\vee^2 \pi(g)$-fixed vector. Let $\langle\,,\,\rangle$ be a G-invariant inner product on V. There is induced a G-invariant Hermitian inner product on $\vee^2 V$ such that $\langle v_1 \vee v_2, w_1 \vee w_2 \rangle = \langle v_1, v_2 \rangle \langle w_1, w_2 \rangle$, and we may define a symmetric bilinear form on V by $B(v, w) = \langle v \vee w, \xi \rangle$. Thus, $\epsilon_\pi = 1$.

Conversely, if $\epsilon_\pi = 1$, let B be a symmetric invariant bilinear form. By the universal property of the symmetric square, there exists a linear form $L : \vee^2 V \longrightarrow \mathbb{C}$ such that $B(v, w) = L(v \vee w)$, and hence a vector $\xi \in \vee^2 V$ such that $B(v, w) = \langle v \vee w, \xi \rangle$, which is a $\vee^2 \pi(g)$-fixed vector.

The case where $\epsilon_\pi = -1$ is identical using the exterior square.   □

Proposition 43.2. *Let* (π, V) *be an irreducible complex representation of the compact group* G. *Then* π *is the complexification of a real representation if and only if* $\epsilon_\pi = 1$. *If this is true,* $\pi(G)$ *is conjugate to a subgroup of the orthogonal group* $\mathrm{O}(n)$.

Proof. First, suppose that $\pi : G \longrightarrow \mathrm{GL}(V)$ is the complexification of a real representation. This means that there exists a real vector space V_0 and a homomorphism $\pi_0 : G \longrightarrow \mathrm{GL}(V_0)$ such that $V \cong \mathbb{C} \otimes_{\mathbb{R}} V_0$ as G-modules. Every compact subgroup of $\mathrm{GL}(V_0) \cong \mathrm{GL}(n, \mathbb{R})$ is conjugate to a subgroup of $O(n)$. Indeed, if $\langle\!\langle \, , \, \rangle\!\rangle$ is a positive definite symmetric bilinear form on V_0, then averaging it gives another positive definite symmetric bilinear form

$$B_0(v, w) = \int_G \langle\!\langle \pi_0(g)v, \pi_0(g)w \rangle\!\rangle \, dg$$

that is G-invariant. Choosing a basis of V_0 that is orthonormal with respect to this basis, the matrices of $\pi_0(g)$ will all be orthogonal. Extending B_0 by linearity to a symmetric bilinear form on V, which we identify with $\mathbb{C} \otimes V_0$, gives a symmetric bilinear form showing that $\epsilon_\pi = 1$.

Conversely, if $\epsilon_\pi = 1$, there exists a G-invariant symmetric bilinear form B on V. We will make use of both B and a G-invariant inner product $\langle \, , \, \rangle$ on V. They differ in that B is linear in the second variable, while the inner product is conjugate linear. If $w \in V$, consider the linear functional $v \mapsto B(v, w)$. Every linear functional is the inner product with a unique element of V, so there exists $\lambda(w) \in V$ such that $B(v, w) = \langle v, \lambda(w) \rangle$. The map $\lambda : V \longrightarrow V$ is $\mathbb{R}$-linear but not $\mathbb{C}$-linear; in fact, it is complex antilinear. Let $V_0 = \{v \in V \mid \lambda(v) = v\}$. It is a real vector space. We may write every element $v \in V$ as a sum $v = u + iw$, where $u, w \in V_0$, taking $u = \frac{1}{2}(v + \lambda(v))$ and $w = \frac{1}{2i}(v - \lambda(v))$. This decomposition is unique since $\lambda(v) = u - iw$, and we may solve for u and w. Therefore, $V = V_0 \oplus iV_0$ and V is the complexification of V_0. Since B and H are both G-invariant, it is easy to see that $\lambda \circ \pi(g) = \pi(g) \circ \lambda$, so π leaves V_0-invariant and induces a real representation with the complexification π. $\qquad\square$

Theorem 43.2. *Let G be a finite group. Let $\mu : G \longrightarrow \mathbb{C}$ be the sum of the irreducible characters of G.*

(i) *Suppose that $\epsilon_\pi = 1$ for every irreducible representation π. Then, for any $g \in G$, $\mu(g)$ is the number of solutions to the equation $x^2 = g$ in G.*

(ii) *Suppose that $\mu(1)$ is the number of solutions to the equation $x^2 = 1$. Then $\epsilon_\pi = 1$ for all irreducible representations π.*

Proof. If π is an irreducible representation of G, let χ_π be its character. We will show

$$\sum_{\text{irreducible } \pi} \chi_\pi(g)\epsilon_{\hat\pi} = \#\{x \in G \mid x^2 = g\}. \tag{43.2}$$

Indeed, by Theorem 43.1, the left-hand side equals

$$\sum_\chi \chi(g) \frac{1}{|G|} \sum_{x \in G} \overline{\chi(x^2)} = \sum_{x \in G} \left[\frac{1}{|G|} \sum_\chi \chi(g) \overline{\chi(x^2)} \right].$$

Let C be the conjugacy class of g. By Schur orthogonality, the expression in brackets equals $1/|C|$ if x^2 is conjugate to g and zero otherwise. Each element of the conjugacy class will have the same number of square roots, so counting the number of solutions to $x^2 \sim g$ (where $\sim$ denotes conjugation) and then dividing by $|C|$ gives the number of solutions to $x^2 = g$. This proves (43.2).

Now (43.2) clearly implies (i). It also implies (ii) because, taking $g = 1$, each coefficient $\chi_\pi(1)$ is a positive integer, so

$$\sum_{\text{irreducible } \pi} \chi_\pi(1) \, \epsilon_{\hat{\pi}} = \sum_{\text{irreducible } \pi} \chi_\pi(1)$$

is only possible if all ϵ_π are equal to 1. $\qquad\square$

Let K be a field and F a subfield. Let V be a K-vector space. If $\pi : G \longrightarrow GL(V)$ is a representation of a group G over K, we say that π is *defined over* F if there exists an F-vector space V_0 and a representation $\pi_0 : G \longrightarrow GL(V_0)$ over F such that π is isomorphic to the representation of G on the K-vector space $K \otimes_F V_0$. The dimension over K of V must clearly equal the dimension of V_0 as an F-vector space.

Theorem 43.3. *Every irreducible representation of S_k is defined over $\mathbb{Q}$.*

Proof. The construction of Theorem 35.1 contained no reference to the ground field and works just as well over $\mathbb{Q}$. Specifically, our formulation of Mackey theory was valid over an arbitrary field, so if λ and μ are conjugate partitions, the computation of Proposition 35.5 shows that there is a unique intertwining operator $\text{Ind}_{S_\lambda}^{S_k}(\varepsilon) \longrightarrow \text{Ind}_{S_\mu}^{S_k}(1)$, where we are now considering representations over $\mathbb{Q}$. The image of this intertwining operator is a rational representation which has a complexification that is the representation ρ_λ of S_k parametrized by λ. $\qquad\square$

In this chapter, we will call an element $x \in G$ an *involution* if $x^2 = 1$. Thus, the identity element is considered an involution by this definition. If $G = S_k$, then by Theorem 43.3 every irreducible representation is defined over $\mathbb{Q}$, a fortiori over $\mathbb{R}$, and so by Theorem 43.2 we have $\epsilon_\pi = 1$ for all irreducible representations π. Therefore, the number of involutions is equal to the sum of the degrees of the irreducible characters, and moreover the sum of the irreducible characters evaluated at $g \in S_k$ equals the number of solutions to $x^2 = g$. In particular, it is a nonnegative integer.

It is possible to prove that the sum of the degrees of the irreducible representations of G is equal to the number of involutions when $G = S_k$ using the Robinson–Schensted correspondence (see Knuth [109], Sect. 5.1.4, or Stanley [153], Corollary 7.13.9). Indeed, both numbers are equal to the number of standard tableaux.

Let G be a group (such as S_k) having the property that all $\epsilon_\pi = 1$, so the number of involutions of G is the sum of the degrees of the irreducible representations. Let $x_1, \ldots, x_h$ be representatives of the conjugacy classes of involutions. The cardinality of a conjugacy class x is the index of its centralizer $C_G(x)$, so $\sum[G : C_G(x_i)]$ is the number of involutions of G. Since this is the sum of the degrees of the irreducible characters of G, it becomes a natural question to ask whether we may specify characters ψ_i of degree 1 of $C_G(x_i)$ such that the direct sum of the induced characters ψ_i^G contains each irreducible character exactly once. If so, these data comprise an *involution model* for G. Involution models do not always exist, even if all $\epsilon_\pi = 1$.

A complete set of representatives of the conjugacy classes of S_k are 1, (12), (12)(34), ... To describe their centralizers, we first begin with the involution $(12)(34)(56)\cdots(2r-1,2r) \in S_{2r}$. Its centralizer, as described in Proposition 37.1, has order $2^r r!$. It has a normal subgroup of order 2^r generated by the transpositions (12), (34), ..., and the quotient is isomorphic to S_r. We denote this group B_{2r}. It is isomorphic to the Weyl group of Cartan type B_r.

Now consider the centralizer in S_k of $(12)(34)\cdots(2r-1,2r)$ where $2r < k$. It is contained in $S_{2r} \times S_{k-2r}$, where the second S_{k-2r} acts on $\{2r+1, 2r+2, \ldots, k\}$ and equals $B_{2r} \times S_{k-2r}$. The theorem of Klyachko, Inglis, Richardson, and Saxl is that we may specify characters of these groups with inductions to S_k that contain every irreducible character exactly once. There are two ways of doing this: we may put the alternating character on S_{k-2k} and the trivial character on B_{2r}, or conversely we may put the alternating character (restricted from S_{2r}) on B_{2r} and the trivial character on S_{k-2r}.

Let ω_{2r} be the character of S_{2r} induced from the trivial character of B_{2r}.

Proposition 43.3. *The restriction of ω_{2r} to S_{2r-1} is isomorphic to the character of S_{2r-1} induced from the character ω_{2r-2} to S_{2r-1}.*

Proof. First, let us show that $B_{2r} \backslash S_{2r} / S_{2r-1}$ consists of a single double coset. Indeed, S_{2r} acts transitively on $X = \{1, 2, \ldots, 2r\}$, and the stabilizer of $2r$ is S_{2r-1}. Therefore, we can identify S_{2r}/S_{2r-1} with X and $B_{2r} \backslash S_{2r} / S_{2r-1}$ with $B_{2r} \backslash X$. Since B_{2r} acts transitively on X, the claim is proved.

Thus, we can compute the restriction of ω_{2r} to S_{2r-1} by Corollary 32.2 to Theorem 32.2, taking $H_1 = B_{2r}$, $H_2 = S_{2r-1}$, $G = S_{2r}$, $\pi = 1$, with $\gamma = 1$ the only double coset representative. We see that the restriction of $\omega_{2r} = \mathrm{Ind}_{H_1}^G(1)$ is the same as the induction of 1 from $H_\gamma = B_{2r} \cap S_{2r-1} = B_{2r-2}$ to H_2. Inducing in stages first from B_{2r-2} to S_{2r-2} and then to S_{2r-1}, this is the same as the character of S_{2r-1} induced from ω_{2r-2}. $\qquad\square$

We are preparing to compute ω_{2r}. The key observation of Inglis, Richardson, and Saxl is that Proposition 43.3, plus purely combinatorial considerations, contains enough information to do this.

We call a partition $\lambda = (\lambda_1, \lambda_2, \ldots)$ *even* if every λ_i is an even integer.

If λ is a partition, let $R_i\lambda = (\lambda_1, \lambda_2, \ldots, \lambda_{i-1}, \lambda_i + 1, \lambda_{i+1}, \ldots)$ be the result of incrementing the ith part. In applying this *raising operator*, we must always check that the resulting sequence is a partition. For this, we need either $i = 1$ or $\lambda_i < \lambda_{i-1}$. Similarly, we have the *lowering operator* $L_i\lambda = (\lambda_1, \lambda_2, \ldots, \lambda_{i-1}, \lambda_i - 1, \lambda_{i+1}, \ldots)$, which is a partition if $\lambda_i > \lambda_{i+1}$.

Lemma 43.1. *Every partition of $2r - 1$ having exactly one odd part is contained in a unique even partition of $2r$.*

Proof. Let μ be a partition of $2r - 1$ having exactly one odd part μ_i. The unique even partition of $2r$ containing μ is $R_i\mu$. Note that this is a partition since $i = 1$ or $\mu_i < \mu_{i-1}$. (We cannot have μ_i and μ_{i-1} both equal since one is odd and the other is even.) $\square$

Proposition 43.4. *Let S be a set of partitions of $2r$. Assume that:*

(i) *Each partition of $2r - 1$ contained in an element of S has exactly one odd part;*

(ii) *Each partition of $2r - 1$ with exactly one odd part is contained in a unique element of S;*

(iii) *The trivial partition $(2r) \in S$.*

Then S consists of the set S_0 of even partitions of $2r$.

Proof. First, we show that S contains S_0. Assume on the contrary that $\lambda \in S_0$ is not in S. We assume that the counterexample λ is minimal with respect to the partial order, so if $\lambda' \in S_0$ with $\lambda' \prec \lambda$, then $\lambda' \in S$. Let $i = l(\lambda)$. We note that $i > 1$ since if $i = 1$, then λ is the unique partition of $2r$ of length 1, namely $(2r)$, which is impossible since $\lambda \notin S$ while $(2r) \in S$ by assumption (iii).

Let $\mu = L_i\lambda$. It is a partition since we are decrementing the last nonzero part of λ. It has a unique odd part μ_i, so by (ii) there is a unique $\tau \in S$ such that $\mu \subset \tau$. Evidently, $\tau = R_j\mu$ for some j. Let us consider what j can be.

We show first that j cannot be $> i$. If it were, we would have $j = i + 1$ because i is the length of μ and λ. Now assuming $\tau = R_{i+1}\mu = R_{i+1}L_i\lambda$, we can obtain a contradiction as follows. We have $\tau_{i-1} = \lambda_{i-1} \geqslant \lambda_i > \lambda_i - 1 = \tau_i$, so $\nu = L_{i-1}\tau$ is a partition. It has three odd parts, namely ν_{i-1}, ν_i and ν_{i+1}. This contradicts (i) for $\nu \subset \tau \in S$.

Also j cannot equal i. If it did, we would have $\tau = R_iL_i\lambda = \lambda$, a contradiction since $\tau \in S$ while $\lambda \notin S$.

Therefore, $j < i$. Let $\sigma = R_jL_i\tau = R_j^2L_i^2\lambda$. Note that σ is a partition. Indeed, either $j = 1$ or else $\tau_j \neq \tau_{j-1}$ since one is odd and the other one is even, and we are therefore permitted to apply R_j. Furthermore, $\tau_i \neq \tau_{i+1}$ since one is odd and the other one is even, so we are permitted to apply L_i.

Since λ is even, σ is even, and since $j < i$, $\sigma \prec \lambda$. By our induction hypothesis, this implies that $\sigma \in S$. Now let $\theta = L_i\tau = L_j\sigma$. This is easily seen to be a partition with exactly one odd part (namely θ_j), and it is contained in two distinct elements of S, namely τ and σ. This contradicts (ii).

This contradiction shows that $S \supset S_0$. We can now show that $S = S_0$. Otherwise, S contains S_0 and some other partition $\lambda \notin S_0$. Let μ be any partition of $2r - 1$ contained in λ. Then μ has exactly one odd part by (i), so by Lemma 43.1 it is contained in some element $\lambda' \in S_0 \subset S$. Since $\lambda \notin S_0$, λ and λ' are distinct elements of S both containing μ, contradicting (ii). $\square$

Theorem 43.4. *The character ω_{2r} of S_{2r} is multiplicity-free. It is the sum of all irreducible characters $\mathbf{s}_\lambda$ with λ an even partition of $2r$.*

Proof. By induction, we may assume that this is true for S_{2r-2}. The restriction of ω_{2r} to S_{2r-1} is the same as the character induced from w_{2r-2} by Proposition 43.3. Using the branching rule for the symmetric groups, its irreducible constituents consist of all $\mathbf{s}_\mu$, where μ is a partition of S_{2r-1} containing an even partition of $2r-2$, and clearly this is the set of partitions of $2r-1$ having exactly one odd part. There are no repetitions.

We see immediately that ω_{2r} is multiplicity-free since its restriction to S_{2r-1} is multiplicity-free. Let S be the set of partitions λ of $2r$ such that $\mathbf{s}_\lambda$ is contained in ω_{2r}. Again using the branching rule for symmetric groups, we see that this set satisfies conditions (i) and (ii) of Proposition 43.4 and condition (iii) is clear by Frobenius reciprocity. The result now follows from Proposition 43.4. $\qquad\qquad\square$

We may now show that S_k has an involution model. The centralizer of the involution $(12)(34)\cdots(2r-1,r)$ is $B_{2r}\times S_{k-2r}$.

Theorem 43.5 (Klyachko, Inglis, Richardson, and Saxl). *Every irreducible character of S_k occurs with multiplicity 1 in the sum*

$$\bigoplus_{2r\leqslant k} \operatorname{Ind}_{B_{2r}\times S_{k-2r}}^{S_k}(1\otimes\varepsilon),$$

where ε is the alternating character of S_{k-2r}.

Proof. We will show that $\operatorname{Ind}_{B_{2r}\times S_{k-2r}}^{S_k}(1\otimes\varepsilon)$ is the sum of the $\mathbf{s}_\lambda$ as λ runs through the partitions of k having exactly $k-2r$ odd parts. Indeed, it is obvious that if λ is a partition of k, there is a unique even partition μ such that $\lambda\supset\mu$ and $\lambda\backslash\mu$ is a vertical strip; the partition μ is obtained by decrementing each odd part of λ. Since ω_{2r} is the sum of all $\mathbf{s}_\lambda$ where λ is a partition of $2r$ into even parts, it follows from Pieri's formula that the character $\omega_{2r}\mathbf{e}_{k-2r}$ is the sum of all $\mathbf{s}_\lambda$ where λ is a partition of k having exactly $k-2r$ odd parts.

We note that the number of odd parts of any partition λ of k is congruent to k modulo 2 because $k=\sum\lambda_i$. The result follows by summing over r. $\qquad\square$

Exercises

The first exercise generalizes Theorem 43.1 of Frobenius and Schur. Suppose that G is a compact group and $\theta:G\longrightarrow G$ is an involution (i.e,, an automorphism satisfying $\theta^2=1$). Let (π,V) be an irreducible representation of G. If $\pi\cong{}^\theta\pi$, where ${}^\theta\pi:V\longrightarrow V$ is the "twisted" representation ${}^\theta\pi(g)=\pi({}^\theta g)$, then by an obvious variant of Proposition 43.1 there exists a symmetric bilinear form $B:V\times V\longrightarrow\mathbb{C}$ such that

$$B_\theta\big(\pi(g)v, \pi({}^\theta g)w\big) = B_\theta(v, w). \tag{43.3}$$

In this case, the *twisted Frobenius–Schur indicator* $\epsilon_\theta(\pi)$ is defined to be the constant equal to ± 1 such that

$$B(v, w) = \epsilon_\theta(\pi)B(w, v).$$

If $\pi \not\cong {}^\theta\pi$ we define $\epsilon_\theta(\pi) = 0$. The goal is to prove the following theorem.

Theorem (Kawanaka and Matsuyama [96]). *Let G be a compact group and θ an involution of G. Let (π, V) be an irreducible representation with character χ. Then*

$$\epsilon_\theta(\pi) = \int_G \chi(g \cdot {}^\theta g)\, dg. \tag{43.4}$$

Exercise 43.1. Assuming the hypotheses of the stated theorem, define a group H that is the semidirect product of G by a cyclic group $\langle t \rangle$ generated by an element t of order 2 such that $tgt^{-1} = {}^\theta g$ for $g \in G$. Thus, the index $[G : H] = 2$. The idea is to use Theorem 43.1 for the group H to obtain the theorem of Kawanaka and Matsuyama for G. Proceed as follows.

Case 1: Assume that $\pi \cong {}^\theta\pi$. In this case, show that there exists an endomorphism $T : V \longrightarrow V$ such that $T \circ \pi(g) = \pi({}^\theta g) \circ T$ and $T^2 = 1_V$. Extend π to a representation π_H of H such that $\pi_H(t) = T$. Let $B_\theta : V \times V \longrightarrow \mathbb{C}$ satisfy (43.3). Then $B(v, w) = B_\theta(v, Tw)$ satisfies (43.1), as does $B(Tv, Tw) = B_\theta(Tv, w)$. Thus, there exists a constant δ such that $B(Tv, Tw) = \delta B(v, w)$. Show that $\delta^2 = 1$ and that

$$\epsilon_\theta(\pi) = \delta\epsilon(\pi). \tag{43.5}$$

Apply Theorem 43.1 to the representation π_H, bearing in mind that the Haar measure on H restricted to G is only half the Haar measure on G because both measures are normalized to have total volume 1. This gives

$$\epsilon(\pi_H) = \frac{1}{2}\left(\epsilon(\pi) + \int_G \chi(g \cdot {}^\theta g)\, dg\right). \tag{43.6}$$

Now observe that if π_H is self-contragredient, then the nondegenerate form that it stabilizes must be a multiple of B. Deduce that if $\delta = 1$ then π_H is self-contragredient and $\epsilon(\pi_H) = \epsilon(\pi)$, while if $\delta = -1$, then $\epsilon(\pi_H) = 0$. In either case reconcile, (43.5) and (43.6) to prove (43.4).

Case 2: Assume that $\pi \not\cong {}^\theta\pi$. In this case, show that the induced representation $\mathrm{Ind}_G^H(\pi)$ is irreducible and call it π_H. Show that

$$\epsilon(\pi_H) = e(\pi) + \int_G \chi(g \cdot {}^\theta g)\, dg.$$

Show using direct constructions with bilinear forms on V and V^H that if either $\epsilon(\pi)$ or $\epsilon_\theta(\pi)$ is nonzero, then π_H is self-contragredient, while if π_H is self-contragredient, then exactly one of $\epsilon(\pi)$ or $\epsilon_\theta(\pi)$ is nonzero, and whichever one is nonzero equals $\epsilon(\pi_H)$.

Exercise 43.2. Let G be a finite group and let θ be an involution. Let $\mu : G \longrightarrow \mathbb{C}$ be the sum of the irreducible characters of G. If $\mu(1)$ equals the number of solutions to the equation $x \cdot {}^{\theta}x = 1$, then show that $\epsilon_{\theta}(\pi) = 1$ for all irreducible representations π. If this is true, show that $\mu(g)$ equals the number of solutions to $x \cdot {}^{\theta}x = g$ for all $g \in G$.

For example, if $G = \mathrm{GL}(n, \mathbb{F}_q)$, it was shown independently by Gow [57] and Klyachko [103] that the conclusions to Exercise 43.2 are satisfied when $G = \mathrm{GL}(n, \mathbb{F}_q)$ and θ is the automorphism $g \longmapsto {}^{t}g^{-1}$.

For the next group of exercises, the group B_{2k} is a Coxeter group with generators

$$(13)(24), \quad (35)(46), \quad \ldots \quad , (2k-3, 2k-1)(2k-2, 2k)$$

and $(2k-1, 2k)$. It is thus a Weyl group of Cartan type B_k with order $k!2^k$. It has a linear character ξ_{2k} having value -1 on these "simple reflections." This is the character $(-1)^{l(w)}$ of Proposition 20.12. Let $\eta_{2k} = \mathrm{Ind}_{B_{2k}}^{S_{2k}}(\xi_{2k})$ be the character of S_{2k} induced from this linear character of B_{2k}. The goal of this exercise will be to prove analogs of Theorem 43.4 and the other results of this chapter for η_{2k}.

Exercise 43.3. Prove the analog of Proposition 43.3. That is, show that inducing the restriction of η_{2r} to S_{2r-1} is isomorphic to the character of S_{2r-1} induced from the character η_{2r-2} to S_{2r-1}.

Let $\mathcal{S}_{2k}$ be the set of characters s_{λ} of S_{2k} where λ is a partition of $2k$ such that if μ is the conjugate partition, then $\mu_i = \lambda_i + 1$ for all i such that $\lambda_i \geqslant i$. For example, the partition $\lambda = (5, 5, 4, 3, 3, 2)$ has conjugate $(6, 6, 5, 3, 2)$, and the hypothesis is satisfied. Visually, this assumption means that the diagram of λ can be assembled from two congruent pieces, as in Fig. 43.1. We will describe these as the "top piece" and the "bottom piece," respectively.

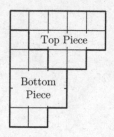

Fig. 43.1. The diagram of a partition of class $\mathcal{S}_{2k}$ when $k = 11$

Let $\mathcal{T}_{2k+1}$ be the set of partitions of $2k+1$ with a diagram that contains an element of $\mathcal{S}_{2k}$.

Exercise 43.4. Prove that if $\lambda \in \mathcal{T}_{2k+1}$, then there are unique partitions $\mu \in \mathcal{S}_{2k}$ and $\nu \in \mathcal{S}_{2k+2}$ such that the diagram of λ contains the diagram of μ and is contained in the diagram of ν. (**Hint:** The diagrams of the skew partitions $\lambda - \mu$ and $\nu - \lambda$, each consisting of a single node, must be corresponding nodes of the top piece and bottom piece.)

Exercise 43.5. Let Σ be a set of partitions of $2k + 2$. Assume that each partition λ of $2k + 1$ is contained in an element of Σ if and only if $\lambda \in \mathcal{T}_{2k+1}$, in which case it is contained in a unique element of Σ. Show that $\Sigma = S_{2k+2}$. [This is an analog of Proposition 43.4. It is not necessary to assume any condition corresponding to (iii) of the proposition.]

Exercise 43.6. Show that η_{2k} is multiplicity-free and that the representations occurring in it are precisely the $\mathbf{s}_\lambda$ with $\lambda \in \mathcal{S}_{2k}$.

44

Some Symmetric Algebras

The results of the last chapter can be translated into statements about the representation theory of $U(n)$. For example, we will see that each irreducible representation of $U(n)$ occurs exactly once in the decomposition of the symmetric algebra of $V \oplus \wedge^2 V$, where $V = \mathbb{C}^n$ is the standard module of $U(n)$. The results of this chapter are also proved in Goodman and Wallach [56], Howe [77], Littlewood [120], and Macdonald [124]. See Theorem 26.6 and the exercises to Chap. 26 for alternative proofs of some of these results.

Let us recall some ideas that already appeared in Chap. 38. If $\rho : G \longrightarrow GL(V)$ is a representation, then $\bigvee V$ and $\bigwedge V$ become modules for G and we may ask for their decomposition into irreducible representations of V. For some representations ρ, this question will have a simple answer, and for others the answer will be complex. The very simplest case is where $G = GL(V)$. In this case, each $\vee^k V$ is itself irreducible, and each $\wedge^k V$ is either irreducible (if $k < \dim(V)$) or zero.

We can encode the solution to this question with generating functions

$$P_\rho^\vee(g;t) = \sum_{k=0}^{\infty} \operatorname{tr}\left(g \mid \vee^k V\right) t^k, \qquad P_\rho^\wedge(g;t) = \sum_{k=0}^{\infty} \operatorname{tr}\left(g \wedge^k V\right) t^k.$$

Proposition 44.1. *Suppose that* $\rho : G \longrightarrow GL(V)$ *is a representation and* $\gamma_1, \ldots, \gamma_d$ *are the eigenvalues of* $\rho(g)$. *Then*

$$P_\rho^\vee(g,t) = \prod_i (1 - t\gamma_i)^{-1}, \qquad P_\rho^\wedge(g,t) = \prod_i (1 + t\gamma_i). \qquad (44.1)$$

Proof. The traces of $\rho(g)$ on $\vee^k V$ and $\wedge^k V$ are

$$h_k(\gamma_1, \ldots, \gamma_d) \qquad \text{and} \qquad e_k(\gamma_1, \ldots, \gamma_d),$$

so this is a restatement of (33.1) and (33.2). $\qquad \square$

D. Bump, *Lie Groups*, Graduate Texts in Mathematics 225,
DOI 10.1007/978-1-4614-8024-2_44, © Springer Science+Business Media New York 2013

We see that for all g, $P_\rho^\vee(g,t)$ is convergent if $t < \max(|\gamma_i|^{-1})$ and has meromorphic continuation in t, while $P_\rho^\wedge(g,t)$ is a polynomial in t of degree equal to the dimension of V. We will denote $P_\rho^\vee(g) = P_\rho^\vee(g,1)$ and $P_\rho^\wedge(g) = P_\rho^\wedge(g,1)$. Then we specialize $t = 1$ in (44.1) and write

$$P_\rho^\vee(g) = \prod_i (1 - \gamma_i)^{-1}, \qquad P_\rho^\wedge(g) = \prod_i (1 + \gamma_i). \qquad (44.2)$$

For the first equation, this is understood to be an analytic continuation since the series defining $P_\rho^\vee$ might not converge when $t = 1$.

Proposition 44.2. *Let* $V = \mathbb{C}^n$ *be regarded as a* $\mathrm{GL}(n,\mathbb{C})$-*module in the usual way. Then*

$$\vee^k(\vee^2 V) \cong \left(\bigotimes^{2k} V \right) \otimes_{\mathbb{C}[S_{2k}]} \omega_{2k}$$

as $\mathrm{GL}(n,\mathbb{C})$-*modules. It is the direct sum of the* π_λ *as* λ *runs through all even partitions of* k.

Proof. Let $\mathbb{C}_{\text{trivial}}$ denote $\mathbb{C}$ denoted as a trivial module of $\mathbb{C}[B_{2k}]$. It is sufficient to prove that

$$\vee^k(\vee^2 V) \cong \left(\bigotimes^{2k} V \right) \otimes_{\mathbb{C}[B_{2k}]} \mathbb{C}_{\text{trivial}} \qquad (44.3)$$

as $\mathrm{GL}(n,\mathbb{C})$-modules. Indeed, assuming this, the right-hand side is isomorphic to

$$\left(\left(\bigotimes^{2k} V \right) \otimes_{\mathbb{C}[S_{2k}]} \mathbb{C}_{[S_{2k}]} \right) \otimes_{\mathbb{C}[B_{2k}]} \mathbb{C}_{\text{trivial}} \cong$$
$$\left(\bigotimes^{2k} V \right) \otimes_{\mathbb{C}[S_{2k}]} \left(\mathbb{C}_{[S_{2k}]} \otimes_{\mathbb{C}[B_{2k}]} \mathbb{C}_{\text{trivial}} \right) \cong \left(\bigotimes^{2k} V \right) \otimes_{\mathbb{C}[S_{2k}]} \omega_{2k}.$$

To prove (44.3), we will use the universal properties of the symmetric power and tensor products to construct inverse maps

$$\vee^k(\vee^2 V) \longleftrightarrow \left(\bigotimes^{2k} V \right) \otimes_{\mathbb{C}[B_{2k}]} \mathbb{C}_{\text{trivial}}.$$

Here $B_{2k} \subset S_{2k}$ acts on $\bigotimes^k V$ on the right by the action (34.1).

First, we note that the map

$$(v_1, \ldots, v_{2k}) \longmapsto (v_1 \vee v_2) \vee \cdots \vee (v_{2k-1} \vee v_{2k})$$

commutes with the right action of B_{2k}. It is $2k$-linear and hence induces a map

$$\alpha : \bigotimes^{2k} V \longrightarrow \vee^k(\vee^2 V), \qquad \alpha(v_1 \otimes \cdots \otimes v_{2k}) = (v_1 \vee v_2) \vee \cdots \vee (v_{2k-1} \vee v_{2k}),$$

and $\alpha(\xi\sigma) = \alpha(\xi)$ for $\sigma \in B_{2k}$. Thus, the map

$$\left(\bigotimes^{2k} V\right) \times \mathbb{C}_{\text{trivial}} \longrightarrow \vee^k(\vee^2 V), \qquad (\xi, t) \mapsto t\alpha(\xi),$$

is $\mathbb{C}[B_{2k}]$-balanced and there is an induced map

$$\left(\bigotimes^{2k} V\right) \otimes_{\mathbb{C}[B_{2k}]} \mathbb{C}_{\text{trivial}} \longrightarrow \vee^k(\vee^2 V),$$
$$(v_1 \otimes \cdots \otimes v_{2k}) \otimes t \mapsto t(v_1 \vee v_2) \vee \cdots \vee (v_{2k-1} \vee v_{2k}).$$

As for the other direction, we first note that for $v_3, v_4, \ldots, v_{2k}$ fixed, using the fact that $\otimes_{\mathbb{C}[B_{2k}]}$ is B_{2k}-balanced, the map

$$(v_1, v_2) \longmapsto (v_1 \otimes v_2 \otimes v_3 \otimes \cdots \otimes v_{2k}) \otimes 1 \in \left(\bigotimes^{2k} V\right) \otimes_{\mathbb{C}[B_{2k}]} \mathbb{C}_{\text{trivial}}$$

is symmetric and bilinear, so there is induced a map

$$\mu_{v_3, v_4, \ldots, v_{2k}} : \vee^2 V \longrightarrow \left(\bigotimes^{2k} V\right) \otimes_{\mathbb{C}[B_{2k}]} \mathbb{C}_{\text{trivial}},$$
$$\mu_{v_3, \ldots, v_4}(v_1 \vee v_2) = (v_1 \otimes v_2 \otimes v_3 \otimes \cdots \otimes v_{2k}) \otimes 1.$$

Now with $\xi_1 \in \vee^2 V$ and $v_5, \ldots, v_{2k}$ fixed, the map

$$(v_3, v_4) \longmapsto \mu_{v_3, v_4, \ldots, v_{2k}}(\xi_1)$$

is symmetric and bilinear, so there is induced a map

$$\nu_{\xi, v_5, \ldots, v_{2k}} : \vee^2 V \longrightarrow \left(\bigotimes^{2k} V\right) \otimes_{\mathbb{C}[B_{2k}]} \mathbb{C}_{\text{trivial}},$$
$$\nu_{\xi, v_5, \ldots, v_{2k}}(v_3 \vee v_4) = \mu_{v_3, v_4, \ldots, v_{2k}}(\xi_1).$$

With $v_5, \ldots, v_{2k}$ fixed, denote by

$$\mu_{v_5, \ldots, v_{2k}} : \vee^2 V \times \vee^2 V \longrightarrow \left(\bigotimes^{2k} V\right) \otimes_{\mathbb{C}[B_{2k}]} \mathbb{C}_{\text{trivial}}$$

the map $\mu_{v_5, \ldots, v_{2k}}(\xi_1, \xi_2) = \nu_{\xi_1, v_5, \ldots, v_{2k}}(\xi_2)$. Continuing in this way, we eventually construct a k-linear map $\mu : \vee^2 V \times \cdots \vee^2 V \longrightarrow \left(\bigotimes^{2k} V\right) \otimes_{\mathbb{C}[B_{2k}]} \mathbb{C}_{\text{trivial}}$ such that

$$\mu(v_1 \vee v_2, \ldots, v_{2k-1} \vee v_{2k}) = (v_1 \otimes \cdots \otimes v_{2k}) \otimes 1.$$

Using the fact that $\otimes_{\mathbb{C}[B_{2k}]}$ is B_{2k}-balanced, the map μ is symmetric and hence induces a map $\vee^k(\vee^2 V) \longrightarrow \left(\bigotimes^{2k} V\right) \otimes_{\mathbb{C}[B_{2k}]} \mathbb{C}_{\text{trivial}}$ that is the inverse of the map previously constructed. We have now proved (44.3). $\qquad\square$

Theorem 44.1. *Let $V = \mathbb{C}^n$ be regarded as a $\mathrm{GL}(n,\mathbb{C})$-module in the usual way. Then*

$$\vee^k \left(\vee^2 V \right) \cong \bigoplus_{\lambda \text{ an even partition of } 2k} \pi_\lambda.$$

Proof. This follows from Proposition 44.2, Theorem 36.4, and the explicit decomposition of Theorem 43.4.   □

Theorem 44.2 (D. E. Littlewood). *Let $\alpha_1, \ldots, \alpha_n$ be complex numbers, $|\alpha_i| < 1$. Then*

$$\prod_{1 \leqslant i \leqslant j \leqslant n} (1 - \alpha_i \alpha_j)^{-1} = \sum_{\lambda \text{ even}} s_\lambda(\alpha_1, \ldots, \alpha_n). \qquad (44.4)$$

The sum is over even partitions.

Proof. This follows on applying (44.2) to the symmetric square representation by using Proposition 44.1 and the explicit decomposition of Theorem 44.2.   □

Theorem 44.3 (D. E. Littlewood). *Let $\alpha_1, \ldots, \alpha_n$ be complex numbers, $|\alpha_i| < 1$. Then*

$$\left[\prod_{1 \leqslant i \leqslant n} (1 + \alpha_i) \right] \left[\prod_{1 \leqslant i \leqslant j \leqslant n} (1 - \alpha_i \alpha_j)^{-1} \right] = \sum_\lambda s_\lambda(\alpha_1, \ldots, \alpha_n).$$

The sum is over all partitions.

Proof. The coefficient of t^k in

$$\left[\prod_{1 \leqslant i \leqslant n} (1 + t\alpha_i) \right] \left[\prod_{1 \leqslant i \leqslant j \leqslant n} (1 - t^2 \alpha_i \alpha_j)^{-1} \right]$$

$$= \left[\sum_k e_k t^k \right] \left[\sum_{\lambda \text{ an even partition of } 2r} s_\lambda t^{2r} \right]$$

is

$$\sum_{2r \leqslant k} e_{k-2r}(\alpha_1, \ldots, \alpha_n) \sum_{\lambda \text{ an even partition of } 2r} s_\lambda.$$

This is the image of $\mathbf{e}_{k-2r}\omega_{2r}$ under the characteristic map, and it equals the sum of the s_λ for all partitions of k by Theorem 43.5. Taking $t = 1$, the result follows.   □

A *polynomial character* of $GL(n, \mathbb{C})$ is one with matrix coefficients that are polynomials in the coordinates functions g_{ij} not involving $\det^{-1}$. As we know, they are exactly the characters of π_λ where $\lambda = (\lambda_1, \ldots, \lambda_n)$ is a partition. We may express Theorem 44.3 as saying that every polynomial character of $GL(n, \mathbb{C})$ occurs exactly once in the algebra $(\bigwedge V) \otimes \bigvee(\bigvee^2 V)$, which is the tensor product of the exterior algebra over V with the symmetric algebra over the exterior square representation.

There are dual forms of these results. Let $\tilde{\omega}_{2k} = \mathrm{Ind}_{B_{2k}}^{S_{2k}}(\varepsilon)$ be the character of S_{2k} obtained by inducing the alternating character ε from B_{2k}.

Proposition 44.3. *The character $\tilde{\omega}_{2k}$ is the sum of the s_λ, where λ runs through all the partitions of k such that the conjugate partition λ^t is even.*

Proof. This may be deduced from Theorem 43.4 as follows. Applying this with $G = S_{2k}$, $H = B_{2k}$, and $\rho = \varepsilon$, we see that $\tilde{\omega}_{2k}$ is the same as ω_{2k} multiplied by the character ε. By Theorem 37.4, this is ${}^t\omega_{2k}$, and by Theorems 43.4, and 35.2, this is the sum of the s_λ with λ^t even. $\square$

Theorem 44.4. *Let $V = \mathbb{C}^n$ be regarded as a $GL(n, \mathbb{C})$-module in the usual way. Then*

$$\bigvee^k(\wedge^2 V) \simeq \left(\bigotimes^{2k} V\right) \otimes_{\mathbb{C}[S_{2k}]} \tilde{\omega}_{2k}$$

as $GL(n, \mathbb{C})$-modules. It is the direct sum of the π_λ as λ runs through all conjugates of even partitions of k.

Proof. Similar to Theorem 44.2. $\square$

Theorem 44.5 (D. E. Littlewood). *Let $\alpha_1, \ldots, \alpha_n$ be complex numbers, $|\alpha_i| < 1$. Then*

$$\prod_{1 \leqslant i < j \leqslant n} (1 - \alpha_i \alpha_j)^{-1} = \sum_{\lambda^t \ even} s_\lambda(\alpha_1, \ldots, \alpha_n). \qquad (44.5)$$

The sum is over even partitions.

Proof. Similar to Theorem 44.2. $\square$

Theorem 44.6 (D. E. Littlewood). *Let $\alpha_1, \ldots, \alpha_n$ be complex numbers, $|\alpha_i| < 1$. Then*

$$\left[\prod_{1 \leqslant i \leqslant n} (1 - \alpha_i)^{-1}\right] \left[\prod_{1 \leqslant i < j \leqslant n} (1 - \alpha_i \alpha_j)^{-1}\right] = \sum_\lambda s_\lambda(\alpha_1, \ldots, \alpha_n).$$

The sum is over all partitions.

Proof. Similar to Theorem 44.3, and actually equivalent to Theorem 44.3 using the identity $(1 + \alpha_i)(1 - \alpha_i^2)^{-1} = (1 - \alpha_i)^{-1}$. $\square$

Exercises

Exercise 44.1. Let η_{2k} be the character of S_{2k} from the exercises of the last chapter, and let $\mathcal{S}_{2k}$ be the set of partitions of $2k$ defined there. Show that

$$\wedge^k(\wedge^2 V) \cong \left(\bigotimes^{2k} V\right) \otimes_{\mathbb{C}[S_{2k}]} \eta_{2k},$$

and deduce that

$$\wedge^k(\wedge^2 V) \cong \bigoplus_{\lambda \in \mathcal{S}_{2k}} \pi_\lambda.$$

Prove also that

$$\wedge^k(\vee^2 V) \cong \bigoplus_{{}^t\lambda \in \mathcal{S}_{2k}} \pi_\lambda.$$

Exercise 44.2. Prove the identities

$$\prod_{1 \leqslant i < j \leqslant n} (1 + \alpha_i \alpha_j) = \sum_k \sum_{\lambda \in \mathcal{S}_{2k}} s_\lambda(\alpha_1, \ldots, \alpha_n),$$

$$\prod_{1 \leqslant i \leqslant j \leqslant n} (1 + \alpha_i \alpha_j) = \sum_k \sum_{{}^t\lambda \in \mathcal{S}_{2k}} s_\lambda(\alpha_1, \ldots, \alpha_n).$$

Explain why, in contrast with (44.4) and (44.5), there are only finitely many nonzero terms on the right-hand side in these identities.

45

Gelfand Pairs

We recall that a representation θ of a compact group G is called *multiplicity-free* if in its decomposition into irreducibles,

$$\theta = \bigoplus_i d_i \pi_i, \tag{45.1}$$

each irreducible representation π_i occurs with multiplicity $d_i = 0$ or 1. A common situation that we have seen already several times is for a group $G \supset H$ to have the property that for some representation τ of H the induced representation $\mathrm{Ind}_H^G(\tau)$ is multiplicity-free.

In this chapter we will see how the question of showing that a representation is multiplicity-free leads to the consideration of a *Hecke algebra*. If the Hecke algebra is commutative, the representation is multiplicity free. If it is not commutative, it may also have an interesting structure, as we will see in the next chapter. Another approach to multiplicity-free representations may be seen in Theorem 26.7

Of course, we have only defined induced representations when H and G are finite. Assuming H and G are finite, saying that $\mathrm{Ind}_H^G(\tau)$ is multiplicity-free means that each irreducible representation π of G, when restricted to H, can contain at most one copy of τ, and formulated this way, the statement makes sense even if H and G are infinite.

The most striking examples we have seen are when $H = S_{k-1}$ and $G = S_k$ and when $H = \mathrm{U}(n-1)$ and $G = \mathrm{U}(n)$. In these examples *every* irreducible representation τ of H has this "multiplicity one" property. Such examples are fairly rare. A far more common circumstance is for a single representation τ of H to have the multiplicity one property. For example, we showed in Theorem 43.4 that inducing the trivial representation from the group B_{2k} of S_{2k} produces a multiplicity-free representation. However, this would not be true for some other irreducible representations.

D. Bump, *Lie Groups*, Graduate Texts in Mathematics 225,
DOI 10.1007/978-1-4614-8024-2_45, © Springer Science+Business Media New York 2013

Proposition 45.1. *Suppose θ is a representation of a finite group G. A necessary and sufficient condition that θ be multiplicity-free is that the ring $\mathrm{End}_G(\theta)$ be commutative.*

Proof. In the decomposition (45.1), we have $\mathrm{End}_G(\theta) = \bigoplus \mathrm{Mat}_{d_i}(\mathbb{C})$. This is commutative if and only if all $d_i \leqslant 1$. $\qquad\square$

Let G be a group, finite for the time being, and H a subgroup. Then (G, H) is called a *Gelfand pair* if the representation of G induced by the trivial representation of H is multiplicity-free. We also refer to H as a *Gelfand subgroup*. More generally, if π is an irreducible representation of H, then (G, H, π) is called a *Gelfand triple* if π^G is multiplicity-free. See Gross [59] for a lively discussion of Gelfand pairs.

From Proposition 45.1, Gelfand pairs are characterized by the commutativity of the endomorphism ring of an induced representation. To study it, we make use of Mackey theory.

Proposition 45.2. *Let G be a finite group, and let H_1, H_2, H_3 be subgroups. Let (π_i, V_i) be complex representations of H_1, H_2, and H_3 and let $L_1 : V_1^G \longrightarrow V_2^G$ and $L_2 : V_2^G \longrightarrow V_3^G$ be intertwining operators. Let $\Delta_1 : G \longrightarrow \mathrm{Hom}(V_1, V_2)$ and $\Delta_2 : G \longrightarrow \mathrm{Hom}(V_2, V_3)$ correspond to L_1 and L_2 as in Theorem 32.1. Then $\Delta_2 * \Delta_1 : G \longrightarrow \mathrm{Hom}(V_1, V_3)$ corresponds to $L_2 \circ L_1 : V_1^G \longrightarrow V_3^G$, where the convolution is*

$$\Delta_2 * \Delta_1(g) = \sum_{\gamma \in H_2 \backslash G} \Delta_2(g\gamma^{-1}) \circ \Delta_1(\gamma).$$

Proof. Note that, using (32.9), the summand $\Delta_2(g\gamma^{-1})\Delta_1(\gamma)$ does not depend on the choice of representative $\gamma \in H_2 \backslash G$. The result is easily checked. $\qquad\square$

Theorem 45.1. *Let H be a subgroup of the finite group G, and let (π, V) be a representation of H. Then (G, H, π) is a Gelfand triple if and only if the convolution algebra $\mathcal{H}$ of functions $\Delta : G \longrightarrow \mathrm{End}_{\mathbb{C}}(V)$ satisfying*

$$\Delta(h_2 g h_1) = \pi(h_2) \circ \Delta(g) \circ \pi(h_1), \qquad h_1, h_2 \in H,$$

is commutative.

We call a convolution ring $\mathcal{H}$ of this type a *Hecke algebra*.

Proof. By Proposition 45.2, this condition is equivalent to the commutativity of the endomorphism ring $\mathrm{End}_G(V^G)$, so this follows from Proposition 45.1. $\qquad\square$

In this chapter, an *involution* of a group G is a map $\iota : G \to G$ of order 2 that is anticommutative:

$${}^{\iota}(g_1 g_2) = {}^{\iota}g_2 \, {}^{\iota}g_1.$$

Similarly, an *involution* of a ring R is an additive map of order 2 that is anticommutative for the ring multiplication.

A common method of proving that such a ring is commutative is to exhibit an involution and then show that this involution reduces to the identity map.

Theorem 45.2. *Let H be a subgroup of the finite group G, and suppose that G admits an involution fixing H such that every double coset of H is invariant: $HgH = H\iota g H$. Then H is a Gelfand subgroup.*

Proof. The ring $\mathcal{H}$ of Theorem 45.1 is just the convolution ring of H-bi-invariant functions on G. We have an involution on this ring:

$$\iota\Delta(g) = \Delta(\iota g).$$

It is easy to check that

$$\iota(\Delta_1 * \Delta_2) = \iota\Delta_2 * \iota\Delta_1.$$

On the other hand, each Δ is constant on each double coset, and these are invariant under ι by hypothesis, so ι is the identity map. This proves that $\mathcal{H}$ is commutative, so (G, H) is a Gelfand pair. $\square$

Let S_n denote the symmetric group. We can embed $S_n \times S_m \to S_{n+m}$ by letting S_n act on the first n elements of the set $\{1, 2, 3, \ldots, n+m\}$ and letting S_m act on the last m elements.

Proposition 45.3. *The subgroup $S_n \times S_m$ is a Gelfand subgroup of S_{n+m}.*

We already know this: the representation of S_{n+m} induced from the trivial character of $S_n \times S_m$ is the product in the ring $\mathcal{R}$ of $\mathbf{h}_n$ by $\mathbf{h}_m$. By Pieri's formula, one computes, assuming without loss of generality that $n > m$,

$$\mathbf{h}_n \mathbf{h}_m = \sum_{k=0}^{m} \mathbf{s}_{(n+m-k,k)}.$$

Thus, the induced representation is multiplicity-free. We prove this again to illustrate Theorem 45.2.

Proof. Let $H = S_n \times S_m$ and $G = S_{n+m}$. We take the involution ι in Theorem 45.2 to be the inverse map $g \longrightarrow g^{-1}$. We must check that each double coset is ι-stable.

It will be convenient to represent elements of S_{n+m} by permutation matrices. We will show that each double coset HgH has a representative of the form

$$\begin{pmatrix} I_r & 0 & 0 & 0 \\ 0 & 0_{n-r} & 0 & I_{n-r} \\ 0 & 0 & I_{m-n+r} & 0 \\ 0 & I_{n-r} & 0 & 0_{n-r} \end{pmatrix}. \tag{45.2}$$

Here I_n and 0_n are the $n \times n$ identity and zero matrices, and the remaining 0 matrices are rectangular blocks.

We start with g in block form,

$$\begin{pmatrix} A & B \\ C & D \end{pmatrix},$$

where A, B, C, and D are *subpermutation matrices*—that is, matrices with only 1's and 0's, and with at most one nonzero entry in each row and column. Here A is $n \times n$ and D is $m \times m$. Let r be the rank of A. Then clearly B and C both must have rank $n - r$, and so D has rank $m - n + r$.

Multiplying A on the left by an element of S_n, we may arrange its rows so that its nonzero entries lie in the first r rows. Then multiplying on the right by an element of S_n, we may put these in the upper left-hand corner. Similarly, we may arrange that D has its nonzero entries in the upper left-hand corner. Now the form of the matrix is

$$\begin{pmatrix} T_r & 0 & 0 & 0 \\ 0 & 0_{n-r} & 0 & U_{n-r} \\ 0 & 0 & V_{m-n+r} & 0 \\ 0 & W_{n-r} & 0 & 0_{n-r} \end{pmatrix},$$

where the sizes of the square blocks are indicated by subscripts. The matrices T, U, V, and W are permutation matrices (invertible). Left multiplication by element of $S_r \times S_{n-r} \times S_{m-n+r} \times S_{n-r}$ can now replace these four matrices by identity matrices. This proves that (45.2) is a complete set of double coset representatives.

Since these double coset representatives are all invariant under the involution, by Theorem 45.2 it follows that $S_n \times S_m$ is a Gelfand subgroup.   □

Proposition 45.4. *Suppose that* (G, H, ψ) *is a Gelfand triple, and let* (π, V) *be an irreducible representation of* G. *Then there exists at most one space* $\mathcal{M}$ *of functions on* G *satisfying*

$$M(hg) = \psi(h)M(g), \qquad (h \in H), \tag{45.3}$$

such that $\mathcal{M}$ *is closed under right translation and such that the representation of* G *on* $\mathcal{M}$ *by right translation is isomorphic to* π.

The space $\mathcal{M}$ is called a *model* of π, meaning a concrete realization of the representation in a space of functions on G.

Proof. This is just the Frobenius reciprocity. The space of functions satisfying (45.3) is $\mathrm{Ind}_H^G(\psi)$, so $\mathcal{M}$, if it exists, is the image of an element of $\mathrm{Hom}_G(V, \mathrm{Ind}_H^G(\psi))$. This is one-dimensional since the induced representation is assumed to be multiplicity-free.   □

We turn now to Gelfand pairs in compact groups. We will obtain a result similar to Theorem 45.1 by a different method.

Let $C(G)$ be the space of continuous functions on the compact group G. It is a ring (without unit) under convolution. If $\phi \in C(G)$, and if (π, V) is a finite-dimensional representation, let $\pi(\phi) : V \longrightarrow V$ denote the endomorphism

$$\pi(\phi) v = \int_G \phi(g) \, \pi(g) \, v \, dg.$$

One checks easily that if $\phi, \psi \in C(G)$, then

$$\pi(\phi * \psi) = \pi(\phi) \circ \pi(\psi).$$

Let H be a closed subgroup of G. Let $\mathcal{H}$ be the subring of $C(G)$ consisting of functions that are both left- and right-invariant under H. If (π, V) is a representation of G, let V^H denote the space of H-fixed vectors.

Theorem 45.3. *Let H be a closed subgroup of the compact group G. Let $\mathcal{H}$ be the subring of $C(G)$ consisting of functions that are both left- and right-invariant under H. If $\mathcal{H}$ is commutative, then V^H is at most one-dimensional for every irreducible representation (π, V) of G.*

In this case, extending the definition from the case of finite groups, we say (G, H) is a *Gelfand pair* or that H is a *Gelfand subgroup* of G.

Proof. Let $\xi, \eta \in V^H$. For $g \in G$, let

$$\phi_{\xi,\eta}(g) = \overline{\langle \pi(g)\xi, \eta \rangle},$$

where $\langle \, , \, \rangle$ is an invariant inner product on V (Proposition 2.1). It is easy to see that $\phi_{\xi,\eta} \in \mathcal{H}$. We will prove that

$$\pi(\phi_{\xi,\eta}) \, v = \tfrac{1}{\dim(V)} \, \langle v, \xi \rangle \, \eta. \tag{45.4}$$

Indeed, taking the inner product of the left-hand side with an arbitrary vector $\theta \in V$, Schur orthogonality (Theorem 2.4) gives

$$\langle \pi(\phi_{\xi,\eta})v, \theta \rangle = \int_G \langle \pi(g)\,v, \theta \rangle \, \overline{\langle \pi(g)\xi, \eta \rangle} \, dg = \tfrac{1}{\dim(V)} \, \langle v, \xi \rangle \, \langle \eta, \theta \rangle,$$

and since this is true for every θ, we have (45.4).

Now we show that the image of $\pi(\phi_{\eta,\xi} * \phi_{\xi,\eta})$ is $\mathbb{C}\eta$. Indeed, applying (45.4) twice, we see that

$$\pi(\phi_{\eta,\xi} * \phi_{\xi,\eta}) \, v = \pi(\phi_{\eta,\xi}) \circ \pi(\phi_{\xi,\eta}) \, v = \tfrac{1}{\dim(V)^2} \, \langle v, \xi \rangle \, \langle \eta, \eta \rangle \, \xi.$$

The image of this is contained in the linear span of η, and taking $v = \xi$ shows that the map is nonzero. Since $\mathcal{H}$ is assumed commutative, this also equals $\pi(\phi_{\xi,\eta} * \phi_{\eta,\xi})$. Hence, its image is also equal to $\mathbb{C}\xi$, and so we see that ξ and η both belong to the same one-dimensional subspace of V. $\square$

To give an example where we can verify the hypotheses of Theorem 45.3, let $G = SO(n+1)$, and let $H = SO(n)$, which we embed into the upper left-hand corner of G:

$$g \longmapsto \begin{pmatrix} g & 0 \\ 0 & 1 \end{pmatrix}.$$

We also embed $K = SO(2)$ into the lower right-hand corner:

$$\begin{pmatrix} a & b \\ -b & a \end{pmatrix} \longmapsto \begin{pmatrix} \begin{array}{c|cc} I_{n-1} & \multicolumn{2}{c}{0} \\ \hline & a & b \\ 0 & -b & a \end{array} \end{pmatrix}. \tag{45.5}$$

Proposition 45.5. *With $G = SO(n+1)$, $H = SO(n)$, and $K = SO(2)$ embedded as explained above, every double coset in $H \backslash G / H$ has a representative in K.*

Proof. Let $g \in G$. Write the last column of g in the form

$$\begin{pmatrix} bv_1 \\ bv_2 \\ \vdots \\ bv_n \\ a \end{pmatrix} = \begin{pmatrix} bv \\ a \end{pmatrix}, \qquad v = \begin{pmatrix} v_1 \\ \vdots \\ v_n \end{pmatrix},$$

where $b^2 + a^2 = 1$ and v has length 1. Complete v to an orthogonal matrix $h \in H$. Then it is simple to check that the last column of $h^{-1}g$ is

$$\begin{pmatrix} 0 \\ \vdots \\ 0 \\ b \\ a \end{pmatrix},$$

so with k the matrix in (45.5), the last column of $k^{-1}h^{-1}g$ is

$$\xi_0 = \begin{pmatrix} 0 \\ \vdots \\ 0 \\ 1 \end{pmatrix}. \tag{45.6}$$

This implies that $k^{-1}h^{-1}g \in O(n)$, so g and k lie in the same double coset. $\square$

Theorem 45.4. *The subgroup $SO(n)$ of $SO(n+1)$ is a Gelfand subgroup.*

Proof. With $G = SO(n+1)$, $H = SO(n)$, and $K = SO(2)$ embedded as explained above, we exhibit an involution of G, namely

$$g \mapsto \begin{pmatrix} I_n & \\ & -1 \end{pmatrix} {}^t g \begin{pmatrix} I_n & \\ & -1 \end{pmatrix}.$$

This involution maps H to itself and is the identity on matrices in $O(2)$. Hence, the involution of $\mathcal{H}$ that it induces is the identity, and $\mathcal{H}$ is therefore commutative. $\qquad\square$

Now let us think a bit about what this means in concrete terms. The quotient G/H may be identified with the sphere S^n. Indeed, thinking of S^n as the unit sphere in $\mathbb{R}^{n+1}$, G acts transitively and H is the stabilizer of a point in S^n.

Consequently, we have an action of G on $L^2(S^n)$, and this may be thought of as the representation induced from the trivial representation of $O(n)$.

Theorem 45.5. *Let (π, V) be an irreducible representation of $O(n+1)$. Then there exists at most one subspace of $L^2(S^n)$ that is invariant under the action of $O(n+1)$ and affords a representation isomorphic to π.*

This gives us a concrete *model* for at least some representations of $O(n+1)$.

Proof. Let $\phi : V \to L^2(S^n)$ be an intertwining operator. It is sufficient to show that ϕ is uniquely determined up to a constant multiple. The $O(n+1)$-equivariance of ϕ amounts to the formula

$$\phi\big(\pi(g)\,v\big)(x) = \phi(v)(g^{-1}x) \tag{45.7}$$

for $g \in O(n+1)$, $v \in V$, and $x \in S^n$.

Let $\langle \cdot, \cdot \rangle$ be an invariant Hermitian form on V. This form is nondegenerate, so each linear functional on V is of the form $v \to \langle v, \eta \rangle$ for some vector η. In particular, with $\xi_0 \in S^n$ as in (45.6), there exists a vector $\eta \in V$ such that

$$\phi(v)(\xi_0) = \langle v, \eta \rangle.$$

By (45.7), we have

$$\phi(v)\big(\pi(g)\,\xi_0\big) = \big\langle \pi(g^{-1})v, \eta \big\rangle = \langle v, \pi(g)\eta \rangle.$$

This makes it clear that ϕ is determined by η, and it also shows that η is $O(n)$-invariant since $\xi_0 \in S^n$ is $O(n)$-fixed. Since the space of $O(n)$-fixed vectors is at most one-dimensional, the theorem is proved. $\qquad\square$

Proposition 45.6. *If $g \in U(n)$, then there exist k_1 and $k_2 \in O(n)$ such that $k_1 g k_2$ is diagonal.*

Proof. Let $x = g\,{}^t g$. This is a unitary symmetric matrix. By Proposition 28.2, there exists $k_1 \in \mathrm{O}(n)$ such that $k_1 x k_1^{-1}$ is diagonal. It is unitary, so its diagonal entries have absolute value 1. Taking their square roots, we find a unitary diagonal matrix d such that $k_1 x k_1^{-1} = d^2$. This means that $(d^{-1} k_1 g)^t (d^{-1} k_1 g) = 1$, so $k_2^{-1} = d^{-1} k_1 g$ is orthogonal and $k_1 g k_2 = d$. $\square$

Theorem 45.6. *The group* $\mathrm{O}(n)$ *is a Gelfand subgroup of* $\mathrm{U}(n)$.

Proof. Let $G = \mathrm{U}(n)$ and $H = \mathrm{O}(n)$, and let $\mathcal{H}$ be the ring of Theorem 45.3. The transpose involution of G preserves H and thus induces an involution of $\mathcal{H}$. By Proposition 45.6, every double coset in $H\backslash G/H$ has a diagonal representative, so this involution is the identity map, and it follows that $\mathcal{H}$ is commutative. Therefore, H is a Gelfand subgroup. $\square$

Exercises

Exercise 45.1. Let G be any compact group. Let $H = G \times G$, and embed G into H diagonally, that is, by the map $g \longmapsto (g, g)$. Use the involution method to prove that G is a Gelfand subgroup of H.

Exercise 45.2. Use the involution method to show that $\mathrm{O}(n)$ is a Gelfand subgroup of $U(n)$.

Exercise 45.3. Show that each irreducible representation of $\mathrm{O}(3)$ has an $\mathrm{O}(2)$-fixed vector, and deduce that $L^2(S^2)$ is the (Hilbert space) direct sum of all irreducible representations of $\mathrm{O}(3)$, each with multiplicity one.

Exercise 45.4 (Gelfand and Graev). Let $G = \mathrm{GL}(n, \mathbb{F}_q)$ and let N be the subgroup of upper triangular unipotent matrices. Let $\psi : \mathbb{F}_q \longrightarrow \mathbb{C}^\times$ be a nontrivial additive character. Define a character ψ_N of N by

$$\psi_N \begin{pmatrix} 1 & x_{12} & x_{13} & \cdots & x_{1n} \\ & 1 & x_{23} & \cdots & x_{2n} \\ & & 1 & & \\ & & & \ddots & \vdots \\ & & & & 1 \end{pmatrix} = \psi(x_{12} + x_{23} + \cdots + x_{n-1,n}).$$

The object of this exercise is to show that $\mathrm{Ind}_G^N(\psi_N)$ is multiplicity-free. This *Gelfand–Graev representation* is important because it contains *most* irreducible representations of the group; those it contains are therefore called *generic*. We will denote by Φ the root system of $\mathrm{GL}(n, \mathbb{F}_q)$ and by Φ^+ the positive roots α_{ij} such that $i < j$. Let Σ be the simple positive roots $\alpha_{i,i+1}$.

(i) Show that each double coset in $N\backslash G/N$ has a representative m that is a *monomial matrix*. In the notation of Chap. 27, this means that $m \in N(T)$, where T is the group of diagonal matrices. (Make use of the Bruhat decomposition.) Let $w \in W = N(T)/T$ be the corresponding Weyl group element.

(ii) Suppose that the double coset of NwN supports an intertwining operator $\mathrm{Ind}(\psi_N) \longrightarrow \mathrm{Ind}(\psi_N)$. (See Remark 32.2.) Show that if $\alpha \in \Sigma$ and $w(\alpha) \in \Phi^+$, then $w(\alpha) \in \Sigma$. (Otherwise, choose x in the unipotent subgroup corresponding to the root α such that $mx = ym$ with $\psi_N(x) \neq 1$ and $\psi_N(y) = 1$, and applying Δ as in Theorem 32.1, obtain a contradiction.)

(iii) Deduce from (ii) that there exist integers $n_1, \ldots, n_r$ such that $\sum n_i = n$ such that

$$
m = \begin{pmatrix} & & M_r \\ & \cdot^{\cdot^{\cdot}} & \\ & M_2 & \\ M_1 & & \end{pmatrix},
$$

where M_i is an $n_i \times n_i$ diagonal matrix.

(iv) Again make use of the assumption that NwN supports an intertwining operator to show that M_i is a scalar matrix.

(v) Define an involution ι of G by

$$
g \longmapsto w_0 \, {}^t g \, w_0, \qquad w_0 = \begin{pmatrix} & & 1 \\ & \cdot^{\cdot^{\cdot}} & \\ 1 & & \end{pmatrix}.
$$

Note that N and its character ψ_N are invariant under ι. Interpret (iv) as showing that every double coset that supports an intertwining operator $\mathrm{Ind}(\psi_N) \longrightarrow \mathrm{Ind}(\psi_N)$ has a representative that is invariant under ι, and deduce that $\mathrm{End}_G(\mathrm{Ind}(\psi_N))$ is commutative and that $\mathrm{Ind}(\psi_N)$ is multiplicity-free.

46
Hecke Algebras

A *Coxeter group* (Chap. 25) is a group W which may be given the following description. The group W has generators s_i $(i = 1, 2, \ldots, r)$ with relations $s_i^2 = 1$ and for each pair of indices i and j the "braid relations"

$$s_i s_j s_i \cdots = s_j s_i s_j \cdots$$

where the number of terms on both sides is the same integer $n(i, j)$. An example is the symmetric group S_k, where s_i is the transposition $(i, i+1)$. In this case $r = k - 1$.

Given a Coxeter group W, we may deform its group algebra as follows. Let $\mathcal{H}(W)$ be the ring with generators t_i satisfying the same braid relations

$$t_i t_j t_i \cdots = t_j t_i t_j \cdots,$$

but we replace the relation $s_i^2 = 1$ by a more general relation

$$t_i^2 = (q - 1)t_i + q.$$

The parameter q may be a complex number or an indeterminate. If $q = 1$, we recover the group algebra of W.

Hecke algebras are ubiquitous. They arise in various seemingly different ways: as endomorphism rings of induced representations for the groups of Lie type such as $GL(k, \mathbb{F}_q)$ (Iwahori [84], Howlett and Lehrer [80]); as convolution rings of functions on p-adic groups (Iwahori and Matsumoto [86]); as rings of operators acting on the equivariant K-theory of flag varieties (Lusztig [122], Kazhdan and Lusztig [98]); as rings of transfer matrices in statistical mechanics and quantum mechanics (Temperley and Lieb [160], Jimbo [90]), in knot theory (Jones [91]), and other areas. It is the context for defining the Kazhdan–Lusztig polynomials, which occur in seemingly unrelated questions in representation theory, geometry and combinatorics [97]. Some of these different occurrences of Hecke algebras may *seem* unrelated to each other, but this can be an illusion when in fact deep and surprising connections exist.

D. Bump, *Lie Groups*, Graduate Texts in Mathematics 225,
DOI 10.1007/978-1-4614-8024-2_46, © Springer Science+Business Media New York 2013

Following Iwahori [84], we will study a certain "Hecke algebra" $\mathcal{H}_k(q)$ that, as we will see, is isomorphic to the Hecke algebra of the symmetric group S_k. The ring $\mathcal{H}_k(q)$ can actually be defined if q is any complex number, but if q is a prime power, it has a representation-theoretic interpretation. We will see that it is the endomorphism ring of the representation of $G = \mathrm{GL}(k, \mathbb{F}_q)$, where $\mathbb{F}_q$ is the finite field with q elements, induced from the trivial representation of the Borel subgroup B of upper triangular matrices in G. The fact that it is a deformation of $\mathbb{C}[S_k]$ amounts to a parametrization of a certain set of irreducible representations of G—the so-called unipotent ones—by partitions.

If instead of $G = \mathrm{GL}(k, \mathbb{F}_q)$ we take $G = \mathrm{GL}(k, \mathbb{Q}_p)$, where $\mathbb{Q}_p$ is the p-adic field, and we take B to be the *Iwahori subgroup* consisting of elements g of $K = \mathrm{GL}(k, \mathbb{Z}_p)$ that are upper triangular modulo p, then one obtains the *affine Hecke algebra*, which is similar to $\mathcal{H}_k(q)$ but infinite-dimensional. It was introduced by Iwahori and Matsumoto [86]. The role of the Bruhat decomposition in the proofs requires a generalization of the Tits' system described in Iwahori [85]. This Hecke algebra contains a copy of $\mathcal{H}_k(p)$. On the other hand, it also contains the ring of K-bi-invariant functions, the so-called *spherical Hecke algebra* (Satake [143], Tamagawa [158]). The spherical Hecke algebra is commutative since K is a Gelfand subgroup of G. The spherical Hecke algebra is (when $k = 2$) essentially the portion corresponding to the prime p of the original Hecke algebra introduced by Hecke [65] to explain the appearance of Euler products as the L-series of automorphic forms. See Howe [76] and Rogawski [137] for the representation theory of the affine Hecke algebra.

Let F be a field. Let $G = \mathrm{GL}(k, F)$ and, as in Chap. 27, let B be the Borel subgroup of upper triangular matrices in G. A subgroup P containing B is called a *standard parabolic subgroup*. (More generally, any conjugate of a standard parabolic subgroup is called *parabolic*.)

Let $k_1, \ldots, k_r$ be positive integers such that $\sum_i k_i = k$. Then S_k has a subgroup isomorphic to $S_{k_1} \times \cdots \times S_{k_r}$ in which the first S_{k_1} acts on $\{1, \ldots, k_1\}$, the second S_{k_2} acts on $\{k_1 + 1, \ldots, k_1 + k_2\}$, and so forth. Let Σ denote the set of $k - 1$ transpositions $\{(1, 2), (2, 3), \ldots, (k - 1, k)\}$.

Lemma 46.1. *Let J be any subset of Σ. Then there exist integers $k_1, \ldots, k_r$ such that the subgroup of S_k generated by J is $S_{k_1} \times \cdots \times S_{k_r}$.*

Proof. If J contains $(1, 2), (2, 3), \ldots, (k_1 - 1, k_1)$, then the subgroup they generate is the symmetric group S_{k_1} acting on $\{1, \ldots, k_1\}$. Taking k_1 as large as possible, assume that J omits $(k_1, k_1 + 1)$. Taking k_2 as large as possible such that J contains $(k_1 + 1, k_1 + 2), \ldots, (k_1 + k_2 - 1, k_1 + k_2)$, the subgroup they generate is the symmetric group S_{k_2} acting on $\{k_1 + 1, \ldots, k_1 + k_2\}$, and so forth. Thus J contains generators of each factor in $S_{k_1} \times \cdots \times S_{k_r}$ and does not contain any element that is not in this product, so this is the group it generates. $\square$

The notations from Chap. 27 will also be followed. Let T be the maximal torus of diagonal elements in G, N the normalizer of T, and $W = N/T$ the

Weyl group. Moreover, Φ will be the set of all roots, Φ^+ the positive roots, and Σ the simple positive roots. Concretely, elements of Φ are the $k^2 - k$ rational characters of T of the form

$$\alpha_{ij} \begin{pmatrix} t_1 & & \\ & \ddots & \\ & & t_n \end{pmatrix} = t_i t_j^{-1},$$

where $1 \leqslant i, j \leqslant n$, Φ^+ consists of $\{\alpha_{ij} \,|\, i < j\}$, and $\Sigma = \{\alpha_{i,i+1}\}$. Identifying W with S_k, the set Σ in Lemma 46.1 is then the set of simple reflections.

Let J be any subset of Σ. Let W_J be the subgroup of W generated by the s_α with $\alpha \in \Sigma$. Then, by Lemma 46.1, we have (for suitable k_i)

$$W_J \cong S_{k_1} \times \cdots \times S_{k_r}. \tag{46.1}$$

Let N_J be the preimage of W_J in N under the canonical projection to W. Let P_J be the group generated by B and N_J. Then

$$P_J = \left\{ \begin{pmatrix} G_{11} & G_{12} & \cdots & G_{1r} \\ 0 & G_{22} & \cdots & G_{2r} \\ \vdots & & \ddots & \vdots \\ 0 & 0 & \cdots & G_{rr} \end{pmatrix} \right\}, \tag{46.2}$$

where each G_{ij} is a $k_i \times k_j$ block. The group P_J is a semidirect product $P_J = M_J U_J = U_J M_J$, where M_J is characterized by the condition that $G_{ij} = 0$ unless $i = j$, and the normal subgroup U_J is characterized by the condition that each G_{ii} is a scalar multiple of the identity matrix in $\mathrm{GL}(k_i)$. The groups P_J with J a proper subset of Σ are called the *standard parabolic subgroups*, and more generally any subgroup conjugate to a P_J is called *parabolic*. The subgroup U_J is the *unipotent radical* of P_J (that is, its maximal normal unipotent subgroup), and M_J is called the standard *Levi subgroup* of P_J. Evidently,

$$M_J \cong \mathrm{GL}(k_1, F) \times \cdots \times \mathrm{GL}(k_r, F). \tag{46.3}$$

Any subgroup conjugate in P_J to M_J (which is not normal) would also be called a Levi subgroup.

As in Chap. 27, we note that a double coset $B\omega B$, or more generally $P_I \omega P_J$ with $I, J \subset \Sigma$, does not depend on the choice $\omega \in N$ of representative for an element $w \in W$, and we will use the notation $BwB = C(w)$ or $P_I w P_J$ for this double coset. Let $B_J = M_J \cap B$. This is the standard "Borel subgroup" of M_J.

Proposition 46.1.

(i) Let $J \subseteq \Sigma$. Then

$$M_J = \bigcup_{w \in W_J} B_J w B_J \quad \text{(disjoint)}.$$

(ii) Let $I, J \subseteq \Sigma$. Then, if $w \in W$, we have

$$BW_I w W_J B = P_I w P_J. \tag{46.4}$$

(iii) The canonical map $w \longmapsto P_I w P_J$ from $W \longrightarrow P_I \backslash G / P_J$ induces a bijection

$$W_I \backslash W / W_J \cong P_I \backslash G / P_J.$$

Proof. For (i), we have (46.3) for suitable k_i. Now B_J is the direct product of the Borel subgroups of these $\mathrm{GL}(k_i, F)$, and W_J is the direct product (46.1). Part (i) follows directly from the Bruhat decomposition for $\mathrm{GL}(k, F)$ as proved in Chap. 27.

As for (ii), since $BW_I \subset P_I$ and $W_J B \subset P_J$, we have $BW_I w W_J B \subseteq P_I w P_J$. To prove the opposite inclusion, we first note that

$$wBW_J \subseteq BwW_J B. \tag{46.5}$$

Indeed, any element of W_J can be written as $s_1 \cdots s_r$, where $s_i = s_{\alpha_i}$, with $\alpha_i \in J$. Using Axiom TS3 from Chap. 27, we have

$$wBs_1 \cdots s_r \subset BwBs_2 \cdots s_r B \cup Bws_1 Bs_2 \cdots s_r B$$

and, by induction on r, both sets on the right are contained in $BwW_J B$. This proves (46.5). A similar argument shows that

$$W_I BwW_J \subseteq BW_I w W_J B. \tag{46.6}$$

Now, using (i),

$$P_I w P_J = U_I M_I w M_J U_J \subset U_I B_I W_I B_I w B_J W_J B_J U_J \subset BW_I BwBW_J B.$$

Applying (46.5) and (46.6), we obtain $BW_I w W_J B \supseteq P_I w P_J$, whence (46.4).

As for (iii), since by the Bruhat decomposition $w \longmapsto BwB$ is a bijection $W \longrightarrow B \backslash G / B$, (46.4) implies that $w \longrightarrow P_I w P_J$ induces a bijection $W_I \backslash W / W_J \longrightarrow P_I \backslash G / P_J$. $\qquad\square$

To proceed further, we will assume that $F = \mathbb{F}_q$ is a finite field. We recall from Chap. 34 that $\mathcal{R}_k$ denotes the free Abelian group generated by the isomorphism classes of irreducible representations of the symmetric group S_k, or, as we sometimes prefer, the additive group of generalized characters. It can be identified with the character ring of S_k. However, we do not need its ring structure, only its additive structure and its inner product, in which the distinct isomorphism classes of irreducible representations form an orthonormal basis.

Similarly, let $\mathcal{R}_k(q)$ be the free Abelian group generated by the isomorphism classes of irreducible representations of $\mathrm{GL}(n, \mathbb{F}_q)$ or equivalently the additive group of generalized characters. Like $\mathcal{R}_k$, we can make $\mathcal{R}_k(q)$ into the k-homogeneous part of a graded ring, a point we will take up in the next chapter.

Proposition 46.2. *Let H be a group, and let M_1 and M_2 be subgroups of H. Then in the character ring of H, the inner product of the characters induced from the trivial characters of M_1 and M_2, respectively, is equal to the number of double cosets in $M_1 \backslash H / M_2$.*

Proof. By the geometric form of Mackey's theorem (Theorem 32.1), the space of intertwining maps from $\mathrm{Ind}_{M_1}^{H}(1)$ to $\mathrm{Ind}_{M_2}^{H}(1)$ is isomorphic to the space of functions $\Delta : H \longrightarrow \mathrm{Hom}(\mathbb{C}, \mathbb{C}) \cong \mathbb{C}$ that satisfy $\Delta(m_2 h m_1) = \Delta(h)$ for $m_i \in M_i$. Of course, a function has this property if and only if it is constant on double cosets, so the dimension of the space of such functions is equal to the number of double cosets. On the other hand, the dimension of the space of intertwining operators equals the inner product in the character ring by (2.7). □

Theorem 46.1. *There is a unique isometry of $\mathcal{R}_k$ into $\mathcal{R}_k(q)$ in which for each subset I of Σ the representation $\mathrm{Ind}_{W_I}^{W}(1)$ maps to the representation $\mathrm{Ind}_{P_I}^{G}(1)$. This mapping takes irreducible representations to irreducible representations.*

Proof. If $I \subseteq \Sigma$, let χ_I denote the character of S_k induced from the trivial character of W_I, and let $\chi_I(q)$ denote the character of G induced from the trivial character of P_I.

We note that the representations χ_I of $\mathcal{R}_k$ span $\mathcal{R}_k$. Indeed, by the definition of the multiplication in $\mathcal{R}$, inducing the trivial representation from $S_{k_1} \times \cdots \times S_{k_r}$ to S_k, where $\sum k_i = k$, gives the representation denoted

$$\mathbf{h}_{k_1} \mathbf{h}_{k_2} \cdots \mathbf{h}_{k_r},$$

which is χ_I. Expanding the right-hand side of (35.10) expresses each $\mathbf{s}_\lambda$ as a linear combination of such representations, and by Theorem 35.1 the $\mathbf{s}_\lambda$ span $\mathcal{R}_k$; hence so do the χ_I.

We would like to define a map $\mathcal{R}_k \longrightarrow \mathcal{R}_k(q)$ by

$$\sum_I n_I \chi_I \longmapsto \sum_I n_I \chi_I(q), \tag{46.7}$$

where the sum is over subsets of Σ. We need to verify that this is well-defined and an isometry.

By Proposition 46.1, if $I, J \subseteq \Sigma$, the cardinality of $W_I \backslash W / W_J$ equals the cardinality of $P_I \backslash G / P_J$. By Proposition 46.2, it follows that

$$\langle \chi_I, \chi_J \rangle_{S_k} = \langle \chi_I(q), \chi_J(q) \rangle_{\mathrm{GL}(k, \mathbb{F}_q)}. \tag{46.8}$$

Now, if $\sum n_I \chi_I(q) = 0$, we have

$$\left\langle \sum_I n_I \chi_I, \sum_I n_I \chi_I \right\rangle_{S_k} = \sum_{I,J} n_I n_J \langle \chi_I, \chi_J \rangle_{S_k}$$

$$= \sum_{I,J} n_I n_J \langle \chi_I(q), \chi_J(q) \rangle_{\mathrm{GL}(k,\mathbb{F}_q)}$$

$$= \left\langle \sum_I n_I \chi_I(q), \sum_I n_I \chi_I(q) \right\rangle_{\mathrm{GL}(k,\mathbb{F}_q)} = 0,$$

so $\sum n_I \chi_I = 0$. Therefore (46.7) is well-defined, and (46.8) shows that it is an isometry.

It remains to be shown that irreducible characters go to irreducible characters. Indeed, if χ is an irreducible character of $W = S_k$, and if $\hat{\chi}$ is the corresponding character of $G = \mathrm{GL}(k,\mathbb{F}_q)$, then $\langle \hat{\chi}, \hat{\chi} \rangle = \langle \chi, \chi \rangle = 1$, so either $\hat{\chi}$ or $-\hat{\chi}$ is an irreducible character, and it is sufficient to show that $\hat{\chi}$ occurs with positive multiplicity in some proper character of G. Indeed, $\chi = \mathbf{s}_\lambda$ for some partition λ, and by (35.10) this means that χ appears with multiplicity one in the character induced from the trivial character of S_λ. Consequently, $\hat{\chi}$ occurs with multiplicity one in $\mathrm{Ind}_{P_I}^G(1)$, where I is any subset of Σ such that $W_I \cong S_\lambda$. This completes the proof. $\square$

If λ is a partition, let $\mathbf{s}_\lambda(q)$, $\mathbf{h}_k(q)$, and $\mathbf{e}_k(q)$ denote the images of the characters $\mathbf{s}_\lambda$, $\mathbf{h}_k$, and $\mathbf{e}_k$, respectively, of S_k under the isomorphism of Theorem 46.1. Thus $\mathbf{h}_k(q)$ is the trivial character. The character $\mathbf{e}_k(q)$ is called the *Steinberg character* of $\mathrm{GL}(k,\mathbb{F}_q)$. The characters $\mathbf{s}_\lambda(q)$ are the *unipotent characters* of $\mathrm{GL}(k,\mathbb{F}_q)$. This is not a proper definition of the term unipotent character because the construction as we have described it depends on the fact that the unipotent characters are precisely those that occur in $\mathrm{Ind}_B^G(1)$. This is true for $G = \mathrm{GL}(n,\mathbb{F})$ but *not* (for example) for $\mathrm{Sp}(4,\mathbb{F}_q)$. See Deligne and Lusztig [41] and Carter [32] for unipotent characters of finite groups of Lie type and Vogan [167] for an extended meditation on unipotent representations.

Proposition 46.3. *As a virtual representation, the alternating character $\mathbf{e}_k$ of S_k admits the following expression:*

$$\mathbf{e}_k = \sum_{J \subseteq \Sigma} (-1)^{|J|} \mathrm{Ind}_{W_J}^W(1).$$

Proof. We recall that $\mathbf{e}_k = \mathbf{s}_\lambda$, where λ is the partition $(1,\ldots,1)$ of K. The right-hand side of (35.10) gives

$$\mathbf{e}_k = \begin{vmatrix} \mathbf{h}_1 & \mathbf{h}_2 & \mathbf{h}_3 & \cdots & \mathbf{h}_k \\ 1 & \mathbf{h}_1 & \mathbf{h}_2 & \cdots & \mathbf{h}_{k-1} \\ 0 & 1 & \mathbf{h}_1 & \cdots & \mathbf{h}_{k-2} \\ \vdots & \vdots & \vdots & & \vdots \\ 0 & 0 & 0 & \cdots & \mathbf{h}_1 \end{vmatrix}.$$

Expanding this gives a sum of exactly 2^{k-1} monomials in the $\mathbf{h}_i$, which are in one-to-one correspondence with the subsets J of Σ. Indeed, let J be given, and let $k_1, k_2, k_3, \ldots$ be as in Lemma 46.1. Then there is a monomial that has $|J|$ 1's taken from below the diagonal; namely, if $\alpha_{i,i+1} \in \Sigma$, then there is a 1 taken from the $i+1, i$ position, and there is an $\mathbf{h}_{k_1}$ taken from the $1, k_1$ position, an $\mathbf{h}_{k_2}$ taken from the $k_1 + 1, k_1 + k_2$ position, and so forth. This monomial equals $(-1)^{|J|} \mathbf{h}_{k_1} \mathbf{h}_{k_2} \cdots$, which is $(-1)^{|J|}$ times the character induced from the trivial representation of $W_J = S_{k_1} \times S_{k_2} \times \cdots$. $\qquad\square$

Theorem 46.2. *As a virtual representation, the Steinberg representation* $\mathbf{e}_k(q)$ *of* $\mathrm{GL}(k, \mathbb{F}_q)$ *admits the following expression:*

$$\mathbf{e}_k(q) = \sum_{J \subseteq \Sigma} (-1)^{|J|} \mathrm{Ind}_{P_J}^P (1).$$

Proof. This follows immediately from Proposition 46.3 on applying the mapping of Theorem 46.1. $\qquad\square$

For our next considerations, there is no reason that F needs to be finite, so we return to the case where $G = \mathrm{GL}(k, F)$ of a general field F. We will denote by U the group of upper triangular unipotent matrices in $\mathrm{GL}(k, F)$.

Proposition 46.4. *Suppose that S is any subset of Φ such that if $\alpha \in S$, then $-\alpha \notin S$, and if $\alpha, \beta \in S$ and $\alpha + \beta \in \Phi$, then $\alpha + \beta \in S$. Let U_S be the set of $g = (g_{ij})$ in $\mathrm{GL}(k, F)$ such that $g_{ii} = 1$, and if $i \neq j$, then $g_{ij} = 0$ unless $\alpha_{ij} \in S$. Then U_S is a group.*

Proof. Let $\tilde{S}$ be the set of (i, j) such that the root $\alpha_{ij} \in S$. Translating the hypothesis on S into a statement about $\tilde{S}$, if $(i, j) \in \tilde{S}$ we have $i < j$, and

$$\text{if both } (i, j) \text{ and } (j, k) \text{ are in } \tilde{S}, \text{ then } i \neq k \text{ and } (i, k) \in \tilde{S}. \qquad (46.9)$$

From this it is easy to see that if g and h are in U_S, then so are g^{-1} and gh. $\qquad\square$

As a particular case, if $w \in W$, then $S = \Phi^+ \cap w\Phi^-$ satisfies the hypothesis of Proposition 46.4, and we denote

$$U_{\Phi^+ \cap w\Phi^-} = U_w^-.$$

Similarly, $S = \Phi^+ \cap w\Phi^+$ meets this hypothesis, and we denote

$$U_{\Phi^+ \cap w\Phi^+} = U_w^+.$$

Finally, let U be the group of all upper triangular unipotent matrices in G, which was denoted N in Chap. 27.

Let $l(w)$ denote the length of the Weyl group element, which (as in Chap. 20) is the smallest k such that w can be written as a product of k simple reflections.

Proposition 46.5. *Let $F = \mathbb{F}_q$ be finite, and let $w \in W$. We have*

$$|U_w^-| = q^{l(w)}.$$

Proof. By Propositions 20.2 and 20.5, the cardinality of $S = \Phi^+ \cap w^{-1}\Phi^-$ is $l(w)$, so this follows from the definition of U_S. $\qquad\square$

Proposition 46.6. *Let $w \in W$. The multiplication map $U_w^+ \times U_w^- \longrightarrow U$ is bijective.*

Proof. We will prove this if F is finite, the only case we need. In this case $U_w^+ \cap U_w^- = \{1\}$ by definition since the sets $\Phi^+ \cap w\Phi^-$ and $\Phi^+ \cap w\Phi^+$ are disjoint. Thus, if $u_1^+ u_1^- = u_2^+ u_2^-$ with $u_i^\pm \in U_w^\pm$, then $(u_2^+)^{-1}u_1^+ = u_2^-(u_1^-)^{-1} \in U_w^+ \cap U_w^-$ so $u_1^\pm = u_2^\pm$. Therefore, the multiplication map $U_w^+ \times U_w^- \longrightarrow U$ is injective. To see that it is surjective, note that

$$|U_w^-| = q^{|\Phi^+ \cap w\Phi^-|}, \qquad |U_w^+| = q^{|\Phi^+ \cap w\Phi^+|},$$

so the order of $U_w^+ \times U_w^-$ is $q^{|\Phi^+|} = |U|$, and the surjectivity is now clear. $\qquad\square$

We are interested in the size of the double coset BwB. In geometric terms, G/B can be identified with the space of F-rational points of a projective algebraic variety, and the closure of BwB/B is an algebraic subvariety in which BwB/B is an open subset; the dimension of this "Schubert cell" turns out to be $l(w)$.

If $F = \mathbb{F}_q$, an equally good measure of the size of BwB is its cardinality. It can of course be decomposed into right cosets of B, and its cardinality will be the order of B times the cardinality of the quotient BwB/B.

Proposition 46.7. *Let $F = \mathbb{F}_q$ be finite, and let $w \in W$. The order of BwB/B is $q^{l(w)}$.*

Proof. We will show that $u^- \longmapsto u^- wB$ is a bijection $U_w^- \longrightarrow BwB/B$. The result then follows from Proposition 46.5.

Note that every right coset in BwB/B is of the form bwB for some $b \in B$. Using Proposition 46.6, we may write $b \in B$ uniquely in the form $u^- u^+ t$ with $u^\pm \in U_w^\pm$ and $t \in T$. Now $w^{-1}u^+ tw = w^{-1}u^+ w.w^{-1}tw \in B$ because $w^{-1}u^+ w \in U$ and $w^{-1}tw \in T$. Therefore $bwB = u^- wB$.

It is now clear that the map $u^- \longmapsto u^- wB$ is surjective. We must show that it is injective; in other words, if $u_1^- wB = u_2^- wB$ for $u_i^- \in U_w^-$, then $u_1^- = u_2^-$. Indeed, if $u^- = (u_1^-)^{-1}u_2^-$ then $w^{-1}u^- w \in B$ from the equality of the double cosets. On the other hand, $w^{-1}u^- w$ is lower triangular by the definition of U_w^-. It is both upper triangular and lower triangular, and unipotent, so $u^- = 1$. $\qquad\square$

With k and q fixed, let $\mathcal{H}$ be the convolution ring of B-bi-invariant functions on G. The dimension of $\mathcal{H}$ equals the cardinality of $B\backslash G/B$, which is

$|W| = k!$ by the Bruhat decomposition. A basis of $\mathcal{H}$ consists of the functions ϕ_w ($w \in W$), where ϕ_w is the characteristic function of the double coset $\mathcal{C}(w) = BwB$. We normalize the convolution as follows:

$$(f_1 * f_2)(g) = \frac{1}{|B|} \sum_{x \in G} f_1(x) f_2(x^{-1}g) = \frac{1}{|B|} \sum_{x \in G} f_1(gx) f_2(x^{-1}).$$

With this normalization, the characteristic function f_1 of B serves as a unit in the ring.

The ring $\mathcal{H}$ is a normed ring with the L^1 norm. That is, we have

$$|f_1 * f_2| \leqslant |f_1| \cdot |f_2|,$$

where

$$|f| = \frac{1}{|B|} \sum_{x \in G} |f(x)|.$$

There is also an *augmentation map*, that is, a $\mathbb{C}$-algebra homomorphism $\epsilon : \mathcal{H} \longrightarrow \mathbb{C}$ given by

$$\epsilon(f) = \frac{1}{|B|} \sum_{x \in G} f(x).$$

By Proposition 46.7, we have

$$\epsilon(\phi_w) = q^{l(w)}. \tag{46.10}$$

Proposition 46.8. *Let* $w, w' \in W$ *such that* $l(ww') = l(w) + l(w')$. *Then*

$$\phi_{ww'} = \phi_w \phi_{w'}.$$

Proof. By Proposition 27.1, we have $\mathcal{C}(ww') = \mathcal{C}(w)\mathcal{C}(w')$. Therefore $\phi_w * \phi_{w'}$ is supported in $\mathcal{C}(ww')$ and is hence a constant multiple of $\phi_{ww'}$. Writing $\phi_w * \phi_{w'} = c\phi_{ww'}$, applying the augmentation ϵ, and using (46.10), we see that $c = 1$. $\square$

Proposition 46.9. *Let* $s \in W$ *be a simple reflection. Then*

$$\phi_s * \phi_s = q\phi_1 + (q-1)\phi_s.$$

Proof. By (27.2), we have $\mathcal{C}(s)\mathcal{C}(s) \subseteq \mathcal{C}(1) \cup \mathcal{C}(s)$. Therefore, there exist constants λ and μ such that $\phi_s * \phi_s = \lambda\phi_1 + \mu\phi_s$. Evaluating both sides at the identity gives $\lambda = q$. Now applying the augmentation and using the special cases $\epsilon(\phi_s) = q$, $\epsilon(f_1) = 1$ of (46.10), we have $q^2 = \lambda \cdot 1 + \mu \cdot q = q + \mu q$, so $\mu = q - 1$. $\square$

Let q be a nonzero element of a field containing $\mathbb{C}$, and let $R = \mathbb{C}[q, q^{-1}]$. Thus q might be a complex number, in which case the ring $R = \mathbb{C}$ or it might be transcendental over $\mathbb{C}$, in which case the ring R will be the ring of Laurent polynomials over $\mathbb{C}$.

We will define a ring $\mathcal{H}_k(q)$ as an algebra over R. Specifically, $\mathcal{H}_k(q)$ is the free $\mathbb{C}[q]$-algebra on generators $f_{s_{\alpha_i}}$ $(i = 1, \ldots, k-1)$ subject to the relations

$$f_{s_{\alpha_i}}^2 = q + (q-1)f_{s_{\alpha_i}}, \tag{46.11}$$

$$f_{s_{\alpha_i}} * f_{s_{\alpha_{i+1}}} * f_{s_{\alpha_i}} = f_{s_{\alpha_{i+1}}} * f_{s_{\alpha_i}} * f_{s_{\alpha_{i+1}}}, \tag{46.12}$$

$$f_{s_{\alpha_i}} * f_{s_{\alpha_j}} = f_{s_{\alpha_i}} * f_{s_{\alpha_j}} \quad \text{if } |i - j| > 1. \tag{46.13}$$

We note that f_{s_α} is invertible, with inverse $q^{-1}f_{\alpha_i} + q^{-1} - 1$, by (46.11).

Although $\mathcal{H}_k(q)$ is thus defined as an abstract ring, its structure reflects that of the Weyl group W of $\mathrm{GL}(k)$, which, as we have seen, is a Coxeter group. We recall what this means. Let $s_{\alpha_1}, \ldots, s_{\alpha_{k-1}}$ be the simple reflections of W. By Theorem 25.1, the group W has a presentation with generators s_{α_i} and relations

$$s_{\alpha_i}^2 = 1,$$

$$s_{\alpha_i} s_{\alpha_{i+1}} s_{\alpha_i} = s_{\alpha_{i+1}} s_{\alpha_i} s_{\alpha_{i+1}}, \quad 1 \leqslant i \leqslant k-2,$$

$$s_{\alpha_i} s_{\alpha_j} = s_{\alpha_j} s_{\alpha_i} \quad \text{if } |i-j| > 1.$$

Of course, since $s_{\alpha_i}^2 = 1$, the relation $s_{\alpha_i} s_{\alpha_{i+1}} s_{\alpha_i} = s_{\alpha_{i+1}} s_{\alpha_i} s_{\alpha_{i+1}}$ is just another way of writing $(s_{\alpha_i} s_{\alpha_{i+1}})^3 = 1$.

Proposition 46.10. *If $q = 1$, the Hecke ring $\mathcal{H}_k(1)$ is isomorphic to the group ring of S_k.*

Proof. This is clear from Theorem 25.1 since if $q = 1$ the defining relations of the ring $\mathcal{H}_k(1)$ coincide with the Coxeter relations presenting S_k. $\quad\square$

Thus $\mathcal{H}_k(q)$ is a deformation of $\mathbb{C}[S_k]$, and its representation theory is the same as the representation theory of the symmetric group, one might therefore ask whether the Frobenius–Schur duality between the representations of S_k and $\mathrm{U}(n)$, which has been a great theme for us, can be extended to representations of this Hecke algebra. The answer is affirmative. The role of $\mathrm{U}(n)$ is played by a "quantum group," which is not actually a group at all but a Hopf algebra. Frobenius–Schur duality in this quantum context is due to Jimbo [89]. See also Zhang [179].

If $w \in W$ is arbitrary, we want to associate an element f_w of $\mathcal{H}_k(q)$ extending the definition of the generators. The next result will make this possible. (Of course, f_w is already defined if w is a simple reflection.)

Proposition 46.11. *Suppose that $w \in W$ with $l(w) = r$, and suppose that $w = s_1 \cdots s_r = s_1' \cdots s_r'$ are distinct decompositions of minimal length into simple reflections. Then*

$$f_{s_1} * \cdots * f_{s_r} = f_{s_1'} * \cdots * f_{s_r'}. \tag{46.14}$$

Proof. Let B be the braid group generated by u_{α_i} parametrized by the simple roots α_i, with $n(u_{\alpha_i}, u_{\alpha_j})$ equal to the order (2 or 3) of $s_{\alpha_i} s_{\alpha_j}$. Let $s_i = s_{\beta_i}$ and $s_i' = s_{\gamma_i}$ with $\beta_i, \gamma_i \in \Sigma$, and let $u_i = u_{\alpha_i}$, $u_i' = u_{\beta_i}$ be the corresponding elements of B. By Theorem 25.2, we have

$$u_1 \cdots u_r = u_1' \cdots u_r'. \tag{46.15}$$

Since the f_{α_i} satisfy the braid relations, there is a homomorphism of B into the group of invertible elements of $\mathcal{H}_k(q)$ such that $u_{\alpha_i} \longmapsto f_{\alpha_i}$. Applying this homomorphism to (46.15), we obtain (46.14). $\square$

If $w \in W$, let $w = s_1 \cdots s_r$ be a decomposition of w into $r = l(w)$ simple reflections, and define

$$f_w = f_{s_1} * \cdots * f_{s_r}.$$

According to Proposition 46.11, this f_w is well-defined.

Theorem 46.3 (Iwahori). *The f_w form a basis of $\mathcal{H}_k(q)$ as a free R-module. Thus, the rank of $\mathcal{H}_k(q)$ is $|W|$.*

Proof. First, assume that q is transcendental, so that R is the ring of Laurent polynomials in q. We will deduce the corresponding statement when $q \in \mathbb{C}$ at the end.

Let us check that

$$\sum_{w \in W} R f_w = \mathcal{H}_k(q). \tag{46.16}$$

It is sufficient to show that this R-submodule is closed under right multiplication by generators f_s of W with s a simple reflection. If $l(ws) = l(w)+1$, then $f_w f_s = f_{ws}$. On the other hand, if $l(ws) = l(w) - 1$, then writing $w' = ws$ we have $f_w f_s = f_{w's} f_s = f_{w'} f_s^2$, which by (46.11) is a linear combination of $f_{w'}$ and $f_{w'} f_s = f_w$.

It remains to be shown that the sum (46.16) is direct. If not, there will be some Laurent polynomials $c_w(q)$, not all zero, such that

$$\sum_w c_w(q) f_w = 0.$$

There exists a rational prime p such that $c_w(p)$ are not all zero. Let $\mathcal{H}$ be the convolution ring of B-bi-invariant functions on $GL(k, \mathbb{F}_p)$. It follows from Propositions 46.8 and 46.9 that (46.11)–(46.13) are all satisfied by the standard generators of $\mathcal{H}$, so we have a homomorphism $\mathcal{H}_k(q) \longrightarrow \mathcal{H}$ mapping each f_w to the corresponding generator ϕ_w of $\mathcal{H}$ and mapping $q \longmapsto p$. The images of the f_w are linearly independent in $\mathcal{H}$, yet since the $c_w(p)$ are not all zero, we obtain a relation of linear independence. This is a contradiction.

The result is proved if q is transcendental. If $0 \neq q_0 \in \mathbb{C}$, then there is a homomorphism $R \longrightarrow \mathbb{C}$, and a compatible homomorphism $\mathcal{H}_k(q) \longrightarrow$

$\mathcal{H}_k(q_0)$, in which $q \longmapsto q_0$. What we must show is that the R-basis elements f_w remain linearly independent when projected to $\mathcal{H}_k(q_0)$. To prove this, we note that in $\mathcal{H}_k(q)$ we have

$$f_w f_{w'} = \sum_{w'' \in W} a_{w,w',w''}(q,q^{-1}) f_{w''},$$

where $a_{w,w',w''}$ is a polynomial in q and q^{-1}. We may construct ring $\tilde{\mathcal{H}}_k(q_0)$ over $\mathbb{C}$ with basis elements $\tilde{f}_w$ indexed by W and specialized ring structure constants $a_{w,w',w''}(q_0, q_0^{-1})$. The associative law in $\mathcal{H}_k(q)$ boils down to a polynomial identity that remains true in this new ring, so this ring exists. Clearly, the identities (46.11)–(46.13) are true in the new ring, so there exists a homomorphism $\mathcal{H}_k(q_0) \longrightarrow \tilde{\mathcal{H}}_k(q_0)$ mapping the f_w to the $\tilde{f}_w$. Since the $\tilde{f}_w$ are linearly independent, so are the f_w in $\mathcal{H}_k(q_0)$.   □

Let us return to the case where q is a prime power.

Theorem 46.4. *Let q be a prime power. Then the Hecke algebra $\mathcal{H}_k(q)$ is isomorphic to the convolution ring of B-bi-invariant functions on $\mathrm{GL}(k, \mathbb{F}_q)$, where B is the Borel subgroup of upper triangular matrices in $\mathrm{GL}(n, \mathbb{F}_q)$. In this isomorphism, the standard basis element f_w ($w \in W$) corresponds to the characteristic function of the double coset BwB.*

Proof. It follows from Propositions 46.8 and 46.9 that (46.11)–(46.13) are all satisfied by the elements ϕ_w in the ring $\mathcal{H}$ of B-bi-invariant functions on $\mathrm{GL}(n, \mathbb{F}_q)$, so there exists a homomorphism $\mathcal{H}_k(q) \longrightarrow \mathcal{H}$ such that $f_w \longmapsto \phi_w$. Since the $\{f_w\}$ are a basis of $\mathcal{H}_k(q)$ and the ϕ_w are a basis of $\mathcal{H}$, this ring homomorphism is an isomorphism.   □

Exercises

Exercise 46.1. Show that any subgroup of $\mathrm{GL}(n, F)$ containing B is of the form (46.2).

Exercise 46.2. For $G = \mathrm{GL}(3)$, describe U_w^+ and U_w^- explicitly for each of the six Weyl group elements.

Exercise 46.3. Let G be a finite group and H a subgroup. Let $\mathcal{H}$ be the "Hecke algebra" of H bi-invariant functions, with multiplication being the convolution product normalized by

$$(f_1 * f_2)(g) = \frac{1}{|H|} \sum_{x \in G} f_1(x) f_2(x^{-1}g).$$

If (π, V) is an irreducible representation of G, let V^H be the subspace of H-fixed vectors. Then V^H becomes a module over $\mathcal{H}$ with the action

$$f \cdot v = |H|^{-1} \sum_{g \in G} f(g)\pi(g)v. \tag{46.17}$$

$f \cdot v = |H|^{-1} \sum_{g \in G} f(g)\pi(g)v$. Show that V^H, if nonzero, is irreducible as an $\mathcal{H}$-module. (**Hint:** If W is a nonzero invariant subspace of V^H, and $v \in V^H$, then since V is irreducible, we have $f_1 \cdot w = v$ for some function f_1 on G, where $f_1 \cdot w$ is defined as in (46.17) even though $f_1 \notin \mathcal{H}$. Show that $f \cdot w = v$, where $f = \varepsilon * f_1 * \varepsilon$ and ε is the characteristic function of H. Observe that $f \in \mathcal{H}$ and conclude that $V^H = W$.)

Exercise 46.4. In the setting of Exercise 46.3, show that $(\pi, V) \longmapsto V^H$ is a bijection between the isomorphism classes of irreducible representations of G with $V^H \neq 0$ and isomorphism classes of irreducible $\mathcal{H}$-modules.

Exercise 46.5. Show that if (π, V) is an irreducible representation of $G = \mathrm{GL}(k, \mathbb{F}_q)$ with character $\mathbf{s}_\lambda(q)$, then the degree of the corresponding representation of $\mathcal{H}_k(q)$ is the degree of the irreducible character $\mathbf{s}_\lambda$ of S_k. (Thus, the degree d_λ of $\mathbf{s}_\lambda$ is the dimension of V^B.) Show that d_λ is the multiplicity of $\mathbf{s}_\lambda(q)$ in $\mathrm{Ind}_B^G(1)$.

Exercise 46.6. Assume that q is a prime. Prove that

$$\mathcal{H}_k(q) \cong \bigoplus_{\lambda \text{ a partition of } k} \mathrm{Mat}_{d_\lambda}(\mathbb{C}) \cong \mathbb{C}[S_k].$$

Exercise 46.7. Prove that the degree of the irreducible character $\mathbf{s}_\lambda(q)$ of $\mathrm{GL}(k, \mathbb{F}_q)$ is a polynomial in q whose value when $q = 1$ is the degree d_λ of the irreducible character $\mathbf{s}_\lambda$ of S_k.

Exercise 46.8. An element of $\mathrm{GL}(k, \mathbb{F}_q)$ is called *semisimple* if it is diagonalizable over the algebraic closure of $\mathbb{F}_q$. A semisimple element is called *regular* if its eigenvalues are distinct. If λ is a partition of k, let c_λ be a regular semisimple element of $\mathrm{GL}(k, \mathbb{F}_q)$ such that

$$c_\lambda = \begin{pmatrix} c_1 & & \\ & \ddots & \\ & & c_r \end{pmatrix}, \qquad c_i \in \mathrm{GL}(\lambda_i, \mathbb{F}_q),$$

and such that the eigenvalues of c_i generate $\mathbb{F}_{q^{\lambda_i}}$. Of course, c_λ isn't completely determined by this description. Such a c_λ will exist (for k fixed) if q is sufficiently large.

(i) Show that, if $k = 2$, then the unipotent characters of $\mathrm{GL}(2, \mathbb{F}_q)$ have the following values:

	$c_{(11)}$	$c_{(2)}$
$\mathbf{s}_{(11)}$	1	1
$\mathbf{s}_{(2)}$	1	-1

Note that this is the character table of S_2.

(ii) More generally, prove that in the notation of Chap. 37, the value of the character $\mathbf{s}_\mu(q)$ on the conjugacy class c_λ of $\mathrm{GL}(k, \mathbb{C})$ equals the value of the character $\mathbf{s}_\mu$ on the conjugacy class $\mathcal{C}_\lambda$ of S_k.

47

The Philosophy of Cusp Forms

There are four theories that deserve to be studied in parallel. These are:

- The representation theory of symmetric groups S_k;
- The representation theory of $GL(k, \mathbb{F}_q)$;
- The representation theory of $GL(k, F)$ where F is a local field;
- The theory of automorphic forms on $GL(k)$.

In this description, a *local field* is $\mathbb{R}$, $\mathbb{C}$, or a field such as the p-adic field $\mathbb{Q}_p$ that is complete with respect to a non-Archimedean valuation. Roughly speaking, each successive theory can be thought of as an elaboration of its predecessor. Both similarities and differences are important. We list some parallels between the four theories in Table 47.1.

The plan of this chapter is to discuss all four theories in general terms, giving proofs only for the second stage in this tower of theories, the representation theory of $GL(n, \mathbb{F}_q)$. (The first stage is already adequately covered.) Although the third and fourth stages are outside the scope of this book, our goal is to prepare the reader for their study by exposing the parallels with the finite field case.

There is one important way in which these four theories are similar: there are certain representations that are the "atoms" from which all other representations are built and a "constructive process" from which the other representations are built. Depending on the context, the "atomic" representations are called *cuspidal* or *discrete series* representations. The constructive process is *parabolic induction* or *Eisenstein series*. The constructive process usually (but not always) produces an irreducible representation.

Harish-Chandra [62] used the term "philosophy of cusp forms" to describe this parallel, which will be the subject of this chapter. One may substitute any reductive group for $GL(k)$ and most of what we have to say will be applicable. But $GL(k)$ is enough to fix the ideas.

In order to explain the philosophy of cusp forms, we will briefly summarize the theory of Eisenstein series before discussing (in a more serious way) a part of the representation theory of $GL(k)$ over a finite field. The reader

D. Bump, *Lie Groups*, Graduate Texts in Mathematics 225,
DOI 10.1007/978-1-4614-8024-2_47, © Springer Science+Business Media New York 2013

only interested in the latter may skip the paragraphs on automorphic forms. When we discuss automorphic forms, we will prove nothing and state exactly what seems relevant in order to see the parallel. For $\mathrm{GL}(k, \mathbb{F}_q)$, we prove more, but mainly what we think is essential to see the parallel. Our treatment is greatly influenced by Howe [74] and Zelevinsky [178]. To go deeper into the representation theory of the finite groups of Lie type, Carter [32] is an exceedingly useful reference.

For the symmetric groups, there is only one "atom"—the trivial representation of S_1. The constructive process is ordinary induction from $S_k \times S_l$ to S_{k+l}, which was the multiplication $\circ$ in the ring $\mathcal{R}$ introduced in Chap. 34. The element that we have identified as atomic was called $\mathbf{h}_1$ there. It does not generate the ring $\mathcal{R}$. However, $\mathbf{h}_1^k$ is the regular representation (or character) of S_k, and it contains every irreducible representation. To construct every irreducible representation of S_k from this single irreducible representation of S_1, the constructive process embodied in the multiplicative structure of the ring $\mathcal{R}$ must be supplemented by a further procedure. This is the extraction of an irreducible from a bigger representation $\mathbf{h}_1^k$ that includes it. This extraction amounts to finding a description for the "Hecke algebra" that is the endomorphism ring of $\mathbf{h}_1^k$. This "Hecke algebra" is isomorphic to the group ring of S_k.

For the groups $\mathrm{GL}(k, \mathbb{F}_q)$, let us construct a graded ring $\mathcal{R}(q)$ analogous to the ring $\mathcal{R}$ in Chap. 34. The homogeneous part $\mathcal{R}_k(q)$ will be the free Abelian group on the set of isomorphism classes of irreducible representations of $\mathrm{GL}(k, \mathbb{F}_q)$, which may be identified with the character ring of this group; the multiplicative structure of the character ring is not used. Instead, there is a multiplication $\mathcal{R}_k(q) \times \mathcal{R}_l(q) \longrightarrow \mathcal{R}_{k+l}(q)$, called *parabolic induction*. Consider the maximal parabolic subgroup $P = MU$ of $\mathrm{GL}(k + l, \mathbb{F}_q)$, where

$$M \cong \mathrm{GL}(k, \mathbb{F}_q) \times \mathrm{GL}(l, \mathbb{F}_q) = \left\{ \begin{pmatrix} g_1 & \\ & g_2 \end{pmatrix} \middle| g_1 \in \mathrm{GL}(k, \mathbb{F}_q),\, g_2 \in \mathrm{GL}(l, \mathbb{F}_q) \right\}$$

and

$$U = \left\{ \begin{pmatrix} I_k & X \\ & I_l \end{pmatrix} \middle| X \in \mathrm{Mat}_{k \times l}(\mathbb{F}_q) \right\}.$$

The group P is a semidirect product, since U is normal, and the composition

$$M \longrightarrow P \longrightarrow P/U$$

is an isomorphism. So given a representation (π_1, V_1) of $\mathrm{GL}(k, \mathbb{F}_q)$ and a representation (π_2, V_2) of $\mathrm{GL}(l, \mathbb{F}_q)$, one may regard the representation $\pi_1 \otimes \pi_2$ of M as a representation of $P/U \cong M$ and pull it back to a representation of P in which U acts trivially. Inducing from P to $\mathrm{GL}(k+l, \mathbb{F}_q)$ gives a representation that we will denote $\pi_1 \circ \pi_2$. By the definition of the induced representation, it acts by right translation on the space $V_1 \circ V_2$ of all functions $f : G \longrightarrow V_1 \otimes V_2$ such that

$$f\left(\begin{pmatrix} g_1 & * \\ & g_2 \end{pmatrix} h\right) = \left(\pi_1(g_1) \otimes \pi_2(g_2)\right) f(h).$$

With this multiplication, $\mathcal{R}(q) = \bigoplus \mathcal{R}_k(q)$ is a graded ring (Exercise 47.1). Inspired by ideas of Philip Hall, Green [58] defined the ring $\mathcal{R}(q)$ and used it systematically by in his description of the irreducible representations of $\mathrm{GL}(k, \mathbb{F}_q)$. Like $\mathcal{R}$, it can be given the structure of a Hopf algebra. See Zelevinsky [178] and Exercise 47.5.

If, imitating the construction with the symmetric group, we start with the trivial representation $\mathbf{h}_1(q)$ of $\mathrm{GL}(1, \mathbb{F}_q)$ and consider all irreducible representations of $\mathrm{GL}(k, \mathbb{F}_q)$ that occur in $\mathbf{h}_1(q)^k$, we get exactly the unipotent representations (i.e., the $\mathbf{s}_k(q)$ of Chap. 46), and this is the content of Theorem 46.1. To get all representations, we need more than this. There is a unique smallest set of irreducible representations of the $\mathrm{GL}(k, \mathbb{F}_q)$—the cuspidal ones—such that we can find every irreducible representation as a constituent of some representation that is a $\circ$ product of cuspidal ones. We will give more precise statements later in this chapter.

At the third stage in the tower of theories, the most important representations are infinite-dimensional, and analysis is important as well as algebra in their understanding. The representation theory of algebraic groups over a local field F is divided into the case where F is Archimedean—that is, $F = \mathbb{R}$ or $\mathbb{C}$—and where F is non-Archimedean.

If F is Archimedean, then an algebraic group over F is a Lie group, more precisely a complex analytic group when $F = \mathbb{C}$. The most important feature in the representation theory of reductive Lie groups is the *Langlands classification* expressing every irreducible representation as a quotient of one that is parabolically induced from discrete series representations. Usually the parabolically induced representation is itself irreducible and there is no need to pass to a quotient. See Knapp [104], Theorem 14.92 on p. 616 for the Langlands classification. Knapp [104] and Wallach [168] are comprehensive accounts of the representation theory of noncompact reductive Lie groups.

For reductive p-adic groups—that is, reductive algebraic groups over a non-Archimedean local field—the situation is similar and in some ways simpler. The most important discrete series representations are the *supercuspidals*. There is again a Langlands classification expressing every irreducible representation as a quotient of one parabolically induced from discrete series. Surveys of the representation theory of p-adic groups can be found in Cartier [33] and Moeglin [130]. Two useful longer articles with foundational material are Casselman [34] and Bernstein and Zelevinsky [16]. The most important foundational paper is Bernstein and Zelevinsky [17]. Chapter 4 of Bump [27] emphasizes $\mathrm{GL}(2)$ but is still useful.

The fourth of the four theories in the tower is the theory of automorphic forms. In developing this theory, Selberg and Langlands realized that certain automorphic forms were basic, and these are called *cusp forms*. The definitive reference for the Selberg–Langlands theory is Moeglin and Waldspurger [131]. Let us consider the basic setup.

Table 47.1. The philosophy of cusp forms

Class of groups	Atoms	Synthetic process	Analytic process	Unexpected symmetry
S_k	h_1	Induction	Restriction	(Trivial)
$GL(k, \mathbb{F}_q)$	Cuspidal representations	Parabolic induction	Unipotent invariants	$\mathcal{R}(q)$ is commutative
$GL(k, F)$ F local	Discrete series	Parabolic induction	Jacquet functors $r_{U,1}$ in [17]	Intertwining integrals such as (47.2)
$GL(k, A)$ A = adele ring of global F	Automorphic cuspidal representations	Eisenstein series	Constant terms	Functional equations

Let $G = \mathrm{GL}(k, \mathbb{R})$. Let Γ be a discrete subgroup of G such that $\Gamma \backslash G$ has finite volume such as $\mathrm{GL}(k, \mathbb{Z})$. An *automorphic form* on G with respect to Γ is a smooth complex-valued function f on G that is K-finite, $\mathcal{Z}$-finite, of moderate growth and automorphic, and has unitary central character. We define these terms now.

The group G acts on functions by right translation: $\rho(g)f(h) = f(hg)$. The group K is the maximal compact subgroup $\mathrm{O}(n)$, and f is K-*finite* if the space of functions $\rho(\kappa)f$ with $\kappa \in K$ spans a finite-dimensional vector space.

The Lie algebra $\mathfrak{g}$ of G also acts by right translation: if $X \in \mathfrak{g}$, then

$$(Xf)(g) = \frac{\mathrm{d}}{\mathrm{d}t} f(ge^{tX}) \Big|_{t=0}.$$

As a consequence, the universal enveloping algebra $U(\mathfrak{g})$ acts on smooth functions. Let $\mathcal{Z}$ be its center. This is a ring of differential operators on G that are invariant under both right and left translation (Exercise 10.2). For example, it contains the Casimir element constructed in Theorem 10.2 (from the trace bilinear form B on $\mathfrak{g}$); in this incarnation, the Casimir element is the *Laplace–Beltrami* operator. The function f is called $\mathcal{Z}$-*finite* if the image of f under $\mathcal{Z}$ is a finite-dimensional vector space.

Embed G into $2k^2$-dimensional Euclidean space $\mathrm{Mat}_k(\mathbb{R}) \oplus \mathrm{Mat}_k(\mathbb{R}) = \mathbb{R}^{2k^2}$ by

$$g \longmapsto (g, g^{-1}).$$

Let $\| \ \|$ denote the Euclidean norm in $\mathbb{R}^{2k^2}$ restricted to G. The function f is said to be of *moderate growth* if $f(g) < C\|g\|^N$ for suitable C and N.

The function f is called *automorphic* with respect to Γ if $f(\gamma g) = f(g)$ for all $\gamma \in \Gamma$.

We will consider functions f such that for some character ω of $\mathbb{R}_+^\times$ we have

$$f\left(\begin{pmatrix} z & & \\ & \ddots & \\ & & z \end{pmatrix} g \right) = \omega(z)\, f(g)$$

for all $z \in \mathbb{R}_+^\times$. The character ω is the *central character*. It is fixed throughout the discussion and is assumed unitary; that is, $|\omega(z)| = 1$.

Let V be a vector space on which K and $\mathfrak{g}$ both act. The actions are assumed to be compatible in the sense that both induce the same representation of $\mathrm{Lie}(K)$. We ask that V decomposes into a direct sum of finite-dimensional irreducible subspaces under K. Then V is called a $(\mathfrak{g}, K)$-*module*. If every irreducible representation of K appears with only finite multiplicity, then we say that V is *admissible*. For example, let (π, H) be an irreducible unitary representation of G on a Hilbert space H, and let V be the space of K-finite vectors in H. It is a dense subspace and is closed under actions of both $\mathfrak{g}$ and K, so it is a $(\mathfrak{g}, K)$-module. The $(\mathfrak{g}, K)$-modules form a category that can be studied by purely algebraic methods, which captures the essence of the representations.

The space $\mathcal{A}(\Gamma \backslash G)$ of automorphic forms is not closed under ρ because K-finiteness is not preserved by $\rho(g)$ unless $g \in K$. Still, both K and $\mathfrak{g}$ preserve the space $\mathcal{A}(\Gamma \backslash G)$. A subspace that is invariant under these actions and irreducible in the obvious sense is called an *automorphic representation*. It is a $(\mathfrak{g}, K)$-module.

Given an automorphic form f on $G = \mathrm{GL}(k, \mathbb{R})$ with respect to $\Gamma = \mathrm{GL}(k, \mathbb{Z})$, if $k = r + t$ we can consider the *constant term* along the parabolic subgroup P with Levi factor $\mathrm{GL}(r) \times \mathrm{GL}(t)$. This is the function

$$\int_{\mathrm{Mat}_{r \times t}(\mathbb{Z}) \backslash \mathrm{Mat}_{r \times t}(\mathbb{R})} f \left(\begin{pmatrix} I & X \\ & I \end{pmatrix} \begin{pmatrix} g_1 & \\ & g_2 \end{pmatrix} \right) \mathrm{d}X$$

for $(g_1, g_2) \in \mathrm{GL}(r, \mathbb{R}) \times \mathrm{GL}(t, \mathbb{R})$. If the constant term of f along every maximal parabolic subgroup vanishes then f is called a *cusp form*. An automorphic representation is called *automorphic cuspidal* if its elements are cusp forms.

Let $L^2(\Gamma \backslash G, \omega)$ be the space of measurable functions on g that are automorphic and have central character ω and such that

$$\int_{\Gamma Z \backslash G} |f(g)|^2 \, \mathrm{d}g < \infty.$$

The integral is well-defined modulo Z because ω is assumed to be unitary. Cusp forms are always square-integrable—an automorphic cuspidal representation embeds as a direct summand in $L^2(\Gamma \backslash G, \omega)$. In particular, it is unitary.

There is a construction that is dual to the constant term in the Selberg–Langlands theory, namely the construction of *Eisenstein series*. Let (π_1, V_1) and (π_2, V_2) be automorphic cuspidal representations of $\mathrm{GL}(r, \mathbb{R})$ and $\mathrm{GL}(t, \mathbb{R})$, where $r + t = k$. Let $P = MU$ be the maximal parabolic subgroup with Levi factor $M = \mathrm{GL}(r, \mathbb{R}) \times \mathrm{GL}(t, \mathbb{R})$. The modular quasicharacter $\delta_P : P \longrightarrow \mathbb{R}_+^\times$ is

$$\delta_P \begin{pmatrix} g_1 & * \\ & g_2 \end{pmatrix} = \frac{|\det(g_1)|^t}{|\det(g_2)|^r}$$

by Exercise 1.2. The space of the $(\mathfrak{g}, K)$-module of the induced representation $\mathrm{Ind}(\pi_1 \otimes \pi_2 \otimes \delta_P^s)$ of G consists of K-finite functions $f_s : G \longrightarrow \mathbb{C}$ such that any element f'_s of the $(\mathfrak{g}, K)$-submodule of $C^\infty(G)$ generated by f_s satisfies the condition that

$$f'_s \begin{pmatrix} g_1 & X \\ & g_2 \end{pmatrix}$$

is independent of X and equals $\delta_P^{s+1/2}$ times a finite linear combination of functions of the form $f_1(g_1)f_2(g_2)$, where $f_i \in V_i$. Due to the extra factor $\delta_P^{1/2}$, this induction is called *normalized induction*, and it has the property that if s is purely imaginary (so that $\pi_1 \otimes \pi_2 \otimes \delta_P^s$ is unitary), then the induced representation is unitary.

Then, for $\mathrm{re}(s)$ sufficiently large and for $f_s \in \mathrm{Ind}(\pi_1 \otimes \pi_2 \otimes \delta_P^s)$, the series

$$E(g, f_s, s) = \sum_{P(\mathbb{Z}) \backslash \mathrm{GL}(k, \mathbb{Z})} f_s(\gamma g)$$

is absolutely convergent. Here $P(\mathbb{Z})$ is the group of integer matrices in P with determinant ± 1.

Unlike cusp forms, the Eisenstein series are not square-integrable. Nevertheless, they are needed for the spectral decomposition of $\mathrm{GL}(k, \mathbb{Z}) \backslash \mathrm{GL}(k, \mathbb{R})$. This is analogous to the fact that the characters $x \longmapsto e^{2\pi i \alpha x}$ of $\mathbb{R}$ are not square-integrable, but as eigenfunctions of the Laplacian, a self-adjoint operator, they are needed for its spectral theory and comprise its continuous spectrum. The spectral problem for $\mathrm{GL}(k, \mathbb{Z}) \backslash \mathrm{GL}(k, \mathbb{R})$ has both a discrete spectrum (comprised of the cusp forms and residues of Eisenstein series) and a continuous spectrum. The Eisenstein series (analytically continued in s and restricted to the unitary principal series) are needed for the analysis of the continuous spectrum.

For the purpose of analytic continuation, we call a family of functions $f_s \in \mathrm{Ind}(\pi_1 \otimes \pi_2 \otimes \delta_P^s)$ a *standard section* if the restriction of the functions f_s to K is independent of s.

Theorem 47.1 (Selberg, Langlands). *Let $r + t = k$. Let P and Q be the parabolic subgroups of $\mathrm{GL}(k)$ with Levi factors $\mathrm{GL}(r) \times \mathrm{GL}(t)$ and $\mathrm{GL}(t) \times \mathrm{GL}(r)$, respectively. Suppose that $f_s \in \mathrm{Ind}(\pi_1 \otimes \pi_2 \otimes \delta_P^s)$ is a standard section. Then $E(g, f_s, s)$ has meromorphic continuation to all s. There exists an intertwining operator*

$$M(s) : \mathrm{Ind}(\pi_1 \otimes \pi_2 \otimes \delta_P^s) \longrightarrow \mathrm{Ind}(\pi_2 \otimes \pi_1 \otimes \delta_Q^{-s})$$

such that the functional equation

$$E(g, f_s, s) = E(g, M(s)f_s, -s) \tag{47.1}$$

is true.

The intertwining operator $M(s)$ is given by an integral formula

$$M(s)f(g) = \int_{\mathrm{Mat}_{t \times r}(\mathbb{R})} f\left(\begin{pmatrix} & -I_t \\ I_r & \end{pmatrix}\begin{pmatrix} I & X \\ & I \end{pmatrix} g\right) dX. \qquad (47.2)$$

This integral may be shown to be convergent if $\mathrm{re}(s) > \frac{1}{2}$. For other values of s, it has analytic continuation. This integral emerges when one looks at the constant term of the Eisenstein series with respect to Q. We will not explain this further but mention it because these *intertwining integrals* are extremely important and will reappear in the finite field case in the proof of Proposition 47.3.

The two constructions—constant term and Eisenstein series—have parallels in the representation theory of $\mathrm{GL}(k, F)$, where F is a local field including $F = \mathbb{R}$, $\mathbb{C}$, or a p-adic field. These constructions are functors between representations of $\mathrm{GL}(k, F)$ and those of the Levi factor of any parabolic subgroup. They are the *Jacquet functors* in one direction and parabolic induction in the other. (We will not define the Jacquet Functors, but they are the functors $r_{U,1}$ in Bernstein and Zelevinsky [17].) Moreover, these constructions also descend to the case of representation theory of $\mathrm{GL}(n, \mathbb{F}_q)$, which we look at next.

An irreducible representation (π, V) of $\mathrm{GL}(k, \mathbb{F}_q)$ is called *cuspidal* if there are no fixed vectors for the unipotent radical of any (standard) parabolic subgroup. If $P \supseteq Q$ are parabolic subgroups and U_P and U_Q are their unipotent radicals, then $U_P \subseteq U_Q$, and it follows that a representation is cuspidal if and only if it has no fixed vectors for the unipotent radical of any (standard) *maximal* parabolic subgroup; these are the subgroups of the form

$$\left\{ \begin{pmatrix} I_r & X \\ & I_t \end{pmatrix} \;\middle|\; X \in \mathrm{Mat}_{r \times t}(\mathbb{F}_q) \right\}, \qquad r + t = k. \qquad (47.3)$$

Proposition 47.1. *Let (π, V) be a cuspidal representation of $\mathrm{GL}(k, \mathbb{F}_q)$. If U is the unipotent radical of a standard maximal parabolic subgroup of $\mathrm{GL}(k, \mathbb{F}_q)$ and if $\eta : V \longrightarrow \mathbb{C}$ is any linear functional such that $\eta(\pi(u)v) = \eta(v)$ for all $u \in U$ and all $v \in V$, then η is zero.*

This means that the contragredient of a cuspidal representation is cuspidal.

Proof. Choose an invariant inner product $\langle\,,\,\rangle$ on V. There exists a vector $y \in V$ such that $\eta(v) = \langle v, y \rangle$. Then

$$\langle v, \pi(u)y \rangle = \langle \pi(u)^{-1}v, y \rangle = \eta(\pi(u)^{-1}v) = \eta(v) = \langle v, y \rangle$$

for all $u \in U$ and $v \in V$, so $\pi(u)y = y$. Since π is cuspidal, $y = 0$, whence $\eta = 0$. $\qquad\square$

Proposition 47.2. *Every irreducible representation (π, V) of $\mathrm{GL}(k, \mathbb{F}_q)$ is a constituent in some representation $\pi_1 \circ \cdots \circ \pi_m$ with the π_i cuspidal.*

Proof. If π is cuspidal, then we may take $m = 1$ and $\pi_1 = \pi$. There is nothing to prove in this case.

If π is not cuspidal, then there exists a decomposition $k = r + t$ such that the space V^U of U-fixed vectors is nonzero, where U is the group (47.3). Let $P = MU$ be the parabolic subgroup with Levi factor $M = \mathrm{GL}(r, \mathbb{F}_q) \times \mathrm{GL}(t, \mathbb{F}_q)$ and unipotent radical U. Then V^G is an M-module since M normalizes U. Let $\rho \otimes \tau$ be an irreducible constituent of M, where ρ and τ are representations of $\mathrm{GL}(r, \mathbb{F}_q)$ and $\mathrm{GL}(t, \mathbb{F}_q)$. By induction, we may embed ρ into $\pi_1 \circ \cdots \circ \pi_h$ and σ into $\pi_{h+1} \circ \cdots \circ \pi_m$ for some cuspidals π_i. Thus, we get a nonzero M-module homomorphism

$$V^U \longrightarrow \rho \otimes \tau \longrightarrow (\pi_1 \circ \cdots \circ \pi_h) \otimes (\pi_{h+1} \circ \cdots \circ \pi_m).$$

By Frobenius reciprocity (Exercise 47.2), there is thus a nonzero $\mathrm{GL}(k, \mathbb{F}_q)$-module homomorphism

$$V \longrightarrow (\pi_1 \circ \cdots \circ \pi_h) \circ (\pi_{h+1} \circ \cdots \circ \pi_m) = \pi_1 \circ \cdots \circ \pi_m.$$

Since π is irreducible, this is an embedding. $\square$

The notion of a cuspidal representation can be extended to Levi factors of parabolic subgroups. Let $\lambda = (\lambda_1, \ldots, \lambda_r)$, where the λ_i are positive integers whose sum is k. We do not assume $\lambda_i \geqslant \lambda_{i+1}$. Such a decomposition we call an *ordered partition of k*. Let

$$P_\lambda = \left\{ \begin{pmatrix} g_{11} & * & \cdots & * \\ & g_{22} & \cdots & * \\ & & \ddots & \vdots \\ & & & g_{rr} \end{pmatrix} \;\middle|\; g_{ii} \in \mathrm{GL}(\lambda_i, \mathbb{F}_q) \right\}.$$

This parabolic subgroup has Levi factor

$$M_\lambda = \mathrm{GL}(\lambda_1, \mathbb{F}_q) \times \cdots \times \mathrm{GL}(\lambda_r, \mathbb{F}_q)$$

and unipotent radical U_λ characterized by $g_{ii} = I_{\lambda_i}$. Any irreducible representation π_λ of M_λ is of the form $\otimes \pi_i$, where π_i is a representation of $\mathrm{GL}(\lambda_i, \mathbb{F}_q)$. We say that π is *cuspidal* if each of the π_i is cuspidal.

Let B_k be the standard Borel subgroup of $\mathrm{GL}(k, \mathbb{F}_q)$, consisting of upper triangular matrices, and let $B_\lambda = \prod B_{\lambda_i}$. We regard this as the Borel subgroup of M_λ. A *standard parabolic subgroup* of M_λ is a proper subgroup Q containing B_λ. Such a subgroup has the form $\prod Q_i$, where each Q_i is either $\mathrm{GL}(\lambda_i, \mathbb{F}_q)$ or a parabolic subgroup of $\mathrm{GL}(\lambda_i, \mathbb{F}_q)$ and at least one Q_i is proper. The parabolic subgroup is maximal if exactly one Q_i is a proper subgroup of $\mathrm{GL}(\lambda_i, \mathbb{F}_q)$ and that Q_i is a maximal parabolic subgroup of $\mathrm{GL}(\lambda_i, \mathbb{F}_q)$. A parabolic subgroup of M_λ has a Levi subgroup and a unipotent radical; if Q is a maximal parabolic subgroup of M_λ, then the unipotent radical of Q is the unipotent radical of the unique Q_i that is a proper subgroup of $\mathrm{GL}(\lambda_i, \mathbb{F}_q)$, and it follows that $\pi = \otimes \pi_i$ is cuspidal if and only if it has no fixed vector with respect to the unipotent radical of any maximal parabolic subgroup of M_λ.

Parabolic induction is as we have already described it for maximal parabolic subgroups. The group $P_\lambda = M_\lambda U_\lambda$ is a semidirect product with the subgroup U_λ normal, and so the composition

$$M_\lambda \longrightarrow P_\lambda \longrightarrow P_\lambda / U_\lambda$$

is an isomorphism, where the first map is inclusion and the second projection. This means that the representation π_λ of M_λ may be regarded as a representation of P_λ in which U_λ acts trivially. Then $\pi_1 \circ \cdots \circ \pi_r$ is the representation induced from P_λ.

Theorem 47.2. *The multiplication in $\mathcal{R}(q)$ is commutative.*

Proof. We will frame our proof in terms of characters rather than representations, so in this proof elements of $\mathcal{R}_k(q)$ are generalized characters of $\mathrm{GL}(k, \mathbb{F}_q)$.

We make use of the involution $\iota : \mathrm{GL}(k, \mathbb{F}_q) \longrightarrow \mathrm{GL}(k, \mathbb{F}_q)$ defined by

$$^\iota g = w_k \cdot {}^t g^{-1} \cdot w_k, \qquad w_k = \begin{pmatrix} & & 1 \\ & \cdot^{\cdot^{\cdot}} & \\ 1 & & \end{pmatrix}.$$

Let $r + t = k$. The involution takes the standard parabolic subgroup P with Levi factor $M = \mathrm{GL}(r, \mathbb{F}_q) \times \mathrm{GL}(t, \mathbb{F}_q)$ to the standard parabolic subgroup $^\iota P$ with Levi factor $^\iota M = \mathrm{GL}(t, \mathbb{F}_q) \times \mathrm{GL}(r, \mathbb{F}_q)$. It induces the map $M \longrightarrow {}^\iota M$ given by

$$\begin{pmatrix} g_1 & \\ & g_2 \end{pmatrix} \longmapsto \begin{pmatrix} {}^\iota g_2 & \\ & {}^\iota g_1 \end{pmatrix}, \qquad g_1 \in \mathrm{GL}(r, \mathbb{F}_q), g_2 \in \mathrm{GL}(t, \mathbb{F}_q),$$

where $^\iota g_1 = w_r \cdot {}^t g_1^{-1} \cdot w_r$ and $^\iota g_2 = w_t \cdot {}^t g_2^{-1} \cdot w_t$. Now since every element of $\mathrm{GL}(n, \mathbb{F}_q)$ is conjugate to its transpose, if μ is the character of an irreducible representation of $\mathrm{GL}(n, \mathbb{F}_q)$ with $n = k, r$, or t, we have $\mu(^\iota g) = \overline{\mu(g)}$. Let μ_1 and μ_2 be the characters of representations of $\mathrm{GL}(r, \mathbb{F}_q)$ and $\mathrm{GL}(t, \mathbb{F}_q)$. Composing the character $\bar\mu_2 \otimes \bar\mu_1$ of $^\iota M$ with $\iota : M \longrightarrow {}^\iota M$ and then parabolically inducing from P to $\mathrm{GL}(k, \mathbb{F}_q)$ will give the same result as parabolically inducing the character directly from $^\iota P$ and then composing with ι. The first way gives $\mu_1 \circ \mu_2$, and the second gives the conjugate of $\bar\mu_2 \circ \bar\mu_1$ (that is, $\mu_2 \circ \mu_1$), and so these are equal. $\qquad \square$

Unfortunately, the method of proof in Theorem 47.2 is rather limited. We next prove a strictly weaker result by a different method based on an analog of the intertwining integrals (47.2). These intertwining integrals are very powerful tools in the representation theory of Lie and p-adic groups, and they are closely connected with the constant terms of the Eisenstein series and with the functional equations. It is for this reason that we give a second, longer proof of a weaker statement.

Proposition 47.3. *Let (π_1, V_1) and (π_2, V_2) be representations of $\mathrm{GL}(r, \mathbb{F}_q)$ and $\mathrm{GL}(t, \mathbb{F}_q)$. Then there exists a nonzero intertwining map between the representations $\pi_1 \circ \pi_2$ and $\pi_2 \circ \pi_1$.*

Proof. Let $f \in V_1 \circ V_2$. Thus $f : G \longrightarrow V_1 \otimes V_2$ satisfies

$$f\left(\begin{pmatrix} g_1 & * \\ & g_2 \end{pmatrix} h\right) = (\pi_1(g_1) \otimes \pi_2(g_2)) f(h), \qquad g_1 \in \mathrm{GL}(r, \mathbb{F}_q), g_2 \in \mathrm{GL}(t, \mathbb{F}_q).$$
$$(47.4)$$

Now define $Mf : G \longrightarrow V_2 \otimes V_1$ by

$$Mf(h) = \tau \sum_{X \in \mathrm{Mat}_{r \times t}(\mathbb{F}_q)} f\left(\begin{pmatrix} & -I_r \\ I_t & \end{pmatrix}\begin{pmatrix} I & X \\ & I \end{pmatrix} h\right),$$

where $\tau : V_1 \otimes V_2 \longrightarrow V_2 \otimes V_1$ is defined by $\tau(v_1 \otimes v_2) = v_2 \otimes v_1$. Let us show that $Mf \in V_2 \circ V_1$. A change of variables $X \longmapsto X - Y$ in the definition of Mf shows that

$$Mf\left(\begin{pmatrix} I_r & Y \\ & I_t \end{pmatrix} h\right) = Mf(h).$$

Also, if $g_1 \in \mathrm{GL}(r, \mathbb{F}_q)$ and $g_2 \in \mathrm{GL}(t, \mathbb{F}_q)$, we have

$$Mf\left(\begin{pmatrix} g_2 & \\ & g_1 \end{pmatrix} h\right) = \tau \sum_{X \in \mathrm{Mat}_{r \times t}(\mathbb{F}_q)} f\left(\begin{pmatrix} g_1 & \\ & g_2 \end{pmatrix}\begin{pmatrix} & -I_r \\ I_t & \end{pmatrix}\begin{pmatrix} I & g_2^{-1} X g_1 \\ & I \end{pmatrix} h\right).$$

Making the variable change $X \longmapsto g_2 X g_1^{-1}$ and then using (47.4) and the fact that $\tau \circ (\pi_1(g_1) \otimes \pi_2(g_2)) = (\pi_2(g_2) \otimes \pi_1(g_1)) \circ \tau$ shows that

$$Mf\left(\begin{pmatrix} g_2 & \\ & g_1 \end{pmatrix} h\right) = (\pi_2(g_2) \otimes \pi_1(g_1)) Mf(h).$$

Thus $Mf \in V_2 \circ V_1$.

The map M is an intertwining operator since G acts on both the spaces of $\pi_1 \circ \pi_2$ and $\pi_2 \circ \pi_1$ by right translation, and $f \longmapsto Mf$ obviously commutes with right translation. We must show that it is nonzero. Choose a nonzero vector $\xi \in V_1 \otimes V_2$. Define

$$f\begin{pmatrix} A & B \\ C & D \end{pmatrix} = \begin{cases} (\pi_1(A) \otimes \pi_2(D))\xi & \text{if } C = 0, \\ 0 & \text{otherwise,} \end{cases}$$

where A, B, C and D are blocks, A being $r \times r$ and D being $t \times t$, etc. It is clear that $f \in V_1 \circ V_2$. Now

$$Mf \begin{pmatrix} & I_t \\ -I_r & \end{pmatrix} = \tau \sum_{X \in \text{Mat}_{r \times t}} f \left(\begin{pmatrix} & -I_r \\ I_t & \end{pmatrix} \begin{pmatrix} I & X \\ & I \end{pmatrix} \begin{pmatrix} & -I_t \\ I_r & \end{pmatrix} \right),$$

and the term is zero unless $X = 0$, so this equals $\tau(\xi) \neq 0$. This proves that the intertwining operator M is nonzero. □

Returning momentarily to automorphic forms, the functional equation (47.1) extends to Eisenstein series in several complex variables attached to cusp forms for general parabolic subgroups. We will not try to formulate a precise theorem, but suffice it to say that if π_i are automorphic cuspidal representations of $\text{GL}(\lambda_i, \mathbb{R})$ and $s = (s_1, \ldots, s_r) \in \mathbb{C}^r$, and if $d_s : P_\lambda(\mathbb{R}) \longrightarrow \mathbb{C}$ is the quasicharacter

$$d_s \begin{pmatrix} g_1 & & \\ & \ddots & \\ & & g_r \end{pmatrix} = |\det(g_1)|^{s_1} \cdots |\det(g_r)|^{s_r},$$

then there is a representation $\text{Ind}(\pi_1 \otimes \cdots \otimes \pi_r \otimes d_s)$ of $\text{GL}(k, \mathbb{R})$ induced parabolically from the representation $\pi_1 \otimes \cdots \otimes \pi_r \otimes d_s$ of M_λ. One may form an Eisenstein series by a series that is absolutely convergent if $\text{re}(s_i - s_j)$ are sufficiently large and that has meromorphic continuation to all s_i. There are functional equations that permute the constituents $|\det|^{s_i} \otimes \pi_i$.

If some of the π_i are equal, the Eisenstein series will have poles. The polar divisor maps out the places where the representations $\text{Ind}(\pi_1 \otimes \cdots \otimes \pi_r \otimes d_s)$ are reducible. Restricting ourselves to the subspace of $\mathbb{C}^r$ where $\sum s_i = 0$, the following picture emerges. If all of the π_i are equal, then the polar divisor will consist of $r(r - 1)$ hyperplanes in parallel pairs. There will be $r!$ points where $r - 1$ hyperplanes meet in pairs. These are the points where the induced representation $\text{Ind}(\pi_1 \otimes \cdots \otimes \pi_r \otimes d_s)$ is maximally reducible. Regarding the reducibility of representations, we will see that there are both similarities and dissimilarities with the finite field case.

Returning to the case of a finite field, we will denote by T the subgroup of diagonal matrices in $\text{GL}(k, \mathbb{F}_q)$. If α is a root, we will denote by U_α the one-dimensional unipotent of $\text{GL}(k, \mathbb{F}_q)$ corresponding to α. Thus, if $\alpha = \alpha_{ij}$ in the notation (27.7), then X_α consists of the matrices of the form $I + tE_{ij}$, where E_{ij} has a 1 in the i, jth position and zeros elsewhere.

If $\lambda = (\lambda_1, \ldots, \lambda_r)$ is an ordered partition of k, π_i are representations of $\text{GL}(\lambda_i, \mathbb{F}_q)$, and $\pi_\lambda = \pi_1 \otimes \cdots \otimes \pi_r$ is the corresponding representation of M_λ, we will use $\text{Ind}(\pi_\lambda)$ as an alternative notation for $\pi_1 \circ \cdots \circ \pi_r$.

Theorem 47.3 (Harish-Chandra). *Suppose that* $\lambda = (\lambda_1, \ldots, \lambda_r)$ *and* $\mu = (\mu_1, \ldots, \mu_t)$ *are ordered partitions of* k, *and let* $\pi_\lambda = \otimes \pi_i$ *and* $\pi'_\mu = \otimes \pi'_j$ *be cuspidal representations of* M_λ *and* M_μ, *respectively. Then*

$$\dim \text{Hom}_{\text{GL}(k, \mathbb{F}_q)}\big(\text{Ind}(\pi_\lambda), \text{Ind}(\pi'_\mu)\big)$$

is zero unless $r = t$. *If* $r = t$, *it is the number of permutations* σ *of* $\{1, 2, \ldots, r\}$ *such that* $\lambda_{\sigma(i)} = \mu_i$ *and* $\pi_{\sigma(i)} \cong \pi'_i$.

See also Harish-Chandra [62], Howe [74] and Springer [151].

Proof. Let V_i be the space of π_i and let V_i' be the space of π_i', so π_λ acts on $V = \otimes V_i$ and π_μ acts on $V' = \otimes V_i'$. By Mackey's theorem in the geometric form of Theorem 32.1, the dimension of this space of intertwining operators is the dimension of the space of functions $\Delta : \mathrm{GL}(k, \mathbb{F}_q) \longrightarrow \mathrm{Hom}_{\mathbb{C}}(V, V')$ such that for $p \in P_\lambda$ and $p' \in P_\mu$ we have

$$\Delta(p'gp) = \pi_\mu'(p')\,\Delta(g)\,\pi_\lambda(p).$$

Of course, Δ is determined by its values on a set of coset representatives for $P_\mu \backslash G / P_\lambda$, and by Proposition 46.1, these may be taken to be a set of representatives of $W_\mu \backslash W / W_\mu$, where if T is the maximal torus of diagonal elements of $\mathrm{GL}(k, \mathbb{F}_q)$, then $W = N(T)/T$, while $W_\lambda = N_{M_\lambda}(T)/T$ and $W_\mu = N_{M_\mu}(T)/T$. Thus W_{P_λ} is isomorphic to $S_{\lambda_1} \times \cdots \times S_{\lambda_r}$ and W_μ is isomorphic to $S_{\mu_1} \times \cdots \times S_{\mu_t}$.

In the terminology of Remark 32.2, let us ask under what circumstances the double coset $P_\mu w P_\lambda$ can support an intertwining operator. We assume that $\Delta(w) \neq 0$.

We will show that $w M_\lambda w^{-1} \supseteq M_\mu$. We first note that $M_\mu \cap w B_k w^{-1}$ is a (not necessarily standard) Borel subgroup of M_μ. This is because it contains T, and if α is any root of M_μ, then exactly one of U_α or $U_{-\alpha}$ is contained in $M_\mu \cap w B_k w^{-1}$ (Exercise 47.3). Now $M_\mu \cap w P_\lambda w^{-1}$ contains $M_\mu \cap w B_k w^{-1}$ and hence is either M_μ or a (not necessarily standard) parabolic subgroup of M_μ. We will show that it must be all of $M_\mu \cap w P_\lambda w^{-1}$ since otherwise its unipotent radical is $M_\mu \cap w U_\lambda w^{-1}$. Now, if $u \in M_\mu \cap w U_\lambda w^{-1}$, then $w^{-1}uw \in U_\lambda$, so

$$\Delta(w) = \Delta(u^{-1} \cdot w \cdot w^{-1}uw) = \pi_\mu'(u^{-1}) \circ \Delta(w). \tag{47.5}$$

This means that any element of the image of $\Delta(w)$ is invariant under $\pi_\mu(u)$ and hence zero by the cuspidality of π_μ. We are assuming that $\Delta(w)$ is nonzero, so this contradiction shows that $M_\mu = M_\mu \cap w P_\lambda w^{-1}$. Thus $M_\mu \subseteq w P_\lambda w^{-1}$. This actually implies that $M_\mu \subseteq w M_\lambda w^{-1}$ because if α is any root of M_μ, then P_λ contains both $w^{-1}U_\alpha w$ and $w^{-1}U_{-\alpha}w$, which implies that M_λ contains $w^{-1}U_\alpha w$, so $U_\alpha \subseteq w M_\lambda w^{-1}$. Therefore $w M_\lambda w^{-1} \supseteq M_\mu$.

Next let us show that $w M_\lambda w^{-1} \subseteq M_\mu$. As in the previous case, $M_\lambda \cap w^{-1} P_\mu w$ contains the (not necessarily standard) Borel subgroup $M_\lambda \cap w^{-1} B_k w$ of M_λ, so either it is all of M_λ or a parabolic subgroup of M_λ. If it is a parabolic subgroup, its unipotent radical is $M_\lambda \cap w^{-1}U_\mu w$. If $u \in M_\lambda \cap w^{-1}U_\mu w$, then by (47.5) we have

$$\Delta(w) = \Delta(wuw^{-1} \cdot w \cdot u^{-1}) = \Delta(w) \circ \pi_\lambda(u^{-1}).$$

By Proposition 47.1, this implies that $\Delta(w) = 0$; this contradiction implies that $M_\lambda = M_\lambda \cap w^{-1} P_\mu w$, and reasoning as before gives $M_\lambda \subseteq w^{-1} M_\mu w$.

Combining the two inclusions, we have proved that if the double coset $P_\mu w P_\lambda$ supports an intertwining operator, then $M_\mu = w M_\lambda w^{-1}$. This means $r = t$.

Now, since the representative w is only determined modulo left and right multiplication by M_μ and M_λ, respectively, we may assume that w takes positive roots of M_λ to positive roots of M_μ. Thus, a representative of w is a "block permutation matrix" of the form

$$w = \begin{pmatrix} w_{11} & \cdots & w_{1r} \\ \vdots & & \vdots \\ w_{t1} & \cdots & w_{tr} \end{pmatrix},$$

where each w_{ij} is a $\mu_i \times \lambda_j$ block, and either $w_{ij} = 0$ or $\mu_i = \lambda_j$ and w_{ij} is an identity matrix of this size, and there is exactly one nonzero w_{ij} in each row and column. Let σ be the permutation of $\{1, 2, \ldots, r\}$ such that $w_{i,\sigma(i)}$ is not zero. Thus $\lambda_{\sigma(i)} = \mu_i$, and if $g_j \in \mathrm{GL}(\lambda_j, \mathbb{F}_q)$, then we can write

$$w \begin{pmatrix} g_1 & & \\ & \ddots & \\ & & g_r \end{pmatrix} = \begin{pmatrix} g_{\sigma(1)} & & \\ & \ddots & \\ & & g_{\sigma(r)} \end{pmatrix} w.$$

Thus

$$\Delta(w) \circ \pi_\lambda \begin{pmatrix} g_1 & & \\ & \ddots & \\ & & g_r \end{pmatrix} = \pi'_\mu \begin{pmatrix} g_{\sigma(1)} & & \\ & \ddots & \\ & & g_{\sigma(r)} \end{pmatrix} \circ \Delta(w),$$

so

$$\Delta(w) \circ \left(\pi_1(g_1) \otimes \cdots \otimes \pi_r(g_r) \right) = \left(\pi'_1(g_{\sigma(1)}) \otimes \cdots \otimes \pi'_r(g_{\sigma(r)}) \right) \circ \Delta(w).$$

Since the representations π and π' of M_λ and M_μ are irreducible, Schur's lemma implies that $\Delta(w)$ is determined up to a scalar multiple, and moreover $\pi'_i \cong \pi_{\sigma(i)}$ as a representation of $\mathrm{GL}(\mu_i, \mathbb{F}_q) = \mathrm{GL}(\lambda_{\sigma(i)}, \mathbb{F}_q)$.

We see that the double cosets that can support an intertwining operator are in bijection with the permutations of $\{1, 2, \ldots, r\}$ such that $\lambda_{\sigma(i)} = \mu_i$ and $\pi_{\sigma(i)} \cong \pi'_i$ and that the dimension of the space of intertwining operators that are supported on a single such coset is 1. The theorem follows. $\square$

This theorem has some important consequences.

Theorem 47.4. *Suppose that $\lambda = (\lambda_1, \ldots, \lambda_r)$ is an ordered partition of k, and let $\pi_\lambda = \otimes \pi_i$ be a cuspidal representation of M_λ. Suppose that no $\pi_i \cong \pi_j$. Then $\pi_1 \circ \cdots \circ \pi_r$ is irreducible. Its isomorphism class is unchanged if the λ_i and π_i are permuted. If $(\mu_1, \ldots, \mu_t)$ is another ordered partition of k, and $\pi'_\mu = \pi'_1 \circ \cdots \circ \pi'_t$ is a cuspidal representation of M_μ, with the π'_i also distinct, then $\pi_1 \circ \cdots \circ \pi_r \cong \pi'_1 \circ \cdots \circ \pi'_t$ if and only if $r = t$ and there is a permutation σ of $\{1, \ldots, r\}$ such that $\mu_i = \lambda_{\sigma(i)}$ and $\pi'_i \cong \pi_{\sigma(i)}$.*

Remark 47.1. This is the *usual* case. If q is large, the probability that there is a repetition among a list of randomly chosen cuspidal representations is small.

Remark 47.2. The statement that the isomorphism class is unchanged if the λ_i and π_i are permuted is the analog of the functional equations of the Eisenstein series.

Proof. By Theorem 47.3, the dimension of the space of intertwining operators of $\text{Ind}(\pi_\lambda)$ to itself is one, and it follows that this space is irreducible. The last statement is also clear from Theorem 47.3. $\square$

Suppose now that l is a divisor of k and that $k = lt$. Let π_0 be a cuspidal representation of $\text{GL}(l, \mathbb{F}_q)$. Let us denote by $\pi_0^{\circ t}$ the representation $\pi_0 \circ \cdots \circ \pi_0$ (t copies). We call any irreducible constituent of $\pi_0^{\circ t}$ a π_0-*monatomic* irreducible representation. As a special case, if π_0 is the trivial representation of $\text{GL}(1, \mathbb{F}_q)$, this is the Hecke algebra identified in Iwahori's Theorem 46.3. There, we saw that the endomorphism ring of $\pi_0^{\circ t}$ was the Hecke algebra $\mathcal{H}_t(q)$, a deformation of the group algebra of the symmetric group S_t, and thereby obtained a parametrization of its irreducible constituents by the irreducible representations of S_t or by partitions of t. The following result generalizes Theorem 46.3.

Theorem 47.5 (Howlett and Lehrer). *Let π_0 be a cuspidal representation of $\text{GL}(l, \mathbb{F}_q)$. Then the endomorphism ring $\text{End}(\pi_0^{\circ t})$ is naturally isomorphic to $\mathcal{H}_t(q^l)$.*

Proof. Proofs may be found in Howlett and Lehrer [80] and Howe [74]. $\square$

Corollary 47.1. *There exists a natural bijection between the set of partitions λ of t and the irreducible constituents $\sigma_{\lambda(\pi)}$ of $\pi_0^{\circ t}$. The multiplicity of $\sigma_{\lambda(\pi)}$ in $\pi_0^{\circ t}$ equals the degree of the irreducible character $\mathbf{s}_\lambda$ of the symmetric group S_t parametrized by λ.*

Proof. The multiplicity of $\sigma_{\lambda(\pi)}$ in $\pi_0^{\circ t}$ equals the multiplicity of the corresponding module of $\mathcal{H}_t(q^l)$. By Exercise 46.5, this is the degree of $\mathbf{s}_\lambda$. $\square$

Although we will not make use of the multiplicative structure that is contained in this theorem of Howlett and Lehrer, we may at least see immediately that

$$\dim\left(\text{End}(\pi_0^{\circ t})\right) = t!, \tag{47.6}$$

by Theorem 47.3, taking $\mu = \lambda$ and all π_i, π_i' to be π_0. This is enough for the following result.

Theorem 47.6. *Let $(\lambda_1, \ldots, \lambda_r)$ be an ordered partition of k, and let $\lambda_i = l_i t_i$. Let π_i be a cuspidal representation of $\text{GL}(l_i, \mathbb{F}_q)$, with no two π_i isomorphic. Let θ_i be a π_i-monatomic irreducible representation of $\text{GL}(\lambda_i, \mathbb{F}_q)$. Let $\theta_\lambda = \otimes \theta_i$. Then $\text{Ind}(\theta_\lambda)$ is irreducible, and every irreducible representation* of $\mathrm{GL}(k, \mathbb{F}_q)$ *is of this type. If* $(\mu_1, \ldots, \mu_t)$ *is another ordered partition of* k, *and* θ_i' *be a family of monatomic representations of* $\mathrm{GL}(\mu_i, \mathbb{F}_q)$ *with respect to another set of distinct cuspidals, and let* $\theta_\mu' = \otimes \theta_i'$. *Then* $\mathrm{Ind}(\theta_\lambda) \cong \mathrm{Ind}(\theta_\mu')$ *if and only if* $r = t$, *and there exists a permutation* σ *of* $\{1, \ldots, r\}$ *such that* $\mu_i = \lambda_{\sigma(i)}$ *and* $\theta_i' \cong \theta_{\sigma(i)}$.

Proof. We note the following *general principle*: χ is a character of any group, and if $\chi = \sum d_i \chi_i$ is a decomposition into subrepresentations such that

$$\langle \chi, \chi \rangle = \sum d_i^2,$$

then the χ_i are irreducible and mutually nonisomorphic. Indeed, we have

$$\sum d_i^2 = \langle \chi, \chi \rangle = \sum d_i^2 \langle \chi_i, \chi_i \rangle + \sum_{i \neq j} d_i d_j \langle \chi_i, \chi_j \rangle.$$

All the inner products $\langle \chi_i, \chi_i \rangle \geqslant 1$ and all the $\langle \chi_i, \chi_j \rangle \geqslant 0$, so this implies that the $\langle \chi_i, \chi_i \rangle = 1$ and all the $\langle \chi_i, \chi_j \rangle = 0$.

Decompose each $\pi_i^{\circ t_i}$ into a direct sum $\sum_j d_{ij} \theta_{ij}$ of distinct irreducibles θ_{ij} with multiplicities d_{ij}. The representation θ_i is among the θ_{ij}. We have

$$\pi_1^{\circ t_1} \circ \cdots \circ \pi_r^{\circ t_r} = \sum_{j_1} \cdots \sum_{j_r} (d_{1j_1} \cdots d_{rj_r}) \, \theta_{1j_1} \circ \cdots \circ \theta_{rj_r}.$$

The dimension of the endomorphism ring of this module is computed by Theorem 47.3. The number of permutations of the advertised type is $t_1! \cdots t_r!$ because each permutation must map the d_i copies of π_i among themselves.

On the other hand, by (47.6), we have

$$\sum_{j_1} \cdots \sum_{j_r} (d_{1j_1} \cdots d_{rj_r})^2 = t_1! \cdots t_r!$$

also. By the "general principle" stated at the beginning of this proof, it follows that the representations $\theta_{1j_1} \circ \cdots \circ \theta_{rj_r}$ are irreducible and mutually nonisomorphic.

Next we show that every irreducible representation π is of the form $\mathrm{Ind}(\theta_\lambda)$. If π is cuspidal, then π is monatomic, and so we can just take $r = t_1 = 1$, $\theta_1 = \pi_1$. We assume that π is not cuspidal. Then by Proposition 47.2 we may embed π into $\pi_1 \circ \cdots \circ \pi_m$ for some cuspidal representations π_i. By Proposition 47.4, we may order these so that isomorphic π_i are adjacent, so π is embedded in a representation of the form $\pi_1^{\circ t_1} \circ \cdots \circ \pi_r^{\circ t_r}$, where π_i are nonisomorphic cuspidal representations. We have determined the irreducible constituents of such a representation, and they are of the form $\mathrm{Ind}(\theta_\lambda)$, where θ_i is π_i-monatomic. Hence π is of this form.

We leave the final uniqueness assertion for the reader to deduce from Theorem 47.3. $\qquad\qquad\square$

The great paper of Green [58] constructs all the irreducible representations of $GL(k, \mathbb{F}_q)$. Systematic use is made of the ring $\mathcal{R}(q)$. However, Green does not start with the cuspidal representations. Instead, Green takes as his basic building blocks certain generalized characters that are "lifts" of modular characters, described in the following theorem.

Theorem 47.7 (Green). *Let G be a finite group, and let $\rho : G \longrightarrow GL(k, \mathbb{F}_q)$ be a representation. Let $f \in \mathbb{Z}[X_1, \ldots, X_k]$ be a symmetric polynomial with integer coefficients. Let $\theta : \bar{\mathbb{F}}_q^\times \longrightarrow \mathbb{C}^\times$ be any character. Let $\chi : G \longrightarrow \mathbb{C}$ be the function*

$$\chi(g) = f\big(\theta(\alpha_1), \ldots, \theta(\alpha_k)\big).$$

Then θ is a generalized character.

Proof. First, we reduce to the following case: $\theta : \bar{\mathbb{F}}_q^\times \longrightarrow \mathbb{C}^\times$ is injective and $f(X_1, \ldots, X_k) = \sum X_i$. If this case is known, then by replacing ρ by its exterior powers we get the same result for the elementary symmetric polynomials, and hence for all symmetric polynomials. Then we can take $f(X_1, \ldots, X_k) = \sum X_i^r$, effectively replacing θ by θ^r. We may choose r to match any given character on a finite field containing all eigenvalues of all g, obtaining the result in full generality.

We recall that if l is a prime, a group is l-*elementary* if it is the direct product of a cyclic group and an l-group. According to Brauer's characterization of characters (Theorem 8.4(a) on p. 127 of Isaacs [83]), a class function is a generalized character if and only if its restriction to every l-elementary subgroup H (for all l) is a generalized character. Thus, we may assume that G is l-elementary. If p is the characteristic of $\mathbb{F}_q$, whether $l = p$ or not, we may write $G = P \times Q$ where P is a p-group and $p \nmid |Q|$. The restriction of χ to Q is a character by Isaacs, [83], Theorem 15.13 on p. 268. The result will follow if we show that $\chi(g_p q) = \chi(q)$ for $g_p \in P$, $q \in Q$. Since g_p and q commute, using the Jordan canonical form, we may find a basis for the representation space of ρ over $\bar{\mathbb{F}}_q$ such that $\rho(q)$ is diagonal and $\rho(g_p)$ is upper triangular. Because the order of g_p is a power of p, its diagonal entries are 1's, so q and $g_p q$ have the same eigenvalues, whence $\chi(g_p q) = \chi(q)$. $\square$

Since the proof of this theorem of Green is purely character-theoretic, it does not directly produce irreducible representations. And the characters that it produces are not irreducible. (We will look more closely at them later.) However, Green's generalized characters have two important advantages. First, their values are easily described. By contrast, the values of cuspidal representations are easily described on the semisimple conjugacy classes, but at other classes require knowledge of "degeneracy rules" which we will not describe. Second, Green's generalized character can be extended to a generalized character of $GL(n, \mathbb{F}_{q^r})$ for any r, a property that ordinary characters do not have.

Still, the cuspidal characters have a satisfactory direct description, which we turn to next. Choosing a basis for $\mathbb{F}_{q^k}$ as a k-dimensional vector space over $\mathbb{F}_q$ and letting $\mathbb{F}_{q^k}^\times$ act by multiplication gives an embedding $\mathbb{F}_{q^k}^\times \longrightarrow$ $GL(k, \mathbb{F}_q)$. Call the image of this embedding $T_{(k)}$. More generally, if $\lambda = (\lambda_1, \ldots, \lambda_r)$ is a partition of k, then T_λ is the group $\mathbb{F}_{q^{\lambda_1}}^\times \times \cdots \times \mathbb{F}_{q^{\lambda_r}}^\times$ embedded in $GL(k, \mathbb{F}_q)$ the same way. We will call any T_λ—or any conjugate of such a group—a *torus*. An element of $GL(k, \mathbb{F}_q)$ is called *semisimple* if it is diagonalizable over the algebraic closure of $\mathbb{F}_q$. This is equivalent to assuming that it is contained in some torus. It is called *regular semisimple* if its eigenvalues are distinct. This is equivalent to assuming that it is contained in a *unique* torus.

There is a very precise duality between the conjugacy classes of $GL(k, \mathbb{F}_q)$ and its irreducible representations. Some aspects of this duality are shown in Table 47.2. In each case, there is an exact numerical equivalence. For example, the number of unipotent conjugacy classes is the number of partitions of k, and this is also the number of unipotent representations, as we saw in Theorem 46.1. Again, the number of cuspidal representations equals the number of regular semisimple conjugacy classes whose eigenvalues generate $\mathbb{F}_{q^k}$. We will prove this in Theorem 47.8.

Table 47.2. The duality between conjugacy classes and representations

Class type	Representation type
Central conjugacy classes	One-dimensional representations
Regular semisimple conjugacy classes	Induced from distinct cuspidals
Regular semisimple conjugacy classes whose eigenvalues generate $\mathbb{F}_{q^k}$	Cuspidal representations
Unipotent conjugacy classes	Unipotent representations
Conjugacy classes whose characteristic polynomial is a power of an irreducible	Monatomic representations

To formalize this duality, and to exploit it in order to count the irreducible cuspidal representations, we will divide the conjugacy classes of $GL(k, \mathbb{F}_q)$ into "types." Roughly, two conjugacy classes have the same type if their rational canonical forms have the same shape. For example, $GL(2, \mathbb{F}_q)$ has four distinct types of conjugacy classes. They are

$$\left\{ \begin{pmatrix} a & \\ & b \end{pmatrix} \Big| a \neq b \right\}, \qquad \left\{ \begin{pmatrix} a & \\ & a \end{pmatrix} \right\},$$
$$\left\{ \begin{pmatrix} a & 1 \\ & a \end{pmatrix} \right\}, \qquad \left\{ \begin{pmatrix} & 1 \\ -\nu^{1+q} & \nu + \nu^q \end{pmatrix} \right\},$$

where the last consists of the conjugacy classes of matrices whose eigenvalues are ν and ν^q, where $\nu \in \mathbb{F}_{q^2} - \mathbb{F}_q$. In the duality, these four types of conjugacy classes correspond to the four types of irreducible representations: the $q + 1$-dimensional principal series, induced from a pair of distinct characters of $\mathrm{GL}(1)$; the one-dimensional representations $\chi \circ \det$, where χ is a character of $\mathbb{F}_q^\times$; the q-dimensional representations obtained by tensoring the Steinberg representation with a one-dimensional character; and the $q - 1$-dimensional cuspidal representations.

Let $f(X) = X^d + a_{d-1}X^{d-1} + \cdots + a_0$ be a monic irreducible polynomial over $\mathbb{F}_q$ of degree d. Let

$$U(f) = \begin{pmatrix} 0 & 1 & 0 & \cdots & 0 \\ 0 & 0 & 1 & & 0 \\ \vdots & & & \vdots & \vdots \\ 0 & 0 & 0 & \cdots & 1 \\ -a_0 & -a_1 & -a_2 & \cdots & -a_{d-1} \end{pmatrix}$$

be the rational canonical form. Let

$$U_r(f) = \begin{pmatrix} U(f) & I_d & 0 & \cdots & 0 \\ 0 & U(f) & I_d & & \\ 0 & 0 & U(f) & & \vdots \\ \vdots & & & \ddots & \\ 0 & & \cdots & & U(f) \end{pmatrix},$$

an array of $r \times r$ blocks, each of size $d \times d$. If $\lambda = (\lambda_1, \ldots, \lambda_t)$ is a partition of r, so that $\lambda_1 \geqslant \cdots \geqslant \lambda_t$ are nonnegative integers with $|\lambda| = \sum_i \lambda_i = r$, let

$$U_\lambda(f) = \begin{pmatrix} U_{\lambda_1}(f) & & \\ & \ddots & \\ & & U_{\lambda_t}(f) \end{pmatrix}.$$

Then every conjugacy class of $\mathrm{GL}(k, \mathbb{F}_q)$ has a representative of the form

$$\begin{pmatrix} U_{\lambda^1}(f_1) & & \\ & \ddots & \\ & & U_{\lambda^m(f_m)} \end{pmatrix}, \tag{47.7}$$

where the f_i are distinct monic irreducible polynomials, and each $\lambda^i = (\lambda_1^i, \lambda_2^i, \ldots)$ is a partition. The conjugacy class is unchanged if the f_i and λ^i are permuted, but otherwise, they are uniquely determined.

Thus the conjugacy class is determined by the following data: a pair of sequences $r_1, \ldots, r_m$ and $d_1, \ldots, d_m$ of integers, and for each $1 \leqslant i \leqslant m$ a partition λ^i of r_i and a monic irreducible polynomial $f_i \in \mathbb{F}_q[X]$ of degree d_i, such that no $f_i = f_j$ if $i \neq j$. The data $(\{r_i\}, \{d_i\}, \{\lambda^i\}, \{f_i\})$ and

$(\{r'_i\}, \{d'_i\}, \{(\lambda')^i\}, \{f'_i\})$ parametrize the same conjugacy class if and only if they both have the same length m and there exists a permutation $\sigma \in S_m$ such that $r'_i = r_{\sigma(i)}$, $d'_i = d_{\sigma(i)}$, $(\lambda')^i = \lambda^{\sigma(i)}$ and $f'_i = f_{\sigma(i)}$.

We say two conjugacy classes are of the same *type* if the parametrizing data have the same length m and there exists a permutation $\sigma \in S_m$ such that $r'_i = r_{\sigma(i)}$, $d'_i = d_{\sigma(i)}$, $(\lambda')^i = \lambda^{\sigma(i)}$. (The f_i and f'_i are allowed to differ.) The set of types of conjugacy classes depends on k, but is independent of q (though if q is too small, some types might be empty).

Lemma 47.1. *Let $\{N_1, N_2, \ldots\}$ be a sequence of numbers, and for each N_k let X_k be a set of cardinality N_k (X_k disjoint). Let Σ_k be the following set. An element of Σ_k consists of a 4-tuple $(\{r_i\}, \{d_i\}, \{\lambda^i\}, \{x_i\})$, where $\{r_i\} = \{r_1, \ldots, r_m\}$ and $\{d_i\} = \{d_1, \ldots, d_m\}$ are sequences of positive integers, such that $\sum r_i d_i = k$, together with a sequence $\{\lambda^i\}$ of partitions of r_i and an element $x_i \in X_{d_i}$, such that no x_i are equal. Define an equivalence relation $\sim$ on Σ_k in which two elements are considered equivalent if they can be obtained by permuting the data, that is, if $\sigma \in S_m$ then*

$$(\{r_i\}, \{d_i\}, \{\lambda^i\}, \{x_i\}) \sim (\{r_{\sigma(i)}\}, \{d_{\sigma(i)}\}, \{\lambda^{\sigma(i)}\}, \{x_{\sigma(i)}\}).$$

Let M_k be the number of equivalence classes. Then the sequence of numbers N_k is determined by the sequence of numbers M_k.

Proof. By induction on k, we may assume that the cardinalities $N_1, \ldots, N_{k-1}$ are determined by the M_k. Let M'_k be the cardinality of the set of equivalence classes of $(\{r_i\}, \{d_i\}, \{\lambda^i\}, \{x_i\}) \in \Sigma_k$ in which no $x_i \in X_k$. Clearly M'_k depends only on the cardinalities $N_1, \ldots, N_{k-1}$ of the sets $X_1, \ldots, X_{k-1}$ from which the x_i are to be drawn, so (by induction) it is determined by the M_i. Now we claim that $N_k = M_k - M'_k$. Indeed, if given $(\{r_i\}, \{d_i\}, \{\lambda^i\}, \{x_i\}) \in \Sigma_k$ of length m, if any $x_i \in X_k$, then since $\sum_{i=1}^m r_i d_i = k$, we must have $m = 1$, $r_1 = 1$, $d_1 = k$, and the number of such elements is exactly N_k. $\qquad \square$

Theorem 47.8. *The number of cuspidal representations of $\mathrm{GL}(k, \mathbb{F}_q)$ equals the number of irreducible monic polynomials of degree k over $\mathbb{F}_q$.*

Proof. We can apply the lemma with X_k either the set of cuspidal representations of S_k or with the set of monic irreducible polynomials of degree k over $\mathbb{F}_q$. We will show that in the first case, M_k is the number of irreducible representations of $\mathrm{GL}(k, \mathbb{F}_q)$, while in the second, M_k is the number of conjugacy classes. Since these are equal, the result follows.

If X_k is the set of cuspidal representations of $\mathrm{GL}(k, \mathbb{F}_q)$, from each element $(\{r_i\}, \{d_i\}, \{\lambda^i\}, \{x_i\}) \in \Sigma_k$ we can build an irreducible representation of $\mathrm{GL}(k, \mathbb{F}_q)$ as follows. First, since x_i is a cuspidal representation of $\mathrm{GL}(d_i, \mathbb{F}_q)$ we can build the x_i-monatomic representations of $\mathrm{GL}(d_i r_i, \mathbb{F}_q)$ by decomposing $x_i^{\circ r_i}$. By Corollary 47.1, the irreducible constituents of $x_i^{\circ r_i}$ are parametrized by partitions of r_i, so x_i and λ^i parametrize an x_i-monatomic representation π_i of $\mathrm{GL}(r_i d_i, \mathbb{F}_q)$. Let $\pi = \pi_1 \circ \cdots \circ \pi_m$. By Theorem 47.4, every irreducible representation of $GL(k, \mathbb{F}_q)$ is constructed uniquely (up to permutation of the π_i) in this way.

On the other, take X_k to be the set of monic irreducible polynomials of degree k over $\mathbb{F}_q$. We have explained above how the conjugacy classes of $GL(k, \mathbb{F}_q)$ are parametrized by such data.   □

Deligne and Lusztig [41] gave a parametrization of characters of any reductive group over a finite field by characters of tori. Carter [32] is a basic reference for Deligne–Lusztig characters. Many important formulae, such as a generalization of Mackey theory to cohomologically induced representations and an extension of Green's "degeneracy rules," are obtained. This theory is very satisfactory but the construction requires l-adic cohomology. For $GL(k, \mathbb{F}_q)$, the parametrization of irreducible characters by characters of tori can be described without resorting to such deep methods. The key point is the parametrization of the cuspidal characters by characters of $T_{(k)} \cong \mathbb{F}_{q^k}$. Combining this with parabolic induction gives the parametrization of more general characters by characters of other tori.

Thus let $\theta : T_{(k)} \cong \mathbb{F}_{q^k} \longrightarrow \mathbb{C}^\times$ be a character such that the orbit of θ under $\mathrm{Gal}(\mathbb{F}_{q^k}/\mathbb{F}_q)$ has cardinality k. The number of $\mathrm{Gal}(\mathbb{F}_{q^k}/\mathbb{F}_q)$-orbits of such characters is

$$\frac{1}{n} \sum_{d|n} \mu\left(\frac{n}{d}\right) q^d, \tag{47.8}$$

where μ is the Möbius function—the same as the number of semisimple conjugacy classes. Then exists a cuspidal character $\sigma_k = \sigma_{k,\theta}$ of $GL(k, \mathbb{F}_q)$ whose value on a regular semisimple conjugacy class g is zero unless g conjugate to an element of $T_{(k)}$, that is, unless the eigenvalues of g are the roots $\alpha, \alpha^q, \ldots, \alpha^{q^{k-1}}$ of an irreducible monic polynomial of degree k in $\mathbb{F}_q[X]$, so that $\mathbb{F}_{q^k} = \mathbb{F}_q[\alpha]$. In this case,

$$\sigma_k(g) = (-1)^{k-1} \sum_{j=0}^{k-1} \theta(\alpha^{q^j}).$$

By Theorem 47.8, the number of $\sigma_{k,\theta}$ is the total number of cuspidal representations, so this is a complete list.

We will first construct σ_k under the assumption that θ, regarded as a character of $\mathbb{F}_{q^k}^\times$, can be extended to a character $\theta : \bar{\mathbb{F}}_q^\times \longrightarrow \mathbb{C}^\times$ that is injective. This is assumption is too restrictive, and we will later relax it. We will also postpone showing that that σ_k is independent of the extension of θ to $\bar{\mathbb{F}}_q^\times$. Eventually we will settle these points completely in the special case where k is a prime.

Let

$$\chi_k(g) = \sum_{i=1}^{k} \theta(\alpha_i), \tag{47.9}$$

where α_i are the eigenvalues of $g \in \mathrm{GL}(k, \mathbb{F}_q)$. By Green's theorem, χ_k is a generalized character.

Proposition 47.4. *Assume that θ can be extended to a character $\theta : \bar{\mathbb{F}}_q^\times \longrightarrow \mathbb{C}^\times$ that is injective. Then the inner product $\langle \chi_k, \chi_k \rangle = k$.*

Proof. We will first prove that this is true for q sufficiently large, then show that it is true for all q. We will use "big O" notation, and denote by $O(q^{-1})$ any term that is bounded by a factor independent of q times q^{-1}. The idea of the proof is to show that as a function of q, the inner product is $k + O(q^{-1})$. Since it is an integer, it must equal k when q is sufficiently large.

The number of elements of $G = \mathrm{GL}(k, \mathbb{F}_q)$ is $q^{k^2} + O(q^{k^2-1})$. This is clear since G is the complement of the determinant locus in $\mathrm{Mat}_k(\mathbb{F}_q) \cong \mathbb{F}_q^{k^2}$. The set G_{reg} of regular semisimple elements also has order $q^{k^2} + O(q^{k^2-1})$ since it is the complement of the discriminant locus. Since $|\chi_k(g)| \leqslant k$ for all g,

$$\langle \chi_k, \chi_k \rangle = \frac{1}{|G|} \sum_{g \in G_{\mathrm{reg}}} |\chi_k(g)|^2 + O(q^{-1}).$$

Because every regular element is contained in a unique conjugate of some T_λ, which has exactly $[G : N_G(T_\lambda)]$ such conjugates, this equals

$$\frac{1}{|G|} \sum_{\lambda \text{ a partition of } k} [G : N_G(T_\lambda)] \sum_{g \in T_\lambda^{\mathrm{reg}}} |\chi_k(g)|^2 + O(q^{-1})$$

$$= \frac{1}{|G|} \sum_\lambda [G : N_G(T_\lambda)] \sum_{g \in T_\lambda} |\chi_k(g)|^2 + O(q^{-1}),$$

the last step using the fact that the complement of the T_λ^{reg} in T_λ is of codimension one. We note that the restriction of χ_k to T_λ is the sum of k distinct characters, so

$$\sum_{g \in T_\lambda} |\chi_k(g)|^2 = k|T_\lambda|.$$

Thus the inner product is

$$k \times \frac{1}{|G|} \sum_\lambda [G : N_G(T_\lambda)] |T_\lambda| + O(q^{-1}).$$

We have

$$\frac{1}{|G|} \sum_\lambda [G : N_G(T_\lambda)] |T_\lambda| = \frac{1}{|G|} \sum_\lambda [G : N_G(T_\lambda)] |T_\lambda^{\mathrm{reg}}| + O(q^{-1})$$

$$= \frac{1}{|G|} |G_{\mathrm{reg}}| + O(q^{-1})$$

$$= 1 + O(q^{-1}).$$

The result is now proved for q sufficiently large.

To prove the result for all q, we will show that the inner product $\langle \chi_k, \chi_k \rangle$ is a polynomial in q. This will follow if we can show that if S is the subset of G consisting of the union of conjugacy classes of a single type, then $[G : C_G(g)]$ is constant for $g \in S$ and

$$\sum_{g \in S} |\chi_k(g)|^2 \tag{47.10}$$

is a polynomial in q. We note that for each type, the index of the centralizer of (47.7) is the same for all such matrices, and that this index is polynomial in q. Thus it is sufficient to show that the sum over the representatives (47.7) is a polynomial in q. Moreover, the value of χ_k is unchanged if every instance of a $U_r(f)$ is replaced with r blocks of $U(f)$, so we may restrict ourselves to semisimple conjugacy classes in confirming this. Thus if $k = \sum d_i r_i$, we consider the sum (47.10), where the sum is over all matrices

$$\begin{pmatrix} U_{(r_1)}(f_1) & & \\ & \ddots & \\ & & U_{(r_m)}(f_m) \end{pmatrix},$$

where f_i are *distinct* irreducible polynomials, each of size d_i, and $U_{(r)}(f)$ is the sum of r blocks of $U(f)$. It is useful to conjugate these matrices so that they are all elements of the same torus T_λ for some λ. The set S is then a subset of T_λ characterized by exclusion from certain (non-maximal) subtori.

Let us look at an example. Suppose that $\lambda = (2, 2, 2)$ and $k = 6$. Then S consists of elements of T_λ, which may be regarded as $(\mathbb{F}_{q^2})^\times$ of the form (α, β, γ), where α, β and γ are distinct elements of $\mathbb{F}_{q^2}^\times - \mathbb{F}_q^\times$. Now if we sum (47.10) over *all* of T_λ we get a polynomial in q, namely $6(q^2 - 1)^3$. On the other hand, we must subtract from this three contributions when one of α, β and γ is in $\mathbb{F}_q^\times$. These are subtori of the form $T_{(2,2,1)}$. We must also subtract three contributions from subgroups of the form $T_{(2,2)}$ in which two of α, β, and γ are equal. Then we must add back contributions that have been subtracted twice, and so on.

In general, the set S will consist of the set T_λ minus subtori $T_1, \ldots, T_N$. If I is a subset of $\{1, \ldots, N\}$ let $T_I = \bigcap_{i \in I} T_i$. We now use the inclusion–exclusion principle in the form

$$\sum_{g \in S} |\chi_k(g)|^2 = \sum_{g \subset T_\lambda} |\chi_k(g)|^2 + \sum_{\varnothing \neq I \subseteq \{1,\ldots,N\}} (-1)^{|I|} \sum_{g \in T_I} |\chi_k(g)|^2.$$

Each of the sums on the right is easily seen to be a polynomial in q, and so is (47.10). □

Theorem 47.9. *Assume that θ is an injective character $\theta : \bar{\mathbb{F}}_q^\times \longrightarrow \mathbb{C}^\times$. For each k there exists a cuspidal $\sigma_k = \sigma_{k,\theta}$ of $\mathrm{GL}(k, \mathbb{F}_q)$ such that if g is a regular semisimple element of $\mathrm{GL}(k, \mathbb{F}_q)$ with eigenvalues that are the Galois conjugates of $\nu \in \mathbb{F}_{q^k}^\times$ such that $\mathbb{F}_{q^k} = \mathbb{F}_q(\nu)$, then*

$$\sigma_{k,\theta}(g) = (-1)^{k-1} \sum_{i=0}^{k-1} \theta(\nu^{q^i}). \tag{47.11}$$

If 1_k denotes the trivial character of $\mathrm{GL}(k, \mathbb{F}_q)$, then

$$\chi_n = \sum_{k=1}^{n} (-1)^{k-1} \sigma_k \circ 1_{n-k}.$$

Note that $\sigma_k \circ 1_{n-k}$ is an irreducible character of $\mathrm{GL}(n, \mathbb{F}_q)$ by Theorem 47.4. So this gives the expression of χ_n in terms of irreducibles.

Proof. By induction, we assume the existence of σ_k and the decomposition of χ_k as stated for $k < n$, and we deduce them for $k = n$.

We will show first that

$$\langle \chi_n, \sigma_k \circ 1_{n-k} \rangle = (-1)^{k-1}. \tag{47.12}$$

Let $P = MU$ be the standard parabolic subgroup with Levi factor $M = \mathrm{GL}(k, \mathbb{F}_q) \times \mathrm{GL}(n-k, \mathbb{F}_q)$ and unipotent radical U. If $m \in M$ and $u \in U$, then as matrices in $\mathrm{GL}(n, \mathbb{F}_q)$, m and mu have the same characteristic polynomials, so $\chi_n(mu) = \chi_n(m)$. Thus, in the notation of Exercise 47.2(ii), with $\chi = \chi_n$, we have $\chi_U = \chi$ restricted to M. Therefore,

$$\langle \chi_n, \sigma_k \circ 1_{n-k} \rangle_G = \langle \chi_n, \sigma_k \otimes 1_{n-k} \rangle_M.$$

Let

$$m = \begin{pmatrix} m_1 & \\ & m_2 \end{pmatrix} \in M, \qquad m_1 \in \mathrm{GL}(k, \mathbb{F}_q), m_2 \in \mathrm{GL}(n-k, \mathbb{F}_q).$$

Clearly, $\chi_n(m) = \chi_k(m_1) + \chi_{n-k}(m_2)$. Now using the induction hypothesis, χ_{n-k} does not contain the trivial character of $\mathrm{GL}(n-k, \mathbb{F}_q)$, hence it is orthogonal to 1_{n-k} on $\mathrm{GL}(n-k, \mathbb{F}_q)$; so we can ignore $\chi_{n-k}(m_2)$. Thus,

$$\langle \chi_n, \sigma_k \circ 1_{n-k} \rangle_G = \langle \chi_k, \sigma_k \rangle_{\mathrm{GL}(k, \mathbb{F}_q)}.$$

By the induction hypothesis, χ_k contains σ_k with multiplicity $(-1)^{k-1}$, and so (47.12) is proved.

Now $\sigma_k \circ 1_{n-1}$ is an irreducible representation of $\mathrm{GL}(n, \mathbb{F}_q)$, by Theorem 47.4, and so we have exhibited $n - 1$ irreducible characters, each of which occurs in χ_n with multiplicity ± 1. Since $\langle \chi_n, \chi_n \rangle = n$, there must be one remaining irreducible character σ_n such that

$$\chi_n = \sum_{k=1}^{n-1} (-1)^{k-1} \sigma_k \circ 1_{n-k} \pm \sigma_n. \tag{47.13}$$

We show now that σ_n must be cuspidal. It is sufficient to show that if U is the unipotent radical of the standard parabolic subgroup with Levi factor

$M = \mathrm{GL}(k, \mathbb{F}_q) \times \mathrm{GL}(n-k, \mathbb{F}_q)$, and if $m_1 \in \mathrm{GL}(k, \mathbb{F}_q)$ and $m_2 \in \mathrm{GL}(n-k, \mathbb{F}_q)$ then

$$\frac{1}{|U|} \sum_{u \in U} \chi_n \left(u \begin{pmatrix} m_1 & \\ & m_2 \end{pmatrix} \right) = \frac{1}{|U|} \sum_{r=1}^{n-1} (-1)^{r-1} (\sigma_r \circ 1_{n-r}) \left(u \begin{pmatrix} m_1 & \\ & m_2 \end{pmatrix} \right),$$

since by Exercise 47.2(ii), this will show that the representation affording the character σ_n has no U-invariants, the definition of cuspidality. The summand on the left-hand side is independent of u, and by the definition of χ_n the left-hand side is just $\chi_k(m_1) + \chi_{n-k}(m_2)$. By Exercise 47.4, the right-hand side can also be evaluated. Using (47.11), which we have assumed inductively for σ_r with $r < n$, the terms $r = k$ and $r = n - k$ contribute $\chi_k(m_1)$ and $\chi_{n-k}(m_2)$ and all other terms are zero.

To evaluate the sign in (47.13), we compare the values at the identity to get the relation

$$n = \sum_{k=1}^{n-1} (-1)^{k-1} \binom{n}{k}_{(q)} \prod_{j=1}^{k-1} (q^j - 1) \pm \prod_{j=1}^{n-1} (q^j - 1),$$

where

$$\binom{n}{k}_{(q)} = \frac{\prod_{j=1}^{n} (q^j - 1)}{\left(\prod_{j=1}^{k} (q^j - 1) \right) \left(\prod_{j=1}^{n-k} (q^j - 1) \right)}$$

is the Gaussian binomial coefficient, which is the index of the parabolic subgroup with Levi factor $\mathrm{GL}(k) \times \mathrm{GL}(n - k)$. Substituting $q = 0$ in this identity shows that the missing sign must be $(-1)^{n-1}$.

If g is a regular element of $T_{(k)}$, then the value of σ_k on a regular element of $T_{(k)}$ is now given by (47.11) since if $k < n$ then $\sigma_k \circ 1_{n-k}$ vanishes on g, which is not conjugate to any element of the parabolic subgroup from which $\sigma_k \circ 1_{n-k}$ is induced.   □

See Exercise 47.9 for an example showing that the cuspidal characters that we have constructed are not enough because of our assumption that θ is injective. Without attempting a completely general result, we will now give a variation of Theorem 47.9 that is sufficient to construct all cuspidal representations of $\mathrm{GL}(k, \mathbb{F}_q)$ when k is prime.

Proposition 47.5. *Let* $\theta : \overline{\mathbb{F}}_q^\times \longrightarrow \mathbb{C}^\times$ *be a character. Assume that the restriction of* θ *to* $\mathbb{F}_q^\times$ *is trivial, but that for any* $0 < d \leqslant k$, *the restriction of* θ *to* $\mathbb{F}_{q^d}^\times$ *does not factor through the norm map* $\mathbb{F}_{q^d}^\times \longrightarrow \mathbb{F}_{q^r}^\times$ *for any proper divisor* r *of* d. *Then*

$$\langle \chi_k, \chi_k \rangle = k + 1.$$

Proof. The proof is similar to Proposition 47.4. It is sufficient to show this for sufficiently large q. As in that proposition, the sum is

$$\frac{1}{|G|} \sum_{\lambda \text{ a partition of } k} [G : N_G(T_\lambda)] \sum_{g \in T_\lambda} |\chi_k(g)|^2 + O(q^{-1}).$$

We note that $[N_G(T_\lambda) : T_\lambda] = z_\lambda$, defined in (37.1). With our assumptions if the partition λ contains r parts of size 1, the restriction of χ_k to T_λ consists of r copies of the trivial character, and $k - r$ copies of other characters, all distinct. (Exercise 47.8.) The inner product of χ_k with itself on T_λ is thus $k - r + r^2$. The sum is thus

$$\sum_\lambda \frac{1}{z_\lambda}(k + r^2 - r) + O(q^{-1}).$$

We can interpret this as a sum over the symmetric group. If $\sigma \in S_k$, let $r(\sigma)$ be the number of fixed points of σ. In the conjugacy class of shape λ, there are $k!/z_\lambda$ elements, and so

$$\sum_\lambda \frac{1}{z_\lambda}(k + r^2 - r) = \frac{1}{k!} \sum_{\sigma \in S_k} (k + r(\sigma)^2 - r(\sigma)).$$

Now $r(\sigma) = \mathbf{h}_{(k-1,1)} = \mathbf{s}_{(k-1,1)} + \mathbf{h}_k$ in the notation of Chap. 37. Here, of course, $\mathbf{h}_k = \mathbf{s}_{(k)}$ is the trivial character of S_k and $\mathbf{s}_{(k-1,1)}$ is an irreducible character of degree $k-1$. We note that $r(\sigma)^2 - r(\sigma)$ is the value of the character $\mathbf{s}_{(k-1,1)}^2 + \mathbf{s}_{(k-1,1)}$, so the sum is

$$\left\langle k\mathbf{h}_k + \mathbf{s}_{(k-1,1)}^2 + \mathbf{s}_{(k-1,1)}, \mathbf{h}_k \right\rangle = k \left\langle \mathbf{h}_k, \mathbf{h}_k \right\rangle + \left\langle \mathbf{s}_{(k-1,1)}^2, \mathbf{h}_k \right\rangle + \left\langle \mathbf{s}_{(k-1,1)}, \mathbf{h}_k \right\rangle$$

where the inner product is now over the symmetric group. Clearly $\langle \mathbf{h}_k, \mathbf{h}_k \rangle = 1$ and $\langle \mathbf{s}_{(k-1,1)}, \mathbf{h}_k \rangle = 0$. Since the character $\mathbf{s}_{(k-1,1)}$ is real and $\mathbf{h}_k$ is the constant function equal to 1,

$$\left\langle \mathbf{s}_{(k-1,1)}^2, \mathbf{h}_k \right\rangle = \left\langle \mathbf{s}_{(k-1,1)}, \mathbf{s}_{(k-1,1)} \right\rangle = 1,$$

and the result follows. $\qquad\square$

Theorem 47.10. *Suppose that n is a prime, and let $\theta : \mathbb{F}_{q^n}^\times \longrightarrow \mathbb{C}^\times$ be a character that does not factor through the norm map $\mathbb{F}_{q^n}^\times \longrightarrow \mathbb{F}_{q^r}^\times$ for any proper divisor r of n. Then there exists a cuspidal character $\sigma_{n,\theta}$ of $\mathrm{GL}(n, \mathbb{F}_q)$ such that if g is a regular semisimple element with eigenvalues $\nu, \nu^q, \ldots \in \mathbb{F}_{q^n}$ then*

$$\sigma_{n,\theta}(g) = (-1)^{n-1} \sum_{i=0}^{n-1} \theta(\nu^{q^i}). \tag{47.14}$$

This gives a complete list of the cuspidal characters of $\mathbb{F}_{q^n}$.

The assumption that n is prime is unnecessary.

Proof. By Exercise 47.11, we can extend θ to a character of $\bar{\mathbb{F}}_q$ without enlarging the kernel. Thus the kernel of θ is contained in $\mathbb{F}_{q^n}^\times$ and does not contain the kernel of any norm map $\mathbb{F}_{q^n}^\times \longrightarrow \mathbb{F}_{q^r}^\times$ for any proper divisor r of n. There are now two cases.

If χ is nontrivial on $\mathbb{F}_q^\times$, then we may proceed as in Theorem 47.9. We are not in the case of that theorem, since we have not assumed that the kernel of θ is trivial, and we do not guarantee that the sequence of cuspidals σ_k that we construct can be extended to all k. However, if $d \leqslant k$, our assumptions guarantee that the restriction of θ to $\mathbb{F}_{q^d}^\times$ does not factor through the norm map to $\mathbb{F}_{q^r}$ for any proper divisor of d, since the kernel of θ is contained in $\mathbb{F}_{q^n}$, whose intersection with $\mathbb{F}_{q^d}$ is just $\mathbb{F}_q$ since n is prime and $d < n$. In particular, the kernel of θ cannot contain the kernel of $N : \mathbb{F}_{q^d}^\times \longrightarrow \mathbb{F}_{q^r}^\times$. We get $\langle \chi_k, \chi_k \rangle = k$ for $k \leqslant n$, and proceeding as in Theorem 47.9 we get a sequence of cuspidal representations σ_k of $\mathrm{GL}(k, \mathbb{F}_q)$ with $k \leqslant n$ such that

$$\chi_k = \sum_{r=1}^{k} (-1)^{r-1} \sigma_r \circ 1_{k-r}.$$

If θ is trivial on $\mathbb{F}_q^\times$, it is still true that the restriction of θ to $\mathbb{F}_{q^d}$ does not factor through the norm map to $\mathbb{F}_{q^r}$ for any proper divisor of d whenever $k \leqslant n$. So $\langle \chi_k, \chi_k \rangle = k + 1$ by Theorem 47.5. Now, we can proceed as before, except that $\sigma_1 = 1_1$, so $\sigma_1 \circ 1_{k-1}$ is not irreducible—it is the sum of two irreducible representations $\mathbf{s}_{(k-1,1)}(q)$ and $\mathbf{s}_{(k)}(q)$ of $\mathrm{GL}(k, \mathbb{F}_q)$, in the notation of Chap. 46. Of course, $\mathbf{s}_{(k)}(q)$ is the same as 1_k in the notation we have been using. The rest of the argument goes through as in Theorem 47.9. In particular the inner product formula $\langle \chi_k, \chi_k \rangle = k + 1$ together with fact that $1_1 \circ 1_{k-1}$ accounts for two representations in the decomposition of χ_k guarantees that σ_k, defined to be $\chi_k - \sum_{r<k} (-1)^r \sigma_r \circ 1_{k-r}$ is irreducible.

The cuspidal characters we have constructed are linearly independent by (47.14). They are equal in number to the total number of cuspidal representations, and so we have constructed all of them. $\qquad\square$

Let us consider next representations of reductive groups over local fields. The problem is to parametrize irreducible representations of Lie and p-adic groups such as $\mathrm{GL}(k, F)$, where $F = \mathbb{R}$, $\mathbb{C}$ or a non-Archimedean local field.

The parametrization of irreducible representations by characters of tori, which we have already seen for finite fields, extends to representations of Lie and p-adic groups such as $\mathrm{GL}(k, F)$, where $F = \mathbb{R}$, $\mathbb{C}$ or a non-Archimedean local field. If T is a maximal torus of $G = \mathrm{GL}(k, F)$, then the characters of T parametrize certain representations of G. As we will explain, not all admissible representations can be parametrized by characters of tori, though (as we will explain) in some sense *most* are so parametrized. Moreover, if we expand the parametrization we can get a bijection. This is the *local Langlands correspondence*, which we will now discuss (though without formulating a precise statement).

In this context, a *torus* is the group of rational points of an algebraic group that, over the algebraic closure of F, is isomorphic to a product of r copies of the multiplicative group G_m. (See Chap. 24.) The torus is called *anisotropic* if it has no subtori isomorphic to G_m over F. If $F = \mathbb{R}$, an anisotropic torus is compact. For example, $\mathrm{SL}(2,\mathbb{R})$ contains two conjugacy classes of maximal tori—the diagonal torus, and the compact torus $\mathrm{SO}(2)$. Over the complex numbers, the group $\mathrm{SO}(2,\mathbb{C})$ is conjugate by the Cayley transform to the diagonal subgroup, since if $a^2 + b^2 = 1$, then

$$c \begin{pmatrix} a & b \\ -b & a \end{pmatrix} c^{-1} = \begin{pmatrix} a + bi & \\ & a - bi \end{pmatrix}, \qquad c = \frac{1}{\sqrt{2i}} \begin{pmatrix} 1 & i \\ 1 & -i \end{pmatrix}.$$

Thus, $\mathrm{SO}(2)$ is an anisotropic torus. If G is semisimple, then G has an anisotropic maximal torus if and only if its maximal compact subgroup K has the same rank as G. An examination of Table 28.1 shows that this is sometimes true and sometimes not. For example, by Proposition 28.3, this will be the case if G/K is a Hermitian symmetric space. The group $\mathrm{SO}(n,1)$ has anisotropic maximal tori if n is even, but not if n is odd. $\mathrm{SL}(k,\mathbb{R})$ does only if $k = 2$.

If F is a local field and E/F is an extension of degree k, then, as in the case of a finite field, we may embed $E^\times \longrightarrow \mathrm{GL}(k,F)$, and the norm one elements will be an anisotropic torus of $\mathrm{SL}(k,F)$. From this point of view, we see why $\mathrm{SL}(2,\mathbb{R})$ is the only special linear group over $\mathbb{R}$ that has an anisotropic maximal torus—the algebraic closure $\mathbb{C}$ of $\mathbb{R}$ is too small.

Let G be a locally compact group and Z its center. Let (π, V) be an irreducible unitary representation of G. By Schur's lemma, $\pi(z)$ acts by a scalar $\omega(z)$ of absolute value 1 for $z \in Z$. Let $L^2(G, \omega)$ be the space of all functions f on G such that $f(zg) = \omega(z)f(g)$ and

$$\int_{G/Z} |f(g)|^2 \, dg < \infty.$$

The group G acts on $L^2(G, \omega)$ by right translation. The representation π is said to be in the *discrete series* if it can be embedded as a direct summand in $L^2(G, \omega)$. If G is a reductive group over a local field, the irreducible representations of G can be built up from discrete series representations of Levi factors of parabolic subgroups by parabolic induction.

Let F be a local field, and let E/F be a finite extension. Then the (relative) Weil group $W_{E/F}$ is a certain finite extension of $E^\times$. It fits in an exact sequence:

$$1 \longrightarrow E^\times \longrightarrow W_{E/F} \longrightarrow \mathrm{Gal}(E/F) \longrightarrow 1.$$

If $E' \supset E$ is a bigger field, there is a canonical map $W_{E'/F} \longrightarrow W_{E/F}$ inducing the norm map $E' \longrightarrow E$, and the absolute Weil group W_F is the inverse limit of the $W_{E/F}$. The discrete series representations of $\mathrm{GL}(k,F)$ are then parametrized by the irreducible k-dimensional complex representations of $W_{E/F}$. This is a slight oversimplification—we are neglecting the Steinberg representation and a few other discrete series that can be parametrized by replacing $W_{E/F}$ by the slightly larger Weil–Deligne group.

This parametrization of irreducible representations of $\mathrm{GL}(k, F)$ by *local Langlands correspondence*. Borel [19] is still a useful reference for the Langlands correspondences, though the correspondence must be made more precise than the formulation in this paper, written before many of the results were proved. Henniart's ICM talk [68] is a good more recent reference. The local Langlands conjectures for $\mathrm{GL}(k)$ over non-Archimedean local fields of characteristic zero were proved by Harris and Taylor [63]. The p-adic case had been proved earlier by Laumon, Rappoport, and Stuhler, and another proof was given soon after Harris and Taylor by Henniart [67].

Assume that $G = \mathrm{GL}(k)$ over a local field F. We now explain why most but not all discrete series representations correspond to characters of anisotropic tori. If T is a maximal torus of G, then T/Z is anisotropic if $T \cong E^\times$ where E/F is an extension of degree k. If θ is a character of $E^\times$ then inducing θ to $W_{E/F}$ gives a representation of $W_{E/F}$ of degree k. This gives a parametrization of many—even most—discrete series representations by characters of tori. In fact, if F is non-Archimedean and the residue characteristic is prime to k, then every irreducible representation is of this form. This is proved in Tate [159] (2.2.5.3). A simple proof when $k = 2$ is given in Bump [27], Proposition 4.9.3.

Although the parametrization of the discrete series representations by characters of tori is thus a more complex story for local fields than for finite fields, the construction of the irreducible representations by parabolic induction still follows the same pattern as in the finite field case. An analog of Theorem 47.3 is true, and the method of proof extends—the function Δ becomes a distribution, and the corresponding analog of Mackey theory is due to Bruhat [26]. Some differences occur because of measure considerations. There are *important* differences between the finite field case and the local field case when reducibility occurs. The finite field statement Corollary 47.1 is both *suggestive* and *misleading* when looking at the local field case. See Zelevinsky [177]. Zelevinsky's complete results are reviewed in Harris and Taylor [63].

Turning at last to automorphic forms, characters of tori still parametrize automorphic representations, and characters of anisotropic tori parametrize automorphic cuspidal representations. Thus, if E/F is an extension of number fields with $[E : F] = k$ and A_E is the adele ring of E, and if θ is a character of $A_E^\times/E^\times$, then there should exist an automorphic representation of $\mathrm{GL}(k, F)$ whose L-function is the same as the L-function of θ. If E/F is cyclic, this is a theorem of Arthur and Clozel [8], Sect. 3.6. In contrast with the situation over local fields, however, where "most" discrete series are parametrized by characters of tori, the cuspidal representations obtained this way are rare. A few more are obtained if we allow parametrizations by the global Weil group, but even these are in the minority. The literature on this topic is too vast to survey here, but we mention one result: in characteristic p, the Langlands parametrization of global automorphic forms on $GL(n)$ was proved by Lafforgue [114].

Exercises

Exercise 47.1 (Transitivity of parabolic induction).

(i) Let P be a parabolic subgroup of $\mathrm{GL}(k)$ with Levi factor M and unipotent radical U, so $P = MU$. Suppose that Q is a parabolic subgroup of M with Levi factor M_Q and unipotent radical U_Q. Show that M_Q is the Levi factor of a parabolic subgroup R of $\mathrm{GL}(k)$ with unipotent radical $U_Q U$.

(ii) In the setting of (i), show that parabolic induction from M_Q directly to $\mathrm{GL}(k)$ gives the same result as parabolically inducing first to M, and then from M to $\mathrm{GL}(k)$.

(iii) Show that the multiplication $\circ$ is associative and that $\mathcal{R}(q)$ is a ring.

Exercise 47.2 (Frobenius reciprocity for parabolic induction). Let $P = MU$ be a parabolic subgroup of $G = \mathrm{GL}(n, \mathbb{F}_q)$.

(i) Let (π, V) be a representation of G and let (σ, W) be a representation of M. Let V^U be the space of U-invariants in V. Since M normalizes U, V^U is an M-module. On the other hand, we may parabolically induce W to a representation $\mathrm{Ind}(\sigma)$ of G. Show that

$$\mathrm{Hom}_G\big(V, \mathrm{Ind}(\sigma)\big) \cong \mathrm{Hom}_M(V^U, W).$$

(**Hint:** Make use of Theorem 32.2. We need to show that

$$\mathrm{Hom}_P(V, W) \cong \mathrm{Hom}_M(V^U, W).$$

Let V_0 be the span of elements of the form $w - \pi(u)w$ with $u \in U$. Show that $V = V^U \oplus V_0$, as M-modules, and that any P-equivariant map $V \longrightarrow W$ factors through $V/V_0 \cong V^U$.)

(ii) Let χ be a character of G, and let σ be a character of M. Let $\mathrm{Ind}(\sigma)$ be the character of the representation of G parabolically induced from σ, and let χ_U be the function on M defined by

$$\chi_U(m) = \frac{1}{|U|} \sum_{u \in U} \chi(mu).$$

Show that χ_U is a class function on M, and that

$$\langle \chi, \mathrm{Ind}(\sigma)\rangle_G = \langle \chi_U, \sigma\rangle_M.$$

Conclude that χ_U is a character of M. [**Note:** Although this statement is closely related to (i), and may be deduced from it, this may also be proved using (32.16) and Frobenius reciprocity for characters, avoiding use of (i).]

Exercise 47.3. Suppose that H is a subgroup of $\mathrm{GL}(k, \mathbb{F}_q)$ containing T such that for each $\alpha \in \Phi$ the group H contains either X_α or $X_{-\alpha}$. Show that H is a (not necessarily standard) parabolic subgroup. If H contains exactly one of X_α or $X_{-\alpha}$ for each $\alpha \in S$, show that H is a (not necessarily standard) Borel subgroup. (See Exercise 20.1.)

The next exercise is very analogous to the computation of the constant terms of Eisenstein series. For example, the computation around pages 39–40 of Langlands [117] is a near exact analog.

Exercise 47.4. Let $1 \leqslant k, r < n$. Let σ_1, σ_2 be monatomic characters of $\mathrm{GL}(r, \mathbb{F}_q)$ and $\mathrm{GL}(n - r, \mathbb{F}_q)$ with respect to a pair of distinct cuspidal representations. Let σ denote the character of the representation $\sigma_1 \circ \sigma_2$ of $\mathrm{GL}(n, \mathbb{F}_q)$, which is irreducible by Theorem 47.6. Let $m_1 \in \mathrm{GL}(k, \mathbb{F}_q)$ and $m_2 \in \mathrm{GL}(n - k, \mathbb{F}_q)$. Let U be the unipotent radical of the standard parabolic subgroup P of $\mathrm{GL}(n, \mathbb{F}_q)$ with Levi factor $M = \mathrm{GL}(k, \mathbb{F}_q) \times \mathrm{GL}(n - k, \mathbb{F}_q)$. if $k = r, k \neq n - r$,

$$\frac{1}{|U|} \cdot \sum_{u \in U} \sigma \left(u \begin{pmatrix} m_1 & \\ & m_2 \end{pmatrix} \right) = \begin{cases} \sigma_1(m_1)\, \sigma_2(m_2) & \text{if } k = r, k \neq n - r, \\ \sigma_1(m_2)\, \sigma_2(m_1) & \text{if } k = n - r, k \neq r, \\ \sigma_1(m_1)\, \sigma_2(m_2) + \sigma_1(m_2)\, \sigma_2(m_1) & \text{if } k = r = n - r. \end{cases}$$

[**Hint:** Both sides are class functions, so it is sufficient to compare the inner products with $\rho_1 \otimes \rho_2$ where ρ_1 and ρ_2 are irreducible representations of $\mathrm{GL}(k, \mathbb{F}_q)$ and $\mathrm{GL}(n - k, \mathbb{F}_q)$, respectively. Using Exercise 47.2 this amounts to comparing $\sigma_1 \circ \sigma_2$ and $\rho_1 \circ \rho_2$. To do this, explain why in the last statement in Theorem 47.6 the assumption that the θ_i' are monatomic with respect to distinct cuspidals may be omitted provided this assumption is made for the θ_i.]

Exercise 47.5. If $k + l = m$, and if $P = MU$ is the standard parabolic of $\mathrm{GL}(m, \mathbb{F}_q)$ with Levi factor $M = \mathrm{GL}(k, \mathbb{F}_q) \times \mathrm{GL}(l, \mathbb{F}_q)$, then the space of U-invariants of any representation (π, V) of $\mathrm{GL}(m, \mathbb{F}_q)$ is an M-module. Show that this functor from representations of $\mathrm{GL}(m, \mathbb{F}_q)$ to representations of $\mathrm{GL}(k, \mathbb{F}_q) \times \mathrm{GL}(l, \mathbb{F}_q)$ can be made the basis of a comultiplication in $\mathcal{R}(q)$ and that $\mathcal{R}(q)$ is a Hopf algebra.

Exercise 47.6. Let $G = \mathrm{GL}(k, \mathbb{F}_q)$. As in Exercise 45.4, let N be the subgroup of upper triangular unipotent matrices. Let $\psi : \mathbb{F}_q \longrightarrow \mathbb{C}^\times$ be a nontrivial additive character, and let ψ_N be the character of N defined by

$$\psi_N \begin{pmatrix} 1 & x_{12} & x_{13} & \cdots & x_{1k} \\ & 1 & x_{23} & \cdots & x_{2k} \\ & & 1 & & \\ & & & \ddots & \vdots \\ & & & & 1 \end{pmatrix} = \psi(x_{12} + x_{23} + \cdots + x_{k-1,k}).$$

Let P be the "mirabolic" subgroup of $g \in G$ where the bottom row is $(0, \ldots, 0, 1)$. (Note that P is not a parabolic subgroup.) Call an irreducible representation of P *cuspidal* if it has no U-fixed vector for the unipotent radical U of any standard parabolic subgroup of G. Note that U is contained in P for each such U. If $1 \leqslant r < k$ let G_r be $\mathrm{GL}(r, \mathbb{F}_q)$ embedded in G in the upper left-hand corner, and let N^r be the subgroup of $x \in N$ in which $x_{ij} = 0$ if $i < j \leqslant r$.

(i) Show that the representation $\kappa = \mathrm{Ind}_N^P(\psi)$ is irreducible. [**Hint:** Use Mackey theory to compute $\mathrm{Hom}_P(\kappa, \kappa)$.]

(ii) Let (π, V) be a cuspidal representation of P. Let L_r be the set of all linear functionals λ on V such that $\lambda(\pi(x)v) = \psi_N(x)v$ for $v \in V$ and $x \in L_r$. Show that if $\lambda \in L_r$ and $r > 1$ then there exists $\gamma \in G_{r-1}$ such that $\lambda' \in L_{r-1}$, where $\lambda'(v) = \lambda(\pi(\gamma)v)$.

(iii) Show that the restriction of an cuspidal representation π of $\mathrm{GL}(k, \mathbb{F}_q)$ to P is a direct sum of copies of κ. Then use Exercise 45.4 to show that at most one copy can occur, so $\pi|_P = \kappa$.

(iv) Show that each irreducible cuspidal representation of $GL(k, \mathbb{F}_q)$ has dimension $(q - 1)(q^2 - 1) \cdots (q^{k-1} - 1)$.

Exercise 47.7. Let $\theta : \mathbb{F}_{q^k}^\times \longrightarrow \mathbb{C}^\times$ be a character.

(i) Show that the following are equivalent.
  (a) The character θ does not factor through the norm map $\mathbb{F}_{q^k} \longrightarrow \mathbb{F}_{q^d}$ for any proper divisor d of k.
  (b) The character θ has k distinct conjugates under $\mathrm{Gal}(\mathbb{F}_{q^k}/\mathbb{F}_q)$.
  (c) We have $\theta^{q^r - 1} \neq 1$ for all divisors r of k.
(ii) Show that the number of such θ satisfying these equivalent conditions given by (47.8), and that this is also the number of monic irreducible polynomials of degree k over $\mathbb{F}_q$.

Exercise 47.8. Suppose that $\theta : \bar{\mathbb{F}}_q^\times \longrightarrow \mathbb{C}^\times$ is a character. Suppose that for all $d \leqslant k$, the restriction of θ to $\mathbb{F}_{q^d}^\times$ does not factor through the norm map $\mathbb{F}_{q^d}^\times \longrightarrow \mathbb{F}_{q^r}^\times$ for any proper divisor r of d. Let λ be a partition of k. Show that the restriction of θ to T_λ contains the trivial character with multiplicity r, equal to the number of parts of λ of size 1, and $k - r$ other characters that are all distinct from one another.

Exercise 47.9. Obtain a character table of $GL(2, \mathbb{F}_3)$, a group of order 48. Show that there are three distinct characters θ of $\mathbb{F}_9^\times$ such that θ does not factor through the norm map $\mathbb{F}_{q^k} \longrightarrow \mathbb{F}_{q^d}$ for any proper divisor of d. Of these, two (of order eight) can be extended to an injective homomorphism $\bar{\mathbb{F}}_3^\times \longrightarrow \mathbb{C}^\times$, but the third (of order four) cannot. If θ is this third character, then χ_2 defined by (47.9) defines a character that splits as $\chi_{\mathrm{triv}} + \chi_{\mathrm{steinberg}} - \sigma_2$, where χ_{triv} and $\chi_{\mathrm{steinberg}}$ are the trivial and Steinberg characters, and σ_2 is the character of a cuspidal representation. Show also that σ_2 differs from the sum of the two one-dimensional characters of $GL(2, \mathbb{F}_3)$ only on the two non-semisimple conjugacy classes, of elements of orders 3 and 6.

Exercise 47.10. Suppose that χ is an irreducible representation of $GL(k, \mathbb{F}_q)$. Let g be a regular semisimple element with eigenvalues that generate $\mathbb{F}_{q^k}$. If $\chi(g) \neq 0$, show that χ is monatomic.

Exercise 47.11. Let θ be a character of $\mathbb{F}_q$. Show that there exists a character $\bar{\theta}$ of $\bar{\mathbb{F}}_q$ extending θ, whose kernel is the same as that of θ.

Exercise 47.12. Let θ be an injective character of $\bar{\mathbb{F}}_q$. Prove the following result.

Theorem. *Let λ be a partition of n and let $t \in T_\lambda$. Then $\sigma_{k,\theta}(t) = 0$ unless $\lambda = (n)$.*

Hint: Assume by induction that the statement is true for all $k < n$. Write $t = (t_1, \ldots, t_r)$ where $t_i \in GL(\lambda_i, \mathbb{F}_q)$ has distinct eigenvalues in $\mathbb{F}_{q^{\lambda_i}}$. Show that

$$(\sigma_k \circ 1_{n-k})(t) = \sum_{\lambda_i} \sigma_k(t_i).$$

Cohomology of Grassmannians

In this chapter, we will deviate from our usual policy of giving complete proofs in order to explain some important matters. Among other things, we will see that the ring $\mathcal{R}$ introduced in Chap. 34 has yet another interpretation in terms of the cohomology of Grassmannians.

References for this chapter are Fulton [53], Hiller [70], Hodge and Pedoe [71], Kleiman [102], and Manivel [126].

We recall the notion of a *CW-complex*. Intuitively, this is just a space decomposed into open cells, the closure of each cell being contained in the union of cells of lower dimension—for example, a simplicial complex. (See Dold [44], Chap. 5, and the appendix in Milnor and Stasheff [129].) Let $\mathbb{B}_n$ be the closed unit ball in Euclidean n-space. Let $\mathbb{B}_n^\circ$ be its interior, the unit disk, and let $\mathbb{S}_{n-1}$ be its boundary, the $n-1$ sphere. We are given a Hausdorff topological space X together with set $\mathcal{S}$ of subspaces of X. It is assumed that X is the disjoint union of the $C_i \in \mathcal{S}$, which are called *cells*. Each space $C_i \in \mathcal{S}$ is homeomorphic to $\mathbb{B}_{d(i)}^\circ$ for some $d(i)$ by a homeomorphism $\varepsilon_i : \mathbb{B}_{d(i)}^\circ \longrightarrow C_i$ that extends to a continuous map $\varepsilon_i : \mathbb{B}_{d(i)} \longrightarrow X$. The image of $\mathbb{S}_{d(i)-1}$ under ε_i lies in the union of cells C_i of strictly lower dimension. Thus, if we define the *n-skeleton*

$$ X_n = \bigcup_{d(i) \leqslant n} C_i, $$

the image of $\mathbb{S}_{d(i)-1}$ under ε_i is contained in $X_{d(i)-1}$. It is assumed that its image is contained in only finitely many C_i and that X is given the *Whitehead topology*, in which a subset of X is closed if and only if its intersection with each C_i is closed.

Let K be a compact Lie group, T a maximal compact subgroup, and X the flag manifold K/T. We recall from Theorem 26.4 that X is naturally a complex analytic manifold. The reason (we recall) is that we can identify $X = G/B$ where G is the complexification of K and B is its Borel subgroup.

The Lefschetz fixed-point formula can be used to show that the Euler characteristic of X is equal to the order of the Weyl group W. Suppose that

D. Bump, *Lie Groups*, Graduate Texts in Mathematics 225,
DOI 10.1007/978-1-4614-8024-2_48, © Springer Science+Business Media New York 2013

M is a manifold of dimension n and $f : M \longrightarrow M$ a map. We define the *Lefschetz number* of f to be

$$\Lambda(f) = \sum_{d=0}^{n} (-1)^d \operatorname{tr}\big(f | H^d(M, \mathbb{Q})\big).$$

A *fixed point* of f is a solution to the equation $f(x) = x$. The fixed point x is *isolated* if it is the only fixed point in some neighborhood of x. According to the "Lefschetz fixed-point formula," if M is a compact manifold and f has only isolated fixed points, the Lefschetz number is the number of fixed points *counted with multiplicity*; see Dold [43].

Let $g \in K$, and let $f = f_g : X \to X$ be translation by g. If g is the identity, then f induces the identity map on X and hence on its cohomology in every dimension. Therefore, the Euler characteristic is $\Lambda(f)$. On the other hand, $\Lambda(f)$ is unchanged if f is replaced by a homotopic map, so we may compute it by moving g to a generator of T. (We are now thinking of X as K/T.) Then $f(hT) = hT$ if and only if g is in the normalizer of T, so there is one fixed point for each Weyl group element. The local Lefschetz number, which is the multiplicity of the fixed point in the fixed point formula, may also be computed for each fixed point (see Adams [2]) and equals 1. So $\Lambda(f) = |W|$, and this is the Euler characteristic of X.

It is possible to be a bit more precise than this: $H^i(X) = 0$ unless i is even and $\sum_i \dim H^{2i}(X) = |W|$. We will explain the reason for this now.

We may give a cell decomposition making X into a CW-complex as follows. If $w \in W$, then BwB/B is homeomorphic to $\mathbb{C}^{l(w)}$, where l is the length function on W. The proof is the same as Proposition 46.7: the unipotent subgroup U_w^- which has the Lie algebra is

$$\bigoplus_{\alpha \in \Phi^+ \cap w\Phi^-} \mathfrak{X}_\alpha$$

is homeomorphic to $\mathbb{C}^{l(w)}$, and $u \longmapsto uwB$ is a homeomorphism of U_w^- onto BwB/B. The closure $\mathcal{C}(w)$ of BwB/B—known as a "closed Schubert cell"— is a union of cells of smaller dimension, so G/B becomes a CW complex. Since the homology of a CW-complex is the same as the cellular homology of its skeleton (Dold [44], Chap. 5), and all the cells in this complex have even dimension—the real dimension of BwB/B is $2l(w)$—it follows that the homology of X is all even-dimensional.

Since X is a compact complex analytic manifold (Theorem 26.4), it is an orientable manifold, and by Poincaré duality we may associate with $\mathcal{C}(w)$ a cohomology class, and these classes span the cohomology ring $H^*(X)$ as a vector space.

This description can be recast in the language of algebraic geometry. A substitute for the cohomology ring was defined by Chow [37]. See Hartshorne [64], Appendix A, for a convenient synopsis of the Chow ring, and see Fulton [54] for a modern treatment. In the graded Chow ring of a nonsingular variety X, the homogeneous elements of degree r are rational equivalence classes of algebraic cycles. Here an *algebraic cycle* of codimension r is an element of the free Abelian group generated by the irreducible subvarieties of codimension r. *Rational equivalence* of cycles is an equivalence relation of algebraic deformation. For divisors, which are cycles of codimension 1, it coincides with the familiar relation of *linear equivalence*. We recall that two divisors D_1 and D_2 are *linearly equivalent* if $D_1 - D_2$ is the divisor of a function f in the function field of X.

The multiplication in the Chow ring is the intersection of cycles. If two subvarieties Y and Z (of codimensions m and n) are given, we say that Y and Z *intersect properly* if every irreducible component of $Y \cap Z$ has codimension $m + n$. (If $m + n$ exceeds the dimension of X, this means that Y and Z have an empty intersection.) *Chow's lemma* asserts that Y and Z may be deformed to intersect properly. That is, there exist Y' and Z' rationally equivalent to Y and Z, respectively, such that Y and Z' intersect properly. The intersection $X \cap Z$ is then a union of cycles of codimension $m + n$, whose sum in the Chow ring is $Y \cap Z$. (They must be counted with a certain *intersection multiplicity*.)

The "moving" process embodied by Chow's lemma will be an issue for us when we consider the intersection pairing in Grassmannians, so let us contemplate a simple case of intersections in $\mathbb{P}^n$. Hartshorne [64], I.7, gives a beautiful and complete treatment of intersection theory in $\mathbb{P}^n$.

The space $\mathbb{P}^n(\mathbb{C})$, which we will come to presently, resembles flag manifolds and Grassmannians in that the Chow ring and the cohomology ring coincide. (Indeed, $\mathbb{P}^n(\mathbb{C})$ is a Grassmannian.) The homology of $\mathbb{P}^n(\mathbb{C})$ can be computed very simply since it has a cell decomposition in which each cell is an affine space $\mathbb{A}^i \cong \mathbb{C}^i$.

$$\mathbb{P}^n(\mathbb{C}) = \mathbb{C}^n \cup \mathbb{C}^{n-1} \cup \cdots \cup \mathbb{C}^0, \qquad \dim(\mathbb{C}^i) = 2i. \tag{48.1}$$

Each cell contributes to the homology in exactly one dimension, so

$$H_i\big(\mathbb{P}^n(\mathbb{C})\big) \cong \begin{cases} \mathbb{Z} & \text{if } i \leqslant 2n \text{ is even,} \\ 0 & \text{otherwise.} \end{cases}$$

The cohomology is the same by Poincaré duality. The multiplicative structure in the ring $H^*\big(\mathbb{P}^n(\mathbb{C})\big)$ is that of a truncated polynomial ring. The cohomology class of a hyperplane $[\mathbb{C}^{n-1}$ in the decomposition (48.1)] is a generator.

Let us consider the intersection of two curves Y and Z in $\mathbb{P}^2(\mathbb{C})$. The intersection $Y \cdot Z$, which is the product in the Chow ring, 1, is a cycle of degree zero, that is, just a sum of points. The rational equivalence class of a cycle of degree zero is completely determined by the number of points, and intersection theory on $\mathbb{P}^2$ is fully described if we know how to compute this number.

Each curve is the locus of a homogeneous polynomial in three variables, and the degree of this polynomial is the *degrees* of the curves, $d(Y)$ and $d(Z)$.

According to *Bezout's theorem*, the number of points in the intersection of Y and Z equals $d(Y)\,d(Z)$.

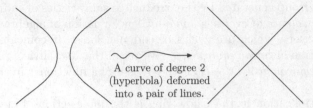

A curve of degree 2 (hyperbola) deformed into a pair of lines.

Fig. 48.1. A curve of degree d in $\mathbb{P}^2$ is linearly equivalent to d lines

Bezout's theorem can be used to illustrate Chow's lemma. First, note that a curve of degree d is rationally equivalent to a sum of d lines (Fig. 48.1), so Y is linearly equivalent to a sum of $d(Y)$ lines, and Z is linearly equivalent to a sum of $d(Z)$ lines. Since two lines have a unique point of intersection, the first set of $d(Y)$ lines will intersect the second set of $d(Z)$ lines in exactly $d(Y)\,d(Z)$ points, which is Bezout's theorem for $\mathbb{P}^2$ (Fig. 48.2).

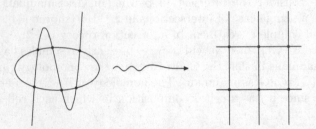

Fig. 48.2. Bezout's theorem via Chow's lemma

It is understood that a point of transversal intersection is counted once, but a point where Y and Z are tangent is counted with a multiplicity that can be defined in different ways.

The intersection $Y \cdot Z$ must be defined even when the cycles Y and Z are equal. For this, one may replace Z by a rationally equivalent cycle before taking the intersection. The self-intersection $Y \cdot Y$ is computed using Chow's lemma, which allows one copy of Y to be deformed so that its intersection with the undeformed Y is transversal (Fig. 48.3). Thus, replacing Y by a rationally equivalent cycle, one may count the intersections straightforwardly (Fig. 48.2).

The Chow ring often misses much of the cohomology. For example, if X is a curve of genus $g > 1$, then $H^1(X) \cong \mathbb{Z}^{2g}$ is nontrivial, yet the cohomology of an algebraic cycle of codimension d lies in $H^{2d}(X)$, and is never odd-dimensional. However, if X is a flag variety, projective space, or Grassmannian, the Chow ring and the cohomology ring are isomorphic. The cup product corresponds to the intersection of algebraic cycles.

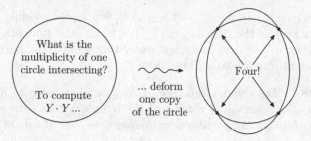

Fig. 48.3. The self-intersection multiplicity of a cycle in $\mathbb{P}^2$

Let us now consider intersection theory on G/P, where P is a parabolic subgroup, that is, a proper subgroup of G containing B. For such a variety, the story is much the same as for the flag manifold—the Chow ring and the cohomology ring can be identified, and the Bruhat decomposition gives a decomposition of the space as a CW-complex. We can write

$$B \backslash G / P \cong W / W_P,$$

where W_P is the Weyl group of the Levi factor of P. If $G = \mathrm{GL}(n)$, this is Proposition 46.1(iii). If $w \in W$, let $\mathcal{C}(w)^\circ$ be the *open Schubert cell* BwP/P, and let $\mathcal{C}(w)$ be its closure, which is the union of $\mathcal{C}(w)^\circ$ and open Schubert cells of lower dimension. The *closed Schubert cells* $\mathcal{C}(w)$ give a basis of the cohomology.

We will discuss the particular case where $G = \mathrm{GL}(r + s, \mathbb{C})$ and P is the maximal parabolic subgroup

$$\left\{ \begin{pmatrix} g_1 & * \\ & g_2 \end{pmatrix} \middle| g_1 \in \mathrm{GL}(r, \mathbb{C}), g_2 \in \mathrm{GL}(s, \mathbb{C}) \right\}$$

with Levi factor $M = \mathrm{GL}(r, \mathbb{C}) \times \mathrm{GL}(s, \mathbb{C})$. The quotient $X_{r,s} = G/P$ is then the *Grassmannian*, a compact complex manifold of dimension rs. In this case, the cohomology ring $H^*(X_{r,s})$ is closely related to the ring $\mathcal{R}$ introduced in Chap. 34.

To explain this point, let us explain how to "truncate" the ring $\mathcal{R}$ and obtain a finite-dimensional algebra that will be isomorphic to $H^*(X_{r,s})$. Suppose that $\mathcal{J}_r$ is the linear span of all $\mathbf{s}_\lambda$ such that the length of λ is $> r$. Then $\mathcal{J}_r$ is an ideal, and the quotient $\mathcal{R}/\mathcal{J}_r \cong \Lambda^{(r)}$ by the characteristic map. Indeed, it follows from Proposition 36.5 that $\mathcal{J}_r$ is the kernel of the homomorphism $\mathrm{ch}^{(n)} : \mathcal{R} \longrightarrow \Lambda^{(n)}$.

We can also consider the ideal ${}^\iota \mathcal{J}_s$, where ι is the involution of Theorem 34.3. By Proposition 35.2, this is the span of the $\mathbf{s}_\lambda$ in which the length of λ^t is greater than s—in other words, in which $\lambda_1 > s$. So $\mathcal{J}_r + {}^\iota \mathcal{J}_s$ is the span of all $\mathbf{s}_\lambda$ such that the diagram of λ does not fit in an $r \times s$ box. Therefore, the ring $\mathcal{R}_{r,s} = \mathcal{R}/(\mathcal{J}_r + {}^\iota \mathcal{J}_s)$ is spanned by the images of $\mathbf{s}_\lambda$ where the diagram of λ *does* fit in an $r \times s$ box. For example, $\mathcal{R}_{3,2}$ is spanned by $\mathbf{s}_{()}$, $\mathbf{s}_{(1)}$, $\mathbf{s}_{(2)}$, $\mathbf{s}_{(11)}$,

$s_{(21)}$, $s_{(22)}$, $s_{(111)}$, $s_{(211)}$, $s_{(221)}$, and $s_{(222)}$. It is a free $\mathbb{Z}$-module of rank 10. In general the rank of the ring $\mathcal{R}_{r,s}$ is $\binom{r+s}{r}$, which is the number of partitions of $r + s$ of length $\leqslant r$ into parts not exceeding s—that is, partitions with diagrams that fit into a box of dimensions $r \times s$.

Theorem 48.1. *The cohomology ring of $X_{r,s}$ is isomorphic to $\mathcal{R}_{r,s}$. In this isomorphism, the cohomology classes of the Schubert cells correspond to the s_λ, as λ runs through the partitions with diagrams that fit into an $r \times s$ box.*

We will not prove this. Proofs (all rather similar and based on a method of Hodge) may be found in Fulton [53], Hiller [70], Hodge and Pedoe [71], and Manivel [126]. We will instead give an informal discussion of the result, including a precise description of the isomorphism and an example.

Let us explain how to associate a partition λ with a diagram that is contained in the $r \times s$ box with a Schubert cell of codimension equal to $|\lambda|$. In fact, to every coset wW_P in W/W_P we will associate such a partition.

Right multiplication by an element of $W_P \cong S_r \times S_s$ consists of reordering the first r columns and the last s columns. Hence, the representative w of the given coset in W/W_P may be chosen to be a permutation matrix such that the entries in the first r columns are in ascending order, and so that the entries in the last s columns are in ascending order. In other words, if σ is the permutation such that $w_{\sigma(j),j} \neq 0$, then

$$\sigma(1) < \sigma(2) < \cdots < \sigma(r), \qquad \sigma(r+1) < \sigma(r+2) < \cdots < \sigma(r+s). \quad (48.2)$$

With this choice, we associate a partition λ as follows. We *mark* some of the zero entries of the permutation matrix w as follows. If $1 \leqslant j \leqslant r$, if the 1 in the ith row is in the last s columns, and if the 1 in the jth column is above (i, j), then we mark the (i, j)th entry. For example, if $r = 3$ and $s = 2$, here are a some examples of a marked matrix:

$$\begin{pmatrix} 1 & & & & \\ & 1 & & & \\ \bullet & \bullet & 1 & & \\ & & & 1 & \\ \bullet & \bullet & & & 1 \end{pmatrix}, \quad \begin{pmatrix} 1 & & & & \\ \bullet & & 1 & & \\ & & 1 & & \\ & & & 1 & \\ \bullet & \bullet & \bullet & & 1 \end{pmatrix}, \quad \begin{pmatrix} 1 & & & & \\ \bullet & & & 1 & \\ \bullet & & & & 1 \\ & & 1 & & \\ & & & 1 & \end{pmatrix} \quad (48.3)$$

Now, we collect the marked columns and read off the permutation. For each row containing marks, there will be a part of the permutation equal to the number of marks in that row. In the three examples above, the respective permutations λ are:

$$(2, 2, 1), \qquad (2, 1, 1), \qquad (2).$$

Their diagrams fit into a 2×3 box. We will write $\mathcal{C}_\lambda$ for the closed Schubert cell $\mathcal{C}(w)$ when λ and w are related this way.

Let F_i be the vector subspace of $\mathbb{C}^{r+s}$ consisting of vectors of the form ${}^t(x_1, \ldots, x_i, 0, \ldots, 0)$. The group G acts on the *Grassmannian* $\mathcal{G}_{r,s}$ of r-dimensional subspaces of $\mathbb{C}^{r+s}$. The stabilizer of F_r is precisely the parabolic subgroup P, so there is a bijection $X_{r,s} \longrightarrow \mathcal{G}_{r,s}$ in which the coset $gP \longmapsto gF_r$. We topologize $\mathcal{G}_{r,s}$ by asking that this map be a homeomorphism.

We can characterize the Schubert cells in terms of this parametrization by means of integer sequences. Given a sequence $(d) = (d_0, d_1, \ldots, d_{r+s})$ with

$$0 \leqslant d_0 \leqslant d_1 \leqslant \cdots \leqslant d_{r+s} = r, \qquad 0 \leqslant d_i \leqslant 1, \tag{48.4}$$

we can consider the set $\mathfrak{C}^\circ_{(d)}$ of V in $\mathcal{G}_{r,s}$ such that

$$\dim(V \cap F_i) = d_i. \tag{48.5}$$

Let $\mathfrak{C}_{(d)}$ be the set of V in $\mathcal{G}_{r,s}$ such that

$$\dim(V \cap F_i) \geqslant d_i. \tag{48.6}$$

The function $V \longmapsto \dim(V \cap F_i)$ is upper semicontinuous on $\mathcal{G}_{r,s}$, that is, for any integer n, $\{V \mid \dim(V \cap F_i) \geqslant n\}$ is closed. Therefore, $\mathfrak{C}_{(d)}$ is closed, and in fact it is the closure of $\mathfrak{C}^\circ_{(d)}$.

Lemma 48.1. *In the characterization of $\mathfrak{C}_{(d)}$ it is only necessary to impose the condition (48.6) at integers $0 < i < r + s$ such that $d_{i+1} = d_i > d_{i-1}$.*

Proof. If $d_{i+1} > d_i$ and $\dim(V \cap F_{i+1}) \geqslant d_{i+1}$, then since $V \cap F_i$ has codimension at most 1 in $V \cap F_{i+1}$ we do not need to assume $\dim(V \cap F_i) \geqslant d_i$. If $d_i = d_{i-1}$ and $\dim(V \cap F_{i-1}) \geqslant d_{i-1}$ then $\dim(V \cap F_{i-1}) \geqslant d_{i-1}$. $\square$

We will show $\mathfrak{C}^\circ_{(d)}$ is the image in $\mathcal{G}_{r,s}$ of an open Schubert cell. For example, with $r = 3$ and $s = 2$, taking w to be the first matrix in (48.3), we consider the Schubert cell BwP/P, which has the image in $\mathcal{G}_{3,2}$ that consists of all bwF_3, where $b \in B$. A one-dimensional unipotent subspace of B is sufficient to produce all of these elements, and a typical such space consists of all matrices of the form

$$\begin{pmatrix} 1 & & & & \\ & 1 & & & \\ & & 1 & \alpha & \\ & & & 1 & \\ & & & & 1 \end{pmatrix} \begin{pmatrix} 1 & & & & \\ & 1 & & & \\ & & & 1 & \\ & & 1 & & \\ & & & & 1 \end{pmatrix} \begin{pmatrix} x_1 \\ x_2 \\ x_3 \\ 0 \\ 0 \end{pmatrix} = \begin{pmatrix} x_1 \\ x_2 \\ \alpha x_3 \\ x_3 \\ 0 \end{pmatrix}$$

with α fixed. These may be characterized by the conditions (48.5) with

$$(d_0, \ldots, d_5) = (0, 1, 2, 2, 3, 3).$$

Proposition 48.1. *The image in $\mathcal{G}_{r,s}$ of the Schubert cell $C(w)$ corresponding to the partition λ (the diagram of which, we have noted, must fit in an $r \times s$ box) is $\mathfrak{C}_{(d)}$, where the integer sequence $(d_0, d_1, \ldots, d_{r+s})$ where*

$$d_k = i \qquad \Longleftrightarrow \qquad s + i - \lambda_i \leqslant k \leqslant s + i - \lambda_{i+1}. \tag{48.7}$$

Similarly, the image of $C(w)^\circ$ is $\mathfrak{C}_{(d)}^\circ$.

We note that, by Lemma 48.1, if (d) is the sequence in (48.7), the closed Schubert cell $\mathfrak{C}_{(d)}$ is characterized by the conditions

$$\dim(V \cap F_{s+i-\lambda_i}) \geqslant i. \tag{48.8}$$

Also, by Lemma 48.1, this only needs to be checked when $\lambda_i > \lambda_{i+1}$. [The characterization of the *open* Schubert cell still requires $\dim(V \cap F_k)$ to be specified for *all* k, not just those of the form $s + i - \lambda_i$.]

Proof. We will prove this for the open cell. The image of $C(w)^\circ$ in $\mathcal{G}_{r,s}$ consists of all spaces bwF_r with $b \in B$, so we must show that, with d_i as in (48.7), we have

$$\dim(bwF_r \cap F_i) = d_i.$$

Since b stabilizes F_i, we may apply b^{-1}, and we are reduced to showing that

$$\dim(wF_r \cap F_i) = d_i.$$

If σ is the permutation such that $w_{\sigma(j),j} \neq 0$, then the number of entries below the nonzero element in the ith column, where $1 \leqslant i \leqslant r$, is $r + s - \sigma(i)$. However, $r - i$ of these are not "marked." Therefore, $\lambda_i = (r + s - \sigma(i)) - (r - i)$, that is,

$$\sigma(i) = s + i - \lambda_i. \tag{48.9}$$

Now wF_r is the space of vectors that have arbitrary values in the $\sigma(1)$, $\sigma(2), \ldots, \sigma(r)$ positions, and all other entries are zero. So the dimension of $wF_r \cap F_i$ is the number of k such that $1 \leqslant j \leqslant r$ and $\sigma(k) \leqslant i$. Using (48.2),

$$\dim(wF_r \cap F_i) = k \qquad \Longleftrightarrow \qquad \sigma(i) \leqslant k < \sigma(i+1),$$

which by (48.9) is equivalent to (48.7). $\qquad \Box$

When (d) and λ are related as in (48.7), we will also denote the Schubert cell $\mathfrak{C}_{(d)}$ by $\mathfrak{C}_\lambda$.

As we asserted earlier, the cohomology ring $X_{r,s}$ is isomorphic to the quotient $\mathcal{R}_{r,s}$ of the ring $\mathcal{R}$, which has played such a role in this last part of the book. To get some intuition for this, let us consider the identity in $\mathcal{R}$

$$\mathbf{s}_{(1)} \cdot \mathbf{s}_{(1)} = \mathbf{s}_{(2)} + \mathbf{s}_{(11)}.$$

By the parametrization we have given, s_λ corresponds to the Schubert cell $\mathfrak{C}_\lambda$. In the case at hand, the relevant cells are characterized by the following conditions:

$$\mathfrak{C}_{(1)} = \{V \mid \dim(V \cap F_s) \geqslant 1\},$$
$$\mathfrak{C}_{(2)} = \{V \mid \dim(V \cap F_{s-1}) \geqslant 1\},$$
$$\mathfrak{C}_{(11)} = \{V \mid \dim(V \cap F_{s+1}) \geqslant 2\}.$$

So our expectation is that if we deform $\mathfrak{C}_{(1)}$ into two copies $\mathfrak{C}'_{(1)}$ and $\mathfrak{C}''_{(1)}$ that intersect properly, the intersection will be rationally equivalent to the sum of $\mathfrak{C}_{(2)}$ and $\mathfrak{C}_{(11)}$. We may choose spaces G_s and H_s of codimension s such that $G_s \cap H_s = F_{s-1}$ and $G_s + H_s = F_{s+1}$. Now let us consider the intersection of

$$\mathfrak{C}'_{(1)} = \{V \mid \dim(V \cap G_s) \geqslant 1\}, \qquad \mathfrak{C}''_{(1)} = \{V \mid \dim(V \cap H_s) \geqslant 1\}.$$

If V lies in both $\mathfrak{C}'_{(1)}$ and $\mathfrak{C}''_{(1)}$, then let v' and v'' be nonzero vectors in $V \cap G_s$ and $V \cap H_s$, respectively. There are two possibilities. Either v' and v'' are proportional, in which case they lie in $V \cap F_{s-1}$, so $V \in \mathfrak{C}_{(2)}$, or they are linearly independent. In the second case, both lie in F_{s+1}, so $V \in \mathfrak{C}_{(11)}$.

The intersection theory of flag manifolds is very similar to that of Grassmannians. The difference is that while the cohomology of Grassmannians for $\mathrm{GL}(r)$ is modeled on the ring $\mathcal{R}$, which can be identified as in Chap. 34 with the ring Λ of *symmetric* polynomials, the cohomology of flag manifolds is modeled on a polynomial ring. Specifically, if B is the Borel subgroup of $G = \mathrm{GL}(r, \mathbb{C})$, then the cohomology ring of G/B is a quotient of the polynomial ring $\mathbb{Z}[x_1, \ldots, x_r]$, where each x_i is homogeneous of degree 2. Lascoux and Schützenberger defined elements of the polynomial ring $\mathbb{Z}[x_1, \ldots, x_r]$ called *Schubert polynomials* which play a role analogous to that of the Schur polynomials (See Fulton [53] and Manivel [126]).

A minor problem is that $H^*(G/B)$ is not precisely the polynomial ring $\mathbb{Z}[x_1, \ldots, x_r]$ but a quotient, just as $H^*(\mathcal{G}_{r,s})$ is not precisely $\mathcal{R}$ or even its quotient $\mathcal{R}/\mathcal{J}_r$, which is isomorphic to the ring of symmetric polynomials in $\mathbb{Z}[x_1, \ldots, x_r]$.

The ring $\mathbb{Z}[x_1, \ldots, x_r]$ should be more properly regarded as the cohomology ring of an infinite CW-complex, which is the cohomology ring of the space $\mathcal{F}_r$ of r-flags in $\mathbb{C}^\infty$. That is, let $\mathcal{F}_{r,s}$ be the space of r-flags in $\mathbb{C}^{r+s}$:

$$\{0\} = F_0 \subset F_1 \subset F_2 \subset \cdots \subset F_r \subset \mathbb{C}^{r+s}, \qquad \dim(F_i) = i. \qquad (48.10)$$

We can regard $\mathcal{F}_{r,s}$ as G/P, where P is the parabolic subgroup

$$\left\{ \begin{pmatrix} b & * \\ & g \end{pmatrix} \mid b \in B, \, g \in \mathrm{GL}(r, \mathbb{C}) \right\}. \qquad (48.11)$$

We may embed $\mathcal{F}_{r,s} \hookrightarrow \mathcal{F}_{r,s+1}$, and the union of the $\mathcal{F}_{r,s}$ (topologized as the direct limit) is $\mathcal{F}_r$. The open Schubert cells in $\mathcal{F}_{r,s}$ correspond to double cosets $B\backslash G/P$ parametrized by elements $w \in S_{r+s}/S_s$. As we increase s, the CW-complex $\mathcal{F}_{r,s}$ is obtained by adding new cells, but only in higher dimension. The n-skeleton stabilizes when s is sufficiently large, and so $H^n(\mathcal{F}_r) \cong H^n(F_{r,s})$ if s is sufficiently large. The ring $H^*(\mathcal{F}_r) \cong \mathbb{Z}[x_1, \ldots, x_r]$ is perhaps the natural domain of the Schubert polynomials.

The cohomology of Grassmannians (and flag manifolds) provided some of the original evidence for the famous conjectures of Weil [170] on the number of points on a variety over a finite field. Let us count the number of points of $X_{r,s}$ over the field $\mathbb{F}_q$ with field elements. Representing the space as $\mathrm{GL}(n, \mathbb{F}_q)/P(\mathbb{F}_q)$, where $n = r + s$, its cardinality is

$$\frac{|\mathrm{GL}(n, \mathbb{F}_q)|}{|P(\mathbb{F}_a)|} = \frac{(q^n-1)(q^n-q)\cdots(q^n-q^{n-1})}{(q^r-1)(q^r-q)\cdots(q^r-q^{r-1})\cdot(q^s-1)(q^s-q)\cdots(q^s-q^{s-1})\cdot q^{rs}}.$$

In the denominator, we have used the Levi decomposition of $P = MU$, where the Levi factor $M = \mathrm{GL}(r) \times \mathrm{GL}(s)$ and the unipotent radical U has dimension rs. This is a Gaussian binomial coefficient $\binom{n}{r}_q$. *It is a generating function for the cohomology ring $H^*(X_{r,s})$.*

Motivated by these examples and other similar ones, as well as the examples of projective nonsingular curves (for which there is cohomology in dimension 1, so that the Chow ring and the cohomology ring are definitely distinct), Weil proposed a more precise relationship between the complex cohomology of a nonsingular projective variety and the number of solutions over a finite field. Proving the Weil conjectures required a new cohomology theory that was eventually supplied by Grothendieck. This is the l-adic cohomology. Let $\bar{F}_q$ be the algebraic closure of $\mathbb{F}_q$, and let $\phi : X \longrightarrow X$ be the geometric Frobenius map, which raises the coordinates of a point in X to the qth power. The fixed points of ϕ are then the elements of $X(\mathbb{F}_q)$, and they may be counted by means of a generalization of the Lefschetz fixed-point formula:

$$|X(F_q)| = \sum_{k=0}^{2n}(-1)^k \operatorname{tr}(\phi|H^k).$$

The dimensions of the l-adic cohomology groups are the same as the complex cohomology, and in these examples (since all the cohomology comes from algebraic cycles) the odd-dimensional cohomology vanishes while on $H^{2i}(X)$ the Frobenius endomorphism acts by the scalar q^i. Thus,

$$|X(F_q)| = \sum_{k=0}^{n} \dim H^{2k}(X)\, q^k.$$

The l-adic cohomology groups have the same dimensions as the complex ones. Hence, the Grothendieck–Lefschetz fixed-point formula explains the extraordinary fact that the number of points over a finite field of the Grassmannian or flag varieties is a generating function for the complex cohomology.

Exercises

Exercise 48.1. Consider the space $\mathcal{F}_{r,s}(\mathbb{F}_q)$ of r-flags in $\mathbb{F}^{r+s}$. Compute the cardinality by representing it as $\mathrm{GL}(n, \mathbb{F}_q)/P(\mathbb{F}_q)$, where P is the parabolic subgroup (48.11). Show that $|\mathcal{F}_{r,s}(\mathbb{F}_q)| = \sum_i d_i(r,s)\, q^i$, where for fixed s, we have $d_i(r,s) = \binom{r+i-1}{i}$.

Exercise 48.2. Prove that $H^*(\mathcal{F}_r)$ is a polynomial ring in r generators, with generators in $H^2(\mathcal{F}_r)$ being the cohomology classes of the canonical line bundles ξ_i; here x_i associates with a flag (48.10) the one-dimensional vector space F_i/F_{i-1}.

Appendix: Sage

Sage is a system of free mathematical software that is under active development. Although it was created to do number theory calculations, it contains considerable code for combinatorics, and other areas. For Lie groups, it can compute tensor products of representations, symmetric and exterior powers, and branching rules. It knows the roots, fundamental dominant weights, Weyl group actions, etc. There is also excellent support for symmetric functions and crystals. Many other things are in Sage: Iwahori Hecke algebras, Bruhat order, Kazhdan–Lusztig polynomials, and so on.

This appendix is not a tutorial, but rather a quick introduction to a few of the problems Sage can solve. For a systematic tutorial, you should go through the *Lie Methods and Related Combinatorics* thematic tutorial available at:

> http://www.sagemath.org/doc/thematic_tutorials/lie.html

Other Sage tutorials may be found at http://www.sagemath.org/help.html. You should learn Sage's systems of on-line documentation, tab completion, and so forth.

You should try to run the most recent version of Sage you can because there are continual improvements. Important speedups were added to the Lie group code in both versions 5.4 and 5.5. For simple tasks Sage can be treated as a command-line calculator (or you can use a notebook interface) but for more complicated tasks you can write programs using Python. You can contribute to Sage development: if you want a feature and it doesn't exist, you can make it yourself, and if it is something others might want, eventually get it into the distributed version.

For computations with representations it is convenient to work in a WeylCharacterRing. There are two notations for these. In the default notation, a representation is represented by its highest weight vector, as an element of its ambient space in the notation of the appendices in Bourbaki [23]. Let us give a brief example of a dialog using the standard notation.

D. Bump, *Lie Groups*, Graduate Texts in Mathematics 225,
DOI 10.1007/978-1-4614-8024-2, © Springer Science+Business Media New York 2013

```
sage: B3=WeylCharacterRing("B3"); B3
The Weyl Character Ring of Type ['B', 3] with Integer Ring
    coefficients
sage: B3.fundamental_weights()
Finite family {1: (1,0,0), 2: (1,1,0), 3: (1/2,1/2,1/2)}
sage: [B3(f) for f in B3.fundamental_weights()]
[B3(1,0,0), B3(1,1,0), B3(1/2,1/2,1/2)]
sage: [B3(f).degree() for f in B3.fundamental_weights()]
[7, 21, 8]
sage: B3(1,1,0).symmetric_power(3)
B3(1,0,0) + 2*B3(1,1,0) + B3(2,1,1) + B3(2,2,1) + B3(3,1,0)
    + B3(3,3,0)
sage: [f1,f2,f3]=B3.fundamental_weights()
sage: B3(f3)
B3(1/2,1/2,1/2)
sage: B3(f3)^3
4*B3(1/2,1/2,1/2) + 3*B3(3/2,1/2,1/2) + 2*B3(3/2,3/2,1/2)
    + B3(3/2,3/2,3/2)
```

This illustrates different ways of interacting with Sage as a command-line interpreter. I prefer to run Sage from within an Emacs buffer; others prefer the notebook. For complicated tasks, such as loading some Python code, you may write your commands in a file and load or attach it. Whatever your method of interacting with the program, you can have a dialog with this one. Sage provides a prompt ("sage:") after which you type a command. Sage will sometimes produce some output, sometimes not; in any case, when it is done it will give you another prompt.

The first line contains two commands, separated by a semicolon. The first command creates the WeylCharacterRing B3 but produces no output. The second command "B3" prints the name of the ring you have just created. Elements of the ring are virtual representations of the Lie group Spin (7) having Cartan type B_3. Addition corresponds to direct sum, multiplication to tensor product.

The ring B3 is a Python class, and like every Python class it has methods and attributes which you can use to perform various tasks. If at the sage: prompt you type B3 then hit the tab key, you will get a list of Python methods and attributes that B3 has. For example, you will notice methods dynkin_diagram and extended_dynkin_diagram. If you want more information about one of them, you may access the on-line documentation, with examples of how to use it, by typing B3.dynkin_diagram?

Turning to the next command, the Python class B3 has a method called fundamental_weights. This returns a Python dictionary with elements that are the fundamental weights. The third command gives the irreducible representations with these highest weights, as a Python list. After that, we compute the degrees of these, the symmetric cube of a representation, the spin representation B3(1/2,1/2,1/2) and its square.

We can alternatively create the WeylCharacterRing with an alternative syntax when you create it with the option style="coroots". This is most appropriate for semisimple Lie groups: for a semisimple Lie group, we recall that the *fundamental dominant weights* are defined to be the dual basis to the coroots. Assuming that the group is simply connected the fundamental dominant weights are weights, that is, characters of a maximal torus. For such a case, where G is semisimple and simply connected, every dominant weight may be uniquely expressed as a linear combination, with nonzero integer coefficients, of the fundamental dominant weights. This gives an alternative notation for the representations, as the following example shows:

```
sage: B3=WeylCharacterRing("B3",style="coroots"); B3
The Weyl Character Ring of Type ['B', 3] with Integer Ring
    coefficients
sage: B3.fundamental_weights()
Finite family {1: (1,0,0), 2: (1,1,0), 3: (1/2,1/2,1/2)}
sage: [B3(f) for f in B3.fundamental_weights()]
[B3(1,0,0), B3(0,1,0), B3(0,0,1)]
sage: [B3(f).degree() for f in B3.fundamental_weights()]
[7, 21, 8]
sage: B3(0,1,0).symmetric_power(3)
B3(1,0,0) + 2*B3(0,1,0) + B3(1,0,2) + B3(0,1,2) + B3(2,1,0)
    + B3(0,3,0)
sage: [f1,f2,f3]=B3.fundamental_weights()
sage: B3(f3)
B3(0,0,1) sage:
B3(f3)^3
4*B3(0,0,1) + 3*B3(1,0,1) + 2*B3(0,1,1) + B3(0,0,3)
```

This is the same series of computations as before, just in a different notation.

For Cartan Type A_r, if you use style="coroots", you are effectively working with the group $SL(r+1,\mathbb{C})$. There is no way to represent the determinant in this notation. On the other hand, if you use the default style, the determinant is represented by A2(1,1,1) (in the case $r = 2$) so if you want to do computations for $GL(r+1,\mathbb{C})$, do not use coroot style.

Sage knows many branching rules. For example, here is how to calculate the restriction of a representation from $SL(4)$ to $Sp(4)$.

```
sage: A3=WeylCharacterRing("A3",style="coroots")
sage: C2=WeylCharacterRing("C2",style="coroots")
sage: r=A3(6,4,1)
sage: r.degree()
6860
sage: r.branch(C2,rule="symmetric")
C2(5,1) + C2(7,0) + C2(5,2) + C2(7,1) + C2(5,3)
    + C2(7,2) + C2(5,4) + C2(7,3) + C2(5,5) + C2(7,4)
```

To get documentation about Sage's branching rules, either see the thematic tutorial or enter the command:

```
sage: get_branching_rule?
```

As another example, let us compute

$$\int_{SU(2)} |\mathrm{tr}(g)|^{20}\, dg.$$

There are different ways of doing this computation. An efficient way is just to compute decompose the tenth power of the standard character into irreducibles: if $\mathrm{tr}(g)^{10} = \sum a_\lambda \chi_\lambda$ then its modulus squared is $\sum a_\lambda^2$.

```
sage: A1=WeylCharacterRing("A1")
sage: A1(1)
A1(0,0)
sage: A1([1])
A1(1,0)
sage: A1([1])^10
42*A1(5,5) + 90*A1(6,4) + 75*A1(7,3) + 35*A1(8,2) + 9*A1(9,1)
    + A1(10,0)
sage: (A1([1])^10).monomial_coefficients()
{(8, 2): 35, (10, 0): 1, (9, 1): 9, (5, 5): 42,
    (7, 3): 75, (6, 4): 90}
sage: sum(v^2 for v in (A1([1])^10).monomial_coefficients().values())
16796
```

Alternatively, $|\mathrm{tr}(g)|^{20}$ is itself a character. We can compute this character, then apply the method monomial_coefficients. This gives a dictionary with entries that are these coefficients. We can extract the value of 0, which we implement as A1.space().zero().

```
sage: z = A1.space().zero(); z
(0, 0)
sage: ((A1([1])^10*A1([0,-1])^10)).monomial_coefficients()[z]
16796
```

Let us check that the moments of the trace are Catalan numbers:

```
sage: [sum(v^2 for v in
    (A1([1])^k).monomial_coefficients().values()) for k in [0..10]]
[1, 1, 2, 5, 14, 42, 132, 429, 1430, 4862, 16796]
sage: [catalan_number(k) for k in [0..10]]
[1, 1, 2, 5, 14, 42, 132, 429, 1430, 4862, 16796]
```

You may also use the method weight_multiplicities of a Weyl character to get a dictionary of weight multiplicities indexed by weight.

```
sage: A2=WeylCharacterRing("A2")
sage: d=A2(6,2,0).weight_multiplicities(); d
{(0, 6, 2): 1, (5, 0, 3): 1, (3, 5, 0): 1, ...
```

(Output suppressed.) Here is how to extract a single multiplicity. The `space` method of the WeylCharacterRing returns the ambient vector space of the weight lattice, and we may use this to generate a key.

```
sage: L=A2.space(); L
Ambient space of the Root system of type ['A', 2]
sage: k=L((3,3,2)); k
(3, 3, 2)
sage: type(k)
<class 'sage.combinat.root_system.ambient_space.
        AmbientSpace_with_category.element_class'>
sage: d[k]
3
```

In addition to the Lie group code in Sage, the symmetric function code in Sage will be useful to readers of this book. You may convert between different bases of the ring of symmetric functions (such as the Schur basis **s** and the power sum basis **p**) and calculate important symmetric functions such as the Hall–Littlewood symmetric functions. Moreover, Sage knows about the Hopf algebra structure on the ring of symmetric functions. For example:

```
sage: Sym = SymmetricFunctions(QQ)
sage: s = Sym.schur()
sage: s[2]^2
s[2, 2] + s[3, 1] + s[4]
sage: (s[2]^2).coproduct()
s[] # s[2,2] + s[] # s[3,1] + s[] # s[4] + 2*s[1] # s[2,1]
  + 2*s[1] # s[3] + s[1, 1] # s[1, 1] + s[1, 1] # s[2]
  + s[2] # s[1, 1] + 3*s[2] # s[2] + 2*s[2, 1] # s[1]
  + s[2, 2] # s[] + 2*s[3] # s[1] + s[3, 1] # s[]
  + s[4] # s[]
sage: def f(a,b): return a*b.antipode()
sage: (s[2]^2).coproduct().apply_multilinear_morphism(f)
0
```

We've computed (in the notation introduced in the exercises to Chapter 35) $m \circ (1 \otimes S) \circ \Delta$ applied to $s_{(2)}^2$. This computation may of course also be done using the defining property of the antipode.

You can get a command line tutorial for the symmetric function code with:

```
sage: SymmetricFunctions?
```

References

1. Peter Abramenko and Kenneth S. Brown. *Buildings*, volume 248 of *Graduate Texts in Mathematics*. Springer, New York, 2008. Theory and applications.
2. J. Adams. *Lectures on Lie Groups*. W. A. Benjamin, Inc., New York-Amsterdam, 1969.
3. J. F. Adams. *Lectures on exceptional Lie groups*. Chicago Lectures in Mathematics. University of Chicago Press, Chicago, IL, 1996. With a foreword by J. Peter May, Edited by Zafer Mahmud and Mamoru Mimura.
4. Gernot Akemann, Jinho Baik, and Philippe Di Francesco, editors. *The Oxford handbook of random matrix theory*. Oxford University Press, Oxford, 2011.
5. A. Albert. *Structure of Algebras*. American Mathematical Society Colloquium Publications, vol. 24. American Mathematical Society, New York, 1939.
6. B. N. Allison. Tensor products of composition algebras, Albert forms and some exceptional simple Lie algebras. *Trans. Amer. Math. Soc.*, 306(2):667–695, 1988.
7. Greg W. Anderson, Alice Guionnet, and Ofer Zeitouni. *An introduction to random matrices*, volume 118 of *Cambridge Studies in Advanced Mathematics*. Cambridge University Press, Cambridge, 2010.
8. J. Arthur and L. Clozel. *Simple Algebras, Base Change, and the Advanced Theory of the Trace Formula*, volume 120 of *Annals of Mathematics Studies*. Princeton University Press, Princeton, NJ, 1989.
9. E. Artin. *Geometric Algebra*. Interscience Publishers, Inc., New York and London, 1957.
10. M. Artin, J. E. Bertin, M. Demazure, P. Gabriel, A. Grothendieck, M. Raynaud, and J.-P. Serre. *Schémas en groupes. Fasc. 5b: Exposés 17 et 18*, volume 1963/64 of *Séminaire de Gémétrie Algébrique de l'Institut des Hautes Études Scientifiques*. Institut des Hautes Études Scientifiques, Paris, 1964/1966 (http://www.math.jussieu.fr/~polo/SGA3/).
11. A. Ash, D. Mumford, M. Rapoport, and Y. Tai. *Smooth Compactification of Locally Symmetric Varieties*. Math. Sci. Press, Brookline, Mass., 1975. Lie Groups: History, Frontiers and Applications, Vol. IV.
12. John C. Baez. The octonions. *Bull. Amer. Math. Soc. (N.S.)*, 39(2):145–205, 2002.
13. W. Baily. *Introductory Lectures on Automorphic Forms*. Iwanami Shoten, Publishers, Tokyo, 1973. Kano Memorial Lectures, No. 2, Publications of the Mathematical Society of Japan, No. 12.

536   References

14. W. Baily, and A. Borel. Compactification of arithmetic quotients of bounded symmetric domains. *Ann. of Math. (2)*, 84:442–528, 1966.

15. A. Berele and J. B. Remmel. Hook flag characters and their combinatorics. *J. Pure Appl. Algebra*, 35(3):225–245, 1985.

16. J. Bernstein and A. Zelevinsky. Representations of the group $GL(n, F)$ where F is a local nonarchimedean field. *Russian Mathematical Surveys*, 3:1–68, 1976.

17. I. Bernstein and A. Zelevinsky. Induced representations of reductive p-adic groups. I. *Ann. Sci. École Norm. Sup. (4)*, 10(4):441–472, 1977.

18. P. Billingsley. *Probability and Measure*. Wiley Series in Probability and Mathematical Statistics. John Wiley & Sons Inc., New York, third edition, 1995. A Wiley-Interscience Publication.

19. A. Borel. Automorphic L-functions. In *Automorphic Forms, Representations and L-Functions (Proc. Sympos. Pure Math., Oregon State Univ., Corvallis, Ore., 1977), Part 2*, Proc. Sympos. Pure Math., XXXIII, pages 27–61. Amer. Math. Soc., Providence, R.I., 1979.

20. A. Borel. *Linear Algebraic Groups*, volume 126 of *Graduate Texts in Mathematics*. Springer-Verlag, New York, second edition, 1991.

21. A. Borel and J. Tits. Groupes réductifs. *Inst. Hautes Études Sci. Publ. Math.*, 27:55–150, 1965.

22. A. Böttcher and B. Silbermann. *Introduction to Large Truncated Toeplitz Matrices*. Universitext. Springer-Verlag, New York, 1999.

23. Nicolas Bourbaki. *Lie groups and Lie algebras. Chapters 4–6*. Elements of Mathematics (Berlin). Springer-Verlag, Berlin, 2002. Translated from the 1968 French original by Andrew Pressley.

24. Nicolas Bourbaki. *Lie groups and Lie algebras. Chapters 7–9*. Elements of Mathematics (Berlin). Springer-Verlag, Berlin, 2005. Translated from the 1975 and 1982 French originals by Andrew Pressley.

25. T. Bröcker and T. tom Dieck. *Representations of Compact Lie Groups*, volume 98 of *Graduate Texts in Mathematics*. Springer-Verlag, New York, 1985.

26. F. Bruhat. Sur les représentations induites des groupes de Lie. *Bull. Soc. Math. France*, 84:97–205, 1956.

27. D. Bump. *Automorphic Forms and Representations*, volume 55 of *Cambridge Studies in Advanced Mathematics*. Cambridge University Press, Cambridge, 1997.

28. D. Bump and P. Diaconis. Toeplitz minors. *J. Combin. Theory Ser. A*, 97(2):252–271, 2002.

29. D. Bump and A. Gamburd. On the averages of characteristic polynomials from classical groups. *Comm. Math. Phys.*, 265(1):227–274, 2006.

30. D. Bump, P. Diaconis, and J. Keller. Unitary correlations and the Fejér kernel. *Math. Phys. Anal. Geom.*, 5(2):101–123, 2002.

31. E. Cartan. Sur une classe remarquable d'espaces de Riemann. *Bull. Soc. Math. France*, 54, 55:214–264, 114–134, 1926, 1927.

32. R. Carter. *Finite Groups of Lie Type, Conjugacy classes and complex characters*. Pure and Applied Mathematics. John Wiley & Sons Inc., New York, 1985. A Wiley-Interscience Publication.

33. P. Cartier. Representations of p-adic groups: a survey. In *Automorphic forms, representations and L-functions (Proc. Sympos. Pure Math., Oregon State Univ., Corvallis, Ore., 1977), Part 1*, Proc. Sympos. Pure Math., XXXIII, pages 111–155. Amer. Math. Soc., Providence, R.I., 1979.

34. W. Casselman. Introduction to the Theory of Admissible Representations of Reductive p-adic Groups. Widely circulated preprint. Available at http://www.math.ubc.ca/~cass/research.html, 1974.

35. C. Chevalley. *Theory of Lie Groups. I.* Princeton Mathematical Series, vol. 8. Princeton University Press, Princeton, N. J., 1946.

36. C. Chevalley. *The Algebraic Theory of Spinors and Clifford Algebras.* Springer-Verlag, Berlin, 1997. Collected works. Vol. 2, edited and with a foreword by Pierre Cartier and Catherine Chevalley, with a postface by J.-P. Bourguignon.

37. W. Chow. On equivalence classes of cycles in an algebraic variety. *Ann. of Math. (2),* 64:450–479, 1956.

38. B. Conrey. L-functions and random matrices. In *Mathematics unlimited—2001 and beyond,* pages 331–352. Springer, Berlin, 2001.

39. C. Curtis. *Pioneers of Representation Theory: Frobenius, Burnside, Schur, and Brauer,* volume 15 of *History of Mathematics.* American Mathematical Society, Providence, RI, 1999.

40. P. A. Deift. *Orthogonal polynomials and random matrices: a Riemann-Hilbert approach,* volume 3 of *Courant Lecture Notes in Mathematics.* New York University Courant Institute of Mathematical Sciences, New York, 1999.

41. P. Deligne and G. Lusztig. Representations of Reductive Groups over Finite Fields. *Ann. of Math. (2),* 103(1):103–161, 1976.

42. P. Diaconis and M. Shahshahani. On the eigenvalues of random matrices. *J. Appl. Probab.,* 31A:49–62, 1994. Studies in applied probability.

43. A. Dold. Fixed point index and fixed point theorem for Euclidean neighborhood retracts. *Topology,* 4:1–8, 1965.

44. A. Dold. *Lectures on Algebraic Topology.* Springer-Verlag, New York, 1972. Die Grundlehren der mathematischen Wissenschaften, Band 200.

45. E. Dynkin. Maximal subgroups of semi-simple Lie groups and the classification of primitive groups of transformations. *Doklady Akad. Nauk SSSR (N.S.),* 75:333–336, 1950.

46. E. Dynkin. Maximal subgroups of the classical groups. *Trudy Moskov. Mat. Obšč.,* 1:39–166, 1952.

47. E. Dynkin. Semisimple subalgebras of semisimple Lie algebras. *Mat. Sbornik N.S.,* 30(72):349–462, 1952.

48. F. Dyson. Statistical theory of the energy levels of complex systems, I, II, III. *J. Mathematical Phys.,* 3:140–156, 157–165, 166–175, 1962.

49. Freeman Dyson. *Selected papers of Freeman Dyson with commentary,* volume 5 of *Collected Works.* American Mathematical Society, Providence, RI, 1996. With a foreword by Elliott H. Lieb.

50. H. Freudenthal. Lie groups in the foundations of geometry. *Advances in Math.,* 1:145–190 (1964), 1964.

51. G. Frobenius. Über die charakterisischen Einheiten der symmetrischen Gruppe. *S'ber. Akad. Wiss. Berlin,* 504–537, 1903.

52. G. Frobenius and I. Schur. Über die rellen Darstellungen der endlichen Gruppen. *S'ber. Akad. Wiss. Berlin,* 186–208, 1906.

53. W. Fulton. *Young Tableaux, with applications to representation theory and geometry,* volume 35 of *London Mathematical Society Student Texts.* Cambridge University Press, Cambridge, 1997.

54. W. Fulton. *Intersection Theory,* volume 2 of *Ergebnisse der Mathematik und ihrer Grenzgebiete.* Springer-Verlag, Berlin, second edition, 1998.

55. I. Gelfand, M. Graev, and I. Piatetski-Shapiro. *Representation Theory and Automorphic Functions.* Academic Press Inc., 1990. Translated from the Russian by K. A. Hirsch, Reprint of the 1969 edition.

56. R. Goodman and N. Wallach. *Representations and Invariants of the Classical Groups*, volume 68 of *Encyclopedia of Mathematics and its Applications.* Cambridge University Press, Cambridge, 1998.

57. R. Gow. Properties of the characters of the finite general linear group related to the transpose-inverse involution. *Proc. London Math. Soc. (3)*, 47(3):493–506, 1983.

58. J. Green. The characters of the finite general linear groups. *Trans. Amer. Math. Soc.*, 80:402–447, 1955.

59. B. Gross. Some applications of Gelfand pairs to number theory. *Bull. Amer. Math. Soc. (N.S.)*, 24:277–301, 1991.

60. R. Gunning and H. Rossi. *Analytic functions of several complex variables.* Prentice-Hall Inc., Englewood Cliffs, N.J., 1965.

61. P. Halmos. *Measure Theory.* D. Van Nostrand Company, Inc., New York, N. Y., 1950.

62. Harish-Chandra. Eisenstein series over finite fields. In *Functional analysis and related fields (Proc. Conf. M. Stone, Univ. Chicago, Chicago, Ill., 1968)*, pages 76–88. Springer, New York, 1970.

63. M. Harris and R. Taylor. *The Geometry and Cohomology of Some Simple Shimura Varieties*, volume 151 of *Annals of Mathematics Studies.* Princeton University Press, Princeton, NJ, 2001. With an appendix by Vladimir G. Berkovich.

64. R. Hartshorne. *Algebraic Geometry.* Springer-Verlag, New York, 1977. Graduate Texts in Mathematics, No. 52.

65. E. Hecke. Über Modulfunktionen und die Dirichletschen Reihen mit Eulerscher Produktentwicklungen, I and II. *Math. Ann.*, 114:1–28, 316–351, 1937.

66. S. Helgason. *Differential Geometry, Lie Groups, and Symmetric Spaces*, volume 80 of *Pure and Applied Mathematics.* Academic Press Inc. [Harcourt Brace Jovanovich Publishers], New York, 1978.

67. Guy Henniart. Une preuve simple des conjectures de Langlands pour $GL(n)$ sur un corps p-adique. *Invent. Math.*, 139(2):439–455, 2000.

68. Guy Henniart. On the local Langlands and Jacquet-Langlands correspondences. In *International Congress of Mathematicians. Vol. II*, pages 1171–1182. Eur. Math. Soc., Zürich, 2006.

69. E. Hewitt and K. Ross. *Abstract Harmonic Analysis. Vol. I, Structure of topological groups, integration theory, group representations*, volume 115 of *Grundlehren der Mathematischen Wissenschaften [Fundamental Principles of Mathematical Sciences].* Springer-Verlag, Berlin, second edition, 1979.

70. H. Hiller. *Geometry of Coxeter Groups*, volume 54 of *Research Notes in Mathematics.* Pitman (Advanced Publishing Program), Boston, Mass., 1982.

71. W. Hodge and D. Pedoe. *Methods of Algebraic Geometry. Vol. II.* Cambridge Mathematical Library. Cambridge University Press, Cambridge, 1994. Book III: General theory of algebraic varieties in projective space, Book IV: Quadrics and Grassmann varieties, Reprint of the 1952 original.

72. Jin Hong and Seok-Jin Kang. *Introduction to quantum groups and crystal bases*, volume 42 of *Graduate Studies in Mathematics.* American Mathematical Society, Providence, RI, 2002.

73. R. Howe. θ-series and invariant theory. In *Automorphic Forms, Representations and L-Functions (Proc. Sympos. Pure Math., Oregon State Univ., Corvallis, Ore., 1977), Part 1*, Proc. Sympos. Pure Math., XXXIII, pages 275–285. Amer. Math. Soc., Providence, R.I., 1979.

74. R. Howe. *Harish-Chandra Homomorphisms for $\mathfrak{p}$-adic Groups*, volume 59 of *CBMS Regional Conference Series in Mathematics*. Published for the Conference Board of the Mathematical Sciences, Washington, DC, 1985. With the collaboration of Allen Moy.

75. Roger Howe. Remarks on classical invariant theory. *Trans. Amer. Math. Soc.*, 313(2):539–570, 1989.

76. R. Howe. Hecke algebras and p-adic GL_n. In *Representation theory and analysis on homogeneous spaces (New Brunswick, NJ, 1993)*, volume 177 of *Contemp. Math.*, pages 65–100. Amer. Math. Soc., Providence, RI, 1994.

77. Roger Howe. Perspectives on invariant theory: Schur duality, multiplicity-free actions and beyond. In *The Schur lectures (1992) (Tel Aviv)*, volume 8 of *Israel Math. Conf. Proc.*, pages 1–182. Bar-Ilan Univ., Ramat Gan, 1995.

78. R. Howe and E.-C. Tan. *Nonabelian Harmonic Analysis*. Universitext. Springer-Verlag, New York, 1992. Applications of $SL(2, \mathbf{R})$.

79. Roger Howe, Eng-Chye Tan, and Jeb F. Willenbring. Stable branching rules for classical symmetric pairs. *Trans. Amer. Math. Soc.*, 357(4):1601–1626, 2005.

80. R. Howlett and G. Lehrer. Induced cuspidal representations and generalised Hecke rings. *Invent. Math.*, 58(1):37–64, 1980.

81. E. Ince. *Ordinary Differential Equations*. Dover Publications, New York, 1944.

82. N. Inglis, R. Richardson, and J. Saxl. An explicit model for the complex representations of s_n. *Arch. Math. (Basel)*, 54:258–259, 1990.

83. I. M. Isaacs. *Character Theory of Finite Groups*. Dover Publications Inc., New York, 1994. Corrected reprint of the 1976 original [Academic Press, New York; MR **57** #417].

84. N. Iwahori. On the structure of a Hecke ring of a Chevalley group over a finite field. *J. Fac. Sci. Univ. Tokyo Sect. I*, 10:215–236, 1964.

85. N. Iwahori. Generalized Tits system (Bruhat decompostition) on p-adic semisimple groups. In *Algebraic Groups and Discontinuous Subgroups (Proc. Sympos. Pure Math., Boulder, Colo., 1965)*, pages 71–83. Amer. Math. Soc., Providence, R.I., 1966.

86. N. Iwahori and H. Matsumoto. On some Bruhat decomposition and the structure of the Hecke rings of $\mathfrak{p}$-adic Chevalley groups. *Inst. Hautes Études Sci. Publ. Math.*, 25:5–48, 1965.

87. N. Jacobson. Cayley numbers and normal simple Lie algebras of type G. *Duke Math. J.*, 5:775–783, 1939.

88. N. Jacobson. *Exceptional Lie Algebras*, volume 1 of *Lecture Notes in Pure and Applied Mathematics*. Marcel Dekker Inc., New York, 1971.

89. M. Jimbo. A q-analogue of $U(\mathfrak{gl}(N + 1))$, Hecke algebra, and the Yang-Baxter equation. *Lett. Math. Phys.*, 11(3):247–252, 1986.

90. Michio Jimbo. Introduction to the Yang-Baxter equation. *Internat. J. Modern Phys. A*, 4(15):3759–3777, 1989.

91. V. Jones. Hecke algebra representations of braid groups and link polynomials. *Ann. of Math. (2)*, 126:335–388, 1987.

92. Victor G. Kac. *Infinite-dimensional Lie algebras*. Cambridge University Press, Cambridge, third edition, 1990.

93. Masaki Kashiwara. On crystal bases. In *Representations of groups (Banff, AB, 1994)*, volume 16 of *CMS Conf. Proc.*, pages 155–197. Amer. Math. Soc., Providence, RI, 1995.

94. N. Katz and P. Sarnak. Zeroes of zeta functions and symmetry. *Bull. Amer. Math. Soc. (N.S.)*, 36(1):1–26, 1999.

95. Nicholas M. Katz and Peter Sarnak. *Random matrices, Frobenius eigenvalues, and monodromy*, volume 45 of *American Mathematical Society Colloquium Publications*. American Mathematical Society, Providence, RI, 1999.

96. N. Kawanaka and H. Matsuyama. A twisted version of the Frobenius-Schur indicator and multiplicity-free permutation representations. *Hokkaido Math. J.*, 19(3):495–508, 1990.

97. David Kazhdan and George Lusztig. Representations of Coxeter groups and Hecke algebras. *Invent. Math.*, 53(2):165–184, 1979.

98. David Kazhdan and George Lusztig. Proof of the Deligne-Langlands conjecture for Hecke algebras. *Invent. Math.*, 87(1):153–215, 1987.

99. J. Keating and N. Snaith. Random matrix theory and $\zeta(1/2 + it)$. *Comm. Math. Phys.*, 214(1):57–89, 2000.

100. A. Kerber. *Representations of permutation groups. I.* Lecture Notes in Mathematics, Vol. 240. Springer-Verlag, Berlin, 1971.

101. R. King. Branching rules for classical Lie groups using tensor and spinor methods. *J. Phys. A*, 8:429–449, 1975.

102. S. Kleiman. Problem 15: rigorous foundation of Schubert's enumerative calculus. In *Mathematical Developments Arising from Hilbert Problems (Proc. Sympos. Pure Math., Northern Illinois Univ., De Kalb, Ill., 1974)*, pages 445–482. Proc. Sympos. Pure Math., Vol. XXVIII. Amer. Math. Soc., Providence, R. I., 1976.

103. A. Klyachko. Models for complex representations of groups $GL(n, q)$. *Mat. Sb. (N.S.)*, 120(162)(3):371–386, 1983.

104. A. Knapp. *Representation Theory of Semisimple Groups, an overview based on examples*, volume 36 of *Princeton Mathematical Series*. Princeton University Press, Princeton, NJ, 1986.

105. A. Knapp. *Lie groups, Lie algebras, and Chomology*, volume 34 of *Mathematical Notes*. Princeton University Press, Princeton, NJ, 1988.

106. A. Knapp. *Lie Groups Beyond an Introduction*, volume 140 of *Progress in Mathematics*. Birkhäuser Boston Inc., Boston, MA, second edition, 2002.

107. M.-A. Knus, A. Merkurjev, M. Rost, and J.-P. Tignol. *The Book of Involutions*, volume 44 of *American Mathematical Society Colloquium Publications*. American Mathematical Society, Providence, RI, 1998. With a preface in French by J. Tits.

108. Donald E. Knuth. Permutations, matrices, and generalized Young tableaux. *Pacific J. Math.*, 34:709–727, 1970.

109. D. Knuth. *The Art of Computer Programming. Volume 3, Sorting and Searching.* Addison-Wesley Publishing Co., Reading, Mass.-London-Don Mills, Ont., 1973. Addison-Wesley Series in Computer Science and Information Processing.

110. S. Kobayashi and K. Nomizu. *Foundations of Differential Geometry. Vol I.* Interscience Publishers, a division of John Wiley & Sons, New York-London, 1963.

111. A. Korányi and J. Wolf. Generalized Cayley transformations of bounded symmetric domains. *Amer. J. Math.*, 87:899–939, 1965.

112. A. Korányi and J. Wolf. Realization of hermitian symmetric spaces as generalized half-planes. *Ann. of Math. (2)*, 81:265–288, 1965.

113. S. Kudla. Seesaw dual reductive pairs. In *Automorphic forms of several variables (Katata, 1983)*, volume 46 of *Progr. Math.*, pages 244–268. Birkhäuser Boston, Boston, MA, 1984.

114. Laurent Lafforgue. Chtoucas de Drinfeld et correspondance de Langlands. *Invent. Math.*, 147(1):1–241, 2002.

115. J. Landsberg and L. Manivel. The projective geometry of Freudenthal's magic square. *J. Algebra*, 239(2):477–512, 2001.

116. S. Lang. *Algebra*, volume 211 of *Graduate Texts in Mathematics*. Springer-Verlag, New York, third edition, 2002.

117. R. Langlands. *Euler Products*. Yale University Press, New Haven, Conn., 1971. A James K. Whittemore Lecture in Mathematics given at Yale University, 1967, Yale Mathematical Monographs, 1.

118. H. B. Lawson and M.-L. Michelsohn. *Spin Geometry*, volume 38 of *Princeton Mathematical Series*. Princeton University Press, Princeton, NJ, 1989.

119. G. Lion and M. Vergne. *The Weil representation, Maslov index and theta series*, volume 6 of *Progress in Mathematics*. Birkhäuser Boston, Mass., 1980.

120. D. Littlewood. *The Theory of Group Characters and Matrix Representations of Groups*. Oxford University Press, New York, 1940.

121. L. Loomis. *An Introduction to Abstract Harmonic Analysis*. D. Van Nostrand Company, Inc., Toronto-New York-London, 1953.

122. George Lusztig. Equivariant K-theory and representations of Hecke algebras. *Proc. Amer. Math. Soc.*, 94(2):337–342, 1985.

123. I. G. Macdonald. Schur functions: theme and variations. In *Séminaire Lotharingien de Combinatoire (Saint-Nabor, 1992)*, volume 498 of *Publ. Inst. Rech. Math. Av.*, pages 5–39. Univ. Louis Pasteur, Strasbourg, 1992.

124. I. Macdonald. *Symmetric Functions and Hall Polynomials*. Oxford Mathematical Monographs. The Clarendon Press Oxford University Press, New York, second edition, 1995. With contributions by A. Zelevinsky, Oxford Science Publications.

125. S. Majid. *A quantum groups primer*, volume 292 of *London Mathematical Society Lecture Note Series*. Cambridge University Press, Cambridge, 2002.

126. L. Manivel. *Symmetric Functions, Schubert Polynomials and Degeneracy Loci*, volume 6 of *SMF/AMS Texts and Monographs*. American Mathematical Society, Providence, RI, 2001. Translated from the 1998 French original by John R. Swallow, Cours Spécialisés [Specialized Courses], 3.

127. H. Matsumoto. Générateurs et relations des groupes de Weyl généralisés. *C. R. Acad. Sci. Paris*, 258:3419–3422, 1964.

128. M. Mehta. *Random Matrices*. Academic Press Inc., Boston, MA, second edition, 1991.

129. J. Milnor and J. Stasheff. *Characteristic Classes*. Princeton University Press, Princeton, N. J., 1974. Annals of Mathematics Studies, No. 76.

130. C. Moeglin. Representations of GL(n) over the real field. In *Representation theory and automorphic forms (Edinburgh, 1996)*, volume 61 of *Proc. Sympos. Pure Math.*, pages 157–166. Amer. Math. Soc., Providence, RI, 1997.

131. C. Mœglin and J.-L. Waldspurger. *Spectral Decomposition and Eisenstein Series, Une paraphrase de l'Écriture [A paraphrase of Scripture]*, volume 113 of *Cambridge Tracts in Mathematics*. Cambridge University Press, Cambridge, 1995.

542   References

132. D. Mumford, J. Fogarty, and F. Kirwan. *Geometric invariant theory*, volume 34 of *Ergebnisse der Mathematik und ihrer Grenzgebiete (2) [Results in Mathematics and Related Areas (2)]*. Springer-Verlag, Berlin, third edition, 1994.

133. I. Pyateskii-Shapiro. *Automorphic Functions and the Geometry of Classical Domains*. Translated from the Russian. Mathematics and Its Applications, Vol. 8. Gordon and Breach Science Publishers, New York, 1969.

134. Martin Raussen and Christian Skau. Interview with John G. Thompson and Jacques Tits. *Notices Amer. Math. Soc.*, 56(4):471–478, 2009.

135. N. Reshetikhin and V. G. Turaev. Invariants of 3-manifolds via link polynomials and quantum groups. *Invent. Math.*, 103(3):547–597, 1991.

136. G. de B. Robinson. On the Representations of the Symmetric Group. *Amer. J. Math.*, 60(3):745–760, 1938.

137. J. Rogawski. On modules over the Hecke algebra of a *p*-adic group. *Invent. Math.*, 79:443–465, 1985.

138. H. Rubenthaler. Les paires duales dans les algèbres de Lie réductives. *Astérisque*, 219, 1994.

139. Michael Rubinstein. Computational methods and experiments in analytic number theory. In *Recent perspectives in random matrix theory and number theory*, volume 322 of *London Math. Soc. Lecture Note Ser.*, pages 425–506. Cambridge Univ. Press, Cambridge, 2005.

140. W. Rudin. *Fourier Analysis on Groups*. Interscience Tracts in Pure and Applied Mathematics, No. 12. Interscience Publishers (a division of John Wiley and Sons), New York-London, 1962.

141. B. Sagan. *The Symmetric Group, representations, combinatorial algorithms, and symmetric functions*, volume 203 of *Graduate Texts in Mathematics*. Springer-Verlag, New York, second edition, 2001.

142. I. Satake. On representations and compactifications of symmetric Riemannian spaces. *Ann. of Math. (2)*, 71:77–110, 1960.

143. I. Satake. Theory of spherical functions on reductive algebraic groups over p-adic fields. *Inst. Hautes Études Sci. Publ. Math.*, 18:5–69, 1963.

144. I. Satake. *Classification Theory of Semi-simple Algebraic Groups*. Marcel Dekker Inc., New York, 1971. With an appendix by M. Sugiura, Notes prepared by Doris Schattschneider, Lecture Notes in Pure and Applied Mathematics, 3.

145. I. Satake. *Algebraic Structures of Symmetric Domains*, volume 4 of *Kano Memorial Lectures*. Iwanami Shoten and Princeton University Press, Tokyo, 1980.

146. R. Schafer. *An Introduction to Nonassociative Algebras*. Pure and Applied Mathematics, Vol. 22. Academic Press, New York, 1966.

147. C. Schensted. Longest increasing and decreasing subsequences. *Canad. J. Math.*, 13:179–191, 1961.

148. J.-P. Serre. *Galois Cohomology*. Springer-Verlag, Berlin, 1997. Translated from the French by Patrick Ion and revised by the author.

149. E. Spanier. *Algebraic Topology*. McGraw-Hill Book Co., New York, 1966.

150. T. Springer. Galois cohomology of linear algebraic groups. In *Algebraic Groups and Discontinuous Subgroups (Proc. Sympos. Pure Math., Boulder, Colo., 1965)*, pages 149–158. Amer. Math. Soc., Providence, R.I., 1966.

151. T. Springer. Cusp Forms for Finite Groups. In *Seminar on Algebraic Groups and Related Finite Groups (The Institute for Advanced Study, Princeton, N.J., 1968/69)*, Lecture Notes in Mathematics, Vol. 131, pages 97–120. Springer, Berlin, 1970.

152. T. Springer. Reductive groups. In *Automorphic forms, representations and L-functions (Proc. Sympos. Pure Math., Oregon State Univ., Corvallis, Ore., 1977), Part 1*, Proc. Sympos. Pure Math., XXXIII, pages 3–27. Amer. Math. Soc., Providence, R.I., 1979.

153. R. Stanley. *Enumerative Combinatorics. Vol. 2*, volume 62 of *Cambridge Studies in Advanced Mathematics*. Cambridge University Press, Cambridge, 1999. With a foreword by Gian-Carlo Rota and appendix 1 by Sergey Fomin.

154. Robert Steinberg. A general Clebsch-Gordan theorem. *Bull. Amer. Math. Soc.*, 67:406–407, 1961.

155. Robert Steinberg. *Lectures on Chevalley groups*. Yale University, New Haven, Conn. (http://www.math.ucla.edu/ rst/), 1968. Notes prepared by John Faulkner and Robert Wilson.

156. E. Stiefel. Kristallographische Bestimmung der Charaktere der geschlossenen Lie'schen Gruppen. *Comment. Math. Helv.*, 17:165–200, 1945.

157. G. Szegö. On certain Hermitian forms associated with the Fourier series of a positive function. *Comm. Sém. Math. Univ. Lund [Medd. Lunds Univ. Mat. Sem.]*, 1952(Tome Supplementaire):228–238, 1952.

158. T. Tamagawa. On the ζ-functions of a division algebra. *Ann. of Math. (2)*, 77:387–405, 1963.

159. J. Tate. Number theoretic background. In *Automorphic forms, representations and L-functions (Proc. Sympos. Pure Math., Oregon State Univ., Corvallis, Ore., 1977), Part 2*, Proc. Sympos. Pure Math., XXXIII, pages 3–26. Amer. Math. Soc., Providence, R.I., 1979.

160. H. N. V. Temperley and E. H. Lieb. Relations between the "percolation" and "colouring" problem and other graph-theoretical problems associated with regular planar lattices: some exact results for the "percolation" problem. *Proc. Roy. Soc. London Ser. A*, 322(1549):251–280, 1971.

161. J. Tits. Algèbres alternatives, algèbres de Jordan et algèbres de Lie exceptionnelles. I. Construction. *Nederl. Akad. Wetensch. Proc. Ser. A 69 = Indag. Math.*, 28:223–237, 1966.

162. J. Tits. Classification of algebraic semisimple groups. In *Algebraic Groups and Discontinuous Subgroups (Proc. Sympos. Pure Math., Boulder, Colo., 1965)*, pages 33–62, Providence, R.I., 1966, 1966. Amer. Math. Soc.

163. Jacques Tits. *Buildings of spherical type and finite BN-pairs*. Lecture Notes in Mathematics, Vol. 386. Springer-Verlag, Berlin, 1974.

164. Marc A. A. van Leeuwen. The Robinson-Schensted and Schützenberger algorithms, an elementary approach. *Electron. J. Combin.*, 3(2):Research Paper 15, approx. 32 pp. (electronic), 1996. The Foata Festschrift.

165. V. Varadarajan. *An Introduction to Harmonic Analysis on Semisimple Lie Groups*, volume 16 of *Cambridge Studies in Advanced Mathematics*. Cambridge University Press, Cambridge, 1989.

166. È. Vinberg, editor. *Lie Groups and Lie Algebras, III*, volume 41 of *Encyclopaedia of Mathematical Sciences*. Springer-Verlag, Berlin, 1994. Structure of Lie groups and Lie algebras, A translation of *Current problems in mathematics. Fundamental directions. Vol. 41* (Russian), Akad. Nauk SSSR, Vsesoyuz. Inst. Nauchn. i Tekhn. Inform., Moscow, 1990 [MR 91b:22001], Translation by V. Minachin [V. V. Minakhin], Translation edited by A. L. Onishchik and È. B. Vinberg.

544   References

167. D. Vogan. *Unitary Representations of Reductive Lie Groups*, volume 118 of *Annals of Mathematics Studies*. Princeton University Press, Princeton, NJ, 1987.

168. N. Wallach. *Real Reductive Groups. I*, volume 132 of *Pure and Applied Mathematics*. Academic Press Inc., Boston, MA, 1988.

169. A. Weil. *L'intégration dans les Groupes Topologiques et ses Applications.* Actual. Sci. Ind., no. 869. Hermann et Cie., Paris, 1940. [This book has been republished by the author at Princeton, N. J., 1941.].

170. A. Weil. Numbers of solutions of equations in finite fields. *Bull. Amer. Math. Soc.*, 55:497–508, 1949.

171. A. Weil. Algebras with involutions and the classical groups. *J. Indian Math. Soc. (N.S.)*, 24:589–623 (1961), 1960.

172. A. Weil. Sur certains groupes d'opérateurs unitaires. *Acta Math.*, 111:143–211, 1964.

173. A. Weil. Sur la formule de Siegel dans la théorie des groupes classiques. *Acta Math.*, 113:1–87, 1965.

174. H. Weyl. Theorie der Darstellung kontinuierlicher halb-einfacher Gruppen durch lineare Transformationen, i, ii and iii. *Math. Zeitschrift*, 23:271–309, 24:328–395, 1925, 1926.

175. J. Wolf. Complex homogeneous contact manifolds and quaternionic symmetric spaces. *J. Math. Mech.*, 14:1033–1047, 1965.

176. J. Wolf. *Spaces of Constant Curvature*. McGraw-Hill Book Co., New York, 1967.

177. A. Zelevinsky. Induced representations of reductive p-adic groups. II. On irreducible representations of $GL(n)$. *Ann. Sci. École Norm. Sup. (4)*, 13(2): 165–210, 1980.

178. A. Zelevinsky. *Representations of Finite Classical Groups, A Hopf algebra approach*, volume 869 of *Lecture Notes in Mathematics*. Springer-Verlag, Berlin, 1981.

179. R. B. Zhang. Howe duality and the quantum general linear group. *Proc. Amer. Math. Soc.*, 131(9):2681–2692 (electronic), 2003.

Index

Abelian subspace, 283
absolute root system, 281, 282
Adams operations, 189, 353
adjoint group, 145
adjoint representation, 54
admissible path, 110
affine Hecke algebra, 472
affine ring, 405
affine root, 307
affine Weyl group, 191, 195, 221
algebraic character, 349
algebraic complexification, 208
algebraic cycle, 519
algebraic representation, 209, 349
alternating map, 59, 356
anisotropic kernel, 265, 284
anisotropic torus, 511
antipodal map, 43
arclength, 110
Ascoli-Arzela Lemma, 21, 22, 24
atlas, 39
augmentation map, 479
automorphic cuspidal representation, 489
automorphic form, 487, 488
automorphic representation, 489

balanced map, 345
base point, 81
Bergman-Shilov boundary, 274
Bezout's Theorem, 520
bialgebra, 375
big Bruhat cell, 254

bilinear form
  invariant, 63
bimodule, 345
Borel subgroup, 227, 232
  standard, 232
boundary
  Bergman-Shilov, 274
boundary component, 272, 274
boundary of a symmetric space, 269
bounded operator, 19
bracket
  Lie, 32
braid group, 216
braid relation, 216
branching rule, 399, 419
Brauer-Klimyk method, 185
Bruhat decomposition, 243, 300
building
  Tits', 195, 214, 243, 276

Cartan decomposition, 89
Cartan involution, 257
Cartan type, 145
  classical, 145
Casimir element, 62, 64, 75, 488
Catalan numbers, 128
Cauchy identity, 241, 395, 415
  dual, 398, 416
  supersymmetric, 406
Cayley numbers, 276, 313
Cayley transform, 37, 268–270
center, 201
central character, 489

Printed in the United States
By Bookmasters